Simons Symposia

Series Editor
Yuri Tschinkel, Courant Institute of Mathematical Sciences and Simons Foundation, New York University, New York, NY, USA

Working to foster communication and enable interactions between experts, volumes in the Simons Symposia series bring together leading researchers to demonstrate the powerful connection of ideas, methods, and goals shared by mathematicians, theoretical physicists, and theoretical computer scientists.

Symposia volumes feature a blend of original research papers and comprehensive surveys from international teams of leading researchers in thriving fields of study. This blend of approaches helps to ensure each volume will serve not only as an introduction for graduate students and researchers interested in entering the field, but also as the premier reference for experts working on related problems.

The Simons Foundation at its core exists to support basic, discovery-driven research in mathematics and the basic sciences, undertaken in pursuit of understanding the phenomena of our world without specific applications in mind. The foundation seeks to advance the frontiers of research in mathematics and the basic sciences by creating strong collaborations and fostering cross-pollination of ideas between investigators, leading to unexpected breakthroughs and a deeper understanding of the world around us.

More information about this series at https://link.springer.com/bookseries/15045

Jennifer S. Balakrishnan • Noam Elkies
Brendan Hassett • Bjorn Poonen
Andrew V. Sutherland • John Voight
Editors

Arithmetic Geometry, Number Theory, and Computation

Springer

Editors
Jennifer S. Balakrishnan
Department of Mathematics and Statistics
Boston University
Boston, MA, USA

Noam Elkies
Department of Mathematics
Harvard University
Cambridge, MA, USA

Brendan Hassett
Institute for Computational and
Experimental Research in Mathematics
Brown University
Providence, RI, USA

Bjorn Poonen
Department of Mathematics
Massachusetts Institute of Technology
Cambridge, MA, USA

Andrew V. Sutherland
Department of Mathematics
Massachusetts Institute of Technology
Cambridge, MA, USA

John Voight
Department of Mathematics
Dartmouth College
Hanover, NH, USA

ISSN 2365-9564 ISSN 2365-9572 (electronic)
Simons Symposia
ISBN 978-3-030-80913-3 ISBN 978-3-030-80914-0 (eBook)
https://doi.org/10.1007/978-3-030-80914-0

Mathematics Subject Classification: 11G40, 11G10, 11G30, 14H40, 11G05

This Springer imprint is published by the registered company Springer Nature Switzerland AG
The registered company address is: Gewerbestrasse 11, 6330 Cham, Switzerland

Foreword

In fall 2017, the collaboration "Arithmetic Geometry, Number Theory, and Computation" was created through awards from the Simons Foundation to our institutions. We are grateful for the support of Jim Simons, Yuri Tschinkel, and other leaders at the Foundation that made this possible.

Our common perspective is that advances in computational techniques accelerate research in arithmetic geometry and number theory, both as a source of data and examples and as an impetus for effective results. The dynamic interplay between experiment, theory, and computation has historically played a pivotal role in the development of number theory. In the eighteenth and nineteenth centuries, Euler and Gauss undertook extensive calculations by hand in the pursuit of data to help formulate and refine conjectures and to produce a source of counterexamples. In the twentieth century, systematic computations of elliptic curves and their L-functions led to the formulation of the Sato–Tate and modularity conjectures, both of which have now been proved, and the conjecture of Birch and Swinnerton-Dyer, which remains open but has been proved in some special cases.

In the twenty-first century, the frontier of research in arithmetic geometry has moved on to curves of higher genus, abelian varieties, and K3 surfaces. Although available computational resources have grown dramatically, the development and implementation of practical algorithms have lagged behind the theory; we seek to correct this imbalance. In contrast to the situation with elliptic curves, in higher dimensions, brute force computation yields very little. To obtain practical algorithms, one must exploit the theoretical infrastructure of modern arithmetic geometry.

We pursue these goals by recruiting graduate students and research scientists with expertise in both computation and fundamental mathematics. Members of our group prove theorems, develop algorithms, explore guiding examples, collect data on arithmetic objects, and support scholarly resources like the *L-functions and Modular Forms Database (LMFDB)* http://www.lmfdb.org.

This volume offers a sampling of our work over the last three years. In addition to the six of us, many authors were directly supported by the collaboration: Eran Assaf, Alex Best, Raymond van Bommel, Edgar Costa, Alex Cowan, Lassina

Dembélé, Maarten Derickx, Sachi Hashimoto, Dohyeong Kim, David Lowry-Duda, Benjamin Matschke, David Roe, and Ciaran Schembri. We are also grateful for the contributions of senior affiliated scientists Andrew Booker, John Cremona, and Kiran Kedlaya.

A number of manuscripts address fundamental questions informed by computation or inspired by the need for effective algorithms. Cowan's contribution "Conjecture: 100% of Elliptic Surfaces over $\mathbb{Q}$ Have Rank Zero" builds on previous work of Nagao and Rosen–Silverman about quantitative measures of growth of ranks of fibers of elliptic surfaces. "Congruent Number Triangles with the Same Hypotenuse" by Lowry-Duda explores coincidences among solutions to the congruent number problem; motivated by observations on Dirichlet series—and encouraged by numerical experiments—the author uncovers some interesting Diophantine geometry. Poonen offers a new proof of Siegel's 1926 theorem on finiteness of integral points in "The S-Integral Points on the Projective Line Minus Three Points via étale Covers and Skolem's Method." His work was inspired by, but is different from, M. Kim's proof via a nonabelian analogue of the Skolem–Chabauty method.

Several manuscripts focus on effective approaches to S-unit equations. "A Robust Implementation for Solving the S-Unit Equation and Several Applications"—by Alvarado, Koutsianas, Malmskog, Rasmussen, Vincent, and West—presents their new algorithms for these problems. Best and Matschke use a reduction to Mordell equations to enumerate "Elliptic Curves with Good Reduction Outside of the First Six Primes."

The development of databases of low genus curves has demanded improved techniques for computing their invariants. "Efficient Computation of BSD Invariants in Genus 2" by van Bommel discusses quantities associated with the Birch and Swinnerton-Dyer conjecture. Picard groups and other invariants of higher-dimensional varieties are also of interest. A p-adic variational approach is developed in "Effective Obstructions to Lifting Tate Classes from Positive Characteristic" by Costa and E. Sertöz.

The work of our collaboration tends to focus attention on examples with unusual properties. Fontaine and Abrashkin showed there are no nonzero abelian varieties over $\mathbb{Q}$ with good reduction everywhere, and we can enumerate elliptic curves with just a few small primes of bad reduction. Mascot, Sijsling, and Voight offer a nice example in this vein in "A Prym Variety with Everywhere Good Reduction over $\mathbb{Q}(\sqrt{61})$." The manuscript "On Rational Bianchi Newforms and Abelian Surfaces with Quaternionic Multiplication"—by Cremona, Dembélé, Pacetti, Schembri, and Voight—presents beautiful modular forms described in the title and the associated very special abelian surfaces.

Modular forms are a major focus of the collaboration, given their historical importance for number theory and across mathematics. Assaf focuses on algorithms for nonstandard levels in "Computing Classical Modular Forms for Arbitrary Congruence Subgroups." We offer a comprehensive survey of computations of modular forms by Best, Bober, Booker, Costa, Cremona, Derickx, Lowry-Duda, Lee, Roe, Sutherland, and Voight. D. Kim lays the groundwork for new algorithms in his contribution "Linear Dependence Among Hecke Eigenvalues." "A Database of Hilbert Modular Forms" by Donnelly and Voight presents what is available on Hilbert modular forms in the LMFDB.

One of the most important tools for effective computation of rational points on curves is the Chabauty–Coleman method, the prototype for nonabelian extensions developed over the last decade. We are especially interested in implementing these techniques for large numbers of examples. Contributions in this direction include "Curves with Sharp Chabauty–Coleman Bound" by Gajović, "Square Root Time Coleman Integration on Superelliptic Curves" by Best, "Computing Rational Points on Genus 3 Hyperelliptic Curves" by de Frutos Fernández and Hashimoto, and "Chabauty–Coleman Computations on Rank 1 Picard Curves" by Hashimoto and Morrison.

Classification of algebraic varieties over finite fields and their Weil polynomials is an emerging focus of number-theoretic databases. "Restrictions on Weil Polynomials of Jacobians of Hyperelliptic Curves"—by Costa, Donepudi, Fernando, Karemaker, Springer, and West—develops results that emerged from experiments, building on theorems in genus two by Howe, Nart, and Ritzenthaler. Dupuy, Kedlaya, Roe, and Vincent present recent progress on building comprehensive databases in "Isogeny Classes of Abelian Varieties over Finite Fields in the LMFDB."

Finally, we include two manuscripts directly documenting recent innovations to the LMFDB. "Zen and the Art of Database Maintenance" by Costa and Roe describe their efforts to modernize the internal architecture—how objects are represented in the system. This work was crucial to recent expansions to the LMFDB and also formed the foundation for the new online directory of research talks and conferences http://researchseminars.org supported by the collaboration. "Visualizing Modular Forms" by Lowry-Duda is a product of serendipity with the program *Illustrating Mathematics* held at the Institute for Computational and Experimental Research in Mathematics (ICERM) in Fall 2019. The "badges" used to represent modular forms on the LMFDB have been updated based on these ideas. We expect these will lead to new insights linking modular forms to geometric and probabilistic processes.

Acknowledgments Balakrishnan is supported by NSF grant DMS-1702196, the Clare Boothe Luce Professorship (Henry Luce Foundation), Simons Foundation grant #550023, and a Sloan Research Fellowship. Elkies is supported by NSF grant DMS-1502161 and Simons Foundation grant #550031. Hassett is supported by NSF grant DMS-1701659 and Simons Foundation grant #546235. Poonen is supported by NSF grant DMS-1601946 and Simons Foundation grants #402472 and #550033. Sutherland also is supported by Simons Foundation grant #550033. Voight is supported by Simons Foundation grant #550029.

Boston University — Jennifer S. Balakrishnan
Harvard University — Noam Elkies
Brown University/ICERM — Brendan Hassett
MIT — Bjorn Poonen
MIT — Andrew V. Sutherland
Dartmouth College — John Voight
December 28, 2020

Contents

A Robust Implementation for Solving the S-Unit Equation and Several Applications

Alejandra Alvarado, Angelos Koutsianas, Beth Malmskog, Christopher Rasmussen, Christelle Vincent, and Mckenzie West

Abstract Let K be a number field, and S a finite set of places in K containing all infinite places. We present an implementation for solving the S-unit equation $x + y = 1$, $x, y \in \mathcal{O}_{K,S}^{\times}$ in the computer algebra package SageMath. This paper outlines the mathematical basis for the implementation. We discuss and reference the results of extensive computations, including exponent bounds for solutions in many fields of small degree for small sets S. As an application, we prove an asymptotic version of Fermat's Last Theorem for totally real cubic number fields with bounded discriminant where 2 is totally ramified. In addition, we use the implementation to find all solutions to some cubic Ramanujan-Nagell equations.

A. Alvarado
Department of Mathematics and Computer Science, Eastern Illinois University, Charleston, IL, USA
e-mail: aalvarado2@eiu.edu

A. Koutsianas
Department of Mathematics, University of British Columbia, Vancouver, BC, Canada
e-mail: akoutsianas@math.ubc.ca

B. Malmskog
Department of Mathematics and Computer Science, Colorado College, Colorado Springs, CO, USA

C. Rasmussen (✉)
Department of Mathematics and Computer Science, Wesleyan University, Middletown, CT, USA
e-mail: crasmussen@wesleyan.edu

C. Vincent
Department of Mathematics and Statistics, University of Vermont, Burlington, VT, USA
e-mail: christelle.vincent@uvm.edu

M. West
Department of Mathematics, University of Wisconsin-Eau Claire, Eau Claire, WI, USA
e-mail: westmr@uwec.edu

J. S. Balakrishnan et al. (eds.), *Arithmetic Geometry, Number Theory, and Computation*, Simons Symposia, https://doi.org/10.1007/978-3-030-80914-0_1

1 Introduction

In 1909, Thue proved there are only finitely many integral solutions to what we now call the Thue equation; i.e, that for any $\mathbb{Q}$-irreducible binary form $F(X, Y)$ of degree at least 3, defined over the integers, there are only finitely many solutions $(x, y) \in \mathbb{Z}^2$ to the equation

$$F(x, y) = c,$$

where c is any non-zero integer [38]. Thue accomplished this by formally factoring F into linear terms of the form $(x - \alpha y)$, where α is algebraic, then bounding the quality of rational approximations of α in terms of the size of x and y. Thus bounds on integer solutions to the Thue equation arose out of the theory of approximating algebraic numbers by rationals. Thue's theorem was generalized by Siegel [33][1] and then Mahler [25]. These generalizations gave rise to a central fact of modern computational number theory: if K is a number field, and S a finite list of places of K including all infinite places, then there are only finitely many solutions (x, y) to the equation

$$x + y = 1, \qquad x, y \in \mathcal{O}_{K,S}^{\times}. \tag{1}$$

Here, $\mathcal{O}_{K,S}^{\times}$ is the unit group of the ring $\mathcal{O}_{K,S}$ of S-integers in K. We refer to (1) as the *S-unit equation*. In this paper, we describe an algorithm to determine the complete set of solutions to the S-unit equation for general K and S. More generally, for fixed $a, b \in \mathcal{O}_{K,S}$, we can see that the equation $ax + by = 1$ will also have only finitely many solutions by expanding the set S to include all primes dividing a and b and searching for solutions to (1). Thus it suffices to solve (1) to address the more general case, and we focus on (1) here (though it should be remarked that this is not the most efficient way to solve $ax + by = 1$).

The work of Gelfond and Schneider, resolving Hilbert's seventh problem in the affirmative (all irrational algebraic powers of algebraic numbers are transcendental once trivial cases are ignored), determined lower bounds on the absolute value of a $\mathbb{Q}$-linear combination of two $\mathbb{Q}$-linearly independent logarithms of algebraic numbers. Alan Baker's 1967 theorem [1] generalized these results to the case of many logarithms. Baker, Wüstholz, and many others continued to improve these bounds. Naturally, one should ask if similar results are available over local fields, and indeed such results began to appear quickly. In 1968, Brumer proved the first analogue of Baker's work for p-adic logarithms [8], followed by many improvements and generalizations, such as the results of Yu [44, 45, 46]. Improvements in both the archimedean and nonarchimedean cases continue to appear, such as in [20, 4, 47, 19].

[1] See also the recent translation [17] by Fuchs.

For any choice of K and S, $\mathcal{O}_{K,S}^\times$ is a finitely generated $\mathbb{Z}$-module. Fixing a basis $\rho_1, \ldots, \rho_t$ for the torsion free part, we can express any $x \in \mathcal{O}_{K,S}^\times$ as $x = \xi \cdot \prod_{i=1}^t \rho_i^{a_i}$ for some root of unity $\xi \in K$ and some $a_i \in \mathbb{Z}$. Building on the lower bounds for linear combinations of logarithms, Győry [18] determined effectively computable bounds for the exponents a_i. This was a great victory for computational number theory, as this provably restricted all solutions to (1) to a finite search space. Unfortunately, the demonstrated bounds were enormous and as a matter of practice, it was computationally infeasible to conduct an exhaustive search for solutions, even in the very simplest cases. Baker and Davenport devised a clever method of reducing the bounds in special cases in [2]. However, in [14], de Weger built on the ideas of Baker-Davenport to develop a powerful general method of algorithmically reducing the bounds to a manageable size, relying on the lattice basis reduction algorithm of Lenstra, Lenstra, and Lovász [24] (henceforth referred to as the "LLL algorithm"). Though it is has not been proven that de Weger's method will always reduce the bounds coming from the results in linear forms of logarithms, this is the rule in practice. In many cases, de Weger's approach provides sufficient improvements that, with careful sieving (or sometimes even with only brute force), the entire search space can be exhausted and complete lists of solutions can be enumerated.

Beyond the improvements provided by LLL-based reduction, many mathematicians have developed further algorithms for efficiently searching below the "LLL bounds" provided by de Weger's work. Two powerful examples are reported in [43] and [37]. Increasingly, the theoretical improvements (assisted by technological improvements) have pushed ambitious and interesting computational problems within reach. For example, Smart determined the entire set of all genus 2 curves over $\mathbb{Q}$ with good reduction away from 2, based in part on solving (1) for a family of number fields unramified away from 2 [35].

We have written a package of Python functions for inclusion in the computer algebra system SageMath [31], which solves the S-unit equation (1) over any number field K and for any finite set S of finite places. As experienced readers may expect, the package is not practical when either $[K : \mathbb{Q}]$ or $|S|$ is too large, although there is no theoretical obstruction. While this package is the independent creation of the authors, it is based in part on the descriptions of algorithms implemented by Smart [34, 35, 36]. Specifically, we follow Smart's development in determining initial large bounds, including the numbering of constants, in [34], with some adjustments and small corrections. In reducing the bounds, we follow [36], again with some adjustments. The sieving step is based on ideas cited by Smart [35] as due to others (as noted in Sect. 6) but has been redeveloped in new notation and style. We include proofs of our versions of results when we made adjustments to versions in the literature. To the authors' knowledge, our package is the first publicly available implementation for solving the S-unit equation over any field other than $\mathbb{Q}$; the present article describes the algorithm and its implementation. The implementation was a highly non-trivial undertaking, involving efforts spreading over more than 7 years on the parts of individuals and the entire team.

We also provide new results facilitated by our implementation. In particular, we first provide a discussion of and link to explicit exponent bounds for solutions of the S-unit equation in all cases (K, S) where $K/\mathbb{Q}$ is ramified only at primes above some subset of $\{2, 3\}$ and

$$[K : \mathbb{Q}] \leq 5, \qquad S \subseteq \{\mathfrak{p} \subseteq \mathcal{O}_K : \mathfrak{p} \mid 6\}.$$

We solve the S-unit equation in the 13 totally real cubic number fields K in which 2 is totally ramified and the absolute discriminant of K, Δ_K, satisfies $|\Delta_K| \leq 2000$, and we use these results to verify that an asymptotic version of Fermat's Last Theorem holds over these fields. Finally, we find all solutions to certain cubic Ramanujan-Nagell equations.

1.1 Overview

The organization of the paper proceeds as follows. We introduce certain notations in Sect. 2. In Sect. 3, we review the relevant work of Baker-Wüstholz and Yu. This is used in Sect. 4 to establish a "pre-LLL" exponent bound for each place in S. In Sect. 5, we explain the process of using LLL to reduce these exponent bounds—the approach is different for archimedean and nonarchimedean places. In Sect. 6, we describe the sieve for further constraining the final search space. We devote Sect. 7 to a discussion of our experimental observations, having now executed our algorithm in several dozen cases. We highlight a special condition (S contains only one finite place) under which a significant improvement in the search space can be obtained. Although narrow in scope, the special condition is sufficiently natural, and the savings sufficiently nontrivial, as to warrant its discussion. Finally, Sect. 8 introduces two applications: an asymptotic version of Fermat's Last Theorem over totally real cubic fields and a solution to a cubic variant of the Ramanujan-Nagell equation.

Acknowledgments We are delighted to recognize the Institute for Computational and Experimental Research in Mathematics for both funding and hosting a 2017 collaboration during which a great deal of this project was completed. Part of this work began at the 2014 workshop SageDays 62, and we would like to thank Anna Haensch and Lola Thompson for organizing that workshop and Microsoft Research and The Beatrice Yormark Fund for Women in Mathematics for funding. Some of the work was supported by the van Vleck fund at Wesleyan University. The authors would like to thank many people for helpful conversations that led to improvements in the code and gave direction to this project, including Bjorn Poonen, Andrew Sutherland, and Norman Danner. We would also like especially to thank David Roe for his contributions to refining and reviewing the code for inclusion in SageMath.

The third author was partially supported in this work by NSA Grant #H98230-16-1-0300. We are very grateful to the anonymous referees for their careful reading of this work and their many helpful comments which have improved the quality of this paper.

2 Notation

2.1 S-Units in Number Fields

Throughout this paper, we let $\bar{\mathbb{Q}}$ denote the algebraic closure of $\mathbb{Q}$ inside $\mathbb{C}$, the field of complex numbers. Unless stated otherwise, we fix the following notation throughout:

K	a number field (assumed to be a subfield of $\bar{\mathbb{Q}}$),
d_K	the absolute degree $[K : \mathbb{Q}]$
w	the number of distinct roots of unity in K
Δ_K	the absolute discriminant of $K/\mathbb{Q}$
$\mathcal{O}_K$	the ring of integers of K
$e_{\mathfrak{p}}$	the ramification index of $\mathfrak{p}$ in $K/\mathbb{Q}$
$f_{\mathfrak{p}}$	the inertial degree of the prime $\mathfrak{p} \subseteq \mathcal{O}_K$ over the rational prime $\mathfrak{p} \cap \mathbb{Z}$
r	the rank of $\mathcal{O}_K^\times$ as a $\mathbb{Z}$-module
S_{fin}	a set $\{\mathfrak{p}_1, \ldots, \mathfrak{p}_s\}$ of s finite places of K
S_∞	the set $\{\mathfrak{p}_{s+1}, \ldots, \mathfrak{p}_{r+s+1}\}$ of all infinite places of K
S	$S_{\text{fin}} \cup S_\infty = \{\mathfrak{p}_1, \ldots, \mathfrak{p}_{r+s+1}\}$
$S_{\mathbb{Q}}$	the set of places of $\mathbb{Q}$ which extend to places of K in S
$\mathcal{O}_{K,S}$	the ring of S-integers in K
$\mathcal{O}_{K,S}^\times$	the group of S-units in K
t	the rank of $\mathcal{O}_{K,S}^\times$ as a $\mathbb{Z}$-module (so $t = r + s$)
ρ_0	a root of unity generating the torsion part of $\mathcal{O}_{K,S}^\times$
$\rho_1, \ldots, \rho_t$	an ordered basis for the torsion-free part of the $\mathbb{Z}$-module $\mathcal{O}_{K,S}^\times$
$\boldsymbol{\rho}$	the ordered list $[\rho_0, \rho_1, \ldots, \rho_t]$

If $f(x) \in \mathbb{Z}[x]$ is a monic and irreducible polynomial, we let K_f denote the number field $\mathbb{Q}(\xi)$, where ξ is a root of $f(x)$. Always, log denotes the principal branch of the complex logarithm function, with argument in $(-\pi, \pi]$.

2.2 Absolute Values and Completions

Each place of K determines an associated absolute value, $|\cdot|_{\mathfrak{p}}$, which we now describe.

Let $|\cdot|$ denote the usual absolute value on $\mathbb{C}$. If $\mathfrak{p}$ is an infinite place, choose $\sigma_{\mathfrak{p}}\colon K \to \mathbb{C}$, an embedding corresponding to $\mathfrak{p}$. The associated absolute value depends on whether $\mathfrak{p}$ is a real or complex (meaning non-real) place of K:

$$|\alpha|_{\mathfrak{p}} := \begin{cases} |\sigma_{\mathfrak{p}}(\alpha)| & \mathfrak{p} \text{ is real,} \\ |\sigma_{\mathfrak{p}}(\alpha)^2| & \mathfrak{p} \text{ is complex.} \end{cases}$$

Now suppose $\mathfrak{p}$ is a finite place. View $\mathfrak{p}$ as a prime ideal of $\mathcal{O}_K$, and let p be the characteristic of the residue field $\mathcal{O}_K/\mathfrak{p}$. Let $\mathrm{ord}_{\mathfrak{p}}$ denote the ordinal function for $\mathfrak{p}$. On $\mathcal{O}_K^\times$ this is defined by

$$\mathrm{ord}_{\mathfrak{p}}(\beta) = m \quad \text{if } \beta \in \mathfrak{p}^m - \mathfrak{p}^{m+1},$$

and it extends to $K^\times$ in the obvious way. We let $|\cdot|_p$ denote the usual absolute value of the p-adic field $\mathbb{Q}_p$. The absolute value associated to $\mathfrak{p}$ on K is

$$|\alpha|_{\mathfrak{p}} := p^{-f_{\mathfrak{p}} \,\mathrm{ord}_{\mathfrak{p}}(\alpha)}.$$

Let $K_{\mathfrak{p}}$ be the $\mathfrak{p}$-adic completion of K with respect to $|\cdot|_{\mathfrak{p}}$; we also use $|\cdot|_{\mathfrak{p}}$ for the absolute value on $K_{\mathfrak{p}}$.

We fix once and for all an algebraic closure $\bar{\mathbb{Q}}_p$ of $\mathbb{Q}_p$, and let $\mathbb{C}_p$ denote the completion of $\bar{\mathbb{Q}}_p$. We use $|\cdot|_p$ to denote the natural extension of $|\cdot|_p$ to all of $\mathbb{C}_p$. We define ord_p on $\mathbb{C}_p^\times$ to satisfy

$$|\alpha|_p = p^{-\mathrm{ord}_p \alpha}, \qquad \alpha \in \mathbb{C}_p^\times.$$

As $p\mathcal{O}_K$ may split into several prime ideals, the absolute value $|\cdot|_p$ on $\mathbb{Q}$ may have several inequivalent extensions to K, of which $|\cdot|_{\mathfrak{p}}$ is just one; so we must take care when viewing $K_{\mathfrak{p}}$ as a subfield of $\bar{\mathbb{Q}}_p$.

For any embedding $\vartheta\colon K \to \bar{\mathbb{Q}}_p$, we obtain a subfield of $\bar{\mathbb{Q}}_p$ as the composite $\vartheta(K)\cdot\mathbb{Q}_p$. By the Prolongation Theorem [21, §18.5], there exists a choice of ϑ such that $(K_{\mathfrak{p}}, |\cdot|_{\mathfrak{p}})$ is value-isomorphic to $(\vartheta(K)\mathbb{Q}_p, |\cdot|_p)$. Henceforth, we always use this isomorphism to view $K_{\mathfrak{p}}$ as a subfield of $\bar{\mathbb{Q}}_p$. As the isomorphism respects the valuations, we know ord_p and $\mathrm{ord}_{\mathfrak{p}}$ satisfy

$$\mathrm{ord}_{\mathfrak{p}} \beta = e_{\mathfrak{p}} \,\mathrm{ord}_p \beta, \qquad \beta \in K_{\mathfrak{p}}. \tag{2}$$

2.3 Height Functions

Suppose $n \geq 1$. We let h denote the standard logarithmic Weil height on $\mathbb{P}^n(K)$. This is defined as follows: for any $\mathbf{x} = (x_0 : \cdots : x_n) \in \mathbb{P}^n(K)$,

$$h(\mathbf{x}) = \frac{1}{d_K} \sum_{\mathfrak{p}} \log\left(\max_j \{|x_j|_{\mathfrak{p}}\}\right),$$

where the sum runs over all places of K. It is a consequence of the product formula [21, Ch. 20, pgs. 326–327] that $h(\mathbf{x})$ is independent of the choice of coordinates for $\mathbf{x}$. For any $\alpha \in K$, set $h(\alpha) = h((1 : \alpha))$. Note that this height is *absolute* in the sense that it is not dependent on which field extension K containing the coordinates of $\mathbf{x}$ is considered.

We introduce a modified version of this height function, used in Sect. 3. Suppose $\alpha_1, \ldots, \alpha_n \in K$, and let $K' = \mathbb{Q}(\alpha_1, \ldots, \alpha_n) \subseteq K$. For any nonzero element $\beta \in K'$, we define the function h' by

$$h'(\beta) = \frac{1}{d_{K'}} \max\left\{d_{K'} \cdot h(\beta), |\log \beta|, 1\right\}.$$

The definition of another height function, $h_{\mathfrak{p}}$, is slightly more technical and will be introduced when needed in Sect. 3.

2.4 p-Adic Logarithms

Inside $\mathbb{C}_p$, consider the open disk

$$\Delta_1 := \{z \in \mathbb{C}_p : |z - 1|_p < 1\}.$$

On Δ_1, we define the p-adic logarithm by the series

$$\log_p z = -\sum_{n \geq 1} \frac{(1-z)^n}{n}. \tag{3}$$

The series is convergent on Δ_1; moreover, on Δ_1 it satisfies the identity

$$\log_p(xy) = \log_p x + \log_p y. \tag{4}$$

If $|z|_p < p^{-\frac{1}{p-1}}$ we have

$$\operatorname{ord}_p\left(\log_p(1+z)\right) = \operatorname{ord}_p z. \tag{5}$$

Based on an idea due to Iwasawa, the p-adic logarithm can be extended to any $z \in \mathbb{C}_p$ such that $|z|_p = 1$; this extension continues to satisfy (4) (see [36, II.2.4]).

2.5 *Solutions to the S-Unit Equation*

We let $A_{K,S}$ denote the additive $\mathbb{Z}$-module $(\mathbb{Z}/w\mathbb{Z}) \times \mathbb{Z}^t$. This is isomorphic to $\mathcal{O}_{K,S}^\times$, and the list of generators $\boldsymbol{\rho}$ determines an isomorphism

$$\Phi_{\boldsymbol{\rho}} \colon A_{K,S} \longrightarrow \mathcal{O}_{K,S}^\times, \qquad \mathbf{a} := (a_0, a_1, \ldots, a_t) \mapsto \prod_{i=0}^{t} \rho_i^{a_i}.$$

We use the shorthand $\boldsymbol{\rho}^{\mathbf{a}} := \Phi_{\boldsymbol{\rho}}(\mathbf{a})$. For obvious reasons, we call the elements of $A_{K,S}$ *exponent vectors*. Much of our discussion will focus on bounds for the entries of an exponent vector. For $\mathbf{a} \in A_{K,S}$, we use the notation $|\mathbf{a}| \leq B$ to signify

$$\max_{0<i\leq t} |a_i| \leq B.$$

Within $\mathcal{O}_{K,S}^\times$, we wish to determine

$$X_{K,S} := \{\tau \in \mathcal{O}_{K,S}^\times : 1 - \tau \in \mathcal{O}_{K,S}^\times\}.$$

Solving the S-unit equation is equivalent to determining the set $X_{K,S}$. We let $E_{K,S}$ denote the corresponding subset $\Phi_{\boldsymbol{\rho}}^{-1}(X_{K,S})$ of $A_{K,S}$.

3 The Bounds of Baker-Wüstholz and Yu

Suppose $\tau_1, \tau_2 \in \mathcal{O}_{K,S}^\times$ provide a solution to the S-unit equation, so that $\tau_1 + \tau_2 = 1$. With respect to the ordered generating set $\boldsymbol{\rho}$, there are unique vectors $\mathbf{b}_i = (b_{i,0}, \ldots, b_{i,t}) \in A_{K,S}$ such that

$$\tau_i = \boldsymbol{\rho}^{\mathbf{b}_i} = \prod_{j=0}^{t} \rho_j^{b_{i,j}}, \qquad i = 1, 2. \tag{6}$$

The techniques of lattice reduction discussed in Sect. 5 will not produce an absolute bound for $|b_{i,j}|$ on their own; they can only be used to improve a known bound. So in this section, we recall bounds established by Baker-Wüstholz [3] and Kunrui Yu [47]. An excellent treatment of the background material appears in [15].

3.1 Statement of Yu's Bound

Let $\mathfrak{p}$ be a finite place of K, and let p denote the rational prime below $\mathfrak{p}$. We let q be the smallest rational prime distinct from p (so $q = 2$ unless $p = 2$, in which case $q = 3$). Let $\zeta_m := \exp(2\pi i/m)$. We say K *satisfies Yu's auxiliary condition* if any of the following hold:

1. $q = 2$ and $p^{f_\mathfrak{p}} \equiv 1 \bmod 4$,
2. $q = 2$ and $\zeta_4 \in K$,
3. $q = 3$ and $\zeta_3 \in K$.

At the end of this section, we explain how the algorithm finds a bound in cases where K does not satisfy Yu's auxiliary condition.

Theorem 3.1 (Yu, [47, pg. 190]) *Suppose $n \geq 1$ and $\mathfrak{p}$ is a prime of $\mathcal{O}_K$. Suppose K is a number field satisfying Yu's auxiliary condition and $\mu_0, \mu_1, \ldots, \mu_{n-1} \in K^\times$ are chosen which satisfy*

$$\operatorname{ord}_\mathfrak{p} \mu_j = 0, \qquad 0 \leq j \leq n-1. \tag{7}$$

Suppose $b_j \in \mathbb{Z}$ and $\Theta := \prod_{j=0}^{n-1} \mu_j^{b_j} \neq 1$. Finally, suppose B satisfies

$$B \geq \max\{|b_0|, \ldots, |b_{n-1}|, 3\}.$$

Then there exist explicit constants C_1^* and Ω, given below, such that

$$\operatorname{ord}_\mathfrak{p} (\Theta - 1) < C_1^* \Omega \log B.$$

3.2 The Constants Ω and Ω'

We first discuss the constant Ω, and the variant Ω' used in the algorithm. In Theorem 3.1, Ω is roughly a product of the logarithmic heights of the μ_j. More precisely, decompose the set $\{\mu_j\}_j$ into a disjoint union $\mathfrak{a} \cup \mathfrak{b}$, where $\mathfrak{a}$ is a maximal subset of $\{\mu_j\}_j$ which is multiplicatively independent. Such a decomposition need not be unique. Because of the possible dependence among the μ_j, Yu requires a modified height function:

$$h_\mathfrak{p}(\mu) := \max\left\{h(\mu), \frac{f_\mathfrak{p}}{\kappa_1(n+4)d_K}\right\}.$$

(The value κ_1 is explained in the following subsection.) The constant Ω, which depends on n, d_K, $\mathfrak{p}$ as well as the μ_j, is then defined by

$$\Omega(n, d_K, \mathfrak{p}) := \prod_{\mu \in \mathfrak{a}} h(\mu) \cdot \prod_{\mu \in \mathfrak{b}} h_{\mathfrak{p}}(\mu).$$

As shown in [47], one may choose any maximal independent set $\mathfrak{a}$ for the computation of Ω. If optimization of the bound is critical, one may search over all possible $\mathfrak{a}$ and take the smallest possible bound. This observation is moot in our use, however.

Corollary 3.2 *Keeping the hypotheses of the previous theorem, suppose also that* $\mu_1, \ldots, \mu_{n-1}$ *are multiplicatively independent. Set*

$$\Omega'(n, d_K, \mathfrak{p}) := h_{\mathfrak{p}}(\mu_0) \prod_{j=1}^{n-1} h(\mu_j).$$

Then $\operatorname{ord}_{\mathfrak{p}}(\Theta - 1) < C_1^* \Omega' \log B$.

Proof In this case, $\mathfrak{b}$ is unique; either $\mathfrak{b} = \{\mu_0\}$ or $\mathfrak{b} = \varnothing$. In either case, $\Omega \leq \Omega'$ and the result follows immediately. □

In the algorithm, we are always in the situation of the Corollary. Rather than decide the question of independence between μ_0 and the other μ_j, we just use the constant Ω'.

3.3 The Constant C_1^*

The value of $C_1^* := C_1^*(n, d_K, \mathfrak{p})$ is dependent on n, d_K, and $\mathfrak{p}$, as follows. Let $u := \operatorname{ord}_q w$, so that q^u is the q-part of w. Set

$$k_2 := c^{(1)} a^{(1)} \cdot \frac{n^n \cdot (n+1)^{n+1}}{n!},$$

$$k_3 := \frac{p^{f_{\mathfrak{p}}}}{q^u} \left(\frac{d_K}{f_{\mathfrak{p}} \log p} \right)^{n+2} \cdot \log \max\{d_K, e\},$$

$$k_4 := \max\left\{ \log\left(e^4 (n+1) d_K\right), e_{\mathfrak{p}}, f_{\mathfrak{p}} \log p \right\}.$$

Here, e denotes the base of the natural logarithm. The constants $a^{(1)}$, κ_1, and $c^{(1)}$ are given in Tables 1 and 2. Finally,

$$C_1^*(n, d_K, \mathfrak{p}) := (n+1) k_2 k_3 k_4. \tag{8}$$

Table 1 The constants $a^{(1)}$ and κ_1

Case	$a^{(1)}$	κ_1
$p = 2$	32	40
$p = 3$	16	20
$p > 3$ and $e_\mathfrak{p} \geq 2$	16	20
$p > 3$ and $e_\mathfrak{p} = 1$	$\frac{8(p-1)}{p-2}$	10

Table 2 The constant $c^{(1)}$

$p \leq 5$		$p > 5$	
Case	$c^{(1)}$	Case	$c^{(1)}$
$p = 2$	160	$p \equiv 1$ (4) and $e_\mathfrak{p} = 1$	1473
$p = 3$ and $d_K = 1$	537	$p \equiv 1$ (4) and $e_\mathfrak{p} \geq 2$	1502
$p = 3$ and $d_K \geq 2$	759	$p \equiv 3$ (4), $e_\mathfrak{p} = 1, d_K = 1$	1288
$p = 5$ and $e_\mathfrak{p} = 1$	1473	$p \equiv 3$ (4), $e_\mathfrak{p} = 1, d_K \geq 2$	1282
$p = 5$ and $e_\mathfrak{p} \geq 2$	319	$p \equiv 3$ (4), $e_\mathfrak{p} \geq 2$	2190

3.4 A Remark About Implementation

For this subsection only, suppose all hypotheses in Theorem 3.1 are satisfied, except K does **not** satisfy Yu's auxiliary condition. Set

$$K' := \begin{cases} K(\zeta_4) & q = 2, \\ K(\zeta_3) & q = 3. \end{cases}$$

Let $\mathfrak{P}$ be a prime of $\mathcal{O}_{K'}$ above $\mathfrak{p}$. Let $e_{\mathfrak{P}|\mathfrak{p}}$ be the ramification index of $\mathfrak{P}$ over $\mathfrak{p}$. Because

$$\operatorname{ord}_\mathfrak{p} \alpha = e_{\mathfrak{P}|\mathfrak{p}} \operatorname{ord}_\mathfrak{P} \alpha, \qquad \alpha \in K^\times,$$

we see $\operatorname{ord}_\mathfrak{P} \mu_j = 0$ for all j. Now Theorem 3.1 applies with K' and $\mathfrak{P}$ in place of K and $\mathfrak{p}$, respectively.

Corollary 3.3 *Under the conditions of this subsection,*

$$\operatorname{ord}_\mathfrak{p} (\Theta - 1) < e_{\mathfrak{P}|\mathfrak{p}} \cdot C_1^*(n, d_{K'}, \mathfrak{P}) \cdot \Omega'(n, d_{K'}, \mathfrak{P}) \cdot \log B.$$

Note that even if $\mathfrak{p}$ splits as $\mathfrak{P}\mathfrak{P}'$ in K', the choice of $\mathfrak{P}$ is irrelevant; both give the exact same bound in the Corollary.

3.5 Bound of Baker-Wüstholz

We now give an effective version of Baker's theorem. (Notations are as in Sects. 2.2 and 2.3.)

Theorem 3.4 (Baker-Wüstholz, [3, pg. 20]) *Let L be a linear form in $t+1$ indeterminates,*

$$L(z_0, \ldots, z_t) = b_0 z_0 + \cdots + b_t z_t, \qquad b_i \in \mathbb{Z}.$$

Let $B = \max\{|b_0|, \ldots, |b_t|\}$, and let $\rho_0, \ldots, \rho_t \in \overline{\mathbb{Q}} - \{0, 1\}$. Let K' be the subfield of $\overline{\mathbb{Q}}$ generated by the ρ_i. If $B > 3$ and

$$\Lambda = L(\log \rho_0, \log \rho_1, \ldots, \log \rho_t) \neq 0,$$

then

$$\log |\Lambda| > -C(t, d_{K'}) \log(B) \prod_{j=0}^{t} h'(\rho_j),$$

where the constant $C(t, d_{K'})$ is defined by

$$C(t, d_{K'}) = 18(t+2)!(t+1)^{(t+2)}(32 d_{K'})^{(t+3)} \log\left(2(t+1)d_{K'}\right).$$

Note that we may be sure $\Lambda \neq 0$ if the set $\{\log \rho_i\}$ is linearly independent over $\mathbb{Q}$.

3.6 Obtaining the Initial Bound

The theorems of Baker-Wüstholz and Yu both provide inequalities of the form "a polynomial function of B is bounded by a polynomial function of $\log(B)$," which in turn guarantee an absolute bound on B. The analysis to determine such a bound explicitly is standard; we will use the following result of Pethő and de Weger for this purpose.

Lemma 3.5 (Pethő and de Weger [29, Lemma 2.2]) *Suppose the real numbers a, b, h satisfy $a \geq 0$, $h \geq 1$, $b > \left(\frac{e^2}{h}\right)^h$, and let $x \in \mathbb{R}$ be the largest solution to the equation*

$$x = a + b(\log x)^h.$$

Then

$$x < 2^h \left(a^{\frac{1}{h}} + b^{\frac{1}{h}} \log\left(h^h b\right)\right)^h.$$

4 Initial Exponent Bounds

4.1 An Upper Bound at the Extremal Place

Suppose (τ_1, τ_2) is a solution to the S-unit equation, with τ_i specified as in (6). We set $B = \max_{i,j} |b_{i,j}|$, and assume $B \geq \max\{4, w\}$. Relabeling τ_1 and τ_2 if necessary, we assume $B = |b_{1,j}|$ for some $1 \leq j \leq t$. Recall that S contains precisely $t+1$ places, $\mathfrak{p}_1, \ldots, \mathfrak{p}_{t+1}$. We choose the indices $k, \ell \in \{1, 2, \ldots, t+1\}$ so that

$$\left|\log|\tau_1|_{\mathfrak{p}_k}\right| = \max_{\mathfrak{p}\in S} \left|\log|\tau_1|_{\mathfrak{p}}\right|, \qquad |\tau_1|_{\mathfrak{p}_\ell} = \min_{\mathfrak{p}\in S} |\tau_1|_{\mathfrak{p}}.$$

Remark 4.1 In the sequel, we number our constants in an effort to stay consistent with the enumeration given in Smart's paper [34]. There, Smart considers a more general unit equation, and so introduces certain constants $c_4(i)$, $c_6(i)$, $c_7(i), \ldots$ whose values are trivial in the present application. So while the alert reader may notice gaps in the enumeration of constants, this is intentional. (Adjusting our implementation to the more general setting is not difficult, but we are satisfied to limit the discussion to match the current state of the implementation.)

For any choice of $U := \{\mathfrak{u}_1, \ldots, \mathfrak{u}_t\} \subseteq S$ define the $t \times t$ matrix

$$M = (m_{i,j}), \qquad m_{i,j} = \log|\rho_j|_{\mathfrak{u}_i}.$$

One may always choose U so that M is invertible (see [15, §5.1]), and so we assume this is the case. We have

$$\begin{pmatrix} b_{1,1} \\ b_{1,2} \\ \vdots \\ b_{1,t} \end{pmatrix} = M^{-1} \begin{pmatrix} \log|\tau_1|_{\mathfrak{u}_1} \\ \log|\tau_1|_{\mathfrak{u}_2} \\ \vdots \\ \log|\tau_1|_{\mathfrak{u}_t} \end{pmatrix}.$$

Let $\|M\|$ be the row norm of M^{-1}, i.e. $\|M\| = \max_i \sum_{j=1}^{t} |m_{i,j}|$, and set

$$c_1 := \max\left\{1, \max_{U\subseteq S}\{||M|| : M \text{ is invertible }\}\right\}.$$

Note that this differs slightly from Smart's definition, to ensure that $c_1 \geq 1$. Then $B \leq c_1 \left|\log|\tau_1|_{\mathfrak{p}_k}\right|$. We define

$$c_2 := \frac{1}{c_1} \qquad c_3 := \frac{0.9999999c_2}{r+s}.$$

By Smart [34, Lemma 2], we have

$$|\tau_1|_{\mathfrak{p}_\ell} \le e^{-c_3 B}. \tag{9}$$

We now have an upper bound on $|\tau_1|_{\mathfrak{p}_\ell}$ in terms of B. We next establish a lower bound, also involving B, which will force a limit on the size of B. The precise argument depends on whether $\mathfrak{p}_\ell$ is a finite or infinite place. For the purposes of the algorithm, we must compute this bound on B for each possible index $1 \le \ell \le t$; we have no choice but to take the largest possible bound, i.e., the larger of the two values K_0 and K_1 determined in the remainder of this section.

4.2 *Case I: $\mathfrak{p}_\ell$ Is Finite*

If $\mathfrak{p}_\ell$ is finite, then let $\mathfrak{p}_\ell$ also denote the associated prime ideal in $\mathcal{O}_K$. Let p be the prime of $\mathbb{Z}$ lying below $\mathfrak{p}_\ell$, and let e_ℓ and f_ℓ denote the ramification index and inertial degree of $\mathfrak{p}_\ell$ over p, respectively. From (9) we have

$$N_{K/\mathbb{Q}}(\mathfrak{p}_\ell)^{-\operatorname{ord}_{\mathfrak{p}_\ell}(\tau_1)} \le e^{-c_3 B}. \tag{10}$$

Setting

$$c_5(\ell) := \frac{c_3}{e_\ell \log N_{K/\mathbb{Q}}(\mathfrak{p}_\ell)},$$

the inequality (10) yields

$$\operatorname{ord}_{\mathfrak{p}_\ell} \tau_1 \ge \frac{c_3 B}{\log N_{K/\mathbb{Q}}(\mathfrak{p}_\ell)} = e_\ell c_5(\ell) B > 0, \tag{11}$$

and so $\operatorname{ord}_{\mathfrak{p}_\ell} \tau_2 = 0$. We would like to apply Yu's Theorem to $\Theta = \tau_2$, but unfortunately the generators ρ_i may have nonzero order with respect to $\mathfrak{p}_\ell$. So we now replace the ρ_i with a different set of generators, as in [34, pp. 824–825]. First, set $n_i := \operatorname{ord}_{\mathfrak{p}_\ell} \rho_i$. Necessarily, there exist indices i for which $n_i \ne 0$. Choose i_0 so that

$$|n_{i_0}| = \min\{|n_i| : n_i \ne 0\},$$

and now relabel so that $i_0 = t$. For $1 \le i \le t-1$, define

$$\mu_i = \rho_i^{n_t} \rho_t^{-n_i},$$

so that $\operatorname{ord}_{\mathfrak{p}_\ell} \mu_i = 0$. Next, for each i with $1 \le i \le t-1$, choose integers d_i, r_i such that

$$0 \le r_i < |n_t| \qquad \text{and} \qquad b_{2,i} = n_t d_i + r_i.$$

Necessarily, $|d_i| \leq B$. Set $N := \sum_{i=1}^{t-1} n_i r_i$.

Lemma 4.1 *We have* $N \equiv 0 \pmod{n_t}$.

Proof Since $\mathrm{ord}_{\mathfrak{p}_\ell}(\tau_2 \rho_0^{-b_{2,0}}) = 0$, we know $\sum_{i=1}^{t} n_i b_{2,i} = 0$. Thus,

$$n_t b_{2,t} = -\sum_{i=1}^{t-1} n_i b_{2,i} = -n_t \sum_{i=1}^{t-1} n_i d_i - \sum_{i=1}^{t-1} n_i r_i,$$

proving the claim. □

Setting $N_0 = \frac{N}{n_t}$ and $\mu_0 := \rho_0^{b_{2,0}} \rho_t^{-N_0} \cdot \prod_{i=1}^{t-1} \rho_i^{r_i}$, we have arranged that

$$\tau_2 = \mu_0 \prod_{i=1}^{t-1} \mu_i^{d_i}, \quad |d_i| \leq B, \quad \mathrm{ord}_{\mathfrak{p}_\ell} \mu_i = 0. \tag{12}$$

Since $0 \leq b_{2,0} < w$ and $0 \leq r_i < |n_t|$, there are only finitely many possible values for μ_0, and this finite set can be determined without any knowledge of B or the $b_{2,i}$. For each μ_0, we may apply Corollary 3.2 or 3.3 as appropriate, and obtain a constant $c_8'(\ell, \mu_0)$ such that

$$\mathrm{ord}_{\mathfrak{p}_\ell} \tau_1 = \mathrm{ord}_{\mathfrak{p}_\ell}(\tau_2 - 1) < c_8'(\ell, \mu_0) \log B.$$

Setting

$$c_8(\ell) := \max\left\{\frac{e^2}{\log 2}, \max_{\mu_0}\{c_8'(\ell, \mu_0)\}\right\},$$

we may be sure every S-unit solution satisfies

$$\mathrm{ord}_{\mathfrak{p}_\ell} \tau_1 < c_8(\ell) \log B. \tag{13}$$

Combining inequalities (11) and (13), we have

$$B < \frac{c_8(\ell)}{e_\ell c_5(\ell)} \log B.$$

Since $c_1 \geq 1$ and $c_8(\ell) \geq e^2 (\log 2)^{-1}$, it follows that

$$\frac{c_8(\ell)}{e_\ell c_5} \geq e^2.$$

Applying Lemma 3.5 with $a = 0$, $b = c_8(\ell)/e_\ell c_5(\ell)$, and $h = 1$, we may conclude

$$B \leq K_0(\ell) := \frac{2c_8(\ell)}{e_\ell c_5(\ell)} \log\left(\frac{c_8(\ell)}{e_\ell c_5}\right).$$

Set

$$K_0 := \max\{K_0(\ell) : \mathfrak{p}_\ell \in S_{\text{fin}}\}.$$

If ℓ corresponds to a finite place, then $B \leq K_0$.

In our implementation, the functions `mus` and `possible_mu0s` are used to recover the μ_i for each finite place $\mathfrak{p}_\ell$. The constants $c_8(\ell)$ determined from Yu's Theorem are computed in `Yu_bound`, while the constant K_0, which may be of independent interest, is computed by `K0_func`.

4.3 Case II: $\mathfrak{p}_\ell$ Is Infinite

We now assume $\mathfrak{p}_\ell$ is infinite. As in §2.3, we let $\sigma_{\mathfrak{p}_\ell}$ denote the embedding of K into $\mathbb{C}$ such that

$$|\alpha|_{\mathfrak{p}_\ell} = \left|\sigma_{\mathfrak{p}_\ell}(\alpha)\right|^{\delta(\ell)}, \quad \text{where } \delta(\ell) = \begin{cases} 1 & \mathfrak{p}_\ell \text{ is real,} \\ 2 & \mathfrak{p}_\ell \text{ is complex.} \end{cases}$$

We let $\alpha^{(\ell)}$ denote $\sigma_{\mathfrak{p}_\ell}(\alpha)$ for any $\alpha \in K$, and we define

$$c_{11}(\ell) := \frac{\delta(\ell)\log 4}{c_3}, \qquad c_{13}(\ell) := \frac{c_3}{\delta(\ell)}.$$

The condition (9) can now be expressed as

$$\left|\tau_1^{(\ell)}\right| \leq e^{-c_{13}(\ell)B}.$$

The choices of $c_{11}(\ell)$ and $c_{13}(\ell)$ guarantee that

$$B \geq c_{11}(\ell) \quad \Longrightarrow \quad \left|\tau_1^{(\ell)}\right| \leq \frac{1}{4}.$$

Set $\Lambda := \log \tau_2^{(\ell)}$. The estimate $|\log z| \leq 2|z-1|$ holds for $|z-1| \leq \frac{1}{4}$, and so

$$|\Lambda| \leq 2\left|\tau_2^{(\ell)} - 1\right| = 2\left|\tau_1^{(\ell)}\right| \leq 2e^{-c_{13}(\ell)B}. \tag{14}$$

The next step is to view Λ as a linear form in logarithms and apply the theorem of Baker and Wüstholz. Set $\zeta := \exp \frac{2\pi\sqrt{-1}}{w} \in \mathbb{C}$. Since ρ_0 is a wth root of unity, there exists $0 \le k < w$ such that $(\rho_0^{(\ell)})^{b_{2,0}} = \zeta^k$. By (6), we have

$$\begin{aligned}\Lambda &= \log\left(\left(\rho_0^{(\ell)}\right)^{b_{2,0}} \cdot \prod_{j=1}^{t}\left(\rho_j^{(\ell)}\right)^{b_{2,j}}\right)\\ &= \log \zeta^k + \sum_{j=1}^{t} b_{2,j} \log \rho_j^{(\ell)} + A \cdot 2\pi\sqrt{-1}\\ &= k \log \zeta + \sum_{j=1}^{t} b_{2,j} \log \rho_j^{(\ell)} + Aw \log \zeta\\ &= (Aw + k) \log \zeta + \sum_{j=1}^{t} b_{2,j} \log \rho_j^{(\ell)},\end{aligned} \tag{15}$$

where we have introduced $A \in \mathbb{Z}$ to adjust for the principal branch of the logarithm. Certainly $|A| \le tB$, and so $|Aw + k| \le (t+1)Bw$. Set

$$b'_{2,j} := \begin{cases} Aw + k & j = 0 \\ b_{2,j} & j > 0 \end{cases}$$

and $L'(z_0, \ldots, z_t) := \sum_{j=0}^{t} b'_{2,j} z_j$. We now have

$$|\Lambda| = \left|L'(\log \zeta, \log \rho_1^{(\ell)}, \ldots, \log \rho_t^{(\ell)})\right|.$$

Taking $K' = \mathbb{Q}(\rho_0, \ldots, \rho_t) \cong \mathbb{Q}(\zeta, \rho_1^{(\ell)}, \ldots, \rho_t^{(\ell)})$, we define

$$c_{14}(\ell) := C(t, d_{K'}) \prod_{j=0}^{t} h'(\rho_j).$$

(Recall that $C(t, d_{K'})$ is defined in Theorem 3.4.) We have $|b'_{2,j}| \le B' := (t+1)Bw$. Applying Theorem 3.4 to Λ, we obtain

$$\log |\Lambda| > -c_{14}(\ell) \cdot \log B' = -c_{14}(\ell) \log\big((t+1)wB\big).$$

Combining this inequality with (14), we obtain

$$2e^{-c_{13}(\ell)B} \ge |\Lambda| \ge e^{-c_{14}(\ell) \log B'}. \tag{16}$$

This yields the inequality

$$B < a(\ell) + b(\ell)\log B,$$

where

$$a(\ell) := \frac{1}{c_{13}(\ell)}\left(\log 2 + c_{14}(\ell)\log\big((t+1)w\big)\right), \quad b(\ell) := \frac{c_{14}(\ell)}{c_{13}(\ell)}.$$

As $c_{13}(\ell) \leq \frac{1}{t}$ and $c_{14}(\ell) \geq 32^3$, we have $a(\ell) \geq 0$ and $b(\ell) \geq e^2$. So by Lemma 3.5, $B < c_{15}(\ell)$ (provided $B \geq c_{11}(\ell)$), where

$$c_{15}(\ell) := 2\big(a(\ell) + b(\ell)\log b(\ell)\big).$$

Thus, setting

$$K_1(\ell) := \max\{c_{11}(\ell), c_{15}(\ell)\},$$
$$K_1 := \max\{K_1(\ell) : \mathfrak{p}_\ell \text{ is infinite}\},$$

we may be sure $B \leq K_1$. In our implementation, the constant K_1 is computed in the function `K1_func`.

Combining all the results of this section, we obtain the following.

Lemma 4.2 *The constant B satisfies $B \leq \max\{4, w, K_0, K_1\}$.*

5 LLL Reduction

In this section we explain how we can reduce the upper bound we have computed in Sect. 4. This is necessary, because in practice the size of the initial bound is extremely large and cannot be used for practical computations. The idea of the method we will present here has its origin in de Weger's thesis [13, 12, 14] where he develops a method based on multi-dimensional approximation lattices of linear forms of p-adic numbers to solve (among many other equations) S-unit equations[2] over $\mathbb{Q}$. These ideas of de Weger have been extended by himself and others to apply over any number field K, and have also been used for the solution of other exponential Diophantine equations [39, 40, 41, 34].

In the reduction step we use the LLL reduction algorithm on lattices generated by integer matrices. So instead of the classical LLL algorithm [24], we use the algorithm in [12]. If $\mathscr{L}$ is a lattice in $\mathbb{R}^n$, let $\mathscr{L}^* = \mathscr{L} - \{\mathbf{0}\}$. For $\mathbf{y} \in \mathbb{R}^n$, we define

[2]It is worth mentioning the recent results of von Känel and Matschke [42], who solve S-unit equations using modularity.

$$\ell(\mathcal{L}, \mathbf{y}) = \begin{cases} \min\limits_{\mathbf{x}\in\mathcal{L}^*} \|\mathbf{x}\|, & \text{if } \mathbf{y} \in \mathcal{L}, \\ \min\limits_{\mathbf{x}\in\mathcal{L}} \|\mathbf{x} - \mathbf{y}\|, & \text{otherwise.} \end{cases}$$

Computing the exact value of $\ell(\mathcal{L}, \mathbf{y})$ is a very challenging problem in general. Instead, the function `minimal_vector` computes a lower bound using standard properties of a reduced basis of a lattice and the LLL algorithm (see [36, Chapter V]). As in the previous section, we follow Smart's notation in [34]. Most of the material we present in this section can also be found in [36, 15].

We preserve the meaning of $\mathfrak{p}_\ell$ from Sect. 4. When $\mathfrak{p}_\ell$ is a finite place, we let p denote the prime of $\mathbb{Z}$ lying below $\mathfrak{p}_\ell$. We continue to assume $B \geq \max\{4, w\}$ in this section.

5.1 Finite Places

Suppose $\mathfrak{p}_\ell$ is a finite place. Set

$$c_{16}(\ell) := 1 + \frac{1}{c_5(\ell)},$$

and suppose that $B \geq c_{16}(\ell)$. Define $\Delta_2 \in K_{\mathfrak{p}_\ell}$ as $\Delta_2 := \log_p \tau_2$. Combining (11), (2), and $B \geq c_{16}(\ell)$, shows that $\operatorname{ord}_p \tau_1 > 1$. Thus $|\tau_1|_p < p^{-\frac{1}{p-1}}$, and by (5),

$$\operatorname{ord}_p \Delta_2 = \operatorname{ord}_p \log_p \tau_2 = \operatorname{ord}_p \log_p(1 - \tau_1) = \operatorname{ord}_p \tau_1 > 1.$$

Let μ_i, d_i be as given in (12), so that we have

$$\Delta_2 = \log_p \tau_2 = \log_p \mu_0 + \sum_{i=1}^{t-1} d_i \log_p \mu_i.$$

Choose $\theta \in K_{\mathfrak{p}_\ell}$ such that $K_{\mathfrak{p}_\ell} = \mathbb{Q}_p(\theta)$, and let $\operatorname{Disc}(\theta)$ denote the discriminant of θ. Set $D_p(\theta) = \operatorname{ord}_p \operatorname{Disc}(\theta)$ and $n = [K_{\mathfrak{p}_\ell} : \mathbb{Q}_p]$, so that $n = e_\ell f_\ell$. Expressing Δ_2 with respect to the power basis, we obtain $\Delta_{2,k} \in \mathbb{Q}_p$ such that $\Delta_2 = \sum_{k=0}^{n-1} \Delta_{2,k}\theta^k$. Further, we may express

$$\Delta_{2,k} = a_{0,k} + \sum_{j=1}^{t-1} d_j a_{j,k}, \qquad a_{j,k} \in \mathbb{Q}_p,\ 0 \leq k \leq n-1. \tag{17}$$

Using an idea due to Evertse [41, p. 257], we have

$$\operatorname{ord}_p \Delta_{2,k} \geq c_5(\ell)B - \frac{D_p(\theta)}{2}.$$

Define

$$c_{17}(\ell) := \min\{\operatorname{ord}_p a_{j,k} : 1 \leq j \leq t-1,\ 0 \leq k \leq n-1\},$$
$$c_{18}(\ell) := c_{17}(\ell) + \frac{D_p(\theta)}{2},$$

and choose $\lambda \in \mathbb{Q}_p$ such that $\operatorname{ord}_p \lambda = c_{17}(\ell)$.

Should there be some index k such that $c_{17}(\ell) > \operatorname{ord}_p(a_{0,k})$, then $\operatorname{ord}_p \Delta_{2,k} = \operatorname{ord}_p a_{0,k} < c_{17}(\ell)$, and consequently

$$B < \frac{c_{18}(\ell)}{c_5(\ell)}.$$

For the remainder, then, we assume

$$c_{17}(\ell) \leq \min\{\operatorname{ord}_p a_{0,k} : 0 \leq k \leq n-1\}.$$

By the choice of λ, $\kappa_{j,k} := a_{j,k}/\lambda$ is a p-adic integer for all j, k, and we may rewrite (17) as

$$\frac{\Delta_{2,k}}{\lambda} = \kappa_{0,k} + \sum_{j=1}^{t-1} d_j \kappa_{j,k}, \quad \text{with} \quad \operatorname{ord}_p\left(\frac{\Delta_{2,k}}{\lambda}\right) \geq c_5(\ell)B - c_{18}(\ell).$$

For any $a \in \mathbb{Z}_p$ and a positive integer z, let $a^{(z)}$ denote the unique integer between 0 and p^z such that $a \equiv a^{(z)} \pmod{p^z}$. For a positive integer u, let $\mathscr{L}$ be the lattice generated by the columns of the matrix

$$\begin{pmatrix} 1 & & 0 & 0 & \cdots & 0 \\ & \ddots & & \vdots & & \vdots \\ 0 & & 1 & 0 & \cdots & 0 \\ \kappa_{1,0}^{(u)} & \cdots & \kappa_{t-1,0}^{(u)} & p^u & & 0 \\ \vdots & & \vdots & & \ddots & \\ \kappa_{1,n-1}^{(u)} & \cdots & \kappa_{t-1,n-1}^{(u)} & 0 & & p^u \end{pmatrix} \in \mathbb{Z}^{(t+n-1)\times(t+n-1)}.$$

Define

$$\mathbf{y} = \left(0 \cdots 0 \ -\kappa_{0,0}^{(u)} \ \cdots \ -\kappa_{0,n-1}^{(u)}\right)^{\mathsf{T}} \in \mathbb{Z}^{t+n-1}.$$

Also set

$$K_0^{\mathrm{LLL}}(\ell) := \max\left\{4, w, \frac{u + c_{18}(\ell)}{c_5(\ell)}, c_{16}(\ell)\right\}.$$

The following lemma is a restatement of [34, Lemma 5], and provides an opportunity to improve the bound on B.[3]

Lemma 5.1 *If $\ell(\mathscr{L}, \mathbf{y}) > \sqrt{t-1} \cdot K_0$, then $B < K_0^{\mathrm{LLL}}(\ell)$.*

In the function `p_adic_LLL_bound` we have implemented the above analysis. In more detail, the functions `log_p` and `embedding_to_Kp` are used to compute the constants $a_{j,k} \in \mathbb{Q}_p$ up to a given precision. If this precision is M, i.e., if the $a_{j,k}$ are stored as pairs of integers modulo p^M, we clearly require $M > u$ for the algorithm to be meaningful. However, the shift by λ requires an additional $c_{17}(\ell)$ p-adic digits of precision. So the algorithm checks that $M > u + c_{17}(k)$; if this fails, then the p-adic logarithms are computed to higher precision and the process is repeated.

The function `log_p` is based on an algorithm of Smart in [36, p. 30]. However, our implementation also resolves a crucial computational problem in the evaluation of $\log_p$ that to our knowledge has not been mentioned in the literature. To understand the issue, we must describe carefully what "computing the logarithm" means in a p-adic setting. Let us view K as a subfield of $K_{\mathfrak{p}}$, and specify $\omega_1, \ldots, \omega_n \in K$, a $\mathbb{Q}_p$-basis for $K_{\mathfrak{p}}$. Suppose $\alpha \in K$ and set $\beta := \log_p \alpha \in K_{\mathfrak{p}}$. Necessarily, there exist $b_j \in \mathbb{Q}_p$ such that $\beta = \sum_j b_j \omega_j$.

As a practical matter, no algorithm can return the true value β; it can only return $\tilde{\beta} \in K$, an approximation to β with $|\beta - \tilde{\beta}|_{\mathfrak{p}}$ very small. In practice however, we require something more specific. We want to find $\tilde{b}_j \in \mathbb{Q}$ such that $|b_j - \tilde{b}_j|_p$ is small for each j. When p splits in K, the original algorithm is not guaranteed to do this.

It can happen that at some other prime $\mathfrak{p}'$ above p, α has a negative valuation. Consequently, the sum (3) used to compute $\tilde{\beta}$ will *not* converge $\mathfrak{p}'$-adically, and the approximations $\tilde{b}_j$ are not guaranteed to be p-adically close to the b_j. We resolve this problem by choosing a suitable element $\eta \in K$ with $\operatorname{ord}_{\mathfrak{p}'} \eta \geq 0$ for all $\mathfrak{p}' \mid p$, such that η also satisfies

$$\operatorname{ord}_{\mathfrak{p}'}(\eta\alpha) \geq 0, \qquad \operatorname{ord}_{\mathfrak{p}}(\eta\alpha) = \operatorname{ord}_{\mathfrak{p}} \eta = 0.$$

Then it holds that

$$\log_p \alpha = \log_p(\eta\alpha) - \log_p \eta.$$

[3] Note s_d in [34] has the value $t - 1$ in our notation.

By evaluating the difference on the right hand side, these $\mathfrak{p}'$-adic divergence issues are avoided, and we may be sure that the individual coefficients $\tilde{b}_j$ approximate the b_j p-adically.

In `p_adic_LLL_bound_one_prime`, we attempt to find a value u such that Lemma 5.1 applies. If successful, we record the improved bound $K_0^{\mathrm{LLL}}(\ell)$. The improvement offered by Lemma 5.1 depends only on the assumption that $\mathfrak{p}_\ell$ is the extremal place, and that some bound $K_0 \geq c_{16}(\ell)$ on the exponents is known. So we may replace K_0 by $K_0^{\mathrm{LLL}}(\ell)$ and attempt to apply Lemma 5.1 again, possibly improving the bound further. Because the application of LLL is very fast compared to the sieving step described in Sect. 6, the algorithm repeats this process until no further improvements can be made to $K_0^{\mathrm{LLL}}(\ell)$. Once each $K_0^{\mathrm{LLL}}(\ell)$ has been optimized in this way, the function `p_adic_LLL_bound` returns

$$K_0^{\mathrm{LLL}} := \max\{K_0^{\mathrm{LLL}}(\ell) : \mathfrak{p}_\ell \in S_{\mathrm{fin}}\}.$$

5.2 *Complex Places*

We now consider the case where $\mathfrak{p}_\ell$ is an infinite complex place. The reduction is quite analogous to the p-adic case; again the standard references are [34, 36, 15]. We keep the notations from Sect. 4.3. For $0 \leq j \leq t$ we define the complex numbers

$$\kappa_j := \begin{cases} \log \zeta & j = 0, \\ \log \rho_j^{(\ell)} & j > 0. \end{cases}$$

As $\mathfrak{p}_\ell$ is an infinite place, we have established already that the $b'_{2,j}$ in

$$\Lambda = \log \tau_2^{(\ell)} = \sum_{j=0}^{t} b'_{2,j} \kappa_j$$

satisfy the bounds

$$|b'_{2,0}| \leq (t+1)Bw, \qquad |b'_{2,j}| \leq B \text{ for } 1 \leq j \leq t.$$

We now attempt to use lattice reduction to improve the bound; the choice of lattice and certain constants will depend slightly on whether the κ_j are all purely imaginary. So we define

$$\sigma := \begin{cases} 1 & \text{every } \kappa_j \text{ is pure imaginary}, \\ 0 & \text{otherwise}, \end{cases}$$

and define

$$S := \big(t - 1 + \sigma\big)K_1^2, \qquad T := \left(\frac{1}{\sqrt{2}}\right)^{1+\sigma} (t + w + tw)K_1.$$

If $\kappa_1, \ldots, \kappa_t$ are not all pure imaginary, relabel $\kappa_1, \ldots, \kappa_t$ so that $\operatorname{Re}\kappa_t \neq 0$. Let $[\cdot]$ denote the nearest integer function. Now define

$$a_{tt} := \begin{cases} [C \operatorname{Re}\kappa_t] & \operatorname{Re}\kappa_t \neq 0, \\ 1 & \operatorname{Re}\kappa_t = 0. \end{cases}$$

Let A be the $(t + 1) \times (t + 1)$ integer matrix

$$A = \begin{pmatrix} 1 & & 0 & 0 & 0 \\ & \ddots & & \vdots & \vdots \\ 0 & & 1 & 0 & 0 \\ [C \cdot \operatorname{Re}\kappa_1] & \cdots & [C \cdot \operatorname{Re}\kappa_{t-1}] & a_{tt} & 0 \\ [C \cdot \operatorname{Im}\kappa_1] & \cdots & [C \cdot \operatorname{Im}\kappa_{t-1}] & [C \cdot \operatorname{Im}\kappa_t] & [C \cdot \frac{2\pi}{w}] \end{pmatrix}. \tag{18}$$

(By design, the upper left $t \times t$ block of A is the identity matrix in case the κ_j are all pure imaginary.) Now, let $\mathscr{L}$ be the lattice generated by the columns of A, and suppose $m_{\mathscr{L}}$ is a positive lower bound for $\ell(\mathscr{L}, \mathbf{0})$. When $\mathfrak{p}_\ell$ is a infinite non-real place, we define:

$$K_1^{\mathrm{LLL}}(\ell) := \max\left\{4, w, \frac{1}{c_{13}(\ell)} \log\left(\frac{2C}{(m_{\mathscr{L}}^2 - S)^{\frac{1}{2}} - T}\right)\right\}.$$

Similar to [36, Lemma VI.2], we have

Lemma 5.2 *Suppose $\mathfrak{p}_\ell$ is a non-real infinite place. With notation as above, suppose C is chosen such that $m_{\mathscr{L}}^2 > T^2 + S$. Then $B \leq K_1^{\mathrm{LLL}}(\ell)$.*

Proof There are two cases to consider, as $\sigma = 0$ or $\sigma = 1$. In each case, our goal is to establish the inequality

$$\sqrt{m_{\mathscr{L}}^2 - S} - T \leq 2Ce^{-c_{13}(\ell)B}, \tag{19}$$

for the result follows by isolating B in the inequality (19).

If the κ_j are not all pure imaginary, we define

$$\Phi_1 := \sum_{j=1}^{t} b'_{2,j} \left[C \cdot \operatorname{Re}(\kappa_j)\right],$$

$$\Phi_2 := b'_{2,0}\left[C \cdot \tfrac{2\pi}{w}\right] + \sum_{j=1}^{t} b'_{2,j}\left[C \cdot \mathrm{Im}(\kappa_j)\right].$$

Then note that

$$|C\Lambda - (\Phi_1 + \Phi_2\sqrt{-1})| \leq T.$$

Therefore

$$|(\Phi_1 + \Phi_2\sqrt{-1})| \leq T + |C\Lambda|.$$

We know from (16) that $2e^{-c_{13}(\ell)B} \geq |\Lambda|$, so

$$|(\Phi_1 + \Phi_2\sqrt{-1})| \leq T + 2Ce^{-c_{13}(\ell)B}.$$

Now notice that the vector

$$\mathbf{y} = (b'_{2,1}, b'_{2,2}, \ldots, b'_{2,t-1}, \Phi_1, \Phi_2)^{\mathsf{T}}$$

is in the lattice $\mathcal{L}$, so $|\mathbf{y}| \geq m_{\mathcal{L}}$. Further,

$$\begin{aligned} m_{\mathcal{L}}^2 \leq |\mathbf{y}|^2 &= \sum_{j=1}^{t-1} (b'_{2,j})^2 + \Phi_1^2 + \Phi_2^2 \\ &\leq (t-1)K_1^2 + \left|\Phi_1 + \Phi_2\sqrt{-1}\right|^2 \\ &\leq S + (T + 2Ce^{-c_{13}(\ell)B})^2, \end{aligned}$$

which implies (19) and the result follows.

In case the κ_j are all pure imaginary, the approach is similar. Set

$$\Phi = b'_{2,0}[C \cdot \frac{2\pi}{w}] + \sum_{j=1}^{t} b'_{2,j}[C \cdot \mathrm{Im}(\kappa_j)].$$

Similar to the other case, we have $\left|C\Lambda - \left(\Phi\sqrt{-1}\right)\right| \leq T$, and therefore $|\Phi| \leq T + |C\Lambda|$. Again applying (16) we obtain

$$|\Phi| \leq T + 2Ce^{-c_{13}(\ell)B}.$$

Now notice that the vector

$$\mathbf{y} = (b'_{2,1}, b'_{2,2}, \ldots, b'_{2,t}, \Phi)^{\mathsf{T}}$$

is in the lattice $\mathscr{L}$, so $|\mathbf{y}| \geq m_{\mathscr{L}}$. Further,

$$\begin{aligned} m_{\mathscr{L}}^2 \leq |\mathbf{y}|^2 &= \sum_{i=1}^{t} (b'_{2,i})^2 + \Phi^2 \\ &\leq tK_1^2 + |\Phi|^2 \\ &\leq S + (T + 2Ce^{-c_{13}(\ell)B})^2. \end{aligned}$$

Again this implies (19). □

5.3 *Real Places*

Now suppose that $\mathfrak{p}_\ell$ is a real infinite place. Although the arguments in Sect. 5.2 apply to $\mathfrak{p}_\ell$, we can obtain a stronger improvement by analyzing this case separately. Replacing ρ_j by $-\rho_j$ as necessary, we may assume $\rho_j^{(\ell)} > 0$ for all j with $1 \leq j \leq t$.

The mere existence of a real place forces $w = 2$ and $0 \leq b_{2,0} \leq 1$. We set $\kappa_0 := \pi\sqrt{-1}$ and define the real numbers

$$\kappa_j := \log \rho_j^{(\ell)}, \qquad 1 \leq j \leq t.$$

As the $\kappa_j \in \mathbb{R}$ for $j > 0$, we may revisit (15); this time we obtain

$$\Lambda = \log \tau_2^{(\ell)} = \sum_{j=0}^{t} b_{2,j}\kappa_j,$$

as no adjustments are required to accommodate the branch cut of the logarithm. Set

$$S := (t-1)K_1^2, \qquad T := \tfrac{1}{2}(tK_1 + 1),$$

and again let $\mathscr{L}$ be generated by the columns of the matrix A in (18). When $\mathfrak{p}_\ell$ is an infinite real place, we define:

$$K_1^{\mathrm{LLL}}(\ell) := \max\left\{4, w, \frac{1}{c_{13}(\ell)} \log\left(\frac{2C}{(m_{\mathscr{L}}^2 - S)^{\frac{1}{2}} - T}\right)\right\}.$$

Lemma 5.3 *Suppose that $\mathfrak{p}_\ell$ is a real infinite place. With notation and definitions as above, suppose C is chosen so that $m_{\mathscr{L}}^2 > T^2 + S$. Then $B \leq K_1^{\mathrm{LLL}}(\ell)$.*

Proof If we define

$$\Phi_1 := \sum_{j=1}^{t} b_{2,j}[C \cdot \kappa_j], \quad \Phi_2 := b_{2,0}[C \cdot \pi],$$

then we obtain

$$|C\Lambda - \Phi_1 - \Phi_2\sqrt{-1}| \le T.$$

Observing that the vector

$$\mathbf{y} = (b_{2,1},\ b_{2,2},\ \cdots,\ b_{2,t-1},\ \Phi_1,\ \Phi_2)^\mathsf{T} \in \mathscr{L},$$

and that $|b_{2,j}| \le B$ for $j > 0$, $|b_{2,0}| \le 1$, the remainder of the proof now follows the logic of Lemma 5.2 exactly. □

5.4 Implementation

The function `minimal_vector` is used in the implementation to compute a value for $m^2_{\mathscr{L}}$. In `cx_LLL_bound`, we have implemented the reduction step for the infinite places applying the above idea. As in the finite case, the parameter C is chosen inside the function and changed as necessary to meet the bound $m^2_{\mathscr{L}} > T^2 + S$ (keeping in mind, of course, that the definitions of S and T depend on the particular place $\mathfrak{p}_\ell$). Notice that the proof of Lemmas 5.2 and 5.3 depend on obtaining true rounding in obtaining the coefficients of the matrix A. In our implementation, we increase precision until this is assured. Similar to the case where $\mathfrak{p}_\ell$ is finite, the improvement of Lemma 5.2 needs only the assumption that $\mathfrak{p}_\ell$ is the extremal place and that some bound K_1 on the exponents is known. So we apply Lemma 5.2 repeatedly until no further improvement to $K_1^{\mathrm{LLL}}(\ell)$ is possible. Once this has been done for each infinite place, we set

$$K_1^{\mathrm{LLL}} := \max\{K_1^{\mathrm{LLL}}(\ell) : \mathfrak{p}_\ell \in S_\infty\}.$$

Consequently, we have the following bound which may be passed to the sieve in the next section.

Lemma 5.4 *Assume that for each ℓ, a value $u = u(\ell)$ or $C = C(\ell)$ exists for which the hypotheses of one of the Lemmas 5.1, 5.2, or 5.3 are met. Then the maximum exponent B appearing in any solution (τ_1, τ_2) of the S-unit equation (1) satisfies*

$$B \le K^{\mathrm{LLL}} := \max\left\{K_0^{\mathrm{LLL}}, K_1^{\mathrm{LLL}}\right\}. \tag{20}$$

We use the proof as an opportunity to summarize the algorithm, up to the sieving step of the next section.

Proof There is nothing to show if the solution set is empty, so let us assume otherwise. We know the S-unit equation has only finitely many solutions. Keeping the notation of (6), let (τ_1, τ_2) be a solution where $B = |\mathbf{a}_i|$ is maximized. One of the places in S, say $\mathfrak{p}_\ell$, is extremal. If $\mathfrak{p}_\ell$ is finite, then the work in Sect. 4.2 demonstrates $B \leq \max\{4, w, K_0(\ell)\}$ by applying one of the corollaries deduced from Yu's bound. If $\mathfrak{p}_\ell$ is infinite, then the work in Sect. 4.3 demonstrates $B \leq \max\{4, w, K_1(\ell)\}$ by applying the theorem of Baker-Wüstholz. This establishes an absolute bound on B.

For each possible ℓ, the techniques of this section attempt to replace this absolute bound with a smaller bound. There is no mathematical proof that the lattice reduction techniques will succeed, i.e., that there will exist appropriate values u and C for which Lemmas 5.1, 5.2, or 5.3 apply. However, when they do exist, the improved bound is provably correct by the same lemmas. Here, such success is presumed for every ℓ, and (20) holds. □

In practice, if the hypotheses of Lemma 5.4 are not established, then one only has the weaker bounds coming from linear forms of logarithms—these are simply too large to allow for a provably complete search. However, the sieve described in the next section can still be used up to any prescribed bound B_0; it will find all solutions satisfying $|\mathbf{a}_i| \leq B_0$.

6 Further Reducing the Search Space: Sieving

The approach taken here, for sieving against primes outside of S, is based on an algorithm described by Smart in [34]. Smart credits Tzanakis and de Weger with this approach [40]; Tzanakis reports that these ideas date back to Andrew Bremner.

6.1 Setup for the Sieve

Recalling the notations of Sect. 2.5, we define for any $m > 0$,

$$A_{K,S,m} := (\mathbb{Z}/w\mathbb{Z}) \times (\mathbb{Z}/m\mathbb{Z})^t.$$

This finite set will provide a useful search space for exponent vectors in a way we will make more precise below. There is an obvious surjective map $\pi_m \colon A_{K,S} \to A_{K,S,m}$. Despite the fact that this map is the identity (and not a reduction map) in the 0th coordinate, we will refer to this as the *reduction modulo* m map, and call an element $\mathbf{a} \in A_{K,S,m}$ an *exponent vector modulo* m.

Let $\tau \in \mathcal{O}_{K,S}^\times$. The *exponent vector* for τ (relative to $\boldsymbol{\rho}$) is $\Phi_{\boldsymbol{\rho}}^{-1}(\tau)$. That is, it is the unique $\mathbf{a} \in A_{K,S}$ such that $\tau = \boldsymbol{\rho}^{\mathbf{a}}$. Given any bound B for the exponent vector of a $\tau \in X_{K,S}$, we obtain a finite subset of $\mathcal{O}_{K,S}^{\times}$ that contains every solution of the S-unit equation. Unfortunately, this is usually still too large of a search space to be practical (see Sect. 7), so we must sieve this finite set (or rather, the equivalent finite set of exponent vectors) prior to the exhaustive search. The sieve attempts to provide an efficient solution to the following problem:

Problem 6.1 Find a small set $Y_{K,S}$ satisfying $E_{K,S} \subseteq Y_{K,S} \subseteq A_{K,S}$.

If we can find a small enough superset $Y_{K,S}$ in a fast enough way, the S-unit equation solutions can then be found by brute force search over $Y_{K,S}$.

Suppose $\mathbf{a} \in A_{K,S}$. We call $\mathbf{b} \in A_{K,S}$ a *complement vector* for $\mathbf{a}$ if $\boldsymbol{\rho}^{\mathbf{a}} + \boldsymbol{\rho}^{\mathbf{b}} = 1$. If a complement vector exists, it must be unique; the existence of a complement vector is equivalent to $\mathbf{a} \in E_{K,S}$, and a pair of complement exponent vectors correspond to a solution of the S-unit equation.

Suppose $q \in \mathbb{Z}$ is a prime number. We say q *avoids* S if $q \notin \mathfrak{p}$ for all ideals $\mathfrak{p} \in S$. If q splits completely in $\mathcal{O}_K$, then there are d_K prime ideals above q in $\mathcal{O}_K$, say $\mathfrak{q}_0, \ldots, \mathfrak{q}_{d_K-1}$. We let $\mathbb{F}_{\mathfrak{q}_j}$ denote the residue field of $\mathfrak{q}_j$. Since q is completely split, we of course have $\mathbb{F}_{\mathfrak{q}_j} \cong \mathbb{F}_q$ for all j.

Suppose $\tau \in \mathcal{O}_{K,S}$, and q is a rational prime number which splits completely in $\mathcal{O}_K$ and which avoids S. The *residue field vector* for τ (with respect to q) is

$$\operatorname{rfv}_q(\tau) := (\tau + \mathfrak{q}_0, \tau + \mathfrak{q}_1, \ldots, \tau + \mathfrak{q}_{d_K-1}) \in \prod_{i=0}^{d_K-1} \mathbb{F}_{\mathfrak{q}_i},$$

where $\tau + \mathfrak{q}_j \in \mathbb{F}_{\mathfrak{q}_j}$ is the reduction of τ modulo $\mathfrak{q}_j$. The residue field vector depends on the ordering of the primes $\mathfrak{q}_j$ above q; we fix one ordering once and for all whenever we consider residue field vectors with respect to q.

Notice that we have the following commutative diagram, whose horizontal rows are exact.

$$\begin{array}{ccccccccc}
 & & & & & & \prod_i \mathbb{F}_{\mathfrak{q}_i}^{\times} & & \\
 & & & & & & \cup\!\uparrow & & \\
1 & \longrightarrow & \mathcal{O}_{K,S}^{\times} \cap (1 + q\mathcal{O}_{K,S}) & \longrightarrow & \mathcal{O}_{K,S}^{\times} & \xrightarrow{\operatorname{rfv}_q} & \operatorname{rfv}_q(\mathcal{O}_{K,S}^{\times}) & \longrightarrow & 1 \\
 & & \cup\!\uparrow & & \uparrow & & \uparrow & & \\
 & & (\mathcal{O}_{K,S}^{\times})^{q-1} & & \Phi_{\boldsymbol{\rho}} \,\Big|\, \cong & & \exists\, \vdots & & \\
 & & \Phi_{\boldsymbol{\rho}} \uparrow & & & & & & \\
0 & \longrightarrow & \{0\} \times ((q-1)\mathbb{Z})^t & \longrightarrow & A_{K,S} & \xrightarrow{\pi_{q-1}} & A_{K,S,q-1} & \longrightarrow & 0
\end{array}$$

Suppose $\mathbf{a} \in A_{K,S,q-1}$. Since any two lifts $\mathbf{a}'$, $\mathbf{a}''$ of $\mathbf{a}$ to $A_{K,S}$ differ by a multiple of $(q-1)$, we see that $\Phi_\rho(\mathbf{a}')$ and $\Phi_\rho(\mathbf{a}'')$ differ by a perfect $(q-1)$th power, and so determine the same residue field vector. In other words, the dashed arrow in the diagram corresponds to a well-defined map $A_{K,S,q-1} \to \prod \mathbb{F}_{\mathfrak{q}_i}^\times$, and so the notion of a residue field vector for $\mathbf{a}$ is well-defined. With this in mind, we abuse notation slightly and also write $\mathrm{rfv}_q \mathbf{a} := \mathrm{rfv}_q \Phi_\rho(\mathbf{a}')$, where $\mathbf{a}' \in A_{K,S}$ is any lift of $\mathbf{a}$.

Lemma 6.1 *Suppose $\tau \in X_{K,S}$ and set $\eta = 1 - \tau$. Then*

(a) $\mathrm{rfv}_q \tau + \mathrm{rfv}_q \eta = (1, 1, \ldots, 1) \in \prod_i \mathbb{F}_{\mathfrak{q}_i}$.
(b) $\mathrm{rfv}_q \tau \in \prod_i \mathbb{F}_{\mathfrak{q}_i}^\times$.
(c) no entry of $\mathrm{rfv}_q \tau$ *is* 1.

Proof Since $\tau + \eta = 1$, it follows that for any j, $\tau + \eta \equiv 1 \pmod{\mathfrak{q}_j}$, verifying (a). As q avoids S, $\tau \notin \mathfrak{q}_j$ for every j. This proves (b). Since (b) holds for both η and τ, (c) follows from (a). □

Suppose $\mathbf{a}$ is an exponent vector modulo $q-1$; i.e., $\mathbf{a} \in A_{K,S,q-1}$. We call $\mathbf{b} \in A_{K,S,q-1}$ a $(q-1)$-*complement vector* for $\mathbf{a}$ if

$$\mathrm{rfv}_q(\mathbf{a}) + \mathrm{rfv}_q(\mathbf{b}) = (1, 1, \ldots, 1) \in \prod_i \mathbb{F}_{\mathfrak{q}_i}.$$

Existence of a $(q-1)$-complement vector is a necessary, but not sufficient, condition for $\mathbf{a}$ to lift to the exponent vector of a unit in a solution to the S-unit equation. Further, any particular $\mathbf{a}$ may have more than one $(q-1)$-complement vector associated to it. We set

$$E_{K,S}(q-1) := \{\mathbf{a} \in A_{K,S,q-1} : \mathbf{a} \text{ has a } (q-1)\text{-complement vector}\}.$$

6.2 *Execution of the Sieve*

The strategy for the sieve is to play the sets $E_{K,S}(q-1)$ off of one another for multiple values of q. Choose a finite list Q of rational prime numbers

$$Q = [q_0, q_1, \ldots, q_{k-1}],$$

each of which splits completely in K and avoids S, and such that

$$\mathrm{lcm}(q_0 - 1, q_1 - 1, \ldots, q_{k-1} - 1) \geq 2B + 1.$$

Any true solution to the S-unit equation corresponds to exponent vectors found in the set $E_{K,S}$, and such vectors must reduce modulo $(q_j - 1)$ to vectors in $E_{K,S}(q_j - 1)$ for each $q_j \in Q$. Conversely, given a choice $\mathbf{a}_i \in E_{K,S}(q_i - 1)$

for each $0 \leq i < k$, there is at most one vector $\mathbf{a} \in A_{K,S}$ such that $\pi_{q_i-1}(\mathbf{a}) = \mathbf{a}_i$ for each i, while also satisfying $|\mathbf{a}| \leq B$. Define π_Q to be the product of the maps π_{q_i-1}:

$$\pi_Q \colon A_{K,S} \longrightarrow \prod_i A_{K,S,q_i-1}.$$

Certainly we have

$$E_{K,S} \subseteq \pi_Q^{-1}\Big(\prod_i E_{K,S}(q_i - 1)\Big).$$

Because lifts from $\prod_i E_{K,S}(q_i - 1)$ to $E_{K,S}$ are unique when they exist, $\prod_i E_{K,S}(q_i - 1)$ provides a reasonable proxy for the search space. We seek to replace each $E_{K,S}(q_i - 1)$ with a subset $Y_i \subseteq E_{K,S}(q_i - 1)$ such that we still have

$$E_{K,S} \subseteq \pi_Q^{-1}\Big(\prod_i Y_i\Big). \tag{21}$$

Suppose q_i, q_j are distinct primes in Q, and suppose $\mathbf{a}_i \in Y_i$, $\mathbf{a}_j \in Y_j$. We say $\mathbf{a}_i$ and $\mathbf{a}_j$ are *compatible* if there exists $\mathbf{a} \in A_{K,S}$ such that $\pi_{q_i-1}(\mathbf{a}) = \mathbf{a}_i$ and $\pi_{q_j-1}(\mathbf{a}) = \mathbf{a}_j$. Notice that for any $i \neq j$, an element $\hat{\mathbf{a}} \in E_{K,S}$ reduces modulo $q_i - 1$ and $q_j - 1$ to produce a compatible pair of exponent vectors.

When $\mathbf{a}_i$ and $\mathbf{a}_j$ are compatible, we further call the pair *complement compatible* if there exist $\mathbf{b}_i \in Y_i$ and $\mathbf{b}_j \in Y_j$ such that

- $\mathbf{b}_i$ is $(q_i - 1)$-complementary to $\mathbf{a}_i$,
- $\mathbf{b}_j$ is $(q_j - 1)$-complementary to $\mathbf{a}_j$,
- $\mathbf{b}_i$ and $\mathbf{b}_j$ are compatible.

Lemma 6.2 *Suppose the sets $Y_i \subseteq E_{K,S}(q_i - 1)$ satisfy condition (21). Further, suppose $\mathbf{a}_i \in Y_i$, and set*

$$Y'_j := \begin{cases} Y_j & j \neq i \\ Y_i - \{\mathbf{a}_i\} & j = i \end{cases}.$$

If there exists $j \neq i$ such that Y_j contains no vectors which are complement compatible to $\mathbf{a}_i$, then

$$E_{K,S} \subseteq \pi_Q^{-1}\Big(\prod_i Y'_i\Big).$$

In other words, under the given condition, we will lose no true solutions by removing $\mathbf{a}_i$ from Y_i.

Proof Towards a contradiction, suppose $\mathbf{a} \in E_{K,S}$ satisfies

$$\pi_{q_i-1}(\mathbf{a}) = \mathbf{a}_i\,.$$

There is a unique $\mathbf{b} \in E_{K,S}$ satisfying $\Phi_\rho(\mathbf{a}) + \Phi_\rho(\mathbf{b}) = 1$. Set

$$\mathbf{a}_j = \pi_{q_j-1}(\mathbf{a}), \quad \mathbf{b}_i = \pi_{q_i-1}(\mathbf{b}), \quad \mathbf{b}_j = \pi_{q_j-1}(\mathbf{b}).$$

Then $\mathbf{a}_i$ and $\mathbf{a}_j$ are compatible by definition. But since $\mathbf{a}_i$ and $\mathbf{a}_j$ cannot be complement compatible, the vectors $\mathbf{b}_i$ and $\mathbf{b}_j$ cannot be compatible. This is impossible, since $\mathbf{b} \in E_{K,S}$. Thus, no such $\mathbf{a}$ exists and the claim holds. □

The algorithm based on this lemma is the following.

Algorithm 6.2 (Sieve) Assume that K, S are fixed and a representation of $\mathcal{O}_{K,S}^\times$ has been computed.

INPUT: $Q = [q_0, q_1, \dots, q_{k-1}]$
OUTPUT: $Y_0, Y_1, \dots, Y_{k-1}$ satisfying (21).

1. Set $Y_i \longleftarrow E_{K,S}(q_i - 1)$ for each i.
2. Loop over $i \in \{0, 1, \dots, k-1\}$:

(a) Loop over $\mathbf{a}_i \in Y_i$:

i. If Y_i contains no $(q_i - 1)$-complement vector for $\mathbf{a}_i$, remove $\mathbf{a}_i$ from Y_i
ii. Loop over $j \in \{0, 1, \dots, i-1, i+1, \dots, k-1\}$:

- If there are no $\mathbf{a}_j \in Y_j$ which are complement compatible with $\mathbf{a}_i$, then remove $\mathbf{a}_i$ from Y_i.

3. Did Step 2 remove any elements from any set Y_i?

- If **YES**, return to Step **2**.
- If **NO**, then **STOP**.

Once the sieve has been completed, we may find all solutions to the S-unit equation by doing an exhaustive search over $\pi_Q^{-1}(\prod_i Y_i)$.

7 Experimental Observations and Computational Choices

In developing this code and in pursuit of applications, we have computed a very large number of examples. Some observations and discussion may be enlightening to a reader who wishes to solve the S-unit equation for their own application.

Our implementation provides the function `sieve_below_bound(K, S, B)`, which returns all solutions to the S-unit equation in $\mathcal{O}_{K,S}^\times$ up to a specified bound B (the maximum absolute value of an entry in an exponent vector). This may be useful in settings where an exhaustive list of solutions is not needed. For example, in the field K_g with $g(x) = x^3 - 3x + 1$, and $S_{\text{fin}} = \{\mathfrak{p} : \mathfrak{p} \mid 2\}$, the provable LLL-reduced exponent bound is 101. However, all solutions actually satisfy the exponent bound 5, and the command `sieve_below_bound(K, S, 5)` executes in under 2 seconds.

7.1 Sieving vs. Simple Exhaustion

Once a bound has been reduced as much as possible by LLL, this search space must be somehow exhausted. This general problem can be solved in multiple ways. Those appearing in the literature can be generally described by the following three ideas:

(1) simple (non-number theory-based) exhaustion,
(2) sieve by reducing the problem modulo primes not in S, and
(3) sieve by reducing the problem modulo powers of primes in S.

Idea (1) could be looked at through the more general lens of efficient programming, and a good programmer may be able to develop their own code to exhaust the search space effectively. The current implementation uses idea (2), inspired by Smart's earlier exposition in [34] and is described here in Sect. 6. Item (3) paraphrases an interesting idea which is first due to de Weger in the case $K = \mathbb{Q}$ [12] and which was generalized to arbitrary number fields K for S-unit equations arising from Thue and Thue-Mahler equations by Tzanakis and de Weger [39, 40, 41]. Wildanger [43] and Smart [37] worked out the details of the full generalization, which was later simplified by Evertse and Győry [15]. This is an extremely promising and potentially effective method of reducing the search space, and has been implemented recently in special cases by several people, including Koutsianas [23], Bennett, Gherga, and Rechnitzer [6], von Känel and Matschke [42], and others. Future work will certainly focus on including this sieving technique for our functions.

In all these methods, we begin with the same search space as in (1), and the computational complexity of a brute force search is easy to estimate. Let B be a bound for the maximum absolute value of an exponent in a solution to the S-unit equation. Since we are searching for a pair $(\tau_1, \tau_2) \in \left(\mathcal{O}_{K,S}^\times\right)^2$, the size of our search space is given by

$$|A_{K,S}|^2 = w^2(2B+1)^{2t}.$$

Thus a naïve brute force search has complexity $O\left(w^2(2B+1)^{2t}\right)$. In practice, a simple exhaustive search can be carried out by checking, for each element τ_1 of $A_{K,S}$, whether $1-\tau_1$ is an S-unit. Assuming this check has constant time for a fixed K and S, we get the less extreme complexity of $O\left(w(2B+1)^t\right)$.

In carrying out computations, we find that the resources required to sieve a search space vary greatly, even for number fields of the same degree and S-unit groups of

Table 3 Runtimes for `sieve_below_bound(K, S, 40)`

$g(x)$	t	w	N	Runtime (in seconds)
$x^4 - x^2 + 1$	2	12	16	1.16
$x^4 + 9$	2	4	0	2.06
$x^4 + 12x^2 + 18$	2	2	0	64

the same rank. For example, we give the run time for three fields K_g, where S_f is the set of primes above 3 in K_g, in Table 3. The column N gives the total number of distinct solutions found. In each case, the LLL-reduced bound is below 40, so complete sets of solutions are found in each case. Computations were performed in a paid account on the CoCalc platform in late 2018.

The resources required depend on the size of the search space, but also can vary greatly based on the particular list of primes Q chosen for the sieve, and even the order of those primes! In many cases, the sieve greatly reduces the time required to exhaust the space. In others, a brute force search of the reduced search space can actually be a better choice, as the sieving computation can take a mysteriously long time. Finding a way to understand and predict these difficulties is a priority for future work. The implementation of idea (3) could also make this unnecessary. In all cases, it is worthwhile to find the smallest reduced bound possible, whether as input for the built-in sieve or for use in a brute force search.

7.2 *Finite Place vs. Infinite Place Bounds*

In general, we find that the LLL-reduced bounds corresponding to $\mathfrak{p}_\ell$ infinite are smaller than the bounds for $\mathfrak{p}_\ell$ finite. To illustrate this, let $\mathscr{K}$ be the set of 85 number fields K satisfying

$$1 \leq [K : \mathbb{Q}] \leq 5, \qquad \Delta_K = \pm 2^a 3^b.$$

If $N \in \mathbb{Z}$, we set

$$S_{\text{fin},K,N} := \{\mathfrak{p} \subseteq \mathcal{O}_K : \mathfrak{p} \mid N\},$$
$$S_{K,N} := S_{\text{fin},K,N} \cup S_\infty.$$

For any choice of $K \in \mathscr{K}$ and $S = S_{K,N}$ where $N \in \{2, 3, 6\}$, we have computed the LLL-reduced bounds under the assumption that $\mathfrak{p}_\ell$ is finite and under the assumption that $\mathfrak{p}_\ell$ is infinite. Complete bound data is available by email request to authors Malmskog or Rasmussen. Here we will consider only the case $S = S_{K,2}$. Now, let $B_1(K)$ and $B_2(K)$ be the bounds obtained in Sect. 5 under the assumption that $\mathfrak{p}_\ell$ is a finite or infinite place, respectively. In Fig. 1, we plot both $B_1(K)$ and $B_2(K)$ against the root discriminant of K (which ranges from 1.74 to 26.56 in $\mathscr{K}$). The bound $B_1(K)$ usually exceeds $B_2(K)$, on average by a factor of ≈ 3.00.

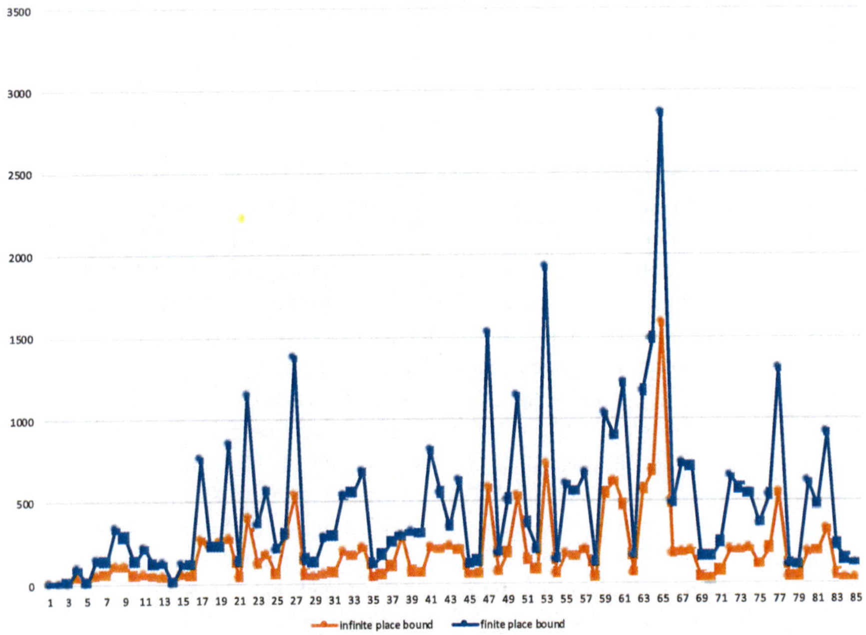

Fig. 1 Bounds $B_1(K)$, $B_2(K)$ (vertical axis) for $S = S_{K,2}$ and $K \in \mathscr{K}$, plotted in order of increasing Δ_K^{1/d_K}

Because the disparity between these bounds is so large, we would prefer to use $B_2(K)$. Generally, we have no control over whether $\mathfrak{p}_\ell$ is finite or infinite. However, if S contains only one finite place, a small trick allows us to use $B_2(K)$. If $(\tau_1, \tau_2) \in \mathscr{O}_{K,S}^\times$ is a solution to the S-unit equation, note that $(\frac{1}{\tau_1}, \frac{-\tau_2}{\tau_1})$ and $(\frac{1}{\tau_2}, \frac{-\tau_1}{\tau_2})$ are also S-unit equation solutions. We define the *solution cycle* of τ_1 to be

$$C(\tau_1) := \left\{\tau_1, 1 - \tau_1, \frac{1}{\tau_1}, 1 - \frac{1}{\tau_1}, \frac{1}{1-\tau_1}, 1 - \frac{1}{1-\tau_1}\right\}.$$

The following result is a restatement of [26, Lemma 6.3].

Lemma 7.1 *Let K be a number field, and suppose S is a finite set of places of K containing all infinite places and at most one finite place, (i.e. $|S_{\text{fin}}| = 1$). Let (τ_1, τ_2) be a solution to the S-unit equation over K. Then at least one element of $C(\tau_1)$ belongs to a solution with $\mathfrak{p}_\ell$ corresponding to an infinite place.*

This implies that under the hypothesis of the lemma, some representative of each solution cycle has an exponent vector bounded by $B_2(K)$; recovering the entire solution cycle from one representative is trivial. Thus, we can determine all solutions to the S-unit equation.

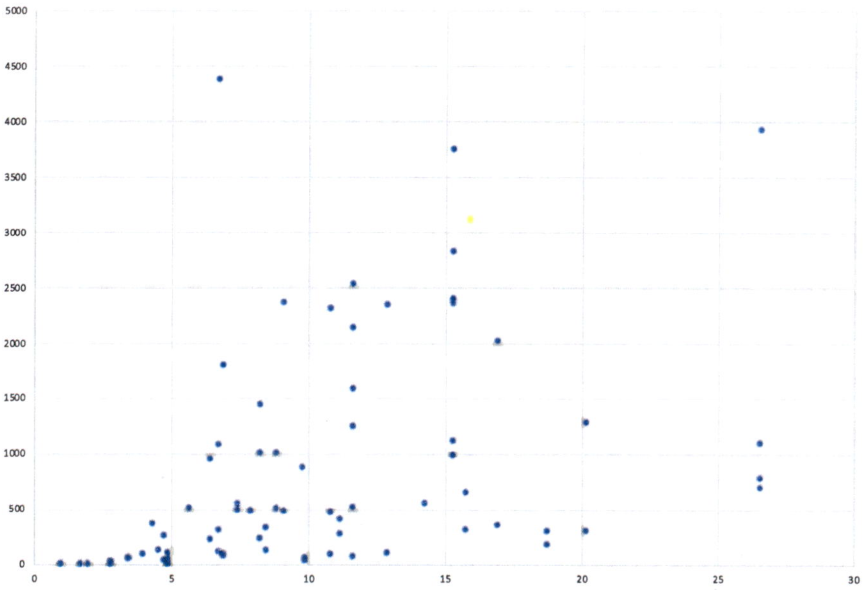

Fig. 2 $R(K)$ (vertical axis) versus Δ_K^{1/d_K} (horizontal axis) for $K \in \mathcal{K}$, $S = S_{K,2}$, and $|S_{\text{fin}}| = 1$

It may seem that the hypothesis of Lemma 7.1—that there is only one finite place in S—is a rather specialized condition. However, many interesting arithmetic applications involve searching for objects with "good" behavior away from one prime p. In such cases, we take $S = S_{K,p}$. Should p ramify in K, the condition $|S_{\text{fin}}| = 1$ is equivalent to p being totally ramified, and this is not so uncommon when $[K : \mathbb{Q}]$ is small. Here, with $S = S_{K,2}$, the lemma applies for 72 of the 85 number fields in $\mathcal{K}$.

To illustrate the utility of Lemma 7.1, consider the ratio of the sizes of the search spaces for two bounds $B_1(K)$ and $B_2(K)$, given by

$$R(K) = \frac{w^2(2B_1(K)+1)^{2t}}{w^2(2B_2(K)+1)^{2t}} \approx \left(\frac{B_1(K)}{B_2(K)}\right)^{2t}.$$

This quantifies the potential savings when the better bound may be used. For $S = S_{K,2}$, Fig. 2 plots the savings $R(K)$ against the root discriminant of K for the 72 fields K in $\mathcal{K}$ for which $|S_{\text{fin}}| = 1$.

8 Applications

A major application of solving S-unit equations is in enumerating solutions to Shafarevich-type problems, for example finding complete lists of curves of a given type with particular reduction properties. The blueprint for this implementation came from Smart's 1997 enumeration of all genus 2 curves over $\mathbb{Q}$ with good reduction away from $p = 2$ [35], building off earlier work with Merriman [27]. In 2017, Malmskog and Rasmussen used these methods to determine all Picard curves defined over $\mathbb{Q}$ with good reduction away from $p = 3$ [26]. The same year, Koutsianas produced a new algorithm that uses solutions to the S-unit equation to find all elliptic curves over an arbitrary number field having good reduction outside S [23]. In the remainder of this article, we provide some new applications of the implementation.

8.1 Asymptotic Fermat

Let $K/\mathbb{Q}$ be a number field. We consider the nontrivial solutions $(a, b, c) \in K^3$ to the Fermat equation:

$$\mathcal{C}_p : a^p + b^p + c^p = 0, \qquad abc \neq 0, \quad p > 3 \text{ a prime.}$$

For fixed p, it follows from the work of Faltings that $\mathcal{C}_p(K)$ is finite, but it is reasonable to ask whether $\bigcup_p \mathcal{C}_p(K)$ is finite or infinite. Finiteness is equivalent to the condition that $\mathcal{C}_p(K) = \varnothing$ for sufficiently large p. We say K satisfies *asymptotic Fermat* if there exists a bound B_K such that $p > B_K$ implies $\mathcal{C}_p(K) = \varnothing$.

There are several number fields K known to satisfy asymptotic Fermat: Jarvis-Meekin [22] demonstrate that $K = \mathbb{Q}(\sqrt{2})$ satisfies asymptotic Fermat with $B_K = 4$. Freitas-Siksek give an explicit family of real quadratic fields of density $\geq \frac{5}{6}$ which satisfy asymptotic Fermat. They also report that the real quartic field, $K = \mathbb{Q}(\sqrt{2+\sqrt{2}})$ satisfies asymptotic Fermat.

In [16], Freitas and Siksek find a condition on a totally real field K which guarantees that K satisfies asymptotic Fermat. For the remainder, suppose K is totally real. Define

$$S = \{\mathfrak{p} : \mathfrak{p} \text{ a nonzero prime ideal of } \mathcal{O}_K \text{ which divides } 2\},$$

$$T = \{\mathfrak{p} \in S : f_{\mathfrak{p}} = 1\}.$$

Theorem 8.1 (Freitas-Siksek) *Let $K/\mathbb{Q}$ be a totally real number field, with either $[K : \mathbb{Q}]$ odd or T nonempty. Suppose that for every solution (τ_1, τ_2) to the S-unit equation, there is some $\mathfrak{p} \in T$ such that* $\max\{|\operatorname{ord}_{\mathfrak{p}}(\tau_1)|, |\operatorname{ord}_{\mathfrak{p}}(\tau_2)|\} \leq 4 \operatorname{ord}_{\mathfrak{p}}(2)$. *Then K satisfies asymptotic Fermat.*

Remark 8.1 We note that Freitas-Siksek's result is actually stronger, and they provide additional conditions under which K must satisfy asymptotic Fermat. Also, more recent work of Şengün-Siksek [32] provides similar criteria for arbitrary number fields. However, the above formulation is sufficient for our application.

The reader may recall that Wiles's classic proof of Fermat's Last Theorem proceeds by taking a hypothetical solution (a, b, c) and noting that the associated

Frey elliptic curve is forced to satisfy an impossible set of constraints (that the curve is not modular). Freitas and Siksek's approach is similar. Given a solution to C_p over K, they produce an elliptic curve E/K (related to, but distinct from, the Frey curve) whose j-invariant is arithmetically constrained. However, the j-invariant is determined by the λ-invariants of E; these λ are guaranteed to arise as solutions to the S-unit equation over K. The result above follows from a delicate analysis of how these constraints interact.

We report a new list of cubic number fields $K/\mathbb{Q}$ which satisfy asymptotic Fermat. Using the implementation of the algorithm described in this paper, we find all solutions to the S-unit equation (S as above), and verify the condition of Freitas-Siksek (this last step is trivial once all solutions have been determined).

Let $\mathscr{K}_X$ denote the set of totally real cubic number fields in which 2 is totally ramified and which have absolute discriminant Δ_K satisfying $|\Delta_K| \leq X$. Table 4 lists all the fields of $\mathscr{K}_X$ for $X = 2000$. For each $K \in \mathscr{K}_X$, we solved the appropriate S-unit equation, and by applying Theorem 8.1, verified that K satisfies asymptotic Fermat. Our results are not effective, as Theorem 8.1 does not provide the bound B_K.

For each $K \in \mathscr{K}_{2000}$, f_K denotes a minimal polynomial for $K/\mathbb{Q}$; Δ_K is the absolute discriminant of K. Because 2 is totally ramified, Lemma 7.1 guarantees that every solution cycle will contain a solution with the extremal place $\mathfrak{p}_\ell$ infinite. Consequently, each solution cycle will contain at least one solution (τ_1, τ_2) satisfying

$$\tau_i = \rho^{\mathbf{a}_i}, \qquad |\mathbf{a}_i| \leq K_1^{\mathrm{LLL}}.$$

(Finding the remaining solutions in the solution cycle is trivial even if they do not satisfy this bound.)

Finally, $N(S, K)$ indicates the number of distinct solutions (τ_1, τ_2) to the S-unit equation found. (These are *unordered* solutions, so that (τ_1, τ_2) and (τ_2, τ_1) are not

Table 4 Fields in $\mathscr{K}_{2000}$ and number of S-unit equation solutions

f_K	Δ_K	K_1^{LLL}	$N(S,K)$
$x^3 - x^2 - 3x + 1$	$2^2 \cdot 37$	225	53
$x^3 - x^2 - 5x - 1$	$2^2 \cdot 101$	175	11
$x^3 - x^2 - 5x + 3$	$2^2 \cdot 3 \cdot 47$	156	5
$x^3 - 6x - 2$	$2^2 \cdot 3^3 \cdot 7$	161	5
$x^3 - x^2 - 7x - 3$	$2^2 \cdot 197$	156	8
$x^3 - 8x - 6$	$2^2 \cdot 269$	176	8
$x^3 - 10x - 10$	$2^2 \cdot 5^2 \cdot 13$	156	8
$x^3 - x^2 - 7x + 5$	$2^2 \cdot 349$	199	8
$x^3 - x^2 - 9x - 5$	$2^2 \cdot 373$	162	8
$x^3 - x^2 - 7x + 1$	$2^2 \cdot 3 \cdot 127$	180	2
$x^3 - x^2 - 9x + 11$	$2^2 \cdot 389$	198	8
$x^3 - 12x - 14$	$2^2 \cdot 3^4 \cdot 5$	164	2
$x^3 - 8x - 2$	$2^2 \cdot 5 \cdot 97$	176	5

Table 5 Solutions to (22) with $3 < q \leq 500$

q	x	k	n	q	x	k	n
11	2	1	1	73	4	2	1
17	−4	4	1	89	2	4	1
17	2	2	1	179	−4	5	1
19	−2	3	1	251	2	5	1
67	4	1	1	307	4	5	1
73	−2	4	1				

considered distinct.) The reader should note that the two trivial solutions over $\mathbb{Q}$, $(-1, 2)$ and $(\frac{1}{2}, \frac{1}{2})$, are counted in each field K.

8.2 *Cubic Ramanujan-Nagell Equations*

In 1913, Ramanujan conjectured that the only solutions of the Diophantine equation $x^2 + 7 = 2^n$ over the natural numbers satisfy $x \in \{1, 3, 5, 11, 181\}$ [30]. This was settled in 1948 by Nagell [28]. The more general family of equations,

$$Ax^2 + B = C^n, \qquad A, B, C \in \mathbb{Z}$$

are called *Ramanujan-Nagell equations*, and the literature for solving such equations is very rich (see for example [10, 9, 11, 7]). Very recently *cubic* Ramanujan-Nagell equations, have attracted the attention of mathematicians [5]. These are equations of the form

$$f(x) = C^n, \qquad f(x) \in \mathbb{Z}[x], \deg f = 3, C \in \mathbb{Z}.$$

We consider the particular example

$$x^3 + 3^k = q^n, \qquad q > 3 \text{ prime}, \quad n, k > 0. \tag{22}$$

If $q = 2$, a more general version of (22) is solved in [5]. Here, we prove the following theorem.

Theorem 8.2 *Let q be a prime with $3 < q \leq 500$. All integer solutions of the cubic Ramanujan-Nagell equation (22) with $k, n > 0$ are listed in Table 5.*

Our method also works for the equation $x^3 + p^k = q^n$, where p, q are different odd primes, and the proof is similar to the case $p = 3$.

Proof Let K be the splitting field for $f(x) = x^3 + 3$. We observe K is unramified outside $\{3, \infty\}$. In fact, K has class number 1 and is totally ramified at 3. Let $\mathfrak{p} = \pi\mathcal{O}_K$ be the unique prime in K above 3. Let S be the set of all places of K above $3, q$, or ∞.

Suppose (q, x, k, n) is a solution to (22). Let β be a root of $f(x)$, and let ζ denote a primitive cube root of unity. Define

$$\alpha_i := (x + \zeta^i \beta^k), \quad \mathfrak{a}_i := \alpha_i \mathcal{O}_K, \quad 0 \leq i \leq 2.$$

Then we must have $\alpha_0\alpha_1\alpha_2 = q^n$ and $\mathfrak{a}_0\mathfrak{a}_1\mathfrak{a}_2 = q^n\mathcal{O}_K$. For $i \neq j$,

$$\alpha_i - \alpha_j = \zeta^i(1 - \zeta^{i-j})\beta^k \in \mathfrak{p}^{2k+3}.$$

Since $(3, q) = 1$, we see $\operatorname{ord}_{\mathfrak{p}} \alpha_i = 0$ for each i. Also, it follows that the $\mathfrak{a}_i$ are pairwise coprime. Thus, if $\mathfrak{q} \mid q\mathcal{O}_K$, then exactly one $\mathfrak{a}_i$ is divisible by $\mathfrak{q}$, and $\operatorname{ord}_{\mathfrak{q}} \alpha_i = n$. Now fix $i' \in \{0, 1, 2\}$ so that $\operatorname{ord}_{\mathfrak{q}} \alpha_{i'} = n$ for at least one $\mathfrak{q} \mid q$. Choose $j' \neq i'$ and set

$$\tau_1 := \frac{\alpha_{i'}}{\alpha_{i'} - \alpha_{j'}}, \quad \tau_2 := \frac{-\alpha_{j'}}{\alpha_{i'} - \alpha_{j'}}.$$

Then (τ_1, τ_2) is a solution to the S-unit equation and $\operatorname{ord}_{\mathfrak{q}} \tau_1 = n$ for some $\mathfrak{q} \mid q$. Choose a root of unity ρ_0 and a basis $\rho_1, \ldots, \rho_t$ for the torsion-free part of $\mathcal{O}_{K,S}^\times$. Choose $b_{i,j} \in \mathbb{Z}$ such that

$$\tau_i = \prod_{j=0}^{t} \rho_j^{b_{i,j}}.$$

There exists B such that $|b_{i,j}| \leq B$. Define

$$c_3 := \sum_{j=1}^{t} \left|\operatorname{ord}_{\mathfrak{p}} \rho_j\right|, \qquad c_{\mathfrak{q}} := \sum_{j=1}^{t} \left|\operatorname{ord}_{\mathfrak{q}} \rho_j\right|,$$

and set $c_q := \max\{c_{\mathfrak{q}} : \mathfrak{q} \mid q\}$. By design,

$$|2k + 3| \leq c_3 B, \qquad |n| \leq c_q.$$

With these bounds established, the solutions to (22) may now be determined by exhaustion. □

As a final remark, we observe that we may choose $\boldsymbol{\rho}$ so that $c_3 = c_q = 1$. Let $\mathfrak{q}_1, \ldots, \mathfrak{q}_g$ be the prime ideals in K above q. As $\mathcal{O}_K$ is a PID, we may choose $\lambda_i \in \mathcal{O}_K$ such that $\mathfrak{q}_i = \lambda_i \mathcal{O}_K$. Let ξ_1, ξ_2 generate the torsion-free part of $\mathcal{O}_K^\times$. The choice $\boldsymbol{\rho} = [\rho_0, \xi_1, \xi_2, \pi, \lambda_1, \ldots, \lambda_g]$ now gives $c_3 = c_q = 1$.

References

1. A. Baker. Linear forms in the logarithms of algebraic numbers. I, II, III. *Mathematika 13 (1966), 204-216; ibid. 14 (1967), 102-107; ibid.*, 14:220–228, 1967.
2. A. Baker and H. Davenport. The equations $3x^2 - 2 = y^2$ and $8x^2 - 7 = z^2$. *Quart. J. Math. Oxford*, 20(2):129-137, 1969.
3. A. Baker and G. Wüstholz. Logarithmic forms and group varieties. *J. Reine Angew. Math.*, 442:19–62, 1993.
4. A. Baker and G. Wüstholz. *Logarithmic forms and Diophantine geometry*, volume 9 of *New Mathematical Monographs*. Cambridge University Press, Cambridge, 2007.
5. M. Bauer and M. A. Bennett. Ramanujan-Nagell cubics. *Rocky Mountain J. Math.*, 48(2):385–412, 2018.
6. M. A. Bennett, A. Gherga, and A. Rechnitzer. Computing elliptic curves over $\mathbb{Q}$. *Math. Comp.*, 88(317):1341–1390, 2019.
7. M. A. Bennett and C. M. Skinner. Ternary Diophantine equations via Galois representations and modular forms. *Canad. J. Math.*, 56(1):23–54, 2004.
8. A. Brumer. On the units of algebraic number fields. *Mathematika*, 14:121–124, 1967.
9. Y. Bugeaud, M. Mignotte, and S. Siksek. Classical and modular approaches to exponential Diophantine equations. II. The Lebesgue-Nagell equation. *Compos. Math.*, 142(1):31–62, 2006.
10. Y. Bugeaud and T. N. Shorey. On the number of solutions of the generalized Ramanujan-Nagell equation. *J. Reine Angew. Math.*, 539:55–74, 2001.
11. J. H. E. Cohn. The Diophantine equation $x^2 + C = y^n$. *Acta Arith.*, 65(4):367–381, 1993.
12. B. M. M. de Weger. Solving exponential Diophantine equations using lattice basis reduction algorithms. *J. Number Theory*, 26(3):325–367, 1987.
13. B. M. M. de Weger. *Algorithms for Diophantine Equations*. PhD thesis, Universiteit Leiden, 1988.
14. B. M. M. de Weger. *Algorithms for Diophantine equations*, volume 65 of *CWI Tract*. Stichting Mathematisch Centrum, Centrum voor Wiskunde en Informatica, Amsterdam, 1989.
15. J.-H. Evertse and K. Győry. *Unit equations in Diophantine number theory*, volume 146 of *Cambridge Studies in Advanced Mathematics*. Cambridge University Press, Cambridge, 2015.
16. N. Freitas and S. Siksek. The asymptotic Fermat's last theorem for five-sixths of real quadratic fields. *Compos. Math.*, 151(8):1395–1415, 2015.
17. C. Fuchs. *On some applications of Diophantine approximations*, volume 2 of *Quaderni/Monographs*. Edizioni della Normale, Pisa, 2014. A translation of Carl Ludwig Siegel's "Über einige Anwendungen diophantischer Approximationen" by Clemens Fuchs, With a commentary and the article "Integral points on curves: Siegel's theorem after Siegel's proof" by Fuchs and Umberto Zannier, Edited by Zannier.
18. K. Győry. On the number of solutions of linear equations in units of an algebraic number field. *Comment. Math. Helv.*, 54(4):583–600, 1979.
19. K. Győry. Bounds for the solutions of S-unit equations and decomposable form equations II. arXiv:1901.11289, January 2019.
20. K. Győry and K. Yu. Bounds for the solutions of S-unit equations and decomposable form equations. *Acta Arithmetica*, 123(1):9–41, 2006.
21. H. Hasse. *Number theory*. Classics in Mathematics. Springer-Verlag, Berlin, german edition, 2002. Reprint of the 1980 English edition [Springer, Berlin; MR0562104 (81c:12001b)], Edited and with a preface by Horst Günter Zimmer.
22. F. Jarvis and P. Meekin. The Fermat equation over $\mathbb{Q}(\sqrt{2})$. *J. Number Theory*, 109(1):182–196, 2004.
23. A. Koutsianas. Computing all elliptic curves over an arbitrary number field with prescribed primes of bad reduction. *Experimental Mathematics*, 2017.
24. A. K. Lenstra, H. W. Lenstra, Jr., and L. Lovász. Factoring polynomials with rational coefficients. *Math. Ann.*, 261(4):515–534, 1982.

25. K. Mahler. Zur Approximation algebraischer Zahlen. I. *Math. Ann.*, 107(1):691–730, 1933.
26. B. Malmskog and C. Rasmussen. Picard curves over $\mathbb{Q}$ with good reduction away from 3. *LMS J. Comput. Math.*, 19(2):382–408, 2016.
27. J. R. Merriman and N. P. Smart. Curves of genus 2 with good reduction away from 2 with a rational Weierstrass point. *Math. Proc. Cambridge Philos. Soc.*, 114(2):203–214, 1993.
28. T. Nagell. The Diophantine equation $x^2 + 7 = 2^n$. *Ark. Mat.*, 4:185–187, 1961.
29. A. Pethö and B. M. M. de Weger. Products of prime powers in binary recurrence sequences. I. The hyperbolic case, with an application to the generalized Ramanujan-Nagell equation. *Math. Comp.*, 47(176):713–727, 1986.
30. S. Ramanujan. Question #464. *J. Indian Math. Soc.*, 5:120, 1913.
31. Sage Developers. *SageMath, the Sage Mathematics Software System (Version 8.4)*, 2018.
32. M. H. Şengün and S. Siksek. On the asymptotic Fermat's last theorem over number fields. *Comment. Math. Helv.*, 93(2):359–375, 2018.
33. C. L. Siegel. Über einige anwendungen diophantischer approximationen. *Abh. der Preuss. Akad. der Wissenschaften Phys. Math. Kl.*, 1:209–266, 1929.
34. N. P. Smart. The solution of triangularly connected decomposable form equations. *Math. Comp.*, 64(210):819–840, 1995.
35. N. P. Smart. S-unit equations, binary forms and curves of genus 2. *Proc. London Math. Soc. (3)*, 75(2):271–307, 1997.
36. N. P. Smart. *The algorithmic resolution of Diophantine equations*, volume 41 of *London Mathematical Society Student Texts*. Cambridge University Press, Cambridge, 1998.
37. N. P. Smart. Determining the small solutions to S-unit equations. *Math. Comp.*, 68(228):1687–1699, 1999.
38. A. Thue. Über Annäherungswerte algebraischer Zahlen. *J. Reine Angew. Math.*, 135:284–305, 1909.
39. N. Tzanakis and B. M. M. de Weger. On the practical solution of the Thue equation. *J. Number Theory*, 31(2):99–132, 1989.
40. N. Tzanakis and B. M. M. de Weger. Solving a specific Thue-Mahler equation. *Math. Comp.*, 57(196):799–815, 1991.
41. N. Tzanakis and B. M. M. de Weger. How to explicitly solve a Thue-Mahler equation. *Compositio Math.*, 84(3):223–288, 1992.
42. R. von Känel and B. Matschke. Solving S-unit, Mordell, Thue, Thue-Mahler and generalized Ramanujan-Nagell equations via Shimura-Taniyama conjecture. Preprint, arXiv:1605.06079, 2016.
43. K. Wildanger. Über das Lösen von Einheiten- und Indexformgleichungen in algebraischen Zahlkörpern. *J. Number Theory*, 82(2):188–224, 2000.
44. K. Yu. Linear forms in p-adic logarithms. *Acta Arith.*, 53(2):107–186, 1989.
45. K. Yu. Linear forms in p-adic logarithms. II. *Compositio Math.*, 74(1):15–113, 1990.
46. K. Yu. Linear forms in p-adic logarithms. III. *Compositio Math.*, 91(3):241–276, 1994.
47. K. Yu. p-adic logarithmic forms and group varieties. III. *Forum Math.*, 19(2):187–280, 2007.

Computing Classical Modular Forms for Arbitrary Congruence Subgroups

Eran Assaf

Abstract In this paper, we prove the existence of an efficient algorithm for the computation of systems of Hecke eigenvalues of modular forms of weight k and level Γ, where $\Gamma \subseteq SL_2(\mathbb{Z})$ is an arbitrary congruence subgroup. We also discuss some practical aspects and provide the necessary theoretical background.

1 Introduction

1.1 Motivation

The absolute Galois group of the rational numbers $G_{\mathbb{Q}} = \mathrm{Gal}(\bar{\mathbb{Q}}/\mathbb{Q})$ is an object of central importance in number theory. It is studied through its action on geometric objects, among which elliptic curves and their torsion points have been playing a dominant role. Let E be an elliptic curve defined over $\mathbb{Q}$, and let p be a prime. Then $G_{\mathbb{Q}}$ acts on its p-torsion points $E[p]$. Therefore, we obtain a representation $\bar{\rho}_{E,p} : G_{\mathbb{Q}} \to GL_2(\mathbb{F}_p)$.

In [28], Serre proved that the index $[GL_2(\mathbb{F}_p) : \bar{\rho}_{E,p}(G_{\mathbb{Q}})]$ is bounded by a constant C_E depending only on E. Serre went on to ask the following question, which remains open to this day, after half a century.

Conjecture 1.1.1 (Serre's Uniformity Problem over $\mathbb{Q}$, [28]) *Is there a constant $C > 0$ such that for any prime $p > C$ and any elliptic curve E over $\mathbb{Q}$ without complex multiplication, the mod p Galois representation $\bar{\rho}_{E,p} : G_{\mathbb{Q}} \to PGL_2(\mathbb{F}_p)$ is surjective?*

There has been much progress in working on the conjecture. The remaining and most difficult case is to exclude the possibility that $\bar{\rho}_{E,p}$ has image contained in the normalizer of a non-split Cartan subgroup of $GL_2(\mathbb{F}_p)$ for large p. These elliptic

E. Assaf (✉)
Dartmouth College, Hanover, NH, USA
e-mail: Eran.Assaf@dartmouth.edu

J. S. Balakrishnan et al. (eds.), *Arithmetic Geometry, Number Theory, and Computation*, Simons Symposia, https://doi.org/10.1007/978-3-030-80914-0_2

curves are classified by a modular curve $X_{ns}^{+}(p)$. Equivalently, one has to show that for all large enough primes p, the only rational points of the modular curve $X_{ns}^{+}(p)$ are CM points.

Therefore, there is great interest in finding explicit equations for $X_{ns}^{+}(p)$ over $\mathbb{Q}$.

More generally, Mazur's Program B [20] suggests, given an open subgroup $G \subseteq GL_2(\hat{\mathbb{Z}})$ to classify all elliptic curves E such that the image of ρ_E is contained in G. To this end, one would like to have explicit equations for the modular curve describing those, X_G, over $\mathbb{Q}$, in order to search for rational points.

The general method to find explicit equations for the modular curves X_G is to compute q-expansions of a basis of cusp forms in $S_k(\Gamma)$ where $\Gamma \subseteq SL_2(\mathbb{Z})$ is the preimage of G in $SL_2(\mathbb{Z})$, and find polynomial relations between them. The complexity of this computation is dominated by the complexity of computing the Hecke operators $\{T_n\}$ on the space of modular forms $S_k(\Gamma)$.

Thus, a main cornerstone of many efforts related to Serre's uniformity conjecture and Mazur's Program B is the computation of Hecke operators and q-expansions of modular forms with arbitrary level Γ.

1.2 Main Results

Let $G \subseteq GL_2(\mathbb{Z}/N\mathbb{Z})$ be a subgroup, and let $k \geq 2$ be an integer. Then G pulls back to a congruence subgroup $\Gamma = \Gamma_G \subseteq SL_2(\mathbb{Z})$ of level N. This paper is concerned mainly with the computation of the space of modular forms of weight k and level Γ_G, which we denote by $M_k(\Gamma_G)$.

Our main goal is to construct an explicit model for $M_k(\Gamma_G)$ over $\mathbb{Q}$, which includes a finite dimensional $\mathbb{Q}$-vector space equipped with a subspace of cusp forms $S_k(\Gamma_G)$, and a family of Hecke operators $\{T_n\}$ acting on it.

After constructing such a model, we proceed to evaluate the systems of eigenvalues of the eigenforms in this space, leading to computation of the zeta functions of the distinct components of the associated Jacobian. We then proceed to compute q-expansions. However, as we shall discuss later, these computations do not bridge the gap in the computational aspect of the efforts for computing q-expansions.

This is achieved most efficiently, under some mild assumptions, using Merel's results [23]. Let us presently describe these assumptions.

We shall make use of $\eta := \begin{pmatrix} -1 & 0 \\ 0 & 1 \end{pmatrix} \in GL_2(\mathbb{Z})$, and denote by $\lambda_N : GL_2(\mathbb{Z}) \to GL_2(\mathbb{Z}/N\mathbb{Z})$ the natural reduction map.

Definition 1.2.1 We say that $G \subseteq GL_2(\mathbb{Z}/N\mathbb{Z})$ is of *real type* if $\lambda_N(\eta) G \lambda_N(\eta)^{-1} = G$. We say that a group $\Gamma \subseteq SL_2(\mathbb{Z})$ is of *real type* if $\eta \Gamma \eta^{-1} = \Gamma$.

Remark 1.2.2 The assumption that G is of real type is essential for the Hecke operators to commute with the star involution (see Lemma 4.1.8). The slightly

Table 1 Summary of complexities for computing Hecke operators

Algorithm	$T_p, p \nmid N$	$T_\alpha, (D(\alpha), N) = 1$	T_α
Complexity	$O(Cdp \log p)$	$O(CdI_{\alpha,\Gamma_G} \log(ND(\alpha)))$	$O\left(d\left(CI_{\alpha,\Gamma_G} \log(ND(\alpha)) + I_G^2 \ln\right)\right)$
$\Gamma_0(N)$	$O(dp \log p)$	$O(dp \log(Np))$	$O(dp \log(Np) + N^2 \log^2 N)$
$\Gamma_{ns}(N)$	$O(dp \log p)$	$O(dp \log(Np))$	$O(dp \log(Np) + N^2)$

weaker assumption, that Γ_G is of real type, is necessary for $Y_\Gamma := \Gamma\backslash\mathcal{H}$ to be defined over $\mathbb{Q}$.

In the description of complexities below, we denote by $I_G := [SL_2(\mathbb{Z}) : \Gamma_G]$ the index of Γ_G, and by $d := \dim S_k(\Gamma_G)$ the dimension of the space of cusp forms. We shall describe the complexity of our algorithms in terms of field operations.

Remark 1.2.3 Since the size of our matrix entries is bounded linearly by the level N, these are essentially logarithmic in N. However, when computing the eigenforms, some of the arithmetic is done over number fields, and the complexity depends on the algorithms used to implement the arithmetic over those fields.

In what follows, we assume the existence of a function CosetIndex(x, R, Γ) that for an element $x \in SL_2(\mathbb{Z})$, a group Γ and a list of coset representatives R for $\Gamma\backslash SL_2(\mathbb{Z})$, returns the unique index i, such that $x \in \Gamma \cdot R[i]$. We will denote by C the complexity of CosetIndex.

Remark 1.2.4 The function CosetIndex could be implemented in general, by noting that $\Gamma_G\backslash SL_2(\mathbb{Z}) \cong G_0\backslash SL_2(\mathbb{Z}/N\mathbb{Z})$, where $G_0 := G \cap SL_2(\mathbb{Z}/N\mathbb{Z})$, and using the Todd-Coxeter algorithm [35], with complexity of $O(I_G^{1/2})$ field operations. However, for certain groups Γ, there are more efficient implementations. For example, if $\Gamma = \Gamma_0(N)$, this can be done in $O(1)$ field operations using the fact that $\Gamma\backslash SL_2(\mathbb{Z}) \cong \mathbb{P}^1(\mathbb{Z}/N\mathbb{Z})$. Moreover, by allowing pre-computation of $O(N^3)$ field operations and memory of $O(N^3)$ field elements, we can use a table and access it using $O(\log N)$ field operations.

For the convenience of the reader, a table is included in which the complexities of the different algorithms for computing Hecke operators are compared to one another, and specialized to $\Gamma_0(N)$, $\Gamma_{ns}(N)$ (Table 1).

Theorem 1.2.5 (Corollary 4.5.11) *There exists an algorithm that given a group of real type $G \subseteq GL_2(\mathbb{Z}/N\mathbb{Z})$, a prime $p \nmid N$ such that $p \bmod N \in \det(G)$ and an integer $k \geq 2$, computes the Hecke operator T_p on the space of modular forms $S_k(\Gamma_G)$ using $O(C \cdot dp \log p)$ field operations.*

Example 1.2.6

(1) If $\Gamma = \Gamma_0(N)$, then $C = O(1)$. Thus we obtain $O(dp \log p)$, the standard complexity.

(2) If $\Gamma = \Gamma_{ns}(N)$ is the non-split Cartan subgroup, then $C = O(1)$. Again we obtain $O(dp \log p)$. Note that here $d = \dim S_k(\Gamma)$, where standard methods cost $O(d_0 p \log p)$, with $d_0 = \dim S_k(\Gamma_0(N^2)) > d$.

We also provide an algorithm for computing Hecke operators corresponding to arbitrary double cosets. For a matrix $\alpha \in GL_2^+(\mathbb{Q})$, we let $D(\alpha) = \det(\alpha)/d_1(\alpha)^2 \in \mathbb{Z}$, with $d_1(\alpha)$ the greatest common divisor of all the entries of α. Denote also $I_{\alpha,\Gamma} = [\Gamma : \alpha^{-1}\Gamma\alpha \cap \Gamma]$.

Theorem 1.2.7 (Corollary 4.3.3) *There exists an algorithm that given a group of real type $G \subseteq GL_2(\mathbb{Z}/N\mathbb{Z})$, an element $\alpha \in GL_2^+(\mathbb{Q})$ such that $\eta^{-1}\alpha\eta \in \Gamma_G\alpha\Gamma_G$ and $\gcd(D(\alpha), N) = 1$, and an integer $k \geq 2$, computes the Hecke operator T_α, corresponding to the double coset $\Gamma_G\alpha\Gamma_G$, on the space of modular forms $S_k(\Gamma_G)$, using $O(CdI_{\alpha,\Gamma_G} \log(ND(\alpha)))$ field operations.*

Example 1.2.8 If $\Gamma = \Gamma_0(N)$ and $\alpha = \begin{pmatrix} 1 & 0 \\ 0 & p \end{pmatrix}$ with $p \nmid N$, then $I_{\alpha,\Gamma} = p+1$, $C = O(1)$ and $D(\alpha) = p$. Thus we obtain $O(dp \log(Np))$, slower than the algorithm described in Theorem 1.2.5. However, this algorithm works in greater generality.

Remark 1.2.9 If G is not of real type, the above algorithm still computes the Hecke operator T_α on the space $S_k(\Gamma_G) \oplus \overline{S_k(\Gamma_G)}$. This, in turn, gives rise to a system of eigenvalues, from which one can deduce the zeta function. An example is (group $8E^1$ in [13])

$$G = \left\langle \begin{pmatrix} 2 & 3 \\ 3 & 5 \end{pmatrix}, \begin{pmatrix} 3 & 0 \\ 0 & 3 \end{pmatrix}, \begin{pmatrix} 7 & 0 \\ 0 & 7 \end{pmatrix}, \begin{pmatrix} 7 & 4 \\ 3 & 1 \end{pmatrix}, \begin{pmatrix} 4 & 1 \\ 3 & 4 \end{pmatrix} \right\rangle \subseteq GL_2(\mathbb{Z}/8\mathbb{Z})$$

which is not conjugate to any subgroup of real type. In this case, $\dim S_2(G) = 1$, and one may compute the Hecke operators using the code as follows.

```
> gens := [[ 7, 0, 0, 7 ],[ 2, 3, 3, 5 ],[ 0, 7, 7, 7 ],
        [ 3, 0, 0, 3 ],[ 4, 7, 7, 3 ]];
> N := 8;
> H_N := sub<GL(2,Integers(N)) | gens>;
> H := PSL2Subgroup(H_N);
> M := ModularSymbols(H, 2, Rationals(), 0);
> S := CuspidalSubspace(M);
> HeckeOperator(S,97);
[18  0]
[ 0 18]
```

We describe a general algorithm to compute the Hecke operators, which works for the case $\gcd(D(\alpha), N) > 1$ as well. To describe its complexity, denote the complexity of membership testing in G by In.

Theorem 1.2.10 (Theorem 4.2.10) *There exists an algorithm that given a group of real type $G \subseteq GL_2(\mathbb{Z}/N\mathbb{Z})$, an element $\alpha \in GL_2^+(\mathbb{Q})$ such that $\eta^{-1}\alpha\eta \in \Gamma_G\alpha\Gamma_G$*

and an integer $k \geq 2$, computes the Hecke operator T_α corresponding to the double coset $\Gamma_G \alpha \Gamma_G$, on the space of modular forms $S_k(\Gamma_G)$, using

$$O\left(d\left(C I_{\alpha,\Gamma_G} \log(ND(\alpha)) + I_G^2 In\right)\right)$$

field operations.

Example 1.2.11 If $\Gamma = \Gamma_0(N)$ and $\alpha = \begin{pmatrix} 1 & 0 \\ 0 & p \end{pmatrix}$ with $p \mid N$, then $I_G = O(N \log N)$, yielding $O(dp \log(Np) + N^2 \log^2 N)$. One may notice that the second term we are picking up, due to the generality, is quite large.

These algorithms allow us to compute zeta functions efficiently. For a complete zeta function, we need also the values of the Hecke operators at primes dividing the level. Therefore, we need to assume an additional hypothesis.

Definition 1.2.12 Let $G \subseteq GL_2(\mathbb{Z}/N\mathbb{Z})$, and let $p \mid N$ be a prime. Let $\{\alpha_{p,i}\}_{i=1}^{r_p}$ be elements in $GL_2^+(\mathbb{Q})$ such that T_p is a linear combination of the $T_{\alpha_{p,i}}$. We say that the Hecke operator T_p on $S_k(\Gamma_G)$ is *effectively computable* if there exists an algorithm to compute a set of such elements $\{\alpha_{p,i}\}$ using $O(Cd(p \log p + kI_G \log(kI_G)) + d^3)$ field operations.

Example 1.2.13

(1) When $\Gamma = \Gamma_0(N)$, T_p is effectively computable for all $p \mid N$, as $\alpha_p = \begin{pmatrix} 1 & 0 \\ 0 & p \end{pmatrix}$ can be computed in $O(1)$ field operations.
(2) When the order generated by Γ contains no elements of determinant p, $T_p = 0$ is effectively computable.

We further denote by In the complexity, in field operations, of membership test in G.

Corollary 1.2.14 (Corollary 6.3.4) *There exists an algorithm that given a group of real type $G \subseteq GL_2(\mathbb{Z}/N\mathbb{Z})$ with surjective determinant such that for all $p \mid N$, T_p is effectively computable, and a positive integer L, returns the zeta functions associated to each of the factors of $Jac(X(\Gamma_G))$ up to precision L^{-s} using*

$$O(d(C(L \log L + N \log N) + N I_G^2 \,\mathrm{In} + I_G \log(I_G)) + d^3)$$

field operations.

This further allows one to compute q-expansions as follows. Consider the inclusion $\iota : S_k(\Gamma_G) \subseteq S_k(\Gamma(N))$, where N is the level of Γ_G. This inclusion is not Hecke equivariant in general. Therefore, our eigenvalues are not the q-expansions of forms in $S_k(\Gamma_G)$.

However, we have an action of $GL_2(\mathbb{Z}/N\mathbb{Z})$ on $S_k(\Gamma(N))$. Write $S_k(\Gamma(N)) = \bigoplus_{i \in I} V_i$, with each V_i an irreducible $GL_2(\mathbb{Z}/N\mathbb{Z})$-representation. Since $V :=$

$\iota(S_k(\Gamma_G)) = S_k(\Gamma(N))^G$, we have $V = \bigoplus_{i\in I} V \cap V_i$. But in $\Gamma(N)$ the q-expansions of newforms do correspond to the zeta functions of the appropriate factors. Therefore, the q-expansions we have found for a basis of $V \cap V_i$ are the q-expansions of newforms in V_i, denote them by g_{ij}. By Atkin et al. [3], for each i, the space V_i is generated by twists of the $\{g_{ij}\}$. Thus, we are now led to finding the linear combinations that land in $V \cap V_i$. This can be done either by computing the $GL(2, \mathbb{Z}/N\mathbb{Z})$-action on cusp forms, as described in [38], or by using modular symbols for $\Gamma(N)$, and solving linear equations. This can be done in several ways, and is currently in the process of being implemented by Josha Box (see [9]) and by the author.

Thus, obtaining the q-expansions remains highly inefficient for two reasons.

(1) In order to obtain the q-expansions one has to work in the larger space $S_k(\Gamma(N))$.
(2) The linear algebra required to obtain the q-expansions has to be done over large fields, as the twisting operators introduce the Gauss sums of the characters, which are elements in the large cyclotomic field $\mathbb{Q}(\zeta_N, \zeta_{\phi(N)})$.

The method illustrated in this paper has been implemented in the MAGMA computational algebra system, by building on the existing package of modular symbols, implemented in MAGMA [8] by William Stein, with contributions by Steve Donnelly and Mark Watkins. The package is publicly available in the github repository [2].

1.3 Recovering Known Results

As an application of our algorithm we recover the equation for the canonical embedding of $X_{S_4}(13)$ in $\mathbb{P}^2_{\mathbb{Q}}$, as was done in [4]. Effectively, this repeats the work done in the original paper, only in an automated fashion. The main difference is that the computation of the eigenform in step 3 of Sect. 4 is done directly in the smaller space $S_2(\Gamma_{S_4})$, instead of inside $S_2(\Gamma_0(169))$.

Corollary 1.3.1 ([4, Theorem 1.8]) *The modular curve $X_{S_4}(13)$ is a genus 3 curve whose canonical embedding in $\mathbb{P}^2_{\mathbb{Q}}$ has the model*

$$4x^3y - 3x^2y^2 + 3xy^3 - x^3z + 16x^2yz - 11xy^2z$$
$$+5y^3z + 3x^2z^2 + 9xyz^2 + y^2z^2 + xz^3 + 2yz^3 = 0.$$

Another application recovers the main result of [6], namely explicit equations for the canonical embeddings of $X_{ns}^+(13)$ and $X_s(13)$ in $\mathbb{P}^2_{\mathbb{Q}}$. Again, this basically repeats the work done in the original paper, only obtaining the result of Sect. 4 in a matter of seconds.

Corollary 1.3.2 ([6]) *The modular curves* $X_{ns}^{+}(13)$ *and* $X_s(13)$ *are defined by the equation*

$$(-y-z)x^3+(2y^2+zy)x^2+(-y^3+zy^2-2z^2y+z^3)x+(2z^2y^2-3z^3y)=0.$$

Similarly, we are also able to reproduce the results of [21] for $X_{ns}^{+}(17)$, $X_{ns}^{+}(19)$ and $X_{ns}^{+}(23)$.

1.4 New Results

An interesting problem that requires the computation of such Hecke operators is the question of decomposition of Jacobians of modular curves. The factors in these decompositions should have interesting arithmetic. We could perform the following using our code in 31 min. Note that using Chen's isogeny, one would need to compute and decompose the space of modular symbols $S_2(\Gamma_0(97^2))$.

Corollary 1.4.1 *The Jacobian of the modular curve* $X_{ns}^{+}(97)$ *decomposes over* $\mathbb{Q}$ *as the direct sum of 13 Hecke irreducible subspaces, of dimensions*

$$3, 4, 4, 6, 7, 7, 12, 14, 24, 24, 24, 56, 168.$$

In particular, it has no elliptic curve factor.

Remark 1.4.2 Any method of computing modular forms of weight $k=1$ relies on computation of modular forms of higher integral weights (e.g. [10], [27]), so that our results apply equally well.

1.5 Related Literature

The theory of modular forms for GL_2 has been established for quite some time. However, the theory usually restricts to the Iwahori level subgroups, $\Gamma_0(N)$ and $\Gamma_1(N)$. There are a few exceptions: Shimura, in [29], explores a mild generalization, and much work has been devoted to some special cases, such as those groups induced by maximal subgroups of $GL_2(\mathbb{Z}/N\mathbb{Z})$. In this vein, computation has also been restricted to the subgroups $\Gamma_0(N)$ and $\Gamma_1(N)$, as in [32] and in [11].

Nevertheless, many new applications arise [4, 5, 16, 34] in which description of the space of modular forms $M_k(\Gamma)$ is needed, when $\Gamma \subseteq SL_2(\mathbb{Z})$ is an arbitrary congruence subgroup.

1.6 Organization

The paper is structured as follows.

In Sect. 2, we introduce the definitions and the statement of the problem.

In Sect. 3, we explain the construction of an explicit model for $M_k(\Gamma)$ using modular symbols, how to construct a boundary map ∂, and through it the subspace of cusp forms $S_k(\Gamma)$. This description of $S_k(\Gamma)$ was already known to Merel, see [23], but we present a general algorithm for the computation of ∂. So far, the literature has only discussed the computation of ∂ for Iwahori level subgroups. (see [12, 32, 37]).

Section 4 contains the core of this work, the computation of the Hecke operators. In Sect. 4.1, we begin by defining the Hecke operators corresponding to double cosets. Then, in Sect. 4.2, we present a general algorithm for computing them, thus proving Theorem 1.2.10. We then present a variant of this algorithm which is more efficient, but applicable only to Hecke operators away from the level, thus yielding Theorem 1.2.7. In order to define the Hecke operators T_n for integers n that are coprime to the level of Γ, i.e. $(n, N) = 1$, in Sect. 4.4, we need to view the modular curve adelically. We start from the adelic Hecke operators and show that they correspond to certain double coset operators that were already defined. In Sect. 4.5, we use the techniques of Merel from [23] to prove Theorem 1.2.5.

Section 5 describes the degeneracy maps, and the old and new subspaces, which will be useful when decomposing our space of cusp forms. Section 6 contains miscellaneous algorithms needed to compute the zeta functions associated to the eigenforms, given the Hecke operators. In Sect. 7 we describe several applications of our result to contemporary research.

1.7 Acknowledgements

It is a pleasure to thank John Voight for many insightful discussions on this project. The author would like to thank deeply Jeremy Rouse for his very useful comments on an earlier draft of this manuscript and for pointing to several interesting references in the literature. The author is also grateful to David Zywina and Elisa Lorenzo Garciá for their remarks on the code package, and to the anonymous referees for their comments on the manuscript and the code. This research was conducted as part of the *Simons Collaboration on Arithmetic Geometry, Number Theory and Computation,* with the support of Simons Collaboration Grant (550029, to John Voight).

2 Setup and Notation

In this section, we present the basic setup and introduce some notations that will be used throughout the paper.

2.1 Congruence Subgroups

Let N be a positive integer, $G \subseteq GL_2(\mathbb{Z}/N\mathbb{Z})$ a subgroup, and $G_0 = G \cap SL_2(\mathbb{Z}/N\mathbb{Z})$. Let $\lambda_N : M_2(\mathbb{Z}) \to M_2(\mathbb{Z}/N\mathbb{Z})$ be the natural reduction map, and let $\lambda_{N,0} : SL_2(\mathbb{Z}) \to SL_2(\mathbb{Z}/N\mathbb{Z})$ be its restriction to $SL_2(\mathbb{Z})$.

Then $\Gamma_G := \lambda_{N,0}^{-1}(G_0) \subseteq SL_2(\mathbb{Z})$ contains $\Gamma(N) := \ker(\lambda_{N,0})$, hence it is a congruence subgroup. Moreover, every congruence subgroup arises this way. In general, we will denote by Γ a congruence subgroup of $SL_2(\mathbb{Z})$.

We denote by $\mathcal{H}$ the complex upper half plane $\mathcal{H} := \{z \in \mathbb{C} \mid \mathrm{Im}(z) > 0\}$. Then $\mathcal{H}$ admits a natural action of $GL_2^+(\mathbb{R})$ via Möbius transformations and we let $Y_\Gamma := \Gamma\backslash\mathcal{H}$ be the affine modular curve of level Γ.

Definition 2.1.1 Denote $\mathcal{H}^* = \mathcal{H} \cup \mathbb{P}^1(\mathbb{Q})$. The group $GL_2^+(\mathbb{Q})$ also acts on $\mathbb{P}^1(\mathbb{Q})$ via Möbius transformations, and we let $X_\Gamma := \Gamma\backslash\mathcal{H}^*$ be the *modular curve of level* Γ. The points of $\Gamma\backslash\mathbb{P}^1(\mathbb{Q}) = X_\Gamma - Y_\Gamma$ are called the *cusps* of X_Γ.

Definition 2.1.2 We say that Γ_G is the congruence subgroup *induced* by G.

Here are several useful examples.

Example 2.1.3 The following are standard examples of congruence subgroups, which will be used throughout.

(1) Let $G = \left\{\begin{pmatrix} a & 0 \\ 0 & 1 \end{pmatrix} \mid a \in (\mathbb{Z}/N\mathbb{Z})^\times\right\}$. Then $\Gamma_G = \Gamma(N)$.

(2) Let $G = \left\{\begin{pmatrix} a & b \\ 0 & 1 \end{pmatrix} \mid a \in (\mathbb{Z}/N\mathbb{Z})^\times, b \in \mathbb{Z}/N\mathbb{Z}\right\}$. Then $\Gamma_G = \Gamma_1(N)$.

(3) Let $G = \left\{\begin{pmatrix} a & b \\ 0 & d \end{pmatrix} \mid a, d \in (\mathbb{Z}/N\mathbb{Z})^\times, b \in \mathbb{Z}/N\mathbb{Z}\right\}$. Then $\Gamma_G = \Gamma_0(N)$.

(4) Let $\mathfrak{h}$ be a subgroup of $(\mathbb{Z}/N\mathbb{Z})^\times$, and let t be a divisor of N. Let
$G = \left\{\begin{pmatrix} a & b \\ 0 & d \end{pmatrix} \mid a \in \mathfrak{h}, d \in (\mathbb{Z}/N\mathbb{Z})^\times, b \in t\mathbb{Z}/N\mathbb{Z}\right\}$. Then $\Gamma_G = \Gamma(\mathfrak{h}, t)$.

Before presenting a non-trivial interesting example, we define the notion of a non-split Cartan subgroup.

Definition 2.1.4 Let A be a finite free commutative $\mathbb{Z}/N\mathbb{Z}$-algebra of rank 2 with unit discriminant, such that for every prime p dividing N, the $\mathbb{F}_p$-algebra A/pA is isomorphic to $\mathbb{F}_{p^2}$. Then the unit group $A^\times$ acts on A by multiplication, and a choice of basis for A induces an embedding $A^\times \hookrightarrow GL_2(\mathbb{Z}/N\mathbb{Z})$. The image of $A^\times$ is called a *non-split Cartan subgroup* of $GL_2(\mathbb{Z}/N\mathbb{Z})$.

Example 2.1.5 Let G be a non-split Cartan subgroup of $GL_2(\mathbb{Z}/N\mathbb{Z})$. Then $\Gamma_G = \Gamma_{ns}(N)$. Similarly, if $\mathfrak{N}(G)$ is its normalizer in $GL_2(\mathbb{Z}/N\mathbb{Z})$ then $\Gamma_{\mathfrak{N}(G)} = \Gamma_{ns}^+(N)$.

2.2 *Modular Forms*

Definition 2.2.1 Let $k \geq 2$ be an integer. Let $g = \begin{pmatrix} a & b \\ c & d \end{pmatrix} \in GL_2^+(\mathbb{Q})$. Let $f : \mathcal{H} \to \mathbb{C}$, where $\mathcal{H} := \{z \in \mathbb{C} \mid \operatorname{Im}(z) > 0\}$. We define for any $z \in \mathcal{H}$

$$f|_g(z) = (\det g)^{k-1} \cdot (cz + d)^{-k} f(gz). \tag{2.2.2}$$

Denote by $M_k(\Gamma)$ the space of holomorphic modular forms of weight k with respect to Γ (i.e. holomorphic functions $f : \mathcal{H} \to \mathbb{C}$ such that $f|_\gamma = f$ for all $\gamma \in \Gamma$ and f is holomorphic at the cusps of $\Gamma\backslash\mathcal{H}$). Denote by $S_k(\Gamma)$ the subspace of cusp forms (those vanishing at the cusps).

Since $\Gamma(N) \subseteq \Gamma$, we know that $\begin{pmatrix} 1 & N \\ 0 & 1 \end{pmatrix} \in \Gamma$, hence $f(z + N) = f(z)$ for all $f \in M_k(\Gamma)$. It follows that f admits a Fourier expansion $f(z) = \sum_{n=0}^{\infty} a_n q^n$, with $a_n \in \mathbb{C}$ and $q := e^{\frac{2\pi i z}{N}}$.

This is now enough to state a computational problem:

Problem 2.2.3 Given a group $G \subseteq GL_2(\mathbb{Z}/N\mathbb{Z})$, an integer $k \geq 2$ and a positive integer L, find the q-expansion of a basis for $S_k(\Gamma_G)$ up to precision q^L.

This problem has been answered extensively in the literature for $\Gamma_G = \Gamma_1(N)$ and $\Gamma_G = \Gamma_0(N)$ (see [31, 32, 17, 11]). In these cases, the solution to Problem 2.2.3 is through the following steps:

(1) Construct an explicit vector space representing $S_k(\Gamma_G)$.
(2) Compute the action of Hecke operators on this space.
(3) Decompose the action into irreducible subspaces.
(4) Write down the q-expansion from the systems of eigenvalues in the above decomposition.

Among these steps, step (3) and (4) follow from (2) by linear algebra techniques. Moreover, the Hecke algebra of $S_k(\Gamma_G)$ is generated by the double coset Hecke operators $\{T_\alpha\}$ (see Corollary 4.1.4). Therefore, we concentrate our efforts on the following problem.

Problem 2.2.4 Given a group $G \subseteq GL_2(\mathbb{Z}/N\mathbb{Z})$, an integer $k \geq 2$ and an element $\alpha \in GL_2^+(\mathbb{Q})$ compute the matrix of the Hecke operator T_α with respect to a basis of $S_k(\Gamma_G)$.

We address this problem in full in Sect. 4, and present several solutions—one general solution with no additional assumptions, and under some mild assumptions, we have a more efficient solution.

Unfortunately, in the general case step (4) does not follow from step (3) in the same way. The reason is that the Hecke operators acting on $S_k(\Gamma_G)$ have no longer a nice description in terms of the Fourier coefficients. For an example of such a description, see [38].

We can and will show that the Hecke operators we construct correspond to the Frobenius action on the curves, hence the system of eigenvalues of any eigenform f does yield the zeta function of A_f, the corresponding factor of the Jacobian.

3 Explicit Computation of $S_k(\Gamma)$

First, we have to construct a model for the vector space $S_k(\Gamma)$. We will do so through the means of modular symbols. For the results of this section, it suffices to assume that Γ is a finite index subgroup of $SL_2(\mathbb{Z})$. Therefore, in this section, we will suppress the notation of the group G in Γ_G.

3.1 Modular Symbols

Let $\mathbb{M}_2$ be the free abelian group generated by the expressions $\{a, b\}$, for $(a, b) \in \mathbb{P}^1(\mathbb{Q}) \times \mathbb{P}^1(\mathbb{Q})$, subject to the relations

$$\{a, b\} + \{b, c\} + \{c, a\} = 0 \quad \forall a, b, c \in \mathbb{P}^1(\mathbb{Q})$$

and modulo any torsion. Let $\mathbb{M}_k := \mathrm{Sym}^{k-2}\,\mathbb{Z}^2 \otimes \mathbb{M}_2$, where Sym^{k-2} is the representation of GL_2 of highest weight $k-2$.

There is a natural left action of $GL_2(\mathbb{Q})$ on $\mathbb{M}_k$ by

$$g(v \otimes \{a, b\}) = gv \otimes \{ga, gb\}.$$

Definition 3.1.1 Let $\mathbb{M}_k(\Gamma) = (\mathbb{M}_k)_\Gamma := \mathbb{M}_k / \langle \gamma x - x \rangle$ be the space of Γ-coinvariants, the space of *modular symbols of weight k for* Γ (over $\mathbb{Z}$). The space of *modular symbols of weight k for* Γ *over a ring R is*

$$\mathbb{M}_k(\Gamma; R) := \mathbb{M}_k(\Gamma) \otimes_{\mathbb{Z}} R.$$

The reason we are interested in the space of modular symbols $\mathbb{M}_k(\Gamma)$ is the following theorem proved by Manin:

Theorem 3.1.2 ([19, Theorem 1.9]) *The natural homomorphism*

$$\varphi : \mathbb{M}_2(\Gamma) \to H_1(X_\Gamma, cusps, \mathbb{Z}),$$

sending the symbol $\{a, b\}$ *to the geodesic path in* X_Γ *between* Γa *and* Γb*, is an isomorphism.*

This will give us the connection to modular forms. We now turn to the reason that modular symbols are useful for computation.

3.2 Manin Symbols

Definition 3.2.1 Let $g \in SL_2(\mathbb{Z})$. The *Manin symbol* $[v, g] \in \mathbb{M}_k(\Gamma)$ is defined as

$$[v, g] := g(v \otimes \{0, \infty\}).$$

Let

$$\sigma = \begin{pmatrix} 0 & -1 \\ 1 & 0 \end{pmatrix}, \quad \tau = \begin{pmatrix} 0 & -1 \\ 1 & -1 \end{pmatrix}, \quad J = \begin{pmatrix} -1 & 0 \\ 0 & -1 \end{pmatrix}.$$

We have a right action of $SL_2(\mathbb{Z})$ on Manin symbols as follows:

$$[v, g]h = [h^{-1}v, gh].$$

The following theorem gives an explicit recipe for constructing a concrete realization of $\mathbb{M}_k(\Gamma)$ as a free module of finite rank.

Theorem 3.2.2 ([23, Propositions 1,3]) *The Manin symbol* $[v, g]$ *depends only on the class* $\Gamma g \in \Gamma \backslash SL_2(\mathbb{Z})$ *and on* v*. The Manin symbols* $\{[v, g]\}_{g \in SL_2(\mathbb{Z}), v \in Sym^{k-2}\mathbb{Z}^2}$ *generate* $\mathbb{M}_k(\Gamma)$*. Furthermore, if* x *is a Manin symbol, then*

$$x + x\sigma = 0,$$

$$x + x\tau + x\tau^2 = 0,$$

$$x - xJ = 0.$$

Moreover, these are all the relations between Manin symbols, i.e. if B *is a basis for* $Sym^{k-2}\mathbb{Z}^2$ *then*

$$\mathbb{M}_k(\Gamma) \cong \left(\bigoplus_{b \in B, g \in \Gamma \backslash SL_2(\mathbb{Z})} \mathbb{Z} \cdot [b, g] \right) / R$$

where R are the above relations, and torsion.

This allows one to compute a model for $\mathbb{M}_k(\Gamma)$ efficiently.

Corollary 3.2.3 *There exists an algorithm that given a finite index subgroup* $\Gamma \subseteq SL_2(\mathbb{Z})$ *and an integer* $k \geq 2$, *computes a basis of* $\mathbb{M}_k(\Gamma)$ *in* $O([SL_2(\mathbb{Z}) : \Gamma])$ *basic* CosetIndex *operations.*

Proof This was implemented by Stein for $\Gamma = \Gamma_0(N)$ in [32], and the general case is similar. □

Remark 3.2.4 The implementation in [32] makes use of the fact that $\Gamma_0(N)\backslash SL_2(\mathbb{Z}) \cong \mathbb{P}^1(\mathbb{Z}/N\mathbb{Z})$. This allows him to perform CosetIndex operations in $O(1)$ field operations. This is the only difference from the general case.

Definition 3.2.5 The module $\mathrm{Sym}^{k-2}\mathbb{Z}^2$ has a basis consisting of homogeneous elements, namely $\{x^w y^{k-2-w}\}_{w=0}^{k-2}$. It follows that $\mathbb{M}_k(\Gamma)$ has a basis consisting of elements of the form $[x^w y^{k-2-w}, g]$. From now on, we assume all given bases to be of this form, and say that the *weight* of a basis element $[x^w y^{k-2-w}, g]$ is w.

3.3 Modular Symbols with Character

Suppose $\Gamma \subseteq \Gamma' \subseteq SL_2(\mathbb{Z})$ are finite index subgroups such that Γ is a normal subgroup of Γ' and let $Q := \Gamma'/\Gamma$. Then Q acts on $\mathbb{M}_k(\Gamma)$ via

$$(\gamma'\Gamma) \cdot [v, g] = [v, \gamma' g]. \tag{3.3.1}$$

Let $\varepsilon : Q \to \mathbb{Q}(\zeta)^\times$ be a character, where $\zeta = \zeta_n$ is an n-th root of unity, and n is the order of ε. Abusing notation, we will denote the induced character on Γ' by ε as well.

Let $\mathbb{M}_k(\Gamma, \varepsilon)$ be the quotient of $\mathbb{M}_k(\Gamma; \mathbb{Z}[\zeta])$ by the relations

$$\gamma' \cdot [v, g] - \varepsilon(\gamma') \cdot [v, g] = 0$$

for all $x \in \mathbb{M}_k(\Gamma; \mathbb{Z}[\zeta])$, $\gamma' \in Q$, and by any $\mathbb{Z}$-torsion.

Note that $\mathbb{M}_k(\Gamma, \varepsilon)$ has a basis consisting solely of elements of the form $[v, g]$ where $g \in \Gamma'\backslash SL_2(\mathbb{Z})$. Then if Q is abelian, this action factorizes to the isotypic components:

$$\mathbb{M}_k(\Gamma) \cong \bigoplus_{\varepsilon \in \hat{Q}} \mathbb{M}_k(\Gamma, \varepsilon). \tag{3.3.2}$$

This decomposition allows for faster computation by implementing $\mathbb{M}_k(\Gamma)$ as the direct sum of these smaller spaces.

3.4 Cuspidal Modular Symbols

Let $\mathbb{B}_2$ be the abelian group generated by the elements $\{a\}$, $a \in \mathbb{P}^1(\mathbb{Q})$. Let $\mathbb{B}_k = \mathrm{Sym}^{k-2}\mathbb{Z}^2 \otimes \mathbb{B}_2$. We have a left action of $GL_2(\mathbb{Q})$ on $\mathbb{B}_k$ via

$$g(v \otimes \{a\}) = gv \otimes \{ga\}.$$

Let $\mathbb{B}_k(\Gamma) := (\mathbb{B}_k)_\Gamma$. This can be viewed naturally as a space of "boundary Manin symbols", using the following boundary map.

Definition 3.4.1 The *boundary map* $\partial : \mathbb{M}_k(\Gamma) \to \mathbb{B}_k(\Gamma)$ is the natural map extending linearly the map

$$\partial(v \otimes \{a, b\}) = v \otimes a - v \otimes b.$$

Again we have a theorem that allows us to find a nice description of this module as a free module of finite rank.

Theorem 3.4.2 ([23, Propositions 4,5]) *Let $\mathcal{R}$ be the equivalence relation on $\Gamma\backslash\mathbb{Q}^2$ given by*

$$[\Gamma(\lambda u, \lambda v)] \sim sign(\lambda)^k[\Gamma(u, v)]$$

for any $\lambda \in \mathbb{Q}^\times$. Let $\mu : \mathbb{B}_k(\Gamma) \to \mathbb{Q}[(\Gamma\backslash\mathbb{Q}^2)/\mathcal{R}]$ be the natural map given by

$$\mu\left(P \otimes \left\{\frac{u}{v}\right\}\right) = P(u, v) \cdot [\Gamma(u, v)]$$

where $P \in Sym^{k-2}\mathbb{Z}^2$ is considered as a homogeneous polynomial in two variables of degree $k-2$ and u, v are coprime.

Then μ is well defined and injective. Moreover,

$$\mu \circ \partial([P, g]) = P(1, 0)[\Gamma g(1, 0)] - P(0, 1)[\Gamma g(0, 1)].$$

The space $\mathbb{S}_k(\Gamma) = \ker \partial$ is called the space of *cuspidal modular symbols*. As in Sect. 3.3, we can define $\mathbb{B}_k(\Gamma, \varepsilon)$, a boundary map $\partial : \mathbb{M}_k(\Gamma, \varepsilon) \to \mathbb{B}_k(\Gamma, \varepsilon)$ and its kernel will be denoted by $\mathbb{S}_k(\Gamma, \varepsilon)$.

In order to have a concrete realization of $\mathbb{S}_k(\Gamma, \varepsilon)$, it remains to efficiently compute the boundary map.

3.5 Efficient Computation of the Boundary Map

In this subsection, we show how to compute the boundary map efficiently. For finite index subgroups $\Gamma \subseteq \Gamma' \subseteq SL_2(\mathbb{Z})$ such that Γ is a normal subgroup of Γ', and a character $\varepsilon : \Gamma'/\Gamma \to \mathbb{Q}(\zeta)^\times$, we denote $c_\varepsilon(\Gamma) = \dim \mathbb{B}_2(\Gamma, \varepsilon)$. We will proceed to prove the following result.

Theorem 3.5.1 *There exists an algorithm that given groups $\Gamma \subseteq \Gamma' \subseteq SL_2(\mathbb{Z})$ such that Γ is a normal subgroup of Γ', a character $\varepsilon : \Gamma'/\Gamma \to \mathbb{Q}(\zeta)^\times$, and an integer $k \geq 2$, computes a basis for $\mathbb{S}_k(\Gamma, \varepsilon)$ in $O([SL_2(\mathbb{Z}) : \Gamma'] \cdot c_\varepsilon(\Gamma) + [SL_2(\mathbb{Z}) : \Gamma])$ basic* CosetIndex *operations.*

The proof of Theorem 3.5.1 will occupy this entire subsection. We will obtain the basis by computing the kernel of the boundary map. Thus, given $\mathbb{M}_k(\Gamma, \varepsilon)$, we would like to compute the boundary map $\mu \circ \partial : \mathbb{M}_k(\Gamma, \varepsilon) \to \mathbb{B}_k(\Gamma, \varepsilon)$.

The idea follows Cremona and Stein [12, Prop. 2.2.3, Lemma 3.2, Prop. 8.13], [32, Alg. 8.12, Prop. 8.13, Alg. 8.14] in that we do not have to compute the space of cusps for Γ a priori. Instead, for every cusp in $\mathbb{P}^1(\mathbb{Q})$ that we encounter, we check whether it is equivalent to the ones previously encountered.

Following that route, we should be able to check whether a cusp is equivalent to 0 (note that this is not the cusp 0, but rather the neutral element of the module $\mathbb{B}_k(\Gamma, \varepsilon)$) or to a previously encountered cusp. We will begin by describing the algorithms for testing cusp equivalence and cusp vanishing, and conclude by presenting the algorithm for computing the boundary map.

These algorithms are implemented in [32] and in [12] for $\Gamma = \Gamma_0(N), \Gamma_1(N)$.

In order to treat the general case, we first prove a criterion for the equivalence.

Proposition 3.5.2 *Let $a, b \in \mathbb{P}^1(\mathbb{Q})$. Let $\Gamma \subseteq SL_2(\mathbb{Z})$ be a subgroup. Let $g_a, g_b \in SL_2(\mathbb{Z})$ be such that $g_a(\infty) = a$ and $g_b(\infty) = b$. Let $T = \begin{pmatrix} 1 & 1 \\ 0 & 1 \end{pmatrix}$. Then $\Gamma \backslash SL_2(\mathbb{Z})$ admits a right action by T, via multiplication, and $\Gamma a = \Gamma b$ if and only if $\Gamma g_a, \Gamma g_b$ lie in the same $\langle T \rangle$-orbit of $\Gamma \backslash SL_2(\mathbb{Z})$.*

Proof If there is some $\gamma \in \Gamma$ such that $\gamma a = b$, then $\gamma g_a(\infty) = g_b(\infty)$, and so

$$g_b^{-1} \gamma g_a \in \text{Stab}_{SL_2(\mathbb{Z})}(\infty) = \langle T \rangle .$$

Therefore we get $g_b^{-1} \gamma g_a = T^n$ for some integer $n \in \mathbb{Z}$. Thus, there exists $n \in \mathbb{Z}$ for which $g_b T^n \in \Gamma g_a$. The converse follows similarly. □

Thus, in order to test for equivalence, we have to compute the T action. This can be done a priori for all cosets, as we proceed to show.

Definition 3.5.3 Let $\Gamma \subseteq SL_2(\mathbb{Z})$ be a finite index subgroup, and let $R = (r_1, \ldots, r_h)$ be coset representatives for $\Gamma \backslash SL_2(\mathbb{Z})$. Let $S = ((o_1, e_1), \ldots, (o_h, e_h))$ be pairs of integers. We say that S is a *$\langle T \rangle$-orbit table* of Γ with respect to R if the following two assumptions hold.

(1) The cosets Γr_i and Γr_j lie in the same $\langle T \rangle$-orbit iff $o_i = o_j$.
(2) For all i, j such that $o_i = o_j$, we have $r_j \cdot T^{e_i - e_j} \cdot r_i^{-1} \in \Gamma$.

Lemma 3.5.4 *There exists an algorithm that given a finite index subgroup* $\Gamma \subseteq SL_2(\mathbb{Z})$*, and a list of coset representatives for* $\Gamma \backslash SL_2(\mathbb{Z})$*, R, computes a* $\langle T \rangle$*-orbit table of* Γ *with respect to R in* $O([SL_2(\mathbb{Z}) : \Gamma])$ *basic* CosetIndex *operations.*

Proof Apply Algorithm 3.5.5. It just goes over the cycles, hence has linear running time. □

Algorithm 3.5.5 OrbitTable(Γ, R). *Constructing the* $\langle T \rangle$*-orbit table.*

Input :

- a finite index subgroup $\Gamma \subseteq SL_2(\mathbb{Z})$.
- a list of coset representatives $R = (r_1, \dots, r_h)$ for $\Gamma \backslash SL_2(\mathbb{Z})$.

Output : a list S, which is a $\langle T \rangle$-orbit table of Γ with respect to R.

(1) for $i \in \{1, \dots, h\}$ do
   (a) $s_i := (0, 0)$
   (b) $t_i :=$ CosetIndex$(r_i \cdot T, R, \Gamma)$
(2) $i := 1, o := 1, e := 0$
(3) while $i \leq h$ do
   (a) $s_i := (o, e)$
   (b) $i := t_i$
   (c) if $s_i \neq (0, 0)$ then
      (i) $o := o + 1$;
      (ii) $e := 0$;
      (iii) while $i \leq h$ and $s_i \neq (0, 0)$ do
         (A) $i := i + 1$;
   (d) else
      (i) $e := e + 1$
(4) return $(s_1, \dots, s_h)$

And then our algorithm for cusp equivalence testing is rather simple.

Algorithm 3.5.6 CuspEquiv(Γ, R, S, a, b). *Test for cusp equivalence.*

Input :

- a finite index subgroup $\Gamma \subset SL_2(\mathbb{Z})$.
- a list of coset representatives $R = (r_1, \dots, r_h)$ for $\Gamma \backslash SL_2(\mathbb{Z})$.
- an orbit table $S = ((o_1, e_1), \dots, (o_h, e_h))$ (the output of Algorithm 3.5.5).
- $a, b \in \mathbb{P}^1(\mathbb{Q})$.

Output : If there exists $\gamma \in \Gamma$ such that $\gamma a = b$, then $(true, \gamma)$, else $(false, 1)$.

(1) Find g_a, g_b such that $g_a(\infty) = a$, $g_b(\infty) = b$.
(2) $i_a :=$ CosetIndex(g_a, R, Γ), $i_b :=$ CosetIndex(g_b, R, Γ)
(3) if $o_{i_a} \neq o_{i_b}$ then return $(false, 1)$;
(4) return $(true, g_b \cdot T^{e_{i_a} - e_{i_b}} \cdot g_a^{-1})$

Corollary 3.5.7 *There exists an algorithm that given a finite index subgroup $\Gamma \subset SL_2(\mathbb{Z})$, a list of coset representatives R for $\Gamma \backslash SL_2(\mathbb{Z})$, the $\langle T \rangle$-orbit table of Γ with respect to R, and $a, b \in \mathbb{P}^1(\mathbb{Q})$, returns whether $\Gamma a = \Gamma b$, in which case it returns also an element $\gamma \in \Gamma$ such that $\gamma a = b$, in $O(1)$ basic* CosetIndex *operations.*

Proof Apply Algorithm 3.5.6. By Proposition 3.5.2, it returns the correct answer. □

Finally, to test whether a cusp vanishes, we use the following proposition (an analogue of [32, Prop. 8.16]). Recall that if $\Gamma \subseteq \Gamma'$ is a normal subgroup, and $Q = \Gamma'/\Gamma$, then Q acts on $\mathbb{M}_k(\Gamma)$ as in (3.3.1).

Proposition 3.5.8 *Let $\Gamma \subseteq \Gamma' \subseteq SL_2(\mathbb{Z})$ be finite index subgroups such that Γ is a normal subgroup of Γ'. Let $Q := \Gamma'/\Gamma$, and let $\varepsilon : Q \to \mathbb{Q}(\zeta)^\times$ be a character. Suppose $a \in \mathbb{P}^1(\mathbb{Q})$ is a cusp. For $x \in \mathbb{P}^1(\mathbb{Q})$, write $[x]$ for the equivalence class of x in $\Gamma \backslash \mathbb{P}^1(\mathbb{Q})$. Then a vanishes modulo the relations*

$$[\gamma' x] = \varepsilon(\gamma') \cdot [x] \quad \forall \gamma' \in \Gamma', x \in \mathbb{P}^1(\mathbb{Q})$$

if and only if there exists $q \in Q$ with $\varepsilon(q) \neq 1$ such that $[q \cdot a] = [a]$.

Proof First suppose that such a q exists. Then

$$[a] = [q \cdot a] = \varepsilon(q) \cdot [a].$$

But $\varepsilon(q) \neq 1$, hence $[a] = 0$.

Conversely, suppose that $[a] = 0$. Because all relations are two-term relations and the Γ-relations identify Γ-orbits, there must exist $\alpha, \beta \in \Gamma'$ such that

$$[\alpha \cdot a] = [\beta \cdot a]$$

and $\varepsilon(\alpha) \neq \varepsilon(\beta)$. Indeed, if this did not occur, we could mod out by the ε relations by writing each $[\alpha \cdot a]$ in terms of $[a]$ and there would be no further relations to kill $[a]$. Next observe that

$$[\beta^{-1}\alpha \cdot a] = \varepsilon(\beta^{-1}) \cdot [\alpha \cdot a] = \varepsilon(\beta)^{-1}[\beta \cdot a] = [a]$$

so that if $q \in Q$ is the image of $\beta^{-1}\alpha \in \Gamma'$, then $\varepsilon(q) = \varepsilon(\beta)^{-1}\varepsilon(\alpha) \neq 1$. □

This gives rise to an algorithm for checking whether a cusp vanishes.

Algorithm 3.5.9 CuspVanishing$(\Gamma, \varepsilon, s, R, S, a)$. *Test for cusp vanishing.*

Input :

- a finite index subgroup $\Gamma \subset SL_2(\mathbb{Z})$.
- a character $\varepsilon : Q = \Gamma'/\Gamma \to \mathbb{Q}(\zeta)^\times$.
- a section $s : Q \to \Gamma'$ of the quotient map.
- a list of coset representatives R for $\Gamma\backslash SL_2(\mathbb{Z})$.
- an orbit table S (the output of Algorithm 3.5.5).
- $a \in \mathbb{P}^1(\mathbb{Q})$.

Output : If $[a] = 0$ in $\mathbb{B}_k(\Gamma, \varepsilon)$ return $true$, return $false$.

(1) For each $q \in Q$ such that $\varepsilon(q) \neq 1$ do:

(a) $equiv, \gamma :=$ CuspEquiv$(\Gamma, R, S, s(q) \cdot a, a)$.
(b) If $equiv$ then return $true$.

(2) return $false$.

Corollary 3.5.10 *There exists an algorithm that given a finite index subgroup* $\Gamma \subseteq \Gamma' \subseteq SL_2(\mathbb{Z})$ *such that* Γ *is normal in* Γ', *a character* $\varepsilon : \Gamma'/\Gamma \to \mathbb{Q}(\zeta)^\times$, *a list of coset representatives* R *for* $\Gamma\backslash SL_2(\mathbb{Z})$, *a* $\langle T\rangle$*-orbit table of* Γ *with respect to* R, *and a cusp* $a \in \mathbb{P}^1(\mathbb{Q})$, *returns whether* $[a] = 0$ *in* $\mathbb{B}_k(\Gamma, \varepsilon)$ *in* $O([\Gamma' : \Gamma])$ *basic* CosetIndex *operations.*

Proof Apply Algorithm 3.5.9. The correctness follows from Proposition 3.5.8. □

Relying on Algorithms 3.5.6 and 3.5.9, we may now formulate an algorithm to compute the boundary map $\mathbb{M}_k(\Gamma, \varepsilon) \to \mathbb{B}_k(\Gamma, \varepsilon)$. In the algorithm below, CuspIndex is a function that given a finite index subgroup $\Gamma \subseteq SL_2(\mathbb{Z})$, a list of coset representatives R for $\Gamma\backslash SL_2(\mathbb{Z})$, a $\langle T\rangle$-orbit table of Γ with respect to R, a cusp a and a list of cusps C, returns an index i such that a is equivalent to C_i in $\mathbb{B}_k(\Gamma)$ and an element $\gamma \in \Gamma$ such that $a = \gamma \cdot C_i$. If $a \notin C$, it appends a to C. This function is obtained by using Algorithm 3.5.6 repeatedly.

Algorithm 3.5.11 BoundaryMap$(\Gamma, \Gamma', \varepsilon, k, B, R, R')$. *Computation of the boundary map.*

Input :

- finite index subgroups $\Gamma \subseteq \Gamma' \subseteq SL_2(\mathbb{Z})$ such that Γ is normal in Γ'.
- a character $\varepsilon : \Gamma'/\Gamma \to \mathbb{Q}(\zeta)^\times$.
- an integer $k \geq 2$.
- $B = (b_1, \ldots, b_n)$, an ordered basis of Manin symbols for $\mathbb{M}_k(\Gamma, \varepsilon)$.
- $R = (r_1, \ldots, r_h)$, coset representatives for $\Gamma\backslash SL_2(\mathbb{Z})$.
- $R' = (r'_1, \ldots, r'_{h'})$, coset representatives for $\Gamma'\backslash SL_2(\mathbb{Z})$.

Output :

- a matrix A.
- a list of boundary Manin symbols C.

such that A represents the boundary map $\mu \circ \partial$ with respect to the bases B, C.

(1) $C := \emptyset$
(2) $A := 0 \in \mathbb{Q}(\zeta)^{n \times \mathbb{N}}$
(3) $S :=$ OrbitTable(Γ, R)
(4) $S' :=$ OrbitTable(Γ', R')
(5) for $i \in \{1, 2, \ldots, n\}$ do

   (a) Write $b_i = [P, g]$
   (b) $w := \text{weight}(b_i)$
   (c) if $w = k - 2$:

      (i) $a := g \cdot \infty$
      (ii) if not CuspVanishing$(\Gamma, \varepsilon, s, R, S, a)$:

         (A) $j, \gamma, C :=$ CuspIndex(Γ', R', S', a, C)
         (B) $A_{ij} := A_{ij} + \varepsilon^{-1}(\gamma)$

   (d) if $w = 0$:

      (i) $a := g \cdot 0$
      (ii) if not CuspVanishing$(\Gamma, \varepsilon, s, R, S, a)$:

         (A) $j, \gamma, C :=$ CuspIndex(Γ', R', S', a, C)
         (B) $A_{ij} := A_{ij} - \varepsilon^{-1}(\gamma)$

(6) return $(A_{ij})_{j=1}^{|C|}, C$

Remark 3.5.12 Note that when $k = 2$, we execute both (c) and (d).

We can now proceed to prove Theorem 3.5.1.

Proof (Of Theorem 3.5.1) The computation of $\mathbb{S}_k(\Gamma, \varepsilon)$ is performed by using Corollary 3.2.3 to construct a basis B of modular symbols for $\mathbb{M}_k(\Gamma, \varepsilon)$, and then applying Algorithm 3.5.11.

The computation of the $\langle T \rangle$-orbit tables for both Γ and Γ' using Algorithm 3.5.5 costs $O([SL_2(\mathbb{Z}) : \Gamma])$ basic CosetIndex operations by Lemma 3.5.4.

The function CuspIndex is implemented by applying Algorithm 3.5.6. The check whether a cusp vanishes is done by applying Algorithm 3.5.9. By Corollaries 3.5.7 and 3.5.10, these algorithms return the required output in $O(1)$ basic CosetIndex operations.

Since there are $2 \cdot [SL_2(\mathbb{Z}) : \Gamma']$ Manin symbols of either maximal or minimal weight, and we have to check equivalence of at most each of these against each of the cusps in the list, which has at most $c_\varepsilon(\Gamma)$ elements in it, it follows that the running time for Algorithm 3.5.11 is $O([SL_2(\mathbb{Z}) : \Gamma'] \cdot c_\varepsilon(\Gamma) + [SL_2(\mathbb{Z}) : \Gamma])$. Thus, we have an algorithm for computing $\mathbb{S}_k(\Gamma, \varepsilon)$ in $O([SL_2(\mathbb{Z}) : \Gamma'] \cdot c_\varepsilon(\Gamma) + [SL_2(\mathbb{Z}) : \Gamma])$ basic CosetIndex operations. □

3.6 Pairing Modular Symbols and Modular Forms

Let $\overline{S}_k(\Gamma) = \{\overline{f} : f \in S_k(\Gamma)\}$ be the space of antiholomorphic cusp forms. We then have a pairing

$$\langle \cdot, \cdot \rangle : \left(S_k(\Gamma) \oplus \overline{S}_k(\Gamma)\right) \times \mathbb{M}_k(\Gamma) \to \mathbb{C}$$

given by

$$\langle (f_1, f_2), P \otimes \{\alpha, \beta\} \rangle = \int_\alpha^\beta f_1(z) P(z, 1) dz + \int_\alpha^\beta f_2(z) P(\overline{z}, 1) d\overline{z}. \tag{3.6.1}$$

The following theorem is the final ingredient needed to get a handle on the actual spaces of modular forms.

Theorem 3.6.2 ([30, Theorem 0.2], and [23, Theorem 3]) *The pairing*

$$\langle \cdot, \cdot \rangle : \left(S_k(\Gamma) \oplus \overline{S}_k(\Gamma)\right) \times \mathbb{S}_k(\Gamma, \mathbb{C}) \to \mathbb{C}$$

is a nondegenerate pairing of complex vector spaces.

This is still not completely satisfactory, as we would like to separate the holomorphic forms from their antiholomorphic counterparts.

3.6.1 The Action of Complex Conjugation

Definition 3.6.3 Let $\eta = \begin{pmatrix} -1 & 0 \\ 0 & 1 \end{pmatrix} \in GL_2(\mathbb{Z})$. If η normalizes Γ, i.e. $\eta \Gamma \eta = \Gamma$, we say that Γ is *of real type*.

We then have the following result.

Proposition 3.6.4 (Merel, [23, Proposition 7]) *Let $\Gamma \subseteq SL_2(\mathbb{Z})$ be a subgroup of real type. Then there is a complex linear involution $\iota : S_k(\Gamma) \oplus \overline{S}_k(\Gamma) \to S_k(\Gamma) \oplus \overline{S}_k(\Gamma)$ which exchanges $S_k(\Gamma)$ and $\overline{S}_k(\Gamma)$, given by $\iota(f)(z) = f(-\overline{z})$. The involution $\iota^\vee : \mathbb{M}_k(\Gamma) \to \mathbb{M}_k(\Gamma)$ defined by*

$$\iota^\vee(v \otimes \{\alpha, \beta\}) = -\eta v \otimes \{\eta\alpha, \eta\beta\} \tag{3.6.5}$$

is adjoint to ι with respect to the pairing (3.6.1). Moreover, $\iota^\vee$ acts on Manin symbols via

$$\iota^\vee([v, g]) = -[\eta v, \eta g \eta^{-1}].$$

We may now state the final result about the pairing, which explains how modular symbols and modular forms are related.

Theorem 3.6.6 (Merel, [23, Proposition 8]) *Let $\mathbb{S}_k(\Gamma)^+$ be the $+1$ eigenspace for ι^* on $\mathbb{S}_k(\Gamma)$. The pairing (3.6.1) induces a nondegenerate bilinear pairing*

$$\mathbb{S}_k(\Gamma)^+ \times S_k(\Gamma) \to \mathbb{C}.$$

We denote by $c(\Gamma) = \dim \mathbb{B}_2(\Gamma)$ the number of cusps of Γ. Then it follows that

Corollary 3.6.7 *There exists an algorithm that given a finite index subgroup $\Gamma \subseteq SL_2(\mathbb{Z})$ and an integer $k \geq 2$, computes a basis for $S_k(\Gamma)$ in $O([SL_2(\mathbb{Z}) : \Gamma] \cdot c(\Gamma))$ basic* CosetIndex *operations.*

Proof In order to compute $\mathbb{S}_k(\Gamma)^+$, it is necessary to replace $\mathbb{B}_k(\Gamma)$ by its quotient modulo the additional relation $[(-u, v)] = [(u, v)]$ for all cusps (u, v). Algorithm 3.5.11 can be modified, as in [32], to treat that case by changing CuspEquiv to check equivalence with both (u, v) and $(-u, v)$. Thus, the complexity described in Theorem 3.5.1 is valid also for the computation of $\mathbb{S}_k(\Gamma, \varepsilon)^+$. In particular, taking $\Gamma' = \Gamma$ and ε to be the trivial character, we obtain the result. □

4 Hecke Operators

Now that we have a realization of $S_k(\Gamma_G)$, we would like to be able to compute Hecke Operators on this space. First, let us recall a few basic facts about Hecke operators in general.

4.1 Hecke Operators on $M_k(\Gamma)$

Definition 4.1.1 Let $\Gamma \subseteq SL_2(\mathbb{Z})$ and suppose $\Delta \subseteq GL_2^+(\mathbb{Q})$ is a set such that $\Gamma\Delta = \Delta\Gamma = \Delta$, and $\Gamma\backslash\Delta$ is finite. Let R be a set of representatives for $\Gamma\backslash\Delta$. Let

$$T_\Delta : M_k(\Gamma) \to M_k(\Gamma)$$
$$T_\Delta(f) = \sum_{\alpha \in R} f|_\alpha$$

where $f|_\alpha$ is the usual right action of $GL_2^+(\mathbb{Q})$ given by (2.2.2). The operator T_Δ is called the *Hecke operator associated to* Δ.

It is a standard result that this operator is well-defined and independent of R.

Proposition 4.1.2 ([29, Prop. 3.37], [14, Chapter 5.1]) *The image of T_Δ lies in $M_k(\Gamma)$, and T_Δ does not depend on the choice of R.*

Definition 4.1.3 The algebra generated by the $\{T_\Delta\}_\Delta$ when Δ runs over all subsets of $GL_2^+(\mathbb{Q})$ such that $\Gamma\Delta = \Delta\Gamma = \Delta$ is *the Hecke algebra of* $M_k(\Gamma)$.

For $\alpha \in GL_2^+(\mathbb{Q})$, we denote $T_\alpha = T_{\Gamma\alpha\Gamma}$.

Corollary 4.1.4 *The Hecke algebra of $M_k(\Gamma)$ is generated by the operators $\{T_\alpha\}_{\alpha\in GL_2^+(\mathbb{Q})}$.*

Proof It is a consequence of the definition that Δ must be a union of double cosets $\Gamma\alpha\Gamma$, where $\alpha \in GL_2^+(\mathbb{Q})$. In particular, the Hecke operators are linearly spanned by the $\{T_\alpha\}$. □

Example 4.1.5 Here are two standard choices for Δ.

(1) If $\Gamma = \Gamma_1(N)$, one usually considers for any $n \in \mathbb{Z}$, the set

$$\Delta_n^1 := \left\{ g \in M_2(\mathbb{Z}) \mid \det(g) = n \text{ and } g \equiv \begin{pmatrix} 1 & * \\ 0 & n \end{pmatrix} \bmod N \right\}.$$

One usually denotes $T_n = T_{\Delta_n^1}$. We note that $\Delta_n^1 = \Gamma \begin{pmatrix} 1 & 0 \\ 0 & n \end{pmatrix} \Gamma$, so that $T_n = T_\alpha$, with $\alpha = \begin{pmatrix} 1 & 0 \\ 0 & n \end{pmatrix}$.

(2) Let $\Gamma = \Gamma(N)$. We may take for $n \in \mathbb{Z}$, the set

$$\Delta_n := \left\{ g \in M_2(\mathbb{Z}) \mid \det(g) = n \text{ and } g \equiv \begin{pmatrix} 1 & 0 \\ 0 & n \end{pmatrix} \bmod N \right\}.$$

Note that again we are taking $\Delta_n = \Gamma \begin{pmatrix} 1 & 0 \\ 0 & n \end{pmatrix} \Gamma$.

As our implementation of modular forms is by using the dual space of modular symbols, we need the dual notion.

Definition 4.1.6 Let $\Delta \subseteq GL_2(\mathbb{Q})$ be such that $\Gamma\Delta = \Delta\Gamma = \Delta$ and such that $\Gamma\backslash\Delta$ is finite. Let R be a set of representatives of $\Gamma\backslash\Delta$. Let

$$T_\Delta^\vee : \mathbb{M}_k(\Gamma) \to \mathbb{M}_k(\Gamma)$$

$$v \otimes \{\alpha, \beta\} \mapsto \sum_{\delta\in R} (\delta v) \otimes \{\delta\alpha, \delta\beta\}.$$

Again, this map does not depend on the choice of R. We then have the following result.

Proposition 4.1.7 (Merel, [23, Proposition 10]) *The operators T_Δ and $T_\Delta^\vee$ are adjoint with respect to the bilinear pairing $\langle \cdot, \cdot \rangle$ defined in* (3.6.1).

Therefore, by Theorem 3.6.2, in order to compute the restriction of the Hecke operators T_Δ to $S_k(\Gamma) \oplus \overline{S}_k(\Gamma)$, it will suffice to compute the Hecke operators $T_\Delta^\vee$ on $\mathbb{S}_k(\Gamma)$. Note that in order to compute the restriction to $S_k(\Gamma)$ using Theorem 3.6.6, we would need T_Δ to commute with the map ι, defined in Proposition 3.6.4.

Lemma 4.1.8 *Let $\Gamma \subseteq SL_2(\mathbb{Z})$ be a subgroup of real type (recall Definition 1.2.1). Let $\alpha \in GL_2^+(\mathbb{Q})$ be such that $\eta^{-1}\alpha\eta \in \Gamma\alpha\Gamma$. Then in $\mathbb{M}_k(\Gamma)$ we have $T_\alpha^\vee \circ \iota^\vee = \iota^\vee \circ T_\alpha^\vee$.*

Proof Let R be a set of representatives for $\Gamma\backslash\Gamma\alpha\Gamma$. For each $\delta \in \Gamma\alpha\Gamma$, we have by the assumptions on Γ and α that

$$\eta^{-1}\delta\eta \in \eta^{-1}\Gamma\alpha\Gamma\eta = \left(\eta^{-1}\Gamma\eta\right)\eta^{-1}\alpha\eta\left(\eta^{-1}\Gamma\eta\right) = \Gamma\eta^{-1}\alpha\eta\Gamma = \Gamma\alpha\Gamma.$$

Therefore, we see that

$$\Gamma\alpha\Gamma = \bigsqcup_{\delta\in R}\Gamma\delta = \eta^{-1}\left(\bigsqcup_{\delta\in R}\Gamma\delta\right)\eta = \bigsqcup_{\delta\in R}\eta^{-1}\Gamma\delta\eta = \bigsqcup_{\delta\in R}\left(\eta^{-1}\Gamma\eta\right)\eta^{-1}\delta\eta = \bigsqcup_{\delta\in R}\Gamma\eta^{-1}\delta\eta,$$

so that $R' = \{\eta^{-1}\delta\eta \mid \delta \in R\}$ is also a set of representatives for $\Gamma\backslash\Gamma\alpha\Gamma$. Now

$$\begin{aligned} T_\alpha^\vee(\iota^\vee(v \otimes \{\alpha, \beta\})) &= T_\alpha^\vee(-\eta v \otimes \{\eta\alpha, \eta\beta\}) \\ &= \sum_{\delta\in R}(-\delta\eta v \otimes \{\delta\eta\alpha, \delta\eta\beta\}) \\ &= \sum_{\delta\in R}(-\eta(\eta^{-1}\delta\eta)v \otimes \{\eta(\eta^{-1}\delta\eta)\alpha, \eta(\eta^{-1}\delta\eta)\beta\}) \\ &= \sum_{\delta'\in R'}(-\eta\delta' v \otimes \{\eta\delta'\alpha, \eta\delta'\beta\}) = \iota^\vee(T_\alpha^\vee(v \otimes \{\alpha, \beta\})). \end{aligned}$$

□

Corollary 4.1.9 *Let $G \subseteq GL_2(\mathbb{Z}/N\mathbb{Z})$ be a subgroup of real type (recall Definition 1.2.1). Let $\alpha \in M_2(\mathbb{Z})$ be such that $\lambda_N(\alpha) \in G$. Then in $\mathbb{M}_k(\Gamma_G)$ we have $T_\alpha^\vee \circ \iota^\vee = \iota^\vee \circ T_\alpha^\vee$.*

Proof Since G is of real type, we have

$$\lambda_N(\eta\Gamma_G\eta^{-1}) = \lambda_N(\eta)G\lambda_N(\eta)^{-1} = G.$$

It follows that $\eta\Gamma_G\eta^{-1} \subseteq \Gamma_G$, and hence Γ_G is of real type.

Also, as $\lambda_N(\alpha) \in G$, we see that

$$\lambda_N(\eta^{-1}\alpha\eta) \in \lambda_N(\eta)^{-1} G \lambda_N(\eta) = G.$$

Moreover, since $\det(\alpha) = \det(\eta^{-1}\alpha\eta)$ and $d_1(\alpha) = d_1(\eta^{-1}\alpha\eta)$, where d_1 is the greatest common divisor of all the entries, α and $\eta^{-1}\alpha\eta$ have the same Smith normal form, showing that $SL_2(\mathbb{Z}) \cdot \alpha \cdot SL_2(\mathbb{Z}) = SL_2(\mathbb{Z}) \cdot \eta^{-1}\alpha\eta \cdot SL_2(\mathbb{Z})$. We invoke [29, Lemma 3.29(1)], noting that $\lambda_N(\Gamma_G\alpha) = G = \lambda_N(\alpha\Gamma_G)$, and so, in Shimura's notation, $\alpha \in \Phi$. We deduce that

$$\Gamma_G\alpha\Gamma_G = \{g \in M_2(\mathbb{Z}) \mid SL_2(\mathbb{Z}) \cdot \alpha \cdot SL_2(\mathbb{Z}) = SL_2(\mathbb{Z}) \cdot g \cdot SL_2(\mathbb{Z}) \text{ and } \lambda_N(g) \in G\}.$$

It then follows that $\eta^{-1}\alpha\eta \in \Gamma_G\alpha\Gamma_G$. The result follows from Lemma 4.1.8. □

4.2 Naive Computation of $T_\alpha^\vee$

We will now present a naive algorithm to compute the Hecke operator $T_\alpha^\vee$ on $\mathbb{M}_k(\Gamma)$ for an arbitrary $\alpha \in GL_2^+(\mathbb{Q})$.

Algorithm 4.2.1 HeckeOperator(α, x, B). *Hecke Operator $T_\alpha^\vee$ on $\mathbb{M}_k(\Gamma)$.*

Input :

- $\alpha \in GL_2^+(\mathbb{Q})$.
- an element $x = v \otimes \{\alpha, \beta\} \in \mathbb{M}_k(\Gamma)$.
- a basis B for the space of modular symbols $\mathbb{M}_k(\Gamma)$.

Output : a vector representing $T_\alpha^\vee(x)$ w.r.t. the basis B.

(1) Let $H := \Gamma \cap \alpha^{-1}\Gamma\alpha$.
(2) Let S be a set of representatives for $H\backslash\Gamma$.
(3) Let $R := \alpha \cdot S = \{\alpha \cdot x : x \in S\}$.
(4) Return $\sum_{r \in R} [\pi(r \cdot x)]_B$, where $[\pi(r \cdot x)]_B$ is the vector representing the modular symbol $\Gamma \cdot (rv) \otimes \{r\alpha, r\beta\}$ w.r.t. B.

Remark 4.2.2 Even though this is unclear at first sight, the functions $(\Gamma_1, \Gamma_2) \mapsto \Gamma_1 \cap \Gamma_2$ and $(\Gamma, x) \mapsto x^{-1}\Gamma x$ also require a nontrivial implementation.

Note also that $\Gamma \subseteq SL_2(\mathbb{Z})$ is defined using $G \subseteq GL_2(\mathbb{Z}/N\mathbb{Z})$. Therefore, if $\alpha \in GL_2^+(\mathbb{Q})$ is such that $\alpha^{-1}\Gamma\alpha \not\subseteq SL_2(\mathbb{Z})$, a straightforward computation could be expensive (either find N such that $N \cdot \alpha^{-1}\Gamma\alpha \subseteq SL_2(\mathbb{Z})$ and change computations to fit, or work in $GL_2^+(\mathbb{Q})$ with generators—that requires finding the generators of Γ). Moreover, as $GL_2^+(\mathbb{Q})$ is not finitely generated, we do not have ready made tools for computing the intersection $\Gamma \cap \alpha^{-1}\Gamma\alpha$.

This has brought us to consider the following alternate path:

Strategy for Computation of Algorithm 4.2.1

(a) Implement the function $(\Gamma, x) \mapsto x^{-1}\Gamma x$ for $x \in GL_2^+(\mathbb{Q})$ when $x^{-1}\Gamma x \subseteq SL_2(\mathbb{Z})$.
(b) Implement an intersection function $(\Gamma_1, \Gamma_2) \mapsto \Gamma_1 \cap \Gamma_2$ for $\Gamma_1, \Gamma_2 \subseteq SL_2(\mathbb{Z})$.
(c) Write the set of representatives $\Gamma\backslash\Gamma\alpha\Gamma = \alpha \cdot \left(\left((\alpha^{-1}\Gamma\alpha) \cap \Gamma\right) \backslash\Gamma\right)$ as a sequence of operations of type (a), (b).

Here is an implementation of the intersection in step 1, using (a) and (b). Below, we denote by $\Gamma^0(N)$ the congruence subgroup Γ_G, where

$$G = \left\{\begin{pmatrix} a & 0 \\ c & d \end{pmatrix} \mid a, d \in (\mathbb{Z}/N\mathbb{Z})^\times, c \in \mathbb{Z}/N\mathbb{Z}\right\}.$$

Algorithm 4.2.3 ConjInter(Γ, α). *Conjugation and Intersection.*

Input :

- $\Gamma \subseteq SL_2(\mathbb{Z})$ a congruence subgroup.
- $\alpha \in GL_2^+(\mathbb{Q})$ such that $\alpha^{-1}\Gamma\alpha \subseteq SL_2(\mathbb{Z})$.

Output : $\Gamma \cap \alpha^{-1}\Gamma\alpha$.

(1) Find matrices $x, \gamma \in SL_2(\mathbb{Z})$ such that $x \cdot \alpha^{-1} \cdot \gamma$ is of the form $\begin{pmatrix} a & 0 \\ 0 & b \end{pmatrix}$ with $a^{-1}b \in \mathbb{Z}$ (Smith Normal Form)
(2) Let $H := \Gamma \cap \left(\alpha^{-1} \cdot \left(\Gamma \cap \left(\gamma \cdot \Gamma^0(a^{-1}b) \cdot \gamma^{-1}\right)\right) \cdot \alpha\right)$.

First, let us prove that Algorithm 4.2.3 indeed returns the correct answer.

Proposition 4.2.4 *Algorithm 4.2.3 returns the group* $\Gamma \cap \alpha^{-1}\Gamma\alpha$.

Proof First, note that if $a^{-1}b \in \mathbb{Z}$, then

$$\begin{pmatrix} a & 0 \\ 0 & b \end{pmatrix}^{-1} SL_2(\mathbb{Z}) \begin{pmatrix} a & 0 \\ 0 & b \end{pmatrix} \cap SL_2(\mathbb{Z}) = \Gamma^0(a^{-1}b).$$

Therefore

$$\gamma^{-1}\alpha x^{-1} SL_2(\mathbb{Z}) x \alpha^{-1}\gamma \cap SL_2(\mathbb{Z}) = \Gamma^0(a^{-1}b).$$

But $x \in SL_2(\mathbb{Z})$, so that $x^{-1}SL_2(\mathbb{Z})x = SL_2(\mathbb{Z})$. It follows that

$$\alpha SL_2(\mathbb{Z})\alpha^{-1} \cap \gamma \cdot SL_2(\mathbb{Z})\gamma^{-1} = \gamma\Gamma^0(a^{-1}b)\gamma^{-1}.$$

But $\gamma \in SL_2(\mathbb{Z})$, so $\gamma SL_2(\mathbb{Z})\gamma^{-1} = SL_2(\mathbb{Z})$. It follows that

$$\alpha SL_2(\mathbb{Z})\alpha^{-1} \cap SL_2(\mathbb{Z}) = \gamma\Gamma^0(a^{-1}b)\gamma^{-1}.$$

Therefore

$$\alpha SL_2(\mathbb{Z})\alpha^{-1} \cap \Gamma = \alpha SL_2(\mathbb{Z})\alpha^{-1} \cap SL_2(\mathbb{Z}) \cap \Gamma = \Gamma \cap \gamma\Gamma^0(a^{-1}b)\gamma^{-1},$$

so that

$$SL_2(\mathbb{Z}) \cap \alpha^{-1}\Gamma\alpha = \alpha^{-1}\left(\Gamma \cap \gamma\Gamma^0(a^{-1}b)\gamma^{-1}\right)\alpha. \tag{4.2.5}$$

Finally, intersecting with Γ, one obtains

$$\alpha^{-1}\Gamma\alpha \cap \Gamma = \Gamma \cap \left(\alpha^{-1}\left(\Gamma \cap \gamma\Gamma^0(a^{-1}b)\gamma^{-1}\right)\alpha\right) = H.$$

□

In order to be able to use the strategy 4.2, we should also show that the conjugations we are computing are in $SL_2(\mathbb{Z})$.

Lemma 4.2.6 *We have* $\gamma\Gamma^0(a^{-1}b)\gamma^{-1} \subseteq SL_2(\mathbb{Z})$ *and* $\alpha^{-1}\left(\Gamma \cap \gamma\Gamma^0(a^{-1}b)\gamma^{-1}\right)$ $\alpha \subseteq SL_2(\mathbb{Z})$.

Proof The first is trivial, as $\gamma \in SL_2(\mathbb{Z})$. The second follows from (4.2.5). □

Remark 4.2.7 Algorithm 4.2.1 can be used to compute $[T_\alpha^\vee]_B$ by applying it to each of the basis vectors.

It remains to show how to compute the conjugation in (a), and the intersection in (b). For the conjugation in (a), we compute the generators of Γ using Farey symbols (see [18, Theorem 6.1]). We then conjugate them to obtain the generators of the conjugated subgroup, and can compute the level and the reduction of this conjugated subgroup. For the intersection, we present the following algorithm.

Algorithm 4.2.8 InterInduced(G, H). *Intersection of induced congruence subgroups.*

Input :

- $G \subseteq GL_2(\mathbb{Z}/N_G\mathbb{Z})$.
- $H \subseteq GL_2(\mathbb{Z}/N_H\mathbb{Z})$.

Output : a group $K \subseteq GL_2(\mathbb{Z}/N\mathbb{Z})$ such that $\Gamma_K = \Gamma_G \cap \Gamma_H$. Here $N = \operatorname{lcm}(N_G, N_H)$.

(1) $d := \gcd(N_G, N_H)$, $G_d := \{g \in G \mid g \equiv 1 \bmod d\}$ and $H_d := \{h \in H \mid h \equiv 1 \bmod d\}$.
(2) $K_d := \lambda_d(G) \cap \lambda_d(H)$, and $s_G : K_d \to G$, $s_H : K_d \to H$ sections of λ_d.
(3) Let $A_G \subseteq GL_2(\mathbb{Z}/N\mathbb{Z})$ be a set of matrices a lifting $Generators(G_d)$ such that $a \equiv 1 \bmod N_H$. (Chinese Remainder Theorem)
(4) Let $A_H \subseteq GL_2(\mathbb{Z}/N\mathbb{Z})$ be a set of matrices a lifting $Generators(H_d)$ such that $a \equiv 1 \bmod N_G$. (Chinese Remainder Theorem)

(5) Let $A_d = \{a_z : z \in Generators(K_d)\} \subseteq GL_2(\mathbb{Z}/N\mathbb{Z})$ be such that $a_z \equiv s_G(z) \bmod N_G$ and $a_z \equiv s_H(z) \bmod N_H$. (Chinese Remainder Theorem)
(6) Return $K := \langle A_G \cup A_H \cup A_d \rangle$.

Proposition 4.2.9 *Algorithm 4.2.8 returns a group K such that $\Gamma_K = \Gamma_G \cap \Gamma_H$.*

Proof First note that for any $p \nmid N$, and any e, $\lambda_{p^e}(\Gamma_G \cap \Gamma_H) = GL_2(\mathbb{Z}/p^e\mathbb{Z})$. Then note that

$$\lambda_N(\Gamma_G \cap \Gamma_H) = \left\{x \in GL_2(\mathbb{Z}/N\mathbb{Z}) \mid \lambda_{N_G}(x) \in G, \quad \lambda_{N_H}(x) \in H\right\}.$$

We will show that $K = \lambda_N(\Gamma_G \cap \Gamma_H)$, hence the result. Let $x \in \lambda_N(\Gamma_G \cap \Gamma_H)$. Then $\lambda_d(x) \in K_d$. Then $\lambda_d(x) = \prod z_i^{e_i}$ for some $\{z_i\} \subseteq Generators(K_d)$. Write $a_i := a_{z_i}$, and consider $a = \prod a_i^{e_i}$. It satisfies $\lambda_d(a) = \lambda_d(x)$, hence $\lambda_d(a^{-1}x) = 1$.

In particular, we see that $\lambda_{N_G}(a^{-1}x) \in G_d$ and $\lambda_{N_H}(a^{-1}x) \in H_d$. Therefore there are $\{x_j\} \subseteq Generators(G_d)$ and $\{y_k\} \subseteq Generators(H_d)$ such that $\lambda_{N_G}(a^{-1}x) = \prod x_j^{f_j}$, $\lambda_{N_H}(a^{-1}x) = \prod y_k^{g_k}$. Let $\{b_j\}$ be the lift of the $\{x_j\}$ in A_G and $\{c_k\}$ the lifts of the $\{y_k\}$ in A_H. Then

$$\lambda_{N_G}\left(\prod b_j^{f_j} \cdot \prod c_k^{g_k}\right) = \lambda_{N_G}(a^{-1}x)$$

and

$$\lambda_{N_H}\left(\prod b_j^{f_j} \cdot \prod c_k^{g_k}\right) = \lambda_{N_H}(a^{-1}x)$$

hence

$$x = \prod a_i^{e_i} \cdot \prod b_j^{f_j} \cdot \prod c_k^{g_k} \in K.$$

□

In order to measure the complexity of Algorithm 4.2.1, in addition to the CosetIndex operation, we use In to denote group membership test in G. We will also need some notation, which we now introduce.

For $\alpha \in GL_2^+(\mathbb{Q})$, let $d_1(\alpha) \in \mathbb{Q}_{>0}$ be maximal such that $\alpha \in d_1(\alpha) \cdot M_2(\mathbb{Z})$.

Let $D(\alpha) := \det(\alpha)/d_1(\alpha)^2 \in \mathbb{Z}$, and let $I_{\alpha,\Gamma} := [\Gamma : \alpha^{-1}\Gamma\alpha \cap \Gamma]$.

This leads us to the main result of this subsection.

Theorem 4.2.10 *There exists an algorithm that given a congruence subgroup of real type $\Gamma \subseteq SL_2(\mathbb{Z})$ of level N, an element $\alpha \in GL_2^+(\mathbb{Q})$ such that $\eta^{-1}\alpha\eta \in \Gamma\alpha\Gamma$ and an integer $k \geq 2$, computes the Hecke operator T_α corresponding to the double coset $\Gamma\alpha\Gamma$, on the space of cusp forms $S_k(\Gamma)$, in complexity*

$$O(C \cdot I_{\alpha,\Gamma} \log(N \cdot D(\alpha)) + [SL_2(\mathbb{Z}) : \Gamma]^2 \cdot \mathrm{In}).$$

Proof We apply Algorithm 4.2.1, using Algorithm 4.2.3 to perform step (1). The conjugation function is computed by finding a set of generators for Γ, using Farey symbols (see [18, Theorem 6.1]), and conjugating them. Intersection is computed using Algorithm 4.2.8. This computes the dual Hecke operator $T_\alpha^\vee$ on $\mathbb{M}_k(\Gamma)$. Since Γ is of real type, and $\eta^{-1}\alpha\eta \in \Gamma\alpha\Gamma$, by Corollary 4.1.9 it follows that $T_\alpha^\vee$ commutes with $\iota^\vee$. This shows that it induces an operator on $\mathbb{S}_k(\Gamma)^+$. Therefore, by Theorem 3.6.6 and Proposition 4.1.7, we obtain T_α.

The cost of step (1) is dominated by computing the Farey symbols, which yields the cost $O([SL_2(\mathbb{Z}) : \Gamma]^2)$.

In step (4), for each element in the basis $[v, g] \in B$, we compute $g\{0, \infty\}$, to get the modular symbol $v \otimes \{a, b\}$. Then, for every $r \in R$ apply r to get $rv \otimes \{ra, rb\}$, and then use continued fractions to convert back to Manin symbols. This is done $I_{\alpha,\Gamma}$ times.

Since $g \in \Gamma\backslash SL_2(\mathbb{Z})$, it can be represented by an element in $SL_2(\mathbb{Z})$ whose entries are bounded by N^2. Further, we can choose the representatives $r \in R$ up to a multiplication by a scalar, hence we can bound its entries by $N^2 \cdot D(\alpha)$.

It follows that the entries of $r \cdot g$ are bounded by $N^4 \cdot \det \alpha$, and so computing continued fraction expansion (Euclid's algorithm) has $O(\log(N \cdot D(\alpha)))$ steps, for each of which one has to perform a CosetIndex operation to find the corresponding element in the vector space.

Thus, the algorithm has complexity

$$O(C \cdot I_{\alpha,\Gamma} \log(N \cdot D(\alpha)) + [SL_2(\mathbb{Z}) : \Gamma]^2 \cdot \mathrm{In}).$$

□

4.3 Faster Implementation of the Hecke Operator $T_\alpha^\vee$

Algorithm 4.2.1 is still quite slow in general, since the conjugation function might be very slow, as it depends quadratically on the index. However, under some simplifying assumptions we may use the following algorithm for the conjugation of an induced congruence subgroup.

Here, we assume that $(D(\alpha), N) = 1$.

Algorithm 4.3.1 ConjInd(G, α). *Conjugation of an induced congruence subgroup.*

Input :

- $G \subseteq GL_2(\mathbb{Z}/N\mathbb{Z})$.
- $\alpha \in GL_2^+(\mathbb{Q})$ such that $(D(\alpha), N) = 1$. Write $n := D(\alpha)$.

Output : a group $H \subseteq GL_2(\mathbb{Z}/nN\mathbb{Z})$ such that $\Gamma_H = \alpha^{-1}\Gamma_G\alpha \cap SL_2(\mathbb{Z})$.

(1) Find $x, y \in SL_2(\mathbb{Z})$ such that $x\alpha y = \begin{pmatrix} d_1(\alpha) & 0 \\ 0 & n \cdot d_1(\alpha) \end{pmatrix}$.

(2) Let $H_N := \lambda_N(d_1(\alpha)^{-1}\alpha)^{-1} G \lambda_N(d_1(\alpha)^{-1}\alpha) \subseteq GL_2(\mathbb{Z}/N\mathbb{Z})$.
(3) Let $H_n := \lambda_n(y)\lambda_n(\Gamma^0(n))\lambda_n(y)^{-1} \subseteq GL_2(\mathbb{Z}/n\mathbb{Z})$.
(4) Let $A_N \subseteq GL_2(\mathbb{Z}/nN\mathbb{Z})$ be a set of matrices a_i lifting $Generators(H_N)$ such that $a_i \equiv 1 \bmod n$. (Chinese Remainder Theorem)
(5) Let $B_n \subseteq GL_2(\mathbb{Z}/nN\mathbb{Z})$ be a set of matrices b_j lifting $Generators(H_n)$ such that $b_j \equiv 1 \bmod N$. (Chinese Remainder Theorem)
(6) Return $H := \langle A_N \cup B_n \rangle$.

Proposition 4.3.2 *Algorithm 4.3.1 returns a group H such that $\Gamma_H = \alpha^{-1}\Gamma_G\alpha \cap SL_2(\mathbb{Z})$.*

Proof First note that replacing α by $d_1(\alpha)^{-1}\alpha$, we may assume $\alpha \in M_2(\mathbb{Z})$ and $d_1(\alpha) = 1$. Let $\Gamma' := \alpha^{-1}\Gamma_G\alpha \cap SL_2(\mathbb{Z})$. If $p \nmid nN$, then $\lambda_{p^e}(\Gamma') = SL_2(\mathbb{Z}/p^e\mathbb{Z})$ (for all e). We also have

$$\lambda_N(\Gamma') = \lambda_N(\alpha)^{-1} G \lambda_N(\alpha)$$

and if $x\alpha y = \begin{pmatrix} 1 & 0 \\ 0 & n \end{pmatrix}$, then

$$\lambda_n(\Gamma') = \lambda_n(y)\lambda_n \begin{pmatrix} 1 & 0 \\ 0 & n \end{pmatrix}^{-1} \lambda_n(x^{-1}\Gamma_G x)\lambda_n \begin{pmatrix} 1 & 0 \\ 0 & n \end{pmatrix} \lambda_n(y)^{-1}.$$

But, since $(n, N) = 1$ and $x^{-1}\Gamma_G x \supseteq \Gamma(N)$, we have $\lambda_n(x^{-1}\Gamma_G x) = SL_2(\mathbb{Z}/n\mathbb{Z})$, and so we see that $\Gamma' = \{g \in SL_2(\mathbb{Z}) \mid \lambda_n(g) \in H_n, \lambda_N(g) \in H_N\}$, with H_n and H_N as in Algorithm 4.3.1. Since $(n, N) = 1$, we claim that $\Gamma' = \Gamma_H$, or equivalently that $\lambda_{nN}(\Gamma') = H$.

Indeed, if $g \in \lambda_{nN}(\Gamma')$, then $\lambda_n(g) \in H_n$ and $\lambda_N(g) \in H_N$. Therefore, there exist $\{x_i\} \subseteq Generators(H_n)$ and $\{y_j\} \subseteq Generators(H_N)$ such that $\lambda_n(g) = \prod x_i^{e_i}$ and $\lambda_N(g) = \prod y_j^{f_j}$. Let $\{a_i\}$ be the matrices in A_n lifting the $\{x_i\}$ and $\{b_j\}$ the matrices in B_N lifting the $\{y_j\}$. Then by the Chinese Remainder Theorem, we have $g = \prod a_i^{e_i} \cdot \prod b_j^{f_j}$. □

Next, we would like to understand the running time of Algorithm 4.2.1, with this improvement.

Corollary 4.3.3 *Let $\alpha \in GL_2^+(\mathbb{Q})$ be such that $(D(\alpha), N) = 1$. Then Algorithm 4.2.1 has complexity of*

$$O(I_{\alpha,\Gamma} \cdot \log(N^4 \cdot D(\alpha)))$$

basic CosetIndex *operations.*

Proof Using Algorithms 4.3.1 and 4.2.8 in step (1), we see that the main contribution to the complexity of Algorithm 4.2.1 comes from step (4), and this was computed in the proof of Theorem 4.2.10. □

4.4 The Hecke Operators T_n

The question of computing the Hecke operators T_n, when $(n, N) > 1$, is a difficult one. We have not been able to find in the literature a good reference even for the definition of these operators. Therefore, we restrict ourselves, for the time being, to n such that $(n, N) = 1$.

4.4.1 Adelic Definition

The only definition given in the literature of the Hecke operator for completely general level structure is in adelic terms.

Therefore, before we state the definition, we need to set up some notation.

Definition 4.4.1 Let $G \subseteq GL_2(\mathbb{Z}/N\mathbb{Z})$. We introduce

$$S_G := \left\{ r \cdot g \cdot s \mid r \in \mathbb{Q}^\times, g \in GL_2(\hat{\mathbb{Z}}), s \in GL_2^+(\mathbb{R}), g \bmod N \in G \right\} \subseteq GL_2(\mathbb{A})$$

where we use the identification $\hat{\mathbb{Z}}/N\hat{\mathbb{Z}} \cong \mathbb{Z}/N\mathbb{Z}$.

Now we may define the Hecke correspondences on the corresponding Shimura variety.

Definition 4.4.2 ([29, Section 7.3]) Let $G \subseteq GL_2(\mathbb{Z}/N\mathbb{Z})$. Let $X_G := GL_2^+(\mathbb{Q})\backslash(\mathcal{H} \times GL_2(\hat{\mathbb{Q}}))/S_G$ be the Shimura variety of level S_G. Let $w \in GL_2(\hat{\mathbb{Q}})$. Let $W = S_G \cap wS_Gw^{-1}$. Then w induces a natural correspondence $X_G \to X_G$ via

$$[z, g] \mapsto \sum_{\alpha \in S_G/W} [z, g\alpha w].$$

We denote this correspondence by $X(w)$.

Lemma 4.4.3 *The correspondence $X(w)$, for $w \in GL_2(\hat{\mathbb{Q}})$, is well defined.*

Proof First, the choice of representatives for S_G/W does not matter: if we replace each α by αw_α, we would get

$$\sum_{\alpha \in S_G/W} [z, g\alpha w_\alpha w] = \sum_{\alpha \in S_G/W} [z, g\alpha w \cdot w^{-1} w_\alpha w] = \sum_{\alpha \in S_G/W} [z, g\alpha w]$$

because $w^{-1}Ww \subseteq S_G$. If $\gamma \in GL_2^+(\mathbb{Q})$, then $X(w)[z, g] = X(w)[\gamma z, \gamma g]$. Also, if we consider an element $[z, gs]$, with $s \in S_G$, it will map to

$$\sum_{\alpha \in S_G/W} [z, gs\alpha w] = \sum_{\alpha \in S_G/W} [z, g\alpha w]$$

since $s\alpha$ still runs through a set of representatives of S_G/W. □

We now connect it to our previous definitions of moduli spaces via the following Lemma.

Lemma 4.4.4 ([24, Lemma 5.13, Theorem 5.17]) *Let* $G \subseteq GL_2(\mathbb{Z}/N\mathbb{Z})$. *Let* C *be a set of representatives for*

$$GL_2^+(\mathbb{Q})\backslash GL_2(\hat{\mathbb{Q}})/S_G \cong \hat{\mathbb{Z}}/\det(S_G) \cong (\mathbb{Z}/N\mathbb{Z})^\times / \det(G).$$

Then

$$GL_2^+(\mathbb{Q})\backslash \left(\mathcal{H} \times G(\hat{\mathbb{Q}})\right)/S_G \cong \bigsqcup_{g\in C} \Gamma_g\backslash\mathcal{H}$$

where $\Gamma_g = gS_Gg^{-1} \cap GL_2^+(\mathbb{Q})$.

Under this identification, if we denote by ω the canonical line bundle on X_G, then we may identify $M_k(\Gamma_G)$ with the space of global sections $H^0(X_G, \omega^{\otimes k})$. The correspondence $X(w)$ induces (by pullback) a map on global sections of line bundles $T(w) : H^0(X_G, \omega^{\otimes k}) \to H^0(X_G, \omega^{\otimes k})$ given by $f \mapsto \sum_{\alpha\in S/W}(\alpha w)^* f$. We then get the following corollary.

Corollary 4.4.5 *The correspondences* $X(w)$ *may be identified via this isomorphism (explicitly given as* $z \mapsto [z, 1]$*) as correspondences* $\Gamma_g\backslash\mathcal{H} \to \Gamma_{gw}\backslash\mathcal{H}$. *In particular, if* $\det(w) \in \det(S_G)$, *then we may identify the correspondence* $X(w)$ *as correspondences on* $\Gamma_G\backslash\mathcal{H}$, *and the operator* $T(w)$ *as an operator on* $M_k(\Gamma_G)$.

The reason for using Definition 4.4.2 is the following important theorem.

Theorem 4.4.6 (Shimura, [29, Theorem 7.9]) *Let* $w \in GL_2(\hat{\mathbb{Q}})$ *be such that* $w_l = 1$ *for all* $l \neq p$, *and* $w_p \in M_2(\mathbb{Z}_p)$ *is an element such that* $\det(w_p) = p$. *Then if* $p \nmid N$, *the correspondence* $X(w)$ *on* X_G *is rational over the field of definition of* X_G *and satisfies* $\tilde{X}(w) \equiv Fr_p +^t Fr_p^p \circ \widetilde{[\det(w)^{-1}]}$, *where* $X \mapsto \tilde{X}$ *is the reduction modulo* p *and* Fr_p *is the Frobenius correspondence on* $\tilde{X}_G \times \tilde{X}_G^p$.

This, in turn, relates these Hecke operators to the zeta function of X_G. Before stating the connection, we introduce some notations.

Let $Z_p(s; X_G/\mathbb{Q})$ denote the local factor at p of the zeta function of X_G. Let $w \in GL_2(\hat{\mathbb{Q}})$ be such that $w_l = 1$ for all $l \neq p$ and $w_p \in M_2(\mathbb{Z}_p)$ be such that $\det(w_p) = p$. Let $\sigma \in SL_2(\mathbb{Z})$ be such that $\lambda_N(\sigma) = p \cdot \lambda_N(w)^{-2}$. Via the diagonal embedding $SL_2(\mathbb{Z}) \hookrightarrow GL_2(\hat{\mathbb{Q}})$, the correspondence $X(\sigma)$ is well defined and as $\det(\sigma) = 1 \in \det(S_G)$, it induces an operator $T(\sigma)$ on $M_k(\Gamma_G)$.

Corollary 4.4.7 *Let* w, σ *be as above. Let* $T(w)$, $T(\sigma)$ *be the corresponding operators on* $M_2(\Gamma_G)$. *Then*

$$(1 - p^{-s})(1 - p^{1-2s})Z_p(s; X_G/\mathbb{Q}) = \det\left(1 - T(w) \cdot p^{-s} + T(\sigma) \cdot p^{1-2s}\right). \tag{4.4.8}$$

Proof This is proved as in [29], Corollary 7.10 and Theorem 7.11. □

Although this, in general, does not yield an application to q-expansions, there are some special cases in which it does. We describe these cases, as we shall want to make use of them in some of our applications.

Recall that the Hecke operators commute, hence have common eigenvalues. These are called eigenforms.

Definition 4.4.9 An element $f \in S_k(\Gamma_G)$ is called an *eigenform* if it is an eigenvector for T_α for all $\alpha \in GL_2^+(\mathbb{Q})$. An eigenform $f = \sum_{n=1}^{\infty} a_n q^n$ is *normalized* if $a_1 = 1$.

Corollary 4.4.10 *Let $\Gamma_G = \Gamma(\mathfrak{h}, t)$, and let $f = \sum_{n=1}^{\infty} a_n q^n \in S_k(\Gamma_G)$ be a normalized eigenform. Let $w \in GL_2(\hat{\mathbb{Q}})$ be such that $w_l = 1$ for all $l \neq p$ and $w_p \in M_2(\mathbb{Z}_p)$ be such that* $\det(w_p) = p$. *Then $T(w)f = a_p f$.*

Proof This is proved in [29, Section 3.5]. □

Yet as another consequence, we get the following fact, which could also be shown directly by double coset computation.

Corollary 4.4.11 *If $p \nmid N$, and $w \in GL_2(\hat{\mathbb{Q}})$ is such that $w_l = 1$ for all $l \neq p$, and $w_p \in M_2(\mathbb{Z}_p)$ is an element such that* $\det(w_p) = p$, *then the operator $T(w)$ is independent of the choice of w.*

This motivates the following definition.

Definition 4.4.12 If $p \nmid N$, and $w \in GL_2(\hat{\mathbb{Q}})$ is such that $w_l = 1$ for all $l \neq p$, and $w_p \in M_2(\mathbb{Z}_p)$ is an element such that $\det(w_p) = p$, we define $T_p := T(w)$.

Next, we would like to write this operator in terms of a double coset operator, as in Definition 4.1.1. In order to do that, we show first that the subgroup W in Definition 4.4.2 is also induced from a subgroup of $GL_2(\mathbb{Z}/N\mathbb{Z})$.

Lemma 4.4.13 *Let $G \subseteq GL_2(\mathbb{Z}/N\mathbb{Z})$. Let $p \nmid N$. Let $w \in GL_2(\hat{\mathbb{Q}})$ be such that $w_l = 1$ for all $l \neq p$, and $w_p \in M_2(\mathbb{Z}_p)$ is an element such that* $\det(w_p) = p$. *Let $W = S_G \cap w S_G w^{-1}$. Then one has*

$$p \cdot M_2(\mathbb{Z}_p) \subseteq M_2(\mathbb{Z}_p) \cap w_p M_2(\mathbb{Z}_p) w_p^{-1}.$$

Let R_p be the image of $M_2(\mathbb{Z}_p) \cap w_p M_2(\mathbb{Z}_p) w_p^{-1}$ in $M_2(\mathbb{Z}/p\mathbb{Z})$, and let $G_p := R_p^\times$. Set

$$G_w := G \times G_p \subseteq GL_2(\mathbb{Z}/(Np)\mathbb{Z}).$$

Then $W = S_{G_w}$.

Proof First, let us show that

$$p \cdot M_2(\mathbb{Z}_p) \subseteq M_2(\mathbb{Z}_p) \cap w_p M_2(\mathbb{Z}_p) w_p^{-1}.$$

Indeed, it is enough to prove that $p \cdot M_2(\mathbb{Z}_p) \subseteq w_p M_2(\mathbb{Z}_p) w_p^{-1}$, which is equivalent to

$$w_p^{-1} p \cdot M_2(\mathbb{Z}_p) \cdot w_p \subseteq M_2(\mathbb{Z}_p).$$

However, as $\det(w_p) = p$, we see that $w_p^{-1} \cdot p = \tilde{w}_p \in M_2(\mathbb{Z}_p)$, thus the claim is trivial.

Now, let $x \in W$. Then, as $x \in S_G$, we may write $x = r \cdot g \cdot s$ with $r \in \mathbb{Q}^\times$, $g \in GL_2(\hat{\mathbb{Z}})$, $s \in GL_2^+(\mathbb{R})$, and $g \bmod N \in G$.

Write $g = g^p g_p$, with $g^p \in \prod_{l \neq p} GL_2(\mathbb{Z}_l)$ and $g_p \in GL_2(\mathbb{Z}_p)$. Recall also that $w_l = 1$ for all $l \neq p$. Since $x \in w S_G w^{-1}$, we have

$$r \cdot g^p \cdot w_p^{-1} g_p w_p \cdot s = r \cdot w^{-1} g w \cdot s = w^{-1} x w \in S_G.$$

Thus, there exists $r' \in \mathbb{Q}^\times$ such that

$$r' \cdot g^p \cdot w_p^{-1} g_p w_p \in GL_2(\hat{\mathbb{Z}}).$$

In particular, for all $l \neq p$, we have $r' \cdot g_l \in GL_2(\mathbb{Z}_l)$, and as $g_l \in GL_2(\mathbb{Z}_l)$, we see that $r' \in \mathbb{Z}_l^\times$. Also, as

$$r' \cdot w_p^{-1} g_p w_p \in GL_2(\mathbb{Z}_p)$$

and $g_p \in GL_2(\mathbb{Z}_p)$, we see that $r' \in \{\pm 1\}$, and $w_p^{-1} g_p w_p \in GL_2(\mathbb{Z}_p)$.

Therefore, we obtain

$$g_p \in w_p GL_2(\mathbb{Z}_p) w_p^{-1} \subseteq w_p M_2(\mathbb{Z}_p) w_p^{-1}.$$

As we already know $g_p \in M_2(\mathbb{Z}_p)$, we see that $g_p \bmod p \in R_p$.

Since $g_p \in GL_2(\mathbb{Z}_p)$, it follows that $g \bmod p = g_p \bmod p \in G_p$, and from $g \bmod N \in G$, it follows that $g \bmod Np \in G \times G_p = G_w$. Thus we have shown that $x \in S_{G_W}$, so that $W \subseteq S_{G_w}$.

Conversely, assume $x \in S_{G_w}$. Then $x = r \cdot g \cdot s$ with $r \in \mathbb{Q}^\times$, $g \in GL_2(\hat{\mathbb{Z}})$, $s \in GL_2^+(\mathbb{R})$, and $g \bmod Np \in G_w$.

It follows that $g \bmod N \in G$ and $g \bmod p \in G_p$. In particular, we see immediately that $x \in S_G$. Moreover,

$$w^{-1} x w = r \cdot g^p \cdot w_p^{-1} g_p w_p \cdot s.$$

Since $g_p \bmod p = g \bmod p \in G_p \subseteq R_p$, we have

$$g_p \in M_2(\mathbb{Z}_p) \cap w_p M_2(\mathbb{Z}_p) w_p^{-1}$$

hence $w_p^{-1} g_p w_p \in M_2(\mathbb{Z}_p)$. Looking at the determinant, we see that it actually lies in $GL_2(\mathbb{Z}_p)$. In particular,

$$g' = g^p \cdot w_p^{-1} g_p w_p \in GL_2(\hat{\mathbb{Z}})$$

and $g' \bmod N \in G$. Therefore, $w^{-1} x w \in S_G$, showing equality. □

We now show how to use strong approximation to rewrite the operator $T(w)$ in the classical description of modular curves.

Lemma 4.4.14 *Let $G \subseteq GL_2(\mathbb{Z}/N\mathbb{Z})$ be a subgroup. Let $p \in \det(G)$ be a prime. Let $w \in GL_2(\hat{\mathbb{Q}})$ be such that $w_l = 1$ for all $l \neq p$, and $w_p \in M_2(\mathbb{Z}_p)$ be such that $\det(w_p) = p$. Then the correspondence $X(w)$ can be realized as*

$$\Gamma_G z \mapsto \sum_{\alpha \in \Gamma_G / \Gamma_{G_w}} \Gamma_G \cdot q(w) \cdot \alpha^{-1} z$$

for some $q(w) \in GL_2^+(\mathbb{Q})$, explicitly constructed as a function of w. Therefore T_p is realized as $f \mapsto \sum_{\alpha \in \Gamma_G / \Gamma_{G_w}} f|_{q(w) \cdot \alpha^{-1}}$.

Proof Let $\delta_p \in G$ be an element such that $\det(\delta_p) = p$. Let $w_p \in M_2(\mathbb{Z})$ be such that $\det(w_p) = p$, and let $\alpha \in \Gamma_G$. Then

$$\det(\delta_p^{-1} \cdot \lambda_N(w_p)) = \det(\delta(p))^{-1} \cdot \det(w_p) = p^{-1} \cdot p = 1$$

so that $\delta_p^{-1} \cdot \lambda_N(w_p) \in SL_2(\mathbb{Z}/N\mathbb{Z})$. Since $\lambda_N : SL_2(\mathbb{Z}) \to SL_2(\mathbb{Z}/N\mathbb{Z})$ is surjective, there exists $\beta_p \in SL_2(\mathbb{Z})$ such that

$$\lambda_N(\beta_p) = \delta_p^{-1} \cdot \lambda_N(w_p).$$

Let $q(w) = \beta_p \cdot p w_p^{-1} \in GL_2^+(\mathbb{Q})$. Then we have

$$\alpha \cdot q(w)^{-1} = p^{-1} \cdot \alpha w_p \beta_p^{-1} = p^{-1} \cdot \alpha w \cdot w^{-1} w_p \beta_p^{-1}.$$

Let $s = w^{-1} w_p \beta_p^{-1}$. Then $s_p = w_p^{-1} w_p \beta_p^{-1} = \beta_p^{-1} \in SL_2(\mathbb{Z}) \subseteq GL_2(\mathbb{Z}_p)$. Also, for all $l \neq p$, one has $s_l = w_p \cdot \beta_p^{-1}$, so that $\det(s_l) = \det(w_p) \cdot \det(\beta_p)^{-1} = p$ and $s_p \in M_2(\mathbb{Z}) \subseteq M_2(\mathbb{Z}_l)$, showing that $s_l \in GL_2(\mathbb{Z}_l)$. Moreover,

$$\lambda_N(s) = \lambda_N(w_p) \cdot \lambda_N(\beta_p)^{-1} = \delta_p \in G$$

showing that $s \in S_G$. Therefore $\alpha w = \alpha \cdot q(w)^{-1} \cdot s^{-1}$, so that (recall that $p \in \mathbb{Q}^\times$ acts trivially)

$$[z, \alpha w] = [z, p^{-1} \cdot \alpha w \cdot s] = [z, \alpha \cdot q(w)^{-1}] = [q(w) \cdot \alpha^{-1} \cdot z, 1].$$

By Lemma 4.4.13, $W = S_{G_W}$, hence

$$S_G/W \cong G \times GL_2(\mathbb{Z}/p\mathbb{Z})/G_w \cong \Gamma_G/\Gamma_{G_w}.$$

It follows that under the isomorphism $[z] \mapsto [z, 1]$, the correspondence $X(w)$ above is interpreted as

$$\Gamma_G z \mapsto \sum_{\alpha \in \Gamma_G/\Gamma_{G_w}} \Gamma_G \cdot q(w) \cdot \alpha^{-1} z.$$

□

4.4.2 Classical Definition

We can now use Lemma 4.4.14 to find an equivalent definition in classical terms of the Hecke operator T_p.

Lemma 4.4.15 *Let $\alpha \in M_2(\mathbb{Z})$ be such that* $\det(\alpha) = p$ *and $\lambda_N(\alpha) \in G$. Then $T_\alpha = T_p$.*

Proof Let $w \in GL_2(\hat{\mathbb{Q}})$ be such that $w_l = 1$ for all $l \neq p$, $w_p \in M_2(\mathbb{Z}_p)$ and $\det(w_p) = p$. By Lemma 4.4.14, the operator $T_p = T(w)$ is given by $f \mapsto \sum_{\beta \in \Gamma_G/\Gamma_{G_w}} f|_{q(w)\cdot\beta^{-1}}$, where $q(w) \in GL_2^+(\mathbb{Q})$. By Lemma 4.4.13, we know that $G_w = G \times G_p$, with $G_p = R_p^\times$ where R_p is the image in $M_2(\mathbb{Z}/p\mathbb{Z})$ of $M_2(\mathbb{Z}_p) \cap w_p M_2(\mathbb{Z}_p) w_p^{-1}$, therefore

$$\Gamma_G/\Gamma_{G_w} \cong (G \times GL_2(\mathbb{Z}/p\mathbb{Z}))\,/G_w \cong GL_2(\mathbb{Z}/p\mathbb{Z})/G_p.$$

Recall also that $\det(w_p) = p$, hence $w_p \in SL_2(\mathbb{Z}_p) \cdot \begin{pmatrix} p & 0 \\ 0 & 1 \end{pmatrix} \cdot SL_2(\mathbb{Z}_p)$.

Write $w_p = \gamma_1 \cdot \begin{pmatrix} p & 0 \\ 0 & 1 \end{pmatrix} \cdot \gamma_2$, for some $\gamma_1, \gamma_2 \in SL_2(\mathbb{Z}_p)$. Then

$$\begin{aligned} w_p M_2(\mathbb{Z}_p) w_p^{-1} &= \gamma_1 \cdot \begin{pmatrix} p & 0 \\ 0 & 1 \end{pmatrix} \cdot \gamma_2 M_2(\mathbb{Z}_p) \gamma_2^{-1} \cdot \begin{pmatrix} p & 0 \\ 0 & 1 \end{pmatrix}^{-1} \cdot \gamma_1^{-1} \\ &= \gamma_1 \cdot \begin{pmatrix} p & 0 \\ 0 & 1 \end{pmatrix} \cdot M_2(\mathbb{Z}_p) \cdot \begin{pmatrix} p & 0 \\ 0 & 1 \end{pmatrix}^{-1} \cdot \gamma_1^{-1} \end{aligned}$$

hence

$$M_2(\mathbb{Z}_p) \cap w_p M_2(\mathbb{Z}_p) w_p^{-1} = \gamma_1 \cdot \Delta^0(p) \cdot \gamma_1^{-1}$$

where $\Delta^0(p) = \left\{ \begin{pmatrix} a & b \\ c & d \end{pmatrix} \in M_2(\mathbb{Z}_p) \mid b \in p\mathbb{Z}_p \right\}$, and passing to the image in $M_2(\mathbb{Z}/p\mathbb{Z})$, we get

$$G_p = R_p^\times = \overline{\gamma}_1 \cdot \Gamma^0(\mathbb{Z}/p\mathbb{Z}) \cdot \overline{\gamma}_1^{-1}$$

where $\Gamma^0(\mathbb{Z}/p\mathbb{Z})$ is the Borel subgroup of lower triangular matrices, and $\overline{\gamma}_1 \in SL_2(\mathbb{Z}/p\mathbb{Z})$.

Also, if $\beta_p \in SL_2(\mathbb{Z})$ is such that $\lambda_N(\beta_p) = \lambda_N(p)^{-1} \cdot \lambda_N(\alpha) \cdot \lambda_N(w_p)$, then $\lambda_N(\beta_p \cdot pw_p^{-1}) = \lambda_N(\alpha)$ and $\alpha^{-1}\beta_p \cdot p \cdot w_p^{-1} \in \Gamma(N) \subseteq \Gamma$, so that $\Gamma\beta_p \cdot pw_p^{-1} = \Gamma\alpha$, thus $\Gamma\beta_p \cdot pw_p^{-1}\Gamma = \Gamma\alpha\Gamma$, and so (as $T_\alpha(f) = \sum_{\delta \in \Gamma\backslash\Gamma\alpha\Gamma} f|_\delta$), we may assume that $\beta_p \cdot pw_p^{-1} = \alpha$, hence $q(w) = \alpha$.

Note then that, as $p \cdot \alpha^{-1} = \gamma_1 \cdot \begin{pmatrix} p & 0 \\ 0 & 1 \end{pmatrix} \cdot \gamma_2 \beta_p^{-1}$, we see that

$$\begin{aligned} \lambda_p(\alpha^{-1}\Gamma\alpha) &= \lambda_p(\alpha^{-1} \cdot M_2(\mathbb{Z}) \cdot \alpha)^\times = \lambda_p(\gamma_1 \cdot \Delta^0(p) \cdot \gamma_1^{-1})^\times \\ &= \lambda_p(\gamma_1 \cdot \Gamma^0(p) \cdot \gamma_1^{-1}) = \overline{\gamma}_1 \cdot \Gamma^0(\mathbb{Z}/p\mathbb{Z}) \cdot \overline{\gamma}_1^{-1} = G_p. \end{aligned}$$

Therefore

$$\Gamma/(\Gamma \cap \alpha^{-1}\Gamma\alpha) \cong GL_2(\mathbb{Z}/p\mathbb{Z})/G_p \cong \Gamma_G/\Gamma_{G_w}.$$

But we have bijection

$$\begin{array}{ccccc} \Gamma/(\Gamma \cap \alpha^{-1}\Gamma\alpha) & \to & (\Gamma \cap \alpha^{-1}\Gamma\alpha)\backslash\Gamma & \to & \Gamma\backslash\Gamma\alpha\Gamma \\ \beta & \mapsto & \beta^{-1} & \mapsto & \alpha \cdot \beta^{-1}. \end{array}$$

Thus, the map $\beta \mapsto \alpha \cdot \beta^{-1} : \Gamma_G/\Gamma_{G_w} \to \Gamma\backslash\Gamma\alpha\Gamma$ is a bijection, so that

$$T_\alpha(f) = \sum_{\delta \in \Gamma\backslash\Gamma\alpha\Gamma} f|_\delta = \sum_{\beta \in \Gamma_G/\Gamma_{G_w}} f|_{\alpha \cdot \beta^{-1}} = \sum_{\beta \in \Gamma_G/\Gamma_{G_w}} f|_{q(w)\cdot\beta^{-1}} = T(w)(f) = T_p(f)$$

establishing the result. □

Corollary 4.4.16 *Assume $p \nmid N$. Let $\alpha \in M_2(\mathbb{Z})$ be such that* $\det(\alpha) = p$ *and $\lambda_N(\alpha) \in G$. Then T_α is independent of α. Moreover, if $\Gamma_G = \Gamma(\mathfrak{h}, t)$ and $f = \sum_{n=1}^\infty a_n q^n$ is an eigenform of the Hecke algebra then $T_\alpha f = a_p f$.*

This allows us to define the Hecke operators at primes not dividing the level (and by multiplicativity to all n such that $(n, N) = 1$).

Definition 4.4.17 Let n be such that $(n, N) = 1$ and $n \in \det(G)$. Let $\alpha \in M_2(\mathbb{Z})$ be such that $\det(\alpha) = n$ and $\lambda_N(\alpha) \in G$. Let $T_n := T_\alpha$.

Therefore, Algorithm 4.2.1 can be used to compute the Hecke operators T_n, when n is coprime to N.

Remark 4.4.18 Note that the definition of the Hecke operators $\{T_n\}$, although independent of the choice of representatives, does depend on G, and not only on Γ_G. This is not just an artifact of the proof, as the following example shows.

Example 4.4.19 Let

$$G_1 = \left\{ \pm \begin{pmatrix} 1 & 0 \\ 0 & d \end{pmatrix} \in GL_2(\mathbb{Z}/7\mathbb{Z}) \mid d \in (\mathbb{Z}/7\mathbb{Z})^\times \right\}.$$

Let $G_2 \subseteq GL_2(\mathbb{Z}/7\mathbb{Z})$ be the subgroup

$$G_2 := \left\langle \pm \begin{pmatrix} 1 & 4 \\ 2 & 6 \end{pmatrix} \right\rangle \subseteq GL_2(\mathbb{Z}/7\mathbb{Z})$$

which is an abelian group isomorphic to $\mathbb{Z}/6\mathbb{Z} \oplus \mathbb{Z}/2\mathbb{Z}$.

Then $\Gamma_{G_1} = \Gamma_{G_2} = \Gamma(7) \cdot \{\pm 1\}$. But while $T_2^{G_1}$ on $S_2(\Gamma(7))$ is the familiar Hecke operator T_α for $\alpha = \begin{pmatrix} 1 & 0 \\ 0 & 2 \end{pmatrix}$, $T_2^{G_2}$ on $S_2(\Gamma(7))$ is simply the identity, since $\alpha \notin G_2$.

Remark 4.4.20 As in [29, Proposition 3.31], one can show that the map

$$\Gamma \alpha \Gamma \mapsto SL_2(\mathbb{Z}) \alpha SL_2(\mathbb{Z}),$$

for all $\alpha \in M_2(\mathbb{Z})$ such that $\lambda_N(\alpha) \in \mathfrak{N}(G)$, defines a homomorphism on the Hecke algebras.

However, as the above example shows, this homomorphism is not injective, and elements α in different cosets of $G \backslash \mathfrak{N}(G)$ give rise to different operators.

4.5 *Efficient Implementation of the Hecke Operators $T_n^\vee$, $n \in \det(G)$*

Algorithm 4.2.1 is not as efficient as we would have liked. Specifically, the logarithmic factor obtained from the passage to modular symbols is, in practice, due to the constant, a particularly high cost. We therefore use the ideas of Merel in

[23] to calculate, at least the operators $\{T_n\}$ more efficiently. Following Section 2.1 in [23], we introduce the definition of a Merel pair.

Definition 4.5.1 Let $\Delta \subseteq GL_2(\mathbb{Q})$ be such that $\Gamma\Delta = \Delta\Gamma$ and such that $\Gamma\backslash\Delta$ is finite. Let $\tilde{\Delta} = \{g \in GL_2(\mathbb{Q}) \mid \tilde{g} := g^{-1}\det(g) \in \Delta\}$. Let $\phi : \tilde{\Delta} \cdot SL_2(\mathbb{Z}) \to SL_2(\mathbb{Z})$ be a map such that

1. For all $\gamma \in \tilde{\Delta} \cdot SL_2(\mathbb{Z})$ and $g \in SL_2(\mathbb{Z})$ we have $\Gamma \cdot \phi(\gamma g) = \Gamma \cdot \phi(\gamma) \cdot g$.
2. For all $\gamma \in \tilde{\Delta} \cdot SL_2(\mathbb{Z})$, we have $\gamma \cdot \phi(\gamma)^{-1} \in \tilde{\Delta}$ (or equivalently $\phi(\gamma) \cdot \tilde{\gamma} \in \Delta$).
3. The map $\Gamma\backslash\Delta \to \tilde{\Delta} \cdot SL_2(\mathbb{Z})/SL_2(\mathbb{Z})$ which associates to $\Gamma\delta$ the element $\tilde{\delta}SL_2(\mathbb{Z})$ is injective (it is necessarily surjective).

We say that the pair (Δ, ϕ) is a *Merel pair* for Γ.

We can now state Merel's condition (C_Δ):

Definition 4.5.2 Let $\sum u_M M \in \mathbb{C}[M_2(\mathbb{Z})]$, and let (Δ, ϕ) be a Merel pair for Γ. We will say that $\sum u_M M$ satisfies the condition (C_Δ) if and only if for all $K \in \tilde{\Delta}SL_2(\mathbb{Z})/SL_2(\mathbb{Z})$ we have the following equality in $\mathbb{C}[\mathbb{P}^1(\mathbb{Q})]$:

$$\sum_{M \in K} u_M([M\infty] - [M0]) = [\infty] - [0].$$

Finally, we recall the following extremely useful theorem.

Theorem 4.5.3 ([23, Theorem 4]) *Let $P \in \mathbb{C}_{k-2}[X, Y]$ and $g \in SL_2(\mathbb{Z})$. Let (Δ, ϕ) be a Merel pair for Γ. Let $\sum u_M M \in \mathbb{C}[M_2(\mathbb{Z})]$ satisfy the condition (C_Δ). We have in $\mathbb{M}_k(\Gamma)$*

$$T_\Delta^\vee([P, g]) = \sum_{M, gM \in \tilde{\Delta}SL_2(\mathbb{Z})} u_M[P|_{\tilde{M}}, \phi(gM)].$$

We also recall Merel's condition (C_n).

Definition 4.5.4 Denote by $M_2(\mathbb{Z})_n$ the set of matrices of $M_2(\mathbb{Z})$ of determinant n. We say that an element $\sum_M u_M M \in \mathbb{C}[M_2(\mathbb{Z})_n]$ satisfies the condition (C_n) if for all $K \in M_2(\mathbb{Z})_n/SL_2(\mathbb{Z})$, we have in $\mathbb{C}[\mathbb{P}^1(\mathbb{Q})]$

$$\sum_{M \in K} u_M([M\infty] - [M0]) = [\infty] - [0].$$

Corollary 4.5.5 *Let (Δ_n, ϕ_n) be a Merel pair for Γ with $\Delta_n \subseteq M_2(\mathbb{Z})_n$. Let $\sum_M u_M M$ satisfy the condition (C_n), then in $\mathbb{M}_k(\Gamma)$*

$$T_{\Delta_n}^\vee([P, g]) = \sum_M u_M[P|_{\tilde{M}}, \phi_n(gM)]$$

where the sum is restricted to matrices M such that $gM \in \tilde{\Delta}_n SL_2(\mathbb{Z})$.

Example 4.5.6 Consider the following examples.

(1) Let

$$\Delta_n := \left\{ g = \begin{pmatrix} a & b \\ c & d \end{pmatrix} \in M_2(\mathbb{Z}) \mid \det(g) = n, N \mid c, N \mid a - 1 \right\}$$

and $\phi_n : \tilde{\Delta}_n SL_2(\mathbb{Z}) \to SL_2(\mathbb{Z})$ is a map such that $\pi(\phi_n(g)) = (0 : 1) \cdot \lambda_N(g) \in \mathbb{P}^1(\mathbb{Z}/N\mathbb{Z})$, where $\pi : SL_2(\mathbb{Z}) \to \mathbb{P}^1(\mathbb{Z}/N\mathbb{Z})$ is the natural surjection. Then in [23, Lemma 1], Merel shows that (Δ_n, ϕ_n) is a Merel pair for $\Gamma_1(N)$, which is key to modern efficient implementation of Hecke operators.

(2) Let $p > 2$ be a prime, and let $u \in \mathbb{F}_p^\times$ be a non-square. Let

$$\Delta_n := \left\{ g = \begin{pmatrix} a & b \\ c & d \end{pmatrix} \mid \det(g) = n, p \mid a - d, p \mid b - uc \right\}$$

and $\phi_n : \tilde{\Delta}_n SL_2(\mathbb{Z}) \to SL_2(\mathbb{Z})$ is a map such that $\phi_n(g)^{-1} \cdot \sqrt{u} = \lambda_p(g)^{-1} \cdot \sqrt{u}$, where the action of $\tilde{\Delta}_n SL_2(\mathbb{Z})$ on $\mathbb{F}_{p^2} - \mathbb{F}_p$ is via Mobius transformations. Then similarly (Δ_n, ϕ_n) is a Merel pair for $\Gamma_{ns}(p)$—the nonsplit Cartan subgroup of level p.

Let us write $G_0 := G \cap SL_2(\mathbb{Z}/N\mathbb{Z})$, and let $\pi : SL_2(\mathbb{Z}) \to G_0 \backslash SL_2(\mathbb{Z}/N\mathbb{Z})$ be the natural map $\pi(g) = G_0 \cdot \lambda_N(g)$, inducing an isomorphism $\Gamma \backslash SL_2(\mathbb{Z}) \cong G_0 \backslash SL_2(\mathbb{Z}/N\mathbb{Z})$. Let $s : G_0 \backslash SL_2(\mathbb{Z}/N\mathbb{Z}) \to SL_2(\mathbb{Z})$ be a section of π.

Let $\delta_n \in G$ be an element such that $\det(\delta_n) = n$, and let

$$\Delta_n := \{\alpha \in M_2(\mathbb{Z}) \mid \det(\alpha) = n, \quad \lambda_N(\alpha) \in G\}. \tag{4.5.7}$$

Let $\phi_n : M_2(\mathbb{Z})_n \to SL_2(\mathbb{Z})$ be the map defined by

$$\phi_n(\alpha) = s(G_0 \cdot n^{-1}\delta_n \cdot \lambda_N(\alpha)).$$

Proposition 4.5.8 *(Δ_n, ϕ_n) is a Merel pair.*

Proof First, note that $\alpha \in \tilde{\Delta}_n$ if and only if

$$\det(\alpha) = \det(\tilde{\alpha}) = n \text{ and } n\lambda_N(\alpha)^{-1} = \lambda_N(\tilde{\alpha}) \in G.$$

Therefore

$$\tilde{\Delta}_n = \left\{ \alpha \in M_2(\mathbb{Z}) \mid \det(\alpha) = n, n^{-1}\lambda_N(\alpha) \in G \right\}.$$

Next, we verify the three conditions in Definition 4.5.1:

1. Let $\gamma \in \tilde{\Delta}_n SL_2(\mathbb{Z})$ and $g \in SL_2(\mathbb{Z})$. Then

$$\begin{aligned}\pi(\phi_n(\gamma g)) &= G_0 \cdot n^{-1}\delta_n \cdot \lambda_N(\gamma g) = G_0 \cdot n^{-1}\delta_n \cdot \lambda_N(\gamma) \cdot \lambda_N(g) \\ &= \pi(\phi_n(\gamma)) \cdot \lambda_N(g) = G_0 \cdot \lambda_N(\phi_n(\gamma)) \cdot \lambda_N(g) \\ &= G_0 \cdot \lambda_N(\phi_n(\gamma) g) = \pi(\phi_n(\gamma) g)\end{aligned}$$

hence $\Gamma \cdot \phi_n(\gamma g) = \Gamma \cdot \phi_n(\gamma) g$.

2. Let $g \in \tilde{\Delta}_n SL_2(\mathbb{Z})$. Assume that $n^{-1}\delta_n \cdot \lambda_N(g) \in G_0$. Then $n^{-1}\lambda_N(g) \in G$, hence $g \in \tilde{\Delta}_n$. Thus $g \in \tilde{\Delta}_n$ iff $n^{-1}\delta_n \cdot \lambda_N(g) \in G_0$. Now, note that

$$G_0 \cdot n^{-1}\delta_n \cdot \lambda_N(g) = \pi(\phi_n(g)) = G_0 \cdot \lambda_N(\phi_n(g))$$

hence

$$n^{-1}\delta_n \cdot \lambda_N(g\phi_n(g)^{-1}) \in G_0$$

showing that $g\phi_n(g)^{-1} \in \tilde{\Delta}_n$.

3. Let $\delta_1, \delta_2 \in \Delta_n$ be such that $\delta_1\delta_2^{-1} \in SL_2(\mathbb{Z})$. By definition $\lambda_N(\tilde{\delta}_1), \lambda_N(\tilde{\delta}_2) \in G$, hence

$$\begin{aligned}\lambda_N(\delta_1\delta_2^{-1}) &= \lambda_N(\tilde{\delta}_1^{-1} \cdot n \cdot n^{-1} \cdot \tilde{\delta}_2) = \\ &= \lambda_N(\tilde{\delta}_1^{-1}\tilde{\delta}_2) = \lambda_N(\tilde{\delta}_1)^{-1} \cdot \lambda_N(\tilde{\delta}_2) \in G\end{aligned}$$

hence $\delta_1\delta_2^{-1} \in \Gamma$.

□

Corollary 4.5.9 *Assume $n \in \det(G)$. Let $\sum_M u_M M$ satisfy the condition (C_n). Then, in $\mathbb{M}_k(\Gamma_G)$*

$$T_n^\vee([P, g]) = \sum_M u_M[P|_{\tilde{M}}, \phi_n(gM)]. \tag{4.5.10}$$

Proof By Definition 4.4.17 and (4.5.7), we see that $T_n^\vee = T_\alpha^\vee$ for any $\alpha \in \Delta_n$, i.e. $T_n^\vee = T_{\Delta_n}^\vee$. The corollary now follows from Corollary 4.5.5, once we realize that $M_2(\mathbb{Z})_n = \tilde{\Delta}_n \cdot SL_2(\mathbb{Z})$.

Indeed, for $M \in M_2(\mathbb{Z})_n$, consider

$$g = M \cdot \phi_n(M)^{-1} = M \cdot s(G_0 \cdot n^{-1}\delta_n \cdot \lambda_N(M))^{-1}.$$

Then

$$\lambda_N(g)^{-1} = \lambda_N(\phi_n(M)) \cdot \lambda_N(M)^{-1} \in G_0 \cdot n^{-1}\delta(n) \cdot \lambda_N(M) \cdot \lambda_N(M)^{-1} = G_0 \cdot n^{-1}\delta_n$$

hence

$$n^{-1}\lambda_N(g) \in \delta_n \cdot G_0 \subseteq G$$

showing that $g \in \tilde{\Delta}_n$. □

Corollary 4.5.11 *Computation of the Hecke operator T_p on $S_k(\Gamma_G)$, for $p \in$ $\det(G)$, can be done in $O(p \log p)$ basic* CosetIndex *operations.*

Proof In [22, Proposition 8], Merel shows that one can compute a certain set $\mathcal{S}_n$ of $O(\sigma_1(n) \cdot \log n)$ matrices, which in [23, Proposition 18], he proves to satisfy condition (C_n) (Definition 4.5.4). Therefore, by Corollary 4.5.9, it is enough to compute, for a vector $[P, g] \in \mathbb{M}_k(\Gamma)$, the sum (4.5.10), implying the resulting complexity for computation of the operator $T_p^\vee$ on $\mathbb{M}_k(\Gamma_G)$.

Let $\alpha \in M_2(\mathbb{Z})$ be such that $\det(\alpha) = p$ and $\lambda_N(\alpha) \in G$. Then by definition $T_\alpha = T_p$. Since $\lambda_N(\alpha) \in G$ and G is of real type, by Corollary 4.1.9, one can then compute $T_p^\vee$ on $\mathbb{S}_k(\Gamma_G)$. Finally, from Proposition 4.1.7 and Theorem 3.6.6, the result follows. □

Remark 4.5.12 In general, the computation of the Hecke operators $\{T_n\}$ for $(n, N) = 1$, can be done by using their multiplicativity property, so Corollary 4.5.11 is all that's needed, and gives a slightly better complexity of $O(p_{\max} \cdot \log n)$ where $p_{\max}$ is the largest prime factor of n.

Note also that this gives an improvement by a factor of $\log(N)$ over Algorithm 4.2.1, which is very significant in practice.

We would also like to mention that under some mild assumptions, the decomposition 3.3.2 is Hecke-equivariant, hence one can compute the Hecke operator on each of the isotypic subspaces separately.

Lemma 4.5.13 *Let $G \subseteq G' \subseteq GL_2(\mathbb{Z}/N\mathbb{Z})$ be such that G is normal in G'. Let $\varepsilon : G'/G \to \mathbb{Q}(\zeta)^\times$ be a character, and let $\alpha \in M_2(\mathbb{Z})$ be such that* $\det(\alpha) = p$ *is a prime and $\lambda_N(\alpha) \in G$. Then the subspace $S_k(\Gamma_G, \varepsilon)$ is invariant under T_α.*

Proof Let $f \in S_k(\Gamma_G, \varepsilon)$, and let $\gamma \in \Gamma_{G'}$. Let R be a set of coset representatives for $\Gamma_G \backslash \Gamma_G \alpha \Gamma_G$. Then

$$T_\alpha(f)|_\gamma = \sum_{\delta \in R} f|_{\delta\gamma} = \sum_{\delta \in R} (f|_\gamma)|_{\gamma^{-1}\delta\gamma} = \sum_{\delta \in R} (\varepsilon(\gamma) \cdot f)|_{\gamma^{-1}\delta\gamma} = \varepsilon(\gamma) \sum_{\delta \in R} f|_{\gamma^{-1}\delta\gamma}.$$

Since $\gamma \in \Gamma_{G'}$ and G is normal in G', for any $x \in \Gamma_G$ we have $\lambda_N(\gamma^{-1} x \gamma) \in G$, hence $\gamma^{-1} x \gamma \in \Gamma_G$. Thus $\gamma^{-1} \Gamma_G \gamma = \Gamma_G$.

As $\Gamma_G \alpha \Gamma_G = \bigsqcup_{\delta \in R} \Gamma_G \delta$, it follows that

$$\Gamma_G \gamma^{-1} \alpha \gamma \Gamma_G = \gamma^{-1} \Gamma_G \alpha \Gamma_G \gamma = \bigsqcup_{\delta \in R} \gamma^{-1} \Gamma_G \delta \gamma = \bigsqcup_{\delta \in R} \Gamma_G \gamma^{-1} \delta \gamma.$$

However, since $\gamma \in \Gamma_{G'}$ and $\lambda_N(\alpha) \in G$, it follows that $\lambda_N(\gamma^{-1}\alpha\gamma) \in G$, and so by Shimura [29, Lemma 3.29], we see that

$$\Gamma_G \gamma^{-1} \alpha \gamma \Gamma_G = \{g \in M_2(\mathbb{Z}) \mid \det(g) = p \text{ and } \lambda_N(g) \in G\} = \Gamma_G \alpha \Gamma_G.$$

It follows that $T_\alpha(f) = \sum_{\delta \in R} f|_{\gamma^{-1}\delta\gamma}$, so that $T_\alpha(f)|_\gamma = \varepsilon(\gamma) \cdot T_\alpha(f)$, hence the result. □

Remark 4.5.14 Note that it is not enough to demand that Γ_G will be normal in $\Gamma_{G'}$. For example, if $t^2 \mid N$ and $t \mid 24$, then, as described in [25], there is an element $u_t \in \mathfrak{N}(\Gamma_0(N))$ defined over $\mathbb{Q}(\zeta_t)$ which commutes only with the Hecke operators T_p such that $p \equiv 1 \bmod t$ (equivalently, p splits in $\mathbb{Q}(\zeta_t)$). Indeed, u_t does not normalize the group $G \subseteq GL_2(\mathbb{Z}/N\mathbb{Z})$ appearing in Example 2.1.3 (3).

5 Degeneracy Maps

Our next step would be to decompose our spaces of modular forms to irreducible modules for the Hecke algebra. This should be possible once we apply our algorithms from Sect. 4. The only obstacle in our path to obtain the eigenvectors lie in the fact that we still do not have a multiplicity one result due to the existence of oldforms. We therefore make a small detour, and describe the generalization to our situation of the degeneracy maps, and the decomposition to spaces of newforms.

5.1 *Petersson Inner Product*

Before we describe the degeneracy maps themselves, in order to obtain a good notion of oldforms, we briefly recall the definitions and properties of the Petersson inner product, so that we will consequently be able to define newforms.

Definition 5.1.1 Let $\Gamma \subseteq SL_2(\mathbb{Z})$ be a congruence subgroup. The *Petersson inner product*,

$$\langle , \rangle_\Gamma : S_k(\Gamma) \times S_k(\Gamma) \to \mathbb{C},$$

is given by

$$\langle f, g\rangle_\Gamma = \frac{1}{V_\Gamma} \int_{X_\Gamma} f(\tau)\overline{g(\tau)} (\mathrm{Im}(\tau))^k \, d\mu(\tau) \qquad (5.1.2)$$

where $d\mu\tau = \frac{dxdy}{y^2}$ is the hyperbolic measure on $\mathcal{H}$ and $V_\Gamma = \int_{X_\Gamma} d\mu\tau$ is the volume of X_Γ.

The following standard results then help us to see that it is indeed a well defined inner product, and to compute adjoints of Hecke operators with respect to this inner product.

Proposition 5.1.3 ([14, Section 5.4 and Proposition 5.5.2]) *Let $\Gamma \subseteq SL_2(\mathbb{Z})$ be a congruence subgroup. The integral in the Petersson inner product (5.1.2) is well defined and convergent. The pairing is linear in f, conjugate linear in g, Hermitian-symmetric and positive definite. Let $\alpha \in GL_2^+(\mathbb{Q})$, and set $\alpha' := \det(\alpha) \cdot \alpha^{-1}$. Then*

(1) If $\alpha^{-1}\Gamma\alpha \subseteq SL_2(\mathbb{Z})$ then for all $f \in S_k(\Gamma)$ and $g \in S_k(\alpha^{-1}\Gamma\alpha)$

$$\langle f|_\alpha, g\rangle_{\alpha^{-1}\Gamma\alpha} = \langle f, g|_{\alpha'}\rangle_\Gamma.$$

(2) For all $f, g \in S_k(\Gamma)$,

$$\langle T_\alpha f, g\rangle = \langle f, T_{\alpha'} g\rangle.$$

In particular, if $\alpha^{-1}\Gamma\alpha = \Gamma$ then the adjoint of $f \mapsto f|_\alpha$ is $g \mapsto g|_{\alpha'}$ and in any case $T_\alpha^\star = T_{\alpha'}$.

The correspondence between α and α' forces us to consider some more operators on this space.

Definition 5.1.4 Let $G \subseteq GL_2(\mathbb{Z}/N\mathbb{Z})$, and let $\mathfrak{N}(G)$ be its normalizer in $GL_2(\mathbb{Z}/N\mathbb{Z})$. Then the quotient $Q := \Gamma_{\mathfrak{N}(G)}/\Gamma_G$ acts naturally on $M_k(\Gamma_G)$ via

$$\begin{aligned} \langle q\rangle : M_k(\Gamma_G) &\to M_k(\Gamma_G) \\ f &\mapsto f|_\alpha \text{ for any } \alpha \in \Gamma_{\mathfrak{N}(G)} \text{ with } \alpha \equiv q \bmod \Gamma_G. \end{aligned}$$

This operator is called a *diamond operator*.

For any irreducible representation $\rho : Q \to GL_{n_\rho}(\mathbb{C})$, we let

$$M_k(\Gamma_G, \rho) := \rho \otimes \mathrm{Hom}_Q(\rho, M_k(\Gamma_G)) \hookrightarrow M_k(\Gamma_G)$$

be the ρ-isotypic component of $M_k(\Gamma)$. Then $M_k(\Gamma_G) = \bigoplus_\rho M_k(\Gamma_G, \rho)$ where ρ runs over all irreducible representations of Q.

Lemma 5.1.5 *Let $\alpha \in M_2(\mathbb{Z})$ be such that $\lambda_N(\alpha) \in G$, and let $q \in \Gamma_{\mathfrak{N}(G)}/\Gamma_G$. Then $T_\alpha \circ \langle q\rangle = \langle q\rangle \circ T_\alpha$.*

Proof Denote by $\tilde{q} \in \Gamma_{\mathfrak{N}(G)}$ an element such that $\tilde{q} \equiv q \bmod \Gamma_G$. Since $\lambda_N(\alpha) \in G$ and $\lambda_N(\tilde{q}) \in \mathfrak{N}(G)$, we see that

$$\lambda_N(\tilde{q}^{-1}\alpha\tilde{q}) = \lambda_N(\tilde{q})^{-1}\lambda_N(\alpha)\lambda_N(\tilde{q}) \in \lambda_N(\tilde{q})^{-1}G\lambda_N(\tilde{q}) = G.$$

Moreover, $\det(\tilde{q}^{-1}\alpha\tilde{q}) = \det(\alpha)$, so that $\lambda_N(\tilde{q}^{-1}\alpha\tilde{q}) \in G_0\lambda_N(\alpha)$. By Shimura [29, Lemma 3.29 (a)], it follows that $\tilde{q}^{-1}\alpha\tilde{q} \in \Gamma_G\alpha\Gamma_G$ hence

$$\tilde{q}^{-1}\Gamma_G\alpha\Gamma_G\tilde{q} = \Gamma_G\tilde{q}^{-1}\alpha\tilde{q}\Gamma_G = \Gamma_G\alpha\Gamma_G$$

so that if $\Gamma_G\alpha\Gamma_G = \bigsqcup_i \Gamma_G\alpha_i$ is a disjoint union, then

$$\Gamma_G\alpha\Gamma_G = \tilde{q}^{-1}\Gamma_G\alpha\Gamma_G\tilde{q} = \tilde{q}^{-1}\left(\bigsqcup_i \Gamma_G\alpha_i\right)\tilde{q} = \bigsqcup_i \tilde{q}^{-1}\Gamma_G\alpha_i\tilde{q} = \bigsqcup_i \Gamma_G\tilde{q}^{-1}\alpha\tilde{q}.$$

Therefore, for any $f \in M_k(\Gamma_G)$ we have

$$\langle q\rangle T_\alpha f = \sum_i f|_{\alpha_i\tilde{q}} = \sum_i f|_{\tilde{q}\tilde{q}^{-1}\alpha_i\tilde{q}} = \sum_i (\langle q\rangle f)|_{\tilde{q}^{-1}\alpha_i\tilde{q}} = T_\alpha\langle q\rangle f.$$

□

Before we continue, we need a simple Lemma, which is a variant of [29, Lemma 3.29].

Lemma 5.1.6 *Let $\Gamma_1 \subseteq \Gamma_2$ be congruence subgroups of levels N_1, N_2 respectively. Let*

$$\Phi_i = \{\alpha \in M_2(\mathbb{Z}) \mid \det\alpha > 0, (\det(\alpha), N_i) = 1, \lambda_{N_i}(\Gamma_i\alpha) = \lambda_N(\alpha\Gamma_i)\}, \quad i \in \{1, 2\}.$$

Then the following assertions hold.

(1) $\Gamma_1\alpha\Gamma_1 = \{\xi \in \Gamma_2\alpha\Gamma_2 \mid \lambda_{N_1}(\xi) \in \lambda_{N_1}(\Gamma_1 \cdot \alpha)\}$ *if* $\alpha \in \Phi_1$.
(2) $\Gamma(N_1)\alpha\Gamma(N_1) = \Gamma(N_1)\beta\Gamma(N_1)$ *if and only if* $\Gamma_2\alpha\Gamma_2 = \Gamma_2\beta\Gamma_2$ *and* $\alpha \equiv \beta \mod N_1$ *(for α, β such that* $(\det\alpha, N_1) = (\det\beta, N_1) = 1$*).*
(3) $\Gamma_2\alpha\Gamma_2 = \Gamma_2\alpha\Gamma_1 = \Gamma_1\alpha\Gamma_2$ *if* $\alpha \in \Phi_2$ *and* $(\det\alpha, N_1) = 1$.
(4) $\Gamma_i\alpha\Gamma_i = \Gamma_i\alpha\Gamma(N_i) = \Gamma(N_i)\alpha\Gamma_i$ *if* $\alpha \in \Phi_i$ *for* $i = 1, 2$.
(5) *If* $\alpha \in \Phi_1$ *and* $\Gamma_1\alpha\Gamma_1 = \bigsqcup_i \Gamma_1\alpha_i$, *then* $\Gamma_2\alpha\Gamma_2 = \bigsqcup_i \Gamma_2\alpha_i$.

Proof First, (4) is simply a restatement of [29, Lemma 3.29 (4)]. To show (3), note that $\gcd(N_1, \det\alpha \cdot N_2) = N_2$. By Shimura [29, Lemma 3.28] and [29, Lemma 3.9], we have

$$\Gamma(N_2) = \Gamma(N_2 \cdot \det\alpha) \cdot \Gamma(N_1) \subseteq \alpha^{-1}\Gamma(N_2)\alpha\Gamma(N_1)$$

so that $\alpha^{-1}\Gamma(N_2)\alpha\Gamma(N_2) \subseteq \alpha^{-1}\Gamma(N_2)\alpha\Gamma(N_1)$. Hence $\Gamma(N_2)\alpha\Gamma(N_2) \subseteq \Gamma(N_2)\alpha\Gamma(N_1) \subseteq \Gamma(N_2)\alpha\Gamma_1$. Therefore $\Gamma_2\alpha\Gamma(N_2) \subseteq \Gamma_2\alpha\Gamma_1 \subseteq \Gamma_2\alpha\Gamma_2$. But from (4), as $\alpha \in \Phi_2$, we know that there is an equality $\Gamma_2\alpha\Gamma(N_2) = \Gamma_2\alpha\Gamma_2$. This shows (3). Next, to see (1) note that by Shimura [29, Lemma 3.29 (1)], we have $\Gamma_1\alpha\Gamma_1 = \{\xi \in SL_2(\mathbb{Z})\alpha SL_2(\mathbb{Z}) \mid \lambda_{N_1}(\xi) \in \lambda_{N_1}(\Gamma_1 \cdot \alpha)\}$. Since we also know $\Gamma_1\alpha\Gamma_1 \subseteq \Gamma_2\alpha\Gamma_2$, (1) follows. (2) is a special case of (1). Finally, let $\alpha \in \Phi_1$, and $\Gamma_1\alpha\Gamma_1 = \bigsqcup_i \Gamma_1\alpha_i$. Then $\Gamma_2\alpha\Gamma_2 = \Gamma_2\alpha\Gamma_1 = \bigcup_i \Gamma_2\alpha_i$ by (3). Assume $\Gamma_2\alpha_i = \Gamma_2\alpha_j$. Then $\alpha_i = \gamma\alpha_j$ for some $\gamma \in \Gamma_2$. By (1)

$$\lambda_{N_1}(\alpha_i) \in \lambda_{N_1}(\Gamma_1) \cdot \lambda_{N_1}(\alpha) = \lambda_{N_1}(\Gamma_1) \cdot \lambda_{N_1}(\alpha_j).$$

Thus, there exists some $\delta \in \Gamma_1$ such that $\alpha_i \equiv \delta\alpha_j \mod N_1$. It follows that $\gamma \equiv \delta \mod N_1$. Since $\Gamma(N_1) \subseteq \Gamma_1$, we have $\gamma \in \Gamma_1$. This proves (5). □

We may now state the adjoints of the Hecke operators with respect to the Petersson inner product.

Theorem 5.1.7 *In the inner product spaces $S_k(\Gamma_G)$, the diamond operators $\langle q\rangle$ and the Hecke operator T_p for $p \in \det(G)$ a prime have adjoints*

$$\langle q\rangle^\star = \langle q\rangle^{-1} \quad \textit{and} \quad T_p^\star = \langle\sigma_p\rangle \cdot T_p$$

where $\sigma_p \in \Gamma_{\mathfrak{N}(G)}$ is an element such that $\lambda_N(\sigma_p) = p \cdot \lambda_N(\alpha)^{-2}$, and $\alpha \in M_2(\mathbb{Z})$ is any element with $\det(\alpha) = p$ such that $\lambda_N(\alpha) \in G$. Thus the Hecke operators $\langle q\rangle$ and T_n for $n \in \det(G)$ are normal.

Proof The first assertion is just a restatement of Proposition 5.1.3 (1). For the second assertion, we note that if $\alpha \in M_2(\mathbb{Z})$ is such that $\lambda_N(\alpha) \in G$ and $\det(\alpha) = p$, then $\alpha \in \Phi_{\mathfrak{N}(G)}$. Indeed, if $\nu \in \mathfrak{N}(G)$ then for any $g \in G$ we have

$$(\lambda_N(\alpha)\nu\lambda_N(\alpha)^{-1})g(\lambda_N(\alpha)\nu^{-1}\lambda_N(\alpha)^{-1}) \in \lambda_N(\alpha)\nu G\nu^{-1}\lambda_N(\alpha)^{-1} \subseteq G$$

so that $\lambda_N(\alpha)\nu\lambda_N(\alpha)^{-1} \in \mathfrak{N}(G)$.

Therefore, by Lemma 5.1.6 (3), we see that $\Gamma_{\mathfrak{N}(G)}\alpha\Gamma_{\mathfrak{N}(G)} = \Gamma_G\alpha\Gamma_{\mathfrak{N}(G)}$. Moreover, we have

$$\lambda_N(\alpha') = \det(\alpha)\lambda_N(\alpha)^{-1} \in \mathfrak{N}(G)$$

hence $\Gamma_{\mathfrak{N}(G)}\alpha'\Gamma_{\mathfrak{N}(G)} = \Delta_p^{\mathfrak{N}(G)} = \Gamma_{\mathfrak{N}(G)}\alpha\Gamma_{\mathfrak{N}(G)} = \Gamma_G\alpha\Gamma_{\mathfrak{N}(G)}$.

Let $\gamma \in \Gamma_{\mathfrak{N}(G)}$ be such that $\alpha' \in \Gamma_G\alpha\gamma$. Now, as γ normalizes Γ_G we see that $\Gamma_G\alpha'\Gamma_G = \Gamma_G\alpha\gamma\Gamma_G = \Gamma_G\alpha\Gamma_G\gamma$ and hence if $\Gamma_G\alpha\Gamma_G = \bigsqcup_i \Gamma_G\alpha_i$, then $\Gamma_G\alpha'\Gamma_G = \bigsqcup_i \Gamma_G\alpha_i\gamma$. It then follows from Proposition 5.1.3 (2) that $T_p^\star = T_\alpha^\star = T_{\alpha'} = \langle\gamma\rangle T_\alpha$. Finally, we note that

$$\lambda_N(\gamma) \in \lambda_N(\Gamma_G)\lambda_N(\alpha)^{-1}\lambda_N(\alpha') \subseteq G \cdot p\lambda_N(\alpha)^{-2} = G \cdot \lambda_N(\sigma_p)$$

so that $\gamma \in \Gamma_G \cdot \sigma_p$, showing that $\langle\gamma\rangle = \langle\sigma_p\rangle$. □

Remark 5.1.8 The σ_p actually acts via a character. Indeed, if $A \subseteq \mathfrak{N}(G)$ is such that A/G is abelian, and A contains the center of diagonal matrices, then $\sigma_p \in A$ and all the irreducible representations of A/G are 1-dimensional.

This leads to the following theorem.

Corollary 5.1.9 *The space $S_k(\Gamma_G)$ has an orthogonal basis of simultaneous eigenvectors for the Hecke operators $\{T_n \mid n \in \det(G)\}$ and the diamond operators $\{\langle q \rangle \mid q \in \Gamma_{\mathfrak{N}(G)}/\Gamma_G\}$.*

Proof This is a corollary of the Spectral Theorem for a family of commuting normal operators. □

Definition 5.1.10 We say that a modular form $0 \neq f \in S_k(\Gamma_G)$ is an *eigenform* if for all $n \in \det(G)$, f is an eigenvector of T_n and for all $q \in \Gamma_{\mathfrak{N}_G}/\Gamma_G$, f is an eigenvector of $\langle q \rangle$.

5.2 Degeneracy Maps

We begin by defining degeneracy maps in general on our space of modular symbols.

Definition 5.2.1 Let $\Gamma_1 \trianglelefteq \Gamma_1'$, $\Gamma_2 \trianglelefteq \Gamma_2'$, be two pairs of finite index subgroups in $SL_2(\mathbb{Z})$. Let $\varepsilon_i : \Gamma_i'/\Gamma_i \to \mathbb{Q}(\zeta_i)^\times$ for $i = 1, 2$ be characters. Let $t \in GL_2^+(\mathbb{Q})$ be such that $t^{-1}\Gamma_1' t \subseteq \Gamma_2'$, and $\varepsilon_2 \circ \mathrm{Inn}(t)^{-1} |_{\Gamma_1'} = \varepsilon_1$. Fix a choice R_t of coset representatives for $\Gamma_1' \backslash t \cdot \Gamma_2'$. Let

$$\alpha_t^\vee : \mathbb{M}_k(\Gamma_1, \varepsilon_1) \to \mathbb{M}_k(\Gamma_2, \varepsilon_2), \qquad \beta_t^\vee : \mathbb{M}_k(\Gamma_2, \varepsilon_2) \to \mathbb{M}_k(\Gamma_1, \varepsilon_1)$$

$$\alpha_t^\vee(x) = t^{-1}x \qquad \beta_t^\vee(x) = \sum_{t\gamma_2 \in R_t} \varepsilon_2(\gamma_2)^{-1} t\gamma_2 \cdot x.$$

We show that these operators are well defined.

Lemma 5.2.2 *The operators $\alpha_t^\vee$ and $\beta_t^\vee$ are well defined, and moreover the composition $\alpha_t^\vee \circ \beta_t^\vee$ is multiplication by $[\Gamma_2' : t^{-1}\Gamma_1' t]$.*

Proof First, note that as $t^{-1}\Gamma_1' t \subseteq \Gamma_2'$, for any $\gamma_1 \in \Gamma_1'$, $\gamma_2 \in \Gamma_2'$ one has

$$\gamma_1 t \gamma_2 = t t^{-1} \gamma_1 t \gamma_2 \in t\Gamma_2'$$

so that Γ_1' indeed acts on $t\Gamma_2'$. To show that α_t is well defined, we must show that for each $x \in \mathbb{M}_k(\Gamma_1, \varepsilon_1)$ and $\gamma_1 \in \Gamma_1'$, we have

$$\alpha_t^\vee(\gamma_1 x - \varepsilon_1(\gamma_1)x) = 0 \in \mathbb{M}_k(\Gamma_2, \varepsilon_2).$$

We have, by assumption

$$\alpha_t^\vee(\gamma_1 x) = t^{-1}\gamma_1 x = t^{-1}\gamma_1 t t^{-1} x = \varepsilon_2(t^{-1}\gamma_1 t) t^{-1} x = \varepsilon_1(\gamma_1)\alpha_t^\vee(x)$$

hence the result.

We next verify that β_t is well defined.

Suppose that $\gamma_1 \in \Gamma_1'$ and $\gamma_2 \in \Gamma_2'$, then

$$\gamma_1 t \gamma_2 = t(t^{-1}\gamma_1 t)\gamma_2.$$

Moreover, for any $x \in \mathbb{M}_k(\Gamma_2, \varepsilon_2)$, we have

$$\varepsilon_2((t^{-1}\gamma_1 t)\gamma_2)^{-1}\gamma_1 t\gamma_2 x = \varepsilon_1(\gamma_1)^{-1}\varepsilon_2(\gamma_2)^{-1}\gamma_1 t\gamma_2 x = \varepsilon_1(\gamma_1)^{-1}\gamma_1(\varepsilon_2(\gamma_2)^{-1}t\gamma_2 x). \tag{5.2.3}$$

But as $\varepsilon_2(\gamma_2)^{-1}t\gamma_2 x \in \mathbb{M}_k(\Gamma_1, \varepsilon_1)$, (5.2.3) is simply $\varepsilon_2(\gamma_2)^{-1}t\gamma_2 x$.

Thus, replacing the representative $t\gamma_2$ by $\gamma_1 t\gamma_2$ does not change the result, so that $\beta_t^\vee$ is independent of the choice of the representatives R_t.

Next, we must show that for any $\gamma \in \Gamma_2'$ and any $x \in \mathbb{M}_k(\Gamma_2, \varepsilon_2)$, $\beta_t^\vee(\gamma x) = \varepsilon_2(\gamma)\beta_t^\vee(x)$. However, for $\gamma \in \Gamma_2'$, using the fact that β_t is independent of the choice of representatives, and that Γ_2' acts on $\Gamma_1' \backslash t\Gamma_2'$ by right translations, we get

$$\begin{aligned}\beta_t^\vee(\gamma x) &= \sum_{t\gamma_2 \in R_t} \varepsilon_2(\gamma_2)^{-1} t\gamma_2\gamma x = \sum_{t\gamma_2\gamma^{-1} \in R_t} \varepsilon_2(\gamma_2\gamma^{-1})^{-1} t\gamma_2\gamma^{-1}\gamma x \\ &= \varepsilon_2(\gamma) \sum_{t\gamma_2 \in R_t} \varepsilon_2(\gamma_2)^{-1} t\gamma_2 x = \varepsilon_2(\gamma)\beta_t^\vee(x).\end{aligned}$$

To compute $\alpha_t^\vee \circ \beta_t^\vee$, we use that $\# R_t = [\Gamma_2' : t^{-1}\Gamma_1' t]$:

$$\begin{aligned}\alpha_t^\vee(\beta_t^\vee(x)) &= \alpha_t^\vee\left(\sum_{t\gamma_2 \in R_t} \varepsilon_2(\gamma_2)^{-1} t\gamma_2 x\right) \\ &= \sum_{t\gamma_2 \in R_t} \varepsilon_2(\gamma_2)^{-1}\gamma_2 x = \sum_{t\gamma_2 \in R_t} x = [\Gamma_2' : t^{-1}\Gamma_1' t]x.\end{aligned}$$

□

We also note that by using the pairing (3.6.1) we can get maps on spaces of cusp forms.

Definition 5.2.4 Let $\Gamma_1 \trianglelefteq \Gamma_1'$, $\Gamma_2 \trianglelefteq \Gamma_2'$, be two pairs of finite index subgroups in $SL_2(\mathbb{Z})$. Let $\varepsilon_i : \Gamma_i'/\Gamma_i \to \mathbb{Q}(\zeta_i)^\times$ for $i = 1, 2$ be characters. Let $t \in GL_2^+(\mathbb{Q})$ be such that $t^{-1}\Gamma_1' t \subseteq \Gamma_2'$, and $\varepsilon_2 \circ \mathrm{Inn}(t)^{-1} |_{\Gamma_1'} = \varepsilon_1$. Fix a choice R_t of coset representatives for $\Gamma_1' \backslash t \cdot \Gamma_2'$. Let

$$\alpha_t : S_k(\Gamma_2, \varepsilon_2) \to S_k(\Gamma_1, \varepsilon_1), \qquad \beta_t : S_k(\Gamma_1, \varepsilon_1) \to S_k(\Gamma_2, \varepsilon_2)$$

$$f \mapsto f|_{t^{-1}} \qquad\qquad f \mapsto \sum_{t\gamma_2 \in R_t} \varepsilon_2(\gamma_2)^{-1} f|_{t\gamma_2}.$$

5.3 Degeneracy Maps and Hecke Operators

Next, as in the classical theory for Iwahori level, we would like to show that the degeneracy maps commute with the Hecke operators. Alas, this is not always the case as the following example shows.

Example 5.3.1 Let $\Gamma_1 = \Gamma(7)$ and let $\Gamma_2 = \Gamma_{ns}(7)$ be the non-split Cartan at level 7. Then there is a natural inclusion map $\alpha_1^\vee : \mathbb{M}_2(\Gamma_1) \to \mathbb{M}_2(\Gamma_2)$, but α_1 does not commute with the standard Hecke operators for these spaces. In this case $\dim S_2(\Gamma_{ns}(7)) = 1$ is spanned by a single element f. We have $\alpha_1(f) = f \in S_2(\Gamma(7))$, but f is not an eigenform of the Hecke algebra on $\Gamma(7)$. Indeed, there are 3 such eigenforms, whose q-expansions are all defined over $\mathbb{Q}(\sqrt{-3})$, while the q-expansion of f is only defined over $\mathbb{Q}(\zeta_7)^+$.

In the next proposition we describe the conditions under which the degeneracy maps commute with the Hecke operators.

Proposition 5.3.2 *Let $N_G \mid N_H$ be positive integers. Let $H \subseteq GL_2(\mathbb{Z}/N_H\mathbb{Z})$ and $G \subseteq GL_2(\mathbb{Z}/N_G\mathbb{Z})$ be subgroups such that $\lambda_{N_G}(H) \subseteq G$. Let p be a prime number such that $p \in \det(H)$, and let $t \in M_2(\mathbb{Z})$ be such that $N_G \cdot \det(t) \mid N_H$ and $t^{-1}\Gamma_H t \subseteq \Gamma_G$. Let $T_{p,?}^\vee$ be the Hecke operator at p on the space $\mathbb{M}_k(\Gamma_?)$ for $? \in \{G, H\}$. Then*

$$T_{p,G}^\vee \circ \alpha_t^\vee = \alpha_t^\vee \circ T_{p,H}^\vee.$$

Proof Denote $d := \det(t)$, so that $d \cdot t^{-1} \in M_2(\mathbb{Z})$. Since $t^{-1}\Gamma_H t \subseteq \Gamma_G$, we have also $d \cdot t^{-1}\Gamma_H t \subseteq d \cdot \Gamma_G$ so that

$$\lambda_{N_G \cdot d}(dt^{-1}) \cdot \lambda_{N_G \cdot d}(\Gamma_H) \cdot \lambda_{N_G \cdot d}(t) = \lambda_{N_G \cdot d}\left(dt^{-1}\Gamma_H t\right) \subseteq \lambda_{N_G \cdot d}(d \cdot \Gamma_G) = d \cdot \lambda_{N_G \cdot d}(\Gamma_G). \tag{5.3.3}$$

However, we also assumed $N_G \cdot d \mid N_H$, hence

$$\lambda_{N_G \cdot d}(\Gamma_H) = \lambda_{N_G \cdot d}(\lambda_{N_H}(\Gamma_H)) = \lambda_{N_G \cdot d}(H). \tag{5.3.4}$$

Let

$$\Delta_p^? := \left\{\alpha \in M_2(\mathbb{Z}) \mid \det(\alpha) = p \text{ and } \lambda_{N_?}(\alpha) \in ?\right\}, \quad ? \in \{G, H\}$$

and let $\alpha \in \Delta_p^H$. Then combining (5.3.3) and (5.3.4) we see that

$$\begin{aligned}\lambda_{N_G \cdot d}(dt^{-1}\alpha t) &= \lambda_{N_G \cdot d}(dt^{-1})\lambda_{N_G \cdot d}(\lambda_{N_H}(\alpha))\lambda_{N_G \cdot d}(t)\\ &\in \lambda_{N_G \cdot d}(dt^{-1})\lambda_{N_G \cdot d}(H)\lambda_{N_G \cdot d}(t) \subseteq d \cdot \lambda_{N_G \cdot d}(\Gamma_G).\end{aligned}$$

In particular, reducing modulo d, we see that $dt^{-1}\alpha t \in d \cdot M_2(\mathbb{Z})$, hence $t^{-1}\alpha t \in M_2(\mathbb{Z})$. Therefore $\lambda_{N_G}(t^{-1}\alpha t) \in \lambda_{N_G}(\Gamma_G) = G$. It follows that $t^{-1}\alpha t \in \Delta_p^G$.

Since $p \in \det(H)$, we also get $p \in \det(\lambda_{N_G}(H)) \subseteq \det(G)$. Therefore, we have by Shimura [29, Lemma 3.29] that $\Delta_p^H = \Gamma_H \alpha \Gamma_H$ and $\Delta_p^G = \Gamma_G \cdot t^{-1}\alpha t \cdot \Gamma_G$.

Let $\Delta_p^H = \bigsqcup_i \Gamma_H \alpha_i$ be its coset decomposition. Then

$$t^{-1}\Gamma_H t \cdot t^{-1}\alpha t \cdot t^{-1}\Gamma_H t = t^{-1}\Gamma_H \alpha \Gamma_H t = t^{-1}\left(\bigsqcup_i \Gamma_H \alpha_i\right) t$$
$$= \bigsqcup_i t^{-1}\Gamma_H \alpha_i t = \bigsqcup_i t^{-1}\Gamma_H t \cdot t^{-1}\alpha_i t.$$

Now, applying Lemma 5.1.6 (5) to $t^{-1}\Gamma_H t \subseteq \Gamma_G$ and $t^{-1}\alpha t$, we obtain that

$$\Delta_p^G = \Gamma_G \cdot t^{-1}\alpha t \cdot \Gamma_G = \bigsqcup_i \Gamma_G \cdot t^{-1}\alpha_i t. \tag{5.3.5}$$

Finally, we use (5.3.5) to compute

$$T_{p,G}^{\vee} \circ \alpha_t^{\vee}(x) = \sum_i t^{-1}\alpha_i t \cdot t^{-1} x = t^{-1}\sum_i \alpha_i x = \alpha_t^{\vee} \circ T_{p,H}^{\vee}(x).$$

□

We also note that by using the pairing (3.6.1) we have the same for cusp forms.

Corollary 5.3.6 *Under the assumptions of Corollary 5.3.2,* $\alpha_t \circ T_{p,G} = T_{p,H} \circ \alpha_t$.

Remark 5.3.7 Note that these do not cover all degeneracy maps. Indeed, there might be $t \in M_2(\mathbb{Z})$ such that $N_G \cdot \det(t) \nmid N_H$, but $t^{-1}\Gamma_H t \subseteq \Gamma_G$ and such that α_t commute with all the Hecke operators (though we must have $\det(t) \mid N_H$). As an example, consider $\Gamma_G = \Gamma_0(N)$, $\Gamma_H = \Gamma(N)$ and $t = \begin{pmatrix} N & 0 \\ 0 & 1 \end{pmatrix}$. However, as the proof shows, for α_t to commute with T_p, it is enough to check that $dt^{-1}\alpha_p t \in d \cdot M_2(\mathbb{Z})$ for all p, which is equivalent to verifying that $\lambda_d(dt^{-1})\lambda_d(\alpha_p)\lambda_d(t) = 0$. As this last expression only depends on $\lambda_{N_H}(\alpha_p)$, hence only on p mod N_H, we just have to check commutation with $\phi(N_H)$ elements. Therefore, one may include these degeneracy maps as well.

This motivates the following definition.

Definition 5.3.8 Let N be a positive integer. Let $H \subseteq GL_2(\mathbb{Z}/N\mathbb{Z})$ be a subgroup. For each $G \subseteq GL_2(\mathbb{Z}/N\mathbb{Z})$ such that $H \subseteq G$, we denote by N_G the level of G. Let T_G be a set of representatives for the orbits of Γ_H on the following set

$$T_G := \Gamma_H \backslash \left\{ t \in M_2(\mathbb{Z}) \mid \det(t) \mid N/N_G \text{ and } t^{-1}\Gamma_H t \subseteq \Gamma_G \right\}. \tag{5.3.9}$$

Let i_G be the map

$$i_G : (S_k(\Gamma_G))^{T_G} \to S_k(\Gamma_H)$$
$$(f_t)_{t \in T_G} \mapsto \sum_{t \in T_G} \alpha_t(f_t).$$

Let Over(H) be the set of minimal overgroups of H in $GL_2(\mathbb{Z}/N\mathbb{Z})$. The subspace of *oldforms at level* H is

$$S_k(\Gamma_H)^{\text{old}} = \sum_{G \in \text{Over}(H)} i_G\left((S_k(\Gamma_G))^{T_G}\right)$$

and the subspace of *newforms at level* H is the orthogonal complement with respect to the Petersson inner product,

$$S_k(\Gamma_H)^{\text{new}} = \left(S_k(\Gamma_H)^{\text{old}}\right)^{\perp}.$$

We have to show that the sum in this definition is well defined.

Lemma 5.3.10 *Let N be a positive integer. Let $H \subseteq G \subseteq GL_2(\mathbb{Z}/N\mathbb{Z})$ be subgroups. Denote by N_G the level of G. Let T_G be the set given in (5.3.9). Then T_G is finite.*

Proof We first note that by the condition $\det(t) \mid N/N_G$, the determinant of t assumes finitely many values, hence it is enough to show that the subset $T_{G,d}$ of such t's with $\det(t) = d$ is finite.

Next, we recall that if Δ_d is the set of matrices in $M_2(\mathbb{Z})$ of determinant d, then $SL_2(\mathbb{Z})\backslash\Delta_d$ is finite. Since Γ_H is of finite index in $SL_2(\mathbb{Z})$, the result follows. □

The Hecke operators respect the decomposition of $S_k(\Gamma_G)$ into old and new.

Proposition 5.3.11 *The subspaces $S_k(\Gamma_G)^{\text{old}}$ and $S_k(\Gamma_G)^{\text{new}}$ are stable under the Hecke operators T_n for $n \in \det(G)$ and the diamond operators $\langle\sigma_p\rangle$, for $p \in \det(G)$ prime.*

Proof By Corollary 5.3.6, the Hecke operators T_p for $p \in \det(G)$ commute with the degeneracy maps α_t. By multiplicativity, it extends to T_n such that $n \in \det(G)$.

For the diamond operators, let $G_i \subseteq GL_2(\mathbb{Z}/N_i\mathbb{Z})$ for $i = 1, 2$ be such that $\lambda_{N_2}(G_1) \subseteq G_2$, and let $t \in M_2(\mathbb{Z})$ be such that $\det(t) \mid N_1/N_2$ and $t^{-1}\Gamma_1 t \subseteq \Gamma_2$, where Γ_i is the congruence subgroup induced by G_i. Let $d = \det(t)$. Then $dt^{-1} \in M_2(\mathbb{Z})$ and so if α is such that $\lambda_{N_1}(\alpha) \in G_1$ and $\det(\alpha) = p$, we have

$$\lambda_{N_2 \cdot d}(dt^{-1}\sigma_p t) = \lambda_{N_2 \cdot d}(dt^{-1})\lambda_{N_2 \cdot d}(p \cdot \lambda_{N_2 \cdot d}(\alpha)^{-2}) \cdot \lambda_{N_2 \cdot d}(t) \in p \cdot \lambda_{N_2 \cdot d}(dt^{-1})\lambda_{N_2 \cdot d}(G_1)\lambda_{N_2 \cdot d}(t).$$

However, since $t^{-1}\Gamma_1 t \subseteq \Gamma_2$, we also have

$$\lambda_{N_2 \cdot d}(dt^{-1})\lambda_{N_2 \cdot d}(G_1)\lambda_{N_2 \cdot d}(t) \subseteq d \cdot \lambda_{N_2 \cdot d}(\Gamma_2)$$

hence $t^{-1}\sigma_p t \in M_2(\mathbb{Z})$ and moreover $\lambda_{N_2}(t^{-1}\sigma_p t) \in p \cdot \lambda_{N_2}(\Gamma_2) = p \cdot G_2 \subseteq \mathfrak{N}(G_2)$. It follows that $t^{-1}\sigma_p t \in \Gamma_{\mathfrak{N}(G_2)}$, and so for any $f \in S_k(\Gamma_2)$ we have

$$\left(\langle\sigma_p\rangle \circ \alpha_t\right)(f) = f|_{t^{-1}\sigma_p} = f|_{t^{-1}\sigma_p t t^{-1}} = \left(\alpha_t \circ \langle t^{-1}\sigma_p t\rangle\right)(f).$$

Finally, we note that $\lambda_{N_2}(t^{-1}\sigma_p t) = p \cdot \lambda_{N_2}(t^{-1}\alpha t)^{-2}$, and that by the proof of Corollary 5.3.2, $t^{-1}\alpha t \in M_2(\mathbb{Z})$ is an element such that $\lambda_{N_2}(t^{-1}\alpha t) \in G_2$. It follows that $\langle t^{-1}\sigma_p t\rangle$ is the diamond operator at p for G_2. This concludes the proof. □

6 Zeta Functions and q-Expansions

Using the results of Sects. 4 and 5, we may now compute the zeta functions associated to eigenforms in $S_k(\Gamma_G)$.

All the ideas and methods used in this section are the same as in the case of level $\Gamma_1(N)$ or $\Gamma_0(N)$, and are known (see [33], [32]) and used to compute the q-expansions.

The process for doing so is divided into three steps:

(1) Compute a subspace of $\mathbb{M}_k(\Gamma)^\vee$ that is isomorphic to $\mathbb{S}_k(\Gamma)$ as a Hecke module, denoted by $\mathbb{S}_k(\Gamma)^\vee$.
(2) Decompose the space $\mathbb{S}_k(\Gamma)^\vee$ into irreducible Hecke modules.
(3) For each irreducible Hecke module in the decomposition, find the system of Hecke eigenvalues, hence the zeta function.

6.1 Constructing Dual Vector Spaces

Since our interest lies in $S_k(\Gamma)$, we would like to have a vector space which is isomorphic to $S_k(\Gamma) \oplus \overline{S}_k(\Gamma)$ as a Hecke module. We could do it by using the pairing (3.6.1), but that would involve computation of integrals and approximation if done directly.

Instead, we can use the fact that $S_k(\Gamma)$ and $\mathbb{S}_k(\Gamma)$ are isomorphic as Hecke modules, thus leading to the following algorithm.

Algorithm 6.1.1 DualVectorSpace(M). *Compute the dual vector space.*

Input : $M \subseteq M_k(\Gamma)$ a subspace.
Output : $M^\vee \subseteq M_k(\Gamma)^\vee$ which is isomorphic to M as a Hecke module.

(1) $V := M_k(\Gamma)^\vee$, $p := 2$
(2) While ($\dim V > \dim M$) do

(a) $T_p :=$ HeckeOperator(V, p).
(b) $T_{p,M} :=$ HeckeOperator$(M_k(\Gamma), p)|_M$.
(c) $c_p :=$ CharPoly$(T_{p,M})$.
(d) $V := \ker(c_p(T_p))$.
(e) $p :=$ NextPrime(p).

(3) Return V.

Here, we assume the existence of a function CharPoly, that given an operator on a vector space, computes its characteristic polynomial.

Remark 6.1.2 We have not defined the Hecke operators T_n for $(n, N) > 1$, and these cannot be ignored (see [7, Example 12.2.11]). It is possible to circumvent this problem, as is done in [7, Section 8.10]. Practically when $p \mid N$, we use instead of T_p the double coset Hecke operators T_α with $\det(\alpha) = p$. There are finitely many such operators, and the operator T_p is a linear combination of them. Thus, for the purpose of describing the dual vector space, and later on, decomposition of the space to irreducible Hecke modules, these suffice.

6.2 *Decomposition of* $\mathbb{S}_k(\Gamma)$

The following theorem is a generalization of [32, Theorem 9.23]. It is an application of Sturm's Theorem, which will enable us to decompose the Hecke module $\mathbb{S}_k(\Gamma)$ in a finite (effectively bounded) number of steps.

Theorem 6.2.1 *Suppose* Γ *is a congruence subgroup of level* N*, and let*

$$r = \text{Sturm}(k, \Gamma) := \left\lfloor \frac{km}{12} - \frac{m-1}{N} \right\rfloor \tag{6.2.2}$$

where $m = [SL_2(\mathbb{Z}) : \Gamma]$*. Then the Hecke algebra*

$$\mathbb{T} = \mathbb{Z}[\ldots, T_n, \ldots] \subseteq \text{End}(S_k(\Gamma))$$

is generated as a $\mathbb{Z}$*-module by the Hecke operators* T_n *for* $n \le r$*.*

Proof Same as in [32, Theorem 9.23]. □

This allows us to decompose the space $\mathbb{S}_k(\Gamma)^\vee = S_k(\Gamma) \oplus \overline{S}_k(\Gamma)$ into irreducible Hecke modules, as follows.

Algorithm 6.2.3 Decomposition(M, p). *Decomposition into irreducible Hecke modules.*

Input :

- $M \subseteq \mathbb{M}_k(\Gamma)$ - a vector subspace stable under the Hecke action.
- p - a prime.

Output : D, a list of subspaces of M, such that $M = \bigoplus_{V_d \in D} V_d$ and $V_d^\vee$ is an irreducible Hecke module.

(1) If $p \mid N$, replace by a larger prime such that $p \nmid N$.
(2) $T_p := \text{HeckeOperator}(M^\vee, p)$, $D := \emptyset$.
(3) $f := \text{CharPoly}(T_p)$. Write $f = \prod_{i=1}^{l} f_i^{a_i}$.
(4) for $i \in \{1, 2, \ldots, l\}$ do

  (a) $V := \ker(f_i(T_p)^{a_i}) \subseteq M^\vee$, $W := V^\vee \subseteq M$.
  (b) if IsIrreducible(W)

    (i) $D := D \cup \{W\}$

    else

    (i) if $W = M$ then $q := \text{NextPrime}(p)$ else $q := 2$.
    (ii) $D := D \cup \text{Decompose}(W, q)$.

(5) Return D.

Here Decompose is a recursive call to the function, and IsIrreducible is creating a random linear combination of Hecke operators, and checks that its characteristic polynomial (on the plus subspace) is irreducible. One could eliminate the randomness to provably verify the irreducibility, as in [32, Chapter 7].

Calling Algorithm 6.2.3 with $M = \mathbb{S}_k(\Gamma)$, $p = 2$ we obtain the decomposition we wanted.

Note that the algorithm terminates if there are no eigenvectors with multiplicities, that is there are no oldforms arising as images of forms of lower levels. In that case, the number of steps in the algorithm can be bounded using Theorem 6.2.1.

In order to deal with the images of oldforms, we may begin by identifying the new subspace, and continuing recursively. The new subspace is computed by noting that it is dual to $\mathbb{S}_k(\Gamma_G)^{\text{new}} := \bigcap_{G' \in \text{Over}(G), t \in T_{G'}} \ker(\alpha_t^\vee)$.

Also, in order to get the old subspace one need not compute the Petersson inner product, but instead note that it is generated by the images of the $\beta_t^\vee$.

$$\mathbb{S}_k(\Gamma_G)^{\text{old}} := \left\langle \text{Im}(\beta_t^\vee) \right\rangle_{G' \in \text{Over}(G), t \in T_{G'}}.$$

6.3 Computing the Zeta Functions

Finally, for a subspace $V \subseteq \mathbb{S}_k(\Gamma_G)$ such that $V^\vee$ is an irreducible Hecke module, we can compute the zeta function of an eigenform (and hence of all eigenforms in $V^\vee$).

Algorithm 6.3.1 ZetaFunction(V, L). *Compute the zeta function associated to an eigenform.*

Input :

- $V \subseteq \mathbb{S}_k(\Gamma_G)$ such that $V^\vee$ is an irreducible Hecke module.
- L - a positive integer.

Output : $Z(f, s) = \sum_{n=1}^{\infty} a_n n^{-s}$ such that $T_n f = a_n f$ for all n (set $a_1 = 1$), given to precision L^{-s}.

(1) Find (randomly) a linear combination of Hecke operators which is irreducible on $(V^\vee)^+$, denote it by T.
(2) Let v be an eigenvector of T over the field $\mathbb{Q}[x]/f_T(x)$, where $f_T =$ CharPoly(T).
(3) Write $v = \sum_{j=i}^{d} c_j e_j^\vee$, where e_j are the basis vectors of $\mathbb{M}_k(\Gamma)$ given by Manin symbols, and $c_i \neq 0$.
(4) For primes $p < L$, set $a_p := \frac{1}{c_i}\langle T_p(e_i), v\rangle$. Complete the other a_n's using multiplicativity.
(5) Return $\sum_{n=1}^{L-1} a_n n^{-s} + O(L^{-s})$.

We start by showing that the algorithm returns the correct output, and measure its complexity.

Lemma 6.3.2 *Let $G \subseteq GL_2(\mathbb{Z}/N\mathbb{Z})$ be a group of real type of index I_G with surjective determinant such that T_p is effectively computable for all $p \mid N$. There exists an algorithm that given a subspace V of $\mathbb{S}_k(\Gamma_G)$ of dimension d, such that $V^\vee$ is irreducible as a module for the Hecke algebra, and an integer L, returns the zeta function associated to an eigenform up to precision L^{-s} in $O(C \log N(L \log L + N) + N I_G^2 \cdot \mathrm{In} + d^3)$, where C is the cost of a basic* CosetIndex *operation and* In *is the cost of the group membership test for G.*

Proof We apply Algorithm 6.3.1. Since G is of real type, by Corollary 4.1.9, the Hecke operator commutes with the star involution, so that step (1) makes sense. For any $p < L$ we have

$$\sum_{j=i}^{d} a_p c_j e_j^\vee = a_p v = T_p^\vee v.$$

In particular,

$$\langle T_p(e_i), v\rangle = \langle e_i, T_p^\vee v\rangle = \left\langle e_i, \sum_{j=i}^{d} a_p c_j e_j^\vee \right\rangle = a_p \cdot c_i.$$

For computation of the Hecke operators away from the level, we apply Corollary 4.5.11, while for the Hecke operators at primes $p \mid N$, we apply Theorem 4.2.10. Note that as the T_p are effectively computable for all $p \mid N$, we may do so with no additional cost to our complexity. Therefore, the complexity of computing the Hecke operators is given by

$$\begin{aligned} C \sum_{p<L, p\nmid N} p \log p + C \sum_{p<L, p\mid N} p \log(N^4 p) + \sum_{p<L, p\mid N} I_G^2 \cdot \text{In} \\ = C \sum_{p<L} p \log p + 4C \log N \sum_{p<L, p\mid N} p + N I_G^2 \cdot \text{In} \\ = O(C(L \log L + N) \log N + N I_G^2 \cdot \text{In}). \end{aligned} \tag{6.3.3}$$

Finally, the d^3 contribution comes from the linear algebra operations. □

We can now pack everything together to produce the following corollary.

Corollary 6.3.4 *There exists an algorithm that given a group of real type $G \subseteq GL_2(\mathbb{Z}/N\mathbb{Z})$ with surjective determinant such that for all $p \mid N$ the Hecke operator is effectively computable and a positive integer L, returns the zeta functions of the factors of $Jac(X_G)$ using*

$$O(d(C \log N(L \log L + N) + N I_G^2 \cdot \text{In} + I_G \log I_G) + d^3)$$

field operations, where $d := \dim S_2(\Gamma_G)$, $I_G := [SL_2(\mathbb{Z}) : \Gamma_G]$, C is the cost of a CosetIndex *operation, and* In *is the cost of membership testing in G.*

Proof We first compute a basis for $\mathbb{S}_2(\Gamma)$ using Theorem 3.5.1. Then we can apply Algorithm 6.2.3 to decompose it into irreducible subspaces. Here, we have to compute the matrices of the Hecke operators up to the Sturm bound (6.2.2), which is linear in I_G.

For computation of the Hecke operators away from the level, we apply Corollary 4.5.11, while for the Hecke operators at primes $p \mid N$, we apply Theorem 4.2.10, as in (6.3.3). Since we now compute their action on a basis, we have to multiply by d, and compute them up to I_G to obtain $O(d(C(I_G \log I_G + N) \log N + N I_G^2 \cdot \text{In}))$.

Finally, for each of the irreducible subspaces, we should apply Algorithm 6.3.1, which by Lemma 6.3.2 costs $O(C \log N(L \log L + N) + N I_G^2 \cdot \text{In} + d^3)$. □

7 Applications

In this section, we present a few of the applications that the above result contributed to, and also some future applications. All time measurements were taken on a MacBook with 2.3 GHz 8-Core Intel Core i9 processor, and 16 GB 2400 MHz DDR4 memory.

7.1 Classification of 2-Adic Images of Galois Representations Associated to Elliptic Curves over $\mathbb{Q}$

In [26], the authors compute the maximal tower of 2-power level modular curves containing a non-cuspidal non-CM rational point, and develop special techniques to compute their equations. Using our implementation, one can obtain the q-expansions of a basis for $S_2(\Gamma)$ for each of those groups.

For example, in [26, Example 6.1] the authors present a curious example associated with the group H_{155}, which is of index 24 and level 16, whose image in $GL_2(\mathbb{Z}/16\mathbb{Z})$ is generated by

$$\begin{pmatrix} 1 & 3 \\ 0 & 3 \end{pmatrix}, \begin{pmatrix} 1 & 0 \\ 2 & 3 \end{pmatrix}, \begin{pmatrix} 1 & 3 \\ 12 & 3 \end{pmatrix}, \begin{pmatrix} 1 & 1 \\ 12 & 7 \end{pmatrix}$$

Then a quick calculation of 1.78 s with our implementation yields the following.

```
> tt := Cputime();
> gens := [[1,3,12,3],[1,1,12,7],[1,3,0,3],[1,0,2,3]];
> N := 16;
> H_N := sub<GL(2,Integers(N)) | gens>;
> H := PSL2Subgroup(H_N);
> M := ModularSymbols(H, 2, Rationals(), 0);
> S := CuspidalSubspace(M);
> D := Decomposition(S, HeckeBound(S));
> qEigenform(D[1],100);
q - 4*q^5 - 3*q^9 - 4*q^13 - 2*q^17 + 11*q^25 -
  4*q^29 +
12*q^37 - 10*q^41 + 12*q^45 - 7*q^49 - 4*q^53 +
  12*q^61 +
16*q^65 - 6*q^73 + 9*q^81 + 8*q^85 + 10*q^89 - 18*q^97
  + O(q^100)
> Cputime(tt);
1.780
```

Proceeding to compute the invariants of the elliptic curve associated to this eigenform, we see that the modular curve X_H is isogenous to the elliptic curve labelled 256b1, defined by the equation $y^2 = x^3 - 2x$ over $\mathbb{Q}$, for which $X_H(\mathbb{Q}) \cong \mathbb{Z}/2\mathbb{Z} \times \mathbb{Z}$ is generated by $(0, 0)$ and $(-1, -1)$.

Remark 7.1.1 For this application, if one wants to compute the equation for the elliptic curve up to isomorphism, one has to compute the j-map. This can be done by computing the q-expansion of Eisenstein modular forms. A way to do so (in an even more complicated case) is explained in detail in [26, Example 6.1]. Our methods simply replicate the efforts originally done by the authors. For convenience, we have also supplied in our code package the code to generate Eisenstein series, e.g. for the weight 2 Eisenstein series, we can do the following.

```
> M := ModularForms(H);
> qExpansion(EisensteinSeries(M)[1], 100);
q^16 + 32*q^32 + 244*q^48 + 1024*q^64 + 3126*q^80
  + 7808*q^96 + O(q^100)
```

7.2 *Modular Curves of Prime-Power Level with Infinitely Many Rational Points*

In [34, Corollary 1.6], the authors introduce a finite set of subgroups G of $GL_2(\mathbb{Z}_l)$ that arise as the image of a Galois representation for infinitely many elliptic curves over $\mathbb{Q}$ with distinct j-invariants.

In section 6, they consider the 250 cases where G has genus 1 and show that the Jacobian of X_G, J_G, for each such group is isogenous to one of a certain finite set of elliptic curves. Then they compute point counts of the reduction modulo several primes to pinpoint the correct isogeny class over $\mathbb{Q}$, and show that one needs only consider a certain set of 28 groups.

These are included in the examples computed in [26], and again, using our implementation, it is possible to construct for each of these subgroups G a basis of q-eigenforms for the modular curve X_G.

The following example took 2.120 s.

```
> tt := Cputime();
> gens := [[2,1,3,2],[0,3,5,8],[1,0,0,5],[1,8,0,3]];
> N := 16;
> H_N := sub<GL(2,Integers(N)) | gens>;
> H := PSL2Subgroup(H_N);
> M := ModularSymbols(H, 2, Rationals(), 0);
> S := CuspidalSubspace(M);
> D := Decomposition(S, HeckeBound(S));
> qEigenform(D[1],100);
q - 2*q^3 + q^9 - 6*q^11 - 6*q^17 - 2*q^19 - 5*q^25
 + 4*q^27
     + 12*q^33 + 6*q^41 + 10*q^43 - 7*q^49 + 12*q^51
       + 4*q^57
     - 6*q^59 + 14*q^67 - 2*q^73 + 10*q^75 - 11*q^81
       - 18*q^83
      - 18*q^89 + 10*q^97 - 6*q^99 + O(q^100)
> Cputime(tt);
2.120
```

This is the newform 256.2.2.a. Computing the elliptic invariants, we see that this is isogenous to the curve 256a2 defined by $y^2 = x^3 + x^2 - 13x - 21$.

7.3 Efficient Computation of q-Eigenforms for $X_{ns}(N)$ and $X_{ns}^{+}(N)$

Using the Merel pair introduced in Example 4.5.6, and a similar one for $\Gamma_{ns}^{+}(p)$, its normalizer, one obtains a highly efficient implementation of the Hecke operators T_n for all n. This allows one to compute the q-eigenform and verify the results of [15] for $X_{ns}(11)$, [6] for $X_{ns}^{+}(13)$, and [21] for $X_{ns}^{+}(17)$, $X_{ns}^{+}(19)$ and $X_{ns}^{+}(23)$, as follows.

```
> tt := Cputime();
> G := GammaNSplus(13);
> M := ModularSymbols(G, 2, Rationals(), 0);
> S := CuspidalSubspace(M);
> D := Decomposition(S, HeckeBound(S));
> qEigenform(D[1], 20);
q + a*q^2 + (-a^2 - 2*a)*q^3 + (a^2 - 2)*q^4 +
(a^2 + 2*a - 2)*q^5 + (-a - 1)*q^6 + (a^2 - 3)*q^7 +
(-2*a^2 - 3*a + 1)*q^8 + (a^2 + 3*a - 1)*q^9 + (-a
+ 1)*q^10 + (-a^2 - 2*a - 2)*q^11 + (a^2 + 3*a)*q^12 +
(-2*a^2 - 2*a + 1)*q^14 + (a^2 + a - 2)*q^15 +
(-a^2 - a + 2)*q^16 + (-a^2 + a + 2)*q^17 +
(a^2 + 1)*q^18 + (-2*a^2 - a + 2)*q^19 + O(q^20)
> BaseRing(Parent($1));
Number Field with defining polynomial x^3 + 2*x^2 - x - 1
over the Rational  Field
> Cputime(tt);
0.160
```

This coincides with the result in section 4 in [6], and took merely 0.16 s. Similarly, in only 0.33 s we get the result for $X_{ns}^{+}(17)$.

```
> tt := Cputime();
> G := GammaNSplus(17);
> M := ModularSymbols(G, 2, Rationals(), 0);
> S := CuspidalSubspace(M);
> D := Decomposition(S, HeckeBound(S));
> [*qEigenform(d,20) : d in D*];
[*
q - q^2 - q^4 + 2*q^5 - 4*q^7 + 3*q^8 - 3*q^9 -
 2*q^10 -
2*q^13 + 4*q^14 - q^16 + 3*q^18 - 4*q^19 + O(q^20),
q + a*q^2 + (-a - 1)*q^3 + (-a + 1)*q^4 + a*q^5 -
 3*q^6 +
(-a - 2)*q^7 - 3*q^8 + (a + 1)*q^9 + (-a + 3)*q^10 -
 3*q^11 +
```

```
(-a + 2)*q^12 + (a - 1)*q^13 + (-a - 3)*q^14 - 3*q^15 +
(-a - 2)*q^16 + 3*q^18 + (-3*a - 1)*q^19 + O(q^20),
q + a*q^2 + (-a^2 + 1)*q^3 + (a^2 - 2)*q^4 +
(-a - 2)*q^5 + (-2*a + 1)*q^6 + (a^2 - 2)*q^7 +
 (-a - 1)*q^8 +
(a^2 - a - 2)*q^9 + (-a^2 - 2*a)*q^10 +
 (2*a^2 + 2*a - 6)*q^11 +
(a - 2)*q^12 + (-2*a^2 - 3*a + 6)*q^13 + (a - 1)*q^14 +
(2*a^2 + 2*a - 3)*q^15 + (-3*a^2 - a + 4)*q^16 +
(-a^2 + a -1)*q^18 + a*q^19 + O(q^20)
*]
> BaseRing(Parent(qEigenform(D[2])));
Number Field with defining polynomial x^2 + x - 3
over the Rational Field
> BaseRing(Parent(qEigenform(D[3])));
Number Field with defining polynomial x^3 - 3*x + 1
over the Rational Field
> Cputime(tt);
0.330
```

which coincides with the results of [21].

In [5], section 6, there is a description of coset representatives for $\Gamma_{ns}(N)$ and $\Gamma_{ns}^{+}(N)$ for general N, which again makes explicit the implementation via modular symbols.

7.4 Decomposition of the Jacobian of $X_{ns}^{+}(p)$

Applying Algorithm 6.2.3 to the space $\mathbb{S}_2(\Gamma_{ns}^{+}(p))$, where $\Gamma_{ns}^{+}(p)$ is yields a decomposition of the Jacobian of $X_{ns}^{+}(p)$. This has been computed using our code for all $p \leq 97$ (see [1]).

7.5 Computation of *q*-Eigenforms for $X_G(p)$ When *G* Is Exceptional

Similarly, the code can be used to compute q-eigenforms for the modular curves $X_G(N)$, when G is an exceptional subgroup of $GL_2(\mathbb{F}_p)$ with surjective determinant, e.g. running it for $X_{S_4}(13)$ we recover [4, Theorem 1.8]. We obtain step 3 of section 4 in the original paper in just a few seconds. (We use the embedding of $S_4 \hookrightarrow GL_2(\mathbb{F}_p)$ using the quaternions as described in [36, Section 11.5].)

```
> tt := Cputime();
> p := 13;
```

```
> B<i,j,k> := QuaternionAlgebra(Rationals(), -1, -1);
> O := QuaternionOrder([1,i,j,k]);
> _, mp := pMatrixRing(O,p);
> S4tp := sub<GL(2,p) | [mp(1+s) : s in [i,j,k]]
>  cat [mp(1-s) : s in [i,j,k]] cat [-1]>;
> H_S4 := sub<GL(2,Integers(p)) | Generators(S4tp)>;
> G_S4 := PSL2Subgroup(H_S4);
> M := ModularSymbols(G_S4, 2, Rationals(), 0);
> S := CuspidalSubspace(M);
> D := Decomposition(S, HeckeBound(S));
> [*qEigenform(d,20) : d in D*];
[*
q + a*q^2 + (-a^2 - 2*a)*q^3 + (a^2 - 2)*q^4 +
(a^2 + 2*a - 2)*q^5 + (-a - 1)*q^6 + (a^2 - 3)*q^7 +
(-2*a^2 - 3*a + 1)*q^8 + (a^2 + 3*a - 1)*q^9 +
(-a + 1)*q^10 + (-a^2 - 2*a - 2)*q^11 +
 (a^2 + 3*a)*q^12 +
(-2*a^2 - 2*a + 1)*q^14 + (a^2 + a - 2)*q^15 +
(-a^2 - a + 2)*q^16 + (-a^2 + a + 2)*q^17 +
(a^2 + 1)*q^18 + (-2*a^2 - a + 2)*q^19 + O(q^20)
*]
> BaseRing(Parent(qEigenform(D[1])));
Number Field with defining polynomial x^3
 + 2*x^2 - x - 1
over the Rational  Field
```

7.6 Smooth Plane Models for Modular Curves

Modular curves of genus 0, 1 obviously admit a smooth plane model always. In general, if a curve admits such a model of degree d, then its genus would be $g = \frac{(d-1)(d-2)}{2}$, so the next numbers to check are $g = 3, 6$. When $g = 3$, by González-Jiménez and Oyono [16], the generic case is a smooth plane quartic, so the answer is "almost always".

However, for $g > 3$, a generic curve does not admit a smooth plane model.

The smallest case for which we do not know the answer is $g = 6$. Thus, computing a basis of eigenforms of $\mathbb{S}_2(\Gamma)$ for each of the congruence subgroups Γ of genus 6 (see [13]), will help us to get a canonical model and from it (maybe) decide whether there exists a plane model.

References

1. Eran Assaf. Dimensions of the decomposition of the Jacobians of $X_{ns}^{+}(p)$ for primes $p < 100$ and characteristic polynomials of the Hecke operators T_p. https://github.com/assaferan/ModularSymbols/blob/master/dec_dims_char_polys_ns_cartan.dat. Last accessed: 2020-02-03.
2. Eran Assaf. Modular symbols package for arbitrary congruence subgroups. https://github.com/assaferan/ModularSymbols/. Last accessed: 2020-02-03.
3. A Oliver L Atkin et al. Twists of newforms and pseudo-eigenvalues of W-operators. *Inventiones mathematicae*, 48(3):221–243, 1978.
4. Barinder Banwait and John Cremona. Tetrahedral elliptic curves and the local-global principle for isogenies. *Algebra & Number Theory*, 8(5):1201–1229, 2014.
5. Burcu Baran. Normalizers of non-split Cartan subgroups, modular curves, and the class number one problem. *Journal of Number Theory*, 130(12):2753–2772, 2010.
6. Burcu Baran. An exceptional isomorphism between modular curves of level 13. *Journal of Number Theory*, 145:273–300, 2014.
7. Alex J. Best, Jonathan Bober, Andrew R. Booker, Edgar Costa, John E. Cremona, Maarten Derickx, Min Lee, David Lowry-Duda, David Roe, Andrew V. Sutherland, and John Voight. Computing classical modular forms. In *Arithmetic Geometry, Number Theory, and Computation*, pages 131–214. Springer, Cham, 2021. https://doi.org/10.1007/978-3-030-80914-0_4
8. Wieb Bosma, John Cannon, and Catherine Playoust. The Magma algebra system I: The user language. *Journal of Symbolic Computation*, 24(3–4):235–265, 1997.
9. Josha Box. Computing models for quotients of modular curves. Res. Number Theory 7(3):51, 2021.
10. Kevin Buzzard. Computing weight one modular forms over $\mathbb{C}$ and $\overline{\mho}_p$. In *Computations with modular forms*, pages 129–146. Springer, 2014.
11. Jean-Marc Couveignes and Bas Edixhoven. *Computational aspects of modular forms and Galois representations*. Princeton University Press, 2011.
12. John E Cremona. Algorithms for modular elliptic curves. 2nd edition. Cambridge University Press, Cambridge, 1997.
13. Chris Cummins and Sebastian Pauli. Congruence subgroups of $PSL_2(\mathbb{Z})$. https://mathstats.uncg.edu/sites/pauli/congruence/csg6.html. Last accessed: 2020-02-03.
14. Fred Diamond and Jerry Michael Shurman. *A first course in modular forms*, volume 228. Springer, 2005.
15. Valerio Dose, Julio Fernández, Josep González, and René Schoof. The automorphism group of the non-split Cartan modular curve of level 11. *Journal of Algebra*, 417:95–102, 2014.
16. Enrique González-Jiménez and Roger Oyono. Non-hyperelliptic modular curves of genus 3. *Journal of Number Theory*, 130(4):862–878, 2010.
17. Lloyd JP Kilford. Modular forms: a classical and computational introduction. *Mathematical intelligencer*, 32(3):58–59, 2010.
18. Ravi S Kulkarni. An arithmetic-geometric method in the study of the subgroups of the modular group. *American Journal of mathematics*, 113(6):1053–1133, 1991.
19. Ju I Manin. Parabolic points and zeta-functions of modular curves. *Mathematics of the USSR-Izvestiya*, 6(1):19, 1972.
20. Barry Mazur. Rational points on modular curves. In *Modular functions of one variable V*, pages 107–148. Springer, 1977.
21. Pietro Mercuri and René Schoof. Modular forms invariant under non-split Cartan subgroups. *Mathematics of Computation*, 89(324):1969–1991, 2020.
22. Loïc Merel. Opérateurs de Hecke pour $\Gamma_0(N)$ et fractions continues. In *Annales de l'institut Fourier*, volume 41, pages 519–537, 1991.
23. Loïc Merel. Universal Fourier expansions of modular forms. In *On Artin's Conjecture for Odd 2-Dimensional Representations*, pages 59–94. Springer, 1994.

24. James S Milne. Introduction to Shimura varieties. *Harmonic analysis, the trace formula, and Shimura varieties*, 4:265–378, 2005.
25. Andrew Ogg. *Modular forms and Dirichlet series*. WA Benjamin New York, 1969.
26. Jeremy Rouse and David Zureick-Brown. Elliptic curves over $\mathbb{Q}$ and 2-adic images of galois. *Research in Number Theory*, 1(1):12, 2015.
27. George J Schaeffer. Hecke stability and weight 1 modular forms. *Mathematische Zeitschrift*, 281(1-2):159–191, 2015.
28. Jean-Pierre Serre. Propriétés galoisiennes des points d'ordre fini des courbes elliptiques. *Invent. math*, 15:259–331, 1972.
29. Goro Shimura. *Introduction to the arithmetic theory of automorphic functions*, volume 1. Princeton University Press, 1971.
30. Vyacheslav Vladimirovich Shokurov. The study of the homology of Kuga varieties. *Izvestiya Rossiiskoi Akademii Nauk. Seriya Matematicheskaya*, 44(2):443–464, 1980.
31. William A Stein. An introduction to computing modular forms using modular symbols. In *An MSRI Proceedings*, 2002.
32. William A Stein. *Modular forms, a computational approach*, volume 79. American Mathematical Soc., 2007.
33. William Arthur Stein. *Explicit approaches to modular abelian varieties*. PhD thesis, University of California, Berkeley, 2000.
34. Andrew Sutherland and David Zywina. Modular curves of prime-power level with infinitely many rational points. *Algebra & Number Theory*, 11(5):1199–1229, 2017.
35. John A Todd and Harold SM Coxeter. A practical method for enumerating cosets of a finite abstract group. *Proceedings of the Edinburgh Mathematical Society*, 5(1):26–34, 1936.
36. John Voight. Quaternion Algebras. Graduate Texts in Mathematics, 288. Springer, 2021.
37. Gabor Wiese et al. *Modular forms of weight one over finite fields*. PhD thesis, Stieltjes Institute for Mathematics, Faculty of Mathematics and Natural Sciences, 2005.
38. David Zywina. Computing actions on cusp forms. *arXiv preprint arXiv:2001.07270*, 2020.

Square Root Time Coleman Integration on Superelliptic Curves

Alex J. Best

Abstract Since Kedlaya first introduced a p-adic algorithm for computing zeta functions of hyperelliptic curves, many related algorithms for computing both zeta functions and Coleman integrals on various classes of algebraic curves have been studied. These algorithms compute in the Monsky-Washnitzer cohomology or the rigid cohomology of the curve to determine the action of Frobenius on this cohomology.

We give a new algorithm for explicitly computing Coleman integrals on superelliptic curves over unramified extensions of p-adic fields. The runtime is softly linear with respect to the square root of the size of the residue field, bringing the runtime in line with that of the corresponding zeta function algorithms. We also describe the implementation of this algorithm in Nemo, a new package for the Julia programming language, which adds functionality for computational number theory. We compare Nemo with other systems in use in this area.

1 Introduction

In [18], Kedlaya introduced an algorithm using p-adic cohomology to compute the zeta function of a hyperelliptic curve defined over a finite field. Since then, there have been advances and generalisations of this method in (at least) three different directions. Firstly [13, 14] and [22, 23] have introduced variants that work on more general curves, respectively; superelliptic curves, cyclic covers of $\mathbf{P}^1$ and general curves with a map to $\mathbf{P}^1$. Secondly, Harvey [16] introduced a variant that runs in time quasilinear in $\sqrt{p}$. Finally, Balakrishnan-Bradshaw-Kedlaya [1] gave an algorithm to compute p-adic Coleman integrals on curves over the p-adics. A large part of the runtime of this algorithm is taken up by computing explicit relations in Monsky-Washnitzer cohomology, the same core procedure used in Kedlaya's algorithm.

A. J. Best (✉)
Department of Mathematics and Statistics, Boston University, Boston, MA, USA

J. S. Balakrishnan et al. (eds.), *Arithmetic Geometry, Number Theory, and Computation*, Simons Symposia, https://doi.org/10.1007/978-3-030-80914-0_3

In [20], Minzlaff showed that the first and second of these could be combined; that is, there exists an algorithm that computes the zeta function of superelliptic curves over $\mathbf{F}_{p^n}$ in $\widetilde{O}(p^{1/2})$ time. Additionally the paper [7] gives an algorithm based on the work of Balakrishnan-Bradshaw-Kedlaya and of Harvey that computes Coleman integrals on hyperelliptic curves over $\mathbf{Q}_p$ in time quasilinear in $\sqrt{p}$.

In this paper, we incorporate the work of Minzlaff into that of [7]. Let a, b be coprime integers with $a > 1, b > 2$, let p be a prime and $q = p^n$, take $h \in \mathbf{Z}_q[x]$ a squarefree polynomial of degree b, and consider the curve

$$C/\mathbf{Z}_q \colon y^a = h(x).$$

Let $g = (a-1)(b-1)/2$ be the genus of C. Let M be the matrix of q-power Frobenius acting on $H^1_{\mathrm{dR}}(C \times_{\mathbf{Z}_q} \mathbf{Q}_q)$ (via comparison with $H^1_{\mathrm{cris}}(C \times_{\mathbf{Z}_q} \mathbf{F}_q, \mathbf{Q}_q)$), in terms of the basis $B = \{\omega_{i,j} = x^i \mathrm{d}x/y^j\}_{i=0,\ldots,b-2, j=1,\ldots,a}$ and $N \in \mathbf{N}_{\geq 1}$ be such that both C and points $P, Q \in C(\mathbf{Q}_{p^n})$ are known to precision p^N, and assume $p > (aN-1)b$. Then, if multiplying two $m \times m$ matrices requires $\widetilde{O}(m^\omega)$ ring operations we have:

Theorem 1.1 *The vector of Coleman integrals* $(\int_P^Q \omega)_{\omega \in B}$ *can be computed in time*

$$\widetilde{O}\left(g^\omega \sqrt{p} n N^{5/2} + N^4 g^4 n^2 \log p\right)$$

to absolute p-adic precision $N - v_p(\det(M - I))$.

The work of Balakrishnan-Tuitman [4] has shown that there exists an algorithm to compute Coleman integrals on general curves defined over p-adic fields. This article shows that fast Coleman integration algorithms are not limited just to the hyperelliptic case.

We also take the opportunity to work out the practical details of the theory without restricting to the case of monic $h(x)$ or curves only defined over $\mathbf{Q}_p$; instead we work over arbitrary unramified extensions of $\mathbf{Q}_p$ throughout.

The algorithms described in this paper have been implemented using the Nemo computer algebra system [12], which is a specialised package for number theory and related mathematics for the programming language Julia [8]. This is a relatively new system for computing in commutative algebra, number theory and group theory that is based on several low-level libraries such as MPIR, Flint, Arb and Antic. A secondary goal of this paper is therefore to compare the performance of the various computer algebra systems available to mathematicians for running this type of computation. The implementation also allows for h to be non-monic and the curve C to be defined over an unramified extension of $\mathbf{Q}_p$.

1.1 Acknowledgements

This work was supported by Simons Foundation grant #550023 and a Hariri Institute Graduate Fellowship.

I would like to thank Jennifer Balakrishnan for many valuable pieces of advice regarding this project and her constant encouragement. Tommy Hofman has patiently answered many of my questions about Nemo and Hecke and has also improved the Julia implementation, making it both more idiomatic and faster; I am very grateful for all of his help. I would also like to thank the anonymous referees for their helpful feedback.

2 Set-Up and Notation

The major outline of the algorithm presented in this paper follows that of [7], adapted to the case of a **(tamely) superelliptic curve**: these will be smooth projective curves C of a specific form, which will be defined over the ring of integers of an unramified extension of $\mathbf{Q}_p$ (such extensions will be denoted as $\mathbf{Q}_q$, where $q = p^n$ is the cardinality of the residue field). The present work can be read in conjunction with [7].

We will denote by σ the Witt vector Frobenius on $\mathbf{Z}_q$ and $\mathbf{Q}_q$. The restriction of being superelliptic for us means that the curve should be given by an affine equation of the form

$$y^a = h(x), \ \deg(h) = b, \ h \text{ squarefree}$$

where $\gcd(a, b) = 1$ and $a > 1, b > 2$. Tameness means that, in addition, we assume $p \nmid a$, though this will be implied by our later assumption (2.1). We will write $h = \lambda x^b + \tilde{h}$ with $\deg \tilde{h} < b$. We can view the projective curve C as living inside of the weighted projective space $\mathbf{P}(a, b, 1)$, where there is a unique point at infinity on the curve, denoted ∞. As h might not be monic, this will be of the form $(\lambda^t : \lambda^s : 0)$ for some $s, t \in \mathbf{Z}$ such that $as - bt = 1$. A superelliptic curve of this form will have genus

$$g = \frac{(a-1)(b-1)}{2}.$$

Each set of points $D \subseteq C(\overline{\mathbf{Q}}_q)$ that is the full preimage of some $x \in C_{\overline{\mathbf{F}}_q}(\overline{\mathbf{F}}_q)$ under the reduction map is known as a **residue disk**.

Such a superelliptic curve comes equipped with a natural map to $\mathbf{P}^1$ given by

$$(x : y : z) \mapsto (x : z^a).$$

We will often be interested in the points where this map is branched (i.e. those with $y = 0$), and call them the **branch points** of this curve. We also call any residue disk containing one of these points a **branch disk**.

These play the analogous role to that of the Weierstrass points in the hyperelliptic case, but we prefer to avoid calling them Weierstrass points as the notion of a Weierstrass point in algebraic geometry would include non-branch points in general, see [21, p. 3372].

We will often work with the affine open subset of the curve obtained by removing all branch points (including ∞), writing

$$A = \mathbf{Z}_q[x, y, y^{-1}]/(y^a - h(x)).$$

This affine space is then

$$\operatorname{Spec} A = U \subseteq C.$$

We denote by $A^\dagger$ the weak completion of A; this is the ring of formal power series

$$A^\dagger = \left\{ \sum_{i=-\infty}^{\infty} \frac{R_i(x)}{y^i} : R_i \in \mathbf{Z}_q[x]_{\deg \le b-1}, \liminf_{|i|\to\infty} \frac{v_p(R_i)}{|i|} > 0 \right\}.$$

We will consider the module of 1-forms

$$\Omega^1_{A^\dagger} = A^\dagger \mathrm{d}x \oplus A^\dagger \mathrm{d}y/(ay^{a-1}\mathrm{d}y - h'(x)\mathrm{d}x)),$$

and the exterior derivative map is denoted

$$\mathrm{d}\colon A^\dagger \to \Omega^1_{A^\dagger}.$$

We also use A_{loc} to denote the $\mathbf{Q}_q$-algebra of $\overline{\mathbf{Q}}_q$-valued functions on $C(\overline{\mathbf{Q}}_q)$ that are given by a convergent power series on each residue disk and are $\mathrm{Gal}(\overline{\mathbf{Q}}_q/\mathbf{Q}_q)$-equivariant. Note that $A^\dagger \hookrightarrow A_{\mathrm{loc}}$. The Monsky-Washnitzer cohomology of A is then defined to be $H^1_{\mathrm{MW}}(A) = \Omega^1_{A^\dagger}/\mathrm{d}(A^\dagger)$.

From now on, we make the assumption that

$$p > (aN - 1)b; \tag{2.1}$$

this simplifies the analysis of denominators appearing later in the algorithm. This assumption is likely completely removable without affecting the asymptotic complexity, or at least smaller p may be used by doing a more involved precision analysis to ensure the right amount of extra precision is used.

3 Coleman Integration

Coleman integration is an integration theory that is in particular defined for 1-forms on curves, viewed as p-adic analytic spaces. This theory has been applied to various questions in the arithmetic of curves over number fields and p-adic fields. These applications include provably determining the set of rational points on a curve using Chabauty-Coleman(-Kim) [19, 3] and determining torsion points on curves. For background on Coleman integration, especially the details of the action of Frobenius when $n > 1$, we refer to [5].

Coleman integration for curves is computable in many cases [6, 1, 4]; that is, given a base point and a 1-form to some finite p-adic precision, the corresponding integral can be computed to some (smaller) precision. This can be done in many instances of interest to arithmetic geometers. Moreover these algorithms perform well in practice, returning an answer in examples of interest to mathematicians in an acceptable real-world runtime.

Coleman integration requires the choice of a *lift of Frobenius*, that is a map

$$\phi\colon A^\dagger \to A^\dagger$$

such that the reduction of this map to $\mathbf{F}_q$

$$\bar{\phi}\colon A^\dagger_{\mathbf{F}_q} \to A^\dagger_{\mathbf{F}_q}$$

is the p-power Frobenius map and $\phi(k) = \sigma(k)$ for all $k \in \mathbf{Z}_q$. There are many possibilities for such a lift in general; in this paper, we will make an explicit choice of a lift for superelliptic curves that allows us to analyse its action in detail. It is therefore important to note that the theory is known to be independent of the choice of ϕ.

We define the action of ϕ on $C(\overline{\mathbf{Q}}_q)$ via

$$\phi(x_0, y_0) = (\sigma^{-1}(\phi(x)(x_0, y_0)), \sigma^{-1}(\phi(y)(x_0, y_0))). \tag{3.1}$$

On functions $f\colon C(\overline{\mathbf{Q}}_q) \to \overline{\mathbf{Q}}_q$ the action of ϕ is then

$$\phi(f)(P) = \sigma f(\phi(P)). \tag{3.2}$$

The key theorem proved by Coleman describing this integration theory can be stated in our setting as follows.

Theorem 3.1 (Coleman) *There is a unique (up to a global constant of integration) $\mathbf{Q}_q$-linear integration map $\int\colon \Omega^1_{A^\dagger} \otimes \mathbf{Q}_q \to A_{\mathrm{loc}}$ satisfying the following:*

(1) Frobenius equivariance: $\int(\phi\omega) = \phi\left(\int \omega\right)$
(2) the fundamental theorem of calculus: $\mathrm{d} \circ \int$ is the canonical inclusion

$$\Omega^1_{A^\dagger} \otimes \mathbf{Q}_q \to \Omega^1_{\text{loc}}$$

(3) and $\int \circ d$ *is the natural map* $A^\dagger \to A_{\text{loc}}/(\textit{constant functions})$.

Given points $P, Q \in C(\overline{\mathbf{Q}}_q)$ *the definite integral* $\int_P^Q \omega$ *is then defined as* $\left(\int \omega\right)(Q) - \left(\int \omega\right)(P)$*, which is a well-defined function of* P, Q.

Balakrishnan-Bradshaw-Kedlaya [1] describe how, using these properties, the computation of Coleman integrals of 1-forms on a curve can be broken up into two parts:

(1) The computation of *tiny integrals* between points in the same residue disk, using local coordinates on the curve.
(2) The evaluation of exact forms appearing when reducing Frobenius pullbacks of differentials in Monsky-Washnitzer cohomology.

In the remainder of this section, we give a description of local coordinates that can be used to compute tiny integrals and describe how this method works in our setting in more detail, and in the next section we describe a procedure for evaluating the relevant exact forms.

3.1 Local Coordinates on Superelliptic Curves

In order to compute Coleman integrals locally on a p-adic disk, we need an explicit local parametrisation of such a disk. To do this we apply Newton's method/Hensel's lemma; to find a power series solution $x(t)$ to a polynomial $F(x(t))$, we start with some $x_0(t)$, and iterate $x_i(t) = x_{i-1}(t) - F(x_{i-1}(t))/F'(x_{i-1}(t))$. The limiting power series is the desired solution; applying this directly with $F(x(t), y(t)) = y(t)^a - h(x(t))$ in the x or y variable leads to the next results, giving local coordinates around a finite branch point or non-branch point.

Proposition 3.2 (Local Coordinates Around a Point not in a Branch Disk) *Let* $P \in C(\mathbf{Q}_q)$ *be a point not in a branch disk, represented as* $P = (X, Y)$ *on* $y^a = h(x)$*. Then local coordinates* $(x(t), y(t))$ *around* P *can be given by*

$$\begin{aligned}
x(t) &= X + t,\\
y_0(t) &= Y,\\
y_i(t) &= \frac{1}{a}\left((a-1)y_{i-1}(t) + \frac{h(x(t))}{y_{i-1}(t)^{a-1}}\right), \textit{ and}\\
y(t) &= \lim_{i\to\infty} y_i(t).
\end{aligned}$$

Proof We may simplify

$$y_i(t) = y_{i-1}(t) - \frac{y_{i-1}(t)^a - h(x(t))}{ay_{i-1}(t)^{a-1}} = y_{i-1}(t) - \frac{1}{a}y_{i-1}(t) + \frac{h(x(t))}{ay_{i-1}(t)^{a-1}}$$
$$= \frac{1}{a}\left((a-1)y_{i-1}(t) + \frac{h(x(t))}{y_{i-1}(t)^{a-1}}\right).$$

■

Proposition 3.3 (Local Coordinates Around a Point in a Finite Branch Disk) *Let $P \in C(\mathbf{Q}_q)$ be a point in a non-infinite branch disk, represented as $P = (X, Y)$ on $y^a = h(x)$. Then local coordinates $(x(t), y(t))$ around P can be given by*

$$y(t) = Y + t,$$
$$x_0(t) = X,$$
$$x_i(t) = x_{i-1}(t) + \frac{y(t)^a - h(x_{i-1}(t))}{h'(x_{i-1}(t))}, \text{ and}$$
$$x(t) = \lim_{i\to\infty} x_i(t).$$

Proof Direct application of Newton's method. ■

For the infinite disk the choice of uniformiser is less obvious and several different choices can be made. For instance, one could take a local coordinate t at ∞ to be such that $x = t^{-a}$. Then given a point (x_0, y_0) in the same residue disk as ∞, to find the value of the parameter t that gives (x_0, y_0), we can take each of the possible roots $\sqrt[a]{x_0}$ and check which of these give y_0.

Instead, we take t to be an appropriate monomial in x, y. This has the advantage that, to find the value of the local coordinate for a point (X_0, Y_0), we may compute the corresponding coordinate t_0 using only multiplication of the values X_0, Y_0. Precisely, we do the following:

Proposition 3.4 (Local Coordinates Around ∞) *Let ∞ be the point at infinity on C. Then local coordinates $(x(t), y(t))$ around ∞ can be given as follows. Let ℓ, k be such that $b\ell - ak = 1$. Set*

$$x_0(t) = \lambda^{-\ell}t^{-a},$$
$$y_0(t) = \lambda^{-k}t^{-b},$$
$$J(x, y) = \begin{pmatrix} -h'(x) & ay^{a-1} \\ -kx^{k-1} & \ell t y^{\ell-1} \end{pmatrix},$$
$$\begin{pmatrix} x_i(t) \\ y_i(t) \end{pmatrix} = \begin{pmatrix} x_{i-1}(t) \\ y_{i-1}(t) \end{pmatrix} - J(x_{i-1}(t), y_{i-1}(t))^{-1} \begin{pmatrix} ty_{i-1}(t)^{\ell} - x_{i-1}(t)^k \\ y_{i-1}(t)^a - h(x_{i-1}(t)) \end{pmatrix}.$$

Then

$$x(t) = \lim_{i\to\infty} x_i(t),$$

$$y(t) = \lim_{i\to\infty} y_i(t).$$

Proof Using the assumption that $\gcd(a, b) = 1$, we may find ℓ, k such that $b\ell - ak = 1$. We now find $x(t), y(t)$ such that $t = x(t)^k/y(t)^\ell$. To do this, we solve $ty(t)^\ell = x(t)^k$ and $y(t)^a = h(x(t))$ simultaneously using multivariate Newton's method; this gives the recurrence stated. ■

When working with hyperelliptic curves, one can use the hyperelliptic involution to avoid computing integrals from infinity to points in branch disks, where the exact forms that would need to be evaluated fail to converge. We can also avoid this issue when working with superelliptic curves as follows:

Proposition 3.5 *Fix μ to be the automorphism of the cover*

$$C \xrightarrow{x} \mathbf{P}^1$$

given by

$$\mu : C \to C$$

$$(x, y) \mapsto (x, \zeta_a y)$$

for some fixed primitive a-th root of unity $\zeta_a \in \overline{\mathbf{Q}}_q$. If P, Q are branch points and ω is a differential for which $\mu^*\omega = \zeta_a^\beta \omega$ for some $\beta \in \{1, \ldots, a-1\}$, then

$$\int_Q^P \omega = 0.$$

Proof We have

$$\int_P^Q \omega = \int_{\mu P}^{\mu Q} \omega = \int_Q^P \mu^*\omega = \zeta_a^\beta \int_P^Q \omega.$$

■

Corollary 3.6 *Let P be a point in a branch disk. We have for any branch point Q*

$$\int_P^\infty \omega = \int_P^Q \omega.$$

Choosing Q to be the branch point in the same disk as P we can therefore compute

$$\int_P^\infty \omega$$

by expanding only in local coordinates around Q.

3.2 The Action of q-Power Frobenius

We now assume that we know the action of ϕ on $\Omega^1_{A^\dagger}$ in the sense that we have the following:

(1) A fixed basis $(\omega_i)_i$ of $H^1_{MW}(C)$, thought of as a column vector
(2) a matrix of Frobenius M,
(3) and a vector of primitives $(f_i)_i$ such that,

$$(\phi^*\omega_i)_i = M(\omega_i)_i + (\mathrm{d}f_i)_i \in (\Omega^1_{A^\dagger})^{2g}.$$

This data is what will be returned by (our generalisation of) Kedlaya's algorithm.

The next lemma calculates the action of ϕ^{*n} on $\Omega^1_{A^\dagger}$ from this data; this is consistent with [2, Rmk. 1] but appears to be different from [1, Rmk. 12].

Lemma 3.7 *The action of a power of Frobenius on the basis differentials is given by*

$$\phi^{*n}(\omega_i)_i = \sum_{t=n-1,\ldots,0}\left(\prod_{s=n-1,\ldots,t+1}\phi^s(M)\right)\phi^{*t}(\mathrm{d}f_i)_i + \prod_{s=n-1,\ldots,0}\phi^s(M)(\omega_i)_i.$$

Proof By induction on n, we will apply the relation

$$\phi^*(\omega_i)_i = (\mathrm{d}f_i)_i + M(\omega_i)_i.$$

The base case

$$\phi^{*0}(\omega_i)_i = 0 + 1\cdot(\omega_i)_i$$

holds trivially (as does the $n = 1$ case which is simply the fundamental relation above), and we have

$$\phi^{*n+1}(\omega_i)_i = \phi^*\left(\sum_{t=n-1,\ldots,0}\left(\prod_{s=n-1,\ldots,t+1}\phi^s(M)\right)\phi^{*t}(\mathrm{d}f_i)_i\right) + \phi^*\left(\prod_{s=n-1,\ldots,0}\phi^s(M)(\omega_i)_i\right)$$

$$= \sum_{t=n-1,\dots,0} \left(\prod_{s=n-1,\dots,t+1} \phi^{s+1}(M) \right) \phi^{*t+1}(\mathrm{d}f_i)_i + \prod_{s=n-1,\dots,0} \phi^{s+1}(M)\phi^*((\omega_i)_i)$$

$$= \sum_{t=n-1,\dots,0} \left(\prod_{s=n-1,\dots,t+1} \phi^{s+1}(M) \right) \phi^{*t+1}(\mathrm{d}f_i)_i$$

$$+ \prod_{s=n-1,\dots,0} \phi^{s+1}(M)(\mathrm{d}f_i)_i + \prod_{s=n-1,\dots,0} \phi^{s+1}(M)M(\omega_i)_i$$

$$= \sum_{t=n-1,\dots,0} \left(\prod_{s=n-1,\dots,t+1} \phi^{s+1}(M) \right) \phi^{*t+1}(\mathrm{d}f_i)_i$$

$$+ \prod_{s=n-1,\dots,0} \phi^{s+1}(M)(\mathrm{d}f_i)_i + \prod_{s=n,\dots,0} \phi^{s}(M)(\omega_i)_i$$

$$= \sum_{t=n-1,\dots,0} \left(\prod_{s=n,\dots,t+2} \phi^{s}(M) \right) \phi^{*t+1}(\mathrm{d}f_i)_i + \prod_{s=n-1,\dots,0} \phi^{s+1}(M)(\mathrm{d}f_i)_i + \prod_{s=n,\dots,0} \phi^{s}(M)(\omega_i)_i$$

$$= \sum_{t=n,\dots,0} \left(\prod_{s=n,\dots,t+1} \phi^{s}(M) \right) \phi^{*t}(\mathrm{d}f_i)_i + \prod_{s=n,\dots,0} \phi^{s}(M)(\omega_i)_i.$$

■

We therefore have

$$\left(\int_P^Q \omega_i \right)_i = \left(\int_P^{\phi^n P} \omega_i + \int_{\phi^n P}^{\phi^n Q} \omega_i + \int_{\phi^n Q}^{Q} \omega_i \right)_i$$

$$= \left(\int_P^{\phi^n P} \omega_i + \int_{\phi^n Q}^{Q} \omega_i \right)_i + \sum_{t=n-1,\dots,0} \left(\prod_{s=n-1,\dots,t+1} \phi^{s}(M) \right) \left(\int_P^Q \phi^{*t} \mathrm{d}f_i \right)_i$$

$$+ \prod_{s=n-1,\dots,0} \phi^{s}(M) \left(\int_P^Q \omega_i \right)_i$$

$$= \left(\int_P^{\phi^n P} \omega_i + \int_{\phi^n Q}^{Q} \omega_i \right)_i + \sum_{t=n-1,\dots,0} \left(\prod_{s=n-1,\dots,t+1} \phi^{s}(M) \right) \left(f_i(\phi^t Q) - f_i(\phi^t P) \right)_i$$

$$+ \prod_{s=n-1,\dots,0} \phi^{s}(M) \left(\int_P^Q \omega_i \right)_i ,$$

hence

$$\left(1 - \prod_{s=n-1,\dots,0} \phi^{s}(M) \right) \left(\int_P^Q \omega_i \right)_i = \tag{3.3}$$

$$\left(\int_P^{\phi^n P} \omega_i\right)_i + \left(\int_{\phi^n Q}^{Q} \omega_i\right)_i + \sum_{t=n-1,\dots,0} \left(\prod_{s=n-1,\dots,t+1} \phi^s(M)\right) \left(f_i(\phi^t Q) - f_i(\phi^t P)\right)_i . \tag{3.4}$$

So computing $\left(\int_P^Q \omega_i\right)_i$ reduces to computing (3.4) and then inverting the matrix $1 - \prod_{s=n-1,\dots,0} \phi^s(M)$.

Thus in order to compute the Coleman integrals between points P, Q we need to know M and evaluations of $\phi^t f_i$ at the points P, Q for $t = 0, \dots, n-1$. From (3.1), (3.2) we see that to compute each $(\phi^t f_i)(P)$ we need to compute $f_i(\phi^t P)$ and apply σ^t.

Remark 3.8 This makes the Teichmüller point variant of Coleman integration algorithms more appealing when working over extensions. In that approach, the Teichmüller point is used in each non-branch disk as a base point and all integrals are computed in two parts: an integral to the Teichmüller point and then tiny integrals for reaching the rest of the disk. The advantage is that all of the points $\phi^t P$ will be equal for a Teichmüller point P, hence also the values $f_i(\phi^t P)$, so fewer evaluations of f_i need to be carried out. We do not take this approach in the current implementation, however cf. Remark 6.1.

4 Reductions in Cohomology

We now describe the reduction process. The foundation for this is the work of Minzlaff in [20], which gives explicit maps, reducing elements of $\Omega^1_{A^\dagger}$ to cohomologous ones with smaller x- or y-degree. To compute Coleman primitives, we must, in addition, record the exact forms subtracted to obtain these cohomologous elements and determine a recurrence that computes the evaluation of the sum of these forms at the end of the reduction process.

We consider the following spaces of differentials for $s, t \in \mathbf{Z}$ and $s \geq -1$:

$$W_{s,t} = \{x^i y^{-j} \cdot x^s y^{-at} \mathrm{d}x : 0 \leq i \leq b-2,\ 1 \leq j \leq a-1\},$$

and for $s = -1$ we take only the subspace with $i \geq 1$.

These differentials are such that $W_{-1,0}$ is spanned by the basis chosen in Theorem 1.1.

We may decompose $W_{s,t}$ into eigenspaces under the action of the superelliptic automorphism μ to obtain

$$W_{s,t} = W^1_{s,t} \oplus \cdots \oplus W^{a-1}_{s,t},$$

where

$$W^\beta_{s,t} = \{x^i y^{-\beta} \cdot x^s y^{-at} \mathrm{d}x : 0 \leq i \leq b-2\}.$$

As the subspaces indexed by β are all preserved in everything that follows, we consider each $1 \leq \beta \leq a-1$ independently from now.

We may choose a lift of Frobenius on $\widetilde{C}$ by letting $\sigma\colon \mathbf{Q}_q \to \mathbf{Q}_q$ be the Witt vector Frobenius and setting

$$\sigma(x) = x^p$$

as shown in [20, Sec 4.] this forces us to take

$$\sigma(y) = y^p \sum_{k=0}^{\infty} \binom{\frac{1}{a}}{k} \frac{(\sigma(h) - h^p)^k}{y^{apk}}.$$

Minzlaff also determined an approximation of the action of Frobenius on the basis differentials. Specifically we have

Lemma 4.1 ([20, Prop. 4.1]) *If we let $\lambda_r^{(k)}$ be the coefficient of x^r in h^k and*

$$\mu_{k,r,j} = p\sigma(\lambda_r^{(k)}) \sum_{\ell=k}^{N-1} (-1)^{\ell+k} \binom{-\frac{j}{a}}{\ell} \binom{\ell}{k} \in \mathbf{Z}_q,$$

then the reduction in cohomology of the 1-form

$$\sum_{k=0}^{N-1} \sum_{r=0}^{bk} \mu_{k,r,j} x^{p(i+r+1)-1} y^{-p(ak+j)} \mathrm{d}x$$

is congruent to that of $\sigma(x^i \mathrm{d}x/y^j)$ modulo p^N. Moreover the exact forms that reduce these respective cohomology classes to the cohomologous linear combination of our chosen basis elements are also congruent modulo p^N.

Proof We denote by D_η the divisor $C \setminus U$. Then Minzlaff [20, Lem. 3.4] shows that for any $\omega \in \Omega_{A/\mathbf{Z}_q}$ with pole divisor

$$(\omega)_\infty \leq m D_\eta,$$

we have

$$p^{\lfloor \log_p(m-1) \rfloor} \pi(\omega) \in \Omega_{A/\mathbf{Z}_q}.$$

However the proof of this statement given there shows that in fact if

$$\omega = \pi(\omega) + \mathrm{d}F$$

then we also have

$$p^{\lfloor \log_p (m-1) \rfloor} F \in A.$$

As in the proof of [20, Prop. 4.1], we apply this to the series

$$\sigma(x^i \mathrm{d}x/y^j) = \sum_{k=0}^{\infty} p \binom{-\frac{j}{a}}{k} (\sigma(h) - h^p)^k x^{p(i+1)-1} y^{-p(ak+j)} \mathrm{d}x,$$

and writing S_k for the kth summand of the infinite sum on the right, we need to show that if $k \geq N$ and

$$S_k = \pi(S_k) + \mathrm{d}F_k,$$

then $\pi(S_k) \equiv 0 \pmod{p^N}$ and $F_k \equiv 0 \pmod{p^k}$. Fixing $k \geq N$, we can write

$$S_k = p^{k+1} \sum_{\ell=0}^{(k+1)p-1} f_\ell(x) y^{m_\ell} \mathrm{d}x, \text{ where } m_\ell = -p(ak+j) + a\ell$$

with $f_\ell \in \mathbf{Z}_q[x]_{\deg \leq b-1}$.

So we apply the above integrality statement to each

$$f_\ell(x) y^{m_\ell} \mathrm{d}x.$$

The same power of p used by Minzlaff is therefore sufficient. ■

The reduction proceeds in two stages, first horizontal reduction that reduces the index s and hence the power of x appearing in the 1-forms and leaves us with forms in $W_{-1,t}$ to consider. Then vertical reduction is performed that reduces the power of y appearing and reduces all forms to $W_{-1,0}$.

We present the vertical reduction step first however, as although it takes place after the horizontal steps when the algorithm is executed, knowing the form of the vertical reduction ahead of time allows us to make some simplifications in the description of the horizontal reduction.

4.1 Vertical Reduction

Because h is squarefree and λ is a unit, we can find for each $i = 0, \ldots, b-2$, a pair of polynomials $R_i, S_i \in \mathbf{Z}_q[x]$ with $\deg R_i \leq b-2$ and $\deg S_i \leq b-1$, such that

$$x^i = R_i h + S_i h'.$$

We fix a choice of such polynomials now.

The vertical reduction proceeds by reducing a 1-form in $W^{\beta}_{-1,t}$ to a cohomologous one in $W^{\beta}_{-1,t-1}$ using the following lemma:

Lemma 4.2 (Vertical Reduction) *We have*

$$x^i y^{-at-\beta}\mathrm{d}x - \frac{-a}{at+\beta-a}\mathrm{d}(S_i(x)y^{-at-\beta+a})$$
$$= \frac{(at+\beta-a)R_i(x)+aS_i'(x)}{at+\beta-a}y^{-a(t-1)-\beta}\mathrm{d}x \in W^{\beta}_{-1,t-1}.$$

Proof Via our choice of R_i, S_i we have that

$$x^i y^{-at-\beta}\mathrm{d}x = \left(R_i(x)h(x) + S_i(x)h'(x)\right) y^{-at-\beta}\mathrm{d}x,$$

and also that

$$\mathrm{d}(S_i(x)y^{-at-\beta+a}) = S_i'(x)y^{-at-\beta+a}\mathrm{d}x + (-at-\beta+a)S_i(x)y^{-at-\beta+a-1}\mathrm{d}y$$
$$= S_i'(x)y^{-at-\beta+a}\mathrm{d}x + (-at-\beta+a)\frac{1}{a}S_i(x)y^{-at-\beta}h'(x)\mathrm{d}x$$

so

$$\frac{-a}{at+\beta-a}\mathrm{d}(S_i(x)y^{-at-\beta+a}) = \frac{-a}{at+\beta-a}S_i'(x)y^{-at-\beta+a}\mathrm{d}x + S_i(x)y^{-at-\beta}h'(x)\mathrm{d}x,$$

and the above equality holds. ■

This reduction is a $\mathbf{Q}_q$-linear map

$$W^{\beta}_{-1,t} \to W^{\beta}_{-1,t-1}.$$

To express it in matrix form, we denote the coefficients of R_i by $r_{i,j}$ and likewise $s'_{i,j}$ for those of S_i', each R_i and S_i' is of degree at most $b-2$.

To compute evaluations of the primitives for Coleman integration, we augment these linear maps to also include a vector of length L, that holds the data of evaluations of the exact form subtracted so far at several points $P_i \in C(\mathbf{Q}_q)$, $i = 1, \ldots, L$. We assume that all of these points do not lie in a branch disk (including the infinite disk). This implies that $x(P_i)$ and $y(P_i)$ are both integral for all i and that $y(P_i)$ is a unit in $\mathbf{Z}_q$. So we have linear maps

$$W^{\beta}_{s+1,t} \times \mathbf{Q}_q^L \to W^{\beta}_{s,t} \times \mathbf{Q}_q^L.$$

Doing this as described gives a matrix whose entries are non-linear functions of the index t, due to the presence of the term $y^{-at+a-\beta}$ in the exact form

$$\frac{-a}{at+\beta-a} S_i(x) y^{-at-\beta+a}.$$

To remedy this, we use the approach of [7] and modify the reduction process to obtain matrices with entries linear functions of t, by factoring out the powers of y in the exact form. As we reduce monomials from $W^{\beta}_{s,t} \to W^{\beta}_{-1,t}$, we can let each later reduction step multiply the exact form by x. This results in a total power of x^s when the reduction finishes in $W^{\beta}_{-1,t}$, which is the same power of x as in the exact form obtained from $W^{\beta}_{s+1,t} \to W^{\beta}_{s,t}$.

Let $D^{\beta}_V(t)$ be the scalar $at+\beta-a \in \mathbf{Z}_q[t]$ and define a $(b-1+L)\times(b-1+L)$ matrix

$$R^{\beta}_V(t) = M^{\beta}_V(t) D^{\beta}_V(t)^{-1},$$

where $M^{\beta}_V(t)$ is the matrix

$$\left(\begin{array}{ccc|ccc} D^{\beta}_V(t) r_{0,0} + a s'_{0,0} & \cdots & D^{\beta}_V(t) r_{b-2,0} + a s'_{b-2,0} & & & \\ \vdots & \ddots & \vdots & & & \\ D^{\beta}_V(t) r_{0,b-2} + a s'_{0,b-2} & \cdots & D^{\beta}_V(t) r_{b-2,b-2} + a s'_{b-2,b-2} & & & \\ \hline -aS_0(x(P_1))y(P_1)^{-\beta+a} & \cdots & -aS_{b-2}(x(P_1))y(P_1)^{-\beta+a} & y(P_1)^{-a} D^{\beta}_V(t) & \cdots & 0 \\ \vdots & \ddots & \vdots & \vdots & \ddots & \vdots \\ -aS_0(x(P_L))y(P_L)^{-\beta+a} & \cdots & -aS_{b-2}(x(P_L))y(P_L)^{-\beta+a} & 0 & \cdots & y(P_L)^{-a} D^{\beta}_V(t) \end{array}\right). \tag{4.1}$$

4.2 *Horizontal Reduction*

We now describe the horizontal reduction process.

The horizontal reduction proceeds by reducing a 1-form in $W^{\beta}_{s+1,t}$ to a cohomologous one in $W^{\beta}_{s,t}$ using the following lemma:

Lemma 4.3 (Horizontal Reduction) *We have*

$$x^{b+s-1} y^{-at-\beta} \mathrm{d}x - \mathrm{d}\left(\frac{a}{as\lambda - (at-a+\beta)\lambda b} x^s y^{-at+a-\beta}\right)$$
$$= -\frac{a(s\tilde{h} - \frac{1}{a}(at-a+\beta)x\tilde{h}')}{as\lambda - (at-a+\beta)\lambda\beta} x^{s-1} y^{-at-\beta} \mathrm{d}x \in W^{\beta}_{s,t}.$$

Proof We directly compute

$$\mathrm{d}(x^s y^{-at+a-\beta}) = s x^{s-1} y^{-at+a-\beta} \mathrm{d}x - (at-a+\beta) x^s y^{-at+a-\beta-1} \mathrm{d}y$$

$$= \left(sx^{s-1}h - \frac{1}{a}(at + \beta - a)x^s h'(x) \right) y^{-at-\beta} dx$$

$$= \left(sx^{s-1}(\lambda x^b + \tilde{h}) - \frac{1}{a}(at + \beta - a)x^s(\lambda b x^{b-a} + \tilde{h}') \right) y^{-at-\beta} dx$$

$$= \left(s(\lambda x^b + \tilde{h}) - \frac{1}{a}(at + \beta - a)x(\lambda b x^{b-a} + \tilde{h}') \right) x^{s-1} y^{-at-\beta} dx.$$

Therefore, by subtracting $\frac{a}{as\lambda-(at-a+\beta)b\lambda} d(x^s y^{-at+a-\beta})$ from $x^b x^s y^{-at-\beta} dx$, the remaining terms are all as stated, and of lower degree so that they lie in $W^{\beta}_{s,t}$. ■

As above we augment our linear maps

$$W^{\beta}_{s+1,t} \to W^{\beta}_{s,t}$$

to also include a vector of length L that holds the data of evaluations of the exact form subtracted so far at $P_i \in C(\mathbf{Q}_q)$, $i = 1, \ldots, L$ not in the branch disks:

$$W^{\beta}_{s+1,t} \times \mathbf{Q}_q^L \to W^{\beta}_{s,t} \times \mathbf{Q}_q^L.$$

Doing this gives a matrix whose entries are non-linear functions of the index, due to the presence of the term x^s in the exact form

$$\frac{-a}{\lambda((at - a + \beta)b - as)} x^s y^{-at+a-\beta}.$$

Instead we multiply the reduction evaluation computed so far by $x(P_i)$ at each reduction step which results in a total power of x^s when the reduction finishes in $W^{\beta}_{-1,t}$. This is precisely the power of x in the exact form obtained from $W^{\beta}_{s+1,t} \to W^{\beta}_{s,t}$. We also do not multiply by $y^{-at+a-\beta}$, as this is the same power of y that is multiplied by in the vertical reduction steps that take place after the horizontal ones.

Rephrasing this in terms of a matrix, the reduction process is given by multiplying by a matrix of the form

$$R^{t,\beta}_H(s) = M^{t,\beta}_H(s) D^{t,\beta}_H(s)^{-1},$$

where $D^{t,\beta}_H(s) = \lambda((at - a + \beta)b - as)$, and $p^{t,\beta}_i$ is the linear function of s obtained as the coefficient of x^i in $as\tilde{h}(x) - (at - a + \beta)x\tilde{h}'(x)$ and where $M^{t,\beta}_H(s)$ is the matrix

$$\left(\begin{array}{cccc|cccc} 0 & \cdots & 0 & p_0^{t,\beta} & & & \\ D_H^{t,\beta}(s) & \cdots & 0 & p_1^{t,\beta} & & & \\ \vdots & \ddots & \vdots & \vdots & & & \\ 0 & \cdots & D_H^{t,\beta}(s) & p_{b-1}^{t,\beta} & & & \\ \hline 0 & \cdots & 0 & -a & x(P_1)D_H^{t,\beta}(s) & \cdots & 0 \\ \vdots & \ddots & \vdots & \vdots & \vdots & \ddots & \vdots \\ 0 & \cdots & 0 & -a & 0 & \cdots & x(P_L)D_H^{t,\beta}(s) \end{array}\right). \tag{4.2}$$

This is the linear recurrence to which we may now apply the modified Bostan-Gaudry-Schost algorithm [7, Thm. 5.2], as compute the result more efficiently than the step-by-step approach alluded to above.

Remark 4.4 This is a slight modification of the approach used in [7, Sec. 4]. Here we make use of the commonality between the factors appearing in the vertical and horizontal stages to give a cleaner recurrence without the "correction factors" appearing in [7, Thm. 4.4].

Remark 4.5 (Regularity) For some applications (such as Chabauty-Coleman) one is only interested in invariant differentials on the Jacobian, or, equivalently, regular 1-forms on the curve, as these are the forms whose integrals define abelian integrals and provide logarithm-type maps to $\mathbf{Q}_q$. In [21, Prop. 2] it is shown that of the $2g$ basis differentials $x^i \mathrm{d}x/y^j$, as above, the g regular 1-forms are those for which

$$i < \left\lceil \frac{b}{a} \right\rceil j - 1 - \left\lfloor j \left(\left\lceil \frac{b}{a} \right\rceil - \frac{b}{a} \right) \right\rfloor.$$

It is interesting to note that the algorithm developed here computes the integral of all $2g$ basis differentials simultaneously and there seems to be no way of computing integrals of only a subset of them via this approach.

5 Runtime and Precision

We now analyse the runtime of the above algorithm, proving Theorem 1.1. We only analyse the steps which differ from that occurring in [20, Sec. 6]. All preparatory steps remain the same. It is only the main reduction steps where we now have larger matrices to compute the evaluations of exact forms.

We may apply [7, Thm. 5.2] as the reduction matrices above are $(m+n)\times(m+n)$ matrices with an $m \times n$ block which is zero and an $n \times n$ block which is diagonal. This uses

$$O\left((\mathrm{MM}(m) + \mathrm{MM}(m,n))\sqrt{K} + \left(m^2 + mn\right)\mathrm{M}(\sqrt{K})\right)$$

ring operations. We apply this for each row of horizontal reductions (with $K = O(pN)$ in the worst case) and for the vertical reductions also (where $K = O(pN)$ again).

The precision loss and gain throughout the algorithm is as in [7, Sec. 7] and [20, Sec. 5.3]. We can therefore compute (3.4) for $Q = \infty$ and L distinct Teichmüller points P via the above reduction procedure to precision N in time

$$\widetilde{O}\left((g^{\omega} + Lg^{\omega-1})\sqrt{p}nN^{5/2} + N^4 g^3 (g+L) n^2 \log p\right)$$

where $\widetilde{O}(t) := O(tf(\log t))$ for some polynomial f. The precision loss from inverting $1 - \prod_{s=n-1,\dots,0} \phi^s(M)$ is precisely

$$v_p\left(\det\left(1 - \prod_{s=n-1,\dots,0} \phi^s(M)\right)\right) = \#\operatorname{Jac} C_{\mathbf{F}_{p^n}}(\mathbf{F}_{p^n}).$$

Applying the above to $L = 2$ gives Theorem 1.1.

6 Implementation

We have implemented the algorithm outlined above in Julia using the functionality provided by the Nemo package [12] and its extension Hecke. This implementation is available online at https://github.com/alexjbest/Coleman.jl.

In this section we describe the implementation and discuss some surrounding issues. To check correctness of the algorithm and implementation, we begin with some examples:

6.1 Examples

6.1.1 A Picard Curve

The curve

$$C\colon y^3 = h(x) = x^4 + 7x^3 + 3x^2 - x$$

was suggested to me by Hashimoto-Morrison [17], it has an algebraic 9-torsion point

$$P = (1, \sqrt[3]{10}),$$

so we take $p = 41$, which splits inside the ring of integers of $\mathbf{Q}(\sqrt[3]{10})$ as a product of a prime of norm 41 and one of norm 41^2, using the latter embedding we can take $P \in C(\mathbf{Q}_{41^2})$. Explicitly if $\mathbf{Q}_{41^2} = \mathbf{Q}_{41}[\alpha]/(\alpha^2 + 38\alpha + 6)$ we have

$$P = (1 + O(41^6), (14 + 30 \cdot 41^1 + 19 \cdot 41^2 + 24 \cdot 41^3 + 35 \cdot 41^5) \cdot \alpha \\ + 11 + 20 \cdot 41^1 + 33 \cdot 41^2 + 23 \cdot 41^3 + 32 \cdot 41^4 + 34 \cdot 41^5 + O(41^6)).$$

As the x-coordinate of this point is 1, it is actually a Teichmüller point (fixed under ϕ), thus we do not have to do any tiny integrals within a disk to compute the Coleman integrals

$$\int_P^\infty \omega$$

for all ω in the basis. Using our package and running the command

```
julia> ColemanIntegrals(3, h, 3, 41, 2, P, :inf)
```

(where respectively the arguments are, degree of the cover, x-polynomial, precision requested, base prime, extension degree, and endpoints) this returns:

$$\begin{pmatrix} 0 \\ (3 + 21 \cdot 41^1 + 32 \cdot 41^2 + O(41^3)) \cdot a + (15 + 37 \cdot 41^1 + 37 \cdot 41^2 + O(41^3)) \\ (32 + 18 \cdot 41^1 + 25 \cdot 41^2 + O(41^3)) \cdot a + (37 + 10 \cdot 41^1 + 9 \cdot 41^2 + O(41^3)) \\ 0 \\ 0 \\ (27 + 10 \cdot 41^1 + 21 \cdot 41^2 + O(41^3)) \cdot a + (30 + 20 \cdot 41^1 + 7 \cdot 41^2 + O(41^3)) \end{pmatrix}.$$

This exactly reflects the fact that $[P - \infty]$ is torsion in the Jacobian and that only the invariant differentials provide group homomorphisms from the Jacobian to $\mathbf{Q}_{41^2}$ (forcing the images of torsion points to be trivial). Note that the basis of differentials is ordered by j and then i here and that the regular differentials are as described in Remark 4.5.

We can also check Galois equivariance for this example, by computing that

$$\left(\int_{P^\sigma}^\infty \omega_i\right)_i = \left(\left(\int_P^\infty \omega_i\right)^\sigma\right)_i,$$

as both evaluate to

$$\begin{pmatrix} 0 \\ (38+19\cdot 41^1+8\cdot 41^2+O(41^3))\cdot a+(24+15\cdot 41^1+32\cdot 41^2+O(41^3)) \\ (9+22\cdot 41^1+15\cdot 41^2+O(41^3))\cdot a+(10+35\cdot 41^1+25\cdot 41^2+O(41^3)) \\ 0 \\ 0 \\ (14+30\cdot 41^1+19\cdot 41^2+O(41^3))\cdot a+(29+25\cdot 41^1+19\cdot 41^2+O(41^3)) \end{pmatrix}.$$

6.2 An Elliptic Curve over a Quartic Field

The elliptic curve with LMFDB label 4.4.725.1-16.1-a1 is defined over the quartic number field $K = \mathbf{Q}(\alpha)$ where α is a root of $x^4 - x^3 - 3x^2 + x + 1$. This curve can be given by the model

$$y^2 = x^3 + (2\alpha^3 + 6\alpha^2 - 9\alpha - 4)x^2 + (32\alpha^2 - 8)x + (16\alpha^3 + 48\alpha^2 - 16\alpha - 16).$$

On this model there is a K-rational 17-torsion point with coordinates

$$P = (0, 4\alpha^2).$$

We work with $p = 43$, which remains inert in K, and fix an embedding

$$K \hookrightarrow \mathbf{Q}_{43^4}.$$

We can then compute the pair of Coleman integrals

$$\int_P^\infty \frac{\mathrm{d}x}{y}, \int_P^\infty x\frac{\mathrm{d}x}{y}$$

as before, obtaining the values 0 and

$$\begin{aligned}&(14+31\cdot 43^1+29\cdot 43^2+32\cdot 43^3+O(43^4))\cdot a^3\\ &+(41+5\cdot 43^1+15\cdot 43^2+12\cdot 43^3+O(43^4))\cdot a^2\\ &+(11+17\cdot 43^1+30\cdot 43^2+11\cdot 43^3+O(43^4))\cdot a\\ &+(26+12\cdot 43^1+28\cdot 43^2+28\cdot 43^3+O(43^4)),\end{aligned}$$

respectively.

6.3 Implementation Details

Minzlaff has made available an open-source implementation of the algorithm in [20]. This includes an implementation of the algorithm of Bostan-Gaudry-Schost and Harvey in the Magma [10] programming language, and computes zeta functions of superelliptic curves over $\mathbf{F}_q$ in $\sqrt{q}$ time. This implementation is included in recent versions of Magma (as `ZetaFunction`) and is separately available online at https://github.com/mminzlaff/superelliptic. Our implementation is built on top of a direct translation of Minzlaff's code into Nemo/Julia.

Remark 6.1 There are two (related) ways to set up a Coleman integration algorithm: integrating to Teichmüller points, or adding the tiny integrals to Frobenius. When using the former, checking that additivity in endpoints holds does not test the implementation for bugs in any serious way as when broken down, the "paths" integrated along form a tree. However with the latter approach, checking additivity in endpoints of the form

$$\int_{P}^{\infty} \omega + \int_{P'}^{P} \omega = \int_{P'}^{\infty} \omega$$

for P, P' in the same residue disk, checks that the local and global integrals are consistent, confirming that there are "non-degenerate triangles" here.

6.3.1 A Comparison of Computer Algebra Systems

In this section, we take the opportunity to compare and contrast programming languages available for working with computational number theory and arithmetic geometry. We discuss some similarities and differences between these computer algebra systems and advantages and disadvantages in the author's experience between the way they behave.

Magma and Julia are similar in several syntactic aspects which made translation from the former to the latter easier. Both operate in a functional way, i.e. `factor(M)` rather than `M.factor()`. This syntax is more similar to the syntax used in mathematics. However it presents difficulties for introspection, that is, it is not always easy to find a function performing a particular task, given only a description of that task or its common name. In the Python-based SageMath, given an object `M`, the user can interactively type `M.` and press `<TAB>` to see a list of all methods associated to the object. It is a common paradigm for a user to search this list to find functions. In languages such as Julia and Magma where the dispatch system makes such functionality difficult, it is important that good searchable documentation exists to help users find the functions they need. For example, a function with a common name like `Normalise` when called as `Normalise;` will show a list of several possibilities of types of objects a Normalise function exists for. The user must then search this list to find the description and signature for the one they wish to call. The Sage user would instead take their object X and type X.norm<TAB> to see if a normalise function exists.

Nemo is a new package in comparison to SageMath and Magma, and so some common general-purpose functionality that would be expected from a general computer algebra system is yet to be implemented. At present this can slow down development of mathematical code using Nemo. However Nemo is built on top of the C libraries Flint, Arb, and Antic, which have been under development for far longer and contain significant functionality. Missing features often simply need to be wrapped from these libraries. Wrapping involves adding only 3–5 lines of code on average as Nemo is tightly coupled to these underlying libraries. It appears that interfacing with external C libraries is simpler when using Julia than in SageMath and Magma.

Magma and SageMath are a lot more stable at present due to the maturity of these systems. This makes continuous integration and testing of code based on Nemo more vital, to prevent code from diverging when the Nemo core is modified in an incompatible way.

The open-source development model and Julia's built-in package manager also mitigates the immaturity problem somewhat, as it is easy to create a public fork of Nemo or to create a package based on it. For instance, at the time of writing there are at least 30 distinct Julia-based repositories on GitHub that use Nemo in some way, and many hundreds of independent Julia packages, some of which also containing functionality relevant to arithmetic geometers. This iterative and modular way of sharing research code is ingrained in the design of Julia and has also been proposed as a more lightweight development model for additional packages on top of the SageMath core https://wiki.sagemath.org/CodeSharingWorkflow. The advantages over more centralised development are a lower barrier for research code to be made available. It is trivial for anyone to host code in a version-controlled repository and create the requisite Julia package files, so that the code may then be installed using the built-in package system. With a more modular system, the notions of ownership and responsibility for individual packages are more clear, users can report bugs and feature requests directly to the developers of smaller modules, rather than all traffic going to a centralised bug tracker, or worse, a closed email list with no public tracking of bugs at all.

For a working mathematician, writing high-level code that is as close as possible to mathematical language is preferable, for speed of development, sharing of research code and identification of bugs. In general, the overhead of using a high-level language can render many potentially practical algorithms too slow for serious use on large scale. For instance, algorithms such as the accumulating remainder tree of [11, 15] can lose their practicality when a large call overhead (such as that in Python) is introduced at each recursive step. This necessitates the use of low-level languages such as C to implement key core functionality, like fundamental arithmetic operations in different rings, operations with polynomials, and linear algebra. When converting from high-level code to low-level, in general more code must be added, to explicitly construct and destruct objects, manage memory, iterate over lists, and handle files. This can often obscure the mathematical process underlying a given algorithm and obstruct understanding and later generalisation.

In SageMath, Cython is used to mitigate such overhead problems somewhat, and to wrap low-level C libraries. Its Python-like syntax makes conversion easier and makes comprehension of code easier, and it does reduce the overhead of Python/ Sage code.

Using domain-specific Julia packages to write high-level code, which is compiled at runtime to a fast lower-level implementation can provide a useful balance for a mathematical user. While there are applications where low-level control is essential: for instance in [9, Sec. 3.2] working at the machine level is necessary to obtain results on such a large scale in as small amount of time (and also therefore cost) as possible. Using a high-level language that is compiled (at runtime or before) into a lower-level language can strike a good balance between speed of the computation and time taken to write the code.

The distinction between compiling at runtime vs. before the software is run is most apparent when experimenting with new code. Anything that can be done at runtime can of course be done before, but the user has more flexibility when they do not have to exit the CAS and recompile to take advantage of compiled code.

6.4 Timings

We provide here some timings of both the Magma implementation of Minzlaff and Julia/Nemo implementation of the underlying zeta function algorithm, which is the only part common to both implementations, and this uses the same algorithm due to Minzlaff. Hence the comparison below is really a test of how the underlying systems handle the operations used by this algorithm (of course further optimizations to both implementations may well be possible). We do not time the Coleman integration code here as the linear recurrence method that underlies the zeta function algorithm is the main component of the runtime for computing Coleman integrals also. Tables 1 and 2 contrast the times taken to compute zeta functions of superelliptic curves of several different types over several medium cardinality fields $\mathbf{F}_q$, between the Magma and Julia/Nemo implementations of the algorithms discussed in this paper.

The dash indicates a parameter range where neither implementation applies due to the standing assumption on p in (2.1).

Table 1 Timings for the Magma implementation

q \ (a,b)	(2,5)	(2,7)	(3,7)	(4,7)
257	0.180	0.647	4.357	10.140
521	0.253	0.963	4.373	10.260
1031	0.290	1.017	6.860	16.543
2053	0.413	1.673	7.080	16.680
4099	0.487	0.960	7.473	21.530
8209	0.750	1.440	8.407	21.767
257^2	0.700	2.743	21.643	-
521^2	1.167	4.060	30.500	78.850

Table 2 Timings for the Nemo implementation

q \ (a,b)	(2,5)	(2,7)	(3,7)	(4,7)
257	0.197	0.651	4.695	15.267
521	0.366	1.27	4.779	17.892
1031	0.415	1.348	13.287	43.691
2053	0.973	3.727	14.135	44.516
4099	1.133	2.332	30.306	114.239
8209	3.261	6.486	36.414	114.61
257^2	0.339	1.487	23.569	-
521^2	0.803	2.992	44.461	111.966

References

1. Balakrishnan, Jennifer S., Robert W. Bradshaw, and Kiran S. Kedlaya. *Explicit Coleman Integration for Hyperelliptic Curves*. In ANTS-IX 2010, LNCS 6197, pp. 16–31, 2010.
2. Balakrishnan, Jennifer S. *Coleman integration for even-degree models of hyperelliptic curves*. LMS Journal of Computation and Mathematics 18.1 (2015): 258–265.
3. Balakrishnan, Jennifer S., Netan Dogra, J. Steffen Müller, Jan Tuitman, and Jan Vonk. *Explicit Chabauty–Kim for the Split Cartan Modular Curve of Level 13*. Annals of Mathematics 189, no. 3 (2019): 885–944. https://www.jstor.org/stable/10.4007/annals.2019.189.3.6.
4. Balakrishnan Jennifer S., and Jan Tuitman. *Explicit Coleman integration for curves*. arXiv preprint arXiv:1710.01673. 2020 Jan 13.
5. Besser, Amnon. *Heidelberg lectures on Coleman integration*. In *The Arithmetic of Fundamental Groups*, pp. 3–52. Springer, Berlin, Heidelberg, 2012.
6. Besser, Amnon, and Rob De Jeu. *$Li^{(p)}$-Service? An Algorithm for Computing p-Adic Polylogarithms*. Mathematics of Computation 77, no. 262 (2008): 1105–34.
7. Best, Alex J. *Explicit Coleman integration in larger characteristic*. Proceedings of the Thirteenth Algorithmic Number Theory Symposium, The Open Book Series, 2(1), 85–102, 2019.
8. Bezanson, Jeff, Alan Edelman, Stefan Karpinski, and Viral B. Shah. *Julia: A Fresh Approach to Numerical Computing*. (2017) SIAM Review, 59: 65–98. https://doi.org/10.1137/141000671. https://julialang.org/research/julia-fresh-approach-BEKS.pdf.
9. Booker, Andrew R., Jeroen Sijsling, Andrew V. Sutherland, John Voight, and Dan Yasaki. *A Database of Genus 2 Curves over the Rational Numbers*. LMS Journal of Computation and Mathematics 19, no. A (2016): 235–54. https://doi.org/10.1112/S146115701600019X.
10. Bosma, Wieb, John Cannon, and Catherine Playoust. *The Magma algebra system. I. The user language*, J. Symbolic Comput., 24 (1997), 235–265.

11. Costa, Edgar, Robert Gerbicz, and David Harvey. *A Search for Wilson Primes*. Mathematics of Computation 83, no. 290 (2014): 3071–91. https://doi.org/10.1090/S0025-5718-2014-02800-7.
12. Fieker, Claus, William Hart, Tommy Hofmann, and Fredrik Johansson. *Nemo/Hecke: Computer Algebra and Number Theory Packages for the Julia Programming Language*. In: *Proceedings of ISSAC '17*, pages 157–164, New York, NY, USA, 2017. ACM, http://doi.acm.org/10.1145/3087604.3087611.
13. Gaudry, Pierrick, and Nicolas Gürel. *An extension of Kedlaya's point-counting algorithm to superelliptic curves*. Advances in Cryptology - ASIACRYPT 2001, Springer, Berlin, Heidelberg, 2001.
14. Gonçalves, Cécile. *A point counting algorithm for cyclic covers of the projective line*. Contemporary mathematics 637 (2015): 145.
15. Harvey, David. *Counting points on hyperelliptic curves in average polynomial time*. Annals of Mathematics 179, no. 2 (2014): 783–803.
16. Harvey, David. *Kedlaya's Algorithm in Larger Characteristic*. IMRN: International Mathematics Research Notices 2007 (2007).
17. Hashimoto, Sachi, and Travis Morrison. *Chabauty-Coleman Computations on Rank 1 Picard Curves*, preprint.
18. Kedlaya, Kiran S. *Counting points on hyperelliptic curves using Monsky-Washnitzer cohomology*, J. Ramanujan Math. Soc. 16 (2001), no. 4, 323–338; errata, ibid. 18 (2003), 417–418.
19. McCallum, William, and Bjorn Poonen. *The Method of Chabauty and Coleman*. Explicit Methods in Number Theory 36 (2012): 99–117.
20. Minzlaff, Moritz. *Computing zeta functions of superelliptic curves in larger characteristic*. Mathematics in Computer Science 3.2 (2010): 209–224.
21. Towse, Christopher. *Weierstrass Points on Cyclic Covers of the Projective Line*. Transactions of the American Mathematical Society 348, no. 8 (1996): 3355–3378.
22. Tuitman, Jan. *Counting points on curves using a map to* $\mathbf{P}^1$. Mathematics of Computation 85.298 (2016): 961–981.
23. Tuitman, Jan. *Counting points on curves using a map to* $\mathbf{P}^1$, *II*. Finite Fields and Their Applications 45 (2017): 301–322.

Computing Classical Modular Forms

Alex J. Best, Jonathan Bober, Andrew R. Booker, Edgar Costa, John E. Cremona, Maarten Derickx, Min Lee, David Lowry-Duda, David Roe, Andrew V. Sutherland, and John Voight

Abstract We discuss practical and some theoretical aspects of computing a database of classical modular forms in the L-functions and modular forms database (LMFDB).

A. J. Best
Department of Mathematics and Statistics, Boston University, Boston, MA, USA

J. Bober
School of Mathematics, University of Bristol, Bristol, UK

The Heilbronn Institute for Mathematical Research, Bristol, UK
e-mail: j.bober@bristol.ac.uk

A. R. Booker · M. Lee
School of Mathematics, University of Bristol, Bristol, UK
e-mail: andrew.booker@bristol.ac.uk; min.lee@bristol.ac.uk

E. Costa · M. Derickx · D. Roe · A. V. Sutherland (✉)
Department of Mathematics, Massachusetts Institute of Technology, Cambridge, MA, USA
e-mail: edgarc@mit.edu; maarten@mderickx.nl; roed@mit.edu; drew@math.mit.edu

J. E. Cremona
Warwick Mathematics Institute, Zeeman Building, University of Warwick, Coventry, UK
e-mail: j.e.cremona@warwick.ac.uk

D. Lowry-Duda
Institute of Computational and Experimental Research in Mathematics, Providence, RI, USA
e-mail: david@lowryduda.com

J. Voight
Department of Mathematics, Dartmouth College, Hanover, NH, USA
e-mail: jvoight@gmail.com

J. S. Balakrishnan et al. (eds.), *Arithmetic Geometry, Number Theory, and Computation*, Simons Symposia, https://doi.org/10.1007/978-3-030-80914-0_4

1 Introduction

1.1 Motivation

Databases of classical modular forms have been used for a variety of mathematical purposes and have almost a 50 year history (see Sect. 2). In this article, we report on a recent effort in this direction in the L-functions and modular forms database (LMFDB [60], https://lmfdb.org); for more on the LMFDB, see the overview by Cremona [32].

1.2 Organization

The paper is organized as follows. In Sect. 2, we begin with a short history, and we follow this in Sect. 3 with a preliminary discussion of Dirichlet characters. Next, in Sect. 4 we make more explicit what we mean by computing (spaces of) modular forms, and then in Sect. 5 we give a short overview of the many existing algorithmic approaches to computing modular forms. We pause in Sect. 6 to prove two technical results. In Sect. 7, we sample the available implementations and make some comparisons. Next, in Sect. 8 we discuss some computational, theoretical, and practical issues that arose in our efforts and in Sect. 9 we explain how we (rigorously) computed the L-functions attached to modular newforms. Turning to our main effort, in Sect. 10 we provide an overview of the computations we performed, make some remarks on the data obtained, and explain some of the features of our database. Finally, in Sects. 11 and 12 we treat twists and issues specific to modular forms of weight 1.

As is clear from this organization, we consider the algorithmic problem of computing modular forms from a variety of perspectives, so this paper need not be read linearly. For the convenience of readers, we draw attention here to a number of highlights:

- In Sect. 2, we survey the rather interesting history of computing databases of modular forms.
- In Sect. 3.2, we exhibit a labeling scheme for Dirichlet characters, due to Conrey.
- In Theorem 4.3.4, we record formulas for the new, old, and total dimensions of spaces of Eisenstein series of arbitrary integer weight $k \geq 2$, level, and character, obtained from work of Cohen–Oesterlé and Buzzard. (Such formulas are not available for weight $k = 1$.)
- In Corollary 6.1.5, we compute an Eichler–Selberg trace formula restricted to the space of newforms; this was used by Belabas–Cohen [4] in their implementation in Pari/GP.
- In Tables 5 and 6, we compare the implementations of Magma and Pari/GP; in Table 7 we note some computationally challenging newspaces.

- In Sect. 8.7, we show that by writing Hecke eigenvalues in terms of an LLL-reduced basis of the Hecke order, we can drastically reduce their total size.
- In Sect. 9.4, we certify analytic ranks of L-functions of modular forms and remark on the ranks occurring in our dataset.
- In Sect. 9.5, we numerically verify a generalization of Chowla's conjecture for central values of non-self-dual modular form L-functions.
- In Sect. 10.2, we present statistics on our data, and in Sect. 10.4 we note some interesting and extreme behavior that we observed in our dataset.
- In Theorems 11.2.4 and 11.2.8, we exhibit simple and effectively computable criteria for rigorously certifying that a modular form has an inner twist.
- In Sect. 12.5, we highlight some interesting and extreme behavior found among weight 1 modular forms in our database.

1.3 Acknowledgments

The authors would like to thank Eran Assaf, Karim Belabas, Henri Cohen, Alan Lauder, David Loeffler, David Platt, Mark Watkins, and the anonymous referees for their comments. This research was undertaken as part of the *Simons Collaboration on Arithmetic Geometry, Number Theory, and Computation*, with the support of Simons Collaboration Grants: 546235, to Brendan Hassett, supporting Lowry-Duda; 550023, to Jennifer Balakrishnan, supporting Best; 550029, to John Voight; and 550033, to Bjorn Poonen and Andrew V. Sutherland, supporting Costa, Derickx, and Roe. Additional support was provided by a Programme Grant from the UK Engineering and Physical Sciences Research Council (EPSRC) *LMF: L-functions and modular forms*, EPSRC reference EP/K034383/1.

2 History

In this section, we survey the history of computing tables of modular forms; for a broader but still computationally-oriented history, see Kilford [54, Section 7.1].

- Perhaps the first systematic tabulation of modular forms was performed by Wada [97, 98]. As early as 1971, he used the Eichler–Selberg trace formula to compute a factorization of the characteristic polynomial of the Hecke operator T_p on $S_2(\Gamma_0(q), \chi)$ for $q \equiv 1 \pmod 4$ prime where χ was either trivial or the quadratic character of conductor q. The total computation time was reported to be about 300 hours on a TOSBAC-3000.
- The next major step was made in the famous *Antwerp IV* tables [71] (published in 1975), motivated by the study of modularity of elliptic curves. Vélu and Stephens–Vélu computed all newforms in $S_2(\Gamma_0(N))$ with $N \leq 200$ using modular symbols [71, Table 3] and these forms were matched with isogeny

classes of elliptic curves over $\mathbb{Q}$ found by Swinnerton-Dyer. Tingley [95] computed the complete splitting into Hecke eigenspaces of $S_2(\Gamma_0(N))$ for $N \leq 300$, extending an earlier table due to Atkin. In particular he found the dimensions of the Atkin–Lehner eigenspaces, and computed the actual eigenvalues as floating point numbers, numerically matching conjugate newforms. By integrating differentials, he also computed elliptic curves from the newforms with integer eigenvalues. In some cases, this computation revealed the existence of elliptic curves not previously found by search. (According to Birch, this was the case for the elliptic curve with Antwerp label 78A and Cremona label 78a1; the curves in its isogeny class have rather large coefficients.)

- Extending the Antwerp IV tables, Cremona [30] (first edition published in 1992) computed a database of newforms in $S_2(\Gamma_0(N))$ with rational coefficients for $N \leq 1000$, providing also a wealth of data on the corresponding (modular) elliptic curves. In the second edition and in later computations, this data was considerably extended. A more recent report [31] was made on the elliptic curve tables to conductor 130,000. By 2016 this database had reached conductor 400,000, and in July 2019 Cremona and Sutherland extended it to conductor 500,000. In this range there are 2,164,260 rational newforms, and the same number of isogeny classes of elliptic curves.
- Miyake [67] published some numerical tables of modular forms as appendices in his book on modular forms; these were computed using the trace formula. These tables included dimensions of $S_k(\Gamma_0(N))$ for $k \geq 2$ even and small values of N, eigenvalues and characteristic polynomials of Hecke operators on $S_2(\Gamma_0(N))$ for small prime values of N, and Fourier coefficients of a primitive form in $S_2(\Gamma_0(N), \chi_N)$ for $N = 29, 37$.
- In the 1990s, Cohen, Skoruppa, and Zagier compiled tables of eigenforms in weights 2 through 12, levels up to 1000 in weight 2 and with a smaller range in higher weight; also some tables of eigenforms with non-trivial character. Their method followed a paper by Skoruppa and Zagier on the trace formula [87], but these tables were not published.
- In the early 2000s, Stein created an online modular forms database [88], computed primarily using a modular symbols package [89] he implemented in Magma [11] starting in the late 1990s. The data was computed using a rack of six custom-built machines and a Sun V480; it was stored in a PostgreSQL database (more than 10 GB), and a (Python-based) web interface to the data was provided. These tables included dimensions, characteristic polynomials, and q-expansions in a variety of weights and levels.
- Using this Magma implementation, Meyer [66] computed a table of newforms for $\Gamma_0(N)$ with rational coefficients: in weight $k = 2$ he went to $N \leq 3000$ and for $k = 4$ to $N \leq 2000$.
- Prior to our work, the LMFDB had a database of classical modular forms computed by Ehlen and Strömberg [42], which used the SageMath [79] implementation of modular symbols. This dataset included partial information on $S_k(\Gamma_0(N))$ for (k, N) in the ranges $[2, 12] \times [1, 100]$ and $[2, 40] \times [1, 25]$, and on $S_k(\Gamma_1(N))$ in the ranges $[2, 10] \times [1, 50]$ and $[2, 20] \times [1, 16]$.

The scope of our modular forms database includes all of the ranges mentioned above (and more), with the exception of Cremona's tables of elliptic curves; see Sect. 10.1 for details.

3 Characters

Our database of modular forms is organized into subspaces identified by a level $N \in \mathbb{Z}_{\geq 1}$, a weight $k \in \mathbb{Z}_{\geq 1}$ and a character $\chi \colon \mathbb{Z} \to \mathbb{C}$ taking values in the cyclotomic field $\mathbb{Q}(\zeta_N)$. In order to identify these subspaces and the modular forms they contain, we adopt a standard convention for identifying Dirichlet characters that is well suited to computation, the Conrey labels recalled in Sect. 3.2 below. We also introduce a convention for identifying Galois orbits of Dirichlet characters that will be used to identify the newform subspaces and newform orbits defined in Sect. 4.

3.1 Definitions

For $N \in \mathbb{Z}_{\geq 1}$, a Dirichlet character of modulus N is a pair (χ, N), where $\chi \colon \mathbb{Z} \to \mathbb{C}$ is a periodic function modulo N that is the extension of a group homomorphism $(\mathbb{Z}/N\mathbb{Z})^\times \to \mathbb{C}^\times$ by zero (defining $\chi(n) = 0$ whenever $\gcd(n, N) \neq 1$)—in particular, χ is totally multiplicative. The degree of a Dirichlet character χ is the degree of the cyclotomic subfield $\mathbb{Q}(\chi) \subseteq \mathbb{C}$ generated by the values of χ.

Given two Dirichlet characters χ, χ' of moduli N, N', we define their product $\chi\chi'$ to be the Dirichlet character of modulus $\operatorname{lcm}(N, N')$ defined by $(\chi\chi')(n) = \chi(n)\chi'(n)$. Under this definition, the set of Dirichlet characters of a fixed modulus N has the structure of a finite abelian group, with identity the principal (or trivial) character with $\chi(n) = 1$ if $\gcd(n, N) = 1$ and $\chi(n) = 0$ otherwise. The order $\operatorname{ord}(\chi)$ of a Dirichlet character χ is its order in this group, i.e., the smallest $m \in \mathbb{Z}_{\geq 1}$ such that χ^m is the principal character.

Let χ be a Dirichlet character of modulus N. Given a multiple N' of N, we may induce χ to a Dirichlet character χ' of modulus N' by $\chi'(n) := \chi(n \bmod N)$ whenever $\gcd(n, N') = 1$ and $\chi'(n) = 0$ otherwise. Consequently, there is a well-defined *minimal* modulus $M := \operatorname{cond}(\chi) \mid N$, called the conductor of χ, such that χ is induced from a Dirichlet character of modulus M. If $\operatorname{cond}(\chi) = N$, i.e., the conductor of χ is equal to its modulus, then we say that χ is a primitive character.

It is sometimes convenient to think about Dirichlet characters *without* a modulus, remembering only a periodic, totally multiplicative arithmetic function χ. In our context, Dirichlet characters arise from modular forms with level structure, so there should be little chance for confusion.

3.2 Conrey Labels

We briefly describe a scheme, due to Brian Conrey, for labeling and computing with Dirichlet characters. Our labeling scheme can be thought of as a choice of an explicit isomorphism between two finite abelian groups: the multiplicative group $(\mathbb{Z}/N\mathbb{Z})^\times$ and the group of Dirichlet characters modulo N. In particular, our Dirichlet characters by definition take values in the complex numbers, so implicit in our choice of labels is a choice of embedding $\mathbb{Q}^{\mathrm{ab}} \hookrightarrow \mathbb{C}$.

For each $N \in \mathbb{Z}_{\geq 1}$, we will construct a function

$$\chi_N \colon (\mathbb{Z}/N\mathbb{Z})^\times \times (\mathbb{Z}/N\mathbb{Z})^\times \to \mathbb{C}^\times \tag{3.2.1}$$

satisfying the following three properties:

- χ_N is multiplicative in each variable (separately);
- χ_N is symmetric (i.e., $\chi_N(m,n) = \chi_N(n,m)$ for all $m, n \in (\mathbb{Z}/N\mathbb{Z})^\times$); and
- χ_N is nondegenerate (i.e., if $\chi_N(m,n) = 1$ for all $m \in (\mathbb{Z}/N\mathbb{Z})^\times$, then $n \equiv 1 \pmod N$).

Moreover, χ_N will be multiplicative in N, and hence it is sufficient to define it for prime powers p^e and then extend $\chi_N(m,n)$ to general N by multiplicativity:

$$\chi_N(m,n) = \prod_{p^e \| N} \chi_{p^e}(m,n).$$

We use the notation $p^e \| N$ to mean that $p^e \mid N$ but $p^{e+1} \nmid N$. On the left side, m and n denote elements of $(\mathbb{Z}/N\mathbb{Z})^\times$, while on the right they denote the images of these in $(\mathbb{Z}/p^e\mathbb{Z})^\times$. We then extend χ_N to a multiplicative, periodic function on $\mathbb{Z} \times \mathbb{Z}$ by setting $\chi_N(m,n) = 0$ whenever $\gcd(mn, N) > 1$.

Under these conditions, fixing one input to χ_N defines a Dirichlet character modulo N and conversely every Dirichlet character arises in this way. Thus each Dirichlet character is given a unique name of the form $\chi_N(m,\cdot)$ for $m \in (\mathbb{Z}/N\mathbb{Z})^\times$. In particular, by symmetry, we see that $\chi_N(1,\cdot)$ is the trivial character modulo N, and $\chi_N(m,\cdot)$ is a quadratic character when $m \not\equiv 1 \pmod N$ but $m^2 \equiv 1 \pmod N$. (More generally, the order of the character $\chi_N(m,\cdot)$ is the multiplicative order of m modulo N.)

We now describe the construction of χ_N.

Odd Prime Powers Let p be an odd prime. Let g be the smallest positive integer that is a primitive root mod p^e for all $e \geq 1$. (This is almost always the same as the smallest primitive root mod p, but may not be; the only odd prime less than a million for which these differ is 40487.) For $m \in (\mathbb{Z}/p^e\mathbb{Z})^\times$, we define $\log_g(m) \in \mathbb{Z}/\phi(p^e)\mathbb{Z}$ by the condition

$$m \equiv g^{\log_g(m)} \pmod{p^e}, \tag{3.2.2}$$

so that $\log_g \colon (\mathbb{Z}/p^e\mathbb{Z})^\times \to \mathbb{Z}/\phi(p^e)\mathbb{Z}$ is an isomorphism of groups.

For $m, n \in (\mathbb{Z}/p^e\mathbb{Z})^\times$, we then define

$$\chi_{p^e}(m,n) := \exp\left(2\pi i \frac{\log_g(m)\log_g(n)}{\varphi(p^e)}\right). \tag{3.2.3}$$

Then χ_{p^e} clearly satisfies the three required conditions (multiplicative, symmetric, and nondegenerate).

Powers of 2 We define χ_2 to be the trivial map (so $\chi_2(1,1) = 1$), and define

$$\chi_4(m,n) = (-1)^{(m-1)(n-1)/2} \tag{3.2.4}$$

for $m, n \in (\mathbb{Z}/4\mathbb{Z})^\times$. Let $e \geq 3$. The group $(\mathbb{Z}/2^e\mathbb{Z})^\times$ is generated by 5 and -1. For $m \in (\mathbb{Z}/2^e\mathbb{Z})^\times$, we define $\epsilon(m) \in \{0, 1\}$ and $\log_5(m) \in \mathbb{Z}/2^{e-2}\mathbb{Z}$ by

$$m \equiv (-1)^{\epsilon(m)} 5^{\log_5(m)} \pmod{2^e} \tag{3.2.5}$$

so that now $(\epsilon, \log_5) \colon (\mathbb{Z}/2^e\mathbb{Z})^\times \to \mathbb{Z}/2\mathbb{Z} \times \mathbb{Z}/2^{e-2}\mathbb{Z}$ is an isomorphism. For $m, n \in (\mathbb{Z}/2^e\mathbb{Z})^\times$, we then define

$$\chi_{2^e}(m,n) := \exp\left(2\pi i \frac{\epsilon(m)\epsilon(n)}{2} + 2\pi i \frac{\log_5(m)\log_5(n)}{2^{e-2}}\right). \tag{3.2.6}$$

As for the case of odd prime power modulus, this function satisfies the required properties.

In this article, as in the LMFDB, the Conrey label of the character $\chi_N(m, \cdot)$ has the form N.m. For example, the Conrey label of $\chi_7(6, \cdot)$, the unique quadratic character of modulus 7, is 7.6.

3.3 *Orbit Labels*

There is an action of the absolute Galois group $\mathrm{Gal}_\mathbb{Q} := \mathrm{Gal}(\mathbb{Q}^{\mathrm{al}} \mid \mathbb{Q})$ of $\mathbb{Q}$ on the set of Dirichlet characters of modulus N, defined by

$$(\sigma\chi)(n) := \sigma(\chi(n)) \tag{3.3.1}$$

for $\sigma \in \mathrm{Gal}_\mathbb{Q}$ and $n \in \mathbb{Z}$.

It is natural to organize characters by Galois orbits, and indeed we will also want to work with modular forms defined without an embedding into the complex numbers, specified up to the action of Galois (see Sect. 4.2). So we also assign an orbit label to each Galois orbit of Dirichlet characters, as follows. To choose this label we lexicographically order the sequences

$$\operatorname{ord}(\chi), \operatorname{Tr}\chi(1), \operatorname{Tr}\chi(2), \operatorname{Tr}\chi(3), \operatorname{Tr}\chi(4), \ldots$$

of integers, where $\operatorname{Tr}\colon \mathbb{Q}(\chi) \to \mathbb{Q}$ is the absolute trace; we then assign the label written in base 26 using the letters of the alphabet, so

a, b, . . . , z, ba, bb, . . . , bz, ca, . . . , zz, baa, ...

For every modulus $N \geq 1$, the Dirichlet character orbit `N.a` is the trivial character, since it is the unique character with (smallest) order 1.

Example 3.3.2 The table below lists the Conrey labels of the eight Dirichlet characters of modulus 20, their values on the generators 11 and 17 of $(\mathbb{Z}/20\mathbb{Z})^\times$, their orders, the absolute traces of their values the first five positive integers coprime to 20 (note $\operatorname{Tr}(\chi(n)) = 0$ if $\gcd(20, n) \neq 1$), and the labels of the six Galois orbits in which they lie.

Conrey label	$\chi(11)$	$\chi(17)$	$\operatorname{ord}(\chi)$	$\operatorname{Tr}(\chi(1))$	$\operatorname{Tr}(\chi(3))$	$\operatorname{Tr}(\chi(7))$	$\operatorname{Tr}(\chi(11))$	$\operatorname{Tr}(\chi(13))$	Orbit label
20.1	1	1	1	1	1	1	1	1	20.a
20.11	−1	1	2	1	−1	−1	1	−1	20.b
20.9	1	−1	2	1	−1	−1	1	1	20.c
20.19	−1	−1	2	1	1	1	1	−1	20.d
20.3	−1	$-i$	4	2	0	0	−2	−2	20.e
20.7	−1	i	4	2	0	0	−2	−2	20.e
20.13	1	$-i$	4	2	0	0	−2	2	20.f
20.17	1	i	4	2	0	0	−2	2	20.f

Remark 3.3.3 The field $\mathbb{Q}(\chi)$ is contained in the coefficient field $\mathbb{Q}(f)$ of a newform f with character χ. When the dimension of $\mathbb{Q}(f)$ is large it may be difficult to compute a complex embedding $\mathbb{Q}(f) \to \mathbb{C}$, and we often need to distinguish embeddings that are compatible with the Hecke action, which means we must know the image of $\mathbb{Q}(\chi)$ under embeddings of $\mathbb{Q}(f)$. Matching up roots of unity of large order can be surprisingly nontrivial! So when computing the coefficient field (as an abstract field, not necessarily embedded in the complex numbers), we compute the values of χ on generators for $(\mathbb{Z}/N\mathbb{Z})^\times$ as elements of the coefficient field. In this way, we may organize embeddings of the coefficient field according to a desired embedding of $\mathbb{Q}(\chi)$.

We could instead keep track of the coefficient field as an extension of $\mathbb{Q}(\chi)$, but that approach creates headaches when comparing results across implementations, it shifts the problem to a different place when working with forms in a Galois orbit, and it does not allow us to represent eigenvalues in terms of a nice LLL-reduced basis (see Sect. 8.7).

4 Computing Modular Forms

In this section, we make precise what it means to *compute modular forms*. For background, we refer to the wealth of references available, for example Cohen–Strömberg [25], Diamond–Shurman [38], Serre [82, Chapter VII], and Stein [89].

4.1 Setup

The group $\mathrm{SL}_2(\mathbb{R})$ acts (on the left) by linear fractional transformations on the upper half-plane $\mathcal{H} := \{z \in \mathbb{C} : \operatorname{Im} z > 0\}$. For $N \in \mathbb{Z}_{\geq 1}$, define the congruence subgroups

$$\begin{aligned}\Gamma_0(N) &:= \left\{\gamma \in \mathrm{SL}_2(\mathbb{Z}) : \gamma \equiv \begin{pmatrix} * & * \\ 0 & * \end{pmatrix} \pmod{N}\right\}, \\ \Gamma_1(N) &:= \left\{\gamma \in \mathrm{SL}_2(\mathbb{Z}) : \gamma \equiv \begin{pmatrix} 1 & * \\ 0 & 1 \end{pmatrix} \pmod{N}\right\}.\end{aligned} \tag{4.1.1}$$

For $\Gamma \leq \mathrm{SL}_2(\mathbb{Z})$ a congruence subgroup, the quotient $Y(\Gamma) := \Gamma\backslash\mathcal{H}$ can be compactified to $X(\Gamma)$ by adding finitely many cusps, identified with the orbits of Γ on $\mathbb{P}^1(\mathbb{Q})$. As usual, we write $X_0(N), X_1(N)$ for the quotients $X(\Gamma)$ with $\Gamma = \Gamma_0(N), \Gamma_1(N)$.

For $k, N \in \mathbb{Z}_{\geq 1}$, a modular form of weight k and level N is a holomorphic function $f : \mathcal{H} \to \mathbb{C}$ that is bounded in vertical strips and satisfies

$$f\left(\frac{az+b}{cz+d}\right) = (cz+d)^k f(z) \tag{4.1.2}$$

for all $\gamma = \begin{pmatrix} a & b \\ c & d \end{pmatrix} \in \Gamma_1(N)$; the $\mathbb{C}$-vector space of such forms is denoted $M_k(\Gamma_1(N))$.

Modular forms are organized by character, as follows. The space $M_k(\Gamma_1(N))$ decomposes according to the action of diamond operators as

$$M_k(\Gamma_1(N)) = \bigoplus_{\chi} M_k(\Gamma_0(N), \chi), \tag{4.1.3}$$

the sum being over all Dirichlet characters $\chi : \mathbb{Z}/N\mathbb{Z} \to \mathbb{C}$ of modulus N, where $M_k(\Gamma_0(N), \chi)$ is the subspace of modular forms with (Nebentypus) character χ consisting of those forms f satisfying

$$f\left(\frac{az+b}{cz+d}\right) = \chi(d)(cz+d)^k f(z) \tag{4.1.4}$$

for all $\gamma \in \Gamma_0(N)$. Throughout, we will abbreviate $M_k(\Gamma_0(N), \chi)$ to $M_k(N, \chi)$ and when χ is trivial, write simply $M_k(N)$.

In order to handle character values with some finesse (as explained above in Sect. 3 and below in Sect. 4.2), we work in the absolute situation (relative to $\mathbb{Q}$) and consider the entire Galois orbit $[\chi]$ of χ, and so we write

$$M_k(\Gamma_0(N), [\chi]) := \bigoplus_{\chi' \in [\chi]} M_k(\Gamma_0(N), \chi'), \tag{4.1.5}$$

so that from (4.1.3) we have

$$M_k(\Gamma_1(N)) = \bigoplus_{[\chi]} M_k(\Gamma_0(N), [\chi]),$$

where the direct sum is now over Galois orbits of characters $[\chi]$. We similarly abbreviate $M_k(\Gamma_0(N), [\chi])$ to just $M_k(N, [\chi])$.

Every such modular form f has a q-expansion (i.e., Fourier expansion at ∞)

$$f(z) = \sum_{n=0}^{\infty} a_n q^n \in \mathbb{C}[[q]] \tag{4.1.6}$$

where $q = \exp(2\pi i z)$ and $z \in \mathcal{H}$. We call $a_n \in \mathbb{C}$ the coefficients of f, and we write $\mathbb{Z}[\{a_n\}_n]$ for the coefficient ring and $\mathbb{Q}(\{a_n\}_n)$ for the coefficient field of f, the subring and subfield of $\mathbb{C}$ generated by its coefficients, respectively.

A modular form f is a cusp form if f vanishes at the cusps of $X_1(N)$. The subspace of cusp forms is denoted $S_k(\Gamma_1(N)) \subseteq M_k(\Gamma_1(N))$, and similarly $S_k(\Gamma_0(N), \chi) \subseteq M_k(\Gamma_0(N), \chi)$. In particular, a cusp form vanishes at the cusp ∞, so that the coefficient a_0 of its q-expansion is zero.

The Petersson inner product provides an orthogonal decomposition

$$M_k(\Gamma_0(N), \chi) = S_k(\Gamma_0(N), \chi) \oplus E_k(\Gamma_0(N), \chi) \tag{4.1.7}$$

where $E_k(\Gamma_0(N), \chi)$ is the space spanned by Eisenstein series, obtained in an explicit way using characters (see Sect. 4.4). Each of the spaces above can further be decomposed into old and new subspaces, and we denote the new subspace by $S_k^{\text{new}}(\Gamma_1(N))$, etc.

The above spaces can be equipped with an action of *Hecke operators* T_n indexed by $n \in \mathbb{Z}_{\geq 1}$. The operators T_n are normal and pairwise commute for $\gcd(n, N) = 1$, so there is a common normalized ($a_1 = 1$) basis for the action of the Hecke operators, called eigenforms; for such forms, $T_n f = a_n f$ for f as in (4.1.6). A new cuspidal eigenform is called an (embedded) newform. The coefficients of a newform are algebraic integers and the coefficient field is a number field. When χ is trivial, this coefficient field is totally real. When χ is trivial, we also have Atkin–Lehner involutions W_p for $p \mid N$, and the Fricke involution $W_N := \prod_{p|N} W_p$. (See Sect. 8.3 below.)

For a subring $A \subseteq \mathbb{C}$, we write $M_k(\Gamma_1(N); A) \subseteq M_k(\Gamma_1(N))$ for the A-submodule of modular forms whose q-expansions have coefficients in A, and similarly with the other decorated spaces.

From now on, we suppose we are given the input of a weight $k \in \mathbb{Z}_{\geq 1}$, a level $N \in \mathbb{Z}_{\geq 1}$, and an orbit of Dirichlet characters χ of modulus N and orbit label N.s; we encode this data of a space of modular forms in the label N.k.s.

Example 4.1.8 For $N = 280$, $k = 2$, and trivial character χ having label 280.a, the space $M_2(280) = M_2(\Gamma_0(280))$ has label 280.2.a.

Remark 4.1.9 We restrict ourselves to integral weight forms in this article. For forms of half-integral weight, the algorithms, applications, and issues that arise are quite different.

4.2 *Galois Digression*

As is usual in Galois theory, it is convenient to work both with abstract objects as well as embedded objects. To this end, we call the Aut($\mathbb{C}$)-orbit of an embedded newform f a newform orbit, and write $[f]$ for this orbit. We call a $\mathbb{Q}$-subspace of $S_k^{\text{new}}(\Gamma_0(N), [\chi]; \mathbb{Q})$ that is irreducible under the action of the Hecke operators a newform subspace.

For an eigenform f in a newform subspace, we obtain an embedded newform by a choice of embedding of its coefficient field into $\mathbb{C}$, and all such embeddings are conjugate under Aut($\mathbb{C}$). Conversely, given an embedded newform $f \in S_k^{\text{new}}(\Gamma_0(N), \chi)$, the $\mathbb{C}$-subspace of $S_k(\Gamma_0(N), [\chi])$ spanned by $\sigma(f) := \sum_n \sigma(a_n(f))q^n$ for $\sigma \in \text{Aut}(\mathbb{C})$ descends to a newform subspace $V_f \subseteq S_k(\Gamma_0(N), [\chi]; \mathbb{Q})$, visibly depending only on the Aut($\mathbb{C}$)-orbit of f. In other words, there is a bijection between newform subspaces $V \subseteq S_k^{\text{new}}(\Gamma_1(N))$ and newform orbits $[f]$ of embedded newforms f of weight k and level N.

The coefficient field K of a newform subspace, defined to be the coefficient field of any eigenform in the subspace, is well-defined as an abstract number field. The expansion (4.1.6) considered in K, is similarly well-defined.

4.3 *Dimensions*

The first thing one may ask to compute for a space of modular forms is just dimensions of the subspaces as defined above: the total dimension $\dim_{\mathbb{C}} M_k(\Gamma_0(N), [\chi])$, the dimension of the Eisenstein subspace $\dim_{\mathbb{C}} E_k(\Gamma_0(N), [\chi])$, and the dimension of the cuspidal subspace $\dim_{\mathbb{C}} S_k(\Gamma_0(N), [\chi])$, as well as the old and new subspaces of each of these. Since these subspaces are naturally vector spaces over $\mathbb{Q}$, we have

$$\dim_{\mathbb{C}} M_k(\Gamma_0(N), [\chi]) = \dim_{\mathbb{Q}} M_k(\Gamma_0(N), [\chi]; \mathbb{Q});$$

moreover, an individual space $M_k(\Gamma_0(N), \chi)$ is a vector space over $\mathbb{Q}(\chi)$ and each summand in (4.1.5) has the same dimension, so these absolute dimensions are the product of their relative dimension by the degree $d = [\mathbb{Q}(\chi) : \mathbb{Q}]$ of χ, i.e., we also have

$$\dim_{\mathbb{C}} M_k(\Gamma_0(N), [\chi]) = \dim_{\mathbb{Q}(\chi)} M_k(\Gamma_0(N), \chi; \mathbb{Q}).$$

Remark 4.3.1 To avoid errors, to compare across packages, and to store data conveniently, we found it essential to compute in the absolute setting (over $\mathbb{Q}$) rather than the relative setting (over $\mathbb{Q}(\chi)$).

For weight $k \geq 2$, these dimensions can be computed using the valence formula, the Riemann–Roch theorem, or the trace formula—they are given explicitly e.g. by Cohen–Strömberg [25, Theorem 7.4.1]. Unfortunately, no formula is known for these dimensions when $k = 1$.

Because they can be understood explicitly in terms of Dirichlet characters, there are separately given formulas for the Eisenstein dimension as well as the dimension of the new and old subspaces in all weights $k \geq 1$: see Cohen–Strömberg [25, Propositions 8.5.15 and 8.5.21] for the full dimension, with the new dimension worked out by Buzzard [17] using a formula of Cohen–Oesterlé [24, Theorem 1] as follows. Lacking a reference for these formulas, we record them here.

For $r, s, p \in \mathbb{Z}$ with p prime and $r > 0$ and $s \leq r$, define

$$\lambda(r, s, p) := \begin{cases} p^{r'} + p^{r'-1}, & \text{if } 2s \leq r = 2r'; \\ 2p^{r'}, & \text{if } 2s \leq r = 2r' + 1; \\ 2p^{r-s}, & \text{if } 2s > r; \end{cases} \tag{4.3.2}$$

and

$$\lambda_{\text{new}}(r, s, p) := \begin{cases} \left.\begin{matrix} 2 \\ 2p-4 \\ 2(p-1)^2 p^{r-s-2} \end{matrix}\right\} & \text{if } 2s > r \text{ and } \begin{cases} r = s; \\ r = s+1; \\ r \geq s+2; \end{cases} \\ \left.\begin{matrix} 0 \\ p-3 \\ (p-2)(p-1)p^{s-2} \end{matrix}\right\} & \text{if } 2s = r \text{ and } \begin{cases} p = 2; \\ r = 2 \text{ and } p \geq 3; \\ r \geq 4; \end{cases} \\ \left.\begin{matrix} 0 \\ p-2 \\ (p-1)^2 p^{r/2-2} \end{matrix}\right\} & \text{if } 2s < r \text{ and } \begin{cases} 2 \nmid r; \\ r = 2; \\ r \geq 4 \text{ and } 2 \mid r. \end{cases} \end{cases} \tag{4.3.3}$$

Theorem 4.3.4 (Cohen–Oesterlé, Buzzard) *Let $N, k \in \mathbb{Z}_{\geq 1}$ and let χ be a character of modulus N and conductor $M \mid N$. Then the following statements hold:*

(a) *If $\chi(-1) \neq (-1)^k$, then* $\dim_{\mathbb{C}} E_k(N, \chi) = \dim_{\mathbb{C}} E_k^{\text{new}}(N, \chi) = 0$.
(b) *For $N = 1$, we have*

$$\dim_{\mathbb{C}} E_k(1) = \dim_{\mathbb{C}} E_k^{\text{new}}(1) = \begin{cases} 1, & \text{if } k \geq 4 \text{ and } 2 \mid k; \\ 0, & \text{otherwise.} \end{cases} \tag{4.3.5}$$

Suppose further that $N \geq 2$ and $\chi(-1) = (-1)^k$, and let

$$\begin{aligned} e &:= \prod_{p \mid N} \lambda(\operatorname{ord}_p(N), \operatorname{ord}_p(M), p) \\ e_{\text{new}} &:= \prod_{p \mid N} \lambda_{\text{new}}(\operatorname{ord}_p(N), \operatorname{ord}_p(M), p). \end{aligned} \tag{4.3.6}$$

Then the following hold:

(c) *We have*

$$\dim_{\mathbb{C}} E_k(N, \chi) = \begin{cases} e - 1 & \text{if } k = 2 \text{ and } \chi \text{ is trivial;} \\ e/2 & \text{if } k = 1; \\ e & \text{otherwise.} \end{cases} \tag{4.3.7}$$

(d) *We have*

$$\dim_{\mathbb{C}} E_k^{\text{new}}(N, \chi) = \begin{cases} e_{\text{new}} + 1 & \text{if } k = 2 \text{ and } \chi \text{ is trivial and } N \text{ is prime;} \\ e_{\text{new}}/2 & \text{if } k = 1; \\ e_{\text{new}} & \text{otherwise.} \end{cases} \tag{4.3.8}$$

(e) *We have*

$$\dim_{\mathbb{C}} E_k(N, [\chi]) = d \dim_{\mathbb{C}} E_k(N, \chi)$$

where $d = [\mathbb{Q}(\chi) : \mathbb{Q}]$ is the degree of χ, and similarly with $\dim_{\mathbb{C}} E_k^{\text{new}}(N, [\chi])$.

Proof The proof is an elaborate and rather tedious exercise in counting characters using the trace formula. □

Table 1 Dimensions for subspaces of $M_3(560, [\chi])$

	Total	New	Old
Modular forms	408	96	312
Cusp forms	360	96	264
Eisenstein series	48	0	48

Example 4.3.9 We consider the space $M_3(560, [\chi])$ with label 560.3.bt; a character χ in this orbit has label 560.bt, order 6, and degree 2. We then compute dimensions as in Table 1.

One can also ask for the full trace form

$$\sum_{n=1}^{\infty} \mathrm{Tr}(T_n \mid S_k(N, [\chi]))q^n \in S_k(N, [\chi]; \mathbb{Z}) \tag{4.3.10}$$

on $S_k(N, [\chi])$ to some (q-adic) precision, with analogous definitions for the other subspaces considered above; see also (4.5.2) below.

4.4 Eisenstein Series

Beyond dimensions, we may next ask for further information about the decomposition of the space $M_k(N, \chi)$. Of course the first step is the decomposition of the Eisenstein subspaces $E_k(N, \chi)$—for this purpose, explicit bases are given by Cohen–Strömberg [25, Theorems 8.5.17, 8.5.22, and 8.5.23].

Remark 4.4.1 We do not currently display an Eisenstein basis in the LMFDB.

4.5 Decomposition of Newspaces into Hecke Orbits

With the Eisenstein subspace described explicitly above, we now turn to the cuspidal subspace. By the newform theory of Atkin–Lehner [1] and Li [59], the multiplicity of the space $S_k^{\mathrm{new}}(M, \chi_M)$ in $S_k(N, \chi)$, is equal to the number of divisors of N/M (so depends only on the conductor and level). While it suffices to study the new subspace, it may be computationally expensive to determine $S_k^{\mathrm{new}}(N, \chi)$ as a subspace of $S_k(N, \chi)$; one way to do this is via projection operators called *degeneracy maps*, one for each prime divisor of N.

At this stage, for each newspace $S_k^{\mathrm{new}}(N, [\chi])$ we may first ask for just the dimensions of its newform subspaces V or Hecke orbits—see Sect. 8.5 below for a discussion of decomposition and irreducibility. When χ is trivial, we may also ask for the decomposition of the space under the Atkin–Lehner involutions and the Fricke involution.

Table 2 Dimensions for subspaces of $S_2^{\text{new}}(3111)$

3	17	61	Fricke	Dimension	Decomposition
+	+	+	+	13	$1+2+3+7$
+	+	−	−	29	29
+	−	+	−	24	$3+21$
+	−	−	+	14	14
−	+	+	−	24	24
−	+	−	+	14	14
−	−	+	+	13	13
−	−	−	−	28	28

Example 4.5.1 The space $S_2^{\text{new}}(3111)$, with trivial character, has dimension 159; it decomposes into newspaces of dimensions $1+2+3+3+7+13+14+14+21+24+28+29=159$, as can be seen in Table 2, which also shows the decomposition into subspaces under the Atkin–Lehner operators.

In practice, one computes this decomposition as follows. We first compute a $\mathbb{Q}(\chi)$-basis for $S_k^{\text{new}}(N,\chi)$ in some manner, and then we compute the matrix of T_p on this basis for $p \nmid N$ for a few primes p in such a way that a small (finite) $\mathbb{Z}$-linear combination $\sum_p c_p T_p$ has squarefree characteristic polynomial. Therefore, the $\mathbb{Q}$-dimension decomposition is simply the degrees of the irreducible factors each multiplied by $[\mathbb{Q}(\chi):\mathbb{Q}]$. (There seems to be no problem in practice finding such a small linear combination, but the best thing that we can say rigorously involves the Sturm bound and appears to be far from optimal.) Already at this point engineering concerns enter: for example, the time to compute such a characteristic polynomial may be faster in certain implementations if done over $\mathbb{Q}$ instead.

With this basic decomposition data in hand, we may continue. For each newform orbit $[f] \leftrightarrow V$ (cf. Sect. 4.2) we wish to compute the following:

(1) The trace form

$$\operatorname{Tr}(f)(q) := \sum_{n=1}^{\infty} \operatorname{Tr}_{K|\mathbb{Q}}(a_n(f))q^n \in S_k(N,[\chi];\mathbb{Z}) \tag{4.5.2}$$

(well-defined on the Galois orbit $[f]$), where K is the coefficient field of f, to precision n up to the *Sturm bound* (see Sect. 8.2). Equivalently, writing $\operatorname{Tr}(f)(q) = \sum_n t_n q^n \in \mathbb{Z}[[q]]$, we have $t_n = \operatorname{Tr}(T_n \mid V)$ as the trace of the Hecke operator T_n restricted to V—see Sect. 8.6 for further discussion.

(2) A minimal polynomial for the coefficient field K of $[f]$.

(3) A finite set of generators for the Hecke kernel for V, the ideal in the Hecke algebra on $S_k^{\text{new}}(N,\chi)$ that vanishes on V; i.e., a finite set of polynomials in T_n such that the ideal generated by these polynomials cuts out exactly V. (We use the Hecke kernel when computing inner twists: see Sect. 11.)

Although it is possible to compute coefficients of the trace form $\operatorname{Tr}(f)$ by computing coefficients of f and taking traces, this is more expensive than other techniques and is not computationally feasible in many cases where it is feasible to compute the trace form (e.g., using the trace formula: see Sect. 5.2). The trace form conveniently records interesting information about the newform orbit, e.g., the coefficient t_1 of the trace form is equal to the dimension of the newform subspace.

Example 4.5.3 Consider the space $S_2(1166, [\chi])$ with label 1166.2.c, the character having order 2 and conductor $53 \mid 1166$. The old subspace decomposes as

$$S_2^{\text{old}}(1166, [\chi]) \simeq S_2^{\text{new}}(53, [\chi])^{\oplus 4} \oplus S_2^{\text{new}}(106, [\chi])^{\oplus 2} \oplus S_2^{\text{new}}(583, [\chi])^{\oplus 2}.$$

The decomposition of the new space $S_2^{\text{new}}(1166, [\chi])$ into irreducibles by $\mathbb{Q}$-dimension is $46 = 2 + 22 + 22$, giving rise to three newform orbits 1166.2.c.a, 1166.2.c.b, 1166.2.c.c with respective trace forms

$$\begin{aligned} \operatorname{Tr}(f_{\mathrm{a}})(q) &= 2q - 2q^4 + 2q^6 - 8q^7 + 4q^9 + O(q^{10}) \\ \operatorname{Tr}(f_{\mathrm{b}})(q) &= 22q - 22q^4 - 6q^6 - 24q^9 + O(q^{10}) \\ \operatorname{Tr}(f_{\mathrm{c}})(q) &= 22q - 22q^4 + 4q^6 + 8q^7 - 34q^9 + O(q^{10}). \end{aligned} \tag{4.5.4}$$

We computed the last two trace forms *without* computing coefficients of a constituent newform belonging to a number field of degree 22, or even determining what this number field is, which would have been much more time consuming. For the newform orbit $[f_{\mathrm{a}}]$, we determined that its coefficient field is $\mathbb{Q}(\sqrt{-1})$, that it can be constructed as the kernel of the linear operator $T_3^2 + 1$ acting on $S_2^{\text{new}}(1166, [\chi])$, and then computed the first 1000 coefficients a_n of its q-expansion $\sum a_n q^n$ as elements of $\mathbb{Q}(\sqrt{-1})$.

4.6 Hecke Eigenvalues

Finally, for a newform f, we can ask for the coefficients of f up to (at least) the Sturm bound. These coefficients can be represented either exactly or as complex numbers (approximately, e.g. using interval arithmetic).

- For exact coefficients, there are issues in representing them compactly: see Sect. 8.7 for our approaches.
- For the numerical (complex) coefficients a_n, the most useful for computing L-functions (see the next section), we ask for these coefficients for each embedded form in the newspace. These coefficients are of size $O(n^{(k-1)/2+\epsilon})$ for all $\epsilon > 0$, so in large weight we prefer to compute the normalized coefficients $a_n/n^{(k-1)/2}$, which by the Ramanujan–Petersson bounds have absolute value of size $O(n^{\epsilon})$.

For large degree coefficient fields, it is often practical to compute numerical coefficients even when storing exact coefficients would be impractical.

Table 3 Embedded newforms for 5355.2.a.bf

Label	$\iota_m(\nu)$	a_2	a_3	a_4	a_5	a_6	a_7	a_8
1.1	2.65109	−1.65109	0	0.726109	1.00000	0	1.00000	2.10331
1.2	−0.273891	1.27389	0	−0.377203	1.00000	0	1.00000	−3.02830
1.3	−1.37720	2.37720	0	3.65109	1.00000	0	1.00000	3.92498

Finally, when the character is trivial, we can also ask for the signs of the Atkin–Lehner involutions (as can be seen in Table 2, for example).

Example 4.6.1 Consider the newform orbit 5355.2.a.bf of dimension 3, with coefficient field $\mathbb{Q}(\nu)$ (LMFDB label 3.3.169.1) where ν is a root of the polynomial $x^3 - x^2 - 4x - 1$. The q-expansion of a newform f in this orbit, with coefficients in $\mathbb{Q}(\nu)$, is

$$f(q) = q + (1 - \beta_1)q^2 + (2 - \beta_1 + \beta_2)q^4 + q^5 + q^7 + (2 - \beta_1 + 2\beta_2)q^8 + O(q^{10})$$

where $\beta_1 = \nu$ and $\beta_2 = \nu^2 - \nu - 3$.

The 3 embedded newforms are labeled 5355.2.a.bf.1.m for $m = 1, 2, 3$ encoding the three embeddings $\iota_m : \mathbb{Q}(\nu) \hookrightarrow \mathbb{C}$; the embedded coefficients to 6 decimal digits are shown in Table 3.

4.7 L-Functions

We can also ask for computations related to (invariants of) L-functions of modular forms, including the sign of the functional equation, the first few zeros, and special values to some precision: see Sect. 9 for more detail.

5 Algorithms

In this section, we give a brief overview of different algorithmic methods to compute modular forms and indicate where they are currently implemented. In our computations for the LMFDB, we only used the first two (modular symbols and the trace formula), but here we also survey the others. Our goal is to give a flavor of what each method entails, referring to the references provided for details. Throughout, we keep notation from the previous section.

5.1 Modular Symbols

The most well-known method to compute modular forms is the method of *modular symbols*, introduced by Birch [6] and developed by Manin [61], Merel [64], Stein [89], and many others. For an extensive history, see Stein [89, 8.10.2], and for a gentle overview see Stein [90]. This method was implemented in Magma [11] by William Stein, with contributions by Steve Donnelly and Mark Watkins, and in SageMath [79] by William Stein, with contributions by David Loeffler, Craig Citro, Peter Bruin, Frédéric Chapoton, Alex Ghitza, and many others.

We now briefly introduce modular symbols. Assume $k \geq 2$. Integration gives a perfect pairing

$$\begin{aligned} S_k(\Gamma_1(N)) \times H_1(X_1(N), \mathbb{R}[x,y]_{k-2}) &\to \mathbb{C} \\ (f, \upsilon \otimes P) &\mapsto \int_\upsilon f(z)\, P(z,1)\, \mathrm{d}z \end{aligned} \tag{5.1.1}$$

where $\mathbb{R}[x,y]_{k-2}$ denotes the $\mathbb{R}$-vector space of homogeneous polynomials of degree $k-2$. In a slogan, (5.1.1) indicates that *the homology of a modular curve is dual to its cusp forms*, and this is formalized as follows. Let $\mathrm{Div}(\mathbb{P}^1(\mathbb{Q}))$ be the free abelian group on symbols $[\alpha]$ for $\alpha \in \mathbb{P}^1(\mathbb{Q})$, and let $\mathrm{Div}^0(\mathbb{P}^1(\mathbb{Q})) \leq \mathrm{Div}(\mathbb{P}^1(\mathbb{Q}))$ be the subgroup of degree zero elements under the natural degree map. Then $\mathrm{Div}^0(\mathbb{P}^1(\mathbb{Q}))$ is generated by elements $\{\alpha, \beta\} := [\alpha] - [\beta]$ for $\alpha, \beta \in \mathbb{P}^1(\mathbb{Q})$, written this way to suggest a path from α to β in $\mathbb{C}$. We define the space of modular symbols of weight k and level N (with $\mathbb{Q}$-coefficients) to be the quotient

$$\mathrm{ModSym}_k(\Gamma_1(N); \mathbb{Q}) := \frac{\mathbb{Q}[x,y]_{k-2} \otimes \mathrm{Div}^0(\mathbb{P}^1(\mathbb{Q}))}{\langle P \otimes \{\alpha,\beta\} - \gamma(P \otimes \{\alpha,\beta\})\rangle_{\alpha,\beta\in\mathbb{P}^1(\mathbb{Q}),\gamma\in\Gamma_1(N)}}$$

under the natural action of $\Gamma_1(N) \leq \mathrm{SL}_2(\mathbb{Q})$. The space $\mathrm{ModSym}_k(\Gamma_1(N); \mathbb{Q})$ of modular symbols has moreover a natural action of Hecke operators and Atkin–Lehner operators.

Theorem 5.1.2 *There is a Hecke-equivariant isomorphism*

$$\mathrm{ModSym}_k(\Gamma_1(N); \mathbb{Q}) \xrightarrow{\sim} M_k(\Gamma_1(N); \mathbb{Q}) \oplus \overline{S}_k(\Gamma_1(N); \mathbb{Q})$$

where $\overline{S}_k(\Gamma_1(N); \mathbb{Q})$ *denotes the space of anti-holomorphic cusp forms, the image of* $S_k(\Gamma_1(n); \mathbb{Q})$ *under complex conjugation.*

Proof See Manin [61], Merel [64], or Stein [89, §8.5]. □

Theorem 5.1.2 has many variants: one may restrict to $\Gamma_0(N)$, work with the (appropriately defined) space of *cuspidal* modular symbols as the kernel of a certain boundary map, carve out just $M_k(\Gamma_1(N); \mathbb{Q})$ as the $+$-space for a natural action of complex conjugation, and so on.

Example 5.1.3 For $\Gamma_0(N)$, the space of modular symbols has a convenient description in terms of *Manin symbols* as follows: $\mathrm{ModSym}_k(\Gamma_0(N); \mathbb{Q})$ is the $\mathbb{Q}$-vector space generated by the set Δ of elements $\delta = (x^i y^{k-2-i}, (c : d))$ for $i = 0, \ldots, k-2$ and $(c : d) \in \mathbb{P}^1(\mathbb{Z}/N\mathbb{Z})$, modulo the subspace

$$\langle \delta + \delta S, \delta + \delta R + \delta R^2 \rangle_{\delta \in \Delta}$$

where $S = \begin{pmatrix} 0 & 1 \\ -1 & 0 \end{pmatrix}$ and $R = \begin{pmatrix} 0 & 1 \\ -1 & 1 \end{pmatrix}$. The Hecke operators do not preserve individual Manin symbols, but there is an efficient procedure (arising from the Euclidean algorithm) for reducing an arbitrary element of $\mathrm{ModSym}_k(\Gamma_0(N); \mathbb{Q})$ to a linear combination of Manin symbols.

One feature of modular symbols is that they are especially well-suited for certain applications, including arithmetic invariants of elliptic curve quotients [31] (and more generally modular abelian varieties) as well as L-values of modular forms (see e.g. Sect. 9.4 below for an application). Moreover, modular symbols can be employed for arbitrary congruence subgroups (see [3] for an example).

In practice, it is quite efficient to compute the space of modular symbols with its Hecke action. It is a matter of sparse linear algebra to compute a basis of modular symbols, a negligible contribution. The number of field operations to compute the action of the Hecke operator T_n on this basis is $\widetilde{O}(nd)$, where d is the $\mathbb{Q}(\chi)$-dimension of the space under consideration: for each of the d basis elements, we sum the action of $\sigma_1(n) := \sum_{d|n} d = \widetilde{O}(n)$ cosets and reduce to the basis in time polynomial in $\log n$ using continued fractions. In this way, we may compute the q-expansions of a basis to precision $O(q^r)$ using $\widetilde{O}(dr^2)$ field operations, and thereby also the trace form.

The most difficult engineering effort that goes into a working implementation of modular symbols is the careful handling of linear algebra aspects: we apply degeneracy operators to obtain precisely the subspace $S_k^{\mathrm{new}}(N, \chi)$, and once the matrices $[T_n]$ representing the Hecke operators are computed on this space, we compute its decomposition into newform subspaces, etc. Indeed, in the preceding paragraphs, the actual *time* complexity of this method may depend on the output desired and the meaning of "arithmetic operation". If we wish for exact results, which is the approach taken by Magma and Pari/GP, then we need to do exact arithmetic with elements of cyclotomic fields, and the larger the order of the corresponding Dirichlet character, the more expensive the computation. Similarly, the coefficients of the newforms themselves may live in a large extension of the field of character values, and the larger this extension is, the harder the computation.

Remark 5.1.4 As alternatives, we may do all of the computations described using floating point approximations to complex numbers, for example using *complex ball arithmetic* to compute rigorous error bounds for all of the output. In this case, the degree of the field of coefficients of the modular form is irrelevant, and the time complexity matches the estimates above; this is particularly attractive if our application is to the computation of Dirichlet coefficients for input into L-function computations. Similar comments apply when performing computations over a finite field, for example, working with coefficients over a finite field with prime cardinality congruent to 1 modulo the order of χ—in this case, we can do all computations over $\mathbb{F}_p$. In both cases, we must do some reconstruction to obtain exact results in characteristic zero.

The above description requires weight $k \geq 2$. For weight 1, there are two approaches that reduce the problem to higher weight. In the approach originated by Buhler [14], further developed by Buzzard [18], and carried out to scale by Buzzard–Lauder [19], we choose nonzero $f \in M_k(\Gamma_1(N))$ and consider $S_1(\Gamma_1(N)) \subseteq f^{-1}M_{k+1}(\Gamma_1(N))$. Intersecting the spaces obtained for many choices of f, we quickly obtain an upper bound for the space $S_1(\Gamma_1(N))$ that can then be matched with a lower bound. Using Buzzard's code, this method was implemented in Magma by Steve Donnelly. (Currently, Magma can provide a basis for the cuspidal subspace, but it does not decompose the space into the old and new subspace and does not provide the action of the Hecke operators; this was implemented by Buzzard–Lauder, but has not yet been incorporated into Magma.) A second related approach is to use the *Hecke stability* method of Schaeffer [80], instead computing the largest subspace of $f^{-1}M_{k+1}(\Gamma_1(N))$ that is stable under the Hecke operators; this has been implemented in SageMath by Schaeffer and Loeffler, and in Pari/GP by Belabas and Cohen [4, §4].

5.2 *Trace Formula*

Perhaps the earliest method to compute modular forms used the *trace formula*. The trace formula is an explicit formula for the trace of a Hecke operator acting on a space of modular forms, and it was pioneered by Selberg [81] and later developed by Eichler [40], Hijikata [47], and Cohen–Oesterlé [24]. A comprehensive treatment with references is the book of Knightly–Li [58], and a tidy presentation is given by Schoof–van der Vlugt [85, Theorem 2.2]. Proofs of the trace formula from different vantage points continue to be developed, see e.g. Popa [74]. This method has been implemented in Pari/GP [72] by Belabas–Cohen [4] and in a standalone implementation by Bober, described in Sect. 7.2.

We again assume $k \geq 2$. An explicit version of the trace formula for $\mathrm{Tr}(T_n \mid S_k(\Gamma_0(N), \chi)) \in \mathbb{Q}(\chi)$ is too complicated to give here. Aside from easily computed terms, it can be naively understood as a weighted sum of (Hurwitz) class numbers of imaginary quadratic fields: for a precise statement, see e.g. Belabas–Cohen [4, Theorem 4]. We obtain $\mathrm{Tr}(T_n \mid S_k^{\mathrm{new}}(N, \chi))$ from a nontrivial application of the Möbius inversion formula, proven in Corollary 6.1.5 below.

Let $d := \dim_{\mathbb{Q}(\chi)} S_k^{\mathrm{new}}(N, \chi) = O(kN)$. The computation of $\mathrm{Tr}(T_n \mid S_k^{\mathrm{new}}(N, \chi))$ requires computing class numbers of imaginary quadratic fields with absolute discriminant up to $O(n)$, and one can compute all of these at once in time complexity $\widetilde{O}(n^{3/2})$. For the purposes of a large-scale computation, these class numbers are cached and may be assumed to be precomputed (their cost amortized over many computations, thereby negligible). Under this assumption, and given factorizations of n and N, to compute $\mathrm{Tr}(T_n \mid S_k^{\mathrm{new}}(N, \chi))$ we sum $O(\sqrt{n})$ terms giving a complexity of $O(\sqrt{n}N^{\epsilon})$ field operations for any $\epsilon > 0$; computing all traces up to $n > d$ then takes $\widetilde{O}(n^{3/2})$ field operations.

In this manner, we compute the relative trace form on the new cuspidal subspace

$$t(q) := \sum_{n=1}^{\infty} \mathrm{Tr}(T_n \mid S_k^{\mathrm{new}}(N, \chi))q^n \in S_k^{\mathrm{new}}(N, \chi; \mathbb{Z}[\chi]), \tag{5.2.1}$$

and from this we quickly compute the full trace form (4.3.10) in $S_k^{\mathrm{new}}(\Gamma_1(N); \mathbb{Z})$. In particular, using the trace formula method we can compute either trace form to precision $O(q^r)$ using $\widetilde{O}(r^{3/2}N^{\epsilon})$ field operations, which for $r > d$ becomes $\widetilde{O}(r^{3/2})$ as in the previous paragraph.

By multiplicity one theorems, and since the Hecke operators act semisimply on the newspace, the images of t under the Hecke operators span $S_k^{\mathrm{new}}(N, \chi)$. Explicitly, applying T_m to t, we obtain

$$(T_m t)(q) = \sum_{n=1}^{\infty} \mathrm{Tr}(T_m T_n \mid S_k^{\mathrm{new}}(N, \chi))q^n, \tag{5.2.2}$$

and the forms $T_1 t, T_2 t, \ldots$ span $S_k^{\mathrm{new}}(N, \chi)$. (We recall that $T_m T_n = T_{mn}$ when $\gcd(m, n) = 1$, and more generally a recursion for the Hecke operators applies. Therefore, these coefficients can again be expressed in terms of traces of Hecke operators.) Once we have a spanning set, we can extract a basis and apply Hecke operators to that basis.

Typically (in practice) we need $O(d)$ forms to span and $O(d)$ coefficients of each form to get a full rank matrix. Thus writing down a basis typically requires the first $O(d^2)$ values of $\mathrm{Tr}(T_n \mid S_k^{\mathrm{new}}(N, \chi))$, which can be computed using $O(d^3)$ field operations. Finding this basis—and the q-expansion to precision $\widetilde{O}(d)$ for each form—is standard linear algebra, accomplished using $\widetilde{O}(d^3)$ field operations. To compute the matrix of the Hecke operator T_n on this basis requires traces up to $O(nd)$ and so $\widetilde{O}(n^{3/2}d^{3/2})$ operations. Finally and similarly, to compute a basis of q-expansions to precision $O(q^r)$ with $r > d$, we compute traces up to $O(rd)$ and apply a change of basis, for a total of $\widetilde{O}(d^{3/2}r^{3/2})$ arithmetic operations.

We summarize the estimated complexity of these two approaches in Table 4, where again d is the $\mathbb{Q}(\chi)$-dimension of the space under consideration and we suppose precision $r > d$.

So although linear algebra eventually dominates both approaches, neither modular symbols nor the trace formula seems to be a winner for all tasks: it seems to be much better to use modular symbols to get information about a small number of Hecke operators, while it is much better to use the trace formula to get a large number of coefficients of a basis of newforms. This heuristic analysis matches our practical experience in the course of our computations.

Table 4 Heuristic complexity of modular form computations

Task	Modular symbols	Trace formula
Full trace form to precision $O(q^r)$, $d = O(r)$	$\widetilde{O}(dr^2)$	$\widetilde{O}(r^{3/2})$
$[T_n]$ on a basis	$\widetilde{O}(dn)$	$\widetilde{O}(d^{3/2}n^{3/2} + d^3)$
Characteristic polynomial of T_n on a basis	$\widetilde{O}(dn + d^3)$	$\widetilde{O}(d^{3/2}n^{3/2} + d^3)$
Basis of q-expansions to precision $O(q^r)$, $d = O(r)$	$\widetilde{O}(dr^2)$	$\widetilde{O}(d^{3/2}r^{3/2})$
Hecke decomposition	$\widetilde{O}(d^3)$	$\widetilde{O}(d^3)$
Minimal polynomials for newspace coefficient fields	$\widetilde{O}(d^3)$	$\widetilde{O}(d^3)$

Similar comments with reference to weight 1 forms apply as in the previous section. The same is true for the issue of time complexity and the coefficient field (see e.g. Remark 5.1.4), with the caveat that the matrices representing Hecke operators using modular symbols tend to be much sparser in comparison to those using the trace formula. In particular, one expects that taking advantage of sparsity will allow a more efficient implementation of the linear algebra aspects for modular symbols.

Remark 5.2.3 In some circumstances, it can be more convenient to work with a basis that is in echelon form with respect to q-expansions (sometimes called a *Victor Miller basis*) in the trace formula method. With such a basis, going back and forth between an action on q-expansions and the matrix form for various linear operators one can see some gains in efficiency.

5.3 Definite Methods

In both of the previous algorithms, we work (either explicitly or implicitly) on the modular curve. In this section, we indicate another class of algorithms that compute systems of Hecke eigenvalues using a different underlying object.

Going back at least to Jacobi, surely the first modular forms studied were theta series. Let $Q(x) = Q(x_1, \ldots, x_d) \in \mathbb{Z}[x_1, \ldots, x_d]$ be a positive definite integral quadratic form in $d = 2k \in 2\mathbb{Z}_{\geq 1}$ variables with discriminant N. Let $P(x)$ be a (nonzero) spherical polynomial of degree $m \geq 0$ with respect to Q, for example $P(x) = 1$. We form the generating series for representations of $n \in \mathbb{Z}_{\geq 0}$ by Q weighted by P, a theta series of Q, by

$$\theta_{Q,P}(q) := \sum_{x \in \mathbb{Z}^d} P(x) q^{Q(x)}. \tag{5.3.1}$$

For example, if $P(x) = 1$, then

$$\theta_{Q,1}(q) = \sum_{n=0}^{\infty} r_Q(n) q^n \in \mathbb{Z}[[q]] \tag{5.3.2}$$

where $r_Q(n) = \#\{x \in \mathbb{Z}^d : Q(x) = n\}$ counts the number of representations of n by Q. By letting $q = e^{2\pi i z}$ for $z \in \mathcal{H}$ as usual, we obtain a holomorphic function $\theta_Q : \mathcal{H} \to \mathbb{C}$. Further, by an application of the Poisson summation formula (see e.g. Miyake [67, Corollary 4.9.5]), we find that $\theta_{P,Q} \in M_{k+m}(\Gamma_0(2N), \chi_N)$ is a classical modular form of weight $k+m$, level $2N$, and character $\chi_N(a) := \left(\dfrac{N}{a}\right)$ of order at most 2.

Turning this around, we can use theta series to compute spaces of classical modular forms. Perhaps the most convenient source of such theta series is to work with quaternary ($d = 4$) quadratic forms of square discriminant coming from quaternion algebras—this method goes by the name *Brandt matrices* as it came about from early work of Brandt. Building on work of Eichler [41], Hijikata–Pizer–Shemanske [48] proved that linear combinations of such theta series span the space of cusp forms, up to twists. (See also Martin [62] for a more recent development.) The coefficients of theta series can then be reformulated in terms of classes of right ideals of specified reduced norm in a quaternion order. This method was first developed in an algorithmic context by Pizer [73]; it has been implemented in Magma by David Kohel and in SageMath by Bober, Alia Hamieh, Victoria de Quehen, William Stein, and Gonzalo Tornaría.

In a little more detail, the method of Brandt matrices runs as follows. Let B be a definite quaternion algebra of discriminant $D := \operatorname{disc} B$, a squarefree product of the primes that ramify in B. Let $\mathcal{O} \subseteq B$ be an Eichler order of level M with $\gcd(D, M) = 1$, and let $N := DM$. Let $\operatorname{Cls} \mathcal{O}$ be the set of locally principal (equivalently, invertible) fractional right $\mathcal{O}$-ideals up to isomorphism (given by left multiplication by an element of $B^\times$). Then $\operatorname{Cls} \mathcal{O}$ is a finite set, so let $\operatorname{Cls} \mathcal{O} = \{[I_1], [I_2], \dots, [I_h]\}$ with $h := \#\operatorname{Cls} \mathcal{O}$. Let $\mathcal{O}_L(I_i)$ be the left order of I_i, and let $\Gamma_i := \mathcal{O}_L(I_i)^\times$ be its unit group with $\#\Gamma_i < \infty$. Let $q_i := \operatorname{nrd}(I_i)$. For $n \in \mathbb{Z}_{\geq 1}$, define

$$\Theta(n)_{i,j} := \Gamma_i \backslash \{\alpha \in I_j I_i^{-1} : \operatorname{nrd}(\alpha) q_i q_j^{-1} = n\}.$$

We have $\alpha \in \Theta(n)_{i,j}$ if and only if $\alpha I_i \subseteq I_j$ with index n^2. To connect this with the previous paragraph, we have

$$\begin{aligned} Q_{ij} &\colon I_j I_i^{-1} \to \mathbb{Z} \\ Q_{ij}(\alpha) &= \operatorname{nrd}(\alpha) \frac{q_i}{q_j} \end{aligned} \tag{5.3.3}$$

is a positive definite integral quaternary quadratic form of discriminant N^2 whose theta series descends to a modular form of level N—in the notation above, we have $r_{Q_{ij}}(n) = \#\Theta(n)_{i,j}$. In this way, we can compute a matrix for the Hecke operator $[T_n]$ acting on $S_k(\Gamma_0(N), \chi)$ by quaternionic arithmetic: for weight $k = 2$, the matrix $[T_n]$ is the adjacency matrix of the directed graph with vertex set $\operatorname{Cls} \mathcal{O}$ and directed edges between $[I_i]$ and $[I_j]$ with multiplicity $\#\Theta(n)_{i,j}$.

The method of Brandt matrices has several advantages. First, the forms computed this way are necessarily new at all primes $p \mid D$, so linear algebra with degeneracy operators can be minimized. Second, the matrices $[T_n]$ of Hecke operators are sparse: for example, in weight $k = 2$ they have nonnegative integer coefficients whose columns sum to $\sigma(n)$. Accordingly, linear algebra steps have an improved complexity both in theory and in practice. Third, Brandt matrices also carry useful arithmetic information about the reduction of modular curves at primes of bad reduction. Fourth, the set $\Theta(n)_{i,j}$ is independent of the weight k and so may be reused. Despite these advantages, the main limitation of Brandt matrices seems to be that it works most simply when there exists a prime p that exactly divides the level N (so that an Eichler order of reduced discriminant N exists); otherwise, we must work with non-Eichler orders. Hence current implementations focus on this case.

The Brandt graph is an expander graph by the Ramanujan–Peterson bound, so with short vector computations one can compute a set of representatives for Cls $\mathcal{O}$ and a spanning set for $S_k(\Gamma_0(N), \chi)$ using $\widetilde{O}(h^2)$ operations; computing a basis from this is a matter of sparse linear algebra and can be considered to be negligible. To compute a single matrix $[T_n]$, in principle we could use Minkowski reduction (together with some awkward corner cases) on $h\sigma(n)$ right ideals using $\widetilde{O}(hn) = \widetilde{O}(dn)$ operations. To compute a basis of q-expansions to precision $O(q^r)$ with $d = O(r)$, for each of the h classes we can enumerate elements of small reduced norm using the Fincke–Pohst algorithm in time proportional to the volume so $\widetilde{O}(dr^2)$, performing reduction with the same complexity. These heuristics match the running time of modular symbols with linear algebra again eventually dominating—however, it is here where sparse linear algebra may ultimately in practice give the Brandt matrix method an edge.

A method that shares much in common with Brandt matrices is the *method of graphs* due to Mestre [65] and Oesterlé. We suppose that $p \parallel N$ and work with the quaternion algebra B of discriminant $D = p$. We recall that there is an equivalence of categories between supersingular elliptic curves over an algebraic closure of $\mathbb{F}_p$ under isogenies and invertible right (or left) $\mathcal{O}$-modules under homomorphisms. So to compute a matrix for the Hecke operator, in place of Cls $\mathcal{O}$ we can compute the set of isomorphism classes of pairs (E, C) where E is a supersingular elliptic curve in characteristic p and C is a cyclic subgroup of order $M = N/p$, and in place of the sets $\Theta(n)_{ij}$ we can enumerate cyclic isogenies between these points up to a natural equivalence.

Finally, a related method of Birch [7] (who sought to generalize the method of graphs beyond discriminant $D = p$) uses *ternary* quadratic forms instead. This method captures all of the advantages above, with an additional feature: work in progress by Hein–Tornaría–Voight shows that one can carve out not just a new subspace but moreover one can specify the Atkin–Lehner eigenvalue, reducing the total dimension and thereby the complexity of linear algebra operations.

5.4 Other Methods

We conclude by briefly indicating two other methods in addition to the above.

- *Multiplying forms of lower weight.* We compute a presentation for the graded ring of modular forms of level N

$$M(\Gamma_1(N)) := \bigoplus_{k=0}^{\infty} M_k(\Gamma_1(N))$$

(or the same for $\Gamma_0(N)$) in terms of a finite set of generators and a Gröbner basis for the ideal of relations among them; see work of Voight–Zureick-Brown [96] for an explicit description of this graded ring in terms of the genus and number of cusps for $\Gamma_1(N)$ (and more generally in terms of the signature of the uniformizing Fuchsian group) as well as further references and discussion. From this, one can compute for each weight k a set of (leading) monomials in the generators that are a $\mathbb{Q}$-basis for $M_k(\Gamma_1(N))$. Using fast Fourier techniques, the multiplication of these q-expansions allows the computation of a basis for large weights k (and fixed level N) quite efficiently in comparison to any of the approaches above.
- *Polynomial-time algorithms.* By work of Edixhoven–Couveignes [39], Bruin [12], and Mascot [63], one can compute coefficients of modular forms of level 1 in polynomial time: for example, for the modular discriminant $\Delta(q) = \sum_n \tau(n) q^n \in S_{12}(1)$, the value $\tau(p)$ for a prime p can be computed in time bounded by a fixed power of $\log p$.

6 Two Technical Ingredients

In this section, we consider two technical results that are needed in the above algorithmic description.

6.1 Eichler–Selberg Trace Formula for Newforms

We first prove a technical result that is used by Belabas–Cohen [4] in the computation of modular forms in Pari/GP [72], as explained above: we describe the trace of Hecke operators on the new subspace in terms of the trace on the total space.

Let χ be a primitive character of conductor $Q \mid N$ and k a positive integer satisfying $\chi(-1) = (-1)^k$; we take these to be fixed and suppress their dependence from the notation.

For any positive integer n, the nth Hecke operator $T_n \colon S_k(N, \chi) \to S_k(N, \chi)$ may be defined by

$$(T_n f)(z) = \frac{1}{n} \sum_{\substack{ad=n \\ \gcd(a,N)=1}} \chi(a)a^k \sum_{b \bmod d} f\left(\frac{az+b}{d}\right).$$

Then

$$(T_n f)(z) = \sum_{m=1}^{\infty} \left(\sum_{\substack{d|(m,n) \\ (d,N)=1}} \chi(d)d^{k-1} a_{\frac{mn}{d^2}} \right) e(mz),$$

where $f(z) = \sum_{m=1}^{\infty} a_m e(mz)$. This operator stabilizes the subspace $S_k^{\mathrm{new}}(N, \chi)$.

Let $\{f_{N,j}\}_{j=1}^{s_N}$ be a basis of normalized newforms for $S_k^{\mathrm{new}}(N, \chi)$ and write

$$f_{N,j}(z) = \sum_{m=1}^{\infty} a_{N,j}(m)e(mz).$$

We assume that each $f_{N,j}$ is an eigenfunction of T_n of eigenvalue $a_{N,j}(n)$ and define

$$g_n = \sum_{j=1}^{s_N} a_{N,j}(n) f_{N,j} = \sum_{m=1}^{\infty} e(mz) \operatorname{Tr}\left(T_n T_m |_{S_k^{\mathrm{new}}(N,\chi)}\right).$$

We parameterize the basis of $S_k(N, \chi)$: for $M_1, M_2 \in \mathbb{Z}_{\geq 1}$ with $Q \mid M_1$ and $M_1 M_2 \mid N$, let

$$f_{M_1,j}^{M_2}(z) := f_{M_1,j}(M_2 z).$$

Then

$$\left\{ f_{M_1,j}^{M_2} \ : \ M_1, M_2 \in \mathbb{Z}_{\geq 1},\ Q \mid M_1,\ M_1 M_2 \mid N \right\}$$

is a basis for $S_k(N, \chi)$. Let us extend the definition of $a_{N,j}(n)$ to $\mathbb{Q}_{>0}$ by writing $a_{N,j}(x) = 0$ if $x \notin \mathbb{Z}_{\geq 1}$.

If $\gcd(n, N) = 1$, then

$$\begin{aligned} T_n f_{M_1,j}^{M_2} &= \sum_{m=1}^{\infty} \sum_{\substack{d|(m,n) \\ (d,N)=1}} \chi(d)d^{k-1} a_{M_1,j}\left(\frac{mn}{d^2 M_2}\right) e(mz) \\ &= \sum_{m=1}^{\infty} a_{M_1,j}\left(\frac{m}{M_2}\right) a_{M_1,j}(n)e(mz) = a_{M_1,j}(n) f_{M_1,j}^{M_2}, \end{aligned}$$

so each $f_{M_1,j}^{M_2}$ is an eigenfunction of T_n. To compute the action of T_n when $\gcd(n, N) > 1$, we need the following theorem.

Theorem 6.1.1 *Let $p \mid N$ be prime, let $\alpha \in \mathbb{Z}_{\geq 0}$, and let $r := \operatorname{ord}_p M_2$. Let χ_{M_1} be the character modulo M_1 induced from χ. Then*

$$T_{p^\alpha} f_{M_1,j}^{M_2} = \begin{cases} f_{M_1,j}^{M_2 p^{-\alpha}}, & \textit{if } \alpha - r \leq 0; \\ a_{M_1,j}(p^{\alpha-r}) f_{M_1,j}^{M_2 p^{-r}} - \chi_{M_1}(p) p^{k-1} a_{M_1,j}(p^{\alpha-r-1}) f_{M_1,j}^{M_2 p^{-r+1}}, & \textit{if } \alpha - r > 0. \end{cases} \tag{6.1.2}$$

Proof By the definition of the Hecke operator, we get

$$T_{p^\alpha} f_{M_1,j}^{M_2} = \sum_{m=1}^{\infty} \sum_{\substack{d \mid (m, p^\alpha), \\ (d,N)=1}} \chi(d) d^{k-1} a_{M_1,j}\left(\frac{mp^\alpha}{M_2 d^2}\right) e(mz) = \sum_{m=1}^{\infty} a_{M_1,j}\left(\frac{mp^\alpha}{M_2}\right) e(mz).$$

If $\alpha - r \leq 0$, then

$$T_{p^\alpha} f_{M_1,j}^{M_2} = \sum_{m=1}^{\infty} a_{M_1,j}\left(\frac{m}{M_2 p^{-r} \cdot p^{-\alpha+r}}\right) e(mz) = f_{M_1,j}^{M_2 p^{-\alpha}}.$$

Assume that $\alpha - r > 0$. Since $f_{M_1,j}$ is a normalized newform for $\Gamma_0(M_1)$, we get

$$\begin{aligned} & a_{M_1,j}\left(\frac{m}{M_2 p^{-r}}\right) \cdot a_{M_1,j}\left(p^{\alpha-r}\right) \\ &= \begin{cases} a_{M_1,j}\left(\frac{mp^\alpha}{M_2}\right), & \text{if } p \mid M_1, \\ \sum_{e=0}^{\min\{\operatorname{ord}_p(m), \alpha-r\}} (\chi(p) p^{(k-1)})^e a_{M_1,j}\left(\frac{mp^{\alpha-2e-r}}{M_2 p^{-r}}\right), & \text{if } p \nmid M_1. \end{cases} \end{aligned} \tag{6.1.3}$$

Then, if $p \mid M_1$, we have

$$T_{p^\alpha} f_{M_1,j}^{M_2} = a_{M_1,j}(p)^{\alpha-r} \cdot f_{M_1,j}^{M_2 p^{-r}}.$$

We now assume that $\alpha - r > 0$ and $p \nmid M_1$. If $\alpha - r = 1$, we have

$$a_{M_1,j}\left(\frac{mp^\alpha}{M_2}\right) = a_{M_1,j}\left(\frac{m}{M_2 p^{-r}}\right) \cdot a_{M_1,j}\left(p^{\alpha-r}\right) - \delta_{\operatorname{ord}_p(m) \geq 1} \chi(p) p^{(k-1)} a_{M_1,j}\left(\frac{mp^{\alpha-2}}{M_2}\right).$$

By taking the summation over $m \in \mathbb{Z}_{\geq 1}$, we get:

$$T_{p^\alpha} f_{M_1,j}^{M_2} = a_{M_1,j}(p^{\alpha-r}) f_{M_1,j}^{M_2 p^{-r}} - \chi(p) p^{k-1} \sum_{m=1}^{\infty} a_{M_1,j}\left(\frac{mp^{\alpha-1}}{M_2}\right) e(mpz)$$

$$= a_{M_1,j}(p^{\alpha-r}) f_{M_1,j}^{M_2 p^{-r}} - \chi(p) p^{k-1} f_{M_1,j}^{M_2 p^{-r+1}}.$$

Note that when $r = 0$ we have $M_1 M_2 p \mid N$, since $p \mid N$.

If $\alpha - r - 2 \geq 0$, by changing α to $\alpha - 2$, we get

$$\chi(p) p^{k-1} a_{M_1,j}\left(\frac{m}{M_2 p^{-r}}\right) \cdot a_{M_1,j}\left(p^{\alpha-2-r}\right) = \sum_{e=1}^{\min\{\mathrm{ord}_p(m), \alpha-2-r\}+1} (\chi(p) p^{(k-1)})^e a_{M_1,j}\left(\frac{m p^{\alpha-2e-r}}{M_2 p^{-r}}\right).$$

By subtracting from (6.1.3), we get

$$\left\{a_{M_1,j}(p^{\alpha-r}) - \chi(p) p^{k-1} a_{M_1,j}(p^{\alpha-r-2})\right\} a_{M_1,j}\left(\frac{m}{M_2 p^{-r}}\right) = a_{M_1,j}\left(\frac{m p^{\alpha}}{M_2}\right)$$

$$+ \begin{cases} -\left(\chi(p) p^{k-1}\right)^{\mathrm{ord}_p(m)+1} a_{M_1,j}\left(\frac{m p^{\alpha-2(\mathrm{ord}_p(m)+1)-r}}{M_2 p^{-r}}\right), & \text{if } 0 \leq \mathrm{ord}_p(m) \leq \alpha - 2 - r, \\ \left(\chi(p) p^{k-1}\right)^{\alpha-r} a_{M_1,j}\left(\frac{m p^{-(\alpha-r)}}{M_2 p^{-r}}\right), & \text{if } \mathrm{ord}_p(m) \geq \alpha - r, \\ 0, & \text{otherwise.} \end{cases}$$

After taking the summation over $m \in \mathbb{Z}_{\geq 1}$ on both sides, we get

$$\sum_{m=1}^{\infty} a_{M_1,j}\left(\frac{m p^{\alpha}}{M_2}\right) e(mz) = T_{\alpha} f_{M_1,j}^{M_2}(z)$$

$$= \left\{a_{M_1,j}(p^{\alpha-r}) - \chi(p) p^{k-1} a_{M_1,j}(p^{\alpha-2-r})\right\} f_{M_1,j}^{M_2 p^{-r}}$$

$$+ \sum_{\ell=0}^{\alpha-2-r} (\chi(p) p^{k-1})^{\ell+1} \sum_{\substack{m=1, \\ p \nmid m}}^{\infty} a_{M_1,j}\left(\frac{m p^{\alpha-2-r-\ell}}{M_2 p^{-r}}\right) e(m p^{\ell} z)$$

$$- (\chi(p) p^{k-1})^{\alpha-r} \sum_{m=1}^{\infty} a_{M_1,j}\left(\frac{m}{M_2 p^{-r}}\right) e(m p^{\alpha-r} z).$$

For the last piece, we have

$$\sum_{m=1}^{\infty} a_{M_1,j}\left(\frac{m}{M_2 p^{-r}}\right) e(m p^{\alpha-r} z) = f_{M_1,j}^{M_2 p^{\alpha-2r}}(z).$$

Now consider

$$\sum_{\substack{m=1, \\ p \nmid m}}^{\infty} a_{M_1,j}\left(\frac{m}{M_2 p^{-r}}\right) e(mz) = f_{M_1,j}^{M_2 p^{-r}}(z) - \sum_{m=1}^{\infty} a_{M_1,j}\left(\frac{m p}{M_2 p^{-r}}\right) e(m p z).$$

Since

$$a_{M_1,j}(p)\cdot f_{M_1,j}^{M_2p^{-r}}(z)=\sum_{m=1}^{\infty}a_{M_1,j}\left(\frac{mp}{M_2p^{-r}}\right)e(mz)+\chi(p)p^{k-1}f_{M_1,j}^{M_2p^{-r+1}}(z),$$

we get

$$\sum_{\substack{m=1,\\ p\nmid m}}^{\infty}a_{M_1,j}\left(\frac{m}{M_2p^{-r}}\right)e(mz)=f_{M_1,j}^{M_2p^{-r}}(z)-a_{M_1,j}(p)\cdot f_{M_1,j}^{M_2p^{-r+1}}(z)+\chi(p)p^{k-1}f_{M_1,j}^{M_2p^{-r+2}}(z).$$

For each $0\le \ell\le \alpha-2-r$, we get

$$\begin{aligned}&\sum_{\substack{m=1,\\ p\nmid m}}^{\infty}a_{M_1,j}\left(\frac{mp^{\alpha-2-r-\ell}}{M_2p^{-r}}\right)e(mp^{\ell}z)\\ &=a_{M_1,j}\left(p^{\alpha-2-r-\ell}\right)\cdot\left\{f_{M_1,j}^{M_2p^{-r+\ell}}(z)-a_{M_1,j}(p)\cdot f_{M_1,j}^{M_2p^{-r+1+\ell}}(z)+\chi(p)p^{k-1}f_{M_1,j}^{M_2p^{-r+2+\ell}}(z)\right\}.\end{aligned}$$

So we finally get

$$\begin{aligned}T_\alpha f_{M_1,j}^{M_2}=&\left\{a_{M_1,j}(p^{\alpha-r})-\chi(p)p^{k-1}a_{M_1,j}(p^{\alpha-2-r})\right\}f_{M_1,j}^{M_2p^{-r}}\\ &+\sum_{\ell=0}^{\alpha-2-r}(\chi(p)p^{k-1})^{\ell+1}a_{M_1,j}\left(p^{\alpha-2-r-\ell}\right)\\ &\qquad\cdot\left\{f_{M_1,j}^{M_2p^{-r+\ell}}-a_{M_1,j}(p)\cdot f_{M_1,j}^{M_2p^{-r+1+\ell}}+\chi(p)p^{k-1}f_{M_1,j}^{M_2p^{-r+2+\ell}}\right\}\\ &-(\chi(p)p^{k-1})^{\alpha-r}f_{M_1,j}^{M_2p^{\alpha-2r}}.\end{aligned}$$

For $s\in\mathbb{Z}_{\ge 0}$, we have

$$a_{M_1,j}(p^s)\cdot a_{M_1,j}(p)=a_{M_1,j}(p^{s+1})+\chi(p)p^{k-1}a_{M_1,j}(p^{s-1}),$$

so we get

$$\begin{aligned}&\sum_{\ell=0}^{\alpha-2-r}(\chi(p)p^{k-1})^{\ell+1}a_{M_1,j}\left(p^{\alpha-2-r-\ell}\right)a_{M_1,j}(p)\cdot f_{M_1,j}^{M_2p^{-r+1+\ell}}\\ &=\sum_{\ell=0}^{\alpha-2-r}(\chi(p)p^{k-1})^{\ell+1}a_{M_1,j}(p^{\alpha-1-r-\ell})\cdot f_{M_1,j}^{M_2p^{-r+1+\ell}}\end{aligned}$$

$$+\sum_{\ell=0}^{\alpha-2-r}(\chi(p)p^{k-1})^{\ell+2}a_{M_1,j}\left(p^{\alpha-3-r-\ell}\right)\cdot f_{M_1,j}^{M_2p^{-r+1+\ell}}$$

$$=\sum_{\ell=1}^{\alpha-1-r}(\chi(p)p^{k-1})^{\ell}a_{M_1,j}(p^{\alpha-r-\ell})\cdot f_{M_1,j}^{M_2p^{-r+\ell}}$$

$$+\sum_{\ell=1}^{\alpha-1-r}(\chi(p)p^{k-1})^{\ell+2}a_{M_1,j}\left(p^{\alpha-2-r-\ell}\right)\cdot f_{M_1,j}^{M_2p^{-r+\ell}}.$$

Then we have

$$\begin{aligned}
T_\alpha f_{M_1,j}^{M_2} &= \left\{a_{M_1,j}(p^{\alpha-r})-\chi(p)p^{k-1}a_{M_1,j}(p^{\alpha-2-r})\right\}f_{M_1,j}^{M_2p^{-r}}\\
&\quad+\sum_{\ell=0}^{\alpha-2-r}(\chi(p)p^{k-1})^{\ell+1}a_{M_1,j}\left(p^{\alpha-2-r-\ell}\right)f_{M_1,j}^{M_2p^{-r+\ell}}\\
&\quad+\sum_{\ell=2}^{\alpha-r}(\chi(p)p^{k-1})^{\ell}a_{M_1,j}\left(p^{\alpha-r-\ell}\right)f_{M_1,j}^{M_2p^{-r+\ell}}\\
&\quad-\sum_{\ell=1}^{\alpha-1-r}(\chi(p)p^{k-1})^{\ell}a_{M_1,j}(p^{\alpha-r-\ell})\cdot f_{M_1,j}^{M_2p^{-r+\ell}}\\
&\quad-\sum_{\ell=1}^{\alpha-1-r}(\chi(p)p^{k-1})^{\ell+2}a_{M_1,j}\left(p^{\alpha-2-r-\ell}\right)\cdot f_{M_1,j}^{M_2p^{-r+\ell}}\\
&\quad-(\chi(p)p^{k-1})^{\alpha-r}f_{M_1,j}^{M_2p^{\alpha-2r}}\\
&= a_{M_1,j}(p^{\alpha-r})f_{M_1,j}^{M_2p^{-r}}-(\chi(p)p^{k-1})a_{M_1,j}(p^{\alpha-r-1})f_{M_1,j}^{M_2p^{-r+1}}.
\end{aligned}$$

Combining, we obtain (6.1.2). □

For $n, N \in \mathbb{Z}_{>0}$, we write $\gcd(n, N^\infty)$ for the largest positive integer d such that $d \mid n$ and $d \mid N^k$ for some $k \in \mathbb{Z}_{\geq 1}$, i.e.,

$$\gcd(n, N^\infty) = \prod_{p\mid\gcd(n,N)} p^{\mathrm{ord}_p(n)}. \tag{6.1.4}$$

The following corollary is then immediate.

Corollary 6.1.5 *With notation as above, we have*

$$\mathrm{Tr}(T_n \mid S_k(N,\chi)) = \sum_{\substack{M\in\mathbb{Z}_{\geq 1}\\ M\mid N\\ \mathrm{cond}(\chi)\mid M}} d\left(\frac{N/M}{\gcd(N/M, n^\infty)}\right)\sum_{\substack{b^2\mid\gcd(n,N^\infty)\\ \gcd(b,M)=1}}\mu(b)\chi(b)b^{k-1}\,\mathrm{Tr}(T_{\frac{n}{b^2}} \mid S_k^{\mathrm{new}}(M,\chi)).$$

6.2 Certifying Generalized Eigenvalues

Second, we show how to certify generalized eigenvalues. Consider the generalized eigensystem

$$Ax = \lambda Bx, \tag{6.2.1}$$

where A and B are real symmetric $n \times n$ matrices, with B positive definite. Choosing R such that $B = R^T R$ and making the change of variables $x = R^{-1}y$, this becomes

$$A'y = \lambda y, \tag{6.2.2}$$

where $A' = (R^{-1})^T AR^{-1}$. Note that A' is again symmetric, so there is an orthonormal basis $\{y_1, \ldots, y_n\}$ with $A'y_j = \lambda_j y_j$. We set $x_j = R^{-1}y_j$, so that the x_j are orthonormal with respect to the inner product defined by B.

Suppose that we have found approximate eigenvalues $\tilde{\lambda}_j$ and eigenvectors $\tilde{x}_j$, i.e. so that $e_j = (A - \tilde{\lambda}_j B)\tilde{x}_j$ is small. Let

$$\tilde{x}_j = \sum_{k=1}^{n} c_{jk} x_k$$

be the expansion of $\tilde{x}_j$ in terms of the eigenbasis. For any $\varepsilon > 0$, define

$$V_{j,\varepsilon} = \operatorname{Span}\{x_k : |\lambda_k - \tilde{\lambda}_j| < \varepsilon\}, \tag{6.2.3}$$

and let

$$v_{j,\varepsilon} = \sum_{\substack{k \\ |\lambda_k - \tilde{\lambda}_j| < \varepsilon}} c_{jk} x_k \tag{6.2.4}$$

be the orthogonal projection (with respect to the inner product defined by B) of $\tilde{x}_j$ onto $V_{j,\varepsilon}$. Then we have

$$\begin{aligned} v_{j,\varepsilon}^T B v_{j,\varepsilon} &= \tilde{x}_j^T B \tilde{x}_j - \sum_{\{k:|\lambda_k - \tilde{\lambda}_j| \geq \varepsilon\}} c_{jk}^2 \geq \tilde{x}_j^T B \tilde{x}_j - \varepsilon^{-2} \sum_{k=1}^{n} c_{jk}^2 (\lambda_k - \tilde{\lambda}_j)^2 \\ &= \tilde{x}_j^T B \tilde{x}_j - \varepsilon^{-2} [(B^{-1}A - \tilde{\lambda}_j)\tilde{x}_j]^T B [(B^{-1}A - \tilde{\lambda}_j)\tilde{x}_j] \\ &= \tilde{x}_j^T B \tilde{x}_j - \varepsilon^{-2} e_j^T B^{-1} e_j \geq \tilde{x}_j^T B \tilde{x}_j - \varepsilon^{-2} b^{-1} |e_j|^2, \end{aligned} \tag{6.2.5}$$

where $b > 0$ is the smallest eigenvalue of B. Note that this is positive if

$$\varepsilon > \varepsilon_j := \frac{|e_j|}{\sqrt{b\tilde{x}_j^T B \tilde{x}_j}},$$

and thus $V_{j,\varepsilon}$ is non-zero. Hence, there is an eigenvalue λ_k in the interval $I_j = [\tilde{\lambda}_j - \varepsilon_j, \tilde{\lambda}_j + \varepsilon_j]$.

Suppose that we are in the favorable situation that the I_j are pairwise disjoint. Then our system has distinct eigenvalues, and we may assume without loss of generality that $\lambda_j \in I_j$. Next, let $\delta_j = \min\{|\lambda - \tilde{\lambda}_j| : \lambda \in \bigcup_{k \neq j} I_k\}$, so that $(\tilde{\lambda}_j - \delta_j, \tilde{\lambda}_j + \delta_j)$ contains λ_j and no other eigenvalues. Put $\Delta_j = \tilde{x}_j - v_{j,\delta_j}$, where $V_{j,\varepsilon}$ and $v_{j,\varepsilon}$ are as above, so that $\tilde{x}_j - \Delta_j$ is an eigenvector with eigenvalue λ_j. To use this in practice, we bound the coordinates of Δ_j from above and add them as small error intervals onto the coordinates of $\tilde{x}_j$. (The resulting vector must then be renormalized in interval arithmetic, according to whatever convention we use, e.g. first Fourier coefficient 1.) To that end, we have

$$\begin{aligned}|R\Delta_j|^2 = \Delta_j^T B \Delta_j = \sum_{\{k : |\lambda_k - \tilde{\lambda}_j| \geq \delta_j\}} c_{jk}^2 \leq \delta_j^{-2} \sum_{k=1}^{n} c_{jk}^2 (\lambda_k - \tilde{\lambda}_j)^2 \\ = \delta_j^{-2} [(B^{-1}A - \tilde{\lambda}_j)\tilde{x}_j]^T B [(B^{-1}A - \tilde{\lambda}_j)\tilde{x}_j] = \delta_j^{-2} e_j^T B^{-1} e_j,\end{aligned} \tag{6.2.6}$$

and thus

$$|\Delta_j| \leq \delta_j^{-1} \sqrt{b^{-1} e_j^T B^{-1} e_j} \leq \frac{|e_j|}{b\delta_j}. \tag{6.2.7}$$

Finally, to estimate b we first compute a double-precision approximation $\tilde{P}$ to the orthogonal matrix which diagonalizes B. We then compute in interval arithmetic the matrices

$$S = (s_{jk}) = \tilde{P}^T B \tilde{P} \quad \text{and} \quad T = (t_{jk}) = \tilde{P}^T \tilde{P}.$$

By Sylvester's law of inertia, we have $b > \lambda$ for any λ such that $S - \lambda T$ is positive definite. In turn, by the Gershgorin circle theorem, this holds if the diagonal entries s_{jj} and t_{jj} are strictly positive and

$$\lambda > \lambda^* := \min_j \frac{2s_{jj} - \sum_k |s_{jk}|}{\sum_k |t_{jk}|}. \tag{6.2.8}$$

Hence $b \geq \lambda^*$.

7 A Sample of the Implementations

7.1 Comparison of Methods

In the course of our computations we made extensive use of the modular forms functionality included in both Pari/GP [72] and Magma [11]. In this section we compare the performance of the two implementations on a small but representative subset of the modular forms we computed: all newforms of weight k and level N with $Nk^2 \leq 1000$ and $k > 1$. We exclude the case $k = 1$ from this comparison because it is not fully supported in Magma and the algorithms used to compute weight one forms are substantially different. For modular forms of weight $k > 1$ the Magma implementation is based on the modular symbols approach, while the Pari/GP implementation uses the explicit trace formula.

For each level N in our chosen range we fix a representative Dirichlet character χ for each Galois orbit $[\chi]$ of modulus N, and for each newspace $S_k^{\text{new}}(N, \chi)$ with $k > 1$ and $Nk^2 \leq 2000$ we carried out the following computations in both Pari/GP and Magma:

(1) Determine the dimensions of the irreducible subspaces of $S_k^{\text{new}}(N, [\chi])$ (the newform orbits).
(2) For each newform orbit $[f]$, compute the first 1000 integer coefficients t_n of the trace form $\text{Tr}(f) = \sum_{n \geq 1} t_n q^n$.
(3) For each newform orbit $[f]$ of (absolute) dimension $d \leq 20$, compute a (reasonably nice) defining polynomial for its coefficient field K and the first 1000 algebraic integer coefficients $a_n(f) \in K$ for a constituent newform f.
(4) For each newform orbit $[f]$ of dimension $d \leq 20$, compute an LLL-optimized basis for its coefficient ring and express the first 1000 coefficients $a_n(f)$ in this basis.

As can be seen in Tables 5 and 6, the explicit trace formula approach used by Pari/GP is faster overall than the modular symbol method implemented in Magma, especially for spaces that consists of a single Galois orbit, but for newspaces that split into multiple Galois orbits it is typically slower, and in general Magma is able to decompose newspaces into Galois orbits more quickly than Pari/GP. The large advantage Pari/GP has on irreducible spaces is due to the fact that in this situation we can use `mfsplit` to determine that the space is irreducible without actually computing any eigenforms, and then use `mftraceform` to compute the trace form for the entire space.

In Table 7 we list the 10 newspaces in our chosen range that were the computationally most difficult for either Magma or Pari/GP. In each case, the 10 most time consuming newspaces accounted for approximately half of the total time to process the 7621 nonzero newspaces in our test range.

Notably, only two newspaces (467.2.c and 497.2.c) were among the computationally most difficult for both methods (these are the two newspaces of largest dimension in our chosen range). Most of the newspaces listed in Table 7 were

Table 5 Magma 2.24-7 vs. Pari/GP 2.12.1 (Intel Xeon W-2155 3.3 GHz); timings for newspaces S of level $N \geq 1$, weight $k > 1$ by Nk^2 range

				Split time (s)		Total time (s)	
Nk^2	Num S	Num f	$\sum \dim(S)$	Magma	Pari/GP	Magma	Pari/GP
[1, 200]	183	214	897	0.4	1.1	73.8	18.2
[201, 400]	453	709	7 560	3.5	17.2	302.4	116.6
[401, 600]	574	1 050	21 452	22.2	50.2	643.4	220.1
[601, 800]	677	1 326	43 515	132.1	70.8	2 444.8	300.6
[801, 1000]	764	1 542	71 358	751.3	322.3	9 216.4	728.2
[1001, 1200]	879	1 805	109 570	2 653.1	1 253.3	36 940.0	2 347.6
[1201, 1400]	905	2 001	152 344	8 889.0	5 517.0	161 327.7	11 855.8
[1401, 1600]	995	2 284	203 492	24 841.1	21 256.5	349 656.8	67 233.4
[1601, 1800]	1 032	2 420	264 506	63 476.2	59 392.6	952 669.0	194 405.8
[1801, 2000]	1 157	2 378	331 348	79 307.2	175 890.1	1 752 685.4	596 779.2
	7 621	15 731	1 206 658	180 089.5	263 771.8	3 266 135.9	874 006.0

Table 6 Magma 2.24-7 vs. Pari/GP 2.12.1 (Intel Xeon W-2155 3.3 GHz); timings for newspaces S of level $N \geq 1$, weight $k > 1$, $Nk^2 \leq 2000$ by $\#S := \#\{f \in S\}$

					Split time (s)		Total time (s)	
$\#S$	Max $\dim(f)$	Num S	Num f	$\sum \dim(S)$	Magma	Pari/GP	Magma	Pari/GP
1	[1, 200]	2 859	2 859	161 375	423.8	529.9	11 967.2	818.0
1	[201, 2000]	1 027	1 027	544 092	26 060.6	55 272.8	701 094.2	53 727.6
1	[2001, ∞]	65	65	215 016	146 044.1	170 751.3	2 226 371.4	163 789.9
2	[1, 200]	1 703	3 406	100 080	278.7	660.9	4 233.8	30 837.0
2	[201, 2000]	145	290	95 704	4 192.1	8188.7	188 745.7	576 764.6
2	[2001, ∞]	4	8	10 870	2 636.5	26 821.1	97 329.8	24 785.1
$\geq$3	[1, 20]	873	4 785	19 282	46.2	64.1	1 596.2	1 197.9
$\geq$3	[21, 200]	235	1 155	23 135	160.8	275.7	1 228.8	5255.3
$\geq$3	[201, ∞]	3	15	1 024	12.0	347.7	1 364.5	357.4
		7 621	15 731	1 206 658	180 089.5	263 771.8	3 266 135.9	874 006.0

computationally much more difficult for one of the two methods: on the largest irreducible spaces in our test range Pari/GP is typically at least ten times as fast as Magma, but for newspaces that split into two large Galois orbits Magma is faster than Pari/GP by a similar (or even larger) factor. This suggests that the optimal approach is to use the explicit trace formula and modular symbol methods in combination. Indeed, a hybrid approach that uses Magma to decompose the space, and then delegates the computation to Pari/GP whenever the newspace contains a Galois orbit of dimension at least 2/3 the dimension of the newspace, takes a total of 264 726 seconds; this is more than 3 times faster than using Pari/GP alone and more than 10 times faster than using Magma alone.

Table 7 Some computationally challenging newspaces

Newspace	$[\mathbb{Q}(\chi) : \mathbb{Q}]$	Decomposition	Magma(s)	Pari/GP(s)
413.2.i	28	420 + 420	68.91	23 711.71
419.2.g	180	6120	79 654.08	5 175.04
424.2.v	24	1248	60 111.92	150.62
431.2.g	168	5880	82 907.51	5 333.59
435.2.bf	12	240 + 240	39.63	25 272.26
443.2.g	192	6912	180 453.61	8 134.21
454.2.c	112	1008 + 1120	1 197.47	44 216.52
467.2.c	232	8816	370 791.77	22 719.24
472.2.l	28	56 + 1568	103 117.42	562.40
478.2.g	96	960 + 960	861.98	87 147.90
479.2.c	238	9282	363 002.59	26 148.67
486.2.i	54	702 + 756	351.60	139 762.27
487.2.k	162	6480	110 903.14	6 766.85
489.2.q	54	702 + 756	202.91	38 345.59
491.2.k	168	6720	12 1405.39	8 558.33
497.2.v	24	408 + 456	99.20	18 438.91
498.2.f	40	560 + 560	269.01	48 844.21
499.2.g	164	6724	119 807.20	12 148.53

7.2 A Trace Formula Implementation with Complex Coefficients

In this section, we describe an implementation of the trace formula using ball arithmetic over the complex numbers due to Bober [8]. This implementation follows the description given in Sect. 5.2. The main focus here is to compute a moderate number of coefficients for all of the newforms in a given space $S_k^{\mathrm{new}}(N, \chi)$ as approximate complex numbers, which enables the computation of modular form L-functions at small height, for example. These computations use Arb [51], a C library which implements arbitrary precision ball arithmetic, so that we can ensure that all of our computations come with rigorous error bounds. There is also some facility for computing with coefficients in a finite field $\mathbb{F}_\ell$, where ℓ is some prime congruent to 1 modulo the order of χ, which is used in the computation of characteristic polynomials of Hecke operators, for example, and in other auxiliary pieces. The package also contains some limited functionality to compute information about spaces of weight one modular forms, which we do not discuss here.

We describe briefly some details of how this implementation works in practice.

To start a computation we first choose a prime and determine a set of trace forms which will give a full rank basis of the space of newforms modulo this prime, avoiding any issues of computing the rank of a matrix with floating point entries. Specifically, we find some matrix of coefficients $(\mathrm{Tr}(T(m_i)T(n_j) \mid S_k^{\mathrm{new}}(N, \chi)))_{1 \le i,j \le d}$ that has full rank, and we also choose our m_i and n_j so that $\gcd(m_i n_j, N) = 1$, which will make later computations easier. Once we know which computation will give us a full rank matrix, we do the computation again over the complex numbers, computing additional coefficients so that we will be able to compute the action of Hecke operators. At this point we find a sum of Hecke operators $T = \sum_n c_n T_n$ such that the characteristic polynomial of T is squarefree (keeping $c_n = 0$ when $\gcd(n, N) \neq 1$).

The diagonalization of T would in general be a computation over the complex numbers, but because we have chosen to only use Hecke operators coprime to the level, we can use knowledge of the arguments of the eigenvalues to turn this into a problem of diagonalizing a real symmetric matrix. This problem is solved by an implementation of Jacobi's algorithm, certifying the result as described in Sect. 6.2. Once we have diagonalized, we obtain a change of basis matrix that takes our trace form basis to the newform basis, and we can compute as many coefficients of newforms as we like by evaluation of the trace formula.

Once we have computed all the embeddings of all of our newforms, we may also wish to compute the decomposition of the space into Hecke-irreducible subspaces. To do this we will compute the characteristic polynomial of a linear combination T of Hecke operators (it is sufficient to find one that is squarefree). If we have enough precision in the Hecke eigenvalues we have computed, we can do this simply by forming the product $\prod_\lambda (x-\lambda)$, where λ ranges over the eigenvalues of T. In general we will find that we do not have enough accuracy to uniquely identify a polynomial with coefficients in $\mathbb{Z}[x]$, however, and we refine the computation by computing this polynomial modulo ℓ for enough small primes ℓ to obtain the polynomial exactly.

The factorization of this Hecke polynomial gives the decomposition of $S_k^{\mathrm{new}}(N, \chi)$ into Hecke irreducible subspaces. However, there is still one more problem we may be faced with: namely, identifying which embeddings correspond to which subspaces. In this problem we have a set of polynomials $f_1, f_2, \ldots, f_n$ and a set of approximations of complex numbers $r_1, r_2, \ldots, r_d$, and all we need to do is determine which complex number is a root of which polynomial. This may seem like a relatively trivial problem, but in fact these polynomials may be enormous and obtaining enough precision in the roots to solve this by simple evaluation may not be feasible.

Example 7.2.1 To give a moderately-sized example, we can consider the space 766.2.c. This space is only 32-dimensional over the field of definition of χ, but there are 190 Galois conjugate characters to consider, so the full degree is 6080. The characteristic polynomial of T_3 acting on the space $S_2^{\mathrm{new}}(766, [\chi])$ is squarefree and factors into 2 irreducible factors of degree 3040; each factor contains more than 1.5 million decimal digits.

To make this problem tractable, we again make use of the arguments of the eigenvalues. Let $f(x)$ be one of these irreducible factors. We know that each root of f can be written as ζt for some root of unity ζ and some real number t, and we find that t is a root of the greatest common divisor of $f(\zeta x)$ and $f(\zeta^{-1} x)$ in $\mathbb{Q}(\zeta)[x]$. In fact, as we prefer to work with real numbers, we compute

$$\gcd(f(\zeta x) + f(\zeta^{-1}x), i(f(\zeta x) - f(\zeta^{-1}x))) \in \mathbb{Q}(\zeta + \zeta^{-1})[x].$$

In principle, computing this greatest common divisor when we have only floating point approximations available could be troublesome, but it is possible because we know what its degree is.

8 Issues: Computational, Theoretical, and Practical

8.1 Analytic Conductor

Earlier efforts to tabulate modular forms have tended to compute all newforms in particular boxes, where the weight and level each vary in a specified range. This approach is easy to describe, but the computational complexity of finding newforms with simultaneously large weight and level ensures that some newforms of interest will be missed (either large weight or large level). Instead of working with boxes, we organized our computation around a single invariant which scales with the complexity of the newform.

Introduced by Iwaniec–Sarnak [50, Eq. (31)] (see also Iwaniec–Kowalski [49, (5.7)]), the analytic conductor of a newform $f \in S_k^{\mathrm{new}}(N, \chi)$ is the positive real number

$$A := N\left(\frac{\exp(\psi(k/2))}{2\pi}\right)^2, \tag{8.1.1}$$

where $\psi(x) := \Gamma'(x)/\Gamma(x)$ is the logarithmic derivative of the Gamma function. The analytic conductor includes a factor that can be thought of as measuring the complexity at infinity. We have $A \sim \dfrac{Nk^2}{16\pi^2}$ as $k \to \infty$, so for simplicity we organized our computations by specifying bounds on the quantity Nk^2.

8.2 Sturm Bound

In this section, we elaborate upon bounds for truncations of q-expansions of modular forms that determine them uniquely. The most well-known of these bounds is due to Hecke (and more generally to Sturm [91, Theorem 1]).

Theorem 8.2.1 (Hecke, Sturm) *Let $\Gamma \leq \mathrm{SL}_2(\mathbb{Z})$ be a congruence subgroup and let f be a modular form of weight k for Γ. Then $f = 0$ if and only if $a_n(f) = 0$ for all $0 \leq n \leq k[\mathrm{SL}_2(\mathbb{Z}) : \Gamma]/12$.*

In fact, for modular forms with character as in our setting, one can apply a sharper bound (as though it was without character) as follows.

Definition 8.2.2 For $k, N \in \mathbb{Z}_{\geq 1}$, define the (Hecke-)Sturm bound

$$\mathrm{Sturm}(k, N) := \frac{k}{12}[\mathrm{SL}_2(\mathbb{Z}) : \Gamma_0(N)] = \frac{Nk}{12}\prod_{p|N}\left(1+\frac{1}{p}\right).$$

Proposition 8.2.3 (Hecke, Sturm) *Let $N, k \geq 1$ and let χ be a character of modulus N. Let $\mathcal{T} \subseteq \mathrm{End}_{\mathbb{C}}(S_k(N, \chi))$ be the $\mathbb{Z}$-subalgebra generated by the Hecke operators T_n for all $n \in \mathbb{Z}_{\geq 1}$, and let $\mathbb{Z}[\chi] \subseteq \mathbb{C}$ be the $\mathbb{Z}$-subalgebra generated by the values of χ. Then there is a natural inclusion $\mathbb{Z}[\chi] \hookrightarrow \mathcal{T}$, and the following statements hold.*

(a) *If $f \in S_k(N, \chi)$ has $a_n(f) = 0$ for all $n \leq \mathrm{Sturm}(k, N)$, then $f = 0$.*
(b) *$\mathcal{T}$ is generated as a $\mathbb{Z}[\chi]$-module by T_n for all $n \leq \mathrm{Sturm}(k, N)$.*
(c) *$\mathcal{T}$ is generated as a $\mathbb{Z}[\chi]$-algebra by T_1 and T_p for all primes $p \leq \mathrm{Sturm}(k, N)$.*

Proof For the inclusion $\mathbb{Z}[\chi] \hookrightarrow \mathcal{T}$, we argue as follows: from the Hecke recursion

$$T_{p^2} - T_p^2 + \chi(p)p^{k-1} = 0 \tag{8.2.4}$$

for $p \nmid N$, we see that $\chi(p)p^{k-1} \in \mathcal{T}$; choosing two distinct primes p congruent modulo N and applying the Chinese Remainder Theorem shows that $\mathbb{Z}[\chi] \subseteq \mathcal{T}$. Consequently, $\mathcal{T}$ is a torsion free $\mathbb{Z}[\chi]$-module. Since $\mathbb{Z}[\chi]$ is a Dedekind domain, $\mathcal{T}$ is locally free.

Abbreviate $S := S_k(N, \chi; \mathbb{Z}[\chi])$. We claim that the pairing

$$\begin{aligned} \mathcal{T} \times S &\to \mathbb{Z}[\chi] \\ (T, f) &\mapsto a_1(Tf) \end{aligned} \tag{8.2.5}$$

is perfect, i.e., the map

$$\begin{aligned} \varphi \colon S &\to \mathrm{Hom}_{\mathbb{Z}[\chi]}(\mathcal{T}, \mathbb{Z}[\chi]) \\ f &\mapsto (T \mapsto a_1(Tf)) \end{aligned} \tag{8.2.6}$$

is an isomorphism. When $\mathbb{Z}[\chi] = \mathbb{Z}$, this is an argument of Ribet [77, Theorem (2.2)], and we only need to make a small modification. The map φ is injective with torsion-free cokernel because $a_1 \circ T_n = a_n$ and the map taking a form to its q-expansion is injective. Since S and $\mathcal{T}$ are locally free $\mathbb{Z}[\chi]$-modules of finite rank, it suffices to show that the rank of $\mathcal{T}$ is at most the rank of S (localizing at primes $\mathfrak{l}$ of $\mathbb{Z}[\chi]$). To this end, consider the other map induced by the pairing, namely,

$$\begin{aligned}\omega\colon \mathcal{T} &\to \mathrm{Hom}_{\mathbb{Z}[\chi]}(S, \mathbb{Z}[\chi]) \\ T &\mapsto (f \mapsto a_1(Tf)).\end{aligned} \tag{8.2.7}$$

We claim that ω is injective. Indeed, if $T \in \ker \omega$, then for all $f \in S$ and all $n \geq 1$ we have

$$0 = \omega(T)(T_n f) = a_1(T T_n f) = a_1(T_n T f) = a_n(Tf)$$

as $\mathcal{T}$ is commutative. Since the q-expansion map is injective, we conclude $Tf = 0$ for all f, so $T = 0$ as an endomorphism, proving the claim. Finally, localizing ω at $\mathfrak{l}$, the injectivity of ω implies the desired rank bound.

We now prove (a) following Buzzard, and suppose that $f \in S_k(N, \chi; \mathbb{Z}[\chi])$ has $a_n(f) = 0$ for all $n \leq \mathrm{Sturm}(k, N)$. Let d be the order of χ, let $s := \mathrm{Sturm}(k, N)$, and consider $f^d \in S_{dk}(\Gamma_0(N); \mathbb{Z}[\chi])$. Since $f = O(q^{s+1})$, we have $f^d = O(q^{d(s+1)})$. Moreover, $\mathrm{Sturm}(kd, N) = ds$, so by the Sturm bound (Theorem 8.2.1) applied to $S_{dk}(\Gamma_0(N); \mathbb{Z}[\chi])$ we conclude $f^d = 0$, which implies $f = 0$.

To prove (b), let $\mathcal{T}_{\leq n} \subseteq \mathcal{T}$ be the $\mathbb{Z}[\chi]$-submodule generated by T_n with $n \leq \mathrm{Sturm}(k, N)$. By the previous paragraph, the pairing (8.2.5) restricted to $\mathcal{T}_{\leq n}$ is still perfect; indeed we can simply argue with $\mathcal{T}_{\leq n}$ in the injectivity of ω in (8.2.7). We conclude that $\mathcal{T}_{\leq n} = \mathcal{T}$.

For part (c), we use multiplicativity to see that T_n for n composite is contained in the algebra generated by the prime power operators T_{p^e}, and then the Hecke recurrence and induction to see that T_{p^e} is contained in the algebra generated by T_1 and T_p.

8.3 Atkin–Lehner Operators and Eigenvalues

Let χ be a Dirichlet character modulo N. For $M \mid N$ with $\gcd(M, N/M) = 1$, there are unique characters $\chi_M \pmod{M}$ and $\chi_{N/M} \pmod{N/M}$ such that $\chi = \chi_M \chi_{N/M}$. The Atkin–Lehner–Li operator W_M [2, §1] maps the space $S_k(N, \chi)$ to $S_k(N, \overline{\chi_M} \chi_{N/M})$. In general $\overline{\chi_M} \chi_{N/M}$ is different from χ, so then this operator cannot be used for splitting up spaces. We have $\overline{\chi_M} \chi_{N/M} = \chi$ when the character χ_M is trivial or quadratic, and in these cases, W_M is an involution on the space $S_k(N, \chi)$, which then splits as the direct sum of ± 1-eigenspaces. Magma only implements Atkin–Lehner operators on spaces with trivial character, where they commute with all Hecke operators and hence map every newform f to $\pm f$, and the sign $\pm$ is the *Atkin–Lehner eigenvalue* of f with respect to M. (By a common abuse of notation and terminology, when M is the power of a prime q not dividing N/M, the operator W_M is often denoted W_q.) In our computations we only compute Atkin–Lehner eigenvalues on newforms with trivial character.

In general, the image of a normalised newform f in $S_k(N, \chi_M \chi_{N/M})$ under W_M is a multiple of a normalised newform in $S_k(N, \overline{\chi_M} \chi_{N/M})$, and the multiple, not necessarily ± 1, is called the *pseudo-eigenvalue* of W_M on f. Atkin–Li [2] do not give a general formula for pseudo-eigenvalues, which are not always easy to compute in practice. See also Belabas–Cohen [4, §§5–6].

When $M = 1$ the operator W_M is trivial, while when $M = N$ it is called the *Fricke involution*. The Fricke involution is the product of all W_q for primes $q \mid N$ (using the convention of the previous paragraph.) For a newform f with trivial character, its Fricke eigenvalue is *minus* the sign of the functional equation of its L-function, and each W_q-eigenvalue is the sign of a certain local functional equation.

8.4 Self-Duality

The coefficient field of a newform $f \in S_k(N, \chi)$ is either totally real or CM [75, Prop 3.2]; we say that f is self-dual in the totally real case. Computing the coefficient field can be time consuming, so we use the following easier criteria when applicable.

Proposition 8.4.1 (Ribet) *Let $f \in S_k(N, \chi)$ have Hecke orbit of dimension d and trace form $\sum_{n=0}^{\infty} t_n q^n$. Then the following statements hold.*

(a) *If χ is trivial or d is odd, then f is self-dual.*
(b) *If χ has order larger than 2, then f is not self-dual.*
(c) *If there exists a prime p so that $t_p \neq 0$ and $\chi(p) \neq 1$, then f is not self-dual.*

Proof See Ribet [75, Propositions 3.2 and 3.3].

8.5 Efficiently Recognizing Irreducibility

Level $N = 2$ is by far the most time-consuming case for Magma (note that this allows for the largest range of k with $N \neq 1$ for any given bound on the analytic conductor). For $k > 400$ with $4 \mid k$, each space takes more than 12 hours of CPU time. However, we observed behavior analogous to the Maeda conjecture in weight 1 up to weight $k \leq 400$, with the additional prediction that the Atkin–Lehner operator splits the space as evenly as possible.

Conjecture 8.5.1 *For all $k \geq 2$, the space $S_k(\Gamma_0(2))$ decomposes under the Atkin–Lehner operator W_2 into Hecke irreducible subspaces of dimensions $\lfloor d/2 \rfloor$ and $\lceil d/2 \rceil$ where $d := \dim_{\mathbb{C}} S_k(\Gamma_0(2))$.*

For general N, the difference $\dim S_2(N)^+ - \dim S_2(N)^-$ can be expressed in terms of class numbers of imaginary quadratic fields (fixed points of Atkin–Lehner involutions), and so this difference is $\ll N^{1/2+\epsilon}$. Perhaps this is explicit enough for $N = 2$ to prove the dimensions above (but not irreducibility), and something similar can be said for N composite.

In practice we should be able to prove that the space is irreducible by working with just one Hecke polynomial as these are often irreducible.

Question 8.5.2 Given an $n \times n$ matrix with entries in $\mathbb{Z}[\zeta_m]$ (typically sparse), is there a fast algorithm that with positive probability correctly determines when its characteristic polynomial is irreducible?

In other words, if you expect that a polynomial is irreducible, can you verify this quickly without factoring the polynomial? Under the expectation that the Galois group of the polynomial is transitive and therefore likely to be S_d, one could succeed in some cases by factoring the polynomial modulo primes. This is different than the typical factorization problems solved in computer algebra systems, which compute a factorization p-adically and then reconstruct the factorization over $\mathbb{Z}$. (See Table 7 for some difficult spaces where this would help.) A natural generalization of this question would be to efficiently determine the degrees of the irreducible factors of its characteristic polynomial (without explicitly computing it).

8.6 Trace Form

As defined in Sect. 4.5, each newform $f \in S_k^{\mathrm{new}}(N, [\chi])$ has an associated trace form $\mathrm{Tr}(f)(q) = \sum_n t_n q^n \in S_k(N, [\chi]; \mathbb{Z})$ equal to the sum of the distinct $\mathrm{Gal}(\mathbb{Q}^{\mathrm{al}} \mid \mathbb{Q})$ conjugates of f and thereby well-defined on its Galois orbit $[f]$. More precisely, we have $\mathrm{Tr}(f) \in S_k^{\mathrm{new}}(\Gamma)$ where

$$\Gamma := \{\left(\begin{smallmatrix} a & b \\ c & d \end{smallmatrix}\right) \in \Gamma_0(N) : \chi(a) = 1\} \supseteq \Gamma_1(N),$$

(but in general $\mathrm{Tr}(f) \notin S_k(N, \chi)$). One can thus apply the Sturm bound (Theorem 8.2.1) for Γ: trace forms of newforms in $S_k(N, \chi)$ with the same Fourier coefficients a_n for $n \leq k[\mathrm{SL}_2(\mathbb{Z}) : \Gamma]/12$ must coincide, but note that this will typically be larger than the Sturm bound $\mathrm{Sturm}(k, N)$. As noted above, we always have $t_1 = [K : \mathbb{Q}]$, where $K = \mathbb{Q}(f)$ is the coefficient field of f.

Trace forms can be efficiently computed using the trace formula. In the common case where $S_k(N, \chi)$ is irreducible, this can be done via the Pari/GP function `mftraceform` [72], which is dramatically faster than computing the coefficients of f as elements of K and taking traces (indeed, it is not even necessary to determine K). More generally, if one knows the decomposition of $S_k(N, [\chi])$ into newform subspaces and has computed trace forms for all but one of them, the remaining trace form can be computed by subtracting corresponding coefficients of the known trace forms from the coefficients given by `mftraceform`, which computes absolute traces of the Hecke operators T_n acting on the entire newspace $S_k(N, [\chi])$. Alternatively, one can sum complex coefficients a_n of the Galois conjugates of f computed to sufficient precision to allow the sum to be identified as a unique integer.

The coefficients t_p of the trace form at primes p are equal to the Dirichlet coefficients of the (typically imprimitive) L-function $L(s) = \sum b_n n^{-s}$ with integer Dirichlet coefficients b_n obtained by taking the product of the L-functions of the Galois conjugates of f. But for nonprime values of n the integer coefficients t_n do not match the integer coefficients b_n unless the newspace has dimension one (in which case $L(s)$ is primitive). Indeed $t_1 = [K : \mathbb{Q}]$ cannot coincide with $b_1 = 1$, and in general the coefficients t_n at nonprime values of n encode different information.

8.7 *Presenting Coefficients Using LLL-Reduction*

One of the most dramatic improvements we saw, both in performance and in display, is in the choice of how to represent coefficient rings. In this section and the next, we explain two such methods.

As explained in Sect. 4.5, one computes a minimal polynomial for the coefficient field by factoring the characteristic polynomial of a Hecke operator. This polynomial may be unwieldy! So we first apply the Pari/GP function `polredbest` which finds an improved minimal polynomial representing the same field by computing an LLL-reduced basis for an order with respect to the Minkowski embedding (whose underlying quadratic form is given by the T_2-norm)—this runs in deterministic polynomial time in the size of the input. When possible, we improve this to `polredabs`, which applies the same technique but to the maximal order (this may require factoring a discriminant, and we frequently encounter situations where this is a bottleneck).

Remark 8.7.1 The function `polredabs` changed in Pari/GP 2.9.5 (Fall 2017); we use the more recent version, described in [28].

We can make significant further improvements by optimizing the $\mathbb{Z}$-basis we use to represent coefficients. Let $f \in S_k^{\text{new}}(N, \chi)$ be a newform. By the Hecke–Sturm bound (Proposition 8.2.3), the coefficient ring of f is generated over $\mathbb{Z}[\chi]$ by the values $a_n(f)$ for $n \leq \text{Sturm}(k, N)$, so by extension we obtain a set of $\mathbb{Z}$-module generators for the ring. We reduce this to an LLL-reduced $\mathbb{Z}$-basis for the coefficient ring, and we rewrite the coefficients in terms of this basis. In our computations, we always use complex precision that is at least as large as the discriminant of the coefficient ring.

We observe the following.

Lemma 8.7.2 *Let F be a number field and let $R \subseteq F$ be a $\mathbb{Z}$-order in F. Then the shortest vectors in R with respect to the T_2-norm are exactly the roots of unity in R.*

Proof The order R contains 1, and $T_2(1) = n := [F : \mathbb{Q}]$. More generally, for any root of unity $\zeta \in R$, we have $T_2(\zeta) = n$. Conversely, let $\alpha \in R$ satisfy $T_2(\alpha) \leq n$. Then by the arithmetic-geometric mean inequality, we have

$$1 \geq \frac{T_2(\alpha)}{n} = \frac{1}{n}\sum_{i=1}^{n} |\alpha_i|^2 \geq \prod_{i=1}^{n} |\alpha_i|^{2/n} = \left|\mathrm{Nm}_{F|\mathbb{Q}}\,\alpha\right|^{2/n}$$

with equality if and only if $|\alpha_1| = \cdots = |\alpha_n| = 1$. But $\alpha \in R$ is integral, so $\left|\mathrm{Nm}_{F|\mathbb{Q}}\,\alpha\right| \geq 1$, so equality holds. By Kronecker's theorem, we conclude that α is a root of unity.

In spite of this lemma, because of nonuniqueness, we may not have 1 as an element of an LLL-reduced basis as there may be more roots of unity than the degree, such as in a cyclotomic field. However, using the above lemma we can recognize the roots of unity in the coefficient ring and thereby recognize when the coefficient ring is cyclotomic itself, where we may institute a canonical basis (see also the next section).

The effect of such a representation is dramatic.

Example 8.7.3 Consider the newform 153.2.e.c. Its coefficient field is $\mathbb{Q}(\nu)$ where ν has minimal polynomial

$$\begin{aligned} x^{20} &- x^{19} + 3x^{18} + 2x^{17} + 13x^{16} - 12x^{15} + 54x^{14} + 27x^{13} + 93x^{12} - 54x^{11} \\ &+ 693x^{10} - 162x^9 + 837x^8 + 729x^7 + 4374x^6 - 2916x^5 + 9477x^4 + 4374x^3 \\ &+ 19683x^2 - 19683x + 59049. \end{aligned}$$

An integral basis written in terms of the powers of ν is too large to record here, and similarly the coefficients of f written in a power basis are enormous!

However, in terms of an LLL-reduced basis $\beta_0 = 1, \cdots, \beta_{19}$, we have coefficients

$$\begin{aligned} a_2(f) &= \beta_{16} \\ a_3(f) &= -\beta_{10} \\ a_4(f) &= -1 - \beta_3 - \beta_5 \\ &\vdots \\ a_{57}(f) &= \beta_2 - \beta_3 - \beta_4 + \beta_5 - 3\beta_7 + 2\beta_9 - \beta_{10} \\ &\qquad + 2\beta_{13} + \beta_{14} + \beta_{15} - \beta_{16} + 2\beta_{17} - \beta_{18} - \beta_{19} \\ &\vdots \end{aligned}$$

that are very small integer linear combinations of the basic elements. Moreover, we have

$$
\begin{aligned}
1 &= \beta_0 \\
\nu &= \beta_1 \\
\nu^2 &= \beta_8 - \beta_7 - \beta_4 + \beta_2 \\
&\vdots \\
\nu^{19} &= 5114\beta_{19} + 2632\beta_{18} + \cdots - 1807\beta_1 - 6756
\end{aligned}
$$

and the powers of ν are reasonably sized $\mathbb{Z}$-linear combinations of our basis elements.

We observe that the matrix that writes the powers of a primitive element in terms of the LLL-reduced basis is noticeably smaller than the other way around. Working with the coefficient ring itself rather than a maximal order containing it is not only more efficient (as it may be prohibitively expensive to compute such a maximal order), but it also seems to give better results.

The intuitive reason that this works is simple: by the Ramanujan–Petersson bounds, the coefficients of a newform are of small size in all complex embeddings, and so it can be expected that writing it in terms of a $\mathbb{Z}$-basis which is LLL-reduced with respect to size provides a small linear combination.

Remark 8.7.4 In the above, we have been concentrating on the case where f is a newform, representing a Galois orbit of newforms, and we write down its q-expansion in terms of a $\mathbb{Z}$-basis for the coefficient ring.

As an alternative, we can consider the $\mathbb{C}$-vector space spanned by f and its conjugates under $\mathrm{Aut}(\mathbb{C})$, making a $\mathbb{C}$-vector space of dimension say d. These conjugates will include conjugates that *do not* preserve the character, so we would either be working implicitly in the direct sum of the spaces over the full Galois orbit of characters, or we need to restrict to quadratic characters, or we only consider conjugates under $\mathrm{Aut}(\mathbb{C} \mid \mathbb{Q}(\chi))$ and get a $\mathbb{Q}(\chi)$-vector space. However, inside this space is a canonical $\mathbb{Q}$-subspace, namely, those forms whose q-expansion belongs to $\mathbb{Q}[[q]]$. So we could instead represent the Galois orbit canonically by an echelonized basis of d individual q-expansions with coefficients in $\mathbb{Q}$. We could then write a representative newform as before as a linear combination of this basis over the coefficient field.

To go from the eigenform to the $\mathbb{Q}$-basis, we apply the operators $\mathrm{Tr}(\beta_i f)$ for β_i any $\mathbb{Q}$-basis for the coefficient field. (To go from the $\mathbb{Q}$-basis to an eigenform one needs to retain sufficiently many eigenvalues to do the linear algebra. In other words, the eigenform contains more information than the $\mathbb{Q}$-basis.) This generalizes the trace form, which is where we take $\beta_i = \beta_0 = 1$.

We could also work integrally and take the $\mathbb{Z}$-module of forms whose q-expansions belong to $\mathbb{Z}$ and then take a LLL-reduced basis which minimizes a (weighted) sum of finitely many coefficients. It is conceivable that in a world where linear algebra over $\mathbb{Q}$ is much faster than linear algebra over number fields that we could succeed in computing a $\mathbb{Q}$-basis in reasonable time but not succeed in computing an eigenform.

8.8 Presenting Coefficients Using a Sparse Cyclotomic Representation

When the coefficient ring of a newform is contained in a cyclotomic field $\mathbb{Q}(\zeta_m)$, writing coefficients in terms of an LLL-optimized basis as described in the previous section does not necessarily give the most compact representation, for two reasons. First, when the coefficient ring is not the maximal order, it may be more compact to express coefficients as elements of the maximal order $\mathbb{Z}[\zeta_m]$. Second, even when the coefficient ring is the maximal order, in which case the LLL-optimized basis will typically be the standard power basis $1, \zeta_m, \zeta_m^2, \ldots, \zeta_m^{\phi(m)-1}$, the eigenvalues a_n can often be written more compactly by expressing them as sparse polynomials in ζ_m rather than integer linear combinations of the power basis. Every integer linear combination of elements of the power basis is of course also a polynomial in ζ_m; the question is whether to allow polynomials of higher degree whose terms involve powers of ζ_m that are not in the power basis (because $m > \phi(m) = [\mathbb{Q}(\zeta_m) : \mathbb{Q}]$), which allows more flexibility and a potentially sparser choice of polynomial.

This added flexibility is particular relevant for weight one newforms, whose coefficients always lie in a cyclotomic field. The correspondence between weight one newforms and (odd irreducible) two-dimensional Artin representations [36] implies that for weight one newforms the eigenvalues a_p can always be written as a sum of at most two roots of unity. For composite values of n the eigenvalues a_n will not be as sparse, but even if one naïvely expands them as products of polynomials in ζ_m for each a_{p^r}, for most a_n we obtain an expression with $O(2^{\log\log n})$ nonzero coefficients (a typical integer n has $\log\log n$ distinct prime factors p and p-adic valuation 1 at all but $O(1)$ of them), which is exponentially sparser than a generic element of $\mathbb{Z}[\zeta_m]$ written in the power basis. For even values of m we can improve on this naïve approach by using the identity $\zeta_m^{m/2} = -1$ to reduce to polynomials of degree at most $m/2 - 1$ in ζ_m; this never increases the number of terms and may reduce it.

For example, the second Fourier coefficient of the newform 3997.1.cz.a is

$$a_2 = -\zeta_{201}^{570} + \zeta_{570}^{244},$$

but when written in terms of the standard power basis $1, \zeta_{570}, \ldots, \zeta_{570}^{143}$ we instead have

$$\begin{aligned} a_2 = 1 &+ \zeta_{570}^{2} + \zeta_{570}^{5} + \zeta_{570}^{11} - \zeta_{570}^{12} - \zeta_{570}^{15} - \zeta_{570}^{17} + \zeta_{570}^{19} - \zeta_{570}^{20} + \zeta_{570}^{21} + \zeta_{570}^{24} + \zeta_{570}^{27} + \zeta_{570}^{30} - \zeta_{570}^{31} \\ &+ \zeta_{570}^{32} - \zeta_{570}^{34} + \zeta_{570}^{35} - \zeta_{570}^{36} - \zeta_{570}^{39} - \zeta_{570}^{42} - \zeta_{570}^{45} + \zeta_{570}^{46} - \zeta_{570}^{47} + \zeta_{570}^{49} - \zeta_{570}^{50} + \zeta_{570}^{51} - \zeta_{570}^{59} \\ &+ \zeta_{570}^{60} - \zeta_{570}^{61} - \zeta_{570}^{64} + \zeta_{570}^{65} - \zeta_{570}^{66} + \zeta_{570}^{74} - \zeta_{570}^{75} - \zeta_{570}^{78} + \zeta_{570}^{79} - \zeta_{570}^{80} + \zeta_{570}^{88} - \zeta_{570}^{89} + \zeta_{570}^{90} \\ &+ \zeta_{570}^{93} - \zeta_{570}^{94} - \zeta_{570}^{97} - \zeta_{570}^{100} - \zeta_{570}^{103} + \zeta_{570}^{104} - \zeta_{570}^{105} - \zeta_{570}^{106} + \zeta_{570}^{107} - \zeta_{570}^{108} + \zeta_{570}^{109} + \zeta_{570}^{112} + \zeta_{570}^{115} \\ &+ \zeta_{570}^{118} - \zeta_{570}^{119} + \zeta_{570}^{120} - \zeta_{570}^{122} + \zeta_{570}^{123} - \zeta_{570}^{124} - \zeta_{570}^{127} - \zeta_{570}^{130} + \zeta_{570}^{134} + \zeta_{570}^{137} + \zeta_{570}^{139} + \zeta_{570}^{142}. \end{aligned}$$

Among the 585 nonzero a_n with $n \leq 2000$, the average number of terms needed to express a_n as a sparse polynomial in ζ_m is 4.1; by contrast, when written in the power basis the average number of nonzero coefficients of a_n is 42.8. This leads to a more than tenfold reduction in storage and a corresponding reduction in the time to transmit or render the q-expansion.

Remark 8.8.1 For modular forms of weight $k > 1$ with cyclotomic coefficient fields $\mathbb{Q}(\zeta_m)$ there is no *a priori* reason to expect the a_p to be expressible as sparse polynomials in ζ_m, and one can see in examples that this is often not the case. One might instead try to apply a more general approach, which, given $\alpha \in \mathbb{Z}[\zeta_m]$ searches for a sparse polynomial $f(\zeta_m)$ of degree less than m with small coefficients that is equivalent to α. We do not know an efficient solution to this problem, but we note that even if one exists, for generic values of α it is unlikely to result in representations that are significantly more compact than using the power basis for purely information theoretic reasons: the number of $\alpha \in \mathbb{Z}[\zeta_m]$ that can be expressed as r-term polynomials in ζ_m using b bits to represent the coefficients must be approximately the same as the number of integer vectors of length $\phi(m)$ that can be encoded in b bits. For this reason we use sparse cyclotomic coefficient representations only for $k = 1$.

8.9 Hecke Kernels

Having determined the decomposition of a newspace $S_k^{\text{new}}(N, [\chi])$ into Hecke orbits V_f corresponding to newforms f, we can compute and store information that will allow us to reconstruct a single Hecke orbit, without having to decompose the entire newspace again. This is particularly useful when the dimension of a particular newform f of interest is much smaller than that of $S_k^{\text{new}}(N, [\chi])$. To achieve this we compute a list of pairs $(p, g_p(X))$, where p is a prime and $g \in \mathbb{Z}[X]$ is the minimal polynomial of the Hecke operator T_p acting on V_f (viewed as a $\mathbb{Q}$-subspace of $S_k^{\text{new}}(N, [\chi])$), such that V_f is equal to the intersection of the kernels of the linear operators $g_p(T_p)$ acting on $S_k^{\text{new}}(N, [\chi])$, in other words, the operators $g_p(T_p)$ generate the Hecke kernel of V_f. Such a list of generators can be used to reconstruct the newform f in Magma via the `Kernel` function.

It is computationally convenient to restrict to primes p not dividing the level N, and to use the same list of primes p for all the newforms in $S_k^{\text{new}}(N, [\chi])$. To this end, for a set of primes $\mathcal{S}$, not dividing N, and a newform f, we let $X_f(\mathcal{S})$ denote the set of pairs (p, g_p), where $g_p \in \mathbb{Z}[X]$ is the minimal polynomial of T_p acting on V_f, and say that $\mathcal{S}$ is a set of distinguishing primes for the newspace $S_k^{\text{new}}(N, [\chi])$ if the sets $X_f(\mathcal{S})$ are distinct as f varies over the newforms of $S_k^{\text{new}}(N, [\chi])$.

We construct a set of distinguishing primes as follows. We start by taking $\mathcal{S}$ to be the empty set. If the newspace $S_k^{\text{new}}(N, [\chi])$ consists of a single Hecke orbit, then $\mathcal{S}$ is a set of distinguishing primes, and otherwise we increase the size of $\mathcal{S}$ by adding the least prime $p \nmid N$ not contained in $\mathcal{S}$ for which

$$\{X_f(\mathcal{S}) : f \in S_k^{\text{new}}(N, [\chi])\} \subsetneq \{X_f(\mathcal{S} \cup \{p\}) : f \in S_k^{\text{new}}(N, [\chi])\}.$$

We observe that the cardinality of the set $\mathcal{S}$ constructed in this fashion is at most one less than the number of Hecke orbits in $S_k^{\text{new}}(N, [\chi])$. This greedy approach to constructing $\mathcal{S}$ does not necessarily minimize its cardinality, but it does minimize the largest p that appears in $\mathcal{S}$, which may be viewed as an invariant of the newspace. For example, we may distinguish the 8 Hecke orbits of the newspace 2608.2.g, where $2608 = 2^4 \cdot 163$, using $p = 3$ and 41. In this case T_3 distinguishes all the forms with the exception of the two CM forms, which both have vanishing a_p for all p split in $\mathbb{Q}(\sqrt{-163})$, hence the smallest prime p such that a_p could possibly distinguish them is 41, and 41 does in fact do so.

Remark 8.9.1 The largest prime p that appears in $\mathcal{S}$ may occasionally exceed the Sturm bound, as in the case of the newforms 66.2.b and 735.2.p, for example. This fact is relevant in the context of Theorem 11.2.8, which we use to determine the group of inner twists of a newform in Sect. 11. This is one reason to compute $a_p(f)$ past the Sturm bound.

9 Computing L-Functions Rigorously

In this section, we describe rigorous methods to compute L-functions of modular forms.

9.1 Embedded Modular Forms

To a newform $f \in S_k^{\text{new}}(N, \chi)$, with q-expansion $\sum a_n q^n$, for each complex embedding of the coefficient field $\iota : \mathbb{Q}(f) \hookrightarrow \mathbb{C}$ we may consider the embedded modular form

$$\iota(f) := \sum \iota(a_n) q^n, \tag{9.1.1}$$

which is a modular form over the complex numbers.

We label such forms by N.k.s.x.c.j, where N.k.s.x is the label of Hecke orbit, N.c is the Conrey label for the character corresponding to the embedding, and j is the index for the embedding within those with the same Dirichlet character; these embeddings are ordered by the vector $\iota(a_n)$, where we order the complex numbers first by their real part and then by their imaginary part.

To such an embedded modular form $\iota(f)$, we may associate a primitive L-function of degree 2

$$L(\iota(f),s) := \sum \iota(a_n)n^{-s} = \prod_p L_p(\iota(f),p^{-s})$$
$$= \prod_{p|N} \left(1-\iota(a_p)p^{-s}\right)^{-1} \prod_{p\nmid N} \left(1-\iota(a_p)p^{-s}+\chi(p)p^{-2s}\right)^{-1}. \tag{9.1.2}$$

Let $\Lambda(\iota(f),s) := N^{s/2}\Gamma_{\mathbb{C}}(s)L(\iota(f),s)$, where $\Gamma_{\mathbb{C}}(s) := 2(2\pi)^{-s}\Gamma(s)$. Then $\Lambda(\iota(f),s)$ continues to an entire function of order 1 and satisfies the functional equation

$$\Lambda(\iota(f),s) = \varepsilon\overline{\Lambda}(\iota(f),k-s), \tag{9.1.3}$$

where ε is the root number of $\Lambda(\iota(f),s)$, a root of unity.

The generalized Riemann hypothesis also predicts that any nontrivial zero of the L-function lies on the line of symmetry of its functional equation $\mathrm{Re}(s) = k/2$, known as the critical line. To study the behavior of $L(\iota(f),s)$ on the critical line, it is natural to introduce the associated Z-function, a smooth real-valued function of a real variable t defined by

$$Z(\iota(f),t) := \overline{\varepsilon}^{1/2}\frac{\gamma(k/2+it)}{|\gamma(k/2+it)|}L(\iota(f),k/2+it), \tag{9.1.4}$$

where $\gamma(s) := N^{s/2}\Gamma_{\mathbb{C}}(s)$ and the square root is chosen so that $Z(t) > 0$ for sufficiently small $t > 0$. By construction, we have $|Z(\iota(f),t)| = |L(\iota(f),k/2+it)|$, the multiset of zeros of $Z(\iota(f),t)$ matches the multiset of zeros of $L(\iota(f),k/2+it)$, and $Z(\iota(f),t)$ changes sign at the zeros of $L(\iota(f),k/2+it)$ of odd multiplicity.

9.2 Computations

Given $\iota(f)$ we would like to compute certain invariants of $L(\iota(f),s)$. For example, the root number ε, the imaginary part of the first few zeros on the critical line, an upper bound on the order of vanishing at $s = k/2$, the leading Taylor coefficient at $s = k/2$, and the plot $Z(\iota(f),t)$ on some interval. Given that a majority of these items cannot be represented exactly, we instead aim to determine a small interval in $\mathbb{R}$ or rectangle in $\mathbb{C}$. Precisely, let b denote the number of bits of target accuracy. We would like to compute the following:

- the root number: $x_\varepsilon, y_\varepsilon \in \mathbb{Z}$ such that $2^{b+1}\mathrm{Re}(z) \in [x_\varepsilon - 1, x_\varepsilon + 1]$ and $2^{b+1}\mathrm{Im}(z) \in [y-1, y+1]$;
- the imaginary part of the first few zeros on the critical line: $t_1, \ldots, t_n \in \mathbb{Z}$ such that $\bigcup_i [t_i - 1, t_i + 1]2^{-b-1}$ covers the first n zeros of $L(\iota(f),k/2+it)$;
- an upper bound on the order of vanishing at $s = k/2$: $r := \max_i\{i : |L^{(i)}(\iota(f),k/2)/i!| < 2^{-b-1}\}$;

- the leading Taylor coefficient at $s = k/2$: $0 \neq s \in \mathbb{Z}$ such that $2^{p+1}L^{(r)}(\iota(f), k/2)/r! \in [s+1, s-1]$;
- an approximation to the plot of $Z(\iota(f), t)$: approximations as doubles of $Z(\iota(f), i\delta)$ for some chosen δ and $i = 0, \ldots, n$.

In order to rigorously compute the items above, we follow an approach that builds on several improvements and extensions of the algorithm from [10] specialized to the motivic case, the details of which will appear in future work. In practice, given the first $C_k\sqrt{N}$ embedded Dirichlet coefficients, with sufficient precision, and while carrying out all floating-point calculations using rigorous error bounds and interval arithmetic [51], one may compute all the items above to the desired bit accuracy. A generic library to carry out such computations, due originally to Dave Platt [27], has been developed.

Example 9.2.1 For an explicit example, we encourage the reader to peruse the source file `examples/cmf_23.1.b.a.cpp` in [27], where the authors show how to use the library to compute all of the items above for the modular form 23.1.b.a, which matches its unique embedding. By running this example, one can compute that

$$\epsilon = (1 \pm 10^{-117}) + (0 \pm 4.7 \times 10^{-59})\, i,$$

(since f is self dual, we must actually have $\epsilon = 1$), and

$$L(f, 1/2) = 0.174036326987934183499504592018 \pm 8.2317 \times 10^{-59},$$

as well as approximate values for the imaginary part of the first ten zeros. Using the notation above, we can represent an approximation to the imaginary part of the first zero

$$5.115683328815117598553356420 38 \pm 3.9443 \times 10^{-31}$$

by the interval $[t_1 - 1, t_1 + 1]2^{-101}$, where

$$t_1 = 12969798084700060914517716069360.$$

The imaginary part of the following nine zeros are approximately 7.15926, 8.88140, 10.2820, 11.4300, 12.9344, 14.6625, 16.4982, 17.1013, and 18.0807.

We carried out this computation with 100 bits of target accuracy for the 14 398 359 embedded newforms in our database with $k \leq 200$. In our computation we observed that it was sufficient to work with 200 bits of precision and $C_k \leq 0.08k\log(k) + 24$. While we did not keep track of CPU time used along the way, by rerunning some of the computations, we extrapolate that we spent at least 11 CPU years on these computations.

9.3 Imprimitive L-Function

Associated to a newform f with coefficient field $\mathbb{Q}(f)$ of degree d, we may also consider the L-function of degree $2d$ associated to its Galois orbit:

$$L(f, s) := \prod_{\iota:\mathbb{Q}(f)\hookrightarrow\mathbb{C}} L(\iota(f), s) = \prod_p L_p(f, p^{-s}). \tag{9.3.1}$$

This gives rise to a $\mathbb{Q}$-primitive L-function with $L_p(f, T) \in 1 + T\mathbb{Z}[T]$, which satisfies the functional equation

$$\Lambda(f, s) := N^{sd/2}\Gamma_{\mathbb{C}}(s)^d L(f, s) = \varepsilon\overline{\Lambda}(f, k - s), \tag{9.3.2}$$

where now we have $\varepsilon = \pm 1$. Using the invariants for each $L(\iota(f), s)$ mentioned above, one can easily deduce the respective invariants for $L(f, s)$.

For these L-functions we would also like to compute the local factors for small p. This is straightforward if one has access to an exact representation of a_p in $\mathbb{Q}(f)$. Otherwise, we relied on Newton identities to compute $L_p(f, T) \in \mathbb{Z}[T]$ from the roots of $L_p(\iota(f), T) \in \mathbb{C}[T]$, while working with interval arithmetic [51]: see L(500.2.e.c) for an example. In some cases, for example when $[\mathbb{Q}(f) : \mathbb{Q}]$ or the weight is large, we were only able to compute the initial coefficients for some local factors—this occurred for L(20.10.e.b), for example.

9.4 Verifying the Analytic Rank

In this section, we discuss methods for rigorously verifying the analytic rank of a modular form L-function. Throughout, let N and k be positive integers and let $f \in S_k(\Gamma_1(N); \mathbb{C})$ be a newform of weight k and level N (with coefficient field embedded in the complex numbers).

Definition 9.4.1 Suppose k is *even*. We define the analytic rank of f to be the order of vanishing of $L(f, s)$ at $k/2$.

When $L^{(n)}(f, k/2) \neq 0$ one can certify such a statement using ball arithmetic by working with enough precision. However, if $L^{(n)}(f, k/2) = 0$, this approach does not work, as there is no known bound ε such that $|L^{(n)}(f, k/2)| < \varepsilon$ implies $L^{(n)}(f, k/2) = 0$. Nonetheless, if the order of vanishing is small, then there are other methods to computationally verify the order of vanishing. Using these methods we were able to provably verify the analytic rank of all modular forms for which the L-functions were computed. The way the analytic computations were verified is detailed below. The strategy used depends on the order of vanishing, and whether the modular form is self-dual or not. The analytic rank zero case is skipped because this can just be done by computing $L(f, k/2)$ to enough precision using interval

Table 8 Number of even weight newforms in the database by analytic rank

	Analytic rank				
	0	1	2	3	≥4
Self-dual	83 338	85 254	2 565	1	0
Not self-dual	63 804	1 798	1	0	0
Total	147 142	87 052	2 566	1	0

arithmetic until 0 is no longer in the computed interval. A summary of analytic ranks of even weight newforms in the database is shown in Table 8.

Self-Dual and Analytic Rank 1 We begin by considering self-dual newforms f whose analytic rank numerically appears to be 1. All such forms in the range of our computation have trivial character. In this case the functional equation takes the form

$$\Lambda(f, s) = \varepsilon' i^k \Lambda(f, k - s), \tag{9.4.2}$$

where $\varepsilon' = \pm 1$ is the eigenvalue of the Atkin–Lehner involution W_N. For such forms in the database, we verified that $\varepsilon' \iota^k = -1$, forcing $\Lambda(f, k/2) = 0$, and the non-vanishing of $N^{s/2}\Gamma_{\mathbb{C}}(k/2)$ then implies that $L(f, k/2) = 0$. The upper bound of 1 on the analytic rank was obtained using interval arithmetic.

Non-self-dual and Analytic Rank 1 Following Stein [89, §8.5], we define a pairing between modular forms and modular symbols

$$S_k(\Gamma_1(N)) \oplus \overline{S}_k(\Gamma_1(N)) \times \mathrm{ModSym}_k(\Gamma_1(N)) \to \mathbb{C}$$

by defining

$$\langle (f, g), P\{a, b\} \rangle := \int_a^b f(z) P(z, 1) dz + \int_a^b g(z) P(\overline{z}, 1) d\overline{z}.$$

This pairing allows one to determine the vanishing of L-functions, because for every integer $1 \le j \le k - 1$ we have

$$L(f, j) = \frac{(-2\pi i)^j}{(j-1)!} \langle (f, 0), X^{j-1} Y^{k-2-(j-1)} \{0, \infty\} \rangle.$$

The pairing is Hecke-equivariant, meaning that $\langle (T_n f, T_n g), x \rangle = \langle (f, g), T_n x \rangle$ for all integers n.

Let $f \in S_k^{\mathrm{new}}(\Gamma_1(N))$ be a newform and $V_f \subseteq S_k^{\mathrm{new}}(\Gamma_1(N))$ the subspace generated by its Galois conjugates. Then by Atkin–Lehner–Li theory, V_f is a simple module over the Hecke algebra $\mathcal{T}$, and there exists a Hecke operator $t_f \in \mathcal{T}$ such that $t_f \colon M_k(\Gamma_1(N)) \to M_k(\Gamma_1(N))$ is a projection onto V_f. Because t_f is a projection we have $\langle (f, 0), x \rangle = \langle (t_f f, t_f 0), x \rangle = \langle (f, 0), t_f x \rangle$ for all $x \in \mathrm{ModSym}_k(\Gamma_1(N))$, and hence in particular this means that if

$$t_f(X^{j-1}Y^{k-2-(j-1)}\{0,\infty\}) = 0$$

then $L(f, j) = 0$.

A map t'_f with the same kernel as t_f can be obtained from $t_f(\mathrm{ModSym}_k(\Gamma_1(N), \mathbb{Q}))$ in Magma using the command `PeriodMapping`. Furthermore, this Magma command only uses exact arithmetic over $\mathbb{Q}$. For all non-self-dual modular forms whose analytic rank numerically seemed to be 1, it was verified that indeed $t'_f(X^{k/2-1}Y^{k/2-1}\{0,\infty\}) = 0$, implying that $L(f, k/2) = 0$. The upper bound of 1 on the analytic rank was again obtained using interval arithmetic.

Self-Dual and Analytic Rank 2 As in the preceding subsection, all newforms in the database whose analytic rank numerically seemed to be 2 have trivial character. This time it was verified that $\varepsilon' \iota^k = 1$ for all these modular forms. In particular, the functional equation then forces all odd derivatives of $\Lambda(f, s)$ to vanish at $k/2$. This forces the order of vanishing of $\Lambda(f, s)$ at $k/2$ to be even, and hence the analytic rank of $L(f, s)$ to be even as well. The techniques of the preceding paragraph were used to prove that for all these modular forms one has that $L(f, k/2) = 0$, which together with the parity argument gives a lower bound of 2 on the analytic rank. The upper bound of 2 was again obtained using interval arithmetic.

Non-self-dual Analytic Rank 2 There is exactly one Galois orbit of non-self-dual newforms in the database whose analytic rank numerically seems to be 2. Let f denote the newform of weight 2 and level 1154 with LMFDB label 1154.2.e.a with coefficient field $\mathbb{Q}(\zeta_3)$. This pair corresponds to an isogeny class of abelian surfaces, and our first goal is to find a representative of this isogeny class. By searching for hyperelliptic curves over $\mathbb{F}_p$ that match the local factors of $L(f, s)$ for small p, and then by lifting their Weierstrass equations to $\mathbb{Z}$ we found the following genus 2 curve:

$$C : y^2 = x^6 - 12x^5 + 34x^4 - 18x^3 - 11x^2 + 6x + 1. \tag{9.4.3}$$

Letting J denote its Jacobian, we find it is of conductor 1154^2 as desired. Our goal is first to show that J really is in the isogeny class of abelian surfaces corresponding to the newform 1154.2.e.a. Using [26] we were able to compute the endomorphism ring of J, and verify that J is of GL_2-type and hence is modular [78, 57]. Thus, J is a good candidate to be a representative of the isogeny class of abelian surfaces corresponding to the newform 1154.2.e.a. Alternatively, one can also verify that J is of GL_2-type by noting that C has an automorphism of order 3 given by $x \mapsto 1-1/x$, $y \mapsto -y/x^3$ and thus showing that its Jacobian is of GL_2-type. Additionally, the Euler factor at 5 of its L-function is

$$1 + 6T + 17T^2 + 30T^3 + 25T^4$$

which is irreducible. Hence its Jacobian is simple, showing that its Jacobian corresponds to a pair of Galois conjugate newforms of level 1154. There is one other pair of Galois conjugate newforms whose coefficient field is $\mathbb{Q}(\zeta_3)$, namely that with LMFDB label 1154.2.c.a. So it remains to show that J does not come from the newform with label 1154.2.c.a. However, the Euler factor of the L-function at 5 for that newform is $1 - 3T + 4T^2 - 15T^3 + 25T^4$ which does not match that of J. This means that Jacobian of C really is in the isogeny class of abelian surfaces corresponding to the newform 1154.2.e.a.

Using the Magma function `RankBounds` one readily computes that J has Mordell-Weil rank 4. In particular, it has rank 2 as a module over $\mathbb{Z}[\zeta_3]$. The generalization by Kato of the work of Kolyvagin and Logachev on the Birch–Swinnerton-Dyer conjecture in the analytic rank 0 and 1 cases to all isogeny factors of $J_1(N)$ (see Kato [53, Corollary 14.3]) shows that the order of vanishing of $L(f, s)$ at 1 cannot be 0 or 1 since this would give J rank 0 or 1 as a $\mathbb{Z}[\zeta_3]$-module. So the order of vanishing is at least 2. An upper bound was again obtained using interval arithmetic.

Self-Dual Analytic Rank 3 The approach here is similar to that in the preceding paragraph and the result was already briefly mentioned in [31, Section 3.4] where the analytic rank is determined for all elliptic curves of conductor $N < 130\,000$, later extended to all elliptic curves of conductor less than 500 000 and rank at most 3. There is only one newform that numerically seems to be of analytic rank 3 in the database, namely 5077.2.a.a of weight 2, level 5077 and trivial character. This modular form corresponds to the elliptic curve $y^2 + y = x^3 - 7x + 6$ which has rank 3 and is the only one in its isogeny class. The verification that its L-function has analytic rank 3 is a famous calculation of Buhler–Gross–Zagier [15], used by Gross–Zagier [45] in their solution to the Gauss class number 1 problem. We confirm it quickly as follows: by known cases of the Birch–Swinnerton-Dyer conjecture, the analytic rank cannot be 0 or 1; by parity of the root number, the analytic rank cannot be 2, so it must be at least 3; and an upper bound on the analytic rank of 3 is obtained by interval arithmetic.

9.5 Chowla's Conjecture

The definition of analytic rank (Definition 9.4.1) as an order of vanishing also makes sense for k odd, and by analogy one might also find it natural to study the central values of $L(f, s)$ at $k/2$ and their derivatives. However, for k odd the central value $s = k/2$ is not a *special* value in the sense of Deligne [37] and thus there is no abelian group whose rank (as a module over an appropriate coefficient ring) is conjecturally related to its leading Taylor coefficient. It would therefore be a stretch to call the order of vanishing at the central point an *analytic rank*. Moreover, one does not expect $L(f, k/2)$ to ever vanish, and this is a generalization of Chowla's conjecture for Dirichlet L-functions [21], as follows.

Let χ be a non-trivial Dirichlet character, then the functional equation associated to $L(\chi, s) := \sum \chi(n)n^{-s}$, similar to Eq. 9.1.3, relates $L(\chi, s)$ to $L(\overline{\chi}, 1 - s)$. The value of $L(\chi, 1)$ is quite well understood. For example, the fact that $L(1, \chi) \neq 0$ gives us Dirichlet's theorem on arithmetic progressions, and for primitive real characters the value $L(\chi, 1)$ gives us Dirichlet's class number formula. As mentioned above, inspired by Definition 9.4.1, one might also find it natural to study the order of vanishing of $L(\chi, s)$ at $s = 1/2$ and its derivatives. However, it is believed that $L(\chi, 1/2) \neq 0$; this was first conjectured by Chowla [21] for primitive real characters and later generalized to other characters. One of the reasons behind such a belief is that for primitive real characters the root number of such L-functions is always 1 [44], and thus there is no simple reason for $L(\chi, 1/2)$ to vanish. Although Chowla's conjecture remains open, it has been numerically verified for all real characters χ of modulus less than 10^{10} [70], and substantial progress towards showing the non-vanishing of $L(\chi, 1/2)$ has also been made, see [44] for a short overview.

A generalization of Chowla's conjecture is that $L(f, k/2) \neq 0$ for k odd. As in the case of Dirichlet L-functions for primitive real characters, we also have that the root number of $L(f, k/2)$ can never be -1 when f is self-dual. This is in stark contrast to the case of self dual even weight modular forms, where the root numbers are split approximately 50–50 between -1 and 1. We verified this generalization of Chowla's conjecture, as we computed $L(f, k/2)$ for every newform in our database with $k \leq 200$, and found that this was nonzero for all the odd weight newforms.

10 An Overview of the Computation

In this section, we provide an overview of the computations we performed, the results of which are now available in the LMFDB [60]. These were accomplished using a combination of Magma, Pari/GP, and SageMath scripts, as well as hand written C code for some of the more computationally-intensive tasks. In aggregate, these computations consumed more than 100 years of CPU time.

10.1 Data Extent

Our database consists of four overlapping sets of newforms described in Table 9. These datasets were chosen both for reasons of mathematical interest, and to ensure that the database included all modular forms contained in existing datasets such as the Stein tables of modular forms [88], the Buzzard-Lauder tables of weight one newforms [19], and the previous database of modular forms contained in the LMFDB. More detailed statistics on the newforms in the database can be found at the statistics page.

For the first dataset (1), we used three independent sources of newform data:

Table 9 Extent of the newform database (only nonzero newspaces are included)

Constraints on $S_k^{\text{new}}(N, \chi)$	Newspaces	Newforms	Embeddings
(1) $Nk^2 \leq 4000$	30 738	67 180	9 966 498
(2) $Nk^2 \leq 40\,000$, $\lvert\chi\rvert = 1$	16 277	170 611	3 092 301
(3) $Nk^2 \leq 40\,000$, $k > 1$, $\dim S_k^{\text{new}}(N, \chi) \leq 100$	30 345	131 540	1 648 617
(4) $Nk^2 \leq 40\,000$, $N \leq 24$ or $Nk^2 \leq 100\,000$, $N \leq 10$ or $N \leq 100$, $k \leq 12$	7 627	12 237	676 574
Union of sets above	62 142	281 219	14 398 359

- Complex eigenvalue data for each embedded newform of weight $k > 1$ computed by the `mflib` software package [8], which uses Arb [51] to rigorously implement the trace formula (as described in [85], for example) to obtain approximate complex values to a precision of 200 decimal digits.
- Exact algebraic eigenvalue data for each newform of weight $k > 1$ and dimension $d \leq 20$ computed using Magma's [11] modular symbols package (originally written by William Stein).
- Exact algebraic eigenvalue data for each newform of weight $k > 1$ and dimension $d \leq 20$ were computed using the modular forms implementation in Pari/GP [72] described in [4], which was also applied to all newforms of weight $k = 1$.

For $k > 1$ and $Nk^2 \leq 4000$ the decomposition of every newspace $S_k^{\text{new}}(N, \chi)$ was computed in all three cases and compared for consistency. Exact algebraic data was computed only for newforms of dimension $d \leq 20$, except for $k = 1$ where exact algebraic data was computed in every case. For newforms of weight $k > 1$ and dimension $d \leq 20$, the algebraic data independently computed by Magma and Pari/GP were checked for consistency (this was not a completely trivial task, as it generally required determining an appropriate automorphism of the coefficient field in order to compare sequences of Fourier coefficients). We also compared the trace forms using all three methods and compared the results for consistency, and for newforms of weight $k = 1$ and level $N \leq 1500$ we compared the Pari/GP computations with the Buzzard-Lauder database [19].

Datasets (2) and (3) were computed entirely in Magma, as was dataset (4), except for 12 spaces of high dimension where complex analytic methods were used. For the portions of these datasets that overlap with the Stein database of modular forms [88], we compared the results for consistency.

For newforms $f = \sum a_n q^n$ of level $N \leq 1000$ we computed 1000 coefficients a_n, while for newforms of level $1001 \leq N \leq 4000$ we computed 2000 coefficients, and for newforms of level $4001 \leq N \leq 10{,}000$ we computed 3000 coefficients. This substantially exceeds the Sturm bound in every case, and also exceeds the bound $30\sqrt{N}$ required for the L-function calculations described in Sect. 9. For every newform in the database we computed complex coefficients to a precision of at least 200 bits. In cases where we compute algebraic coefficient data we computed an optimized representation using an LLL-basis as described in Sect. 8.7, along with a set of generators for the coefficient ring.

For each newform we determined any non-trivial self-twists admitted by the newform (CM, RM, or both), and for newforms with algebraic eigenvalue data available, we computed all inner twists as described in Sect. 11. We also computed the analytic rank of every newform, as described in Sect. 9.4, and for weight one newforms we computed the image of the associated projective Artin representation and a defining polynomial for its kernel, as described in Sect. 12. These computations have now all been rigorously verified.

In addition to the newform database, we computed dimension tables for all newspaces in the range $Nk^2 \leq 40{,}000$ with $k > 1$, and we computed trace forms for all newspaces of level $N \leq 4000$ in this range using the `mftraceform` function in Pari/GP.

10.2 Statistics

In addition to the ability to browse and to search for examples with specific properties, the modular forms database allows for an investigation of arithmetic statistics. The LMFDB [60] includes precomputed tables displaying how various quantities vary across the database, some of which we have duplicated here in Tables 10, 11, 12, and 13.

In addition to these static tables, we have added *dynamic statistics*
https://www.lmfdb.org/ModularForm/GL2/Q/holomorphic/dynamic_stats
which allow users to customize which variables to view and any constraints to impose. For example, a researcher might create a table displaying how the weight and level vary among forms with complex multiplication. We hope that this new feature will enable examination of large-scale patterns, both in the modular form data and elsewhere in the LMFDB.

Remark 10.2.1 The statistics and examples presented in this article reflect the dataset defined in Sect. 10.1, which represents the state of the LMFDB as of January 2020. As new data is added to the LMFDB these statistics may no longer match those displayed in the LMFDB, and the number of newforms returned by some of the example queries listed below may increase.

Table 10 Distribution of analytic ranks

Analytic rank	0	1	2	3
Count	191 520	87 052	2 566	1
Proportion	68.12%	30.96%	0.91%	0.00%
Example	23.1.b.a	37.2.a.a	389.2.a.a	5077.2.a.a

Table 11 Distribution of projective images

Projective image	A_4	S_4	A_5	D_2	D_n
Count	458	1 033	202	1 311	17 613
Proportion	2.37%	5.35%	1.05%	6.79%	91.23%
Example	124.1.i.a	148.1.f.a	1 763.1.p.b	3 600.1.e.a	3 997.1.cz.a

Table 12 Distribution of inner twists

Inner twists	Unknown	1	2	4	6	8	10	12
Count	73 993	129 197	47 492	25 803	24	4 295	6	51
Proportion	26.31%	45.94%	16.89%	9.18%	0.01%	1.53%	0.00%	0.02%
Inner twists	16	20	24	32	40	44	56	
Count	311	3	14	20	7	1	2	
Proportion	0.11%	0.00%	0.00%	0.01%	0.00%	0.00%	0.00%	

Table 13 Distribution of self twist types by weight

	Weight					
	1	2	3	4	5-316	Total
Neither	1 693	174 853	11 117	27 877	40 278	255 818
	8.77%	98.27%	87.85%	98.02%	93.91%	90.97%
CM only	15 841	3 074	1 538	563	2 613	23 629
	82.05%	1.73%	12.15%	1.98%	6.09%	8.40%
RM only	461					461
	2.39%					0.16%
Both	1 311					1 311
	6.79%					0.47%

10.3 Data Reliability

All of our modular form data was computed or verified using rigorous algorithms that do not depend on any unproved assumptions or conjectures.

- Self-twists were either verified via Theorem 11.2.4 and Proposition 11.1.7 using exact algebraic Fourier coefficients a_n or ruled out using complex approximations of sufficient precision to rigorously distinguish zero and nonzero values of a_n and checking for self-twists by all primitive quadratic characters ψ of conductor dividing the level (a newform that admits a self-twist by ψ must have $a_n = 0$ whenever $\psi(a_n) \neq 1$).
- We computed and verified inner twists for all newforms in our dataset that are either of weight one or have dimension at most 20 by computing sufficiently many algebraic Fourier coefficients and applying Theorem 11.2.4 and Proposition 11.1.7.

- Analytic ranks were computed using complex approximations as described in Sect. 9 and then rigorously verified using the symbolic methods described in Sect. 9.4.
- For weight one newforms the classification of projective images as D_n, A_4, S_4, A_5 was rigorously verified by explicitly computing the number field fixed by the kernel of the associated projective Galois representation. As described in Sect. 12, this was accomplished using a combination of the ray class field functionality provided by Pari/GP and Magma, the rigorous tabulation of all A_4, S_4, and A_5 number fields with compatible ramification, and the explicit computation of quotients of ring class fields of orders in imaginary quadratic fields via the theory of complex multiplication.

In addition to using mathematically rigorous algorithms, we performed a variety of consistency checks intended to catch any errors in the software packages used to compute modular forms data, or any errors that might have been introduced during post-processing. The following checks have been performed:

- All newforms of weight $k > 1$ and level N satisfying $Nk^2 \leq 2000$ have been independently computed using Magma and Pari/GP. By comparing the results of these computations we have verified that the decompositions of each newspace $S_k^{\mathrm{new}}(N, \chi)$ into Galois orbits agree (with matching coefficient fields), that the first 1000 coefficients of the trace forms for each Galois orbit agree, and for newforms of dimension $d \leq 20$, that there is an automorphism of the coefficient field that relates the sequences of algebraic eigenvalues $(a_1, \ldots, a_{1000})$ computed by Pari/GP and Magma.
- For all newforms of weight $k > 1$ and level N satisfying $Nk^2 \leq 4000$ we have verified that the trace forms computed by Magma (using modular symbols) agree with the trace forms obtained from complex analytic data computed using the explicit trace formula. This also verifies the dimensions of the coefficient fields.
- For newforms of weight $k = 1$ and level $N \leq 1000$ we have matched the data computed using Pari/GP with the tables computed by Buzzard and Lauder [19].
- For all dihedral newforms of weight $k = 1$ and level $N \leq 4000$ we have matched trace forms with data computed using the explicit trace formula in Pari/GP with data independently computed using the ray class field functionality implemented in Pari/GP and Magma.

As a consistency check for our L-function computations, after computing a provisional list of all non-trivial zeros on the critical line up to a chosen height bound b we confirmed that no zeros are missing, in other words, that the Riemann Hypothesis holds for each L-function up to height b. We use the method described in [16] based on the Weil–Barner explicit formula. If an L-function also arises from another object in the LMFDB for which we already had computed its L-function we verified that these computations match.

10.4 Interesting, Extreme Behavior and Examples from the Literature

When putting modular forms in a database it is easy to view them as an aggregate, but of course each modular form is distinct and many have unique interesting properties.

We take this opportunity to recall the rich history and special properties of several forms in this database. We also provide links between these forms and the literature and note several forms that have naturally arisen in previous work. We focus on weight $k \geq 2$ in this section; see Sect. 12.5 for interesting behavior in weight $k = 1$.

- The most well known, and the prototypical, example of a modular form is the Ramanujan Δ function, of weight 12 and level 1; its label is 1.12.a.a. This is the lowest weight in which a cusp form appears for the full modular group, so many properties of more general newforms were first noticed for Δ. Similarly, Δ has served as a testing ground for techniques and results before they were known more generally. For instance, the Ramanujan–Petersson conjecture was first made by Ramanujan for Δ but later extended to all newforms. Additionally, computation of the q-expansion coefficients of Δ, traditionally denoted by $\tau(n)$ and known as Ramanujan's τ function, is the subject of the monograph [39].
- By the modularity theorem, newforms of weight 2 with rational coefficients correspond to isogeny classes of elliptic curves over $\mathbb{Q}$. The smallest level in which a weight 2 form appears is 11, corresponding to the smallest conductor of an elliptic curve over $\mathbb{Q}$. Here we necessarily have trivial character and the label is 11.2.a.a; this form has q-expansion

$$q \prod_{k \geq 1} (1 - q^k)^2 (1 - q^{11k})^2.$$

- The weight 2 newforms with CM by fields with the largest absolute discriminants in the database are 2169.2.d.a with CM by $\mathbb{Q}(\sqrt{-723})$, 8388.2.e.c and 2097.2.d.a with CM by $\mathbb{Q}(\sqrt{-699})$, 2061.2.c.c with CM by $\mathbb{Q}(\sqrt{-687})$, and 7524.2.l.b with CM by $\mathbb{Q}(\sqrt{-627})$—the last of these has 8 inner twists.
- The weight 2 newform 867.2.i.a with CM by $\mathbb{Q}(\sqrt{-51})$ has 32 inner twists, and the weight 1 newform 3481.1.d.a with CM by $\mathbb{Q}(\sqrt{-59})$ has 56 inner twists.
- The weight 3 newform 7.3.b.a has CM by $\mathbb{Q}(\sqrt{-7})$, making it the first (by analytic conductor) newform of weight ≥ 3 with CM.
- Watkins [99, §9.1.3] discusses several examples of modular forms of analytic rank 2. The query https://www.lmfdb.org/ModularForm/GL2/Q/holomorphic/?weight=4-&analytic_rank=2- returns 130 forms of weight at least 4 and analytic rank at least 2, many of which are mentioned by Watkins, including 2 of weight 8.
- Watkins also discusses modular forms of weight 2 with which are non-self-dual yet have positive analytic rank, particularly examples with quadratic character, such as 122.2.b.a. The query https://www.lmfdb.org/ModularForm/

GL2/Q/holomorphic/?weight=2&char_order=2&is_self_dual=no&analytic_rank=1- produces 567 such examples. In larger weight we have 8.14.b.a which is non-self-dual and has analytic rank 1, as does 162.12.c.i.

- The index of the coefficient ring in the ring of integers of the coefficient field can get quite large, as in the case of the newform 8.21.d.b where the index is at least $2^{153} \cdot 3^{15} \cdot 5^4 \cdot 7^2$. In weight 2, the largest index we computed was $2^{26} \cdot 3^4$ for 2016.2.k.b and 4032.2.k.h.
- Many newforms in our database have very large Hecke orbits. For example, the newform 983.2.c.a has relative dimension 81 over its character field $\mathbb{Q}(\zeta_{491})$ and $\mathbb{Q}$-dimension 39 690.

10.5 Pictures

For every newform f, every nonempty newspace $S_k^{\text{new}}(N, \chi)$, and $S_k^{\text{new}}(\Gamma_1(N))$ for which we have all the newforms, we have created a portrait based on their trace forms, a total of 641 562 portraits. The picture is generated by plotting the absolute value of the trace form in the Poincaré disk, obtained as the image of $(1-iz)/(z-i)$ in $\mathcal{H}$, where the color hue represents the absolute value modulo 1 (with blue being zero, and increasing through purple, red, orange, yellow, ...). For example, as the trace form is always zero at ∞, the top center is always blue, see Fig. 1.

We deviated from the normal approach, used by `complex_plot` in SageMath, of representing magnitude by brightness (with zero being black and infinity being white) and the argument by hue, as this often leads to an overexposed or underexposed picture, see Fig. 2.

Given the number of portraits needed, we limited ourselves to the first 100 Dirichlet coefficients of the trace form, working with 200 bits of precision, evaluating it in a 300×300 grid in $[-1, 1]^2$, and storing the picture as a 184×184 PNG. Overall this consumed about 100 CPU days, and their disk footprint is 45 GB. For aesthetic reasons, the portraits presented here were computed to a higher quality, which creates some discrepancies with the online version, especially in higher weight newforms.

Even though we opted for a plot with less information, it still captures some mathematically interesting features. For example, the behavior on the edge of the disk is a good indicator for level and weight, see Figs. 3 and 4.

The size of the blue spot on top center is inversely correlated with the growth of the trace form away from ∞, thus for fixed weight this is a good indicator for the dimension, see Fig. 5: their dimensions are 1, 4, 33, and 120, respectively.

Finally, one could also be tempted to infer the self-twists of a newform by comparing it with other forms in $S_k^{\text{new}}(\Gamma_1(N))$, see Fig. 6 for $S_1^{\text{new}}(\Gamma_1(164))$.

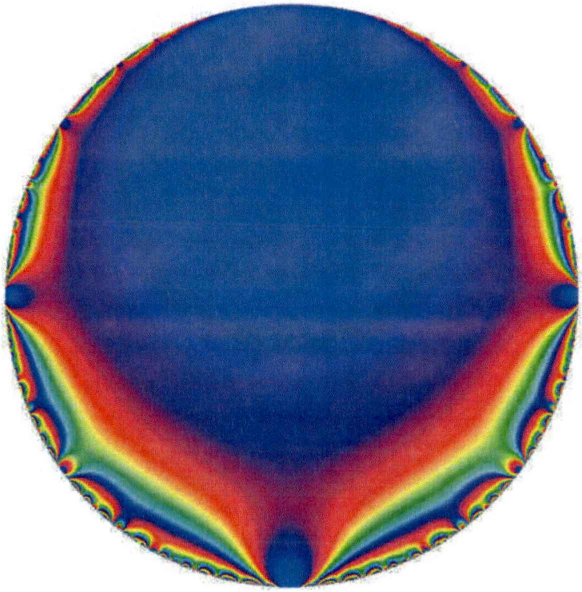

Fig. 1 Portrait of 23.1.b.a

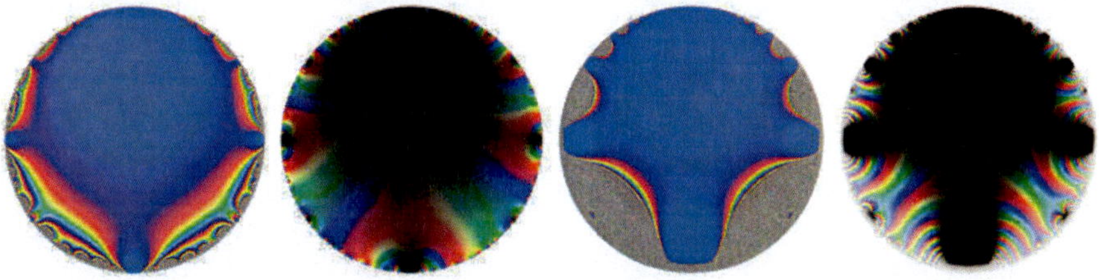

Fig. 2 Portraits of 11.2.a.a and 1.12.a.a and their plots using `complex_plot` in SageMath

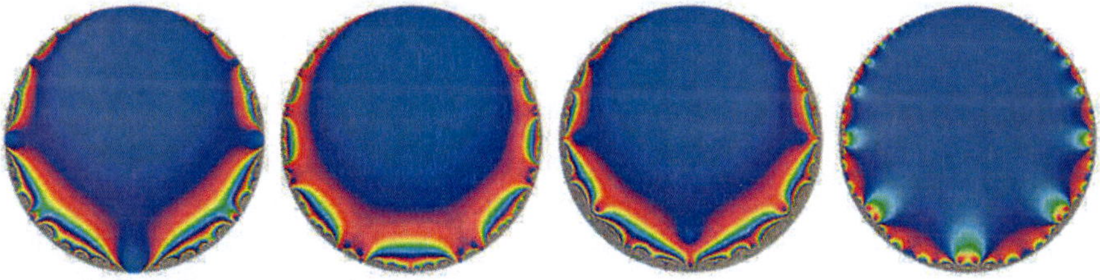

Fig. 3 The portraits for 11.2.a.a, 100.2.a.a, 1001.2.a.a, and 9996.2.a.a

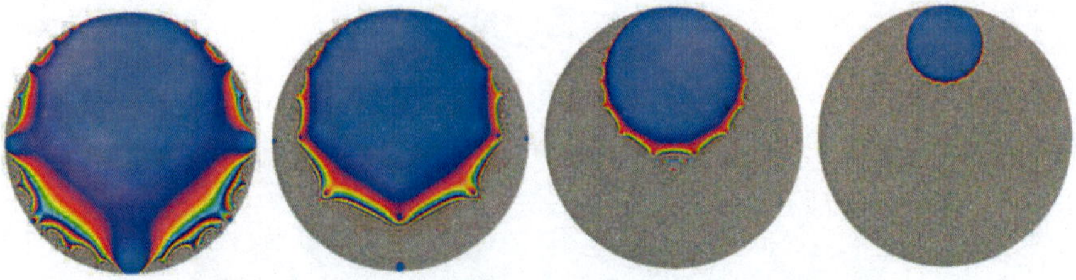

Fig. 4 The portraits for 7.3.b.a, 7.9.b.a, 7.27.b.a, and 7.81.b.a

Fig. 5 The portraits for 9359.2.a.a, 9359.2.a.e, 9359.2.a.k, and 9359.2.a.r

Fig. 6 The portraits for 164.1.d.a, 164.1.d.b, 164.1.j.a, and 164.1.l.a

Remark 10.5.1 These portraits differ from those now used in the LMFDB. Between writing and publishing this article we chose to instead use the pure phase portraits describe in §2.2.5 of [61].

10.6 Features

In parallel to carrying out the computations described elsewhere in this paper, we rewrote the user interface to the database. We highlight some of the more prominent new features in this section, some of which are being extended to other sections of the LMFDB.

The search interface includes multiple modes for viewing results. After entering constraints such as weight, level and dimension, there are four different search buttons available. In addition to the standard list of results, a user can choose to go straight to a randomly chosen newform. Alternatively, there are dimension tables available which display the dimension of the spaces of newforms as a function of weight and level. Finally, a table of traces allows for searching on specific Fourier coefficients, including specifying a particular class modulo an arbitrary integer. This feature can be used to find modular forms matching geometric objects via point-counting.

All of these search modes are also available for newspaces. For newspaces, the list mode shows the dimensions of the corresponding newforms as well as the Atkin–Lehner dimensions in the case of trivial character. For both newforms and newspaces, users can customize the order of the search results.

The homepage for an individual newform has also been completely restructured. Newforms can be downloaded and reconstructed in Magma, allowing for further computations if desired. We include complex eigenvalues for embedded modular forms even when exact Fourier coefficients are not feasible to compute.

One of the key motivations for our extensive computations of (exact or inexact) Fourier coefficients of newforms is to allow their L-functions to be computed. In addition to providing additional mathematical information about the newform, such as its analytic rank and special values, this allows us to automatically connect newforms to other objects in the LMFDB. Examples include:

- The L-function L(256.2.a.e) lists both the Bianchi modular form 2.0.4.1-4096.1-b and the Hilbert modular form 2.2.8.1-1024.1-m as origins (both arise as base changes of 256.2.a.e), as well as the corresponding elliptic curve isogeny classes 2.0.4.1-4096.1-b over $\mathbb{Q}(i)$ and 2.2.8.1-1024.1-m over $\mathbb{Q}(\sqrt{2})$.
- The L-function L(72.2.d.a) has (at least) three additional origins: the Hilbert modular form 2.2.8.1-81.1-b, the elliptic curve isogeny class 2.2.8.1-81.1-b, and the isogeny class 5184.a of the Jacobian of the genus 2 curve 5184.a.46656.1.
- The L-function L(1948.1.b.a) also arises as the L-function of (the Galois orbit of) the icosahedral Artin representation 2.1948.24T576.1. The L-functions home page also lists the four conjugate Artin representations (and four embedded weight one newforms) whose L-functions are primitive factors of this imprimitive L-function of degree 8.

11 Twisting

In this section, we discuss twists of modular forms and related computational issues. For background and further reading, we refer the reader to the foundational articles by Ribet [75, 76].

11.1 Definitions

We begin with definitions, followed by some examples. Throughout this section, let $f \in S_k^{\text{new}}(N, \chi)$ be a newform of weight $k \in \mathbb{Z}_{\geq 1}$, level $N \in \mathbb{Z}_{\geq 1}$, and character χ, and let $K_f := \mathbb{Q}(\{a_n(f)\}_n) \subseteq \mathbb{C}$ be its coefficient field. Let ψ be a Dirichlet character of conductor $\operatorname{cond}(\psi)$, and let ψ_0 be the primitive Dirichlet character inducing ψ (with $\operatorname{cond}(\psi_0) = \operatorname{cond}(\psi)$). Then there is a unique newform $g := f \otimes \psi$ characterized by the property that

$$a_n(g) = \psi_0(n)a_n(f) \quad \text{for all } n \text{ coprime to } N\operatorname{cond}(\psi); \tag{11.1.1}$$

we call g the twist of f by ψ. However, more is true: in fact, we have

$$a_n(g) = \psi_0(n)a_n(f) \quad \text{for all } n \text{ coprime to } \operatorname{cond}(\psi) \tag{11.1.2}$$

including those n that are not necessarily coprime to $N\operatorname{cond}(\psi)$: see Atkin–Li [2, Theorem 3.2]. By the recurrence satisfied by the Hecke operators, (11.1.2) is equivalent to the condition

$$a_p(g) = \psi(p)a_p(f) \quad \text{for all } p \nmid \operatorname{cond}(\psi). \tag{11.1.3}$$

The newform g has character $\chi\psi^2$ (by (11.1.8) below) and level dividing $\operatorname{lcm}(N, \operatorname{cond}(\psi)\operatorname{cond}(\chi\psi))$ (by Lemma 11.2.1 below). We call the newform g the twist of f by ψ and say that g is a twist of f.

As above, the group $\operatorname{Aut}(\mathbb{C})$ acts on the set of newforms in $S_k^{\text{new}}(N, \chi)$, with $a_n(\sigma(f)) = \sigma(a_n(f))$ for all $n \geq 1$. We have $\sigma(f) \in S_k^{\text{new}}(N, \sigma(\chi))$, where $\sigma(\chi)(n) = \sigma(\chi(n))$ for all $n \geq 1$. If $g = f \otimes \psi$, then $\sigma(g) = \sigma(f) \otimes \sigma(\psi)$ for all $\sigma \in \operatorname{Aut}(\mathbb{C})$. Accordingly, the set

$$[f] \otimes [\psi] := \{f' \otimes \psi' : f' \in [f], \psi' \in [\psi]\} \tag{11.1.4}$$

has an action of $\operatorname{Aut}(\mathbb{C})$ and so consists of finitely many $\operatorname{Aut}(\mathbb{C})$-orbits (possibly more than one). Accordingly, we say that $[g]$ is a twist of $[f]$ by $[\psi]$ if there exist $f' \in [f]$, $\psi' \in [\psi]$, $g' \in [g]$ such that $g' = f' \otimes \psi'$, or equivalently, $[g] \subseteq [f] \otimes [\psi]$.

Example 11.1.5 The newform orbits 3380.1.v.e and 3380.1.v.g are both twists of 3380.1.g.c by 13.f (and by 260.bc).

With this Galois digression out of the way, we return to the treatment of twists of (embedded) newforms.

Definition 11.1.6 Let ψ be a Dirichlet character and $\sigma \colon K_f \hookrightarrow \mathbb{C}$ be a field embedding. We say that f admits an inner twist by the pair (ψ, σ) if $f \otimes \psi = \sigma(f)$. In the special case that $\sigma = \operatorname{id}|_{K_f}$, we say that f admits a self-twist by ψ.

Let $\mathrm{InnTw}(f)$ denote the set of inner twists of f and $\mathrm{SelfTw}(f) \subseteq \mathrm{InnTw}(f)$ the subset of self-twists. Then projection onto the first component identifies $\mathrm{SelfTw}(f)$ with a subgroup of Dirichlet characters. By (11.1.2), the form f has an inner twist by (ψ, σ) if and only if $\sigma(a_n) = \psi(n)a_n$ for almost all n. The twist is said to be inner because such twists stay "within" the Galois orbit of f (a nontrivial inner twist is sometimes also referred to as an "extra twist"). Every newform has a trivial self-twist by $(\mathsf{1.a}, \mathrm{id}|_{K_f})$.

Proposition 11.1.7 (Ribet [76], Momose [68]) *The following statements hold.*

(a) *If* $(\psi, \sigma) \in \mathrm{InnTw}(f)$*, then*

$$\sigma(\chi) = \chi\psi^2;$$

so if $\psi \in \mathrm{SelfTw}(f)$ *then* ψ *is quadratic.*

(b) *If* $(\psi, \sigma) \in \mathrm{InnTw}(f)$ *then* $\sigma \in \mathrm{Aut}(K_f)$.

(c) $\mathrm{InnTw}(f)$ *naturally forms a group under*

$$(\psi, \sigma) \cdot (\psi', \sigma') := (\psi\,\sigma(\psi'), \sigma\sigma').$$

(d) *There is an exact sequence of groups*

$$1 \to \mathrm{SelfTw}(f) \to \mathrm{InnTw}(f) \xrightarrow{\pi} \mathrm{Aut}(K_f)$$
$$(\psi, \sigma) \mapsto \sigma.$$

Let $A := \pi(\mathrm{InnTw}(f))$*. Then* $\mathrm{InnTw}(f) \simeq \mathrm{SelfTw}(f) \times A$ *is a* direct *product.*

(e) *The projection* $(\psi, \sigma) \mapsto \psi$ *from* $\mathrm{InnTw}(f)$ *to the set of Dirichlet characters is an injective map of sets.*

(f) *The group* A *is abelian.*

(g) *Suppose* $\mathrm{SelfTw}(f)$ *is trivial. Then* π *is an isomorphism and the assignment* $\sigma \mapsto \psi_\sigma$ *if and only if* $(\psi_\sigma, \sigma) \in \mathrm{InnTw}(f)$ *is a well-defined* 1*-cocycle, i.e.,*

$$\psi_{\sigma\sigma'} = \psi_\sigma\sigma(\psi_{\sigma'}).$$

Proof These results originate with Ribet [76, §3] and Momose [68, Lemma (1.5)], but they work under the hypothesis that f has no self-twists. For clarity, we repeat these arguments to show this hypothesis is unnecessary. Let $f(q) = \sum_n a_n q^n$.

Part (a) follows by looking at (Nebentypus) characters using the Hecke recurrence (or the determinant of the associated Galois representations). Explicitly, on the one hand, the character of $\sigma(f)$ is $\sigma(\chi)$; on the other, if ϵ is the character of $f \otimes \psi$ then for all good primes p the Hecke recurrence reads

$$\begin{aligned} \varepsilon(p)p^{k-1} &= a_p(f \otimes \psi)^2 - a_{p^2}(f \otimes \psi)^2 \\ &= \psi(p)^2(a_p(f)^2 - a_{p^2}(f)) = \psi(p)^2\chi(p)p^{k-1} \end{aligned} \tag{11.1.8}$$

so $\varepsilon = \chi\psi^2$. Consequently, a self-twist by ψ gives $\chi = \chi\psi^2$, so ψ^2 is the trivial character.

For part (b), by (a) we have $\psi^2 = \sigma(\chi)\chi^{-1}$, and we claim ψ takes values in $\mathbb{Q}(\chi)$: indeed, if $\chi(n) = \zeta$ is a primitive dth root of unity, then checking cases based on the parity of d reveals that $\sigma(\zeta)/\zeta \in \langle\zeta^2\rangle$. Since $\mathbb{Q}(\psi) \subseteq K_f$, we conclude $\sigma(a_n) = \psi(n)a_n \in K_f$ for almost all n, so $\sigma(K_f) \subseteq K_f$ as desired.

For part (c), we start with $\sigma'(a_n) = \psi'(n)a_n$ and apply σ to get

$$(\sigma\sigma')(a_n) = \sigma(\psi')(n)\sigma(a_n) = \sigma(\psi')(n)\psi(n)a_n$$

for almost all n, so $(\psi\sigma(\psi'), \sigma\sigma') \in \mathrm{InnTw}(f)$. This product is associative: the identity element in $\mathrm{InnTw}(f)$ is $(1.\mathrm{a}, \mathrm{id}\,|_{K_f})$, and inverses are given by $(\psi, \sigma)^{-1} = (\sigma^{-1}(\psi), \sigma^{-1})$.

In part (d), the exact sequence is evident from (c). The group $\mathrm{InnTw}(f)$ visibly has the structure of a semidirect product $\mathrm{InnTw}(f) \simeq \mathrm{SelfTw}(f) \rtimes A$ via $A \to \mathrm{Aut}(\mathrm{SelfTw}(f))$ by $\sigma \mapsto (\psi \mapsto \sigma(\psi))$. However, by (b) $\mathrm{SelfTw}(f)$ consists only of quadratic characters, so $\sigma(\psi) = \psi$ for all σ so the product is direct.

Part (e) follows from the fact that ψ uniquely determines σ.

Part (f) is claimed by Ribet [76, Proposition (3.3)]: we prove it as follows. As in (a), let $\chi(n) = \zeta$ and $\sigma(\chi)(n) = \zeta^k$. Then again $\psi(n) = \zeta^{(k-1)/2}$ (for some choice of square root of ζ). Write similarly $\sigma'(\chi)(n) = \zeta^{k'}$. Then

$$\frac{\sigma'(\psi)}{\psi}(n) = \frac{\zeta^{k'(k-1)/2}}{\zeta^{(k-1)/2}} = \zeta^{(k-1)(k'-1)/2}$$

is well-defined, and by symmetry this is equal to $(\sigma(\psi)/\psi)(n)$, giving $\psi\,\sigma(\psi') = \psi'\,\sigma'(\psi)$, and similarly $\sigma'(\chi)(n) = \zeta^{k'}$. This calculation shows the projection of the products $(\psi, \sigma)(\psi', \sigma') = (\psi\sigma(\psi'), \sigma\sigma')$ and $(\psi', \sigma')(\psi, \sigma) = (\psi'\sigma'(\psi), \sigma\sigma')$ agree. By part (e), it follows that $\sigma\sigma' = \sigma'\sigma$ and A is abelian.

Finally, part (g) is immediate from (c).

Example 11.1.9 Consider the (embedded) newform 180.1.m.a.107.2; it represents the unique newform orbit in the space 180.1.m of weight 1 and level 180 with character orbit 180.m, whose q-expansion begins

$$f(q) = q - \zeta_8^3 q^2 - \zeta_8^2 q^4 + \zeta_8^3 q^5 - \zeta_8 q^8 + O(q^{10}),$$

where $\zeta_8 = \exp(2\pi i/8) = (1+i)/\sqrt{2}$ is the primitive eighth root of unity in the upper quadrant and $K_f = \mathbb{Q}(\zeta_8)$.

The group $\mathrm{SelfTw}(f)$ of self-twists is of order 2 with nontrivial character 4.3, the quadratic character of conductor 4 associated to the field $\mathbb{Q}(\sqrt{-1})$. The group of inner twists has order $\#\mathrm{InnTw}(f) = 8$, and we compute $\mathrm{InnTw}(f) \simeq (\mathbb{Z}/2\mathbb{Z})^3$, generated by the elements

$$(4.3, \mathrm{id}), (3.2, \zeta_8 \mapsto -\zeta_8), (5.3, \zeta_8 \mapsto \zeta_8^3).$$

The character ψ_5 with label 5.3 has order 4, so letting $\sigma_3 \in \mathrm{Aut}(\mathbb{Q}(\zeta_8))$ by $\sigma_3(\zeta_8) = \zeta_8^3$, we have

$$(\psi_5, \sigma_3)^2 = (\psi_5\,\sigma_3(\psi_5), \sigma_3^2) = (\psi_5\psi_5^{-1}, \mathrm{id}) = 1.$$

The projection of $\mathrm{InnTw}(f)$ onto the set of characters yields characters with conductors 1, 3, 4, 5, 12, 15, 20, 60.

Example 11.1.10 For f with label 361.2.e.d and $K_f = \mathbb{Q}(\zeta_{18})$, we have no nontrivial self-twists and $\pi\colon \mathrm{InnTw}(f) \to \mathrm{Aut}(K_f)$ is an isomorphism onto its image. In fact, we compute that π is surjective, so $\mathrm{InnTw}(f) \simeq \mathbb{Z}/6\mathbb{Z}$. More precisely, the elements of order 3 in $\mathrm{InnTw}(f)$ correspond to the characters 19.7 and 19.11 of order 3, and in the character orbit 19.e there are three characters whose elements match with automorphisms of order 2 and two of order 6.

Example 11.1.11 Among the forms of weight $k = 2$, trivial character, and dimension 2, we can search for forms with inner twist, and we should see a table that matches Cremona [29, Table 3] up to level $N \leq 300$. The lists match with one exception: we found one form 169.2.a.a that was missed by Cremona.

Newforms of weight $k \geq 2$ that admit nontrivial self-twists are commonly said to have *complex multiplication*, for reasons we now explain.

Proposition 11.1.12 (Ribet) *The following statements hold.*

(a) *If $k \geq 2$ and f has nontrivial self-twist by ψ, then ψ is associated to an imaginary quadratic field and is unique, i.e.,* $\mathrm{SelfTw}(f) \simeq \mathbb{Z}/2\mathbb{Z}$.
(b) *If $k = 1$, then f has nontrivial self-twist by ψ if and only if f has dihedral projective image. If so, then ψ may be real or imaginary and* $\mathrm{SelfTw}(f)$ *is a subgroup of* $(\mathbb{Z}/2\mathbb{Z})^2$.

Proof For part (a), see Ribet [75, Theorem (4.5)], a consequence of the theory of complex multiplication.

For part (b), we recall Sect. 12 and observe that f has self-twist by ψ if and only if $a_p(f) = 0$ for all p inert in $\mathbb{Q}(\psi)$ and by classification this happens if and only if the image of the projective Galois representation is dihedral. In this case, let L be the fixed field of the kernel of the projective Galois representation associated to f, so $\mathrm{Gal}(L \mid \mathbb{Q}) \simeq D_n$, the dihedral group of order $2n$. Then for each quadratic subfield $F \subseteq L$, the form f has self-twist by the character associated to F. Accordingly, when $n > 2$ the subfield F and associated self-twist character are unique, and when $n = 2$ (so K is biquadratic) there are three distinct subfields and corresponding characters and there is a real quadratic subfield.

Example 12.5.1 shows that forms in Proposition 11.1.12(b) indeed occur. In light of Proposition 11.1.12, we make the following definition.

Definition 11.1.13 We say f has real multiplication (RM) if f has self-twist by a character attached to a real quadratic field and complex multiplication (CM) if f has self-twist by a character attached to an imaginary quadratic field.

Remark 11.1.14 It is common in the literature to just replace the term *self-twist* by *complex multiplication*. By Proposition 11.1.12(a), there is no harm in this for weight $k \geq 2$, but for weight $k = 1$ we think this is potentially confusing, and we want to avoid saying "f has complex multiplication by $\mathbb{Q}(\sqrt{5})$."

Example 11.1.15 As in the proof of Proposition 11.1.12(b), weight 1 forms can have RM or CM or both. Forms with RM correspond precisely to ray class characters of real quadratic fields that are of mixed signature (i.e., even at one real place and odd at another).

Example 11.1.16 CM modular forms may also have an inner twist that is not a self-twist: the smallest example by analytic conductor is 52.1.j.a, having CM by $\mathbb{Q}(\sqrt{-1})$ and two inner twists that are not self-twists. This phenomenon is not restricted to weight 1, for example the same is true of the form with label 20.2.e.a.

Continuing with the theme of working with newforms that have not yet been embedded, we conclude this section by showing that the inner twist group is well-defined on the Galois orbit.

Lemma 11.1.17 *For all* $\tau \in \mathrm{Aut}(\mathbb{C})$, *we have an isomorphism of groups*

$$\begin{aligned} \mathrm{InnTw}(f) &\xrightarrow{\sim} \mathrm{InnTw}(\tau(f)) \\ (\psi, \sigma) &\mapsto (\tau\psi, \tau\sigma\tau^{-1}). \end{aligned} \tag{11.1.18}$$

Proof From $\sigma(a_n) = a_n\psi(n)$ for almost all n we conclude

$$(\tau\sigma\tau^{-1})(\tau(a_n)) = \tau(a_n)(\tau\psi)(n)$$

for almost all n, and conversely.

11.2 Detecting Inner Twists

With definitions out of the way, we now drill down to precisely understand the level of twists. We keep notation from the previous section, in particular $f(q) = \sum_n a_n(f)q^n \in S_k^{\mathrm{new}}(N, \chi)$ is a newform and ψ is a Dirichlet character of conductor $\mathrm{cond}(\psi)$.

Lemma 11.2.1 *Let M be the level of $f \otimes \psi$, so $f \otimes \psi \in S_k^{\mathrm{new}}(M, \chi\psi^2)$. Then the following statements hold:*

(a) *For all primes p, we have the inequality*

$$\mathrm{ord}_p(M) \le \max\bigl(\mathrm{ord}_p(N), \mathrm{ord}_p(\mathrm{cond}(\psi)\,\mathrm{cond}(\chi\psi))\bigr),$$

with equality if $\mathrm{ord}_p(N) \ne \mathrm{ord}_p(\mathrm{cond}(\psi)\,\mathrm{cond}(\chi\psi))$. *In particular, M divides* $\mathrm{lcm}(N, \mathrm{cond}(\psi)\,\mathrm{cond}(\chi\psi))$.

(b) *For all primes p we have*

$$\mathrm{ord}_p(\mathrm{cond}(\psi)) \le \mathrm{ord}_p(\mathrm{cond}(\psi)\,\mathrm{cond}(\chi\psi)) \le \max(\mathrm{ord}_p(N), \mathrm{ord}_p(M)).$$

In particular, $\mathrm{cond}(\psi)\,\mathrm{cond}(\chi\psi)$ *divides* $\mathrm{lcm}(M, N)$, *and if M divides N, then* $\mathrm{cond}(\psi)\,\mathrm{cond}(\chi\psi)$ *divides N.*

Proof Statement (a) can be found in Booker–Lee–Strömbergsson [9, Lemma 1.4]: this improves the upper bound of Shimura [86, Proposition 3.64] and Atkin–Li [2, Proposition 3.1] that

$$M \mid \mathrm{lcm}(N, \mathrm{cond}(\psi)^2, \mathrm{cond}(\chi)\,\mathrm{cond}(\psi)), \qquad (11.2.2)$$

which can be proven directly.

For statement (b), we prove the contrapositive. Let $p \mid \mathrm{cond}(\psi)\,\mathrm{cond}(\chi\psi)$ and suppose that $\mathrm{ord}_p(\mathrm{cond}(\psi)\,\mathrm{cond}(\chi\psi)) > \mathrm{ord}_p(N)$. Then by (b) we have

$$\mathrm{ord}_p(M) = \mathrm{ord}_p(\mathrm{cond}(\psi)\,\mathrm{cond}(\chi\psi)) > \mathrm{ord}_p(N).$$

Lemma 11.2.3 *If* $a_p(f) \ne 0$ *for some prime number p, then* $\mathrm{ord}_p(N) \in \{1, \mathrm{ord}_p \mathrm{cond}(\chi)\}$.

Proof If $\mathrm{ord}_p(N) = 0$, then $\mathrm{ord}_p(\mathrm{cond}(\chi)) = 0$. If $\mathrm{ord}_p \mathrm{cond}(\chi) \ne \mathrm{ord}_p(N)$, i.e., χ is a character modulo N/p, then $a_p(f) \ne 0$ implies $\mathrm{ord}_p(N) = 1$ by a result of Li [59, Theorem 3].

We recall by Proposition 11.1.7(b) that if $(\psi, \sigma) \in \mathrm{InnTw}(f)$, then $\sigma \in \mathrm{Aut}(K_f)$. But since we do not need this in the proof, we state the following theorem more generally.

Theorem 11.2.4 *Let* $f(q) = \sum_n a_n(f)q^n \in S_k^{\mathrm{new}}(N, \chi)$, *and let* $\sigma \in \mathrm{Gal}(\widetilde{K}_f \mid \mathbb{Q})$ *where* $\widetilde{K}_f \subseteq \mathbb{C}$ *is the Galois closure of* K_f. *Let* ψ *be a primitive Dirichlet character, and let* ψ' *be the primitive character that induces* $\chi\psi$. *Then* $f \otimes \psi = \sigma(f)$ *if and only if all of the following conditions hold:*

(i) $\mathrm{cond}(\psi)\,\mathrm{cond}(\psi') \mid N$*;*
(ii) $\chi\psi^2 = \sigma(\chi)$*; and*
(iii) $\sigma(a_p(f)) \in \bigl\{a_p(f)\psi(p), \overline{a_p(f)}\psi'(p)\bigr\}$ *for all primes* $p \le \mathrm{Sturm}(k, N)$.

Proof Let $\bar{f} \in S_k^{\mathrm{new}}(N, \overline{\chi})$ denote the dual of f, with coefficients $a_n(\bar{f}) = \overline{a_n(f)}$. Thus $\bar{f} = f \otimes \overline{\chi}$ (cf. Atkin–Li [2, Proposition 1.5] or Ribet [75, §1, p. 21]) and consequently $f \otimes \psi = \bar{f} \otimes \psi'$ as

$$a_n(\bar{f})\psi'(n) = a_n(f)\bar{\chi}(n)(\chi\psi)(n) = a_n(f)\psi(n)$$

whenever $\gcd(n, N) = 1$.

First we prove (⇒), and suppose that $f \otimes \psi = \sigma(f)$. By Proposition 11.1.7 we have $\chi\psi^2 = \sigma(\chi)$. Since $\mathrm{cond}(\sigma(f)) = \mathrm{cond}(f) = N$, we have $\mathrm{cond}(\psi)\,\mathrm{cond}(\psi') \mid N$ by Lemma 11.2.1(c). Let $D := \gcd(\mathrm{cond}(\psi), \mathrm{cond}(\psi'))$. Then

$$\mathrm{cond}(\chi) = \mathrm{cond}(\psi'\overline{\psi}) \mid \mathrm{lcm}(\mathrm{cond}(\psi), \mathrm{cond}(\psi')) = (\mathrm{cond}(\psi)\,\mathrm{cond}(\psi')/D) \mid (N/D). \tag{11.2.5}$$

Let p be prime. If $p \nmid \mathrm{cond}(\psi)$ then $\sigma(a_p(f)) = a_p(f \otimes \psi) = a_p(f)\psi(p)$. Similarly, if $p \nmid \mathrm{cond}(\psi')$ then $\sigma(a_p(f)) = a_p(\bar{f} \otimes \psi') = \overline{a_p(f)}\psi'(p)$. Hence we may suppose that $p \mid D$, so by (11.2.5) we have $\mathrm{ord}_p(N) > \max\{1, \mathrm{ord}_p\,\mathrm{cond}(\chi)\}$. By Lemma 11.2.3, it follows that $a_p(f) = 0$, and thus $\sigma(a_p(f)) = a_p(f)\psi(p)$.

Now we prove the converse (⇐), and suppose that conditions (i)–(iii) hold. Let M be the level of $f \otimes \psi$. Let Q denote the product of primes $p \mid N$ such that either

- $p \nmid M$, or
- $a_p(f) = 0$ and $a_p(f \otimes \psi) \neq 0$.

Let ξ denote the trivial character modulo Q, and define

$$g(q) := \sum_{n=1}^{\infty} a_n(f \otimes \psi)\xi(n)q^n. \tag{11.2.6}$$

We claim that conditions (i)–(ii) imply that $g \in S_k(N, \chi\psi^2)$. By Atkin–Li [2, Proposition 3.1] it suffices to show that

$$\mathrm{lcm}\big(M, \mathrm{cond}(\psi\psi')Q, Q^2\big) \mid N. \tag{11.2.7}$$

By Lemma 11.2.1(a) and the fact that $\mathrm{cond}(\psi)\,\mathrm{cond}(\psi') \mid N$, we have

$$\mathrm{cond}(\psi\psi') \mid M \mid \mathrm{lcm}\{N, \mathrm{cond}(\psi)\,\mathrm{cond}(\psi')\} = N,$$

so to prove (11.2.7) it suffices to show that $\mathrm{ord}_p(N) \geq 1 + \max\{1, \mathrm{ord}_p\,\mathrm{cond}(\psi\psi')\}$ for all primes $p \mid Q$.

Let p be such a prime. Then either $p \nmid M$ or $a_p(f) = \overline{a_p(f)} = 0$ and $a_p(f \otimes \psi) = a_p(\bar{f} \otimes \psi') \neq 0$. In either case we must have $p \mid \gcd(\mathrm{cond}(\psi), \mathrm{cond}(\psi'))$ and, by Lemma 11.2.3, $\mathrm{ord}_p(M) \in \{1, \mathrm{ord}_p\,\mathrm{cond}(\psi\psi')\}$. It follows that

$$\max\{1, \mathrm{ord}_p\,\mathrm{cond}(\psi\psi'), \mathrm{ord}_p(M)\} \leq \max\{\mathrm{ord}_p\,\mathrm{cond}(\psi), \mathrm{ord}_p\,\mathrm{cond}(\psi')\}.$$

Since $p \mid \gcd(\mathrm{cond}(\psi), \mathrm{cond}(\psi'))$, we have $\min\{\mathrm{ord}_p\,\mathrm{cond}(\psi), \mathrm{ord}_p\,\mathrm{cond}(\psi')\} \geq 1$, and hence

$$\operatorname{ord}_p(\operatorname{cond}(\psi)\operatorname{cond}(\psi')) \geq 1 + \max\{1, \operatorname{ord}_p \operatorname{cond}(\psi\psi'), \operatorname{ord}_p(M)\}.$$

By Lemma 11.2.1(b) we have

$$\operatorname{ord}_p(N) = \operatorname{ord}_p(\operatorname{cond}(\psi)\operatorname{cond}(\psi')) \geq 1 + \max\{1, \operatorname{ord}_p \operatorname{cond}(\psi\psi')\}.$$

This concludes the proof that $g \in S_k(N, \chi\psi^2)$.

Next, we claim that $a_n(g) = \sigma(a_n(f))$ for all $n \leq \operatorname{Sturm}(k, N)$. Since both sequences are multiplicative and $\chi\psi^2 = \sigma(\chi)$, it suffices to verify this equality at primes, p. There are three cases to consider:

- If $p \nmid N$ then $a_p(f)\psi(p) = \overline{a_p(f)}\psi'(p)$, so that $\sigma(a_p(f)) = a_p(g)$.
- If $p \mid N$ and $a_p(f) = 0$ then $a_p(g) = 0$ by construction, and $\sigma(a_p(f)) = 0$.
- If $p \mid N$ and $a_p(f) \neq 0$ then $0 \neq \sigma(a_p(f)) \in \{a_p(f)\psi(p), \overline{a_p(f)}\psi'(p)\}$.
  - If $\sigma(a_p(f)) = a_p(f)\psi(p)$ then $p \nmid \operatorname{cond}(\psi)$, so $a_p(f)\psi(p) = a_p(f \otimes \psi)$.
  - If $\sigma(a_p(f)) = \overline{a_p(f)}\psi'(p)$ then $p \nmid \operatorname{cond}(\psi')$, so

$$\overline{a_p(f)}\psi'(p) = a_p(\bar{f} \otimes \psi') = a_p(f \otimes \psi).$$

  In either case, we conclude that $\sigma(a_p(f)) = a_p(f \otimes \psi) = a_p(g)$.

By the Hecke–Sturm bound (Proposition 8.2.3), it follows that $g = \sigma(f)$. Finally, since f is a newform, $\sigma(f)$ is as well, and thus $\sigma(f) = g = f \otimes \psi$, by strong multiplicity one. □

We conclude with a variant, similarly useful for algorithmic purposes. We recall the notion of *distinguishing primes* from Sect. 8.9.

Theorem 11.2.8 *With the same hypotheses as in Theorem 11.2.4, we have $f \otimes \psi = \sigma(f)$ if and only if all of the following conditions hold:*

(i) $\operatorname{cond}(\psi)\operatorname{cond}(\chi\psi) \mid N$;
(ii) $\chi\psi^2 = \sigma(\chi)$;
(iii) $\sigma(a_p(f)) = a_p(f)\psi(p)$ *for all primes* $p \leq \operatorname{Sturm}(k, N)$ *with* $p \nmid N$; *and*
(iv) $\sigma(a_p(f)) = a_p(f)\psi(p)$ *for p in a set of distinguishing primes for f.*

Proof The implication ($\Rightarrow$) is clear, so we prove ($\Leftarrow$).

As in the proof of ($\Leftarrow$) of Theorem 11.2.4, we again consider the form g as in (11.2.6) with ξ the trivial character modulo Q. Let N_g be the level of g. Then in the proof we showed that $N_g \mid N$ and $h := g - \sigma(f) \in S_k(N, \sigma(\chi))$. By (iii) and Hecke recursion, we have $a_n(h) = 0$ for all $n \leq \operatorname{Sturm}(k, N)$ coprime to N.

If $N_g = N$, then by (iv), we have $\sigma(f) = f \otimes \psi$. So we may assume that N_g is a proper divisor of N. We now employ degeneracy operators to upgrade (iii). It is convenient to switch from lower-triangular to upper-triangular matrices. Let

$$\Gamma^1(N) := \left\{\gamma \in \mathrm{SL}_2(\mathbb{Z}) : \gamma \equiv \begin{pmatrix} 1 & 0 \\ * & 1 \end{pmatrix} \pmod{N}\right\}$$

and similarly $\Gamma^0(N)$, and define spaces of modular forms on these groups similarly. We refer to Diamond–Shurman [38, §5.7] for the results we need. The groups $\Gamma_1(N)$ and $\Gamma^1(N)$ are conjugate by the matrix $\begin{pmatrix} N & 0 \\ 0 & 1 \end{pmatrix}$, giving an isomorphism $\iota_N := S_k(\Gamma_1(N)) \to S_k(\Gamma^1(N))$ whose effect on Fourier expansions is $\sum_n b_n q^n \mapsto \sum_n b_n q_N^n$ where $q_N := \exp(2\pi i z/N)$. Moreover, this map preserves the Nebentypus character. For any $d \mid N$, the trace operator defines a map

$$\pi_d \colon S_k(\Gamma^1(N)) \to S_k(\Gamma_d) \subseteq S_k(\Gamma^1(N))$$

where $\Gamma_d := \Gamma_1(N) \cap \Gamma^0(N/d)$: its effect on Fourier expansions is

$$\sum_{n=1}^{\infty} b_n q_N^n \mapsto \sum_{\substack{n=1 \\ d|n}}^{\infty} b_n q_N^n.$$

The operator π_d is a projection operator, and for $d, d' \mid N$ with $\gcd(d, d') = 1$ we have $\pi_d \pi_{d'} = \pi_{d'} \pi_d$.

Consider

$$h' := \prod_{p|N} (1 - \pi_p)\iota_N(h) \in S_k(\Gamma^0(N), \chi).$$

By construction, multiplicativity, and (iii), we have $a_n(h') = 0$ for all $n \leq \mathrm{Sturm}(k, N)$. Then by the Hecke–Sturm bound (Proposition 8.2.3), we conclude $h' = 0$. Thus

$$h(q) = \sum_{\substack{n=1 \\ \gcd(n,N)\neq 1}}^{\infty} a_n(h) q^n. \tag{11.2.9}$$

We have realized h as a sum of oldforms. Turning this back to $\Gamma_1(N)$, we conclude that

$$h(q) = \sum_{p|N} h_p(q^p) \tag{11.2.10}$$

with $h_p(q) \in S_k(\Gamma_p, \sigma(\chi)_p)$, as in the oldform theory of Atkin–Lehner [1, Theorem 1] and Li [59, Corollary 1]; moreover, $h_p = 0$ if and only if h is new at p.

We now show that $h = 0$. Let $p \mid N$. If χ is not a character modulo N/p, then $S_k(\Gamma_p, \sigma(\chi)_p) = 0$ so $h_p = 0$. So suppose χ is a character modulo N/p.

- Suppose that $a_p(f) \neq 0$. Then by Lemma 11.2.3, we have $p \parallel N$. Thus $\mathrm{ord}_p(\mathrm{cond}(\chi)) = 0$, so by (i) we have $\mathrm{ord}_p(N) \geq 2\,\mathrm{ord}_p(\mathrm{cond}(\psi))$. If

$\mathrm{ord}_p(\mathrm{cond}(\psi)) = 0$, then we have twisted by a character trivial at p, so $\mathrm{ord}_p(M) = \mathrm{ord}_p(N)$ by Lemma 11.2.1(b). Therefore $f \otimes \psi$ is new at p, so g is new at p and $a_p(g) = a_p(f \otimes \psi)$ so $h_p = 0$. If instead $\mathrm{ord}_p(\mathrm{cond}(\psi)) \geq 1$, then $p^2 \mid N$, a contradiction.
- Suppose $a_p(f) = 0$. If $a_p(f \otimes \psi) \neq 0$, then by construction, $a_p(g) = 0$ so by multiplicativity $a_n(f) = a_n(g)$ for all $p \mid n$; therefore $h_p = 0$.

We have shown that $\sigma(f) = g$. We then conclude as in the end of the proof of Theorem 11.2.4. □

Example 11.2.11 Consider the space 24.2.f.a. There are two Galois-conjugate newforms with the same Nebentypus character. The Sturm bound is 8, but the smallest $p \nmid N$ where the Fourier coefficients differ is 11. In particular, this shows that in the Hecke–Sturm bound (Proposition 8.2.3) we cannot ignore the primes $p \mid N$.

The virtue of Theorems 11.2.4 and 11.2.8 is that they give explicit criteria to certify inner twists, with care taken concerning primes dividing the level.

11.3 Computing Inner Twists

We used Theorem 11.2.8 to compute the complete group of inner twists for all the modular forms in our dataset. Specifically, we enumerate the finite set X of Dirichlet characters ψ satisfying condition (i) of Theorem 11.2.8 for which $\chi\psi^2$ is conjugate to χ. Note that the set X does not depend on f or its coefficient field, only the character χ and level N. We then determine the subset of X that satisfy conditions (iii) and (iv) for some $\sigma \in \mathrm{Gal}(\widetilde{K}_f)$ as follows:

(1) We first remove from X all characters ψ for which there is a prime $p \leq \mathrm{Sturm}(k, N)$ not dividing N such that $a_p(f)\psi(p)$ is not conjugate to $a_p(f)$; this is accomplished by comparing the minimal polynomials of $a_p(f)\psi(p)$ and $a_p(f)$.
(2) For all remaining $\psi \in X$, set $T := \mathrm{Gal}(\widetilde{K}_f)$ and for successive primes $p \leq \mathrm{Sturm}(k, N)$ with $p \nmid N$, replace T with $\{\sigma \in T : \sigma(a_p(f)) = a_p(f)\psi(p)\}$, stopping if T becomes empty. This yields a list of candidate inner twists (ψ, σ) containing $\mathrm{InnTw}(f)$.
(3) Finally, for each candidate (ψ, σ) we check whether (iv) holds; if so then Theorem 11.2.8 implies that (ψ, σ) is an inner twist of f.

As shown by Example 11.2.11, the third step above is potentially necessary, but in our computation we never encountered a case where a candidate inner twist that survived step (2) was discarded in step (3).

Remark 11.3.1 The Magma function `InnerTwists` implements a weaker form of Theorem 11.2.4. It requires checking eigenvalues up to the Sturm bound for level $\mathrm{lcm}(N, \mathrm{cond}(\psi)^2, \mathrm{cond}(\psi)\,\mathrm{cond}(\chi))$, and it performs eigenvalue comparisons using complex approximations that do not guarantee a rigorous result. Indeed, even when the optional parameter `Proof` is set to `True`, Magma version 2.24–7 displays the following message:

```
WARNING: Even if Proof is True, the program does not prove
that every twist returned is in fact an inner twist (though
they are up to precision 0.00001).
```

12 Weight One

Modular forms of weight one are of particular interest due to the connection with Artin representations, provided by a theorem of Deligne and Serre [36]: one can associate to each weight one newform f an odd irreducible two-dimensional Galois representations $\rho_f\colon G_{\mathbb{Q}} \to \mathrm{GL}_2(\mathbb{C})$ for which $L(f, s) = L(\rho_f, s)$ (recall that a Galois representation is odd if complex conjugation has determinant -1). Following the proof of Serre's conjecture by Khare and Wintenberger [57], we now know that the map $f \mapsto \rho_f$ is in fact a bijection. This connection allows one to attach several additional arithmetic invariants to weight one newforms that we would like to compute, including:

- The projective image of ρ_f in $\mathrm{PGL}_2(\mathbb{C})$, which by Klein's classification is isomorphic to either D_n (dihedral of order $2n$, including $D_2 := \mathbb{Z}/2\mathbb{Z} \times \mathbb{Z}/2\mathbb{Z}$), or one of the exceptional groups A_4 (tetrahedral), S_4 (octahedral), or A_5 (icosahedral).
- The projective field of ρ_f: the fixed field of the kernel of $G_{\mathbb{Q}} \xrightarrow{\rho_f} \mathrm{GL}_2(\mathbb{C}) \twoheadrightarrow \mathrm{PGL}_2(\mathbb{C})$.
- The Artin image of ρ_f: the finite group $\rho_f(G_{\mathbb{Q}}) \leq \mathrm{GL}_2(\mathbb{C})$.
- The Artin field of ρ_f: the fixed field of $\ker \rho_f$, with Galois group isomorphic to $\rho_f(G_{\mathbb{Q}})$.

One can also consider the projective representation $\bar{\rho}_f\colon G_{\mathbb{Q}} \to \mathrm{PGL}_2(\mathbb{C})$ induced by ρ_f as an invariant in its own right: it uniquely determines the twist class of f. Two newforms f and g are said to be twist equivalent if $g = f \otimes \psi$ for some Dirichlet character ψ, and in weight 1 this occurs if and only if $\bar{\rho}_f = \bar{\rho}_g$.

12.1 Computational Observations

The Deligne–Serre theorem also has important computational implications. In the typical case where ρ_f is a dihedral representation (meaning that its projective image is dihedral), the Artin L-function $L(\rho_f, s)$ is also the Weber L-function $L(\omega, s)$ of a ray class character ω of the quadratic field K fixed by the preimage of $C_n \subseteq D_n \simeq \bar{\rho}_f(G_{\mathbb{Q}})$. (For $n = 2$ there are three choices for $C_2 \subseteq D_2$; we can use any one of the three.) The quadratic field K and the ray class character ω necessarily satisfy

$$|d_K|\,\mathrm{Nm}(\mathrm{cond}(\omega)) = \mathrm{cond}(\rho_f) = N, \tag{12.1.1}$$

where d_K is the discriminant of K and N is the level of f. In order to obtain an odd representation ρ_f we also require that if K is a real quadratic field then the modulus for ω should include exactly one of the real places of K.

For any given level N, it is straightforward to enumerate all quadratic fields K of discriminant $d_K \mid N$, all $\mathcal{O}_K$-ideals of absolute norm dividing $N/\,|d_K|$, and all ray class characters ω of K for the modulus with finite part I and infinite part compatible with an odd representation. This makes it feasible to explicitly compute Fourier expansions of all dihedral newforms of level N to any desired precision; to compute $a_p(f)$ for $p \nmid N$ this simply amounts to evaluating the corresponding ray class character ω at the prime ideals of $\mathcal{O}_K$ above p.

Pari/GP contains extensive support for computing with ray class characters that are particularly efficient in the case of quadratic fields. We used this to compute all dihedral newforms of level $N \leq 40{,}000$ with Fourier coefficients $a_n(f)$ computed for $n \leq 6000$ (well past the Sturm bound). This yielded a total of 572,462 dihedral newforms, corresponding to 14,634,052 embedded newforms. The largest dimension we found was 2818, which arises for a dihedral newform of level 39473, and the largest projective image we found was D_{2846} for a newform of level 39,851.

These computations go far beyond the extent of our database described in Sect. 10.1, which only covers levels $N \leq 4000$ in weight one. For comparison, the largest dimension arising for $N \leq 4000$ is 232 and the largest projective image is D_{285}. The reason for this discrepancy is that while it is computationally very easy to compute dihedral newforms, to obtain a complete enumeration of all the newforms in a given weight one newspace, one must also enumerate the tetrahedral, octahedral, and icosahedral newforms, which is more difficult—particularly in the icosahedral case. Interestingly, the main difficulty often lies not in enumerating these exceptional newforms, but in verifying that one has actually found them all. In contrast to the case $k > 1$ where there are well known dimension formulas, while there are computational tricks that work well in special cases, to our knowledge no efficient method for computing $\dim S_1^{\mathrm{new}}(N)$ for general N is currently known.

12.2 Classifying the Projective Image

The Pari/GP function `mfgaloistype` can be used to classify the projective image, but given that we actually computed the projective field in every case (which of course determines the projective image), we did not exploit this feature.

Remark 12.2.1 Buzzard–Lauder [19] describe an approach to classifying the projective image by computing projective orders of elements that they applied to all weight one newforms of level up to 1500. They note in their paper that their approach relies on the convenient fact that there are no weight one newforms of level $N \leq 1500$ with projective image A_4 whose coefficient field contains $\mathbb{Q}(\sqrt{5})$. Five such examples arise in our dataset, the first of which is 2299.1.w.a.

12.3 Computing the Projective Field

Our strategy for computing the projective field is to exhaustively compute a complete set of candidates and then rule out all but one. As noted in Sect. 12.1, we can effectively determine all the dihedral forms at each level, so we know in advance exactly which forms are dihedral (and the exact order of the projective image in each of these cases). In cases where a dihedral image has moderate degree—less than 100, say—it is feasible to use the ray class field functionality in Pari/GP to compute the projective field. This notably includes all of the dihedral projective fields whose distinguished quadratic subfield is real: the largest such example in our database is 2605.1.bd.a with projective image D_{40}.

The dihedral fields in which the distinguished subfield is imaginary quadratic can be much larger: the largest example 3997.1.cz.a has projective image D_{285}. In these cases, we exploit the fact that every dihedral field whose distinguished quadratic subfield is imaginary can be realized as a subfield of a ring class field that can be explicitly computed using the theory of complex multiplication. There is a well-developed theory for efficiently computing these ring class fields, even in cases where the degree may be in the millions, motivated by applications to cryptography and elliptic curve primality proving (the CM method for constructing elliptic curves over finite fields).

Given a dihedral weight one newform $f \in S_1^{\text{new}}(N, \chi)$ with dihedral image D_n and distinguished imaginary quadratic field K, there is a finite set of possible suborders $\mathcal{O}$ of $\mathcal{O}_K$ and conductors c such that the projective field of f arises as a cyclic degree-n extension of K of conductor c contained in the ring class field K of $\mathcal{O}$. The enumeration of these dihedral fields was achieved using an algorithm based on the techniques developed by Enge-Sutherland [43] and Sutherland [92, 93] that will be described in a forthcoming paper.

Having enumerated a complete list of candidate fields $L := \mathbb{Q}[x]/(g_L(x))$, for successive primes $p \nmid N$ we can compute the order of $\rho_f(\text{Frob}_p)$ in $\text{PGL}_s(\mathbb{C})$ by determining the positive integer n for which $a_p(f)^2/\chi(p) = \zeta_n + \zeta_n^{-1} + 2$ and compare this to the inertia degree of the primes above p in $\mathcal{O}_L$. This will eventually eliminate all but one candidate field, since the sequence of inertia degrees uniquely determines a Galois number field, and in practice this happens very quickly. To accelerate the computation we precompute defining polynomials for the real cyclotomic fields we may encounter and use p coprime to the discriminants of the defining polynomials g_L so that we can compute the inertia degree as the degree of the irreducible factors of $g_L(x)$ in $\mathbb{F}_p[x]$.

For the non-dihedral projective images we used the methods of Cohen–Diaz y Diaz–Olivier [22, 23] to enumerate all A_4 and S_4 fields unramified outside a given set of primes, and for the A_5 fields we used existing tables of fields in the Jones–Roberts database and the LMFDB combined with a targeted Hunter search for some missing cases, as described by Jones–Roberts [52]. This allowed us to construct complete lists of candidate fields for each non-dihedral weight one form from which we then ruled out all but one candidate by comparing orders of Frobenius elements with inertia degrees as described above.

12.4 Computing the Artin Image, the Artin Field, and the Associated Artin Representation

As of January 2020 the LMFDB contained 5116 odd two-dimensional Artin representations of conductor $N \leq 4000$, all of which we were able to uniquely match to a corresponding newform of weight one. For each of these Artin representations the LMFDB provides the Artin image, the Artin field, and a complete description of the Artin representation given values on each conjugacy class of Frobenius elements. We were also able to compute the Artin image and Artin field for 833 additional weight one newforms that are twists of a weight one newform for which we know the corresponding Artin representation by taking the compositum of the known Artin field with an appropriate cyclotomic field.

There is work in progress to add as many of the Artin representations corresponding to the remaining 14,190 weight one newforms as possible; these will be linked to the corresponding weight 1 newforms as they become available.

12.5 Interesting and Extreme Behavior

Weight one modular forms behave rather differently than those of higher weight. As seen in Sect. 12, one important invariant of weight one forms is the projective image of the associated Galois representation. We will discuss some forms with dihedral projective image first.

Hecke also constructed weight one modular forms starting from imaginary quadratic fields with odd class number at least 3. The first examples of such fields come from $\mathbb{Q}(\sqrt{-23})$, $\mathbb{Q}(\sqrt{-31})$, $\mathbb{Q}(\sqrt{-39})$, and the corresponding modular forms are the three smallest level weight one newforms; these have labels 23.1.b.a, 31.1.b.a and 39.1.d.a, respectively. [46]

Example 12.5.1 The last of these, 39.1.d.a, is the D_2 form of lowest level and has CM by both $\mathbb{Q}(\sqrt{-3})$ and $\mathbb{Q}(\sqrt{-39})$, and RM by $\mathbb{Q}(\sqrt{13})$. This form appears in work of Darmon–Lauder–Rotger [34, Example 2.5].

The first examples of newforms with RM but no CM occur in level 145 with 145.1.f.a (RM by $\mathbb{Q}(\sqrt{5})$, [34, Example 3.3], [33, Example 4.1]) and 145.1.h.a (RM by $\mathbb{Q}(\sqrt{29})$, [35, Example 1.2]).

The problem of constructing weight one forms whose projective image is not dihedral was considered by Tate and Serre in the 1970s. These forms are sometimes called *non-banal* or *exotic*. Such forms divide up into 3 cases based on their projective image, which can be one of A_4, S_4, A_5: the forms are then known as tetrahedral, octahedral and icosahedral, respectively.

Tate together with his students, Flath, Kottwitz, Tunnell, and Weisinger, and additionally Atkin, exhibited a form of level 133, with projective image A_4 described in a letter to Atkin [94, p. 713]; this form is 133.1.m.a in our database. The smallest level example is actually in level 124, given by 124.1.i.a.

In the octahedral case, the smallest level example is in level $4 \cdot 37 = 148$ with label 148.1.f.a; this newform is discussed by Buzzard [18, §2.3] and Darmon–Lauder–Rotger [33, Example 5.6].

Many modular forms previously considered in the literature with interesting Galois representations can now be found in our database. Ogasawara [69] takes the mod-3 Galois representations attached to certain elliptic curves and constructs a $\mathrm{GL}_2(\mathbb{F}_3)$ Artin representation: for example, the elliptic curve of conductor 11 with label 11.a3 is used and the corresponding octahedral modular form of weight one over $\mathbb{Q}(\sqrt{-2})$ is constructed. Using the q-expansion coefficients given there, we can use the trace search functionality to locate a (unique) matching form in our database: 3267.1.b.d. We then verify that it has the right Artin field: a degree 8 extension over which 11.a3 gains 3-torsion.

Buhler [13, 14] constructs the icosahedral Galois representation of level 800, labeled 800.1.bh.a. Kiming–Wang [55] gave several more instances of icosahedral newforms of weight one with characters of order 2, showing their existence in order to verify the Artin conjecture in these cases. The new database now contains all but one of these: 2083.1.b.b, 1948.1.b.a, 3004.1.b.a, 3548.1.d.a, 3676.1.c.a, 2336.1.c (two newforms). The only newspace discussed in loc. cit. with level outside our range would have label 6176.1.b. The database also contains the icosahedral newforms 1376.1.r.a, 2416.1.p.a, 3184.1.t.a, 3556.1.ba.a and 3756.1.q.b which were all shown to satisfy Artin's conjecture by Buzzard–Stein [20]. The proof of Serre's conjecture [57] established Artin's conjecture for all odd irreducible two-dimensional representations, including all of the icosahedral cases. The smallest level example of an icosahedral newform is 633.1.m.b.

Constructing exotic forms of prime level with specific projective image is also a much studied problem. Such forms do not exist in the tetrahedral case [83, Thm. 7, p. 245], leaving only octahedral and icosahedral forms with the possibility of prime level.

In the octahedral case the smallest prime level is 229, and the space of newforms 229.1.d splits into two Galois orbits, (see Serre [83, p. 265]). The second smallest level is 283, where we have the newform 283.1.b.b that appears also in work of Serre [84].

In the icosahedral case, we have seen above the first example of such a form: the one with level 2083 of Kiming–Wang. In fact the query for forms with projective image A_5 shows that there are 4 such forms with prime level ≤ 4000: 2083.1.b.b, 2707.1.b.b, 3203.1.b.a, 3547.1.b.c. It is conjectured that these forms are rare.

Conjecture 12.5.2 *For any $\epsilon > 0$, the number of exotic newforms of prime level N is $O_\epsilon(N^\epsilon)$.*

Bhargava–Ghate [5] have shown an averaged version of this conjecture in the octahedral case.

References

1. A. O. L. Atkin and Joseph Lehner, *Hecke operators on $\Gamma_0(m)$*, Math. Ann. **185** (1970), 134–160.
2. A. O. L. Atkin and Wen-Ch'ing Winnie Li, *Twists of newforms and pseudo-eigenvalues of W-operators*, Invent. Math. **48** (1978), no. 3, 221–243.
3. B. Banwait and J. Cremona, *Tetrahedral elliptic curves and the local-to-global principle for isogenies*, Algebra & Number Theory **8** (2014), no. 5, 1201–1229.
4. Karim Belabas and Henri Cohen, *Modular forms in Pari/GP*, Res. Math. Sci. **5** (2018), no. 3, Paper No. 37, 19 pp.
5. Manjul Bhargava and Eknath Ghate, *On the average number of octahedral newforms of prime level*, Math. Ann. **344** (2009), no. 4, 749–768.
6. B. J. Birch, *Elliptic curves over $\mathbb{Q}$: A progress report*, 1969 Number Theory Institute (State Univ. New York, Stony Brook, N.Y., 1969), Proc. Sympos. Pure Math., vol. 20, Amer. Math. Soc., Providence, 1971, 396–400.
7. B. J. Birch, *Hecke actions on classes of ternary quadratic forms*, Computational number theory (Debrecen, 1989), de Gruyter, Berlin, 1991, 191–212.
8. Jonathan Bober, `mflib` software library, available at https://github.com/jwbober/mflib, 2019.
9. Andrew R. Booker, Min Lee, and Andreas Strömbergsson, *Twist-minimal trace formulas and the Selberg eigenvalue conjecture*, J. Lond. Math. Soc. **102** (2020), no. 3, 1067–1134.
10. Andrew R. Booker, *Artin's conjecture, Turing's method, and the Riemann hypothesis*, Exp. Math. **15** (2006), no. 4, 385–408.
11. Wieb Bosma, John Cannon, and Catherine Playoust, *The Magma algebra system. I. The user language*, J. Symbolic Comput. **24** (1997) no. 3–4, 235–265.
12. Peter Bruin, *Computing coefficients of modular forms*, Actes de la Conférence "Théorie des Nombres et Applications", 19–36, Publ. Math. Besançon Algèbre Théorie Nr., 2011, Presses Univ. Franche-Comté, Besançon, 2011.
13. Joe Buhler, *An icosahedral modular form of weight one*, Modular functions of one variable V, eds. Jean-Pierre Serre and Don Bernard Zagier, Lecture Notes in Math., vol. 601, Springer, Berlin-Heidelberg, 1977, 289–294.
14. Joe P. Buhler, *Icosahedral Galois representations*, Lecture Notes in Math., vol. 654, Springer-Verlag, Berlin-New York, 1978.
15. Joe P. Buhler, Benedict H. Gross, and Don B. Zagier, *On the conjecture of Birch and Swinnerton-Dyer for an elliptic curve of rank 3*, Math. Comp. **44** (1985), 473–481.
16. Jan Büthe, *A method for proving the completeness of a list of zeros of certain L-functions*, Math. Comp. **84** (2015), no. 295, 2413–2431.
17. Kevin Buzzard, *Dimension of spaces of Eisenstein series*, preprint available at http://wwwf.imperial.ac.uk/~buzzard/maths/research/notes/dimension_of_spaces_of_eisenstein_series.pdf, 2012.
18. Kevin Buzzard, *Computing weight one modular forms over $\mathbb{C}$ and $\overline{\mathbb{F}}_p$*, Computations with modular forms, Contrib. Math. Comput. Sci., vol. 6, Springer, Cham, 2014, 129–146.
19. Kevin Buzzard and Alan Lauder, *A computation of modular forms of weight one and small level*, Ann. Math. Qué. **41** (2017), no. 2, 213–219.

20. Kevin Buzzard and William A. Stein, *A mod five approach to modularity of icosahedral Galois representations*, Pacific J. Math. **203** (2002), no. 2, 265–282.
21. Sarvadaman Chowla, *The Riemann hypothesis and Hilbert's tenth problem*, Mathematics and Its Applications, vol. 4, Gordon and Breach, New York, 1965.
22. Henri Cohen, Francisco Diaz y Diaz, and Michel Olivier, *Construction of tables of quartic fields*, Algorithmic number theory (Leiden, 2000), Lecture Notes in Comput. Sci., vol. 1838, Springer, Berlin, 2000, 257-268.
23. Henri Cohen, Francisco Diaz y Diaz, and Michel Olivier, *Constructing complete tables of quartic fields using Kummer theory*, Math. Comp. **72** (2003), no. 242, 941–951.
24. H. Cohen and J. Oesterlé, *Dimensions des espaces de formes modulaires*, Modular functions of one variable VI, eds. Jean-Pierre Serre and Don Zagier, Lecture Notes in Math., vol. 627, Springer, Berlin, 1977, 69–78.
25. Henri Cohen and Fredrik Strömberg, *Modular forms: a classical approach*, Grad. Studies in Math., vol. 179, Amer. Math. Soc., Providence, 2017.
26. Edgar Costa, Nicolas Mascot, Jeroen Sijsling, and John Voight, *Rigorous computation of the endomorphism ring of a Jacobian*, Math. Comp. **88** (2019), 1303–1339.
27. Edgar Costa and David Platt, *A generic L-function calculator for motivic L-functions*, available at https://github.com/edgarcosta/lfunctions, 2019.
28. John E. Cremona, Aurel Page, and Andrew V. Sutherland, *Sorting and labelling integral ideals in a number field*, https://arxiv.org/abs/2005.09491.
29. John E. Cremona, *Abelian varieties with extra twist, cusp forms, and elliptic curves over imaginary quadratic fields*, J. London Math. Soc. (2) **45** (1992), no. 3, 404–416.
30. John E. Cremona, *Algorithms for modular elliptic curves*, 2nd ed., Cambridge University Press, Cambridge, 1997.
31. John E. Cremona, *The elliptic curve database for conductors to 130000*, Algorithmic number theory, eds. Florian Hess, Sebastian Pauli, and Michael Pohst, Lecture Notes in Comput. Sci., vol. 4076, Springer, Berlin, 2006, 11–29.
32. John E. Cremona, *The L-functions and modular forms database project*, Found. Comput. Math. **16** (2016), no. 6, 1541–1553.
33. Henri Darmon, Alan Lauder, and Victor Rotger, *Stark points and p-adic iterated integrals attached to modular forms of weight one*, Forum Math. Pi **3** (2015), e8, 95 pp.
34. Henri Darmon, Alan Lauder, and Victor Rotger, *First order p-adic deformations of weight one newforms*, L-functions and automorphic forms, Contrib. Math. Comput. Sci., vol. 10, Springer, Cham, 2017, 39–80.
35. Henri Darmon, Alan Lauder, and Victor Rotger, *Overconvergent generalised eigenforms of weight one and class fields of real quadratic fields*, Adv. Math. **283** (2015), 130–142.
36. Pierre Deligne and Jean-Pierre Serre, *Formes modulaires de poids 1*, Ann. Sci. Éc. Norm. Sup. (4) **7** (1974), no. 4, 507–530.
37. P. Deligne, *Valeurs de fonctions L et périodes d'intégrales*, appendix by N. Koblitz and A. Ogus, Automorphic forms, representations and L-functions (Oregon State Univ., Corvallis, Ore., 1977), Part 2, eds. A. Borel, W. Casselman, Proc. Sympos. Pure Math., vol. 33, Amer. Math. Soc., Providence, 1979, 313–346.
38. Fred Diamond and Jerry Shurman, *A first course in modular forms*, Grad. Texts in Math., vol. 228, Springer-Verlag, New York, 2005.
39. Bas Edixhoven and Jean-Marc Couveignes, *Computational aspects of modular forms and Galois representations: how one can compute in polynomial time the value of Ramanujan's tau at a prime*, Annals of Math. Studies, vol. 176, Princeton University Press, Princeton, NJ, 2011.
40. M. Eichler, *Einige Anwendungen der Spurformel im Bereich der Modularkorrespondenzen*, Math. Ann. **168** (1967), 128–137.
41. M. Eichler, *The basis problem for modular forms and the traces of the Hecke operators*, Modular functions of one variable I (Proc. Internat. Summer School, Univ. Antwerp, Antwerp, 1972), ed. W. Kuijk, Lecture Notes in Math., vol. 320, Springer, Berlin, 1973, 75–151.

42. Stephan Ehlen and Fredrik Strömberg, `modforms-db` software package, available at https://github.com/sehlen/modforms-db/tree/refactor, 2014.
43. Andreas Enge and Andrew V. Sutherland, *Class invariants by the CRT method*, Algorithmic number theory, Lecture Notes in Comput. Sci., vol. 6197, Springer, Berlin, 2010, 142–156.
44. Daniel Fiorilli, *On the non-vanishing of Dirichlet L-functions at the central point*, Q. J. Math. **66** (2015), no. 2, 517–528.
45. B.H. Gross and D.B. Zagier, *Heegner points and derivatives of L-series*, Invent. Math. **84** (1986), 225–320.
46. Erich Hecke, *Mathematische Werke*, Göttingen, Vandenhoeck & Ruprecht, 1959.
47. Hiroaki Hijikata, *Explicit formula of the traces of Hecke operators for* $\Gamma_0(N)$, J. Math. Soc. Japan **26** (1974), 56–82.
48. Hiroaki Hijikata, Arnold K. Pizer, and Thomas R. Shemanske, *The basis problem for modular forms on* $\Gamma_0(N)$, Mem. Amer. Math. Soc. **82** (1989), no. 418.
49. Henryk Iwaniec and Emmanuel Kowalski, *Analytic number theory*, Amer. Math. Soc. Colloquium Publications, vol. 53, Amer. Math. Soc., Providence, 2004.
50. H. Iwaniec and P. Sarnak, *Perspectives on the analytic theory of L-functions*, GAFA 2000 (Tel Aviv, 1999), Geom. Funct. Anal. 2000, Special Volume, Part II, 705–741.
51. Fredrik Johansson, *Arb: a C library for ball arithmetic*, ACM Communications in Computer Algebra **47** (2013), no. 4, 166–169.
52. John W. Jones and David P. Roberts, *Timing analysis of targeted Hunter searches*, Algorithmic number theory (Portland, OR, 1998), Lecture Notes in Comput. Sci., vol. 1423, Springer, Berlin, 1998, 412–423.
53. Kazuya Kato, *p-adic Hodge theory and values of zeta functions of modular forms*, Astérisque **295** (2004), 117–290.
54. Lloyd J. P. Kilford, *Modular forms: A classical and computational introduction*, 2nd ed., Imperial College Press, London, 2015.
55. Ian Kiming and Xiang Dong Wang, *Examples of 2-dimensional, odd Galois representations of* A_5*-type over* $\mathbb{Q}$ *satisfying the Artin conjecture*, On Artin's conjecture for odd 2-dimensional representations, Lecture Notes in Math., vol. 1585, Springer, Berlin, 1994, 109–121.
56. David Lowry-Duda, *Visualizing modular forms*, J. S. Balakrishnan et al. (eds.), Arithmetic geometry, number theory, and computation, Simons symposia, Springer, Cham, 2021, 539–560.
57. Chandrashekhar Khare and Jean-Pierre Wintenberger, *Serre's modularity conjecture (I)*, Invent. Math. **178** (2009), no. 3, 485–504.
58. Andrew Knightly and Charles Li, *Traces of Hecke operators*, Math. Surveys Monogr., vol. 133, Amer. Math. Soc., Providence, 2006.
59. Wen-Ch'ing Winnie Li, *Newforms and functional equations*, Math. Ann. **212** (1975), no. 4, 285–315.
60. The LMFDB Collaboration, *The L-functions and Modular Forms Database*, http://www.lmfdb.org, 2019.
61. Ju. I. Manin, *Parabolic points and zeta functions of modular curves*, Izv. Akad. Nauk SSSR Ser. Mat., **36** (1972), 19–66, translated in Math.-USSR Izvestija, **6** (1972), no. 1, 19–64.
62. Kimball Martin, *The basis problem revisited*, arXiv:1804.04234v2, 2019.
63. Nicolas Mascot, *Certification of modular Galois representations*, Math. Comp. **87** (2018), 381–423.
64. Loïc Merel, *Universal Fourier expansions of modular forms*, On Artin's conjecture for odd 2-dimensional representations, Lecture Notes in Math., vol. 1585, Springer, Berlin, 1994, 59–94.
65. J.-F. Mestre, *La méthode des graphes. Exemples et applications*, Proceedings of the international conference on class numbers and fundamental units of algebraic number fields (Katata, 1986), Nagoya Univ., Nagoya, 1986, 217–242.
66. Christian Meyer, *Newforms of weight two for* $\Gamma_0(N)$ *with rational coefficients*, http://meyer-idstein.de/weight2.pdf, 2005.
67. Toshitsune Miyake, *Modular forms*, Springer Monographs in Math., Springer-Verlag, Berlin, 2006.

68. Fumiyuki Momose, *On the l-adic representations attached to modular forms*, J. Fac. Sci. Univ. Tokyo Sect. IA Math. **28** (1981), no. 1, 89–109.
69. Takeshi Ogasawara, *Octahedral newforms of weight one associated to three-division points of elliptic curves*, Funct. Approx. Comment. Math. **49** (2013), no. 1, 103–109.
70. Sami Omar, *Non-vanishing of Dirichlet L-functions at the central point*, Algorithmic Number Theory (ANTS 2008), Lecture Notes in Comp. Sci. **5011** (2008) 443–453.
71. *Modular functions of one variable IV*, Proceedings of the International Summer School on Modular Functions of One Variable and Arithmetical Applications, RUCA, University of Antwerp, Antwerp, July 17–August 3, 1972, eds. Bryan J. Birch and Willem Kuyk, Lecture Notes in Math., vol. 476, Springer-Verlag, Berlin, 1975.
72. The PARI-group, *Pari/GP* (versions 2.11 and 2.12), Univ. Bordeaux, available at http://pari.math.u-bordeaux.fr, 2019.
73. Arnold Pizer, *An algorithm for computing modular forms on* $\Gamma_0(N)$, J. Algebra **64** (1980), no. 2, 340–390.
74. Alexandru Popa, *On the trace formula for Hecke operators on congruence subgroups, II*, Res. Math. Sci. **5** (2018), no. 1, Paper No. 3, 24 pp.
75. Kenneth A. Ribet, *Galois representations attached to eigenforms with Nebentypus*, Modular functions of one variable V, eds. Jean-Pierre Serre and Don Bernard Zagier, Lecture Notes in Math., vol. 601, Springer, Berlin-Heidelberg, 1977, 17–51.
76. Kenneth A. Ribet, *Twists of modular forms and endomorphisms of abelian varieties*, Math. Ann. **253** (1980), no. 1, 43–62.
77. Kenneth A. Ribet, *Mod p Hecke operators and congruences between modular forms*, Invent. Math. **71** (1983), no. 1, 193–205.
78. Kenneth A. Ribet, *Abelian varieties over* $\mathbb{Q}$ *and modular forms*, Modular curves and abelian varieties, eds. John E. Cremona, Joan-C. Lario, Jordi Quer, and Kenneth A. Ribet, Progress in Math. **224**, Birkhäuser, Basel, 2004, 241–261.
79. The Sage Developers, *SageMath* (version `8.8`), available at https://www.sagemath.org, 2019.
80. George J. Schaeffer, *Hecke stability and weight 1 modular forms*, Math. Z. **281** (2015), no. 1–2, 159–191.
81. A. Selberg, *Harmonic analysis and discontinuous groups in weakly symmetric Riemannian spaces with applications to Dirichlet series*, J. Indian Math. Soc. (N.S.) **20** (1956), 47–87.
82. Jean-Pierre Serre, *A course in arithmetic*, Grad. Texts in Math. **7**, Springer-Verlag, New York, 1973.
83. J. P. Serre, *Modular forms of weight one and Galois representations*, Algebraic number fields: L-functions and Galois properties (Proc. Sympos., Univ. Durham, Durham, 1975), ed. A. Fröhlich, Academic Press, London, 1977, 193–268.
84. Jean-Pierre Serre, *On a theorem of Jordan*, Bull. Amer. Math. Soc. **40** (2003), no. 4, 429–440.
85. Rene Schoof and Marcel van der Vlugt, *Hecke operators and the weight distribution of certain codes*, J. Combin. Ser. A. **57** (1991) no. 2, 163–186.
86. Goro Shimura, *Introduction to the arithmetic theory of automorphic functions*, Kano Memorial Lectures, No. 1, Publications of the Mathematical Society of Japan, no. 1, Princeton Univ. Press, Princeton, N.J., 1971.
87. Nils-Peter Skoruppa and Don Zagier, *Jacobi forms and a certain space of modular forms*, Invent. Math. **94** (1988), no. 1, 113–146.
88. William Stein, *The Modular Forms Database*, available at http://wstein.org/Tables/.
89. William A. Stein, *Modular forms, a computational approach*, with an appendix by Paul E. Gunnells, Graduate Studies in Math., vol. 79, Amer. Math. Soc., Providence, 2007.
90. William A. Stein, *An introduction to computing modular forms using modular symbols*, Algorithmic number theory: lattices, number fields, curves and cryptography, Math. Sci. Res. Inst. Publ., vol. 44, Cambridge Univ. Press, Cambridge, 2008, 641–652.
91. Jacob Sturm, *On the congruence of modular forms*, Number theory (New York, 1984-1985), Lecture Notes in Math., vol. 1240, Springer, Berlin, 1987, 275–280.
92. Andrew V. Sutherland, *Computing Hilbert class polynomials with the Chinese remainder theorem*, Math. Comp. **80** (2011), no. 273, 501–538.

93. Andrew V. Sutherland, *Accelerating the CM method*, LMS J. Comput. Math. **15** (2012), 172–204.
94. John Tate, *Collected works of John Tate, Part I (1951-1975)*, eds. Barry Mazur and Jean-Pierre Serre, Amer. Math. Soc., Providence, 2016.
95. Dave J. Tingley, *Elliptic curves uniformized by modular functions*, Ph.D. thesis, University of Oxford, 1975.
96. John Voight and David Zureick-Brown, *The canonical ring of a stacky curve*, accepted to Mem. Amer. Math. Soc., preprint available at https://arxiv.org/abs/1501.04657v3, 2019.
97. Hideo Wada, *Tables of Hecke operators. I*, Seminar on Modern Methods in Number Theory (Inst. Statist. Math., Tokyo, 1971), Paper No. 39, Inst. Statist. Math., Tokyo, 1971, 1–10.
98. Hideo Wada, *A table of Hecke operators. II*, Proc. Japan Acad. **49** (1973), no. 6, 380–384.
99. Mark Watkins, *A discursus on 21 as a bound for ranks of elliptic curves over $\mathbb{Q}$, and sundry related topics*, available at https://magma.maths.usyd.edu.au/~watkins/papers/DISCURSUS.pdf, 2015.

Elliptic Curves with Good Reduction Outside of the First Six Primes

Alex J. Best and Benjamin Matschke

Abstract We present a database of rational elliptic curves, up to $\mathbb{Q}$-isomorphism, with good reduction outside $\{2, 3, 5, 7, 11, 13\}$. We provide a heuristic involving the *abc* and BSD conjectures that the database is likely to be the complete set of such curves. Moreover, proving completeness likely needs only more computation time to conclude. We present data on the distribution of various quantities associated to curves in the set. We also discuss the connection to S-unit equations and the existence of rational elliptic curves with maximal conductor.

1 Introduction

Databases or tables of all elliptic curves subject to various constraints have been published since the 1970s, including in the well known Antwerp IV conference proceedings [6]. Such tables are useful both in identifying a given curve appearing in nature, or for proving a curve with certain properties does not exist. Tables can also be used to answer distributional questions about properties of elliptic curves when ordered in different ways.

The most well known such tables are those of elliptic curves over $\mathbb{Q}$ with bounded conductor due to Cremona [12, 11], which now form part of the LMFDB [26].

One may instead, however, construct tables of elliptic curves with bad reduction only at primes in a specified set S of rational primes. These are exactly the primes dividing the conductor. Organising curves by their primes of bad reduction can be quite useful in practise; it is often possible to prove a particular curve has good reduction outside certain places, and then conclude that the curve is contained in such a table for some S.

A. J. Best · B. Matschke (✉)
Department of Mathematics and Statistics, Boston University, Boston, MA, USA
e-mail: matschke@bu.edu

J. S. Balakrishnan et al. (eds.), *Arithmetic Geometry, Number Theory, and Computation*, Simons Symposia, https://doi.org/10.1007/978-3-030-80914-0_5

In particular many classical diophantine equations can be phrased in terms of the existence of elliptic curves with specified places of bad reduction, see Sects. 4.1 and 4.2.

In this paper we compute and study what is conjecturally the complete set of isomorphism classes of elliptic curves over $\mathbb{Q}$ with good reduction away from the first six primes $\{2, 3, 5, 7, 11, 13\}$. This set and the code and auxiliary data used to compute it (including Mordell–Weil bases for almost 100,000 Mordell curves) are available at

https://github.com/elliptic-curve-data/ec-data-S6.

Many of the curves in this set have quite large conductor, but nevertheless, by virtue of having bad reduction at only a few small primes, can be simpler arithmetically than other curves with smaller conductor.

History We now give a non-exhaustive overview of previous work computing databases of elliptic curves over $\mathbb{Q}$.

In the late 1980's, Brumer and McGuinness [8] computed rational elliptic curves of prime discriminant Δ bounded by $|\Delta| \leq 10^8$. Stein and Watkins [36] then extended this database to include almost all curves up to $|\Delta| \leq 10^{12}$ with either conductor $N \leq 10^8$ or prime conductor less than 10^{10}.

To compute the set of elliptic curves with bounded conductor, Tingley [40] used modular symbols to find all elliptic curves with $N \leq 200$. This was greatly extended and improved by Cremona [12, 11], who has currently computed all of these curves up to $N \leq 500000$. Initially this approach was only known to compute modular elliptic curves, and it was only when modularity was proved that it was confirmed [7] that over $\mathbb{Q}$ being modular is not a restriction.

A third natural way to construct a database of elliptic curves is by restricting the set of places of bad reduction, i.e. the primes that divide N (or equivalently, primes that divide the minimal discriminant). For any finite set of rational primes S, let $M(S)$ denote the finite set of elliptic curves over $\mathbb{Q}$ with good reduction outside of S, up to $\mathbb{Q}$-isomorphism, and let

$$N_S := \prod_{p \in S} p.$$

We may then hope to compute the set $M(S)$ for various sets S.

The set $M(\{2, 3\})$ was computed by Coghlan [9] and Stephens [37], and Coghlan's data was republished as Table 4 in [6]. Agrawal, Coates, Hunt and van der Poorten [1] computed $M(\{11\})$ via a reduction to Thue–Mahler equations. Cremona and Lingham [13] computed $M(\{2, p\})$ for $p \leq 23$ via a reduction to the computation of S-integral points on Mordell curves. Koutsianas [25] used a reduction to S-unit equations over number fields to compute $M(\{2, 3, 23\})$, as well as curves $E \in M(S)$ for various other S satisfying certain restrictions on the 2-division field of E. Von Känel and the second author [22] computed $M(\{2, 3, 5, 7, 11\})$ as well as all $M(S)$ with $N_S \leq 1000$ using an elliptic logarithm sieve to compute S-integral points on elliptic curves. Bennett and Rechnitzer [4] and Bennett, Gherga and Rechnitzer [3] computed $M(\{p\})$ for all $p \leq 500000$ using a refinement of the reduction to Thue–Mahler equations and Thue equations. The latter paper also recomputes $M(\{2, 3, 5, 7, 11\})$ using this approach. Moreover, using a heuristic they computed all curves in $M(\{p\})$ for $p \leq 10^{10}$, without guaranteeing completeness.

Finally we mention that there are various extensions to the above methods to compute elliptic curves over number fields with good reduction outside of a given set of places. In particular the aforementioned approaches of Cremona and Lingham [13] and of Koutsianas [25] generalise to the number field setting.

Outline The aim of this paper is to compute the set $M(\{2, 3, 5, 7, 11, 13\})$. We have computed a subset of this that is heuristically the full set, but is not proved to be complete by our method at present.[1] In Sects. 1.1 and 1.2 we give a summary of our data and discuss some statistics of the data. We compare our data to Cremona's database in Sect. 1.3.

Our computation relies on a reduction to solving Mordell equations in S-integers; this is discussed in Sect. 2.1. The main computational bottleneck is computing the Mordell–Weil bases of a large set of Mordell curves; this is elaborated upon in Sect. 2.2.

In Sects. 2.3 and 2.4, we discuss a heuristic that our database should be complete and the possibility of proving completeness via additional computation. In Sect. 3, we show some results suggested by the data regarding the question for which sets S there are elliptic curves with good reduction outside S of maximal possible conductor. In Sect. 4, we discuss connections and applications to solving other classical diophantine equations including S-unit, Thue–Mahler and Ramanujan–Nagell equations.

1.1 Summary of the Database

Let $S(n)$ denote the set of the first n rational primes. According to our computation, the set $M(S(6))$ contains 4576128 curves in total; see Table 1. Here, $j(M(S(n)))$ is the set of distinct j-invariants of curves in $M(S(n))$; the cardinality of this set is therefore the number of $\overline{\mathbb{Q}}$-isomorphism classes of curves in $M(S(n))$.

When $n \geq 2$ we can obtain all of $M(S(n))$ by taking a representative of each $\overline{\mathbb{Q}}$-isomorphism class of curves in $M(S(n))$ and twisting this representative by all integers divisible only by primes in $S(n)$. For $j \neq 0, 1728$, we only have quadratic twists. When $j = 1728$ we have quartic twists, and for $j = 0$ sextic twists (our assumption that $n \geq 2$ implies that $0, 1728 \in j(M(S(n)))$), giving the equation

$$\#M(S(n)) = 2^{n+1}(\#j(M(S(n))) - 2) + 2 \cdot 4^n + 2 \cdot 6^n.$$

[1] However, work in progress by the second author gives the same set of curves using a different method.

Table 1 Numbers of elliptic curves with good reduction outside $S(n)$ up to $\mathbb{Q}$-isomorphism and up to $\overline{\mathbb{Q}}$-isomorphism. The asterisk refers to the possible incompleteness of this paper's table. The case $n = 0$ is the classical result that there is no elliptic curve over $\mathbb{Q}$ with everywhere good reduction

n	$\#M(S(n))$	$\#j(M(S(n)))$	
0	0	0	Tate (cf. Ogg [29])
1	24	5	[9, 37, 29]
2	752	83	[9, 37]
3	7600	442	[22]
4	71,520	2140	[22]
5	592,192	8980	[22, 3]
6	4,576,128*	34,960*	This paper

This holds in all cases above and provides a quick check that nothing that obviously should have been in the database has been missed.

Each curve in $M(S(6))$ has conductor $N \mid 2^8 3^5 5^2 7^2 11^2 13^2$, which gives, a priori, 4374 possibilities for N. It turns out that exactly 4344 of them are indeed attained by curves in our set. The 30 exceptions for which there is no curve with that conductor are

$$\{1, 2, 3, 4, 5, 6, 7, 8, 9, 10, 12, 13, 16, 18, 22, 25, 28, 60, 81, 165,$$
$$169, 351, 945, 1280, 1820, 2673, 2816, 9984, 13365, 362880\}.$$

These exceptions factor as follows:

$$\{1, 2, 3, 2^2, 5, 2 \cdot 3, 7, 2^3, 3^2, 2 \cdot 5, 2^2 \cdot 3, 13, 2^4, 2 \cdot 3^2, 2 \cdot 11, 5^2, 2^2 \cdot 7, 2^2 \cdot 3 \cdot 5,$$
$$3^4, 3 \cdot 5 \cdot 11, 13^2, 3^3 \cdot 13, 3^3 \cdot 5 \cdot 7, 2^8 \cdot 5, 2^2 \cdot 5 \cdot 7 \cdot 13, 3^5 \cdot 11, 2^8 \cdot 11,$$
$$2^8 \cdot 3 \cdot 13, 3^5 \cdot 5 \cdot 11, 2^7 \cdot 3^4 \cdot 5 \cdot 7\}.$$

These (non-)conductors are all within the range of Cremona's database, and we can therefore check that there are indeed no elliptic curves with any of these numbers as their conductor. We note that the largest conductor for which no elliptic curve of that conductor exists is less than the square root of the largest possible conductor of a curve in $M(S(6))$.

Next we consider isogeny classes in $M(S(6))$. This is also a natural partition of curves in the database as $M(S(n))$ is closed under taking isogenies (any two isogenous curves have the same conductor). Our data contains 3688192 disjoint isogeny classes in total: 2966912 classes of cardinality one, 646784 of cardinality two, 4608 of cardinality three, 60928 of cardinality five, 6784 of cardinality six, 2176 of cardinality eight, and no others. An example of a curve in $M(S(6))$ with isogeny class of cardinality 8 is

$$y^2 = x^3 + 827614112325\,x + 276113445805174250.$$

Edgar Costa has computed the analytic ranks of all curves in our table, as well as the leading coefficients and root numbers of the associated L-series. His computations use interval arithmetic, and hence the leading coefficients are given with exact error bounds. The standard problem that remains is that it is impossible to verify numerically that the lower derivatives vanish exactly, and thus the computed analytic rank is actually only an upper bound once the rank is large enough. According to his computations, there are 1884428 curves of analytic rank zero in our data, 2267261 of analytic rank one, 406309 of analytic rank two, 18003 curves of analytic rank three, and the remaining 127 curves are of analytic rank four. We can compare this to the number of rational elliptic curves with conductor bound $N \leq 500000$ of each rank using Cremona's database: for these curves, Cremona computed analytic and algebraic ranks (and checked that they coincide), and found that there are 1632686 curves of rank zero, 2124004 of rank one, 461670 of rank two, 11243 of rank three, and 1 of rank four. In both tables, we observe, similarly, a larger number of rank 1 curves than rank 0 curves. An intriguing difference is the larger number of rank 4 curves in our data, compared to a similar total number of curves when ordered by conductor.

1.2 Distribution of Quantities

In this section we study the distribution of various arithmetical quantities associated to curves in our dataset. As these curves have bad reduction at only the first six primes, they are quite structured, and it is interesting to compare answers to distributional questions to when curves are ordered with respect to conductor or discriminant.

One fundamental quantity is the conductor. We plot the distribution of the logarithm of the conductor for the curves in our database as a histogram in Fig. 1a. We take the logarithm of N due to the multiplicative nature of the conductor. Indeed, if the conductor exponents f_p in $N = \prod_{p \in S} p^{f_p}$ were uniformly and independently distributed (which they are not), then in Fig. 1a we would see an approximately normal distribution with mean 14.037 and standard deviation 4.382. The observed distribution of $\log(N)$ is comparatively lopsided: It appears denser in the larger conductor range. This could be explained by the fact that one can turn good into additive reduction at $p \geq 3$ via twisting by p (as the reduction of E at p will have Kodaira symbol I_0^* by Tate's algorithm), without leaving $M(S(6))$.

The *Szpiro ratio* of an elliptic curve over $\mathbb{Q}$ is defined to be the ratio

$$\sigma = \frac{\log |\Delta_E|}{\log N}$$

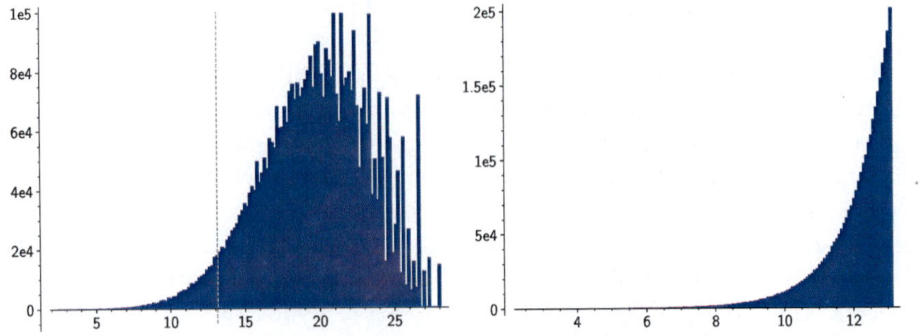

(a) Elliptic curves in our data set.

(b) Elliptic curves with $N \leq 500000$.

Fig. 1 Histograms of logarithms of conductors: (**a**) shows the curves we computed within $M(S(6))$. For a comparison, (**b**) shows all rational elliptic curves with $N \leq 500000$ according to Cremona's database. The bar at $\log(500000) \approx 13.1$ signifies the end of the overlap of both tables

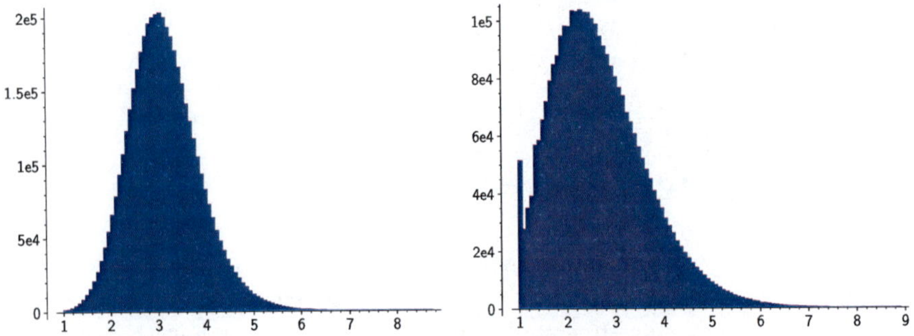

(a) Elliptic curves in $M(S(6))$.

(b) Elliptic curves with $N \leq 500000$.

Fig. 2 Histograms of Szpiro ratios $\sigma = \log(\Delta_E)/\log(N)$. (**a**) shows the curves we computed within $M(S(6))$. For a comparison, (**b**) shows all rational elliptic curves with $N \leq 500000$ according to Cremona's database. We observe three differences: the larger maximal value for σ in (**b**) (namely 8.903700), the larger mean for σ in (**a**), and (**b**) contains a significant number of curves with $\sigma = 1$ (namely 602)

of the logarithms of the minimal discriminant of the curve and its conductor. Figure 2a sketches the distribution of Szpiro ratios for the curves in our database. Szpiro's conjecture states that $|\Delta_E| = O_\varepsilon(N^{6+\varepsilon})$, or equivalently, that for any $\delta > 0$ there are only finitely many elliptic curves over $\mathbb{Q}$ with $\sigma > 6 + \delta$.

Indeed, the largest Szpiro ratios occurring among all curves in our dataset are approximately

$$8.757316,\ 8.371586,\ 8.11481 \text{ and } 8.034917\ldots.$$

The curves for which these ratios occur are all in the LMFDB and have labels *858.k2*, *2574.j2*, *910.e1*, and *9438.m2* respectively. The second and fourth of these are both quadratic twists of the first. It seems that these three have a large Szpiro ratio due to a factor of 3^{21} in each of their discriminants. The third has a factor of 2^{63} in its discriminant. These are the only four curves in our set with $\sigma \geq 8$. There are 123 curves in the database with $\sigma \geq 7$, only 15 of which have conductor larger than 500000. The largest σ from our data with $N > 500000$ has $N = 532350$ and $\sigma \approx 7.161459$.

1.3 Comparison with Cremona's Database

Cremona [12, 11] has computed the set of all rational elliptic curves with conductor less than various bounds, currently up to $N \leq 500{,}000$.

If S is the set of primes of bad reduction of an elliptic curve E of conductor N, then

$$N_S \leq N \leq 1728 N_S^2.$$

Thus, in principle, the problems of computing $M(S)$ and all curves of bounded conductor are equivalent. Both parameters S and N stratify the infinite set of rational elliptic curves. In practise, however, these stratifications differ considerably: for example, $M(S(6))$ contains 14,216 curves of conductor $2^8 3^5 5^2 7^2 11^2 13^2 \approx 10^{12}$, which is considerably larger than 500,000; and on the other hand, $M(S(6))$ does not contain the four curves with conductor 17.

Cremona's database contains at present 1,238,682 distinct j-invariants, whereas the computation we performed resulted in 34,960, because for each j-invariant, our set contains at least 128 distinct twists. On the other hand, Cremona's database contains 3,064,705 $\mathbb{Q}$-isomorphism classes of curves, whereas ours contains 4,576,128. Despite the fact that the two databases contain more or less the same number of curves, there are 4,376,070 curves in our set not contained in the Cremona database, that is, less than 5% of our set overlaps with his. We observe significant differences in the distributions of $\log(N)$ and of σ for both data sets, see Figs. 1 and 2.

Cremona's tables contain more information about each curve present there than our tables currently do, including Manin constants, generators for the Mordell–Weil group, BSD invariants, modular degrees, optimality data, sets of integral points and images of Galois representations. Much of this data would be prohibitively difficult to compute for every curve in our set due in part to the size of the conductors of some of the curves in our table.

2 Computation

In this section, we discuss the reduction of computing $M(S)$ to the problem of solving Mordell equations, the computation of the requisite Mordell–Weil bases, which is then the dominant computational task to be undertaken, and the heuristic completeness of the obtained data.

The code implementing the methods described here and computed data are available online. The repository *https://github.com/elliptic-curve-data/ec-data-S6* contains the majority of the code, and the file `mordell.sage` of *https://github.com/bmatschke/solving-classical-diophantine-equations/* contains an implementation of the algorithm of von Känel and the second author [22].

2.1 Computation Method

Let S denote a finite set of rational primes, let $M(S)$ denote the set of elliptic curves over $\mathbb{Q}$ with good reduction outside of S, up to $\mathbb{Q}$-isomorphism, and let

$$N_S := \prod_{p\in S} p.$$

For this section, we assume that $2, 3 \in S$, which can be achieved by enlarging S if necessary. Let $\mathcal{O}_S = \mathbb{Z}[1/N_S]$ denote the ring of S-integers and $\mathcal{O}_S^*$ the group of S-units.

A theorem of Shafarevich [31] states that for any such S, the set of curves $M(S)$ is finite. This can be seen as follows: for any $E \in M(S)$, choose a minimal Weierstrass model for E and consider the c_4 and c_6 invariants and discriminant Δ_E of this model. These invariants satisfy the equation $c_6^2 = c_4^3 - 1728\Delta_E$ and $\Delta_E \in \mathbb{Z} \cap \mathcal{O}_S^*$. If necessary, we may divide this equation by a power of p^6 for each $p \in S$ to obtain an equality of the form $Y^2 = X^3 + a$, where $X, Y \in \mathcal{O}_S$ and $a = \pm\prod_{p\in S} p^{e_p}$ with $0 \le e_p \le 5$ $(p \in S)$. The pair (X, Y) can then be regarded as an S-integral point on the Mordell curve

$$E_a\colon y^2 = x^3 + a.$$

By a theorem of Siegel [32, 33], the set $E_a(\mathcal{O}_S)$ is finite. From any point in $E_a(\mathcal{O}_S)$, we can recover potential invariants c_4 and c_6 that produce the point, up to any factors of p^6 in c_4^3 and c_6^2 for $p \in S$. This recovers E up to a quadratic twist by a positive S-unit. Moreover, there are exactly $2^{|S|}$ such twists.

We deduce that $M(S)$ is finite, and its computation reduces to the computation of $E_a(\mathcal{O}_S)$ for finitely many values of a. To determine $E_a(\mathcal{O}_S)$, we use the algorithm of von Känel and the second author [22], who gave a method to compute S-integral points on rational elliptic curves E provided that generators of the free part of $E(\mathbb{Q})$ are known. Their implementation uses an elliptic logarithm sieve, which can compute $E_a(\mathcal{O}_S)$ in quite an efficient manner. Thus to compute $M(S)$, it turns out that computing the necessary Mordell–Weil bases of $2 \cdot 6^{|S|}$ Mordell curves is the computational bottleneck. In Sect. 2.2 we discuss this in detail.

2.2 *Computing Mordell–Weil Bases*

We have carried out the approach outlined above for $S = S(6)$. We now discuss the most computationally intensive part of the process, which is finding the generators of the free part of the Mordell–Weil group for a number of Mordell curves, many of which have large discriminant. We will use the term Mordell–Weil basis to refer to these generators. Note that finding the generators of the torsion subgroup is both computationally easier and completely classified for Mordell curves [19], so henceforth we assume it is known.

The curves we consider are those with

$$a \in \{\pm 2^{e_2} 3^{e_3} 5^{e_5} 7^{e_7} 11^{e_{11}} 13^{e_{13}} : 0 \leq e_p \leq 5\}, \tag{2.1}$$

giving us 93,312 curves for which we wish to find the Mordell–Weil bases.

We can reduce the number of curves that we need to consider using the following fact.

Lemma 2.2 *All Mordell curves have a 3-isogeny given by*

$$y^2 = x^3 + a \to y^2 = x^3 - 27a \tag{2.3}$$

$$(x, y) \mapsto \left(\frac{y^2 + 3a}{x^2}, y\frac{y^2 - 9a}{x^3} \right) \tag{2.4}$$

As the composition of two such isogenies is an isomorphism between two models of the same curve, these 3-isogenies partition our set of Mordell curves into pairs. The upshot is that if we can find generators of the Mordell–Weil group of one of each pair, we can easily find generators for the other by pushing the basis forward along the isogeny and saturating, if necessary. Using this we need only compute the Mordell–Weil bases of half the curves, and we may choose which of each pair to consider.

2.2.1 Standard Techniques

Out of the 93,312 Mordell curves, we have computed what should be the analytic rank of those with positive a using Pari/GP's `ellanalyticrank` [39], via Sage [16]. Using the above isogeny, the other half of the curves will have the same Mordell–Weil rank. Of these curves, 20,215 have analytic rank zero, 23,186 have analytic rank one, 3112 have analytic rank two, 142 have analytic rank three, and only 1 curve has analytic rank four which is

$$y^2 = x^3 + 82063881900.$$

We assume that the output of `ellanalyticrank` is correct and that the analytic ranks are as stated above. As Pari does not use interval arithmetic, it is not clear to what extent these computations are guaranteed to be correct (especially for the high rank cases). As we shall see below we have found as many generators as there should be for almost all curves. For many of the curves, once a set of generators is found, descent techniques can be used to prove that the algebraic rank equals what is implied by BSD and that the set of generators is complete.

By the work of Gross–Zagier [20] and Kolyvagin [24], it is known that analytic rank ≤ 1 implies the rank equals the analytic rank. Therefore no further computation is required for the analytic rank 0 curves above. For the analytic rank 1 curves, we need only to find a single non-torsion point which we can then saturate to find a basis.

For many rank 1 and 2 curves in the set and for all curves of rank at least 3, a combination of the built-in Magma and Sage functions and a few other techniques summarized below sufficed to compute the Mordell–Weil bases. These included two and four-descents methods, point-searching with Stoll's `ratpoints` program [38] and Simon's `ellQ` [35] to search for points in some instances.

For the curves of rank at least 2, sometimes it was only possible to find a subset of a set of generators on each curve of each three-isogenous pair. However in this case, it was often possible to map one set of generators via the isogeny to the other curve, and combine the generators to give a basis for the Mordell–Weil group of one (and therefore both) curves. This happened mostly when the height of the found generators grew when mapped to the isogenous curve.

In rank 1, Heegner points are available in addition to the other machinery of point-searching and descent [10]. In theory, computing a Heegner point is guaranteed to terminate and if the found point is non-torsion, then it is known that the curve has algebraic rank 1. However, in order to compute Heegner points, we need to find the images of points under the modular parameterization, and hence we may need to compute a large number of Frobenius eigenvalues to find the image to a large enough precision in order to recover an algebraic point.

Using a combination of all of these techniques, we found bases for all curves except for 16481 of the rank 1 curves, and we found a single generator (but not the full basis) for all but 33 of the rank 2 curves. There was one additional rank 2 curve for which we did not find any infinite-order points with these methods (E_a for $a = 2 \cdot 3 \cdot 5 \cdot 7 \cdot 11^4 \cdot 13^5$). It is likely that a part of the rank 1 cases would be amenable to the techniques mentioned, by using larger search bounds or more time or memory. However it seemed a different approach was needed to find bases of the hard rank 1 curves as well as all remaining rank 2 curves.

2.2.2 12-Descent

To determine the generators on these harder curves, we used the 12-descent routine in Magma designed and implemented by Fisher [17]. This works by combining a 3-cover obtained from a 3-descent procedure with a 4-cover from doing 2-descent and then 4-descent. In our setting, the presence of a 3-isogeny for all of our target curves allows us to use 3-descent by isogeny to obtain the 3-cover. This is more efficient, as the number fields involved are smaller than those involved in a general 3-descent. The implementation for this in Magma is due to Creutz.

Fisher's algorithm then determines a 12-cover and a map to the original curve from each pair of one 3-cover and one 4-cover coming from these lower descents. Therefore to find a generator of the Mordell curve, we loop over all 4-covers and 3-covers of the curve coming from descent and search for points on the corresponding 12-cover. It is expected that if an n-cover has small enough coefficients, the height of a preimage of a point of height h is roughly $h/2n$. Therefore, given an estimate of the canonical height of a generator of the Mordell curve (coming from the regulator estimated via BSD) and a bound for the difference of the naïve and canonical heights on an elliptic curve (such as [28]), we can search for points on the cover which should map to a generator. Because this point should have smaller height, this should substantially reduce the time needed to search for points, compared with simply searching on the original curve. Using this, we reduce the bound for the height search by a factor of up to 24 if the coefficients of the 12-cover are not too large. To search for points on the 12-covers, we use the Magma method `PointSearch`, implemented by Watkins [41]; see also [43]. This approach has previously been used to find generators of large height on single Mordell curves [21].

Due to the fact that we do not know the order $|\text{Ш}|$ of the Shafarevich-Tate group for our curves, the regulator may give an overestimate for the height of a generator, as BSD will only allow us to determine $\sqrt{|\text{Ш}|} \cdot R$ from readily available information.

This procedure was carried out with increasing timeout, up to a maximum of 12 hours, and was broadly successful in finding a generator of the rank 1 and 2 curves for which more standard methods failed.

2.2.3 Remaining Curves

The combination of these methods has generally sufficed to determine the Mordell–Weil bases of the necessary Mordell curves. However, there are 306 rank 1 curves remaining (up to the 3-isogeny above), for which we have so far been unable to find the Mordell–Weil bases.[2] A combination of large conductor and large regulator

[2]Allan MacLeod has communicated to us the Mordell–Weil bases of 10 more of these curves, using his own implementations of similar techniques to those outlined above. Nevertheless a fast general method to find bases for all of our curves remains elusive.

(and hence either large generator height or large |Ш|) has prevented any of the above methods from working in a reasonable time frame.

The Mordell curve with smallest regulator for which we do not know a generator is

$$y^2 = x^3 + 730033053750,$$

with regulator approximately 167.305352.

The largest regulators occurring for the remaining curves arise for

$$y^2 = x^3 \pm 9045090090045009000000,$$

which interestingly are quadratic twists of each other (by -1). Their regulators are 17,550.10 in the $+$ case and 17,628.52 in the $-$ case. However, these curves are somewhat exceptional; not all curves are quite so large. The mean of the remaining regulators is 2622.49.

To attack the remaining curves, several options exist to compute a generator. We have explored these for a few of the remaining curves that we expect to be "easier".

We attempted to make use of Magma's `HeegnerPoint` method. As described in Watkins [42], this allows the user to use 4-descent to construct a 4-cover of the target elliptic curve and then find a Heegner point on the cover, reducing the required precision needed and hence the number of required Frobenius traces. Unfortunately, this Magma method fails on many of our difficult examples, presumably because both the conductor of the curve and the height of the Heegner point are both large enough that the number of Frobenius traces needed becomes unwieldy for Magma. Due to the closed-source nature of Magma (and the `HPInternal2` and `FrobeniusTracesDirect` methods in particular) we have been unable to rectify these problems. It is also unclear whether or not Magma's algorithm for computing all traces of Frobenius for primes below a given bound for one of our curves is optimal. As our curves are Mordell curves they have CM (by $\sqrt{-3}$); therefore, to compute the Frobenius traces, we may make use of Cornacchia's algorithm [10, pp. 597].

The highly optimised `smalljac` package [23] (available from Sutherland's webpage) includes an implementation of this algorithm in the case of j-invariant 0, and we expect that using this will be the most effective way to compute enough Frobenius traces to find a Heegner point on the remaining curves.

Happily, Pari/GP's `ellheegner` method is more reliable on our examples, though it does appear to use the covering method. Thus we expect that it will successfully find a generator on several of the remaining curves given enough time. It is not clear that Cornacchia's algorithm is used to compute all of the Frobenius traces in Pari.

For instance, this Pari function has returned a point successfully for one of the "missing" curves, we have found a point of

$$y^2 = x^3 + 425985404554710000$$

of height 956.2822, and it is possible this case was more tractable due to the fact that this is really the double of a generator which has height 239.07055 instead. As we are not actually missing a generator on this curve, we may check that indeed it does not give any extra elements of $M(S(6))$. However, as it took far longer to find this point, and it required more interactive experimentation with parameters than the descent methods that we used for the vast majority of the curves, we prefer to present it separately from the main data.

In theory, with an increased height bound for point-searching on 12-covers and with enough time, a point should be found on such a cover in the same way as we found the above. There are two potential issues with this. Firstly a lattice reduction algorithm is used in the point-search procedure. It often happens that this method gets stuck if these lattices happen to be ill-conditioned for Magma's algorithm. This can stall the point search, and we are not aware of the true cause or of ways of avoiding this other than restarting and hoping to get lucky. The second is that the coefficients of the 12-cover can be quite large, which can reduce the effectiveness of the height-saving of the algorithm. Thus it is very important to minimise the 12-cover, as described by Fisher, as much as possible to get the most use out of the method. It is plausible that with more work minimizing the 12-covers, the runtime of point-searching can be made more feasible.

We have checked another "missing" example where 12-descent succeeds with more individual care than we were able to take at scale. This was curve E_a for $a = 139413405126996000$, which has regulator 1504.24027. With a height bound of 10^{21} on the 12-covers, the descent finds a point that gives us a generator of height 1504.24027 on E_a. It is interesting that this point is not a multiple of any smaller generator, suggesting that Ш is trivial here.

Higher descents are also a potential avenue to complete the process of finding generators for the remaining curves. The work of Fisher allows one to combine covers of coprime degrees subject to a numerical condition on the degrees. This includes the case of combining an n-cover with an $(n+1)$-cover to obtain an $n(n+1)$-cover. This could conceivably be used to compute $8 \cdot 9 = 72$-covers on Mordell curves by combining 8-descent and 9-descent (as a second p-isogeny descent), both of which have been implemented in Magma. However, it is unknown at present how to make describing and combining such covers practical.

2.3 Completeness of the Data

First, for many Mordell curves, we have computed what should be their rank by computing the analytic rank. This is easier to compute than the algebraic rank in general. According to BSD, these ranks are equal, but this is not known in general. Computing the algebraic rank is more computationally intensive and can be obstructed by non-trivial Ш. However the analogous computation was performed in [22] for $S(5)$.

We have in some cases allowed Magma to assume GRH, which speeds up computation of class groups, and hence, descent machinery. This does not invalidate searching for points on the corresponding covers. Any rational points found are then verified unconditionally to be independent elements of the Mordell–Weil group, but when proving that the algebraic ranks agree with the analytic ones, either GRH or a longer computation time is required.

Secondly, and more seriously, we are missing any S-integral points on 612 Mordell curves E_a of rank 1, because so far we were not able to find the generator of Mordell–Weil for 306 curves (as once one curve from each isogeny class's basis is found, the other may be computed relatively easily). Assuming BSD, we may estimate the regulators of these curves up to a factor of $\sqrt{|\text{Ш}|}$. In the rank 1 case, the regulator is simply the height of a Mordell–Weil generator. So in the missing cases, either the generators are of large height or $\sqrt{|\text{Ш}|}$, is large as their product is at least 150.

To relate this to the S-integral points on these curves, we recall that the *abc* conjecture can be used to prove the weak Hall conjecture, which states that integral points (x, y) on the Mordell curve $E_a\colon y^2 = x + a$ satisfy $x = O(a^{2+\varepsilon})$ for any $\varepsilon > 0$, see Schmidt [30]. The same proof can be used to show (asymptotic) upper height bounds for S-integral points on E_a. These make it seem unlikely that an E_a of rank 1 with a very large Mordell–Weil generator has an S-integral point. These estimates could be made explicit if we assume, for example, Baker's explicit *abc* conjecture [2]. We give more details on this heuristic in Sect. 2.4.

These missing Mordell–Weil generators of curves of rank 1 could be computed via the Heegner point method, which is, for example, implemented in Pari/GP [39], whose complexity to find $P \in E_a(\mathbb{Q})$ is proportional to $\sqrt{N}h(P)$. Thus, together with BSD, we estimate that we can prove completeness of our database in about 50 CPU years. This is probably less than the (quote) "many thousand machine hours on 80 cores" that Bennett, Gherga and Rechnitzer [3] used to recompute the database of [22] for $S(5)$. The original computation of $M(S(5))$ [22] was not timed, but recalling from memory, it took approximately one CPU year.

2.4 An S-Integral Weak Hall Conjecture and the abc Conjecture

In this section, we will discuss an S-integral analogue of the classical Hall conjecture and how it adds to our heuristic for why our database should be complete. As for the classical Hall conjecture, we will show that it is implied by the *abc* conjecture.

For this section, we will use the following terminology. For any finite set of rational primes S, we call a pair of integers (x, y) *S-primitive* if there is no $p \in S$ such that p^6 divides both x^3 and y^2. We formulate an S-integral generalization of the weak Hall conjecture.

Conjecture 2.5 (An S-integral weak Hall conjecture) *Let S be a finite set of rational primes. Let $D \neq 0$ be an integer. For any $\varepsilon > 0$, any S-primitive solution (x, y) of the equation*

$$y^2 = x^3 + aD, \qquad x, y \in \mathbb{Z}, \quad a \in \mathbb{Z} \cap \mathcal{O}_S^\times, \tag{2.6}$$

satisfies

$$\max(|x|^{1/2}, |y|^{1/3}) = O_\varepsilon((N_S D)^{1+\varepsilon}). \tag{2.7}$$

Recall that the abc conjecture states that for any $\varepsilon > 0$ the following holds. If a, b, c are coprime integers with $a + b + c = 0$, then

$$\max(|a|, |b|, |c|) \leq O_\varepsilon(\mathrm{rad}(abc)^{1+\varepsilon}), \tag{2.8}$$

where $\mathrm{rad}(abc) = \prod_{p \mid abc} p$. More explicitly, (2.8) states that $\max(|a|, |b|, |c|) \leq K_\varepsilon \,\mathrm{rad}(abc)^{1+\varepsilon}$, where K_ε is a constant that depends only on ε.

Theorem 2.9 *The abc conjecture implies the S-integral weak Hall conjecture. More explicitly, if the abc conjecture holds for some $0 < \varepsilon \leq 0.1$ with constant K_ε, then any S-primitive solution (x, y) of (2.6) satisfies*

$$\max(|x|^{1/2}, |y|^{1/3}) \leq K_\varepsilon^{1+10\varepsilon}(N_s D)^{1+12\varepsilon}. \tag{2.10}$$

Our proof largely follows Schmidt's proof [30] that abc implies the classical weak Hall conjecture, although the proof below avoids some technicalities by choosing s and t (see proof) in an efficient way.

Proof Suppose (x, y) is an S-primitive solution of (2.6). Let $g = \gcd(x^3, y^2)$. Let $A = x^3/g$, $B = -y^2/g$ and $C = aD/g$, which are coprime integers. As $A + B + C = 0$, the abc conjecture implies that

$$\max(|x|^3/g, |y|^2/g) \leq K_\varepsilon \,\mathrm{rad}(ABC)^{1+\varepsilon}. \tag{2.11}$$

We claim that

$$\mathrm{rad}(ABC) \mid \frac{xyN_S D}{g}. \tag{2.12}$$

To see this, we consider two cases.

Case 1. If some $p \in S$ divides ABC, then by S-primitivity of (x, y) we have $\mathrm{ord}_p(y) \leq 2$ or $\mathrm{ord}_p(x) \leq 1$. In either case, $\mathrm{ord}_p(g) \leq 4$. If $\mathrm{ord}_p(g) = 4$, then $p^2 \mid x$, $p^2 \mid y$, $p \mid N_S$, and thus $p \mid xyN_S D/g$. The cases $\mathrm{ord}_p(g) \in \{0, 2, 3\}$ are similar, and $\mathrm{ord}_p(g) = 1$ is a priori not possible.

Case 2. Suppose some $p \notin S$ divides ABC. If $p \nmid g$ then the obvious $p \mid xyD$ suffices. If $p \mid g$, then $\mathrm{ord}_p(xyD/g) \geq \mathrm{ord}_p(g)(1/3 + 1/2 + 1 - 1) > 0$ and so $p \mid xyD/g$. This finishes the proof of (2.12).

Plugging (2.12) into (2.11) implies that $\max(|x|^3, |y|^2) \leq K_\varepsilon(xyN_SD)^{1+\varepsilon}$ and hence

$$|x|^{3s}|y|^{2t} \leq K_\varepsilon^{s+t}(xyN_SD)^{(s+t)(1+\varepsilon)}.$$

For $s = (1-\varepsilon)/(1-5\varepsilon)$ and $t = (1+\varepsilon)/(1-5\varepsilon)$ we obtain

$$|x| \leq K_\varepsilon^{2/(1-5\varepsilon)}(N_SD)^{(2+2\varepsilon)/(1-5\varepsilon)}.$$

Similarly for $s = (1+\varepsilon)/(1-5\varepsilon)$ and $t = (2-\varepsilon)/(1-5\varepsilon)$ we obtain

$$|y| \leq K_\varepsilon^{3/(1-5\varepsilon)}(N_SD)^{(3+3\varepsilon)/(1-5\varepsilon)}.$$

This yields

$$\max(|x|^{1/2}, |y|^{1/3}) \leq K_\varepsilon^{1/(1-5\varepsilon)}(N_SD)^{(1+\varepsilon)/(1-5\varepsilon)}.$$

For $\varepsilon \leq 0.1$ this reduces to the claimed bounds.

Let us relate this to S-integral points on the above Mordell curves $E_a\colon y^2 = x^3 + a$, where a is as in (2.1) an S-unit with bounded exponents. Suppose $P = (X, Y) \in E_a(\mathcal{O}_S)$. We can clear denominators of X and Y by multiplying X^3 and Y^2 by suitable powers of p^6 for each $p \in S$, and call the resulting integers $\widetilde{X}$ and $\widetilde{Y}$. This yields a relation $\widetilde{Y}^2 = \widetilde{X}^3 + \widetilde{a}$, to which we can apply the S-integral weak Hall conjecture 2.5 (with $D = 1$), or alternatively (2.10) as implied by the *abc* conjecture. We obtain conjectural asymptotic height bounds for $|\widetilde{X}|^3$ and $|\widetilde{Y}|^2$, which imply up to a small explicit constant (depending on S) the same bound on the naïve height of P, which in turn is, up to an explicitly bounded error, the Néron–Tate height $\hat{h}(P)$.

In case $S = S(6)$, we can thus make the following heuristic. First, assume that the *abc* conjecture holds for $\varepsilon = 0.1$ with a constant $K_\varepsilon \leq 1.1 \cdot 10^8$. We checked that this bound indeed holds for all *abc*-triples of the ABC@Home project by de Smit [14] for which we could compute the radical. Using this ε and K_ε and the above reasoning, we would obtain a bound for $\hat{h}(P)$ of approximately $2(2\log K_\varepsilon + 2.2\log N_S) \leq 120$.

3 Attainability of Maximal Conductor by Curves in $M(S)$

In this section, we prove some results suggested by empirical observations of our data.

Specifically we ask the following: Given a set of rational primes S, what is the highest possible conductor of an elliptic curve over $\mathbb{Q}$ with good reduction outside S? An immediate upper bound is $M_S := \prod_{p \in S} p^{f_p}$ where $f_2 = 8$, $f_3 = 5$, and $f_p = 2$ for $p \geq 5$. More specifically we may then ask:

Question 3.1 *Does there exist a curve of conductor of M_S for any set S?*

The answer to this question is no without further conditions on S. For example there does not exist an elliptic curve with good reduction away from 5; however, the answer is positive for a large class of S. Motivated by our data, we have the following sufficient criterion.

Theorem 3.2 *Let S be a finite set of rational primes that contains either 2 or 3 (or both). Then there exists an elliptic curve over $\mathbb{Q}$ with conductor $N = M_S$.*

In order to prove the theorem, we recall the notion of quadratic twists of elliptic curves. For any rational elliptic curve $E\colon y^2 = x^3 + ax + b$ and an integer d, we denote by $E^d\colon y^2 = x^3 + d^2ax + d^3b$ its quadratic twist by d.

The theorem now follows immediately from the following lemma. The proof is constructive.

Lemma 3.3 *Let d be a square-free product of primes $p \geq 5$.*

1. *Let $E_{\{2,3\}}\colon y^2 = x^3 - 18x + 24$. Then $E^d_{\{2,3\}}$ has conductor $N = 2^8 3^5 d^2$ and Kodaira type* III *at 2,* II *at 3, and* I_0^* *at $p \geq 5$ with $p \mid d$.*
2. *Let $E_{\{2\}}\colon y^2 = x^3 + 8x$. Then $E^d_{\{2\}}$ has conductor $N = 2^8 d^2$ and Kodaira type* III^* *at 2 and* I_0^* *at $p \geq 5$ with $p \mid d$.*
3. *Let $E_{\{3\}}\colon y^2 + y = x^3 - 1$. Then $E^d_{\{3\}}$ has conductor $N = 3^5 d^2$ and Kodaira type* II *at 3 and* I_0^* *at $p \geq 5$ with $p \mid d$.*

Proof This is a straightforward computation with Tate's algorithm, which we omit here. For the convenience of the reader, it is available as an appendix of the GitHub and arXiv version of this paper, which can be found at:

https://github.com/elliptic-curve-data/ec-data-S6/blob/master/docs/paper.pdf

□

We remark that in general, twisting an elliptic curve E by a prime $p \geq 5$ may change the reduction type of E at 2 and 3, but this does not happen for the three curves listed in the lemma.

Silverman [34, Exercises 4.52, 4.53] gives two families of elliptic curves defined over $\mathbb{Q}$ that have maximal possible conductor exponent at 3 and at 2, respectively, and also have this property after base changing to a number field. The above curve $E_{\{2\}}$ belongs to Silverman's latter family.

4 Applications

In this section we will briefly discuss some applications of the dataset.

4.1 Solving S-Unit Equations

Let S be a finite set of rational primes. As above, denote by $\mathcal{O}_S$ and $\mathcal{O}_S^*$ the S-integers and S-units, respectively. The S-unit equation is the equation

$$x + y = 1, \qquad x, y \in \mathcal{O}_S^*. \tag{4.1}$$

This classical diophantine equation is intimately related to the *abc* conjecture; this can be seen by clearing denominators to obtain an *abc* equation. Also, more generally, S-unit equations over number fields are known to have only finitely many solutions, as was first shown by Siegel [32] and Mahler [27]. Siegel [32, 33] used this to prove that any hyperelliptic curve of genus at least one has only finitely many S-integral points.

It turns out that solving S-unit equations can be reduced to the computation of $M(S \cup \{2\})$ via Frey–Hellegouarch curves: If (x, y) is a solution of the S-unit equation, then $E_x : Y^2 = X(X-1)(X-x)$ lies in $M(S \cup \{2\})$. Moreover any curve $E \in M(S \cup \{2\})$ can be obtained in this way from at most six different solutions of (4.1), and these can be computed from the six possible modular λ-invariants of E. In our case, (4.1) for $S = S(6)$ is exactly the case that was considered by de Weger [15]. He proved that, up to symmetry, it has exactly 545 solutions. We checked that the curves associated to all of these can be found in our database, which means that our database certainly contains all Frey–Hellegouarch curves with good reduction outside $S(6)$. We remark that (4.1) has been solved for $S = S(16)$, as well as for all S with $N_S \leq 10^7$ [22]. This is far out of reach for the above method of reducing (4.1) to computing $M(S)$.

In the other direction, the computation of $M(S)$ can be reduced to solving S'-unit equations over finitely many number fields, where the number fields are all possible number fields K of degree at most six that are unramified outside $S \cup \{2\}$ and S' being the primes in K above $S \cup \{2\}$. This link was made into an algorithm by Koutsianas [25].

4.2 Other Diophantine Problems

Many other diophantine problems reduce to the computation of $M(S)$, notably cubic Thue–Mahler equations

$$ax^3 + bx^2y + cxy^2 + dy^3 = m \prod_{p \in S} p^{e_p}, \qquad x, y \in \mathbb{Z}, \quad e_p \in \mathbb{Z}_{\geq 0}\ (p \in S),$$

where $a, b, c, d, m \in \mathbb{Z}$ and $m \neq 0$ are given such that the left-hand side has non-vanishing discriminant. Likewise generalized Ramanujan–Nagell equations

$$x^2 + b = y, \qquad x \in \mathcal{O}_S, \ y \in \mathcal{O}_S^*,$$

where $b \neq 0$ is a given integer, can be reduced to computing $M(S)$. In particular, we can find solutions for these equations for $S = S(6)$ via our computation of curves in $M(S)$, which subject to the hypothesis that we have in fact found the whole set $M(S)$ should be the complete sets of solutions of these equations; see the above discussion on completeness in Sect. 2.3.

4.3 *n*-Congruences Between Elliptic Curves

Given $n \in \mathbb{N}$, a pair of elliptic curves $E_1, E_2/\mathbb{Q}$ for which $E_1[n] \simeq E_2[n]$ as Galois modules are called n-congruent. The Frey–Mazur conjecture implies that there should be an absolute bound C such that if $p \geq C$ and E_1, E_2 are p-congruent, then E_1 and E_2 must be isogenous. The only known example of a pair of non-isogenous 17-congruent elliptic curves, found by Cremona and then Billerey [5], occurs for a pair of curves with good reduction outside of 3, 5, 7, 13. Using our database we searched for similar examples of n-congruences between curves for primes $13 \leq n \leq 47$. We found several instances of 13-congruences that were outside the range of existing databases. Fisher has recently found an infinite family of 13-congruent curves [18], of which the examples in our database are all members. We did not find any further examples of 17 (or higher) congruences between curves in our database, other than quadratic twists of the example of Cremona–Billerey mentioned above.

Acknowledgments It is our pleasure to thank Edgar Costa for various useful comments and for computing the analytic ranks of all curves in our database, as well as the leading coefficients and root numbers of the associated L-series. They are available from the same GitHub repository. We would also like to thank the anonymous referees for their helpful comments on earlier versions of this article.

The first author was supported by Simons Foundation grant #550023 and a Hariri Institute Graduate Student Fellowship.

The second author was partially supported by Excellence Initiative of Université de Bordeaux (IdEx project DiGeMANT), and by Simons Foundation grant #550023.

References

1. M. K. Agrawal, J. H. Coates, D. C. Hunt, and A. J. van der Poorten. Elliptic curves of conductor 11. *Math. Comp.*, 35(151):991–1002, 1980.
2. Alan Baker. Experiments on the abc-conjecture. *Publ. Math. Debrecen*, 65(3–4):253–260, 2004.
3. Michael A. Bennett, Adela Gherga, and Andrew Rechnitzer. Computing elliptic curves over $\mathbb{Q}$. *Math. Comp.*, 88(317):1341–1390, 2019.
4. Michael A. Bennett and Andrew Rechnitzer. Computing elliptic curves over $\mathbb{Q}$: bad reduction at one prime. In *Recent progress and modern challenges in applied mathematics, modeling and computational science*, volume 79 of *Fields Inst. Commun.*, pages 387–415. Springer, New York, 2017.
5. Nicolas Billerey. On some remarkable congruences between two elliptic curves, 2016. arXiv:1605.09205.
6. Bryan J. Birch and Willem Kuyk, editors. *Modular functions of one variable. IV.* Lecture Notes in Mathematics, Vol. 476. Springer-Verlag, Berlin-New York, 1975.
7. Christophe Breuil, Brian Conrad, Fred Diamond, and Richard Taylor. On the modularity of elliptic curves over **Q**: wild 3-adic exercises. *J. Amer. Math. Soc.*, 14(4):843–939, 2001.
8. Armand Brumer and Oisín McGuinness. The behavior of the Mordell-Weil group of elliptic curves. *Bull. Am. Math. Soc.*, 23(2):375–382, 1990.
9. Francis Coghlan. Elliptic Curves with Conductor $2^m 3^n$. Ph.D. thesis, Manchester, England, 1967.
10. Henri Cohen. *Number Theory: Volume I: Tools and Diophantine Equations*. Graduate Texts in Mathematics, Number Theory. Springer-Verlag, New York, 2007.
11. John E. Cremona. Elliptic curve data. https://johncremona.github.io/ecdata/.
12. John E. Cremona. *Algorithms for modular elliptic curves*. Cambridge University Press, second edition, 1997.
13. John E. Cremona and Mark P. Lingham. Finding all elliptic curves with good reduction outside a given set of primes. *Experiment. Math.*, 16(3):303–312, 2007.
14. Bart de Smit. ABC@Home. https://www.math.leidenuniv.nl/~desmit/abc/.
15. Benjamin M. M. de Weger. Solving exponential Diophantine equations using lattice basis reduction algorithms. *J. Number Theory*, 26(3):325–367, 1987.
16. The Sage Developers. *SageMath (Version 9.0)*, 2020. http://www.sagemath.org.
17. Tom Fisher. Finding rational points on elliptic curves using 6-descent and 12-descent. arXiv:0711.3774, 2007.
18. Tom Fisher. On families of 13-congruent elliptic curves, 2019. arXiv:1912.10777.
19. Rudolf Fueter. Ueber kubische diophantische Gleichungen. *Comment. Math. Helv.*, 2(1):69–89, 1930.
20. Benedict H. Gross and Don B. Zagier. Heegner points and derivatives of l-series. *Invent. Math.*, 84(2):225–320, 1986.
21. Jamie Weigandt (https://mathoverflow.net/users/4872/jamieweigandt). Rational points on $y^2 = x^3 - 86069^5$. MathOverflow. https://mathoverflow.net/q/105591 (version: 2012-09-25).
22. Rafael von Känel and Benjamin Matschke. Solving S-unit, Mordell, Thue, Thue–Mahler and generalized Ramanujan–Nagell equations via Shimura–Taniyama conjecture. arXiv:1605.06079, https://bmatschke.github.io/solving-classical-diophantine-equations/, 2016.
23. Kiran S. Kedlaya and Andrew V. Sutherland. Computing L-Series of Hyperelliptic Curves. In Alfred J. van der Poorten and Andreas Stein, editors, *Algorithmic Number Theory*, Lecture Notes in Computer Science, pages 312–326, Berlin, Heidelberg, 2008. Springer.
24. Victor A. Kolyvagin. Euler systems. In *The Grothendieck Festschrift*, pages 435–483. Springer, 2007.
25. Angelos Koutsianas. Computing all elliptic curves over an arbitrary number field with prescribed primes of bad reduction. *Exp. Math.*, 28(1):1–15, 2019.

26. The LMFDB Collaboration. The L-functions and modular forms database. https://www.lmfdb.org, 2020.
27. Kurt Mahler. Zur Approximation algebraischer Zahlen. I. *Math. Ann.*, 107(1):691–730, 1933.
28. J. Steffen Müller and Michael Stoll. Computing Canonical Heights on Elliptic Curves in Quasi-Linear Time. *LMS J. Comput. Math.*, 19(A):391–405, 2016. arXiv:1509.08748.
29. Andrew P. Ogg. Abelian curves of 2-power conductor. *Math. Proc. Camb. Philos. Soc.*, 62(2):143–148, 1966.
30. Wolfgang M. Schmidt. *Diophantine approximations and Diophantine equations*, volume 1467 of *Lecture Notes in Mathematics*. Springer-Verlag, Berlin, 1991.
31. Igor R. Shafarevich. Algebraic number fields. In *Proceedings of an International Congress on Mathematics, Stockholm*, pages 163–176, 1962.
32. Carl L. Siegel. Über einige Anwendungen diophantischer Approximationen. *Abh. Preuß. Akad. Wiss., Phys.-Math. Kl.*, 1929(1):70, 1929.
33. Joseph H. Silverman. *Arithmetic of elliptic curves.*, volume 106 of *Graduate Texts in Mathematics*. Springer, 1986.
34. Joseph H. Silverman. *Advanced topics in the arithmetic of elliptic curves*, volume 151 of *Graduate Texts in Mathematics*. Springer, 1994.
35. Denis Simon. Computing the rank of elliptic curves over number fields. *LMS J. Comput. Math.*, 5:7–17, 2002. Program ellQ: https://simond.users.lmno.cnrs.fr/ellQ.gp.
36. William A. Stein and Mark Watkins. A Database of Elliptic Curves — First Report. In Claus Fieker and David R. Kohel, editors, *Algorithmic Number Theory*, Lecture Notes in Computer Science, pages 267–275, Berlin, Heidelberg, 2002. Springer.
37. Nelson M. Stephens. The Birch Swinnerton-Dyer Conjecture for Selmer curves of positive rank. Ph.D. Thesis, Manchester, 1965.
38. Michael Stoll. Documentation for the ratpoints program. arXiv:math/0803.3165, 2014.
39. The PARI Group, Univ. Bordeaux. *PARI/GP version 2.11.2*, 2019. http://pari.math.u-bordeaux.fr/.
40. Dave J. Tingley. Elliptic curves uniformized by modular functions. Ph.D. thesis, University of Oxford, 1975.
41. Mark Watkins. Searching for points using the Elkies ANTS-IV algorithm. http://magma.maths.usyd.edu.au/~watkins/papers/padic.ps.
42. Mark Watkins. Some Remarks on Heegner Point Computations. arXiv:math/0506325, 2006.
43. Thomas Womack. *Explicit Descent on Elliptic Curves*. PhD thesis, University of Nottingham, July 2003.

Efficient Computation of BSD Invariants in Genus 2

Raymond van Bommel

Abstract Recently, all Birch and Swinnerton-Dyer invariants, except for the order of Ш, have been computed for all curves of genus 2 contained in the L-functions and Modular Forms Database [LMFDB]. This report explains the improvements made to the implementation of the algorithm described in [vBom19] that were needed to do the computation of the Tamagawa numbers and the real period in reasonable time. We also explain some of the more technical details of the algorithm, and give a brief overview of the methods used to compute the special value of the L-function and the regulator.

Mathematics Subject Classification (2010) 11G40, 11G10, 11G30, 14H40

1 Introduction

The Birch and Swinnerton-Dyer conjecture has an extensive computational history. The conjecture has originally been conceived by Birch and Swinnerton-Dyer based on computational findings on elliptic curves. Later, it has been generalised by Tate to the case of abelian varieties over general number fields. We formulate the version for principally polarised abelian varieties over $\mathbb{Q}$ here.

Conjecture 1 (BSD for Abelian Varieties over $\mathbb{Q}$, [BiSw65, Tate66]) *Let $A/\mathbb{Q}$ be a principally polarised abelian variety of dimension d and algebraic rank r. Let $L(A, s)$ be its L-function, R its regulator, Ш its Tate-Shafarevich group and Ω its real period. For each prime number p, let c_p be the Tamagawa number of A at p. Then $L(A, s)$ has a zero of order r at $s = 1$ and*

R. van Bommel (✉)
Department of Mathematics, Massachusetts Institute of Technology, Cambridge, MA, USA
e-mail: bommel@mit.edu

J. S. Balakrishnan et al. (eds.), *Arithmetic Geometry, Number Theory, and Computation*, Simons Symposia, https://doi.org/10.1007/978-3-030-80914-0_6

$$\lim_{s\to 1}(s-1)^{-r}L(A,s)=\frac{R\cdot\Omega\cdot|\text{Ш}|\cdot\prod_p c_p}{|A(\mathbb{Q})_{\text{tors}}|^2}.$$

Since the group Ш is very hard to compute in general, it has been common to verify the conjecture "up to squares". For this, one computes the other terms numerically, and checks that the value of |Ш| conjectured by the formula is close to a square or twice a square, according to what the criterion of Poonen and Stoll, [PoSt99], predicts.

For example, this has been done in the paper [FLSSSW] by Flynn, Leprévost, Schaefer, Stein, Stoll and Wetherell for a collection of 32 modular Jacobians of hyperelliptic curves of genus 2 with small conductor. Modularity was needed for the computation of the special value of the L-function. In [vBom19], the author developed and implemented algorithms in Magma to compute Tamagawa numbers and real periods for hyperelliptic curves of genus 2. The algorithms only use the equation of the curve and build upon existing methods implemented by Steve Donnelly to compute regular models.

The current goal is to compute all these BSD-invariants for the curves of genus 2 in the [LMFDB], see also [BSSVY], and the curves of higher genus that will appear in the future in the [LMFDB], see also [Suth19]. The Tamagawa numbers had already been computed for all but 54 of the genus 2 curves in the database. For the other 54 curves, and for most of the real periods, the algorithms developed in [vBom19] did not give a result within a reasonable amount of time.

Given that there are more than 66,000 curves of genus 2 in the [LMFDB], the computation of a Tamagawa number or real period should ideally take at most a few seconds for most of the curves, and only more than an hour in very exceptional cases. In this report, we will explain the improvements that have been made to achieve this, and therefore finish the verification of BSD up to squares for all curves of genus 2 in the [LMFDB]. We also report on the computation of the real periods for both hyperelliptic and non-hyperelliptic curves of genus 3 from [Suth19]. Moreover, this report will also serve as a more extensive documentation of the algorithms, explaining more details of the computation than the description in [vBom19].

Notation Throughout this text, let C over $\mathbb{Q}$ be a smooth projective curve, let p be a prime number and let $\mathcal{C}\,/\,\mathbb{Z}_{(p)}$ be a regular model of C over $\mathbb{Z}_{(p)}$. Let J be the Jacobian of C.

2 Tamagawa Numbers

Classically, Tamagawa numbers are invariants which are associated to the so called Tamagawa measure on an algebraic group. However, in the context of the Birch and Swinnerton-Dyer conjecture, Tamagawa numbers are defined a bit differently. To find more about the relation between the classical definition and the one used in the Birch and Swinnerton-Dyer conjecture, see [Blo80] for example. We will use the following definition.

Definition 2 Let A be an abelian variety over $\mathbb{Q}$. Let p be a prime, and let $\mathcal{A}/\mathbb{Z}_{(p)}$ be a Néron model of $A/\mathbb{Q}$. Let $\varphi = \mathcal{A}_{\mathbb{F}_p}/\mathcal{A}^0_{\mathbb{F}_p}$ be the component group scheme of the special fibre of $\mathcal{A}$. Then the *Tamagawa number* of A at p is defined as $c_p(A) := |\varphi(\mathbb{F}_p)|$.

In practice, if we have a curve of higher genus, it is already infeasible to compute equations for its Jacobian in projective space. The computation of a Néron model is certainly out of scope. The following theorem by Raynaud gives a relation between the component group scheme and a regular model of the curve.

Theorem 3 ([BoLi99, Theorem 1.1, p. 3]) *Let C, and J be as usual. Let $\mathcal{C}^{hens}$ be a regular model of C over the strict henselisation of $\mathbb{Z}_{(p)}$, and let I be the set of components in the special fibre of $\mathcal{C}^{hens}$. Let*

- *$\overline{\alpha}$ be the linear map $\mathbb{Z}^I \to \mathbb{Z}^I$ given by*

$$\Gamma \mapsto \sum_{\Gamma' \in I} \langle \Gamma, \Gamma' \rangle \cdot \Gamma', \qquad \text{for } \Gamma \in I,$$

  where $\langle \cdot, \cdot \rangle$ is the intersection pairing on $\mathcal{C}^{hens}$,
- *$\overline{\beta}$ be the linear map $\mathbb{Z}^I \to \mathbb{Z}$ given by*

$$\Gamma \mapsto \text{multiplicity}(\Gamma), \qquad \text{for } \Gamma \in I,$$

- *φ_J be the component group of the special fibre of a Néron model of J.*

Then $\mathrm{Gal}\left(\overline{\mathbb{F}_p}/\mathbb{F}_p\right)$ *acts in a natural way on I and $\mathbb{Z}^I$, and hence on* $\operatorname{im}\overline{\alpha}$ *and* $\ker\overline{\beta}$, *and there is a canonical exact sequence of* $\mathrm{Gal}\left(\overline{\mathbb{F}_p}/\mathbb{F}_p\right)$*-modules*

$$0 \to \operatorname{im}\overline{\alpha} \to \ker\overline{\beta} \to \varphi_J\left(\overline{\mathbb{F}_p}\right) \to 0.$$

Briefly summarised, our algorithm to compute the Tamagawa numbers (see also [vBom19]) consists of the following steps:

1. Compute a regular model $\mathcal{C}$ for C.
2. Find the set I of components of the special fibre of $\mathcal{C}^{\text{hens}}$.
3. Compute the action of Frobenius on I.
4. Use Theorem 3 to construct $\varphi_J(\overline{\mathbb{F}_p})$ as a $\mathrm{Gal}\left(\overline{\mathbb{F}_p}/\mathbb{F}_p\right)$-module.
5. The Tamagawa number c_p is then the cardinality of

$$\varphi_J(\mathbb{F}_p) = \varphi_J(\overline{\mathbb{F}_p})^{\mathrm{Gal}\,(\overline{\mathbb{F}_p}/\mathbb{F}_p)}.$$

For step 1, we use the method `RegularModel` in `Magma`, which gives us patching data for a regular model. In Sect. 2.1, we explain in more detail how this model is computed and represented.

The rest of this section will be focused on step 3, the computation of the action of Frobenius on the components. Section 2.2 contains an outline of the computation of this action. The further subsections explain some problems we encountered during this computation and how we solved these.

2.1 Computation of a Regular Model in Magma

The existence and construction of a regular model for C relies on the resolution of singularities for arithmetic surfaces. The following result by Lipman implies that such resolutions exist.

Theorem 4 ([Lip78, Art86]) *Suppose $\mathcal{A}/\mathbb{Z}_{(p)}$ is a normal proper flat model of $C/\mathbb{Q}$, i.e. $\mathcal{A}$ is an arithmetic surface over $\mathbb{Z}_{(p)}$ and its generic fibre is isomorphic to C. Define a sequence*

$$\mathcal{A} = \mathcal{A}_0 \leftarrow \mathcal{A}_1 \leftarrow \mathcal{A}_2 \leftarrow \dots$$

of arithmetic surfaces as follows: the morphism $\mathcal{A}_n \leftarrow \mathcal{A}_{n+1}$ is the normalisation of the blow-up of $\mathcal{A}_n$ in the non-regular points of $\mathcal{A}_n$. Then for sufficiently large n, the arithmetic surface $\mathcal{A}_n$ is regular.

In other words, we can obtain a regular model from any flat model by repeatedly normalising it and blowing up the non-regular points. This is not exactly the way in which `Magma` computes a regular model, but it is close. What the algorithm in `Magma` actually does is to construct a sequence of blow-ups

$$\mathcal{C}_0 \leftarrow \mathcal{C}_1 \leftarrow \mathcal{C}_2 \leftarrow \dots \leftarrow \mathcal{C}_N = \mathcal{C}, \tag{1}$$

where $\mathcal{C}_0$ is some proper flat model of C we start with, $\mathcal{C}$ is a regular model we end with, and each morphism $\mathcal{C}_n \leftarrow \mathcal{C}_{n+1}$ is a blow-up in either a point or irreducible component of the special fibre of $\mathcal{C}_n$. If there is a choice between blowing up a point or a component, components will be blown up first (although the latter has only been implemented for curves over $\mathbb{Q}$).

Let us briefly describe the data involved in this production of a regular model by `Magma`. In general, the input of the algorithm is a smooth projective curve defined over a number field K and a prime ideal $\mathfrak{p}$ of $\mathcal{O}_K$, the prime at which the model must become regular. The regular model is stored as a collection of affine patches and transition maps between some of these patches. Each time a blow-up is done inside a certain patch, some new patches with transition maps to and from the original patch are created. Other relevant data that are stored are sets of non-regular points,

points where components in the special fibre of the model intersect, multiplicities of these components, and intersection numbers.

In some cases, because the non-regular points in the special fibre are blown up one at a time, it could happen that such a non-regular point is not defined over $\mathcal{O}_K/\mathfrak{p}$. In this case, the number field K is extended and the algorithm is restarted with this new number field as base field. Also in the end, when we want to compute the set I of geometric components of the regular model (i.e. the set of components in the special fibre of $\mathcal{C}^{\text{hens}}$), the number field may need to be extended.

The reader might wonder at this point why we chose to work over number fields instead of over local fields. Indeed, the theory is cleaner to state in the local setting, but the problem is that some of the computational tools that we need have not been implemented for local fields. For example, multivariate polynomial factorisation has only been implemented for finite fields, number fields, function fields of curves over such fields, $\mathbb{Z}$, and some finitely generated extensions of these rings.

2.2 Outline of Algorithm to Compute the Galois Action

Even though our curve C is defined over $\mathbb{Q}$, we have seen in the previous subsection that it might be necessary to work over a general number field K. We will go through the algorithm that `Magma` uses to create its regular models, and see how we can track the action of Frobenius on the special fibre during this process. Recall the notation from Eq. (1).

At the start, we have a proper flat model $\mathcal{C}_0$ all of whose equations are defined using coefficients in $\mathbb{Z}_{(p)}$. First we compute the irreducible components occurring in the special fibre of $\mathcal{C}_0$ and the action of Frobenius on them. This can simply be done by applying Frobenius on the ideal defining such a component in one of the affine patches containing the component. Then we proceed to look at the non-regular points.

Suppose P is a non-regular point in the special fibre of $\mathcal{C}_0$. Then all Galois conjugates of P are also non-regular in $\mathcal{C}_0$. Suppose P is defined over $\mathbb{F}_{p^\ell}$ for some $\ell \geq 1$. Then the special fibre of the blow up of $\mathcal{C}_0$ in P can be defined over $\mathbb{F}_{p^\ell}$.

Let Comp_P be the set of components in the special fibre of $\mathcal{C}$ that map to $P \in \mathcal{C}_0$, and define Comp_Q analogously for any Galois conjugate Q of P. Then Frobenius sends components in Comp_P to components in $\text{Comp}_{\text{Frob}(P)}$. Of course we have the following properties.

Proposition 5

(a) The bijection $\text{Comp}_P \to \text{Comp}_{\text{Frob}(P)}$ *induced by Frobenius preserves intersection numbers.*
(b) If $Q \neq P$ is a Galois conjugate of P, then components in Comp_P *and* Comp_Q *do not intersect.*

(c) *For any component D of the special fibre of $\mathcal{C}$ and any component E in* $\mathrm{Comp}_{\mathrm{Frob}(P)}$ *we have*

$$\langle D, E\rangle = \langle \mathrm{Frob}^{-1}(D), \mathrm{Frob}^{-1}(E)\rangle.$$

As a consequence, intersection numbers of components in $\mathrm{Comp}_{\mathrm{Frob}(P)}$ with any other component of the special fibre of $\mathcal{C}$ can be reduced to an intersection number of a component in Comp_P with either another component of Comp_P, or a component not lying in any of the Comp_Q, with Q ranging over the Galois conjugates of P.

In particular, it is not of importance for us to compute which component of Comp_P is exactly mapped to which component of $\mathrm{Comp}_{\mathrm{Frob}(P)}$. Instead, our algorithm just creates ℓ copies of Comp_P, and lets Frobenius map a component to the corresponding component in the next copy, for the first $\ell - 1$ copies. For the ℓ-th copy the action of Frobenius is computed differently. For this, we compute the action of Frob^ℓ on Comp_P in a recursive way.

The algorithm backtracks the construction of the regular model, while keeping track of all the intersection numbers and the action of Frobenius. To summarise our recursive algorithm:

Algorithm 6 *Input: a non-regular point P of $\mathcal{C}_n$. Suppose $k(P) = \mathbb{F}_{p^\ell}$.*
Output: combinatorial description of Comp_P, *the set of components in the special fibre of $\mathcal{C}$ that map to P. This combinatorial description includes the action of* Frob^ℓ *and the intersection numbers.*

Step 1 Find equations for the components in the special fibre of $\mathcal{C}_{n+1}$ which are contracted to P, and compute the action of Frob^ℓ on these components, and the intersections of these components with each other.

Step 2 Loop through all Galois orbits of non-regular points in $\mathcal{C}_{n+1}$ mapping to P. Execute steps 3 to 7 for each orbit.

Notation for steps 3 to 7: Q is a point of such an orbit and $k(Q) = \mathbb{F}_{p^{\ell \cdot m}}$.

Step 3 Recursively, call Algorithm 6 with the non-regular point Q of $\mathcal{C}_{n+1}$ as input. In this way, we compute the action of $\mathrm{Frob}^{\ell \cdot m}$ on Comp_Q, and the intersections of these components with each other.

Step 4 Compute the intersection of the components considered in step 1 with the components in Comp_Q. (This is already done by the `RegularModel` command in `Magma`.)

Step 5 Create $m - 1$ more copies of Comp_Q for $\mathrm{Comp}_{\mathrm{Frob}^\ell(Q)}$, $\mathrm{Comp}_{\mathrm{Frob}^{2\ell}(Q)}$, et cetera.

Step 6 The action of Frob^ℓ is the identity $\mathrm{Comp}_{\mathrm{Frob}^{\ell \cdot i}(Q)} \to \mathrm{Comp}_{\mathrm{Frob}^{\ell \cdot (i+1)}(Q)}$ for $i = 0, \ldots, m - 2$. For $i = m - 1$, the action is given by the action computed in step 3.

Step 7 The intersection of components contained in $\mathrm{Comp}_{\mathrm{Frob}^{\ell \cdot i}(Q)}$ and components contained in $\mathrm{Comp}_{\mathrm{Frob}^{\ell \cdot j}(Q)}$ is 0 for distinct $i, j \in \{0, \ldots, m - 1\}$. The intersection number of a component $D \in \mathrm{Comp}_{\mathrm{Frob}^{\ell \cdot i}(Q)}$ for $i =$

$1, \ldots, m-1$ with a component E considered in step 1, equals the intersection number of $\mathrm{Frob}^{-i \cdot \ell}(D)$ and $\mathrm{Frob}^{-i \cdot \ell}(E)$.

Step 8 Combine and output all the combinatorial data that has been collected.

The improvement in comparison with the previous implementation is the more systematic implementation of Steps 5, 6, and 7, which led to a significant speed-up.

2.3 Problem Arising During the Algorithm

One problem that arose regularly in our algorithm is the following. When we blow-up our point P with coordinates in $\mathbb{F}_{p^\ell}$, the special fibre of the new patches are not necessarily defined over $\mathbb{F}_{p^\ell}$ even though they could have been. This is due to the choice we have when parametrising the blow-up and is illustrated in the following example.

Example 7 Suppose $K = \mathbb{Q}[a]/(a^4 + a + 1)$ and $\mathfrak{p} = 2\mathcal{O}_K$. Consider the affine curve given by $y^2 = 2(x^2 + x + 1)$ in $\mathbb{A}^2$ over $\mathcal{O}_{K,\mathfrak{p}}$. It has two Galois conjugate non-regular points $(\bar{a}^5, 0)$ and $(\bar{a}^{10}, 0)$ in the special fibre, with coordinates in $\mathbb{F}_4$. Let us do a blow-up in $(\bar{a}^5, 0)$. We rewrite the equation as

$$y^2 = 2x'^2 + (4a^5 + 2)x' + (4a^3 - 4a),$$

where $x' = x - a^5$. To get one of the charts of the blow-up, we 'substitute' $y = tx'$ and $2 = sx'$. We get the equation

$$t^2x'^2 = s^2x'^3 + (2a^5 + 1)sx'^2 + (a^3 - a)s^2x'^2$$

which after dividing by x'^2 gives the following equations for the blow-up:

$$t^2 = s^2x' + (2a^5 + 1)s + (a^3 - a)s^2, \qquad sx' = 2.$$

Now we see that the special fibre is not defined over $\mathbb{F}_4$, as $\bar{a}^3 - \bar{a} \notin \mathbb{F}_4$.

As this problem occurred quite often, we tried to look for a solution that is efficient to implement. We are aware of two suitable possible solutions:

1. Blow-up Galois conjugated points at the same time (e.g. $(\bar{a}^5, 0)$ and $(\bar{a}^{10}, 0)$ in the previous example). This ensures all schemes stay defined over $\mathbb{Z}_{(p)}$. We choose not to take this route as the blow-up becomes significantly more complicated to represent when blowing up multiple points at the same time.
2. For any number field K that we encounter in our algorithm (e.g. when we start with a curve over $\mathbb{Q}$, but we need to extend the field we are working over several times in order to blow up non-regular points, as described in Sect. 2.1), make sure that for each d dividing $[K : \mathbb{Q}]$ there is a subfield $K_d \subset K$ of absolute degree

d. In this way, if a non-regular point in the special fibre is defined over $\mathbb{F}_{p^d}$, all equations for the blow-up can be taken to lie inside K_d and the special fibre of the blow-up is guaranteed to be defined over $\mathbb{F}_{p^d}$.

In the next section, we explain what should be kept in mind when constructing number fields as in **2.**, and how it has been implemented in our algorithm.

2.4 Construction of a Suitable Extension

For this subsection we say that a number field K has the *subfield property*, if for every divisor d of $[K : \mathbb{Q}]$ there is a subfield of K of absolute degree d. Our goal is now to construct number fields having the subfield property, as described in the previous subsection. Let us first prove that they exist, and that we can extend them.

Proposition 8 *Let ℓ be a positive integer and let p be a prime. Let K be a number field having the subfield property such that p is inert in K. Then there exists an extension $K \subset L$ of degree ℓ such that p is inert in L, and L has the subfield property.*

Proof It suffices to consider the case ℓ is prime. Write $[K : \mathbb{Q}] = \ell^e \cdot m$, where m is relatively prime to ℓ. Let K_{ℓ^e} be a subfield of K of absolute degree ℓ^e, and let K_m be a subfield of absolute degree m. Then construct any extension $K_{\ell^e} \subset L_{\ell^{e+1}}$ such that p is inert in $L_{\ell^{e+1}}$, for example by constructing an equation for the extension of residue fields $\mathbb{F}_{p^{\ell^e}} \subset \mathbb{F}_{p^{\ell^{e+1}}}$ and lifting it to K_{ℓ^e}. Define L as a compositum of the linearly disjoint fields K_m and $L_{\ell^{e+1}}$. Then p is inert in L. Now we show that L has the subfield property.

If d is a divisor of $[L : \mathbb{Q}] = \ell^{e+1} \cdot m$, then either d divides $[K : \mathbb{Q}]$, or it is of the form $\ell^{e+1} \cdot m'$ for a certain divisor m' of m. In the former case, K already has a subfield of degree d. In the latter case, K has a subfield $K_{m'}$ of degree m' and the subfield $L_{\ell^{e+1}} K_{m'}$ of L has degree d.

There are of course many ways to construct fields with the subfield property, but one thing to keep in mind is that we would like to avoid the situation where the coefficients for equations for the number field get too large to do any meaningful computation with them.

Our first attempt was to only consider abelian extensions, i.e. subextensions of cyclotomic fields. They typically have very small defining equations. However, the problem is that we could not answer the following question affirmatively.

Question 9 Is Proposition 8 still true if K and L are required to be abelian?

So we decided to construct the fields following the strategy of the proof of 8. To keep the defining equations for our number fields small, we apply a reduction algorithm every step. This has been implemented in `Magma` under the name `OptimisedRepresentation`. For this function, it is important that the ring of integers of K is computable, which is a property we would like to have anyway for the computation of a regular model. A big bottleneck in the computation of the ring of integers is the factorisation of the discriminant of K, therefore we would like to keep this small.

Hence, we use the following strategy for each extension $K \subset L$ we construct. We first fix the extension modulo p, as we want p to remain inert in L, i.e. we extend K taking the root of some yet to be determined polynomial $f \in K[x]$ and we fix f modulo p. We then try to find a local minimum for the discriminant of f. I.e., we start with any such f, and we repeatedly add or subtract p times a randomly chosen monomial. If the discriminant of f gets larger, then we revert the last change.

In practice, this worked for all curves of genus 2 contained in [LMFDB]. We identified two ways in which this algorithm could still be further improved. First, there might be better ways to find 'small' field extensions. Second, it might be worthwhile to determine in advance for which d the subfield K_d of absolute degree d is actually needed. Then there would be fewer constraints and hence more candidates for the number field K.

2.5 Implementation

The algorithm has been implemented by the author in `Magma`, changing partly the way regular models are constructed. The computation of the Galois action on the component group took a negligible amount of time after a suitable regular model was constructed. Most time was spent on the construction of a regular model.

For the 66,158 curves of genus 2 in the [LMFDB], the construction of suitable regular models for all of the primes of bad reduction, took about 2.02 s on average per curve. For 36 of these curves, the computation took longer than 60 s, the longest one taking about 1600 s.

In principle, the implementation can also be used to compute Tamagawa numbers for the curves of genus 3 from [Suth19]. Currently, this has not been done, and it is not inconceivable that there could be a few tricky cases for which the computation of the ring of integers of a suitable field extension, as described in Sect. 2.4, becomes a bottleneck.

3 Real Periods

Let A be an abelian variety over $\mathbb{Q}$ of dimension g. The real period has been defined in different ways in the past. Traditionally, this has been defined in terms of an integral of a g-form, the so-called Néron differential, along $A(\mathbb{R})$. It can also be defined in terms of integrals of 1-forms along homology cycles of A. For the comparison of the different definitions see for example [Gro82].

Definition 10 Let m be the number of connected components of $A(\mathbb{R})$. Let $\mathcal{A}$ be a Néron model of A over $\mathbb{Z}$, and let $(\gamma_1, \ldots, \gamma_g)$ be a basis for the group $H_1(A(\mathbb{C}), \mathbb{Z})^{\mathrm{Gal}(\mathbb{C}/\mathbb{R})}$ of homology cycles invariant under complex conjugation. Let $(\omega_1, \ldots, \omega_g)$ be a $\mathbb{Z}$-basis of $\Omega^1_{\mathcal{A}/\mathbb{Z}}(\mathcal{A})$. Then the real period of A is defined as

$$\Omega_A = m \cdot \left| \det \left(\int_{\gamma_i} \omega_j \right)_{i,j=1}^{g} \right|.$$

In the case of a Jacobian with Néron model $\mathcal{J}$, the Abel-Jacobi map gives a bijection between $\Omega^1_{\mathcal{J}/\mathbb{Z}}(\mathcal{J})$ and the global sections $\omega_{\mathcal{C}/\mathbb{Z}}(\mathcal{C})$ of the canonical sheaf of a regular model, see for example [vBom19, Sect. 3.2] and [Mil86, Sect. 2]. As a consequence, we get the following result, which gives us a practical way to compute the real period of the Jacobian of a curve.

Proposition 11 *Let C, J and $\mathcal{C}$ be as usual. Then*

$$\Omega_J = m \cdot \left| \det \left(\int_{\gamma_i} \omega_j \right)_{i,j=1}^{g} \right|,$$

where $(\gamma_1, \ldots, \gamma_g)$ is a $\mathbb{Z}$-basis of $H_1(C(\mathbb{C}), \mathbb{Z})^{\mathrm{Gal}(\mathbb{C}/\mathbb{R})}$, and $(\omega_1, \ldots, \omega_g)$ is a $\mathbb{Z}$-basis of $\omega_{\mathcal{C}/\mathbb{Z}}(\mathcal{C})$, and m is the number of connected components of $J(\mathbb{R})$.

An algorithm to compute real periods is described in [vBom19, Algorithm 13]. We repeat it here.

Algorithm 12 *Input: a curve C of genus g over $\mathbb{Q}$.*
Output: the real period Ω_J of its Jacobian J.

Step 1 Compute the big period matrix $(\int_{\gamma_i} \omega_j)_{i=1,\ldots,2g}^{j=1,\ldots g}$. Here $\underline{\omega} = (\omega_1, \ldots, \omega_g)$ is any basis of $\Omega^1_{C/\mathbb{Q}}(C)$ and $(\gamma_1, \ldots, \gamma_{2g})$ is a symplectic basis of $H_1(C(\mathbb{C}), \mathbb{Z})$.

Step 2 For each subset $I \subset \{1, \ldots, 2g\}$ with $|I| = g$, calculate the covolume $P_I := \left| \det \left(\int_{\gamma_i} \omega_j + \overline{\int_{\gamma_i} \omega_j} \right)_{i \in I}^{j=1,\ldots,g} \right|$.

Step 3 Compute a generator P for the lattice inside $\mathbb{R}$ spanned by the P_I.

Step 4 For each bad prime p, construct a regular model $\mathcal{C}^p/\mathbb{Z}_{(p)}$ of C.

Step 5 For each bad prime p and each of the differentials $\omega_1, \ldots, \omega_g$, check if the differential has a pole on any of the irreducible components of the special fibre of $\mathcal{C}^p$. If so, adjust the basis by multiplying the differential having a pole with p to get a new basis $\underline{\omega}'$ and apply Step 5 again (until the basis is not changing anymore).

Step 6 For each bad prime p and each $(c_j)_{j=1}^{g} \in \{0, \ldots, p-1\}^g \setminus \{(0, 0, \ldots, 0)\}$, check if $\sum_j c_j \omega_j$ vanishes on the whole special fibre of $\mathcal{C}^p$. If so, adjust the basis $\underline{\omega}'$ by replacing one of the ω_j such that $c_j \neq 0$ with $\frac{1}{p} \sum_j c_j \omega_j$, then apply Step 6 again (until the basis is not changing anymore).

Step 7 For each bad prime p compute p^{a-b}, where a is the number of basis adjustments done in Step 5, and b is the number of basis adjustments done in Step 6 (this is also the determinant of the change of basis matrix whose columns express $\underline{\omega}'$ in terms of $\underline{\omega}$). Then take the product W over p of these determinants, and output $W \cdot P$.

In order to compute the real periods for the 66, 158 hyperelliptic curves of genus 2 contained in [LMFDB] some optimisations had to be done:

1. In our original implementation, many computations appeared to be done multiple times. To avoid this, new data structures have been constructed to store all rings, ideals, their Gröbner bases and other relevant objects that are needed multiple times during the computation.
2. In our original implementation, we used Van Wamelen's algorithm [vWam06] implemented in `Magma` to compute the big period matrix in Step 1. This only works for hyperelliptic curves with a simplified Weierstraß model of odd degree. We moved to Pascal Molin and Christian Neurohr's implementation that can compute the Riemann surface associated to any curve [MN19, Neu18]. This also had advantages for the hyperelliptic case as it allowed us to take simpler (e.g. even degree) models, which typically give rise to simpler regular models.
3. In Step 6 of the algorithm, for $p^g - 1$ differentials it is tested whether or not they vanish on the special fibre. This should not be necessary and in principle be a linear algebra problem. This is explained in more detail in Sect. 3.3.
4. The Gröbner basis computations over $\mathbb{Z}$ took most of the time in this algorithm. In some cases, even the computation of the function field over $\mathbb{Q}$ of one of the patches of the regular model was taking a lot of time. In Sect. 3.2, it is explained how we circumvented Gröbner basis computations over $\mathbb{Z}$.

Moreover, we will consider some other ideas that could be used to improve this code even further in the future.

3.1 Computation of the Order of Vanishing of a Function

First, we will briefly explain how the order of vanishing of a function is computed. We reduced our computation to one inside a polynomial ring over $\mathbb{Z}$. In the following example, we illustrate how this is done.

Example 13 Consider the scheme given by the equation $y^2 = x^2 - 2x - 2$ in $\mathbb{A}^2_{\mathbb{Z}}$. We take $p = 2$. We want to compute the order of vanishing of the function 2 on the component given by the vanishing of $x + y$ and 2. In other words, we want to compute the multiplicity of that component in the special fibre.

We start with the ideal $I = (x + y,\ 2) \subset \mathbb{Z}[x, y]$ and notice that 2 lies in it, so the order of vanishing is at least 1. Then we compute

$$I_2 = I^2 + (y^2 - x^2 + 2x + 2) = (x^2 + y^2 + 2,\ 2x + 2,\ 2y + 2,\ 4).$$

Even though 2 does not lie in I_2, we see that the ideal quotient

$$(I_2 : (2)) = (x + 1,\ y + 1,\ 2) \not\subset I.$$

In other words: 2 when multiplied with some unit lands in I_2, and 2 vanishes with order at least 2. We proceed to compute

$$\begin{aligned} I_3 &= I^3 + (y^2 - x^2 + 2x + 2) \\ &= (x^2 + 2x + 3y^2 + 6,\ 2xy + 2x + 2y^2 + 2y,\ 4x + 4,\ 4y + 4,\ 8) \end{aligned}$$

and

$$(I_3 : (2)) = (x^2 + y^2 + 2,\ xy + x + y^2 + y,\ 2x + 2,\ 2y + 2,\ 4),$$

which is easily seen to be contained in I. Hence, the vanishing order of the function 2 on the component given by the vanishing of $x + y$ and 2, equals 2.

So in order to compute the order of vanishing, we make use of the following proposition.

Proposition 14 *Let R be a commutative ring. Let J be an ideal of R, and let I be a prime ideal of R containing J. Let n be an integer. Then an element $f \in R$ maps to an element in $I^n \cdot (R/J)_I \subset (R/J)_I$, i.e. f vanishes with order at least n on the component defined by I in* Spec(R/J)*, if and only if the ideal quotient $(I^n + J : (f))$ is not contained in I.*

Proof Suppose $\overline{f} \in R/J$ lands in $I^n \cdot (R/J)_I$ after localisation. This means that there are $g \in I^n$ and $h \notin I$, such that $(\overline{f} - \overline{g})\overline{h} = 0 \in R/J$, or in other words $(f - g)h \in J$. From this, we deduce that $h \in (I^n + J : (f))$, as desired.

Conversely, if $h \in (I^n + J : (f))$ is an element such that $h \notin I$, then we can write hf as $j + g$ for some $j \in J$ and $g \in I^n$. Then we get $\overline{hf} = \overline{g} \in I^n \cdot (R/J)_I$. As $\overline{h}$ is a unit inside $(R/J)_I$ we get that $\overline{f} \in I^n \cdot (R/J)_I$, as desired.

This leads to the following algorithm.

Algorithm 15 *Input: an $f \in \mathbb{Z}[x_1, \ldots, x_m]$ and ideals I and J of $\mathbb{Z}[x_1, \ldots, x_m]$ such that $V(J)$ is a flat curve over $\mathbb{Z}$ on which f does not vanish. We assume I is a prime ideal containing J, defining an irreducible component in the special fibre $V(J)_p$ of this curve. We assume that the local ring $\mathcal{O}_{V(J),I}$ is regular.*
Output: the order of vanishing of f at $V(I)$.

Step 1 If f is not irreducible, find a factorisation for f and run the rest of the algorithm for each of its irreducible factors.
Step 2 Start with $n = 1$ and $I_0 = J$.
Step 3 Compute a Gröbner basis for the ideal $I_n = I_{n-1} \cdot I + J$ of $\mathbb{Z}[x_1, \ldots, x_m]$.
Step 4 Compute generators for $(I_n : (f))$.
Step 5 If any of these generators is not contained in I, then return the value $n - 1$. Else, increase n by 1 and proceed with Step 3.

Proposition 16 *Algorithm 15 terminates and is correct.*

Proof By induction we see that $I_n = I^n + J$. The correctness follows immediately from Proposition 14. As f does not vanish on $V(J)$, the element f reduces to a non-zero element of the discrete valuation ring $\mathcal{O}_{V(J),I}$. In particular, the order of vanishing of f at $V(I)$ is finite, and the algorithm will terminate.

The ideal quotient and membership of ideals can be checked using Gröbner basis machinery. This has been implemented in Magma and other packages in the case R is a polynomial ring over a field or a polynomial ring over $\mathbb{Z}$.

In case we are working over a ring of integers $\mathcal{O}_K$ which is not $\mathbb{Z}$, we can represent $\mathcal{O}_K$ as a $\mathbb{Z}$-algebra of finite type and still use the Gröbner basis machinery over $\mathbb{Z}$. For the computation of the real period, the field extensions that are necessary when computing a regular model, as explained in Sect. 2.1, are not chosen as in Sect. 2.4, but they are chosen in such a way that $\mathcal{O}_K$ is monogenic. In this way, we can proceed with the Gröbner basis calculations at the cost of adding just one extra variable. This is how we get an algorithm to compute the order of vanishing of a polynomial.

In general, a function will be of the form $\frac{f}{g}$, where f and g can be written as polynomials with integer coefficients. The order is then computed as the difference between the order of vanishing of f and g.

The reader might be wondering at this point why we are working over $\mathbb{Z}$ instead of over the ring of integers of a local field. The main reason is the availability of an algorithm to factor multivariate polynomials over $\mathbb{Z}$ in Magma. If f and g are two polynomials, then the computation of the order of vanishing of f and g generally takes way less time than the computation of the order of vanishing of $f \cdot g$ (without using the factorisation).

3.2 *How to Avoid Difficult Computations in Characteristic 0*

3.2.1 Working over $\mathbb{Z}/p^n\mathbb{Z}$ Instead of over $\mathbb{Z}$

In Example 13, all the calculations could have been done over $\mathbb{Z}/4\mathbb{Z}$ instead of $\mathbb{Z}$, as we already know in advance that 4 has a strictly higher order of vanishing than 2. In fact, we replaced the current algorithm to compute the multiplicity of a component with this new algorithm working over $\mathbb{Z}/p^2\mathbb{Z}$, leading to big speed-ups in some difficult cases.

In general, when we consider a polynomial f and we want to determine whether f vanishes with order at least r in a component with multiplicity m, it suffices to do all computations over $\mathbb{Z}/p^{\lfloor r/m \rfloor + 1}\mathbb{Z}$. In this way, we can avoid Gröbner basis computations over $\mathbb{Z}$, which take significantly more time.

Proposition 17 *Let f, r, m be as above. If Algorithm 15 is run over $\mathbb{Z}/p^{\lfloor r/m \rfloor + 1}\mathbb{Z}$ instead of $\mathbb{Z}$ and outputs a value less than r, then the output is correct. If the output is greater than or equal to r, then f vanishes with order at least r.*

Proof Instead of testing whether f lies in $I^n \cdot (R/J)_I$, see also Proposition 14, we are testing whether $f + p^{\lfloor r/m \rfloor + 1} \cdot g$ lies in $I^n \cdot (R/J)_I$, for some element $g \in \mathbb{Z}[x_1, \ldots, x_m]$. Because $p^{\lfloor r/m \rfloor + 1}$ vanishes with order $m \cdot (\lfloor r/m \rfloor + 1) > r$, the strong triangle inequality yields that $f \in I^n \cdot (R/J)_I$ if and only if $f + p^{\lfloor r/m \rfloor + 1} \cdot g \in I^n \cdot (R/J)_I$ in case $n \leq r$. This proves the proposition.

3.2.2 Optimising the Ideals for Fast Computation

Another big improvement came from a slight modification of the ideals I_n, as described in Example 13. A priori we were computing I_n as $I_n = I^n + J$, where J is an ideal inside a polynomial ring $\mathbb{Z}[x_1, \ldots, x_\ell]$ corresponding to an affine patch of C, and then computing a Gröbner basis. In our new implementation we defined ideals I_n^{modified} and J_n as follows:

$$J_n = \begin{cases} I & \text{if } n = 1, \\ I_{n-1}^{\text{modified}} \cdot I + J & \text{else,} \end{cases}$$

$$I_n^{\text{modified}} = \begin{cases} I & \text{if } n = 1, \\ J_n + \langle x \in \text{GB}(I_{n-1}^{\text{modified}}) : (J_n : (x)) \not\subset I \rangle & \text{else,} \end{cases}$$

where GB gives a Gröbner basis of an ideal. Basically, the elements that we added to I_n^{modified} are exactly those generators of $I_{n-1}^{\text{modified}}$ that already appear to vanish with order n at I. The following proposition will formalise this and prove that we still get the correct answer if use the ideal I_n^{modified} instead of I_n.

Proposition 18 *For any positive integer n, there is an equality*

$$(I_n/J)_I = (I_n^{\text{modified}}/J)_I$$

of ideals of $(\mathbb{Z}[x_1, \ldots, x_\ell]/J)_I$.

Proof We prove the statement by induction to n. The case $n = 1$ is trivial.

Let $k \geq 2$ be an integer and suppose the statement is true for all $n < k$. We start with the equality $(I_{k-1}/J)_I = (I_{k-1}^{\text{modified}}/J)_I$. Multiplying both sides with the ideal $(I/J)_I$, gives us that $(I_k/J)_I = (J_k/J)_I$. Now we will prove the equality $(J_k/J)_I = (I_k^{\text{modified}}/J)_I$, which will finish the proof of the proposition.

Let $x \in \mathbb{Z}[x_1, \ldots, x_\ell]$ by any element with $(J_k : (x)) \not\subset I$. This means that x multiplied with a unit in R_I lies inside J_k. In other words, $\overline{x}$ lies in $(J_k/J)_I$. Therefore, the equality $(J_k/J)_I = (I_k^{\text{modified}}/J)_I$ follows by definition of I_k^{modified}.

Corollary 19 *Algorithm 15 is still correct if I_n is replaced by I_n^{modified}.*

One might wonder why we chose to use this rather artificial looking ideal I_n^{modified} instead of I_n. The following example, in which we also use our improvements from the previous Sect. 3.2.1, illustrates the advantage of this approach.

Example 20 Consider the ideal

$$J = (y^{100} - x^{100} + 2x + 2,\ xz - 2)$$

inside the ring $\frac{\mathbb{Z}}{2^{18}\mathbb{Z}}[x, y, z]$ equipped with the graded reverse lexicographic order. Now we look at the component in the special fibre given by $I = (x + y, z, 2)$. Then the direct computation of a Gröbner basis for I_{18} took 43.5 s, while the inductive computation of a Gröbner basis for I_{18}^{modified} only took 2.32 s. Moreover, the former Gröbner basis has 61 elements, while the latter only has 30 elements.

3.2.3 Avoiding Function Field Computations

In our previous implementation, we computed a function field for each patch of the regular model. This function field was used to represent differentials on this patch. The computation of this function field is quite expensive. In the following example, it did not seem to finish within reasonable time.

Example 21 Consider the curve 1328.a.84992.1, given by

$$y^2 + (x + 1)y = 4x^5 + 9x^4 + 16x^3 + 13x^2 + 8x + 1.$$

The regular model that `Magma` computes at 2 for this curve, has a patch given by two equations. One of these equations has 428 terms and is of degree 56. The computation of a function field over $\mathbb{Q}$ for this patch did not finish in 24 h. Although we could not determine why the computation took so long, we suspect that it is due to the exponential time needed to compute a Gröbner basis for an ideal.

However, in practice, we do not need this function field. Except for the first patches, every patch $\mathcal{P}$ is defined using three variables x, y, and z, and two equations $f = g = 0$. Without loss of generality, we can assume that all differentials on $\mathcal{P}$ can be represented as $h \cdot dx$ with $h \in K(x, y, z)$, where K is the base field (this does not need to be $\mathbb{Q}$, as sometimes there is need extend the base field, as explained in Sect. 2.1). The only operation for which we used the function field, was to convert a differential of the shape $j \cdot dy$ or $j \cdot dz$ into a differential of the shape $h \cdot dx$. The following algorithm gives us a way to achieve this in polynomial time.

Algorithm 22 *Input: polynomials $f, g \in K[x, y, z]$ defining the affine patch $\mathcal{P}$. Output: relatively prime $a, b \in K[x, y, z]$ such that $a \cdot dx = b \cdot dy$ holds on $\mathcal{P}$.*

Step 1 The equations $df = dg = 0$, give rise to the following linear relations over $K[x, y, z]$ for dx, dy and dz:

$$\begin{pmatrix} \frac{\partial f}{\partial x} & \frac{\partial f}{\partial y} & \frac{\partial f}{\partial z} \\ \frac{\partial g}{\partial x} & \frac{\partial g}{\partial y} & \frac{\partial g}{\partial z} \end{pmatrix} \cdot \begin{pmatrix} dx \\ dy \\ dz \end{pmatrix} = 0. \tag{2}$$

Step 2 Consider

$$M = \begin{pmatrix} \frac{\partial f}{\partial y} & \frac{\partial f}{\partial z} \\ \frac{\partial g}{\partial y} & \frac{\partial g}{\partial z} \end{pmatrix}.$$

Multiply Eq. (2) on the left side by the adjugate of M to get a relation of the form

$$\begin{pmatrix} F & D & 0 \\ G & 0 & D \end{pmatrix} \cdot \begin{pmatrix} dx \\ dy \\ dz \end{pmatrix} = 0, \tag{3}$$

for some polynomials D, F and G in $K[x, y, z]$.

Step 3 Then we compute $a = \frac{F}{\gcd(D,F)}$ and $b = \frac{D}{\gcd(D,F)}$.

Proposition 23 *Algorithm 22 is correct and the number of field operations in K is polynomial in the input size.*

Proof It is clear that equality (3) always holds. To check that Step 3 is well-defined, we will prove that $D \neq 0$. Suppose $D = 0$. Because dx is a non-vanishing differential by assumption, we get $F = G = 0$. The Jacobian matrix

$$N = \begin{pmatrix} \frac{\partial f}{\partial x} & \frac{\partial f}{\partial y} & \frac{\partial f}{\partial z} \\ \frac{\partial g}{\partial x} & \frac{\partial g}{\partial y} & \frac{\partial g}{\partial z} \end{pmatrix}$$

has full rank, as $f = g = 0$ defines a curve in $\mathbb{A}^3$. Hence $\operatorname{adj}(M)N = 0$ implies that $\operatorname{adj}(M) = 0$ and $M = 0$. But that immediately contradicts the fact that N is of full rank.

Now we will look at the number of field operations, let B be a bound for the number of coefficients of f and g. To compute the six partial derivatives, we need at most $\mathcal{O}(B)$ field operations. For the matrix multiplication in Step 3, we need to multiply polynomials with at most B coefficients with each other, so this can be done using at most $\mathcal{O}(B^2)$ field operations. To see that the number of field operations needed in Step 3 is polynomial, we refer the reader to [Witt04, chap. 6].

3.3 How to Avoid Checking p^g Differentials

In Step 6 of Algorithm 12, a candidate basis $\omega_1, \ldots, \omega_g$ for the global sections of the canonical sheaf $\omega_{\mathcal{C}/\mathbb{Z}_{(p)}}$ is considered. For each linear combination $\sum_j c_j \omega_j$ with the $c_j \in \{0, \ldots, p-1\}$ not all equal to zero, it is tested whether $\sum_j c_j \omega_j$ vanishes on the special fibre. If so, then $\frac{1}{p} \sum_j c_j \omega_j$ can replace one of the current candidate basis elements.

For big values of p, the checking of $p^g - 1$ differentials can take a lot of time. To check whether a differential vanishes on all components of the special fibre of $\mathcal{C}$, quite a few expensive Gröbner basis computations are needed. We would like to limit such computations as much as possible. Of course, it would suffice to actually check just $\frac{p^g-1}{p-1}$ of these differentials, but for $g \geq 2$, this could still be a big number.

If D is a component of the special fibre $\mathcal{C}_p$, then $\mathcal{O}_{\mathcal{C},D}$ is a discrete valuation ring whose residue field $\mathbb{F}_D$ is a function field over a finite field. So in principle, checking whether $\sum_j c_j \omega_j$ is vanishing on D is an $\mathbb{F}_p$-linear algebra problem inside $\mathbb{F}_D$, which could be solved much more efficiently.

If we write $\omega_j = \frac{f_j}{h_j} d$ for polynomials f_j, h_j and d is a local generator on a sufficiently small affine neighbourhood of the generic point of D for the sheaf $\omega_{\mathcal{C}/\mathbb{Z}_{(p)}}$, as in [vBom19, Lemma 12], then we would need to determine the residue class of $\frac{f_j}{h_j}$ inside $\mathbb{F}_D$. The problem that we now encounter is that both f_j and h_j could vanish on D with the same order $n > 0$. We could try to check if $f_j - \alpha \cdot h_j$ vanishes on D with order greater than n, where α ranges over lifts of elements in $\mathbb{F}_D$, but then we still need to check $|\mathbb{F}_D|$ elements, which is exactly what we wanted to avoid. Although the vanishing of h_j on D sometimes happens, it seemed that in most cases there were a bunch of components for which this does not happen. Therefore, we decided to only use a linear algebra approach on those components on which h_j does not vanish. We then check in the end if the differential we found really vanishes on the whole special fibre.

Remark 24 An alternative approach would be to try to repeatedly divide a uniformiser out of f_j and h_j until they both have order 0. However, the difficulty with this approach is, that we are doing our computations over $\mathbb{Z}/p^e\mathbb{Z}$ for some exponent e (see Sect. 3.2.1), and we cannot divide by the nilpotent uniformiser.

Instead of actually doing linear algebra in the different function fields $\mathbb{F}_D$, we opted for a faster approach, which works in almost all of the cases. We precompute a finite set S of (not necessarily rational) random closed points in different components of the special fibre, and specialise to these points. This leads to the following algorithm.

Algorithm 25 *Input: precomputed set S of closed points on the special fibre of $\mathcal{C}$, polynomials $f_1, \ldots, f_g, h_1, \ldots, h_g$ such that $\omega_j = \frac{f_j}{h_j} d$ for $j = 1, \ldots, g$.*
Output: a subspace V of $\mathbb{F}_p^g$ such that all $(c_1, \ldots, c_g) \in \mathbb{F}_p^g$ such that $\sum_j c_j \omega_j$ vanishes on the special fibre of $\mathcal{C}$ are contained in V.

Step 1 For each $j = 1, \ldots, g$ compute the subset $S_j = \{P \in S : h_j(P) = 0\}$. Compute $S' = S \setminus \bigcup_{j=1}^{g} S_j$.

Step 2 Construct the $|S'| \times g$ matrix corresponding to the linear map $\mathbb{F}_p^g \to \mathbb{F}_p^{S'}$ mapping the j-th standard basis vector to $(f_j(P) \cdot h_j(P)^{-1})_{P \in S'}$.

Step 3 Output the kernel of M.

Proposition 26 *Algorithm 25 is correct, i.e. the vector space V satisfies the output conditions, and the number of field operations needed is polynomial in the input size.*

Proof If $\sum_j c_j \omega_j$ vanishes on the special fibre of $\mathcal{C}$, then $\sum_j c_j \frac{f_j}{h_j}$ must vanish on the special fibre. In particular, $\sum_j c_j \frac{f_j}{h_j}$ must vanish in any point of S', and $(c_j)_{j=1}^{g}$ lies in the kernel of the linear map constructed in Step 2.

The evaluation of the polynomials in Step 1 is polynomial in the input size: at most $\mathcal{O}(g \cdot |S| \cdot B \cdot \log D)$ operations are needed, where B is a bound for the number of coefficients of $f_1, \ldots, h_g$, and D is a bound for their degree. For Step 2, we need $\mathcal{O}(g \cdot |S'| \cdot B \cdot \log D)$ operations to construct a matrix for the linear map. Finally, the Gauss elimination to find generators for the kernel can be done using $\mathcal{O}(\max(g, |S'|)^3)$ field operations.

In most cases in our computation, the vector space V appeared to be at most one-dimensional and the full computation to determine whether a differential vanishes, had to be done at most once. In almost all cases for which the space had dimension greater than 1, the prime p appeared to be small.

3.4 Possible Ideas for Improvements

Let D be a component of the special fibre of $\mathcal{C}$. We propose two ideas that could help in the task of finding the order of vanishing of a function f at D.

The first idea is to try to remedy the defect described in Remark 24. For example, if we know that an integer is $4 \mod 12$ and we divide it by 2, we cannot tell its value modulo 12, but we do know for sure that it is $2 \mod 6$. In the same way, we can reduce the modulus when we divide by a uniformiser to solve the problem described in Remark 24.

The second idea comes from a comparison with methods used in real analysis. If you want to see if a C^∞ function which vanishes in a point P, is vanishing twice, you usually compute its derivative and check if that is vanishing. Of course, this does not work in characteristic p, but the Hasse derivative could work, see for example [Gol03, Sect. 1.3]. If there are methods to compute the Hasse derivative more efficiently than the Gröbner basis computations that we are currently doing, this could lead to much smaller runtimes.

Average time per curve (in seconds)	first 200 curves	slowest curve
Computation of big period matrix	1.20	1.31
Construction of regular models	0.22	1.22
Finding a basis of $\omega_{\mathcal{C}/\mathbb{Z}_{(p)}}(\mathcal{C})$	1.03	6665.40
Total runtime	2.46	6668.03

Fig. 1 Average runtime per curve

3.5 Implementation

The algorithm has been implemented by the author in `Magma` and used to compute the real period for all 66, 158 curves of genus 2 in the [LMFDB], all hyperelliptic curves and all but 8 non-hyperelliptic curves of genus 3 in [Suth19]. For the 8 missing curves, we could not compute the real period due to a bug in `Magma`'s machinery to compute the big period matrix of the Jacobian of these curves.

The average runtime was about 1.67 s per curve for the 66,158 curves of genus 2. For 89 of these curves the computation took more than 60 s, the longest runtime for a single curve being 6668 s. For the first 200 curves appearing in the database, and for the slowest curve, we have more detailed information showing how the runtime has been spent between the different parts of the computation (Fig. 1).

Almost all of the time spent finding a basis of $\omega_{\mathcal{C}/\mathbb{Z}_{(p)}}(\mathcal{C})$ was spent on actual Gröbner basis computations.

4 Other BSD Invariants

There have also been recent computations of other Birch and Swinnerton-Dyer invariants of Jacobians of genus 2 curves to the [LMFDB]. In this section, we summarise them.

The leading coefficients for the L-functions of these Jacobians have been computed using methods developed by Bober, Booker, Costa, Lee, Platt and Sutherland, to appear in [BBCLPS], which builds upon work of Booker, see [Boo06]. The code developed by Costa and Platt is available at [CoPl19].

Generators for the Mordell-Weil group of the Jacobians have been computed using code of Stoll. This is an improvement of the old `j-points` code. The algorithm roughly consists of the following three phases:

1. try to find a tight upper bound R for rank of the Mordell-Weil group, i.e. we want R to equal the rank of J, but at this point we can only prove that the rank of J is bounded by R;
2. search for points in $J(\mathbb{Q})$, until they generate a subgroup of rank R;
3. saturate this subgroup to obtain the Mordell-Weil group.

For the upper bound for the rank, the following methods are used:

- the 2-Selmer group of the Jacobian and the 2-Selmer set of the Pic^1 of the curve can be computed and used to obtain bounds for the rank, see [Sto01, Creu18],
- the method of visualisation of elements of Ш can be used to further improve these bounds, see [CrMa00, Bru04, BrFl06],
- 2-power isogenous abelian surfaces are computed and the same methods are used to compute upper bounds for the rank of their Mordell-Weil group, which turned out to be better than the upper bound found for J itself in some cases.

For the search phase, the strategy is to look for points up to a certain height. As the canonical height is hard to control, a naive height function is used in this phase. The difference between the naive and canonical height is bounded by some positive real number δ, that we can actually compute. So to find all points up to canonical height B, one has to enumerate all points up to naive height $B + \delta$. In the new version of `j-points`, a new modified naive height function has been used, which has a smaller difference with the canonical height than the classical naive height function, see [MüSt16, part IV]. This reduces the size of the required search space significantly.

For the saturation phase, it is important that we have found all points up to canonical height ε for some $\varepsilon > 0$. If all generators for the free part of the subgroup we found have canonical height at most H, then the index of the subgroup inside the Mordell-Weil group can only have prime factors that are smaller than $\frac{H}{\varepsilon}$. We can then saturate the subgroup at every prime number $p < \frac{H}{\varepsilon}$ and in this way, we are sure that we found the whole Mordell-Weil group. The saturation has also been made faster using the modified naive height function, see also [MüSt16, part IV].

With this code, we have been able to compute generators for almost all Mordell-Weil groups of genus 2 curves in the [LMFDB]. For the few for which we did not directly find generators, we found generators for an abelian surface isogenous to the Jacobian. The following example is the one for which it was most difficult to find a generator.

Example 27 The Jacobian for the genus 2 curve 900617.a.900617.1 given by

$$y^2 + (x^2 + x)y = x^5 - 65x^4 + 224x^3 + 30x^2 + x$$

is isogenous to the Jacobian of the genus 2 curve given by

$$y^2 = -1110x^6 - 2790x^5 - 9315x^4 + 11160x^3 + 18315x^2 - 9540x - 12765.$$

On the latter Jacobian, we were able to find a generator of the rank 1 Mordell-Weil group of canonical height 16.281246 using the code of Stoll. We then computed the image of this point under the isogeny, and used the modified height bound and improved saturation, to prove that the we actually found a generator of canonical height 65.124982 on the Jacobian of 900617.a.900617.1.

For the computation of the regulators from the set of generators of the Mordell-Weil group, we used an algorithm which is based on an approach to decompose the height in different local contributions using Arakelov theory, see also [Hol12, Mül14, BHM20].

Acknowledgments We would like to thank Michael Stoll for sharing his code for the computation of Mordell-Weil groups, and his extensive explanation of this code. Moreover, we thank Edgar Costa for his explanation of the machinery used to compute special values of L-functions. Several anonymous referees are thanked for their useful comments that led to improvements of this article.

Raymond van Bommel has been supported by the Simons Collaboration on Arithmetic Geometry, Number Theory, and Computation (Simons Foundation grant 550033).

References

[Art86] M. Artin, *Lipmans Proof of Resolution of Singularities for Surfaces.* In: G. Cornell, J. H. Silverman (eds), *Arithmetic Geometry.* Springer, New York, NY, 1986.

[BiSw65] B. J. Birch, H. P. F. Swinnerton-Dyer. Notes on elliptic curves. II. *J. Reine Angew. Math.* **218** (1965), 79–108.

[BBCLPS] Jonathan W. Bober, Andrew R. Booker, Edgar Costa, Min Lee, David J. Platt, Andrew Sutherland, *Computing motivic L-functions*, in preparation.

[vBom19] Raymond van Bommel, Numerical verification of the Birch and Swinnerton-Dyer conjecture for hyperelliptic curves of higher genus over $\mathbb{Q}$ up to squares. *Exp. Math.* (2019), doi:10.1080/10586458.2019.1592035.

[Blo80] Spencer Bloch, A note on height pairings, Tamagawa numbers, and the Birch and Swinnerton-Dyer conjecture. *Invent. math.* **58**, no. 1 (1980): 65–76.

[BHM20] Raymond van Bommel, David Holmes, J. Steffen Müller, Explicit arithmetic intersection theory and computation of Néron-Tate heights. *Math. Comp.* **89** (2020), no. 321, 395–410.

[Boo06] Andrew R. Booker, Artin's conjecture, Turing's method, and the Riemann hypothesis. *Exp. Math.* **15** (2006), no. 4, 385–408.

[BSSVY] A. R. Booker, A. V. Sutherland, J. Voight, D. Yasaki, A database of genus-2 curves over the rational numbers. *LMS Journal of Computation and Mathematics* **19**, no. A (2016), 235–254.

[BoLi99] S. Bosch, Q. Liu, Rational points of the group of components of a Néron model. *Manuscripta Math.* **98** (1999), no. 3, 275–293.

[Bru04] Nils Bruin, Visualising Sha[2] in abelian surfaces. *Math. Comp.* **73** (2004), no. 247, 1459–1476.

[BrFl06] N. Bruin, E. V. Flynn, Exhibiting SHA[2] on hyperelliptic Jacobians. *J. Number Theory* **118** (2006), no. 2, 266–291.

[CoPl19] Edgar Costa and David Platt, *A generic L-function calculator for motivic L-functions*, available at https://github.com/edgarcosta/lfunctions, 2019

[CrMa00] John E. Cremona, Barry Mazur, Visualizing elements in the Shafarevich-Tate group. *Experiment. Math.* **9** (2000), no. 1, 13–28.

[Creu18] Brendan Creutz, Improved rank bounds from 2-descent on hyperelliptic Jacobians. *Int. J. Number Theory* **14** (2018), no. 6, 1709–1713.

[FLSSSW] E. V. Flynn, F. Leprévost, E. F. Schaefer, W. A. Stein, M. Stoll, J. Wetherell, Empirical evidence for the Birch and Swinnerton-Dyer conjectures for modular Jacobians of genus 2 curves. *Math. Comp.* **70** (2001), no. 236, 1675–1697.

[Gol03] David M. Goldschmidt, *Algebraic functions and projective curves.* Graduate Texts in Mathematics, 215. *Springer-Verlag, New York*, 2003.

[Gro82] B. H. Gross, On the conjecture of Birch and Swinnerton-Dyer for elliptic curves with complex multiplication. *Number theory related to Fermat's last theorem* (Cambridge, Mass., 1981), 219–236, Progr. Math., 26, *Birkhäuser. Boston, Mass.*, 1982.

[Hol12] David Holmes, Computing Néron-Tate heights of points on hyperelliptic Jacobians. *J. Number Theory* **132** (2012), no. 6, 1295–1305.

[Lip78] Joseph Lipman, Desingularization of two-dimensional schemes. *Ann. Math.* **107** (1978), no. 1, 151–207.

[LMFDB] The LMFDB Collaboration, *The L-functions and Modular Forms Database.* https://www.lmfdb.org.

[Mil86] J. S. Milne, Jacobian varieties. *Arithmetic geometry (Storrs, Conn., 1984)*, 167–212. Springer, New York, 1986.

[MN19] Pascal Molin, Christian Neurohr, Computing period matrices and the Abel-Jacobi map of superelliptic curves. *Math. Comp.* **88** (2019), no. 316, 847–888.

[Mül14] J. Steffen Müller, Computing canonical heights using arithmetic intersection theory. *Math. Comp.* **83** (2014), no. 285, 311–336.

[MüSt16] Jan Steffen Müller, Michael Stoll, Canonical heights on genus-2 Jacobians. *Algebra Number Theory* **10** (2016), no. 10, 2153–2234.

[Neu18] Christian Neurohr, Efficient integration on Riemann surfaces & applications. *PhD thesis* (2018), https://oops.uni-oldenburg.de/3607/1/neueff18.pdf.

[PoSt99] B. Poonen, M. Stoll, The Cassels-Tate pairing on polarized abelian varieties. *Ann. of Math.* **150** (1999), no. 3, 1109–1149.

[Sto01] Michael Stoll, Implementing 2-descent for Jacobians of hyperelliptic curves. *Acta Arith.* **98** (2001), no. 3, 245–277.

[Suth19] A. V. Sutherland, A database of nonhyperelliptic genus 3 curves over Q, *Thirteenth Algorithmic Number Theory Symposium (ANTS XIII)*, Open Book Series **2** (2019), 443–459.

[Tate66] J. Tate, On the conjectures of Birch and Swinnerton-Dyer and a geometric analog. *Séminaire Bourbaki, Vol. 9* (1964–1966), Exp. No. 306, 415–440, Soc. Math. France, Paris, 1995.

[vWam06] Paul B. van Wamelen, Computing with the analytic Jacobian of a genus 2 curve. *Discovering mathematics with Magma,* 117–135. Algorithms Comput. Math., 19, Springer, Berlin, 2006.

[Witt04] A. Wittkopf, *Algorithms and implementations for differential elimination.* PhD dissertation, 2004.

Restrictions on Weil Polynomials of Jacobians of Hyperelliptic Curves

Edgar Costa, Ravi Donepudi, Ravi Fernando, Valentijn Karemaker, Caleb Springer, and Mckenzie West

Abstract Inspired by experimental data, we investigate which isogeny classes of abelian varieties defined over a finite field of odd characteristic contain the Jacobian of a hyperelliptic curve. We provide a necessary condition by demonstrating that the Weil polynomial of a hyperelliptic Jacobian must have a particular form modulo 2. For fixed $g \geq 1$, the proportion of isogeny classes of g-dimensional abelian varieties defined over $\mathbb{F}_q$ which fail this condition is $1 - Q(2g + 2)/2^g$ as $q \to \infty$ ranges over odd prime powers, where $Q(n)$ denotes the number of partitions of n into odd parts.

E. Costa (✉)
Department of Mathematics, Massachusetts Institute of Technology, Cambridge, MA, USA
e-mail: edgarc@mit.edu

R. Donepudi
Department of Mathematics, University of Illinois at Urbana-Champaign, Urbana, IL, USA
e-mail: donepud2@illinois.edu

R. Fernando
Department of Mathematics, University of California — Berkeley, Berkeley, CA, USA
e-mail: fernando@berkeley.edu

V. Karemaker
Mathematical Institute, Utrecht University, Utrecht, The Netherlands
e-mail: V.Z.Karemaker@uu.nl

C. Springer
Department of Mathematics, The Pennsylvania State University, University Park, PA, USA
e-mail: cks5320@psu.edu

M. West
Department of Mathematics, University of Wisconsin-Eau Claire, Eau Claire, WI, USA
e-mail: westmr@uwec.edu

J. S. Balakrishnan et al. (eds.), *Arithmetic Geometry, Number Theory, and Computation*, Simons Symposia, https://doi.org/10.1007/978-3-030-80914-0_7

1 Introduction

The question of which abelian varieties arise as Jacobians of curves has a long and rich history. It has classically been investigated over the complex numbers as the Schottky Problem, using techniques from differential geometry and Hodge theory. In positive characteristic, some of these tools are no longer available, but instead the Frobenius endomorphism becomes a formidable weapon.

In this article, we study g-dimensional abelian varieties which are defined over a finite field $\mathbb{F}_q$ of odd cardinality. The $\mathbb{F}_q$-isogeny class of such an abelian variety A is uniquely determined by the characteristic polynomial of its Frobenius endomorphism. This polynomial is called the *Weil polynomial* of the abelian variety and has the following form:

$$Z_A(t)=t^{2g}+a_1t^{2g-1}+\cdots+a_{g-1}t^{g+1}+a_gt^g+a_{g-1}qt^{g-1}+\cdots+a_1q^{g-1}t+q^g \in \mathbb{Z}[t].$$

Given $Z_A(t)$, one would like to determine if the isogeny class of A contains the Jacobian of a smooth curve over $\mathbb{F}_q$.

In genus one, the problem is straightforward, as every one-dimensional abelian variety is an elliptic curve, which is isomorphic to its Jacobian. For genus two, the problem was solved by Howe, Nart, and Ritzenthaler [HNR09] who give an explicit classification using only elementary restrictions involving the integers a_1, a_2, and q. Their method relies on three key facts:

(1) the Jacobian of a curve is canonically principally polarized;
(2) a principally polarized abelian surface is isomorphic to exactly one of the following: a Jacobian of a genus two curve, a product of two elliptic curves, or the restriction of scalars of an elliptic curve from a quadratic extension of the ground field;
(3) every genus two curve is hyperelliptic.

In particular, they take advantage of the canonical involution associated with any genus two curve.

Solving this problem in higher genus appears to be significantly more complicated. For example, every curve of genus three is isomorphic to either a hyperelliptic curve or to a smooth plane quartic curve. Curves of the latter type generically do not possess any non-trivial automorphisms, so the arguments of Howe, Nart and Ritzenthaler for genus two cannot easily be extended to non-hyperelliptic genus three curves.

In this paper, we focus on a more accessible question, namely, whether a given isogeny class of abelian varieties contains the Jacobian of a hyperelliptic curve. We do this by studying the geometric configurations of the Weierstrass points of hyperelliptic curves, as is done in the $g = 2$ case, e.g., [MN02, Appendix]. In doing so, we obtain parity conditions on the coefficients of the Weil polynomials which prevent certain isogeny classes from containing the Jacobian of a hyperelliptic curve. For instance, in genus three we obtain:

Theorem 2.8 *Let q be an odd prime power. The isogeny classes of three-dimensional abelian varieties corresponding to Weil polynomials of the form*

$$t^6 + a_1 t^5 + a_2 t^4 + a_3 t^3 + q a_2 t^2 + q^2 a_1 t + q^3$$

with $a_2 \equiv 0 \pmod 2$ and $a_3 \equiv 1 \pmod 2$ do not contain the Jacobian of a hyperelliptic curve over $\mathbb{F}_q$.

In fact, we can perform the same analysis for abelian varieties of any dimension; as far as we know, this is the first non-existence result for Jacobians in isogeny classes when $g > 2$ and q is any odd prime power. Indeed, we demonstrate that the Weil polynomial of the Jacobian of a hyperelliptic curve of genus g over a finite field of odd characteristic must be congruent modulo 2 to a polynomial of the form $\prod_{i=1}^{r}(t^{d_i} - 1)/(t-1)^2 \in \mathbb{F}_2[t]$ where $2g + 2 = d_1 + \cdots + d_r$ is a partition. We call a Weil polynomial *admissible* if it takes this form modulo 2, and *inadmissible* otherwise. Notice that it is easy to test whether a polynomial is admissible by explicitly trying all possible partitions, and an isogeny class is guaranteed to not contain a hyperelliptic Jacobian if its Weil polynomial is inadmissible.

By using asymptotic results on the number of Weil polynomials of a fixed degree whose coefficients lie in prescribed congruence classes modulo an integer, it is possible to determine the asymptotic proportion of isogeny classes with admissible, or inadmissible, Weil polynomials.

Theorem 3.3 *Let $c(q, g)$ be the proportion of isogeny classes of g-dimensional abelian varieties over $\mathbb{F}_q$ with admissible Weil polynomial. For all $g \geq 2$ we have*

$$\lim_{q \to \infty} c(q, g) = \frac{Q(2g+2)}{2^g},$$

as q ranges over odd prime powers, where $Q(2g+2)$ is the number of partitions of $2g+2$ into distinct parts.

For comparison, notice that for fixed $g > 6$, it is possible to see by simply counting the number of isogeny classes of g-dimensional abelian varieties over $\mathbb{F}_q$ versus the number of hyperelliptic curves of genus g over $\mathbb{F}_q$ that actually 0% of g-dimensional abelian varieties over $\mathbb{F}_q$ contain a hyperelliptic Jacobian as $q \to \infty$ ranges over odd prime powers. However, this counting argument does not provide a way to identify which isogeny classes do or do not contain a hyperelliptic Jacobian. In contrast, for any fixed $g \geq 1$, any isogeny class with an inadmissible Weil polynomial is explicitly known to not contain a hyperelliptic Jacobian. Rephrasing the theorem above for fixed $g > 6$, the proportion of isogeny classes of g-dimensional abelian varieties over $\mathbb{F}_q$ which do not contain a hyperelliptic Jacobian but are not found by testing for inadmissible Weil polynomials is $Q(2g+2)/2^g$ as $q \to \infty$ ranges over odd prime powers. As noted in Remark 3.4, this discrepancy approaches zero as g grows.

On the other hand, over the algebraic closure $\overline{\mathbb{F}}_q$, determining which isogeny classes contain Jacobians remains an open question. In order for our methods to shed light in this context, the behavior of admissibility under finite field extensions must first be understood. We leave this question for future work.

The outline of the paper is as follows. In Sect. 2, we recall some results on the geometry of hyperelliptic curves and their Jacobians and prove Theorem 2.8 and its generalizations to higher genus. In Sect. 3, we study asymptotic consequences of these non-existence results, deriving Theorem 3.3. In Sect. 4, we determine restrictions on point counts of curves, allowing us to reprove Theorem 2.8 by elementary methods. Finally, in Sect. 5, we present experimental data on the optimality of our results.

2 Weil Polynomials Mod 2

Let C be a hyperelliptic curve of genus g defined over $\mathbb{F}_q$ (of characteristic $p \neq 2$) with canonical degree two map $\pi : C \to \mathbb{P}^1$. Let $W := \{\alpha_1, \alpha_2, \ldots, \alpha_{2g+2}\}$ be the support of ramification divisor of π, i.e., the geometric Weierstrass points of C. The $\mathbb{F}_q$-Frobenius endomorphism Frob acts on W by permuting its elements; we denote the cardinalities of the orbits of this action by d_i. (Alternatively, W consists of some number, r, of $\mathbb{F}_q$-places with respective degrees d_i, where $d_1 + \cdots + d_r = 2g + 2$.)

The multiset $d_C := \{d_i\}$ forms a partition of the integer $2g + 2$; we call this the *degree set* of the curve C, and by convention we order the d_i so that $d_1 \leq d_2 \leq \cdots \leq d_r$.

Explicitly, C can be given in coordinates by $y^2 = f(x)$, where f is a squarefree polynomial of degree either $2g + 1$ or $2g + 2$. Then W consists of the points $(\alpha, 0)$, where α runs over the roots of f, together with the point at infinity if $\deg(f) = 2g + 1$. Then the d_i are precisely the degrees of the irreducible factors of f, along with an extra 1 in the case $\deg(f) = 2g + 1$.

Remark 2.1 If q is small compared to the genus, not all partitions of $2g + 2$ may arise as degree sets of genus g hyperelliptic curves, due to the finite number of irreducible polynomials of any fixed degree in $\mathbb{F}_q[x]$. For example, if $2g+2 > q+1$, then the partition $2g + 2 = 1 + \cdots + 1$ cannot be realized, as there are not enough $\mathbb{F}_q$-points for C to ramify over.

Let J denote the Jacobian variety of C whose elements are degree-zero divisors on C modulo linear equivalence. Then the group $J[2]$ of (geometric) 2-torsion of J is an $\mathbb{F}_2$-vector space that admits the following explicit description.

Lemma 2.2 *The group $J[2]$ of 2-torsion elements of J forms a $2g$-dimensional vector space over $\mathbb{F}_2$. Explicitly, this group can be expressed as the vector space obtained from $\mathbb{F}_2^{2g+2}$ by considering all vectors with an even number of non-zero entries and forming the quotient by $\langle(1, 1, \ldots, 1)\rangle$.*

Proof The first claim follows since $p \neq 2$. For the second, we argue as follows; cf. [Mum07, Corollary 2.11] over $\mathbb{C}$, which extends to any algebraically closed field of characteristic unequal to two. Every element of $J[2]$ can be explicitly represented either by a divisor

$$e_U := \sum_{P \in U} P - |U|(\infty),$$

if the map $\pi : C \to \mathbb{P}^1$ is ramified at infinity, in which case the symbol ∞ slightly abusively denotes $\pi^{-1}(\infty) \in C$, or by a divisor

$$e_U := \sum_{P \in U} P - \frac{|U|}{2}(\infty_1 + \infty_2),$$

if π is split at infinity, in which case $\pi^{-1}(\infty) = \{\infty_1, \infty_2\}$. Either way, the set $U \subseteq W$ is a subset of even cardinality, and two divisors, e_U and $e_{U'}$, represent the same element of $J[2]$ if either $U = U'$ or $U = W \setminus U'$. In the $\mathbb{F}_2$-vector space V whose standard basis is indexed by W, the above set of representatives determine the subspace of vectors with an even number of non-zero entries. Two vectors $v_1, v_2 \in V$ yield the same element of $J[2]$ precisely when their sum is contained in $\langle(1, 1, \ldots, 1)\rangle$. The second claim follows. □

We will also need the following standard fact about vector spaces, whose proof is omitted.

Lemma 2.3 *Consider an exact sequence of vector spaces*

$$0 \to W_1 \to W_2 \to W_3 \to 0$$

and a linear map $T\colon W_2 \to W_2$ *such that* $T(W_1) \subseteq W_1$. *We will also denote the induced map* $T\colon W_3 \to W_3$. *For* $i \in \{1, 2, 3\}$, *denote the characteristic polynomial of* T *on* W_i *by* $\chi(T, W_i)$. *Then we have*

$$\chi(T, W_2) = \chi(T, W_1)\chi(T, W_3).$$

Denote the base-change of C (resp. J) to $\overline{\mathbb{F}}_q$ by C^{alg} (resp. J^{alg}). Exploiting the action of Frob on the geometric Weierstrass points W, we obtain the following result on the characteristic polynomial of Frob modulo 2.

Proposition 2.4 *Let* C *be a hyperelliptic curve of genus* g *defined over* $\mathbb{F}_q$. *Let* $\{d_i\}_{i=1}^r$ *be the partition of* $2g + 2$ *which records the sizes of the orbits of Frobenius acting on the* $2g + 2$ *geometric Weierstrass points. For any prime* $\ell \neq p$ *we have*

$$\det\Big(1 - \mathrm{Frob}\, t \mid H^1_{\acute{e}t}(C^{\mathrm{alg}}, \mathbb{Q}_\ell)\Big) \equiv \Big(\prod_{i=1}^{r} t^{d_i} - 1\Big)/(t-1)^2 \pmod 2. \qquad (2.5)$$

Proof We have (cf. [Mil86, Corollary 9.6]) that

$$\det\left(1 - \mathrm{Frob}\, t \mid H^1_{\text{ét}}(C^{\mathrm{alg}}, \mathbb{Q}_\ell)\right) = \det\left(1 - \mathrm{Frob}\, t \mid H^1_{\text{ét}}(J^{\mathrm{alg}}, \mathbb{Q}_\ell)\right)$$
$$= \det\left(1 - \mathrm{Frob}\, t \mid V_\ell(J)\right),$$

where $V_\ell(J) = T_\ell(J) \otimes_{\mathbb{Z}_\ell} \mathbb{Q}_\ell$, and where $T_\ell(J)$ is the ℓ-adic Tate module of J. By the Weil conjectures the polynomials above all have integer coefficients. By taking $\ell = 2$, we further have

$$\det\left(1 - \mathrm{Frob}\, t \mid V_2(J)\right) \equiv \det\left(1 - \mathrm{Frob}\, t \mid J[2]\right) \pmod 2.$$

As before, consider the $\mathbb{F}_2$-vector space $W_2 := \mathbb{F}_2^{2g+2}$ whose standard basis is indexed by W and which is acted on by the $\mathbb{F}_q$-Frobenius endomorphism Frob. By assumption, the action of Frob on W_2 can be represented by a block-diagonal matrix with r blocks, whose ith block (of order d_i) is a cyclic permutation of basis vectors. Since the characteristic polynomial of a cyclic permutation of order n is $t^n - 1$, it follows that

$$\chi(\mathrm{Frob}, W_2) = \prod_{i=1}^{r} \left(t^{d_i} - 1\right).$$

Furthermore, let W_1 be the codimension-one subspace of vectors with an even number of non-zero entries. Then W_1 is stable under Frob. Moreover, Frob acts invertibly on W_1 and hence also on the one-dimensional quotient $W_3 := W_2/W_1$, so that $\chi(\mathrm{Frob}, W_3) = t - 1$. Applying Lemma 2.3 to the short exact sequence $0 \to W_1 \to W_2 \to W_2/W_1 \to 0$ yields

$$\chi(\mathrm{Frob}, W_1) = \frac{\prod_{i=1}^{r}(t^{d_i} - 1)}{t - 1}.$$

Now let W_4 be the one-dimensional subspace of $\langle(1, 1, \ldots, 1)\rangle \subseteq W_1$ and define $W_5 := W_1/W_4$. Similarly, W_4 is stable under Frob with $\chi(\mathrm{Frob}, W_4) = t - 1$ since Frob acts invertibly. Applying Lemma 2.3 to the short exact sequence $0 \to W_4 \to W_1 \to W_5 \to 0$, we find

$$\chi(\mathrm{Frob}, W_5) = \frac{\prod_{i=1}^{r}\left(t^{d_i} - 1\right)}{(t - 1)^2}.$$

Since, by Lemma 2.2, Frob acts on W_5 as it does on $J[2]$, the result follows. □

Remark 2.6 It is possible to prove a congruence modulo 2 similar to (2.5) in Proposition 2.4 also when the finite field $\mathbb{F}_q$ is allowed to have characteristic 2. In this case, Frobenius acts on Weierstrass points as it does on the étale part of

Table 2.9 Weil coefficients modulo 2 and corresponding partitions for threefolds

Coefficients $(a_1, a_2, a_3) \pmod 2$	Partition of 8
(0, 1, 1)	{3, 5}
(1, 1, 0)	{1, 1, 1, 1, 1, 3}, {1, 1, 1, 2, 3}, {1, 2, 2, 3}, {1, 3, 4}
(1, 0, 0)	{1, 1, 1, 5}, {1, 2, 5}
(0, 0, 0)	{1, 1, 3, 3}, {1, 1, 6}, {2, 3, 3}, {2, 6}
(0, 1, 0)	{1, 1, 1, 1, 1, 1, 1, 1}, {1, 1, 1, 1, 1, 1, 2}, {1, 1, 1, 1, 2, 2},
	{1, 1, 1, 1, 4}, {1, 1, 2, 2, 2}, {1, 1, 2, 4},
	{2, 2, 2, 2}, {2, 2, 4}, {4, 4}, {8}
(1, 1, 1)	{1, 7}

$J[2]$. Since we require odd characteristic throughout the rest of the paper, notably in Theorem 3.1, we do not pursue this further.

Definition 2.7 We call any polynomial of the form (2.5) an *admissible Weil polynomial modulo 2*. Note that the notion of admissibility is independent of q.

By applying Proposition 2.4 to all possible partitions of 8 we obtain our main theorem.

Theorem 2.8 *Let q be an odd prime power. The isogeny classes of three-dimensional abelian varieties corresponding to Weil polynomials of the form*

$$t^6 + a_1t^5 + a_2t^4 + a_3t^3 + qa_2t^2 + q^2a_1t + q^3$$

with $a_2 \equiv 0 \pmod 2$ and $a_3 \equiv 1 \pmod 2$ do not contain the Jacobian of a hyperelliptic curve over $\mathbb{F}_q$.

Proof For each of the twenty-two partitions $\{d_i\}$ of $2g + 2 = 8$, we compute the corresponding polynomial $(t-1)^{-2}\prod_{i=1}^{r}(t^{d_i} - 1)$. By Proposition 2.4, these are all of the admissible Weil polynomials modulo 2 for Jacobians of hyperelliptic curves of genus three. In Table 2.9, we tabulate the coefficients (a_1, a_2, a_3) of the resulting Weil polynomials modulo 2 with the corresponding partitions $\{d_i\}$. □

Although Theorem 2.8 only considers the case of three-dimensional abelian varieties, Proposition 2.4 applies much more generally. Indeed, for any $g \geq 1$, we can produce a list of admissible Weil polynomials modulo 2 for Jacobians of hyperelliptic curves of genus g, independent of q. The following result counts the number of admissible polynomials in terms of partitions.

Proposition 2.10 *The number of admissible Weil polynomials modulo 2 for Jacobians of hyperelliptic curves of genus g over finite fields of odd characteristic is equal to $Q(2g + 2)$, the number of partitions of $2g + 2$ into distinct parts, or equivalently, the number of partitions of $2g + 2$ into odd parts.*

Proof It is well known that the number of partitions of $2g + 2$ into distinct parts is equal to the number of partitions of $2g + 2$ into odd parts; see [NZM91, Theorem

10.2]. For the remainder of this proof, our perspective will focus on partitions into distinct parts.

Proposition 2.4 shows how to compute admissible Weil polynomials modulo 2 using partitions, although many different partitions can correspond to a single admissible polynomial. For the purposes of this proof, we will call two partitions $\{d_1, \ldots, d_r\}$ and $\{e_1, \ldots, e_s\}$ of an integer n *equivalent* if

$$\prod_{i=1}^{r} (t^{d_i} - 1) \equiv \prod_{j=1}^{s} (t^{e_j} - 1) \pmod 2.$$

Using Eq. (2.5), it suffices to prove that every equivalence class of partitions of $2g + 2$ contains precisely one partition with distinct parts.

Observe that every partition of $2g + 2$ is equivalent to a partition with distinct parts. Indeed, if $\{d_1, \ldots, d_{r+1}\}$ is any partition of $2g + 2$ where $d_r = d_{r+1}$, then we can construct another partition $\{d_1, \ldots, d_{r-1}, d_r'\}$ with $d_r' = 2d_r$. These two partitions are equivalent since $t^{2d_r} - 1 \equiv (t^{d_r} - 1)^2 \mod 2$. Thus by induction on the number of equal parts, we conclude that every admissible Weil polynomial modulo 2 arises as a partition of $2g + 2$ into distinct parts.

Now suppose that $\{d_1, \ldots, d_r\}$ and $\{e_1, \ldots, e_s\}$ are two equivalent partitions of $2g + 2$ into distinct parts. Without loss of generality, we order the parts so that d_1 and e_1 are the smallest parts of their respective partitions. After expanding the polynomials above, we see that the non-constant monomials of smallest degree are t^{d_1} and t^{e_1}, respectively. This implies that $d_1 = e_1$, hence $\{d_2, \ldots, d_r\}$ and $\{e_2, \ldots, e_s\}$ are equivalent partitions of $2g + 2 - d_1$. Thus by induction on r, we conclude that two equivalent partitions of $2g + 2$ into distinct parts are equal. □

Remark 2.11 In contrast with Remark 2.1, every admissible Weil polynomial mod 2 arises as the reduction mod 2 of a Weil polynomial for a hyperelliptic curve over $\mathbb{F}_q$, for each q: Since there is at least one irreducible polynomial of every degree, we can construct a hyperelliptic curve ramified at points of degree d_i, where $\{d_i\}$ is the partition of $2g + 2$ into distinct parts.

In Table 2.12, we tabulate the number of admissible and inadmissible Weil polynomials modulo 2 for Jacobians of hyperelliptic curves of genus $g \leq 7$, i.e., the numbers $Q(2g + 2)$ and $2^g - Q(2g + 2)$.

Table 2.12 The number of admissible and inadmissible Weil polynomials modulo 2 for hyperelliptic curves of small genus

g	1	2	3	4	5	6	7
$Q(2g+2)$	2	4	6	10	15	22	32
	100.00%	100.00%	75.00%	62.50%	46.88%	34.38%	25.00%
$2^g - Q(2g+2)$	0	0	2	6	17	42	96
	0.00%	0.00%	25.00%	37.50%	53.12%	65.62%	75.00%

3 Asymptotics

The results in the previous section show that an isogeny class cannot contain a hyperelliptic Jacobian if the coefficients of the corresponding Weil polynomial satisfy a certain parity condition. Therefore, it is natural to ask how many Weil polynomials satisfy any given parity condition, or more generally, how many Weil polynomials are congruent to a given fixed polynomial of the correct form modulo an integer m. The following theorem answers this question asymptotically for abelian varieties of fixed dimension.

Theorem 3.1 *Let $g \geq 2$ and m be fixed integers, and let $f(t) \in \mathbb{Z}[t]$ be a fixed polynomial of the form*

$$f(t) = t^{2g} + a_1 t^{2g-1} + \cdots + a_{g-1} t^{g+1} + a_g t^g + a_{g-1} q t^{g-1} + \cdots + a_1 q^{g-1} t + q^g.$$

For a prime power q coprime to m, write $e_{f,m}(q, g)$ for the proportion of isogeny classes of g-dimensional abelian varieties over $\mathbb{F}_q$ whose Weil polynomial is congruent to $f(x)$ modulo m. Then

$$\lim_{q \to \infty} e_{f,m}(q, g) = \frac{1}{m^g},$$

where the limit is taken over all prime powers q coprime to m.

Proof This is essentially a theorem of Holden [Hol04, Theorem 5], although we present a slightly more general statement here. Specifically, Holden's theorem only considers the case when $m = \ell$ is a prime, and the limit is taken over $q = p^r$ for a single fixed prime p. However, the more general version of the theorem presented above is obtained immediately from Holden's methods, as follows. Denote by $I(q, g)$ the number of isogeny classes of g-dimensional abelian varieties over $\mathbb{F}_q$ and write $I_{f,m}(q, g)$ for the number of such isogeny classes whose Weil polynomial is congruent to $f(x)$ modulo m. With this notation we may write $e_{f,m}(q, g) = \frac{I_{f,m}(q,g)}{I(q,g)}$. Using lattices, DiPippo and Howe obtained upper and lower bounds for $I(q, g)$; see [DH98, Theorem 1.2] and [DH00]. Holden employed analogous techniques to obtain similar bounds for $I_{f,m}(q, g)$; see [Hol04, Proposition 2.2]. Using these bounds, as in the proof of [Hol04, Theorem 5], we find

$$\begin{aligned}
&\frac{v_g r(q) q^{g(g+1)/4} m^{-g} - 2c(g, m) q^{g(g+1)/4-1/2} m^{1-g}}{v_g r(q) q^{g(g+1)/4} + (v_g + 3c(g, 1)) q^{g(g+1)/4-1/2}} \\
&\qquad \leq \frac{I_{f,m}(q, g)}{I(q, g)} \\
&\qquad \leq \frac{v_g r(q) q^{g(g+1)/4} m^{-g} + (v_g + 3c(g, 1)) q^{g(g+1)/4-1/2} m^{1-g}}{v_g r(q) q^{g(g+1)/4} - 2c(g, m) q^{g(g+1)/4-1/2}},
\end{aligned}$$

where v_g is a constant depending g, and $c(g, m)$ is a constant depending on both g and m, and $r(q) = \varphi(q)/q$, where φ denotes Euler totient function. Letting $q \to \infty$, the theorem follows. □

Recall that, in light of Theorem 2.8 and Proposition 2.10, we say that the Weil polynomial of an isogeny class of g-dimensional abelian varieties is admissible if it is congruent modulo 2 to a polynomial of the form $\prod_{i=1}^{r}(t^{d_i}-1)/(t^2-1) \in \mathbb{F}_2[t]$ for some partition $2g+2 = d_1+\cdots+d_r$, and is inadmissible otherwise. By combining Theorems 2.8 and 3.1, we obtain the following corollary.

Corollary 3.2 *Denote by $c(q)$ the proportion of isogeny classes of abelian threefolds over $\mathbb{F}_q$ with admissible Weil polynomials. Then*

$$\lim_{q\to\infty} c(q) = 75\%,$$

where the limit is taken over all odd prime powers q.

More generally, by combining Theorem 2.8 and Proposition 2.10, we can determine the proportion of isogeny classes of g-dimensional abelian varieties over $\mathbb{F}_q$ with admissible Weil polynomials in terms of the number of partitions of $2g+2$ into distinct parts.

Theorem 3.3 *Let $c(q, g)$ be the proportion of isogeny classes of g-dimensional abelian varieties over $\mathbb{F}_q$ with admissible Weil polynomial. For all $g \geq 2$ we have*

$$\lim_{q\to\infty} c(q, g) = \frac{Q(2g+2)}{2^g},$$

as q ranges over odd prime powers, where $Q(2g+2)$ is the number of partitions of $2g+2$ into distinct parts.

Proof By Proposition 2.10, we know there is a set of polynomials $S = \{f_1, \ldots, f_{Q(2g+2)}\}$ such that the Weil polynomial of the Jacobian of a hyperelliptic curve of genus g over any finite field of odd characteristic is equivalent modulo 2 to some polynomial in S. Thus, the result follows immediately from Theorem 3.1. □

Remark 3.4 For fixed $g \geq 2$, the number of isogeny classes of g-dimensional abelian varieties over $\mathbb{F}_q$ and the number of hyperelliptic curves of genus g over $\mathbb{F}_q$ are asymptotically bounded by $q^{g(g+1)/4}$ and q^{2g-1}, respectively, as $q \to \infty$ ranges over odd prime powers; see [DH98, Theorem 1.1], and [Nar09, Table 1, Corollary 3.4]. For $g > 6$, this means that asymptotically 0% of the isogeny classes of g-dimensional abelian varieties over $\mathbb{F}_q$ contain a hyperelliptic Jacobian as $q \to \infty$ ranges over odd prime powers. Comparing this reality to Theorem 3.3, we see that the proportion of isogeny classes of g-dimensional abelian varieties that have an admissible Weil polynomial but still do not contain a hyperelliptic Jacobian is $\frac{Q(2g+2)}{2^g}$ as $q \to \infty$. Notice that this discrepancy gets smaller as g grows. Indeed, $Q(N) \sim \frac{3^{3/4}}{12N^{3/4}} \exp(\pi\sqrt{N/3})$ as $N \to \infty$ [FS09, Figure 1.9], so it follows that $\frac{Q(2g+2)}{2^g} \to 0$ as $g \to \infty$.

4 Point Counts

In Sect. 2 we determined restrictions modulo 2 for Weil polynomials of hyperelliptic Jacobians. These Weil polynomials govern the point counts of the corresponding hyperelliptic curve $C/\mathbb{F}_q$ over all extensions of $\mathbb{F}_q$. In this section, we determine 2-adic restrictions on the point counts of hyperelliptic curves over extensions of $\mathbb{F}_q$ by more elementary means. Rather than studying the action of Frobenius on $J[2]$, we study its action on the Weierstrass points of the curve directly. In particular, this provides an alternative proof of Theorem 2.8.

4.1 Restrictions on Parity of Point Counts

As before, let q be an odd prime power, $C/\mathbb{F}_q$ be a hyperelliptic curve of genus $g > 1$. Recall that we denote the support of the ramification divisor by W and we denote by $\{d_i\}$ a partition of $2g + 2$, corresponding to the decomposition of W into Frobenius orbits. We begin with the following observations on the point counts of C over extensions of $\mathbb{F}_q$.

Lemma 4.1 *For each $n \geq 1$, we have $\#W(\mathbb{F}_{q^n}) = \sum_{i:d_i|n} d_i$.*

Proof A point of degree d contributes d $\mathbb{F}_{q^n}$-points if $d|n$, and none otherwise. □

Lemma 4.2 *For each $n \geq 1$, we have $\#C(\mathbb{F}_{q^n}) \equiv \#W(\mathbb{F}_{q^n}) \pmod 2$.*

Proof The $\mathbb{F}_{q^n}$-points of W are precisely the $\mathbb{F}_{q^n}$-points of C that are fixed by the hyperelliptic involution. Since all other points appear in pairs, it follows that $\#C(\mathbb{F}_{q^n}) \equiv \#W(\mathbb{F}_{q^n}) \pmod 2$. □

Corollary 4.3 *For each $n \geq 1$, we have $\#C(\mathbb{F}_{q^n}) \equiv \#C(\mathbb{F}_{q^{2n}}) \pmod 2$.*

Proof The numbers n and $2n$ share the same set of odd divisors, therefore by Lemmas 4.1 and 4.2

$$\#C(\mathbb{F}_{q^n}) \equiv \#W(\mathbb{F}_{q^n}) = \sum_{i:d_i|n} d_i \equiv \sum_{i:d_i|2n} d_i = \#W(\mathbb{F}_{q^{2n}}) \equiv \#C(\mathbb{F}_{q^{2n}}) \pmod 2.$$

□

Corollary 4.4 *For each n, we have*

$$n \cdot \#\{i : d_i = n\} \equiv \sum_{d|n} \mu(n/d)\#C(\mathbb{F}_{q^d}) \pmod 2,$$

where μ is the Möbius function.

Proof Using Lemma 4.1 we can restate Lemma 4.2 as

$$\#C(\mathbb{F}_{q^n}) \equiv \sum_{d|n} d \cdot \#\{i : d_i = d\} \pmod 2.$$

The result follows by applying Möbius inversion. □

Remark 4.5 For C as above, consider the binary sequence

$$a_C(n) := \#C(\mathbb{F}_{q^n}) \pmod 2.$$

Lemma 4.2 and Corollary 4.4 imply that a_C determines, and is determined by, the numbers $n \cdot \#\{i : d_i = n\} \pmod 2$ for all n. These numbers are encoded in the degree set d_C; they tell us the parity of $\#(i : d_i = n)$ when n is odd, and give no information when n is even.

More precisely, Lemma 4.2 and Corollary 4.4 give us a dictionary between the sequence a_C of parities of point counts and the set

$$\{d : d \text{ is odd and appears an odd number of times in } d_C\}.$$

This is consistent with Proposition 2.10.

This is already enough to prove that some sequences of point count parities are inconsistent with the identity $\sum d_i = 2g + 2$. For example:

Lemma 4.6 *If the genus of C is 3, then we cannot have $\#C(\mathbb{F}_q) \equiv 0$, $\#C(\mathbb{F}_{q^3}) \equiv 1$, and $\#C(\mathbb{F}_{q^5}) \equiv 0 \pmod 2$.*

Proof If such a curve C existed, Corollary 4.4 implies that $\#\{d_i = 1\}$ is even, $\#\{d_i = 3\}$ is odd, and $\#\{d_i = 5\}$ is even. Thus $\#\{d_i = 3\} = 1$ and $\#\{d_i = 5\} = 0$, which contradicts $\#\{d_i = 1\}$ being even. □

4.2 Restrictions on Point Counts Modulo Powers of 2

We have just seen some restrictions on the point counts modulo 2 of hyperelliptic curves. In this section we will obtain further restrictions modulo higher powers of 2.

Fix an integer $m \geq 1$. Let G be the group $\{\pm 1\} \times \mathrm{Gal}(\mathbb{F}_{q^{2^m}}/\mathbb{F}_q)$ of order 2^{m+1}. This acts on the $\mathbb{F}_{q^{2^m}}$-points of C, where the first factor acts by the hyperelliptic involution and the second by field automorphisms. We will study $\#C(\mathbb{F}_{q^{2^m}}) \pmod{2^{m+1}}$ by examining the orbits of the action of G.

Proposition 4.7 *If C is a genus g hyperelliptic curve over $\mathbb{F}_q$, then*

$$\#C(\mathbb{F}_{q^n}) \equiv 2(q^n + 1) - \#W(\mathbb{F}_{q^n}) \pmod{2^{a+1}},$$

where $n = 2^a \cdot m \geq 1$ with m odd.

Proof We will count the $\mathbb{F}_{q^n}$-points of C by considering the fibers of the hyperelliptic map $\pi : C(\mathbb{F}_{q^n}) \to \mathbb{P}^1(\mathbb{F}_{q^n})$. If $x \in \mathbb{P}^1(\mathbb{F}_{q^n})$, then $\pi^{-1}(x)$ can be computed by extracting the square roots of $f(x)$, where f is a polynomial of degree $2g+1$ or $2g+2$. Accordingly, $\pi^{-1}(x)$ contains either zero, one, or two $\mathbb{F}_{q^n}$-points, and a preimage of size one occurs if and only if $x \in W$. Further, if n is even and x is defined over $\mathbb{F}_{q^{n/2}}$ (equivalently, if its degree is not divisible by 2^a), then the preimage must have size two, since quadratic equations $y^2 = f(x)$ over $\mathbb{F}_{q^{n/2}}$ can be solved over $\mathbb{F}_{q^n}$. Thus, the preimage may have size zero only if the degree of x is divisible by 2^a. In this case, the Galois orbit of x has size divisible by 2^a, so if there are $\mathbb{F}_{q^n}$-points above x, they occur in an orbit whose size is divisible by 2^{a+1}. Hence, to count modulo 2^{a+1}, we may assume that all unramified $\mathbb{F}_{q^n}$-points of $\mathbb{P}^1$ have exactly two preimages defined over $\mathbb{F}_{q^n}$, and subtract the number of $\mathbb{F}_{q^n}$-points of W to obtain the result. □

As with Proposition 2.4, this implies that certain sequences of point counts of hyperelliptic curves (equivalently, certain Weil polynomials of their Jacobians) cannot be realized.

Corollary 4.8 *Assume that the genus of C is 3. If $\#C(\mathbb{F}_q) \equiv 1 \pmod 2$ and $\#C(\mathbb{F}_{q^3}) \equiv 0 \pmod 2$, then we have $\#C(\mathbb{F}_{q^{2^m}}) \equiv -1 \pmod{2^{m+1}}$ for all $m \geq 1$.*

Proof Since $\#C(\mathbb{F}_q)$ is odd, we have that $\#\{d_i = 1\}$ is odd by Lemma 4.2. Analogously, since $\#\{d_i = 1\}$ is odd and $\#C(\mathbb{F}_{q^3})$ is even, it follows that $\#\{d_i = 3\}$ is odd and hence must equal one. Altogether, we have $d_j = 3$ for a unique j and $d_{j'} \in \{1, 2, 4\}$ for $j' \neq j$. Hence $\#\left(\bigcup_{k\geq 1} W(\mathbb{F}_{q^{2^k}})\right) = 5$, and so $\#W(\mathbb{F}_{q^{2^m}}) \equiv 5 \pmod{2^{m+1}}$. The result follows by applying Proposition 4.7 to $n = 2^m$, since we have $q^{2^m} \equiv 1 \pmod{2^{m+1}}$ for q odd. □

4.3 *Reproving Theorem 2.8*

The results in the previous subsections can be used to reprove Theorem 2.8 as follows:

Alternative Proof of Theorem 2.8 Suppose that the statement is false, so that there is some hyperelliptic curve C of genus three whose Jacobian has the prescribed Weil polynomial. By using Newton's formulae and [Mil86, Theorem 11.1], we can write the point counts $\#C(\mathbb{F}_{q^k})$ in terms of the coefficients of the Weil polynomial for all $k \geq 1$, as follows:

$$\#C(\mathbb{F}_q) = q + 1 + a_1;$$

$$\#C(\mathbb{F}_{q^2}) = q^2 + 1 - a_1^2 + 2a_2;$$

$$\#C(\mathbb{F}_{q^3}) = q^3 + 1 + a_1^3 - 3a_1a_2 + 3a_3;$$

$$\#C(\mathbb{F}_{q^4}) = q^4 + 1 - a_1^4 + 4a_1^2a_2 - 4a_1a_3 - 2a_2^2 + 4qa_2;$$
$$\#C(\mathbb{F}_{q^5}) = q^5 + 1 + a_1^5 - 5a_1^3a_2 + 5a_1a_2^2 - 5qa_1a_2 - 5a_2a_3 + 5q^2a_1.$$

We will complete our proof by showing that the restrictions $a_2 \equiv 0 \pmod 2$ and $a_3 \equiv 1 \pmod 2$ on the coefficients force the point counts to fall into the impossible cases listed in the lemmas above. First consider the case when $a_1 \equiv 0 \pmod 2$. Expanding the equations above leads to

$$\#C(\mathbb{F}_q) \equiv 0 \pmod 2, \quad \#C(\mathbb{F}_{q^2}) \equiv 2 \pmod 4,$$
$$\#C(\mathbb{F}_{q^3}) \equiv 1 \pmod 2, \quad \#C(\mathbb{F}_{q^4}) \equiv 2 \pmod 8,$$
$$\#C(\mathbb{F}_{q^5}) \equiv 0 \pmod 2.$$

In particular, Lemma 4.6 applies, and we reach a contradiction.

Now assume that $a_1 \equiv 1 \pmod 2$. The analogous computation gives

$$\#C(\mathbb{F}_q) \equiv 1 \pmod 2, \quad \#C(\mathbb{F}_{q^2}) \equiv 1 \pmod 4,$$
$$\#C(\mathbb{F}_{q^3}) \equiv 0 \pmod 2, \quad \#C(\mathbb{F}_{q^4}) \equiv 5 \pmod 8,$$
$$\#C(\mathbb{F}_{q^5}) \equiv 0 \pmod 2.$$

Corollary 4.8 applies, and we again reach a contradiction. □

Remark 4.9 The argument above can be automated, and by iterating over partitions of $2g + 2$, and applying Proposition 4.7, one can rule out Weil polynomials modulo 2, similarly to Proposition 2.10. We have verified that the sets of unrealizable Weil polynomials obtained by these two different procedures agree up to genus $g \leq 10$ by using the Newton identities. Indeed, we expect both sets to match for any g, although we are unable to prove it in general.

5 Experimental Data

The main results of this paper were inspired by experimental data, which we include here to illustrate the phenomena. By performing an exhaustive search, the number of Jacobians of hyperelliptic curves in each isogeny class over $\mathbb{F}_q$ (up to isomorphism of principally polarized abelian varieties) has been computed for threefolds and prime powers $q \leq 13$ that are either prime or odd. Similarly, by iterating over the isomorphism classes of smooth plane quartic curves, the number of isogeny classes of abelian threefolds which contain the Jacobian of a smooth plane quartic curve has been computed for $q = 2, 3$, and 5. Both searches were done by Andrew Sutherland using the techniques developed in [KS08, KS16], and the data have been incorporated in the *L-functions and Modular Forms Database* [LMF19]; see www.LMFDB.org/Variety/Abelian/Fq/.

Table 5.1 Types of Jacobians per isogeny classes of abelian threefolds

	Contains Jacobian				Total
q	Yes	No	Hyperelliptic	Quartic curve	
2	108	107	59	73	215
	50.23%	49.77%	27.44%	33.95%	
3	479	198	297	389	677
	70.75%	29.25%	43.87%	57.46%	
5	2611	342	1723	2471	2953
	88.42%	11.58%	58.35%	83.68%	

By combining these data sets, one can also deduce which isogeny classes of abelian threefolds do not contain a Jacobian for $q = 2, 3$, and 5, see Table 5.1. These data sets, and more precisely the multisets of virtual point counts modulo 2 and 4 that can be extracted from this data, provide motivation for Theorem 2.8.

In Fig. 5.2, we compare the proportion of isogeny classes of abelian threefolds over $\mathbb{F}_q$ which do not contain a hyperelliptic Jacobian for $q \leq 13$ with the proportion of such isogeny classes which are ruled out via Theorem 2.8, giving us some insight into the efficiency of Theorem 2.8 as q grows. Examples of isogeny classes which do not contain a (hyperelliptic) Jacobian but are not ruled out by Theorem 2.8 are given by the following Weil polynomials, cf. isogeny classes `3.3.a_ab_ac` and `3.5.ac_j_aw` on the LMFDB [LMF19]:

$$Z_1(t) = t^6 - t^4 - 2t^3 - 3t^2 + 27,$$

$$Z_2(t) = t^6 - 2t^5 + 9t^4 - 22t^3 + 45t^2 - 50t + 125.$$

In fact, neither isogeny class contains the Jacobian of any curve. These isogeny classes correspond to the row of Table 2.9 with the most partitions.

In Table 5.3, we present the number of isogeny classes which are ruled out by Theorem 2.8, for $17 \leq q \leq 31$. The data show that the proportion of isogeny classes ruled out by Theorem 2.8 quickly approaches 25%, as expected; see Corollary 3.2. Furthermore, naive extrapolation seems to indicate that the proportion of isogeny classes of abelian threefolds over $\mathbb{F}_q$ which do not contain a hyperelliptic Jacobian approaches 25% from above as $q \to \infty$ ranges over odd prime powers, while the proportion of the isogeny classes that are ruled out via Theorem 2.8 approaches 25% from below. Similarly, in Tables 5.4 and 5.5 we display the number of isogeny classes which cannot contain a hyperelliptic Jacobian by Proposition 2.4, and as before, we observe proportions which are already very close to the ones attained in the q-limit; see Corollary 3.2.

To generate Tables 5.3, 5.4 and 5.5, we enumerated all isogeny classes through their Weil polynomials: First, we enumerated Weil polynomials of degree $2g$, and then we filtered by the Honda–Tate condition on its factors, see [Wat69, Chapter 2], to only keep the ones that correspond to an isogeny class of abelian varieties

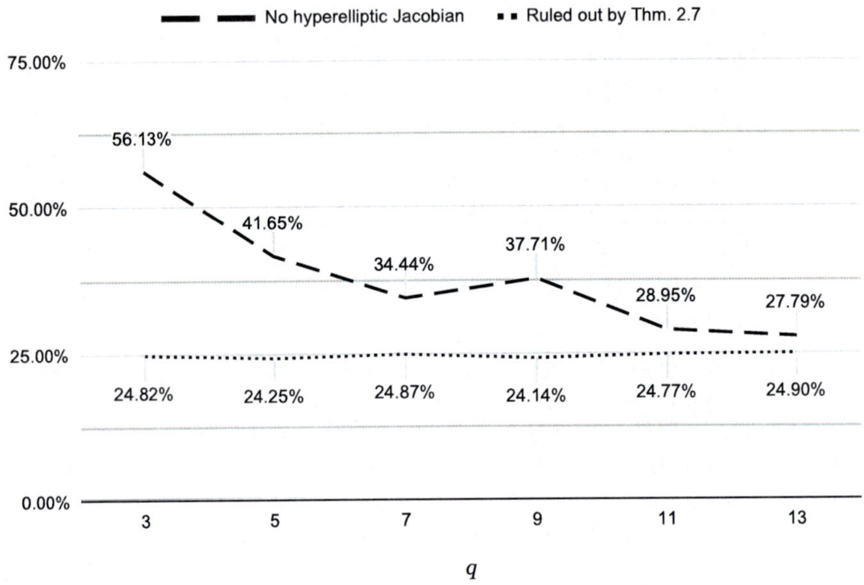

Fig. 5.2 Effectiveness of Theorem 2.8

Table 5.3 Isogeny classes ruled out by Proposition 2.4 for genus 3 and small q

q	17	19	23	25	27	29	31
Number of isogeny classes	112,283	156,589	277,517	332,166	333,695	555,843	678,957
Inadmissible by Theorem 2.8	27,974	39,034	69,268	82,564	83,350	138,730	169,574
	24.91%	24.93%	24.96%	24.86%	24.98%	24.96%	24.98%

Table 5.4 Isogeny classes ruled out by Proposition 2.4 for genus 4 and small q

q	3	5	7	9	11	13	$q \to \infty$
Number of isogeny classes	10,963	132,839	705,593	2,232,114	6,718,947	15,477,119	
Inadmissible by Proposition 2.4	3856	48,910	262,564	829,189	2,513,570	5,794,772	
	35.17%	36.82%	37.21%	37.15%	37.41%	37.44%	37.50%

Table 5.5 Isogeny classes ruled out by Proposition 2.4 in genus 5 and small q

q	3	5	$q \to \infty$
Number of isogeny classes	267,465	11,902,325	
Inadmissible by Proposition 2.4	137,866	6,286,570	
	51.55%	52.82%	53.12%

of dimension g. We enumerated Weil polynomials using *root-unitary*,[1] which implements the techniques introduced in [Ked08].

Acknowledgments This project started at the AMS MRC workshop "Explicit methods in arithmetic geometry in characteristic p" held at Whispering Pines (RI). The authors thank Andrew Sutherland for suggesting this problem, and Jeff Achter, Everett Howe, Bjorn Poonen, Christophe Ritzenthaler, David Roe, Andrew Sutherland, and Christelle Vincent for helpful discussions. They also thank the anonymous referees for careful reading and useful feedback. The first author was supported by the Simons Collaboration in Arithmetic Geometry, Number Theory, and Computation via Simons Foundation grant 550033. The third author was partially supported by NSF RTG grant DMS-1646385. The fourth author was supported by an AMS–Simons Travel Grant. The fifth author was partially supported by National Science Foundation award CNS-1617802.

References

[DH98] Stephen A. DiPippo and Everett W. Howe. Real polynomials with all roots on the unit circle and abelian varieties over finite fields. *J. Number Theory*, 73(2):426–450, 1998.

[DH00] Stephen A. DiPippo and Everett W. Howe. Corrigendum: "Real polynomials with all roots on the unit circle and abelian varieties over finite fields" [J. Number Theory **73** (1998), no. 2, 426–450; MR1657992 (2000c:11101)]. *J. Number Theory*, 83(1):182, 2000.

[FS09] Philippe Flajolet and Robert Sedgewick. *Analytic combinatorics*. Cambridge University Press, Cambridge, 2009.

[HNR09] Everett W. Howe, Enric Nart, and Christophe Ritzenthaler. Jacobians in isogeny classes of abelian surfaces over finite fields. *Ann. Inst. Fourier (Grenoble)*, 59(1):239–289, 2009.

[Hol04] Joshua Holden. Abelian varieties over finite fields with a specified characteristic polynomial modulo l. *J. Théor. Nombres Bordeaux*, 16(1):173–178, 2004.

[Ked08] Kiran S. Kedlaya. Search techniques for root-unitary polynomials. In *Computational arithmetic geometry*, volume 463 of *Contemp. Math.*, pages 71–81. Amer. Math. Soc., Providence, RI, 2008.

[KS08] Kiran S. Kedlaya and Andrew V. Sutherland. Computing L-series of hyperelliptic curves. In *Algorithmic number theory*, volume 5011 of *Lecture Notes in Comput. Sci.*, pages 312–326. Springer, Berlin, 2008.

[KS16] Kiran S. Kedlaya and Andrew V. Sutherland. A census of zeta functions of quartic K3 surfaces over $\mathbb{F}_2$. *LMS J. Comput. Math.*, 19(suppl. A):1–11, 2016.

[LMF19] The LMFDB Collaboration. The L-functions and modular forms database. http://www.lmfdb.org, 2019. [Online; accessed 30 October 2019].

[1] https://github.com/kedlaya/root-unitary.

[Mil86] J. S. Milne. Jacobian varieties. In *Arithmetic geometry (Storrs, Conn., 1984)*, pages 167–212. Springer, New York, 1986.

[MN02] Daniel Maisner and Enric Nart. Abelian surfaces over finite fields as Jacobians. *Experiment. Math.*, 11(3):321–337, 2002. With an appendix by Everett W. Howe.

[Mum07] David Mumford. *Tata lectures on theta. II.* Modern Birkhäuser Classics. Birkhäuser Boston, Inc., Boston, MA, 2007. Jacobian theta functions and differential equations, With the collaboration of C. Musili, M. Nori, E. Previato, M. Stillman and H. Umemura, Reprint of the 1984 original.

[Nar09] Enric Nart. Counting hyperelliptic curves. *Adv. Math.*, 221(3):774–787, 2009.

[NZM91] Ivan Niven, Herbert S. Zuckerman, and Hugh L. Montgomery. *An introduction to the theory of numbers*. John Wiley & Sons, Inc., New York, fifth edition, 1991.

[Wat69] William C. Waterhouse. Abelian varieties over finite fields. *Ann. Sci. École Norm. Sup. (4)*, 2:521–560, 1969.

Zen and the Art of Database Maintenance

Edgar Costa and David Roe

Abstract The last decade has seen a proliferation of online mathematical resources. We discuss some of the technical challenges involved in creating and maintaining a mathematical database. In particular, we report on the transition of the *L-functions and Modular Forms Database* (LMFDB) between two database systems. We also highlight some of the improvements to the LMFDB that we have made as part of this transition.

1 Introduction

Mathematics has a long history of using computations to aid in forming conjectures and searching for counterexamples. In the past few decades, computers have taken on a central role, both in performing many calculations and in hosting the results. As computational and storage capacity has increased, the size of these results has grown to the point where the task of searching and maintaining the data requires specialized knowledge. We discuss some of the challenges involved, as well as the main tools available to address them. We focus on one particular case study: the effort to migrate the database system supporting *L-functions and Modular Forms Database* (LMFDB) [29] from MongoDB [31] to PostgreSQL [38]

We hope that this article will be useful to mathematicians aspiring to create or improve their own databases, to database engineers who are considering switching from MongoDB to PostgreSQL, and to mathematicians participating in large software projects for whom some of the LMFDB's lessons might prove useful.

We begin with a survey of existing mathematical databases to highlight the diversity of approaches used to disseminate mathematical data. While some focus on communicating theorems and propositions, we pay more attention to those which center on examples. Among these, we observe a range of search functionality.

E. Costa · D. Roe (✉)
Department of Mathematics, Massachusetts Institute of Technology, Cambridge, MA, USA
e-mail: edgarc@mit.edu; roed@mit.edu

J. S. Balakrishnan et al. (eds.), *Arithmetic Geometry, Number Theory, and Computation*, Simons Symposia, https://doi.org/10.1007/978-3-030-80914-0_8

Since the LMFDB prioritizes the ability to search by properties of each object, we hope that our experience working with MongoDB and PostgreSQL will help other mathematical resources improve their capabilities in this area.

Next, we consider some of the important choices involved in setting up and maintaining such a database. We divide this discussion into three parts: how to encode and manipulate the data, how the user interacts with the database, and where it is physically stored. In each category, we present several options and a brief analysis of their pros and cons.

The technical core of the article is the comparison of MongoDB and PostgreSQL, along with our experience in switching between them. We were motivated to change systems primarily for performance reasons: many searches that we wanted to support stalled on common inputs. We did not see a way to resolve the issue within MongoDB. Several missing features in MongoDB also pushed us toward shifting: support for arbitrary precision integers and transactions that allow multiple commands to be rolled back if there is a failure.[1] We summarize the major differences, including the distinction between a relational databasc and a document-oriented database, and describe some of the features of each that were important in different stages of the LMFDB's history. We also outline our process for both porting the data and updating the codebase to use a different system.

We finish by highlighting some of the benefits provided by performing such a major structural change. For example, the process of rewriting a large amount of code allowed us to refactor and encapsulate many of the LMFDB's core functions. Moreover, these code changes enabled us to dramatically improve performance and add new features, such as verification suites that check for consistency within the database. Going forward, the ability to make connections between different kinds of mathematics is one of the core aims of the LMFDB, and a relational database will help us achieve this goal.

2 Using Databases in Mathematics

In order to put our work on the LMFDB in context, we give an overview of some other existing mathematical databases.

We will be primarily interested in datasets consisting of information on mathematical objects, such as specific number fields, graphs or curves, rather than on mathematical statements. There are many valuable resources focusing on theorems and propositions, from general encyclopedias such as the Encyclopedia of Mathematics [16] and Mathworld [48] to the preprint server arXiv [5] to numerous domain specific references [33, 14, 1]. However, the design imperatives for such projects favor a different type of database than the ones we are interested in. Since the content of theorem-oriented sites is likely to be textual, the standard tools for

[1] MongoDB 4.0 added support for transactions but was released after we switched to PostgreSQL.

internet searching such as Google tend to work effectively. Moreover, because humans are directly creating the content, the tools for inputting and managing data are different. Wiki [23, 1, 33, 6] and blog [3] software is often used to make it easier for many users to be involved in content creation. Finally, when algorithms are generating data, it is possible to end up with orders of magnitude more content, leading to various challenges of scale not faced by human-created datasets.

One of the main roles that a mathematical database plays is to record the results of interesting computations. Such collections of examples are useful in many areas of mathematics, including group theory [49, 7, 22, 15], Lie theory [2, 8], graph theory [10], knot theory [28], integer sequences [44], Euclidean geometry [26], algebraic geometry [3] and number theory [11, 29]. In each case, much of the utility of the database is shaped by how the results are displayed to a user, whether as a list of integers with annotations, a picture of a knot or a webpage showing various invariants of the object. Some databases [30, 7, 22] are embedded within computer algebra systems such as Sagemath [42], Magma [9], or GAP [18], which allows a user to perform further computations with the objects in the database easily. Others are available online in various formats; we will discuss different methods of data distribution in Sect. 3.

Collecting a large number of examples into one database makes possible a valuable paradigm for experimental mathematics: the ability to search through examples for a few that have particular specific desired properties or are counterexamples to a conjecture. In addition to the LMFDB, a number of other mathematical databases also have search functionality as one of their core features [8, 41, 44, 37, 27, 10, 28]. Searching requires a more substantial infrastructure than is needed for just displaying data; the discussion in Sect. 4 highlights some of the factors to consider when choosing a search engine.

A final role played by many online mathematical databases is to provide domain-specific exposition surrounding the examples that they host [11, 14, 1, 23, 41, 28, 15]. The LMFDB aims to provide enough definitions and background for mathematically literate users to be able to understand the contents. The main tool for doing so is a webpage element called a *knowl*, which is a bit of context-free knowledge. Within the LMFDB, knowls are embedded inside webpages or other knowls, and expand to provide more information when clicked. Such exposition is an important part of the presentation of a database, but is peripheral to our main topic of the functioning of the underlying system.

Another important aspect of a database project that falls beyond this article's scope is the social and financial efforts required to sustain a collaboration. There are numerous examples of projects that are no longer accessible because the authors have moved on or passed away. Projects have various ways of reaching out to potential developers, ranging from academic workshops and conferences, to the standard open source tools such as GitHub [19], to forms allowing users to easily submit data [41, 10, 44]. In this article, we limit our attention to the technical side of the job: from setting up hosting servers, to the different options for database software, to details on how the LMFDB functions.

2.1 The LMFDB Project

The Langlands program, first formulated by Robert Langlands in the 1960s, is a set of widespread conjectures aimed at understanding and explaining the interconnections between a dizzying array of subfields of mathematics, including number theory, representation theory, algebraic geometry, and harmonic analysis—and in the twenty-first century its reach continues to expand. The Langlands program has been called "one of the biggest ideas to come out of mathematics in the last 50 years" and "the Grand Unified Theory of mathematics" [17, p. 3].

To provide compelling visual and computational displays of the Langlands program "in action," a database was created called the *L-functions and Modular Forms Database* (LMFDB), available at the website https://www.lmfdb.org/. The LMFDB was first conceived in 2007 [12] and remains the object of a significant amount of ongoing work by over one hundred mathematicians [29, Acknowledgments].

The LMFDB hosts a variety of databases, including some predating the project [13, 24, 25], and connects them through the Langlands program; we recommend online experimentation or [12] for more details.

3 Nuts and Bolts

When it comes to hosting a mathematical database, there are a wide range of options available. Until the last several decades, the only choice was to use a print format, for example in a book or an article, and rely on a publisher to distribute its content and to make it available to the rest of scientific community. Since the publisher takes care of most of the technical issues, this method offers the benefit of simplicity to the author. Still, distribution in print provides several disadvantages compared to digital distribution via the internet. For instance, the act of publishing through print is a lengthy process, and the data is less immediately accessible than online distribution, putting barriers in the way of the database user. Furthermore, once the data is published, it is challenging to address mistakes, add new data or features. Finally, it can only accommodate small databases, due to the physical requirements to print.

On the other hand, given that the digital distribution of mathematical data is not taken on by publishers, authors have many more choices to make. We break down these choices into the following related categories:

- back end—How to encode and manipulate the database in a machine readable format.
- front end—How the user interacts with the database.
- hosting—Where to physically store the database.

We will overview some of the possible options in each of these categories

3.1 Back End

We organize the options for how to encode the database in a machine readable format by how easy it is to search for content.

Human Readable Files Storing the data set in a human readable format, like a plain text file or a comma-separated values file, is one of the most straightforward possible solutions. It does not require special software to access the data, very similar to a table in print. However, it can be cumbersome and impractical to search or manipulate the data without additional tools. Note that simple searches are possible using tools like grep or the find feature of a text editor. However, these methods do not offer the full search capability of a database.

Application Dependent Files A more advanced solution is to store the dataset in an application dependent format. For example, the Small Groups Library [7] available in the GAP computer algebra system [18] uses a specific method to encode solvable groups. In many cases, such a choice enables direct manipulation of the desired objects. For example, using SageMath pickles makes it easy to load the data into an active SageMath session. However, the dataset will not likely be efficiently searchable, searching for objects with specific properties will usually require to loop over every item in the dataset (a sequential search). Moreover, specialized data formats are vulnerable to backward-incompatible changes in the software used to load them and are more difficult to maintain in the long run.

Database Structure Finally, one may store the dataset in a database format, i.e. in a format optimized for searching. As above, such a solution will require additional software for users to manipulate the data, but the data formats are more stable in practice since database software is widely used beyond mathematics. There are many types of database available, each optimized with different applications; in this article, we will focus on two:

- relational database—a collection of ledger-style tables with a fixed schema, where the structure of each row is constant. This kind of database has been prevalent for decades; major examples include Oracle [35], MySQL [34], PostgreSQL [38], and SQLite [21].
- document oriented database—a schema free database, where each object is stored in a single instance in the database, and every stored object can be different from every other. MongoDB [31] is one of the most famous instantiations of this paradigm.

In Sect. 4, we will contrast PostgreSQL and MongoDB.

3.2 Front End

There are many choices for how to enable a user to interact with data. Different options have varying degrees of accessibility and utility; we describe three possibilities below.

Minimal The author may decide to provide direct access to the database. For example, the author may make the database files available through email or a webpage; or allow direct read access to a SQL server [8].

Computer Algebra System If the database is provided in a computer algebra system dependent files, these might evolve into being part of algebra system itself, either as a standard or an optional package. For example, the Small Groups Library [7] is available in GAP and the K3 Database in Magma [30]. This option is extremely convenient for users of those software systems but limits their availability to others.

Static Website The user may also interact with the database through a collection of web pages displaying the data. These web pages can range widely in complexity. For example, one can manually curate each page individually [23, 3] or automatically generate them as static web pages [26, 14, 45, 28, 15]. Such webpages can provide a very usable interface, but their main weakness is that the html code for displaying the data must be duplicated for each object presented, introducing a large overhead that becomes prohibitive as the size of the database grows.

Dynamic Website Alternatively, one can instead use a web application to generate each page when it is accessed. This is essential for large datasets, as generating and storing a static web page for each object can be infeasible. The LMFDB follows such an approach, where the web page displaying the content is generated on the fly through a dedicated web application based on Flask [40] and HTML templates, and only a small amount of the content is human-curated. Other examples of mathematical databases using such an approach include [6, 10]. The main downside of this approach is a higher level of complexity when compared to the static option.

3.3 Hosting

Finally, we discuss some of the options for where to physically host the front end and the database. We focus on how each possibility varies in terms of cost, ease of maintenance and scalability.

Personal Hard-Drive This is one of the simplest options, as it has no initial set up cost. However, it severely limits the database's accessibility to others due to the lack of a front end.

Personal Web Page If one already maintains a personal web page, a convenient option is to also use it to host the database and the front end. However, the scope might be limited, as many universities will only allow their users to host static content.

Web Framework Another convenient option is to use a web framework service (e.g. GitHub, CoCalc, Wordpress/Wikidot/blogs). For example, some people might dump their database into GitHub, while others might also use it to host the front end. These options offer a lower initial set up cost with a minimal long term maintenance cost, however, the scope is also limited by the framework, since they control what software is used to host content.

On a University/Personal Server This solution provides complete control. However, there is both a high initial cost in buying the servers and setting them up. Moreover, one needs to maintain the server for the duration of its life: installing security updates, dealing with service interruptions, and managing backups. In addition, the database's scope might be limited by the capabilities of the initial hardware choices. The LMFDB uses a couple of servers purchased through grants to provide a development platform for our contributors, and host the back end and the front end for beta.lmfdb.org

Using a Hosting Service Servers in the cloud [4, 20] provide complete control and have no monetary startup costs, but require ongoing payments proportional to the resources used. The main benefit is that one does not need to worry about maintaining the servers themselves; cloud servers also have the ability to easily scale the resources available if necessary. This option is currently used by the LMFDB to host www.lmfdb.org.

4 MongoDB vs PostgreSQL

In 2009, when MongoDB was initially released as an open source project, it presented itself as an exciting new option to easily store mathematical objects. As a document-oriented database, where each document was stored independently in a JSON-like format, it did not require a schema, a boon when creating new collections of mathematical objects where the data requirements only gradually became clear. Moreover, Python bindings were available, a language that many computational number theorists were familiar with due to its use in SageMath. This setup worked well for many years; however, as the LMFDB grew, various shortcomings of using a non-relational database became more evident. In March 2018, we decided to transition from MongoDB to PostgreSQL, a popular open source implementation of the SQL language that grew out of earlier projects founded in the 1970s and 1980s. As a relational database, PostgreSQL stores data in a structured way, where the data is organized in tables, and each table has a specified schema, i.e., each row must have the same layout, with the types of columns constant across all rows.

4.1 Comparing Two Database Systems

MongoDB and PostgreSQL offer two different database paradigms. MongoDB, as a document-oriented database, is meant to facilitate the storage of unstructured data, where the fields and types present in each document can vary within a collection. This flexibility is extremely convenient since the data structure can organically evolve as the project develops. However, it can also easily lead to errors and unnecessary overhead for a structured database. For example, field names were kept short in MongoDB to save space, and multiple typos in the field names were found during the transition to PostgreSQL. Furthermore, differing layouts across documents required more complicated processing code. In contrast, PostgreSQL is a relational database with fixed schemas that enable queries across multiple tables. Since the LMFDB was already a uniform set of fields in MongoDB (so much so that an inventory application was written to extract a schema from the data), creating a formal schema was fairly straightforward. Database-level schema support improves robustness and performance, and PostgreSQL's JSONB type allows for schemaless columns when desired.

One benefit of switching has been that storing the data in PostgreSQL requires substantially less space than in MongoDB. For example, when converting our biggest MongoDB collection `Lfunctions.Lfunctions` to the PostgreSQL table `lfunc_lfunctions`, we observed a space savings of about 42%, going from 194 GiB to 113 GiB. These savings translate both to monetary savings (we were spending a total of about $2000 per year paying for storage space using MongoDB) and to query speed when a sequential search is performed.

Two main factors contribute to the smaller storage requirements. First, we no longer need to (repeatedly) store the name of each field in every document; this was a MongoDB requirement, as fields in a document can vary across a collection. For some of the LMFDB's collections, these names accounted for a substantial fraction of the storage requirement. In addition to the space benefits, the switch to PostgreSQL also relaxes the pressure to minimize the lengths of these names, improving readability.

The second factor contributing to space savings is PostgreSQL's support for arbitrary precision integers. As a mathematical database, the LMFDB frequently uses large integers, which had to be stored as strings in MongoDB. In addition to the storage consequences, various workarounds were required to sort and search the data correctly. The elimination of these hacks has simplified some of the supporting code.

While smaller tables help with query performance, indexes provide a more powerful tool for speeding up searches. Indexes facilitate the location of records with constraints on a specified set of columns by storing additional data. Having such indexes available is key to providing quick search results for typical queries. For example, both MongoDB and PostgreSQL support indexes based on binary trees, which work well for totally ordered data such as integers and strings. However, there is another query type that is not well supported by binary trees: given a column

Table 1 Times for counting the number fields K satisfying the given constraint on set of ramified primes

Query	Result/total	MongoDB[a]	PostgreSQL
$\#\{K : 5 \notin R_K\}$	71%	6h 9min	2.93s
$\#\{K : \{2, 5\} \subset R_K\}$	18%	10min 2s	6.9s
$\#\{K : R_K \subset \{2, 3, 5\}\}$	4%	14min 34s	3.42s
$\#\{K : R_K \subset \{2, 3, 73\}\}$	0.1%	8min 11s	170 ms

[a]We used WiredTiger for comparisons since it outperforms MMAP when using normal amounts of RAM.

Table 2 Times for fetching the objects/rows satisfying the given constraint

Query	MongoDB	PostgreSQL
$\{K : 5 \notin R_K, \lvert \mathrm{disc}_K \rvert \leq 2000\}$	1.95s	0.1s
$\{E : \#\text{Ш} = 4, \text{rank } E = 1, \#E_{\text{torsion}} = 1\}$	1min 58s	1.95s

representing a set of integers, search for rows that contain a particular set, or are contained in a particular set. For example, we store the set of ramified primes R_K for number fields K, and we want to be able to search on number fields by specifying such constraints on R_K. Even though we supported such searches through our web interface, many such searches took several minutes to perform within MongoDB, which caused our web front end to timeout. In PostgreSQL, most of these queries finish in a couple of seconds, both when the number of results is not large enough to justify using an index (the first three rows of Table 1), and when generalized inverted indexes (GIN) [43] apply (the last row of Table 1).

The significant improvement in indexless queries is another considerable advantage of the transition to PostgreSQL. Such queries are relevant, since non-typical queries can be the most interesting mathematically, and since it is impossible to construct all indexes in advance because the LMFDB offers the user a lot of freedom to select search parameters. We present two such examples in Table 2 (the first for number fields and the second for elliptic curves), where the number of results is about 1000 in each case. We also note that successful queries are instantaneous on a second run, as both databases cache the results.

While search performance is vital for website usability, the process of uploading data and creating indexes is also essential to make the development sustainable and to aid in the implementation of new features. PostgreSQL's `COPY FROM` command allows for the rapid importing of data from plain-text files without a Python intermediary, which is useful when the data is generated by some other system such as Magma [9]. For example, we imported a 14,398,359 row table from a 680 GB text file in about 8 h. With MongoDB we estimate that would have taken us several days. Creating indexes in PostgreSQL is also dramatically faster than in MongoDB, approximately by an order of magnitude. Additionally, PostgreSQL offers a more diverse and mature set of open source front ends and administration tools, for example [36, 47], that lower the learning curve for the SQL language. The transition has substantially improved database management.

In addition to performance improvements, PostgreSQL makes new kinds of queries possible. Relational databases allow joining tables, returning values from both tables when they share some specified relationship. The existing schemas, having been designed for MongoDB, do not take advantage of this feature, but new tables on finite groups and on p-adic tori will.

4.2 *Transition*

Once the decision had been made to switch from MongoDB to PostgreSQL, we faced two main tasks: to port the data and to update the LMFDB codebase to use PostgreSQL. The majority of the work was done by the second author between March 2018 and August 2018, with assistance from the first author starting in July.

Since PostgreSQL requires the specification of schemas, the initial step of the data migration involved defining the schema for each table. In practice, most of the tables had a schema already, which were made visible by a custom inventory module within the LMFDB.[2] Unfortunately, the accuracy of these schema varied: some collections had multiple independent kinds of records within the same table, while others had typos in some of the field names. Moreover, the inventory system itself proved to reflect the types of some columns inaccurately, since it inferred the type from finding a single record. For example, it concluded that a specific field was a real number, while some documents had complex values. Eventually, we overcame these issues and created a schema for each PostgreSQL table.

In some cases, there were multiple possible types to choose from for a given column. PostgreSQL 9.4 added support for a binary JSON type, which allows for unstructured data to be stored in a column. It provided an alternative to arrays that could also store an analog of Python's dictionaries, with strings as keys. For simplicity, we initially chose to use this JSONB type instead of arrays, since it simplified some of the supporting code. However, after much of the rest of the transition was complete, we realized that when arrays are possible (namely, when the type is uniform across all entries), they provide some significant advantages. Namely, arrays are more space efficient than the JSONB type, and indexes on arrays support more operations than indexes on a JSONB column. We are gradually transitioning JSONB to array columns where useful.

We also took advantage of the transition to change some data types that had been chosen due to the deficiencies in MongoDB. For example, many large integers had been stored as strings since MongoDB lacked an arbitrary precision data type. With the switch to PostgreSQL, we were able to use the numeric type instead, which simplified code used to sort and compute with the results.

[2]The inventory displayed the kinds of documents within each collection, the fields present on each type of document, the indexes present in each collection, and both machine and human generated information on the types of each field. It was a very useful resource for constructing schema.

Since one of the goals of the switch was to improve performance, we made some changes to the schemas to help speed up queries. In particular, we split some of the largest tables in half, putting columns that were used for searching into a search table and columns that were only used for display into an *extras table*. This significantly reduced the search table's size, improved index efficiency, and sped up indexless queries that had to traverse the whole table. Query performance is also affected by sort order, and the default sorting on many tables uses four or five columns. We added a primary key column to each table, and for large tables we ordered this primary key by the table's standard order. This feature decreased the size of indexes and improved their performance since search results could be ordered by a single numeric column rather than a combination of many columns. This optimization was feasible because the data in the LMFDB changes infrequently, and because each search page comes with a default order.

With the schemas created and export scripts written, the actual changeover of the data went reasonably smoothly. PostgreSQL provides a COPY FROM command to load data from a text file, which runs very quickly. After several false starts, exporting 400 GB of data from MongoDB to text files took three or 4 days; copying it between servers on opposite sides of the Atlantic took less than a day, and loading it into PostgreSQL took less than a day.

Adapting the LMFDB codebase to use PostgreSQL took longer. The backend code for the LMFDB is written in Python, and the Python bindings for MongoDB and PostgreSQL take a very different approach. PyMongo [32] offers a high level interface to the database that integrates python data structures. Queries are performed by constructing dictionaries that specify the values of certain columns and have special keys for adding more complicated constraints such as inequalities or boolean combinations of conditions. Psycopg [46], on the other hand, is a thin interface that offers the user the ability to execute SQL statements. Constructing queries using dictionaries worked well as a model for the LMFDB, since the search pages provided many inputs that could be processed independently into a dictionary. Moreover, we did not want to require LMFDB developers to learn SQL. Therefore, we implemented a high-level interface to PostgreSQL that translated a dictionary-based query system similar to PyMongo's into SQL SELECT statements while avoiding SQL injection attacks. This query interface has already proved useful in rapidly prototyping another project, researchseminars.org, and we plan to make it available as a stand-alone tool.

5 Benefits

5.1 Abstraction

As the LMFDB has grown, the most common paradigm for adding new features has been to take a functional section of the website, copy it to a new folder, and modify the templates and backend code. This approach has the benefit of easily producing valid code since it is iteratively making small changes to an already functional webpage. However, it leads to large quantities of repeated code, which makes fixing bugs and adding features much harder, since the same change must be applied in dozens of locations. The standard solution is to encapsulate tasks into functions that can then be used repeatedly and changed as needed. Of course, the LMFDB has used functions like this since the beginning, but often these functions are not written in a way that allows them to be used in multiple sections of the website.

The task of switching from MongoDB to PostgreSQL was made more difficult by this low level of abstraction, but it also presented an opportunity. Since the database change necessitated changing a large number of files throughout the project, it offered a perfect time to make additional changes. We highlight three in particular to illustrate the kinds of encapsulation that accompanied the database change: a new data-management interface, more robust and uniform handling of search inputs, and common tools for creating statistics pages.

We centralized many of the scripts used to upload data into a suitc of data management tools. There is now a single object available from the Sage command line that has methods for performing a variety of tasks: creating new tables and modifying existing schemas, copying data to and from text files, moving data between the beta and production versions of the database, accessing PostgreSQL's tools for analyzing slow queries, adding indexes and constraints, and tracking statistics as data is changed. The centralization of these tasks allows us to reduce code duplication and use best practices to handle errors.

Each section of the LMFDB has a search page, where users can input constraints they would like their curve, field or group to satisfy. Some input boxes require an integer or list of integers; others need a label string in a particular format; others a rational number. The processing code must process the user provided text, raising appropriate errors on invalid input, and transform the set of inputs into a database query. Before the database change, we had already begun the process of encapsulating this task. Each input box corresponds to a specific field or column in the database, and there is a processing function for each type of input (list of integers, floating point, etc.). As we switched each part of the LMFDB to PostgreSQL, we ensured that it was using these functions.

Once these queries have been constructed, we need to execute the search and pass the results to the Jinja [39] template that actually creates the webpage seen by the user. Various parts of this task are shared among all sections of the LMFDB. For example, we have to be able to handle errors in user input, errors or timeouts in the execution of the query, and jump boxes that allow a user to go straight to an object with a particular label. As part of the switch to PostgreSQL, we created a *search wrapper* that performs all of these jobs, and adopted it across the LMFDB. In addition to simplifying the code overall, this change made the user experience more uniform and allowed for the creation of new features, which will be described in the next section.

Many parts of the LMFDB had a statistics page prior to the database change. These pages described the number of objects with certain attributes; for example, a table of how many number fields were available by signature, or a table giving the fraction of elliptic curves with a specified torsion structure. Along with the database change, we created a class to support the database queries required for these statistics, and a common template to display them nicely. This change has the benefit that it automates the process for updating these counts when developers add data. It also makes it easier to create these statistics pages for more sections of the LMFDB, and makes it easier to implement new features such as user-requested statistics.

5.2 New Features

In addition to the intrinsic benefits of using PostgreSQL, the transition has eased the implementation of a number of new features. While many of these would have been possible using MongoDB, the improved abstraction made implementing them much more manageable. And some, such as dynamic statistics and verifications, rely on PostgreSQL's improved performance and the relational model.

PostgreSQL supports transactions, allowing one to roll back to a functional earlier state if some later action fails. We have built these transactions into our interface layer, making it possible to safely modify rows in place, rather than creating a copy of the whole table when changes are made.

The fact that all interactions with the database pass through our interface layer allows us to add custom logging. For example, all queries that take longer than a threshold are recorded, allowing us to focus on creating the right indexes. We also log high-level actions, such as adding new data to a table or changing a schema, along with a username to provide a record of what changes have been made to the data. These logs help us ensure the quality of our data.

The statistics infrastructure has enabled a new user interface which we refer to as *dynamic statistics*. Rather than presenting only a pre-selected set of views, the user can specify which variables they are interested in and the objects' constraints. The system then generates a display describing the possible values of those variables and the number of objects possessing each possible set of values. For example, a user can create a table to view how the weight and level of modular forms vary among forms with complex multiplication. As of June 2020 we are still in the process of enabling this feature throughout the LMFDB. We hope that this feature will provide researchers with a new method of interacting with the LMFDB, allowing them to look for large scale patterns in these objects in addition to searching for objects with particular properties.

Many browse pages in the LMFDB already had a link to view a random object. Because of the new search wrapper abstraction, it was easy to add the ability to return a random object satisfying a search query. This feature saves time when browsing since it allows you to more quickly reach the homepage of an object satisfying a particular constraint, such as CM elliptic curve or a weight one modular form.

During the transition to PostgreSQL, we modified the search results pages so that they do not always provide the total count. The count is provided only if it is sufficiently small or has already been cached (though it is available in every case if requested). This change dramatically improved response times in some cases, where a large number of results makes finding the first few much easier than counting them all.

Mathematical data differ from data in most other applications because it has a notion of correctness that can be checked, rather than just providing a record of a real-world measurement. Some of these verifications are internal, such as checking that the defining polynomial of a number field is irreducible or that the sum of the complex embeddings of classical modular matches the trace of the exact newform. Others rely on connections with other tables in the database, such as checking that invariants match up between an elliptic curve and its corresponding modular form. These kinds of verifications across different tables are much faster using PostgreSQL, since the relational database model is explicitly designed to search across different tables. The switch to PostgreSQL made it possible to create a framework to write such consistency checks and run them whenever new data is added.

We are excited about the new features that we added to the LMFDB and we believe that the transition to PostgreSQL will aid with the long term maintenance of the project. We look forward to how the abstraction layer we have created around database interactions will help standardize the user experience across LMFDB. We hope others will learn from our experience in their efforts to leverage computation and data towards the study of mathematics.

References

1. Scott Aaronson, Greg Kuperberg, and Christopher Granade, *Complexity Zoo*, 2005–2019. https://complexityzoo.uwaterloo.ca.
2. Jeffrey Adams, Annegret Paul, Ran Tui, Susana Salamanca-Riba, Peter Trapa, Marc van Leeuwen, and David Vogan, *Atlas of Lie Groups and Representations*, 2002–2019. http://www.liegroups.org.
3. Mohammad Akhtar, Tom Coates, Alessio Corti, Sergey Galkin, Vasily Golyshev, and Alexander M. Kasprzyk, *Fano Varieties and Extremal Laurent Polynomials*, 2010–2012. http://fanosearch.net.
4. *Amazon Web Services*, Amazon.com, Inc, 2006–2019. https://aws.amazon.com/.
5. *arXiv*, Cornell University, 1991–2019. https://arxiv.org/.
6. Dror Bar-Natan and Scott Morrison, *The Knot Atlas*, 2006–2019. http://katlas.org.
7. Hans Ulrich Besche, Bettina Eick, and E. A. O'Brien, *The groups of order at most 2000*, Electron. Res. Announc. of the AMS **7** (2001), 1–4.
8. Birne Binegar, *Unipotent Muffin Research Kitchen*, 2009–2019. http://lie.math.okstate.edu/UMRK/UMRK.html.
9. Wieb Bosma, John Cannon, and Catherine Playoust, *The Magma algebra system. I. The user language*, J. Symbolic Comput. **24** (1997), no. 3-4, 235–265. Computational algebra and number theory (London, 1993). MR1484478

10. G. Brinkmann, K. Coolsaet, J. Goedgebeur, and H. Mélot, *House of Graphs: a database of interesting graphs*, Discrete Applied Mathematics **161** (2013), 311–314. http://hog.grinvin.org.
11. Chris Caldwell, *The Prime Pages*, 1994–2019. https://primes.utm.edu.
12. John Cremona, *The L-functions and modular forms database project*, Found. Comput. Math. **16** (2016), no. 6, 1541–1553. MR3579716.
13. John E. Cremona, *Algorithms for modular elliptic curves*, 2nd ed., Cambridge Univ. Press, Cambridge, 1997.
14. H.N. de Ridder et al., *Information System on Graph Classes and their Inclusions*, 2001–2014. http://graphclasses.org.
15. Tim Dokchitser, *GroupNames*, 2017–2019. http://groupnames.org.
16. *Encyclopedia of Mathematics*, Kluwer Academic Publishers, 2002–2012. https://www.encyclopediaofmath.org.
17. Edward Frenkel, *Love and math: The heart of hidden reality*, Basic Books, 2013.
18. *GAP – Groups, Algorithms, and Programming*, The GAP Group, 2019. https://www.gap-system.org.
19. *GitHub*, GitHub, Inc., 2008–2019. https://github.com.
20. *Google Cloud Platform*, Google Inc., 2008–2019. https://cloud.google.com/.
21. Richard Hipp et. al., *SQLite*, 2000–2019. https://www.sqlite.org.
22. Alexander Hulpke, *Constructing transitive permutation groups*, Journal of Symbolic Computation **39** (2005), no. 1, 1–30.
23. Joel David Hamkins, Victoria Gitman, et al., *Cantor's Attic*, 2011–2019. http://cantorsattic.info.
24. John Jones and David Roberts, *A database of local fields*, Journal of Symbolic Computation **41** (2006), no. 1, 80–97.
25. ——, *A database of number fields*, LMS J. Comput. Math. **17** (2014), no. 1, 595–618.
26. Clark Kimberling, *Encyclopedia of Triangle Centers*, 1999–2019. https://faculty.evansville.edu/ck6/encyclopedia/ETC.html.
27. Maximilian Kreuzer and Harald Skarke, *Calabi-Yau data*, 2000–2019. http://hep.itp.tuwien.ac.at/~kreuzer/CY/.
28. Charles Livingston and Jae Choon Cha, *KnotInfo: Table of Knot Invariants*, 2019. https://www.indiana.edu/~knotinfo/.
29. *The L-functions and modular forms database*, The LMFDB Collaboration, 2009–2019. http://www.lmfdb.org.
30. *Magma databases*, Magma, 2001–2018. http://magma.maths.usyd.edu.au/magma/download/db/.
31. *MongoDB*, MongoDB, Inc, 2009–2019. https://www.mongodb.com/.
32. *PyMongo*, MongoDB, Inc, 2009–2019. https://api.mongodb.com/python/current/.
33. Vipul Naik, *Groupprops, The Group Properties Wiki*, 2008–2019. https://groupprops.subwiki.org.
34. *MySQL*, Oracle Corporation, 1995–2019. https://www.mysql.com/.
35. *Oracle Database*, Oracle Corporation, 1979–2019. https://www.oracle.com/database.
36. *pgAdmin4*, The pgAdmin Development Team, 2019. https://www.pgadmin.org.
37. Simon Plouffe, Adam Van Tuyl, Paul Irvine, Loki Jörgenson, David H. Bailey, Peter Borwein, Jonathan Borwein, and Yasumasa Kanada, *The Inverse Symbolic Calculator*, 1986–2019. http://wayback.cecm.sfu.ca/projects/ISC.
38. *PostgreSQL*, The PostgreSQL Global Development Group, 1996–2019. https://www.postgresql.org/.
39. Armin Ronacher, et. al., *Jinja 2*, 2008–2019. http://jinja.pocoo.org/.
40. ——, *Flask: a Python microframework*, 2010–2019. http://flask.pocoo.org/.
41. Martin Rubey, Christian Stump, et al., *FindStat – The combinatorial statistics database*, 2013–2019. http://www.FindStat.org.
42. *Sagemath, the Sage Mathematics Software System*, The Sage Developers, 2019. http://www.sagemath.org.

43. Teodor Sigaev and Oleg Bartunov, *Gin for postgresql*, 2008–2019. http://www.sai.msu.su/~megera/wiki/Gin.
44. N. J. A. Sloane et al., *The On-Line Encyclopedia of Integer Sequences*, 1964–2019. http://oeis.org.
45. *The syzygy database*, The SyzygyData Group, 2017. http://www.syzygydata.com.
46. Daniele Varrazzo, *Psycopg*, 2000–2019. http://initd.org/psycopg/.
47. Jakub Vrána, *Adminer*, 2019. https://www.adminer.org.
48. Eric Weisstein, *Wolfram Mathworld*, 1995–2019. http://mathworld.wolfram.com/.
49. Robert Wilson, Peter Walsh, Jonathan Tripp, Ibrahim Suleiman, Richard Parker, Simon Norton, Simon Nickerson, Steve Linton, John Bray, and Rachel Abbott, *ATLAS of Finite Group Representations*, 1985–2019. http://brauer.maths.qmul.ac.uk/Atlas/v3/.

Effective Obstruction to Lifting Tate Classes from Positive Characteristic

Edgar Costa and Emre Can Sertöz

Abstract We give an algorithm that takes a smooth hypersurface over a number field and computes a p-adic approximation of the obstruction map on the Tate classes of a finite reduction. This gives an upper bound on the "middle Picard number" of the hypersurface. The improvement over existing methods is that our method relies only on a single prime reduction and gives the possibility of cutting down on the dimension of Tate classes by two or more. The obstruction map comes from p-adic variational Hodge conjecture and we rely on the recent advancement by Bloch–Esnault–Kerz to interpret our bounds.

1 Introduction

Let $X_{\mathbb{C}} := Z(f) \subset \mathbb{P}^{2r+1}_{\mathbb{C}}$ be a smooth complex hypersurface of even dimension $2r$ with the polynomial f having algebraic coefficients. The dimension $\rho^r(X_{\mathbb{C}})$ of the space of codimension r algebraic cohomology classes $A^r(X_{\mathbb{C}}) \otimes_{\mathbb{Z}} \mathbb{Q}$ in $\mathrm{H}^{2r}_{\mathrm{Betti}}(X_{\mathbb{C}}, \mathbb{Q})$ is called the *middle Picard number* of $X_{\mathbb{C}}$. Due to the Lefschetz hyperplane theorem, there are no interesting algebraic cycles in codimension different from r. The middle Picard number is a difficult invariant to compute, but it strongly constrains the geometry, arithmetic, and dynamics of the hypersurface. Recently, there have been numerous developments on computing the middle Picard number and we review them in Sect. 1.5.

The main contributions of this paper are Algorithm 4.1 and Proposition 2.22. They give practical means for computing upper bounds for $\rho^r(X_{\mathbb{C}})$ starting with the defining equation f. The key idea is to take a prime reduction of the hypersurface and compute the obstruction to lifting algebraic (or Tate) classes of

E. Costa (✉)
Department of Mathematics, Massachusetts Institute of Technology, Cambridge, MA, USA
e-mail: edgarc@mit.edu

E. C. Sertöz
Max Planck Institute for Mathematics in the Sciences, Leipzig, Germany

J. S. Balakrishnan et al. (eds.), *Arithmetic Geometry, Number Theory, and Computation*, Simons Symposia, https://doi.org/10.1007/978-3-030-80914-0_9

the reduction to characteristic zero. We give an implementation[1] of this algorithm in SageMath [Sag20]. In Sect. 5, this implementation is used to illustrate the strengths and weaknesses of the method. To the best of our knowledge, our method gives a superior performance over any other method that has a computer implementation.

There is a practical method, due to van Luijk [vLui07], for bounding the Picard number of K3 surfaces, which has now become the standard. When the K3 surface X is defined over a number field, the Picard numbers of its prime reductions give upper bounds on the Picard number $\rho^1(X_{\mathbb{C}})$. The method of van Luijk can improve these bounds by at most one, via a comparison of two different reductions. However, some K3 surfaces with real multiplication do not admit a prime reduction where the Picard number jump is less than two [Cha14].

Our approach is different, and we do not suffer from these limitations. For instance, we do not need two distinct prime reductions to improve the bound. We also see in Example 5.9 that our method does not necessarily break down in the presence of real multiplication. Additionally, in Example 5.7, we give an example of a K3 surface, whose Picard number jumps by six at a prime reduction, and show that our method detects this jump.

Let us recall that for simply connected surfaces such as K3s, the Picard group and its image in integral cohomology, namely the Néron–Severi group, coincide. These are free abelian groups whose rank equals the Picard number.

In Example 5.5, we go through the 184,725 smooth quartic K3 surfaces in the database of [LS19] to verify computations and benchmark performance. In Sect. 5, we give many more examples, involving Picard numbers of quintic surfaces and endomorphisms of Jacobian threefolds.

For the rest of the paper we will work with number fields as opposed to the field of complex numbers. In the end we do not lose generality since $\rho^r(X_{\mathbb{C}}) = \rho^r(X_{\overline{\mathbb{Q}}})$, see for instance Lemma 2.5.

1.1 Lifting Algebraic Cycles from Finite Characteristic

We will now sketch the core concept behind our algorithm. To simplify the exposition, let us assume f has rational coefficients so that $X = Z(f) \subset \mathbb{P}^{2r+1}_{\mathbb{Q}}$. For most prime numbers $p \in \mathbb{Z}$ the reduction $X_p := Z(f \bmod p) \subset \mathbb{P}^{2r+1}_{\mathbb{F}_p}$ will be well defined and smooth. Fix one such prime p. The choice of such a prime introduces an infinite sequence of hypersurfaces

$$Z(f \bmod p^n) \subset \mathbb{P}^{2r+1}_{\mathbb{Z}/p^n} \tag{1.1.1}$$

that can be said to approximate X.

[1]The code is available at https://github.com/edgarcosta/crystalline_obstruction.

As we shall soon see, one of the strong points of working with a finite field $\mathbb{F}_p$ is that the determination of algebraic cycles on the variety $X_p/\mathbb{F}_p$ is greatly simplified in comparison to $X/\mathbb{Q}$. After determining algebraic cycles on X_p, one must determine which cycles can be lifted to $Z(f \bmod p^n)$ for all n, and eventually to X.

Hodge theory gives an obstruction to lifting cycles from characteristic p, namely the map (1.1.3) below. The p-adic variational Hodge conjecture states that any unobstructed cycle lifts to X. See [Eme97] and [AMMN20] for more details on this conjecture. Bloch–Esnault–Kerz [BEK14] prove that an intermediate statement is true: an unobstructed cycle can lift compatibly to all the hypersurfaces $Z(f \bmod p^n)$ approximating X. Except when $r = 1$, it is unknown whether this is sufficient for lifting to X.

As an example, suppose X is a K3 surface. Then the obstruction to lifting a curve $C \subset X_p$ can be represented by a p-adic integer $a_1 p + a_2 p^2 + \cdots \in p\mathbb{Z}_p$. The curve C is the restriction of a curve in $Z(f \bmod p^n)$ if and only if $a_1 = a_2 = \cdots = a_{n-1} = 0$, see also [EJ11b].

The crystalline cohomology is the cohomology theory we need in order to compute obstructions of cycles. This theory associates a well behaved cohomology to smooth and proper varieties over finite fields [BO83, Ber74, Cha98]. A remarkable feature of the theory is that the crystalline cohomology of X_p can be identified with the algebraic de Rham cohomology of X with p-adic coefficients (Sect. 3.1),

$$\mathrm{H}^{2r}_{\mathrm{crys}}(X_p/\mathbb{Z}_p) \otimes \mathbb{Q}_p \simeq \mathrm{H}^{2r}_{\mathrm{dR}}(X/\mathbb{Q}) \otimes_{\mathbb{Q}} \mathbb{Q}_p. \tag{1.1.2}$$

The algebraic de Rham cohomology is defined as the hypercohomology of the usual differential complex $\Omega^\bullet_{X/\mathbb{Q}}$, see [Gro66] for the conception of this idea and [HM17, §3] for more details. The resulting cohomology groups have coefficients in the base field and when tensored with real or complex numbers, one recovers the classical de Rham cohomology groups.

The image of the crystalline cycle class map is a group $A^r(X_p) \subset \mathrm{H}^{2r}_{\mathrm{crys}}(X_p/\mathbb{Z}_p)$ representing algebraic cohomology classes of X_p. On the other hand, the cycle class map for de Rham cohomology also gives us the group $A^r(X) \subset \mathrm{H}^{2r}_{\mathrm{dR}}(X/\mathbb{Q})$ of algebraic cycles on X that are defined over $\mathbb{Q}$. Let us denote the $\mathbb{Q}$-spans of $A^r(X)$ and $A^r(X_p)$ with $A^r(X)_{\mathbb{Q}}$ and $A^r(X_p)_{\mathbb{Q}}$ respectively.

There is a containment $\iota\colon A^r(X)_{\mathbb{Q}} \hookrightarrow A^r(X_p)_{\mathbb{Q}}$ obtained by reducing subvarieties of X modulo p, see Sect. 2.1.4. Moreover, comparing with complex Hodge theory one sees that $A^r(X)$ must be contained in the r-th piece of the Hodge filtration $\mathrm{F}^r\mathrm{H}^{2r}_{\mathrm{crys}}(X_p/\mathbb{Z}_p)$ on $\mathrm{H}^{2r}_{\mathrm{crys}}(X_p/\mathbb{Z}_p)$. Consider the quotient map

$$\pi : \mathrm{H}^{2r}_{\mathrm{crys}}(X_p/\mathbb{Z}_p) \to \mathrm{H}^{2r}_{\mathrm{crys}}(X_p/\mathbb{Z}_p)/\mathrm{F}^r\mathrm{H}^{2r}_{\mathrm{crys}}(X_p/\mathbb{Z}_p) \tag{1.1.3}$$

and note that $\pi(A^r(X)) = 0$. The theorem of Bloch–Esnault–Kerz (Theorem 2.7) states that any element in the kernel of

$$\pi_A := \pi|_{A^r(X_p)} \tag{1.1.4}$$

lifts to a compatible sequence of algebraic cycles on $Z(f \bmod p^n)$ for each $n \geq 1$. It is unknown whether every element in $\ker \pi_A$ comes from an element in $A^r(X)$, see *loc. cit.* However, when $r = 1$ we do have $A^1(X) = \ker \pi_A$ due to [Gro63, Theorem 5.1.4].

See Sect. 2.1.6 for more details on the obstruction map π_A. We also recommend the introduction of the paper [BEK14] for the intuition it provides in interpreting their main result.

1.2 Tate Classes As a Substitute for Algebraic Classes

Tate's conjecture gives a hold on the space of algebraic cohomology classes $A^r(X_p)$. The Frobenius action on X_p induces an action on crystalline cohomology by functoriality and, therefore, a map Frob: $\mathrm{H}^{2r}_{\mathrm{crys}}(X_p/\mathbb{Z}_p) \to \mathrm{H}^{2r}_{\mathrm{crys}}(X_p/\mathbb{Z}_p)$. Tate's conjecture states that the $\mathbb{Q}_p$-span of $A^r(X_p)$ coincides with p^r-eigenspace of Frob.

Write $T^r(X_p) := \ker(\mathrm{Frob} - p^r)$. The elements of $T^r(X_p)$ are called Tate classes. The inclusion $A^r(X_p) \subset T^r(X_p)$ is always true and $A^r(X_p)_{\mathbb{Q}} \otimes_{\mathbb{Q}} \mathbb{Q}_p \simeq T^r(X_p)$ is the Tate conjecture. This conjecture is known to hold for K3 surfaces [Cha13, Mad15, KM16].

Consider the restriction of the quotient map (1.1.3) to the space of Tate classes,

$$\pi_T := \pi|_{T^r(X_p)}. \tag{1.2.1}$$

By the discussion above, $A^r(X)$ is contained in $\ker \pi_T$ and one can deduce the inequality

$$\dim_{\mathbb{Q}} A^r(X)_{\mathbb{Q}} \leq \dim_{\mathbb{Q}_p} \ker \pi_T. \tag{1.2.2}$$

Our Algorithm 4.1 describes a practical method to bound $\dim_{\mathbb{Q}_p} \ker \pi_T$. We should point out that the codomain of π_T is typically small when viewed as a $\mathbb{Q}_p$-vector space but infinite dimensional as a $\mathbb{Q}$-vector space. If X is a K3 surface then the codomain of π_T is 1 dimensional. Therefore, as it stands, the bound (1.2.2) can be underwhelming. Ideally, one should seek to recover a $\mathbb{Q}$-structure on $T^r(X_p)$ containing $A^r(X_p)$ to leverage the fact that $\dim_{\mathbb{Q}} \mathbb{Q}_p = \infty$. In this way, the codomain of π_T would have the capacity to obstruct arbitrarily many classes. We do not take this approach in this paper, but we say more about it in Sect. 1.4.

Instead, we make use of the Galois module structure on Tate classes to improve (1.2.2). A slight modification of the discussion above allows one to pass to the algebraic closures of the base fields. This allows us to compute bounds on $\dim_{\mathbb{Q}} A^r(X_{\overline{\mathbb{Q}}}) = \dim_{\mathbb{Q}} A^r(X_{\mathbb{C}})$. Furthermore, these bounds on $\dim_{\mathbb{Q}} A^r(X_{\mathbb{C}})$ can be computed without leaving the comfort of $\mathrm{H}^{2r}_{\mathrm{crys}}(X_p/\mathbb{Z}_p)$. We explain how to enlarge the base field in Sect. 2.3. In the process, we show that algebraic cycles appearing over different base fields can be obstructed independently. This is the content of Proposition 2.22, which improves over (1.2.2).

1.3 A Note on Using Finite Approximations

A p-adic approximation of the Frobenius action on crystalline cohomology Frob: $\mathrm{H}^{2r}_{\mathrm{crys}}(X_p/\mathbb{Z}_p) \to \mathrm{H}^{2r}_{\mathrm{crys}}(X_p/\mathbb{Z}_p)$ can be obtained as in [AKR10, Cos15, CHK19]. Using this approximation, we find a vector space approximating the space of Tate cycles $T^r(X_p)$. We work in a coordinate system that respects the Hodge structure, see Sect. 3.4, and allows us to approximate the obstruction map π_T on $T^r(X_p)$. With careful management of precision, we can compute a rigorous upper bound on the dimension of the kernel of π_T, see Sect. 4.1.10.

1.4 A Limitation and the Need for Integral Structure

There is a shortcoming of our approach. Even when the Tate conjecture holds, and even when $A^r(X)$ is characterized in $A^r(X_p)$ by the kernel of π_A, the inequality $\dim_{\mathbb{Q}} A^r(X)_{\mathbb{Q}} \leq \dim_{\mathbb{Q}_p} \ker \pi_T$ can be strict. This is because $A^r(X)_{\mathbb{Q}}$ is a rational vector space, while the obstruction map can be irrational, see Example 5.8 where Proposition 2.22 cannot help.

This issue can be circumvented to a large extent if one can identify the image of $A^r(X_p)$ inside $T^r(X_p)$. It is a well-known shortcoming of the Tate conjecture that, unlike the Hodge conjecture, there is no description of this image, even conjecturally.

Suppose we can compute (or approximate) any $\mathbb{Z}$-lattice Λ inside $T^r(X_p)$ that would contain $A^r(X_p)$. Using pLLL [IN17], the p-adic version of LLL [LLL82], we can get a good guess on the "integral" kernel of π_A restricted to Λ. This would sacrifice rigor in return for what is most probably the correct value of $\rho^r(X_{\mathbb{C}})$.

This idea can be compared to the one used in [LS19] in the setting of complex periods. There, integral Betti cohomology serves as the $\mathbb{Z}$-lattice Λ inside the complex Betti cohomology.

In the present paper, we do not consider the problem of identifying a lattice $\Lambda \subset T^r(X_p)$. Nevertheless, we tried to set-up the theory in a way that anticipates this development. We addressed issues of torsion in integral crystalline cohomology and the state of knowledge regarding the properties of the obstruction map with integral coefficients, see Sects. 3.3 and 3.4.

1.5 Previous Approaches

Given the importance of $A^r(X)$, several techniques exist in the literature for obtaining information on $A^r(X)$ or $\rho^r(X)$ for a given X.

For example, the authors of [PTvL15] provide an algorithm for surfaces conditional on the computability of the étale cohomology with finite coefficients. In [HKT13], the authors provide an algorithm for K3 surfaces of degree 2 conditional on an effective version of the Kuga–Satake construction. Another algorithm to compute the geometric Picard number of a K3 surface, conditional on the Hodge conjecture for $X \times X$, is presented in [Cha14].

Often, as in the algorithms mentioned above, one obtains a lower bound for $\rho(X)$ by exhibiting divisors explicitly. However, there is no known practical algorithm to do this in general. Nonetheless, if X has some additional structure, practical methods may arise. For example, when X is a product of curves [CMSV19], is a quotient of another variety by finite group [Shi86], or is an elliptic surface [Shi72, Shi90].

The specialization homomorphism of algebraic cycles $\iota\colon A^1(X) \hookrightarrow A^1(X_p)$ is used frequently to compute Picard numbers of surfaces. For instance, one may compare the lattice structure on $A^1(X_p)$ for two different primes to limit the image of ι [vLui07], or use the Artin–Tate conjecture for surfaces [Klo07]. One can also view ι as a morphism of two Galois modules, as was done in [EJ11a], see also Proposition 2.22. Alternatively, when explicit elements of $A^1(X_p)$ are known, one can rely on their geometry to show that some of them cannot be lifted. This becomes a powerful tool in the absence of torsion in the cokernel of ι, see [EJ11b].

The methods outlined in the previous paragraph are strongest when the rank jump between $A^1(X)$ and $A^1(X_p)$ is at most one. However, Charles [Cha14] proved that some K3 surfaces may never admit a prime reduction where $\dim A^1(X_p) - \dim A^1(X) \leq 1$, see Example 5.9.

We should highlight the difficulty in computing Picard numbers of surfaces. For instance, only recently did Schütt [Sch15] obtain the set of values that can be attained as the Picard number of a quintic surface. We still do not know this set for sextic surfaces. See the introduction and Sect. 2 of *loc. cit.* for a comprehensive overview.

The papers [Ser19, LS19] tackle the computation of the middle Picard number of a hypersurface using complex transcendental methods. That method seems very reliable (Example 5.5), but proving the result of its computation can be challenging, see for instance [MS].

1.6 Overview

In Sect. 2, we present the theoretical framework of our method. This includes an overview of the problem of lifting algebraic cycles from positive characteristic. We state the Bloch–Esnault–Kerz theorem (Theorem 2.7) and show how to use it in computations, to constrain the liftability of certain Tate classes. In Sect. 3, we recall how to effectively compute in the crystalline cohomology of a smooth hypersurface. Here, we slightly strengthen the known results to better handle torsion. In Sect. 4, we present the main algorithm of this paper (Algorithm 4.1) and clarify each computational step. In Sect. 5, we explore examples that illustrate the strengths and weaknesses of the method. We also demonstrate the usage of the code `crystalline_obstruction`.

2 Lifting Algebraic Cycles from Positive Characteristic

In this section, we will be working with a smooth projective variety X defined over a number field K. The goal is to constrain the dimension of the span of algebraic cohomology classes on $X_{\overline{K}} := X \times_K \overline{K}$, for an algebraic closure $\overline{K}$ of K.

We begin with a review of the core concepts we will use. In Sect. 2.1 we setup notation for the passage to a finite field from K. We then recall the statement of Tate's conjecture for finite fields in Sect. 2.2.

The later subsections are intended to reposition these concepts to simplify computations. In particular, we do not want to extend the base for crystalline cohomology when computing. Section 2.3 makes the first simplification that allows us to find "eventual Tate classes", see (2.3.2). In Sect. 2.4, we show that the obstruction map can be studied at the level of eventual Tate classes and the obstruction map can be applied seperately on elements that appear at different levels of field extensions. The conclusion, Proposition 2.22, multiplies the extent to which we can obstruct classes.

2.1 Passage from Number Fields to Finite Fields

Let K be a number field with ring of integers $\mathcal{O}_K$. Fix an unramified prime ideal $\mathfrak{p} \subset \mathcal{O}_K$. Localizing and completing $\mathcal{O}_K$ at $\mathfrak{p}$ we get a local field $\mathbb{Q}_\mathfrak{p}$ and its ring of integers $\mathbb{Z}_\mathfrak{p}$. We will continue to write $\mathfrak{p}$ for the maximal ideal of $\mathbb{Z}_\mathfrak{p}$. Let k be the residue field at $\mathfrak{p}$ and p the characteristic of k.

The ring of Witt vectors $W(\mathsf{k})$ of k and $\mathbb{Z}_\mathfrak{p}$ are canonically isomorphic. In fact, for some n there is an isomorphism $\mathsf{k} \simeq \mathbb{F}_{p^n}$ and $W(k)$ is determined uniquely up to isomorphism as the unramified extension of degree n of the ring of p-adic integers $\mathbb{Z}_p$.

2.1.1 Good Reduction

With $X \to \operatorname{Spec} K$ a smooth projective variety, assume that $\mathfrak{p} \subset \mathcal{O}_K$ is chosen so that that X has *good reduction* at $\mathfrak{p}$. That is, we assume that there is a regular scheme $\mathcal{X} \to \operatorname{Spec} \mathbb{Z}_\mathfrak{p}$, smooth over the base, such that $X_{\mathbb{Q}_\mathfrak{p}} := X \times_K \mathbb{Q}_\mathfrak{p}$ is identified with the generic fiber $\mathcal{X} \times_{\mathbb{Z}_\mathfrak{p}} \mathbb{Q}_\mathfrak{p}$. We will write $X_\mathfrak{p}$ for the special fiber of $\mathcal{X}$, and $\overline{X}_\mathfrak{p} := X_\mathfrak{p} \times_k \overline{k}$ for its base change to an algebraic closure $\overline{k}$ of k.

We will also use "thickenings" of $X_\mathfrak{p}$, namely $\mathcal{X}_n := \mathcal{X} \otimes_{\mathbb{Z}_\mathfrak{p}} (\mathbb{Z}_\mathfrak{p}/\mathfrak{p}^n)$ for each $n \geq 1$. Let $\widehat{\mathcal{X}}$ be the formal scheme obtained by completing $\mathcal{X}$ along the special fiber $X_\mathfrak{p}$.

2.1.2 The Specialization Map on Subvarieties

For any $r \geq 0$ and a reduced scheme Y, the algebraic cycle group $\mathcal{Z}^r(Y)$ of Y is the free abelian group generated by codimension r subvarieties of Y. When Y is a formal scheme, we use $\mathcal{Z}^r(Y)$ to denote formal subschemes of codimension r. Let us write $\mathcal{Z}^r(\mathcal{X})' \subset \mathcal{Z}^r(\mathcal{X})$ to be the free abelian group generated by subvarieties in $\mathcal{X}$ that are flat over the base $\mathbb{Z}_\mathfrak{p}$.

There is an isomorphism $\mathcal{Z}^r(X_{\mathbb{Q}_\mathfrak{p}}) \to \mathcal{Z}^r(\mathcal{X})'$ which maps each subvariety $V \subset X$ to its closure in $\mathcal{X}$. The inverse of this map is given by intersecting a flat variety in $\mathcal{X}$ with the generic fiber $X_{\mathbb{Q}_\mathfrak{p}}$. There is also a restriction map $\mathcal{Z}^r(\mathcal{X})' \to \mathcal{Z}^r(X_\mathfrak{p})$ obtained by intersecting a subvariety of $\mathcal{X}$ by the special fiber $X_\mathfrak{p}$. Composing the isomorphism above with the restriction map gives the *specialization map* [Ful98, §20.3]

$$\operatorname{sp}\colon \mathcal{Z}^r(X_{\mathbb{Q}_\mathfrak{p}}) \to \mathcal{Z}^r(X_\mathfrak{p}). \tag{2.1.1}$$

Let us point out that the specialization map factors through $\mathcal{Z}^r(\widehat{\mathcal{X}})$,

$$\mathcal{Z}^r(X_{\mathbb{Q}_\mathfrak{p}}) \to \mathcal{Z}^r(\widehat{\mathcal{X}}) \to \mathcal{Z}^r(X_\mathfrak{p}), \tag{2.1.2}$$

though the first map does not need to be surjective.

2.1.3 A Hodge Filtration on Crystalline Cohomology

For each $r \geq 0$, the de Rham cohomology $\mathrm{H}^r_{\mathrm{dR}}(\mathcal{X}/\mathbb{Z}_\mathfrak{p})$ of $\mathcal{X}$ comes with the Hodge filtration $\mathrm{F}^\bullet \mathrm{H}^r_{\mathrm{dR}}(\mathcal{X}/\mathbb{Z}_\mathfrak{p})$. Using the Berthelot comparison isomorphism [Ber97, Shi02] (see also Sect. 3.1),

$$\Phi\colon \mathrm{H}^r_{\mathrm{dR}}(\mathcal{X}/\mathbb{Z}_\mathfrak{p}) \xrightarrow{\sim} \mathrm{H}^r_{\mathrm{crys}}(X_\mathfrak{p}/\mathbb{Z}_\mathfrak{p}), \tag{2.1.3}$$

we can carry the Hodge filtration over to the crystalline cohomology of $X_\mathfrak{p}$. Let us point out that the resulting filtration is not intrinsic to $X_\mathfrak{p}$ but depends on the model $\mathcal{X}/\mathbb{Z}_\mathfrak{p}$. The filtration modulo $\mathfrak{p}$ is, however, intrinsic to $X_\mathfrak{p}$.

Definition 2.1 Let $\mathrm{F}^{\bullet}_{\mathcal{X}}\mathrm{H}^r_{\mathrm{crys}}(X_{\mathfrak{p}}/\mathbb{Z}_{\mathfrak{p}})$ denote the filtration induced by the Hodge filtration on de Rham cohomology carried over by the comparison isomorphism (2.1.3).

2.1.4 Cycle Class Maps

There are cycle class maps c_{dR} [Har75] and c_{crys} [GM87] (see also [BEK14, §8] for a review of the crystalline cycle map with integral coefficients) making the following diagram commutative:

$$\begin{array}{ccccc} \mathcal{Z}^r(X) & \xrightarrow{\sim} & \mathcal{Z}^r(\mathcal{X})' & \longrightarrow & \mathcal{Z}^r(X_{\mathfrak{p}}) \\ \downarrow c_{\mathrm{dR}} & & \downarrow & & \downarrow c_{\mathrm{crys}} \\ \mathrm{H}^{2r}_{\mathrm{dR}}(X/\mathbb{Q}_{\mathfrak{p}}) & \longleftarrow & \mathrm{H}^{2r}_{\mathrm{dR}}(\mathcal{X}/\mathbb{Z}_{\mathfrak{p}}) & \underset{\Phi}{\xrightarrow{\sim}} & \mathrm{H}^{2r}_{\mathrm{crys}}(X_{\mathfrak{p}}/W) \end{array} \tag{2.1.4}$$

We define the middle vertical arrow so as to make the square on the right commute. The map Φ on the bottom right is the Berthelot comparison map (2.1.3). The map on the bottom left is the composition:

$$\mathrm{H}^{2r}_{\mathrm{dR}}(\mathcal{X}/\mathbb{Z}_{\mathfrak{p}}) \to \mathrm{H}^{2r}_{\mathrm{dR}}(\mathcal{X}/\mathbb{Z}_{\mathfrak{p}}) \otimes_{\mathbb{Z}_{\mathfrak{p}}} \mathbb{Q}_{\mathfrak{p}} \xrightarrow{\sim} \mathrm{H}^{2r}_{\mathrm{dR}}(X/\mathbb{Q}_{\mathfrak{p}}). \tag{2.1.5}$$

Definition 2.2 The image of the cycle class maps, in the appropriate cohomology, will be denoted with $A^r(X)$ and $A^r(X_{\mathfrak{p}})$. Their tensor with $\mathbb{Q}$ are denoted by $A^r(X)_{\mathbb{Q}}$ and $A^r(X_{\mathfrak{p}})_{\mathbb{Q}}$, respectively. The elements in the image of the cycle class maps, or elements in their $\mathbb{Q}$-span, will be called *algebraic cohomology classes.*

Lemma 2.3 *The maps above give an injective homomorphism* $\iota\colon A^r(X)_{\mathbb{Q}} \hookrightarrow A^r(X_{\mathfrak{p}})_{\mathbb{Q}}$.

Proof The cycle class maps are compatible in the sense that the diagram (2.1.4) commutes after tensoring with $\mathbb{Q}$. Furthermore, the horizontal bottom arrows are all isomorphisms once tensored with $\mathbb{Q}$, and thus we have injectivity. □

Remark 2.4 In fact, even without tensoring with $\mathbb{Q}$, we can obtain an injection. The specialisation map $A^r(X_{\overline{K}}) \to A^r(\overline{X}_{\mathfrak{p}})$ preserves the intersection pairing, see [Ful98, Corollary 20.3]. Using that the polarization maps to the polarization, and using the Hodge index theorem on $A^r(X_{\overline{K}})$, we conclude that no element can map to zero.

2.1.5 Dimensions of the Space of Algebraic Cycles Over Different Fields

Let $\rho^r_{\mathrm{Betti}}(X_{\mathbb{C}})$ denote the dimension of the space of codimension r algebraic cycles $A^r(X_{\mathbb{C}})$ in the Betti cohomology $\mathrm{H}^{2r}_{\mathrm{Betti}}(X_{\mathbb{C}}, \mathbb{C})$ of the associated complex manifold.

Let L/K be a field extension of K. Define $\rho^r_{\mathrm{dR}}(X_L)$ as the dimension of the algebraic cycles in the *de Rham* cohomology of $X_L := X \times_K L$.

Let $\overline{K} \subset \mathbb{C}$ be the algebraic closure of $K \hookrightarrow \mathbb{C}$. Let $K \hookrightarrow \mathbb{Q}_{\mathfrak{p}}$ be a localization as in Sect. 2.1 and let $\overline{\mathbb{Q}}_{\mathfrak{p}}$ be an algebraic closure of $\mathbb{Q}_{\mathfrak{p}}$. Then we have the following series of equalities:

$$\rho^r_{\mathrm{Betti}}(X_{\mathbb{C}}) = \rho^r_{\mathrm{dR}}(X_{\mathbb{C}}) = \rho^r_{\mathrm{dR}}(X_{\overline{K}}) = \rho^r_{\mathrm{dR}}(X_{\overline{\mathbb{Q}}_{\mathfrak{p}}}). \tag{2.1.6}$$

The first equality follows from the standard comparison isomorphisms. The other equalities result from the following well-known fact.

Lemma 2.5 *Suppose K is algebraically closed and characteristic zero. Let L/K be a field extension with L algebraically closed. Let X_K/K be a smooth projective variety and X_L its pullback to L. Then for every $r \geq 0$, $\rho^r_{\mathrm{dR}}(X_K) = \rho^r_{\mathrm{dR}}(X_L)$.*

Sketch of Proof In characteristic zero, the cycle class map c_{dR} is well defined, and it factors through the Chow groups CH^r (the algebraic cycle group $\mathcal{Z}^r$ modulo algebraic equivalence). But the Chow groups of X_K and X_L are canonically isomorphic.

The inclusion $\mathrm{CH}^r(X_K) \hookrightarrow \mathrm{CH}^r(X_L)$ is induced by pulling back varieties in X_K to X_L. In the reverse direction, take a subvariety $V \subset X_L$. Without loss of generality, we may assume L is the field of definition of V over K. In particular, L now has finite transcendence degree over K. We can find an affine K-variety Y with function field L over which V extends to a flat family of varieties $\mathcal{V} \to Y$ so that $\mathcal{V} \hookrightarrow X_K \times_K Y$ is a closed immersion. Any fiber of $\mathcal{V} \to Y$ over a K-valued point defines an element in $\mathrm{CH}^r(X_K)$. □

2.1.6 The Obstruction Map

The goal here is to give a partial description of the image of the inclusion $\iota \colon A^r(X)_{\mathbb{Q}} \hookrightarrow A^r(X_{\mathfrak{p}})_{\mathbb{Q}}$. That is the image of the composed map $\mathcal{Z}^r(X) \to \mathcal{Z}^r(X_{\mathfrak{p}}) \to \mathrm{H}^{2r}_{\mathrm{crys}}(X_{\mathfrak{p}}/\mathbb{Z}_{\mathfrak{p}})$. The following theorem instead allows us to describe the image of the composed map $\mathcal{Z}^r(\widehat{\mathfrak{X}}) \to \mathcal{Z}^r(X_{\mathfrak{p}}) \to \mathrm{H}^{2r}_{\mathrm{crys}}(X_{\mathfrak{p}}/\mathbb{Z}_{\mathfrak{p}})$ (see also (2.1.2)).

Definition 2.6 An algebraic cycle $\xi \in A^r(X_{\mathfrak{p}})$ is *unobstructed* (with respect to $\mathfrak{X}$) if $\xi \in \mathrm{F}^r_{\mathfrak{X}}\mathrm{H}^{2r}_{\mathrm{crys}}(X_{\mathfrak{p}}/\mathbb{Z}_{\mathfrak{p}})$, with the filtration defined in Definition 2.1. A cycle is *obstructed* otherwise.

Note that unobstructed cycles form a subgroup. The terminology is justified by the following theorem of Bloch, Esnault, and Kerz, which states that modulo torsion, the group of unobstructed cycles, can be lifted to $\widehat{\mathcal{X}}$.

Theorem 2.7 ([BEK14, Theorem 1.3]) *Suppose $\mathcal{X}/\mathbb{Z}_\mathfrak{p}$ is smooth and projective. For each $r \geq 0$ and $p = \mathrm{char}\, \mathsf{k} > \dim X_\mathfrak{p} + 6$ the image of the composition $\mathcal{Z}^r(\widehat{\mathcal{X}}) \otimes \mathbb{Q} \to \mathcal{Z}^r(X_\mathfrak{p}) \otimes \mathbb{Q} \twoheadrightarrow A^r(X_\mathfrak{p})_\mathbb{Q}$ coincides with the group of unobstructed cycles.*

Definition 2.8 The *obstruction map* of $X_\mathfrak{p}$ with respect to $\mathcal{X}$ is the quotient map

$$\pi_\mathcal{X} \colon \mathrm{H}^{2r}_{\mathrm{crys}}(X_\mathfrak{p}/\mathbb{Z}_\mathfrak{p}) \to \mathrm{H}^{2r}_{\mathrm{crys}}(X_\mathfrak{p}/\mathbb{Z}_\mathfrak{p})/\mathrm{F}^r_\mathcal{X}\mathrm{H}^{2r}_{\mathrm{crys}}(X_\mathfrak{p}/\mathbb{Z}_\mathfrak{p}).$$

If it is clear from context, we may tensor $\pi_\mathcal{X}$ with $\mathbb{Q}_\mathfrak{p}$ without changing notation. Restrictions of $\pi_\mathcal{X}$ to a subspace V of cohomology will be abbriviated to π_V.

Remark 2.9 If we are given $X_\mathfrak{p}$ but not X the model $\mathcal{X}/\mathbb{Z}_\mathfrak{p}$ is not unique. The choice of a model $\mathcal{X}/\mathbb{Z}_\mathfrak{p}$ impacts which cycles are obstructed.

Remark 2.10 A theorem of this nature—in particular, the ability to define the Chern class map for line bundles—was the original motivation for the definition of crystalline cohomology [Gro68, §7.4]. For line bundles, this goal was realized by Berthelot and Ogus in [BO83].

2.2 *Finding the Image of the Chern Class Map via Tate's Conjecture*

Due to its computational complexity, we would like to avoid computing and representing actual subvarieties in X or even $X_\mathfrak{p}$ as much as possible. On the other hand, the $\mathbb{Q}_\mathfrak{p}$-span of algebraic cycles in cohomology has, at least conjecturally, a computationally tractable description as *Tate classes* [Tat66b, Mil07].

Recall that there is the arithmetic Frobenius map $X_\mathfrak{p} \to X_\mathfrak{p}$ over k. As crystalline cohomology is functorial, the relative Frobenius map induces a map on cohomology:

$$\mathrm{Frob}_\mathsf{k} \colon \mathrm{H}^{2r}_{\mathrm{crys}}(X_\mathfrak{p}/\mathbb{Z}_\mathfrak{p}) \to \mathrm{H}^{2r}_{\mathrm{crys}}(X_\mathfrak{p}/\mathbb{Z}_\mathfrak{p}). \tag{2.2.1}$$

Definition 2.11 The *integral Tate classes* are the $q^{-r}\,\mathrm{Frob}_\mathsf{k}$-fixed elements of crystalline cohomology, they form $T^r(X_\mathfrak{p})_{\mathrm{int}} \subset \mathrm{H}^{2r}_{\mathrm{crys}}(X_\mathfrak{p}/\mathbb{Z}_\mathfrak{p})$. The fixed elements of $q^{-r}\,\mathrm{Frob}_\mathsf{k}$ in $\mathrm{H}^{2r}_{\mathrm{crys}}(X_\mathfrak{p}/\mathbb{Z}_\mathfrak{p}) \otimes \mathbb{Q}_\mathfrak{p}$ are *Tate classes* and they form the space $T^r(X_\mathfrak{p})$.

Conjecture 2.12 (Tate Conjecture) *For each r, algebraic cycles span $T^r(X_\mathfrak{p})$; that is; $A^r(X_\mathfrak{p}) \otimes_\mathbb{Z} \mathbb{Q}_\mathfrak{p} \simeq T^r(X_\mathfrak{p})$.*

Remark 2.13 The stronger statement $A^r(X_\mathfrak{p}) \otimes_\mathbb{Z} \mathbb{Z}_\mathfrak{p} = T^r(X_\mathfrak{p})_{\text{int}}$ is called the *integral Tate conjecture*. This version is often false due to torsion, though it may be true modulo torsion. In any case, when $r = 1$, then the usual Tate conjecture implies the integral Tate conjecture [Mil07].

Remark 2.14 Tate himself proved the Tate conjecture for abelian varieties over finite fields [Tat66a]. More recently, the Tate conjecture was proven for K3 surfaces [Cha14, Mad15, KM16]. For an overview of the Tate conjecture over finite fields, as well as recent progress on it, we recommend [Mil07, Tot17].

2.3 Enlarging the Base Field

Given the smooth projective variety X/K we want to compute the dimension of $A^r(X_{\overline{K}})$, with $\overline{K}$ an algebraic closure of K. If $X_\mathfrak{p}$ is our choice of finite reduction, it is likely that $A^r(X_{\overline{K}})$ will not map into $T^r(X_\mathfrak{p})$ and we need to enlarge the residue field k. The dimension of the space of Tate classes will increase as we enlarge k but will eventually stop growing. This terminal dimension, as well as the terminal dimension of the unobstructed Tate cycles, can be computed without ever enlarging the base field. We will explain how to do this here, paying attention to computational limitations.

Let $\chi_{2r} \in \mathbb{Q}[t]$ be the characteristic polynomial of $q^{-r}\,\mathrm{Frob}_\mathsf{k}$, a scaling of the Frobenius map (2.2.1). Factorize this characteristic polynomial over $\mathbb{Q}[t]$ as follows

$$\chi_{2r}(t) = h(t) \prod_i \Phi_i(t)^{\gamma_i}, \tag{2.3.1}$$

where, for each i, the polynomial $\Phi_i(t) \in \mathbb{Z}[t]$ is the i-th cyclotomic polynomial, i.e., the minimal polynomial of any primitive i-th root of unity, the exponents $\gamma_i \geq 0$, and no root of $h(t)$ is a root of unity. Set $u := \mathrm{lcm}\{i \mid \gamma_i \neq 0\}$.

Consider now the following subspace of $\mathrm{H}^{2r}_{\mathrm{crys}}(X_\mathfrak{p}/\mathbb{Z}_\mathfrak{p}) \otimes \mathbb{Q}_\mathfrak{p}$,

$$\widetilde{T}^r(X_\mathfrak{p}) := \ker\left(\mathrm{Frob}_\mathsf{k}^u - q^{ur}\right). \tag{2.3.2}$$

Elements of $\widetilde{T}^r(X_\mathfrak{p})$ may be viewed as "eventual Tate classes" as they will span the Tate classes when the base field k is enlarged, as its dimension will match the rank of $A\big((X_\mathfrak{p})_{\overline{k}}\big)$, as shown below.

Proposition 2.15 *Using the obstruction map (Definition 2.8) we obtain the following bound:*

$$\dim A^r(X_{\overline{K}})_\mathbb{Q} \leq \dim \ker \pi|_{\widetilde{T}^r(X_\mathfrak{p})} \tag{2.3.3}$$

Proof Crystalline cohomology base changes like the de Rham cohomology. However, the natural action of $\mathrm{Frob}_{\mathsf{k}}$ on the extended scalars will not be linear. Instead, if k'/k is a base extension then the natural action of $\mathrm{Frob}_{\mathsf{k}'}$ will correspond to the linear extension of $\mathrm{Frob}_{\mathsf{k}}^{[\mathsf{k}':\mathsf{k}]}$ from $\mathrm{H}^{2r}_{\mathrm{crys}}(X_{\mathfrak{p}}/\mathbb{Z}_{\mathfrak{p}})$. It follows that if $[\mathsf{k}':\mathsf{k}] = u$ then the span of $\widetilde{T}^r(X_{\mathfrak{p}})$ will give the new space of Tate classes. Further extensions of the base field will not increase the dimension of the space of Tate classes. Observing that the obstruction map $\pi_{\mathcal{X}}$ extends linearly with base change, we conclude the proof using Theorem 2.7. □

In practice we will approximate $\mathrm{Frob}_{\mathsf{k}}$ to a few p-adic digits (often 5–20 digits). Increasing the precision is costly and each power of $\mathrm{Frob}_{\mathsf{k}}$ will lose some of that precision. Considering that u maybe quite large, we will compute $\widetilde{T}^r(X_{\mathfrak{p}})$ in the following way which requires taking at most the v-th power of $\mathrm{Frob}_{\mathsf{k}}$, with $v := \max_i\{\deg \Phi_i \mid \gamma_i \neq 0\}$.

Proposition 2.16 *Let* $T_i^r(X_{\mathfrak{p}}) := \ker\left(\Phi_i(q^{-r}\,\mathrm{Frob}_{\mathsf{k}})\right)$. *Then* $\widetilde{T}^r(X_{\mathfrak{p}}) = \bigoplus_{i\geq 0} T_i^r(X_{\mathfrak{p}})$, *where the sum is taken over i with* $\gamma_i \neq 0$.

Proof The restriction of $q^{-r}\,\mathrm{Frob}_{\mathsf{k}}$ to $\widetilde{T}^r(X_{\mathfrak{p}})$ is annihilated by the polynomial $t^u - 1$. Therefore, its minimal polynomial will have only reduced cyclotomic factors. Now apply the primary decomposition theorem from linear algebra [HK71, §6.8, Theorem 12]. □

2.4 Improving the Obstruction Using the Frobenius Decomposition

The space of "eventually Tate classes" $\widetilde{T} := \widetilde{T}^r(X_{\mathfrak{p}})$ from (2.3.2) admits the Frobenius decomposition $\widetilde{T} = \bigoplus_i T_i$ of Proposition 2.16. It is tempting to obstruct each T_i individually in order to improve the upper bounds we can get for the dimension of $A^r(X_{\overline{K}})$. Assuming the full Tate conjecture, as described below, we show that this is possible here.

The full Tate conjecture assumes, in addition to the Tate conjecture, that the intersection product on algebraic cohomology classes is non-degenerate [Mil07]. This additional requirement always holds for surfaces. In particular, K3 surfaces over finite fields continue to satisfy the full Tate conjecture [Cha13, Mad15, KM16].

For this section let $A := A^r(X_{\overline{K}})_{\mathbb{Q}}$, $B := A^r(\overline{X}_{\mathfrak{p}})_{\mathbb{Q}}$ and $H := \mathrm{H}^{2r}_{\mathrm{crys}}(X_{\mathfrak{p}}/\mathbb{Z}_{\mathfrak{p}}) \otimes \mathbb{Q}_p$. We warn the reader that B does not map into H. Pick a finite extension k'/k where all algebraic cycle classes of $\overline{X}_{\mathfrak{p}}$ can be defined. Recall that $W(\mathsf{k}')$ is the ring of Witt vectors of k' and $Q(\mathsf{k}')$ is its fraction field. Then $q^{-r}\,\mathrm{Frob}_{\mathsf{k}}$ maps B equivariantly into

$$H' := \mathrm{H}^{2r}_{\mathrm{crys}}(X_{\mathfrak{p}} \times_{\mathsf{k}} \mathsf{k}'/W(\mathsf{k}')) \otimes_{W(\mathsf{k}')} Q(\mathsf{k}') \simeq H \otimes_{\mathbb{Q}_{\mathfrak{p}}} Q(\mathsf{k}'). \qquad (2.4.1)$$

The natural action of Frob_k on H' is *not* the one that linearly extends $q^{-r}\,\mathrm{Frob}_\mathsf{k}$ from H. Overcoming this non-linearity is the main technical obstacle in this section. Let $\sigma\colon Q(\mathsf{k}') \to Q(\mathsf{k}')$ be the field homomorphism induced by the Frobenius map of k on k'. The natural Frob_k action on H' is the σ-linear extension of the Frob_k action on H.

Lemma 2.17 *The image of the inclusion* $\iota\colon A^r(X_{\overline{K}}) \hookrightarrow A^r(\overline{X}_p)$ *is invariant under the action of* $q^{-r}\,\mathrm{Frob}_\mathsf{k}$.

Proof We need to find an action on the left hand side that commutes with the Frobenius action on the right hand side. Note that there is a finite Galois extension K'/K such that $A^r(X_{\overline{K}}) = A^r(X_{K'})$. We can also choose a prime $\mathfrak{q}$ of K' lying above $\mathfrak{p}$ for the reductions. It is standard that one can lift the Frobenius action into the subgroup of $\mathrm{Gal}(K'/K)$ that fixes $\mathfrak{q}$ [Mil17, § 8]. □

Let $\chi_A, \chi_B \in \mathbb{Q}[t]$ be the characteristic polynomials of the $q^{-r}\,\mathrm{Frob}_\mathsf{k}$ action on A and B respectively. We will write $\chi_{\widetilde{T}} \in \mathbb{Q}[t]$ for the characteristic polynomial of $q^{-r}\,\mathrm{Frob}_\mathsf{k}$ on $\widetilde{T} \subset H$ from (2.3.2). We know that $q^{-r}\,\mathrm{Frob}_\mathsf{k}$ is of finite order on each of these spaces, therefore we can factor these characteristic polynomials as follows:

$$\chi_A = \prod_i \Phi_i^{\alpha_i},\ \chi_B = \prod_i \Phi_i^{\beta_i},\ \text{and}\ \chi_{\widetilde{T}} = \prod_i \Phi_i^{\gamma_i}, \tag{2.4.2}$$

where $\Phi_i \in \mathbb{Z}[t]$ is the cyclotomic polynomial for any primitive i-th root of unity. Note that we want to determine the dimension $\dim A$ which equals $\sum_i \alpha_i \deg \Phi_i$.

Decompose these spaces into $A = \bigoplus_i A_i$, $B = \bigoplus_i B_i$ and $\widetilde{T} = \bigoplus_i T_i$ as in Proposition 2.16, with $A_i = \ker\left(\Phi_i(q^{-r}\,\mathrm{Frob}_\mathsf{k})\right)$ and so on.

Lemma 2.18 *For each i we have* $\alpha_i \le \beta_i \le \gamma_i$.

Proof The first inequality $\alpha_i \le \beta_i$ follows from Lemma 2.17 since $B_i \subset A_i$. The inequality $\beta_i \le \gamma_i$ is more subtle, but standard. Either one can argue as in Proposition 2.19 below, or deduce it from the Weil conjectures in comparison with the étale cohomology with $\mathbb{Q}_\ell$-coefficients. □

For each i, let $T_i' \simeq T_i \otimes_{\mathbb{Q}_\mathsf{p}} Q(\mathsf{k}')$ be the span of T_i in H', see (2.4.1).

Proposition 2.19 *Assuming the full Tate conjecture for X, we have* $T_i' \cap B = B_i$.

Proof For each i, let k_i/k be a degree i extension and let H_i be the crystalline cohomology of $X_{\mathfrak{p}^i} := X_\mathfrak{p} \times_\mathsf{k} \mathsf{k}_i$. The operator $q^{-ri}\,\mathrm{Frob}_\mathsf{k}^i - 1$ is linear on H_i and it annihilates $A^r(X_{\mathfrak{p}^i})$. Moreover, $B_{\le i} = \bigoplus_{j \le i} B_j$ coincides with $A^r(X_{\mathfrak{p}^i})$ because both sides are the fixed elements of $A^r(\overline{X}_\mathfrak{p})$ with respect to the absolute Galois group of k_i.

Having assumed the full Tate conjecture, we may conclude that $B_{\le i}$ is a $\mathbb{Q}$-substructure for $T_{\le i}'$. In particular, elements of B are $Q(\mathsf{k}')$-linearly independent in the ambient space H'. This implies the equality $\widetilde{T}_{\le i}' \cap B = B_{\le i}$. The proposition for $i = 1$ follows immediately. Fix i and assume that the proposition holds for each $j < i$.

Although the $\mathrm{Frob}_{\mathsf{k}}$ action is not $Q(\mathsf{k}')$-linear on T_i', the space T_i' is nevertheless a $\mathbb{Q}$-vector space that is $\mathrm{Frob}_{\mathsf{k}}$ invariant. Thus, the intersection $T_i' \cap B$ is a $\mathrm{Frob}_{\mathsf{k}}$ invariant $\mathbb{Q}$-subspace of $B_{\leq i}$. In particular, it admits a decomposition into a direct sum, with each component lying in one B_j for $j \leq i$. But distinct spaces T_j' are disjoint and each T_j' contains B_j when $j < i$. It follows immediately that the intersection $T_i' \cap B$ must lie in B_i. The equality follows. □

For each $i \in \mathbb{Z}_{>0}$ define the integer r_i as

$$r_i = \begin{cases} 0, & \text{if} \quad \pi_{T_i} = 0 \\ \deg \Phi_i, & \text{if} \quad \pi_{T_i} \neq 0. \end{cases} \tag{2.4.3}$$

Corollary 2.20 *Assuming the full Tate conjecture,* $\dim_{\mathbb{Q}} A^r(X_{\overline{K}}) \leq \dim_{\mathbb{Q}_p} \widetilde{T} - \sum_i r_i$.

Proof With notation as in (2.4.2) we recall $\dim A = \sum_i \alpha_i \deg \Phi_i$, with corresponding sums for the dimensions of B and $\widetilde{T}$. Thus, we need only show $\alpha_i < \gamma_i$ when the obstruction map on T_i is non-zero. If $\beta_i = \gamma_i$ then, in light of Proposition 2.19, the span of B_i and T_i must agree in H'. If $\pi_{T_i} \neq 0$ then the containment $A_i \subset B_i$ must be strict by Theorem 2.7, implying $\alpha_i < \beta_i = \gamma_i$. On the other hand, if $\beta_i < \gamma_i$ then there is nothing more to do. □

Remark 2.21 We can use the bound in Corollary 2.20 even when computing the obstructions with finite precision. Let $r_{i,N}$ to be 0 or $\deg \Phi_i$ depending on whether $\pi_{T_i} \equiv 0 \bmod p^N$. Clearly, $r_{i,N} \leq r_i$ and $\dim_{\mathbb{Q}_p} \widetilde{T} - \sum_i r_{i,N}$ also serves as an upper bound.

The result of Corollary 2.20 is easy to use. However, we can improve these bounds by computing more. Define the following map on T_i

$$\pi_i : v \mapsto \left(\pi(v), \pi \circ q^{-r} \mathrm{Frob}_{\mathsf{k}}(v), \ldots, \pi \circ \left(q^{-r} \mathrm{Frob}_{\mathsf{k}}\right)^{u_i - 1}(v)\right). \tag{2.4.4}$$

Observe that this map is $\mathbb{Q}_p$-linear and let $L_i = \ker \pi_i$. The space L_i is the largest $\mathrm{Frob}_{\mathsf{k}}$-invariant subspace of $\ker \pi_{T_i}$.

Proposition 2.22 *Assuming the full Tate conjecture*

$$\dim_{\mathbb{Q}} A^r(X_{\overline{K}}) \leq \sum_i \dim_{\mathbb{Q}_p} L_i.$$

Proof Let u_i be the degree of Φ_i. We extend the map π_i to the span T_i' of T_i in H' as follows. On each element we use the same definition

$$\pi_i^! : v \mapsto \left(\pi(v), \pi \circ q^{-r}\,\mathrm{Frob}_{\mathsf{k}}(v), \ldots, \pi \circ \left(q^{-r}\,\mathrm{Frob}_{\mathsf{k}}\right)^{u_i - 1}(v)\right). \tag{2.4.5}$$

However, we modify the $Q(\mathsf{k}')$-action on the codomain so that $Q(\mathsf{k}')$ acts σ^i-linearly on the i-th coordinate. As a consequence, $\pi_i^!$ is $Q(\mathsf{k}')$-linear. Therefore, the kernel of $\pi_i^!$ is the span L_i' of L_i in H'.

Since $A_i \subset B_i$ is in the kernel of π and A_i is invariant under Frobenius, A_i is contained in the kernel of $\pi_i^!$. Thus, the $Q(\mathsf{k}')$-span of A_i lies in L_i'. Since elements of B_i are $Q(\mathsf{k}')$-linearly independent, the same holds for A_i. We conclude that $\dim_{\mathbb{Q}} A_i \leq \dim_{Q(\mathsf{k}')} L_i' = \dim_{\mathbb{Q}_{\mathsf{p}}} L_i$. □

2.5 Bounds on the Characteristic Polynomial of the Frobenius

Let $\overline{X}_{\mathsf{p}}$ denote the scheme $X_{\mathsf{p}} \times_{\mathsf{k}} \bar{\mathsf{k}}$ for an algebraic closure $\bar{\mathsf{k}}$ of k. Let $q = p^n$ be the number of elements in k. The Weil conjectures (now a theorem [Del74]) tell us that the Hasse–Weil zeta function of X_{p} has the form

$$Z(X_{\mathsf{p}}, t) := \exp\left(\sum_{m=1}^{\infty} \frac{\#X_{\mathsf{p}}(\mathbb{F}_{q^m})}{m} t^m\right) = \prod_{i=0}^{2\dim(X_{\mathsf{p}})} P_i(X_{\mathsf{p}}, t)^{(-1)^i}, \tag{2.5.1}$$

where $P_i(X_{\mathsf{p}}, t)$ is the reciprocated characteristic polynomial of the Frobenius action on $\mathrm{H}^i_{\mathrm{crys}}(X_{\mathsf{p}}/\mathbb{Z}_{\mathsf{p}})$.

$$P_i(X, t) := \det(1 - t\,\mathrm{Frob}_{\mathsf{k}}) \in 1 + t\mathbb{Z}[t]. \tag{2.5.2}$$

The polynomial $P_i(X_{\mathsf{p}}, t)$ has integer coefficients, has constant term 1, and all of its roots in $\mathbb{C}$ have Euclidean norm $q^{-i/2}$. This information is crucial in deducing P_i from an approximate representation of the Frobenius (see Sect. 4.1.5).

Note that the χ_{2r}, defined in Sect. 2.3, and P_r are related by

$$t^{\deg \chi_{2r}} \chi_{2r}(1/t) = P_{2r}(X, tq^{-r}). \tag{2.5.3}$$

Remark 2.23 The Tate conjecture implies that the number of real roots of $P_{2r}(X_{\mathsf{p}} \times_{\mathsf{k}} \mathsf{k}', q^{-r}t) \in \mathbb{Q}[t]$ for a sufficiently large (but finite) extension k'/k coincides with $\dim A^r(\overline{X}_{\mathsf{p}})$. Thus, we expect the parity of $\dim A^r(\overline{X}_{\mathsf{p}})$ to match the parity of $\dim \mathrm{H}^{2r}_{\mathrm{crys}}(X_{\mathsf{p}}/\mathbb{Z}_{\mathsf{p}}) = \deg P_{2r}$.

3 Computing in Crystalline Cohomology

Computations in crystalline cohomology are made possible by comparing it with two other cohomology theories. Berthelot's comparison theorem relates crystalline cohomology to a characteristic 0 de Rham cohomology. Monsky–Washnitzer cohomology, on the other hand, allows one to represent the Frobenius map explicitly. In this section, we outline this construction.

The de Rham cohomology is amenable to computation in general. For hypersurfaces in particular, Griffiths' explicit description of a basis makes these computations highly practical, see Sect. 3.4. The approximations of the action of the Frobenius are made in terms of the this basis.

This approach to approximating the Frobenius action was first conceptualized by Kedlaya [Ked08], first implemented by Abbott, Kedlaya and Roe [AKR10], and shown to be practical in larger characteristic by Costa and Harvey [Har10a, Har10b, Har10c, Cos15].

We summarize the required statements from [AKR10] and re-frame some using $\mathbb{Z}_p$ coefficients as opposed to $\mathbb{Q}_p$ coefficients (e.g. Proposition 3.8).

3.1 Crystalline to de Rham Cohomology

The de Rham cohomology of an affine variety does not behave well in finite characteristic. On the other hand, in characteristic zero, the cohomology of a hypersurface is made explicit by working with the (affine) complement of the hypersurface. The de Rham cohomology of a *log pair* is what is needed to carry the characteristic zero advantages to positive characteristic.

Let $\mathcal{Y}$ be a smooth proper variety over $\mathbb{Z}_p$ and let $\mathcal{X} \hookrightarrow \mathcal{Y}$ be a subvariety that is a relative normal crossing divisor. Such a pair $(\mathcal{Y}, \mathcal{X})$ is called a *smooth proper pair* over $\mathbb{Z}_p$ [AKR10, Section 2.2]. Denote the complement of $\mathcal{X}$ in $\mathcal{Y}$ by $\mathcal{U} := \mathcal{Y} \setminus \mathcal{X}$. In practice, we will take $\mathcal{Y} = \mathbb{P}^{2r+1}_{\mathbb{Z}_p}$ and $\mathcal{X}$ a smooth hypersurface.

The de Rham cohomology, as well as the crystalline cohomology, of a smooth proper pair $(\mathcal{Y}, \mathcal{X})$ is well defined [Shi02, Chapter 2]. The cohomology of the hypersurface complement $\mathcal{U}$, at least over the generic fiber, can be computed via the pair $(\mathcal{Y}, \mathcal{X})$. Indeed, there exists a natural isomorphism [Kat89, Theorem 6.4]:

$$\mathrm{H}^i_{\mathrm{dR}}((\mathcal{Y}, \mathcal{X})/\mathbb{Z}_p) \otimes \mathbb{Q}_p \simeq \mathrm{H}^i_{\mathrm{dR}}(\mathcal{U}_{\mathbb{Q}_p}/\mathbb{Q}_p). \tag{3.1.1}$$

We may compare the de Rham and crystalline cohomologies of a smooth proper pair $(\mathcal{Y}, \mathcal{X})$ by the following generalization of Berthelot's comparison theorem. We denote by $\mathcal{Y}_p$ and $\mathcal{X}_p$ the special fibers of $\mathcal{Y}$ and $\mathcal{X}$ over $\mathbb{Z}_p$.

Theorem 3.1 (Berthelot Comparison Theorem [Ber97, Shi02]) *For each $i \geq 0$, there is a canonical isomorphism*

$$H^i_{\mathrm{dR}}((\mathcal{Y},\mathcal{X})/\mathbb{Z}_{\mathrm{p}}) \simeq H^i_{\mathrm{crys}}(\mathcal{Y}_{\mathrm{p}},\mathcal{X}_{\mathrm{p}}).$$

The functoriality of crystalline cohomology thus equips the de Rham cohomology of $(\mathcal{Y},\mathcal{X})$ *with a Frobenius action.*

Let us remark that if $\mathcal{X}=\emptyset$ then $\mathrm{H}^i_{\mathrm{dR}}((\mathcal{Y},\mathcal{X})/\mathbb{Z}_{\mathrm{p}})$ is denoted by $\mathrm{H}^i_{\mathrm{dR}}(\mathcal{Y}/\mathbb{Z}_{\mathrm{p}})$ as it coincides with the usual de Rham cohomology of $\mathcal{Y}$. The analogous notational convention holds for the crystalline cohomology.

3.2 *Splitting the Cohomology of a Hypersurface*

We will first recall the characteristic zero statement regarding the splitting of the cohomology of a smooth hypersurface. Then, we will give the equivalent statement over $\mathbb{Z}_{\mathrm{p}}$.

3.2.1 Splitting in Characteristic Zero

Let K be a characteristic zero field and $X \subset \mathbb{P}^{n+1}_K$ a smooth hypersurface with complement U. By the Lefschetz hyperplane theorem and Poincaré duality, the only non-trivial cohomology group of X is in degree n. Moreover, there is a natural splitting of the cohomology of X which is orthogonal with respect to the cup product:

$$\mathrm{H}^n_{\mathrm{dR}}(X/K) \simeq \mathrm{H}^{n+2}_{\mathrm{dR}}(\mathbb{P}^{n+1}_K/K) \oplus^{\perp} \mathrm{H}^{n+1}_{\mathrm{dR}}(U/K). \qquad (3.2.1)$$

The projection onto the first factor, the cohomology of $\mathbb{P}^{n+1}_K$, has the following geometric interpretation. Thinking of the underlying complex analytic manifolds, there is a natural map from the singular homology of $\mathbb{P}^{n+1}_{\mathbb{C}}$ to the homology of $X_{\mathbb{C}}$ obtained by intersecting a homology class in $\mathbb{P}^{n+1}$ with X. The dual of this map gives the projection onto the first factor above. The kernel of this projection is called the *primitive part of the cohomology* and is canonically identified with $\mathrm{H}^{n+1}_{\mathrm{dR}}(U)$. Note that if n is odd then $\mathrm{H}^{n+2}_{\mathrm{dR}}(\mathbb{P}^{n+1}_K)$ is zero.

3.2.2 Splitting Over the Ring of Witt Vectors

We will now work over the Witt vectors $\mathbb{Z}_{\mathrm{p}}$ of the residue field k of characteristic p. We take $\mathcal{X} \hookrightarrow \mathbb{P}^{2r+1}_{\mathbb{Z}_{\mathrm{p}}}$ to be a smooth hypersurface of even dimension, the odd case being much simpler and of less interest for our purposes. Recall that the de Rham cohomology groups of $\mathbb{P}^{2r+1}_{\mathbb{Z}_{\mathrm{p}}}$, $\mathcal{X}$, and $(\mathbb{P}^{n+1}_{\mathbb{Z}_{\mathrm{p}}},\mathcal{X})$ admit canonical Frobenius actions via Theorem 3.1. The goal of this section is to state and prove Proposition 3.2, which is not explicitly stated in [AKR10] but follows from the arguments presented there.

To put Proposition 3.2 in context, let us recall some of the simpler results. From [AKR10, Corollary 3.1.4] we have

$$\mathrm{H}^i_{\mathrm{dR}}(\mathbb{P}^n_{\mathbb{Z}_p}) = \begin{cases} \mathbb{Z}_p & i = \{0, 2, \ldots, 2n\} \\ 0 & \text{otherwise}, \end{cases} \qquad \forall n \geq 0. \tag{3.2.2}$$

Furthermore, by arguing as in [Pan20, Lemma 2.4] or [DI87], we see that the cohomology groups $\mathrm{H}^i_{\mathrm{dR}}(\mathcal{X}/\mathbb{Z}_p)$ are $\mathbb{Z}_p$-torsion-free for all i. Comparing with characteristic zero, we conclude that:

$$\mathrm{H}^i_{\mathrm{dR}}(\mathcal{X}) = \begin{cases} \mathbb{Z}_p & i \in \{0, 2, \ldots, 2n\} \setminus \{2r\} \\ 0 & \text{otherwise}. \end{cases} \tag{3.2.3}$$

Proposition 3.2 *Suppose the degree of $\mathcal{X}$ is not divisible by p. Then, there is a natural Frobenius equivariant splitting of the cohomology of $\mathcal{X}$:*

$$\mathrm{H}^{2r}_{\mathrm{dR}}(\mathcal{X}/\mathbb{Z}_p) \simeq \mathrm{H}^{2r+2}_{\mathrm{dR}}(\mathbb{P}^{2r+1}_{\mathbb{Z}_p}/\mathbb{Z}_p)(1) \oplus \mathrm{H}^{2r+1}_{\mathrm{dR}}((\mathbb{P}^{n+1}_{\mathbb{Z}_p}, \mathcal{X})/\mathbb{Z}_p)(1).$$

Proof Let $\mathcal{Y} = \mathbb{P}^{2r+1}_{\mathbb{Z}_p}$. There is a long exact sequence associated to proper pairs [AKR10, Proposition 2.2.8]. Combining with (3.2.2) and ignoring the Frobenius action we get the following exact sequence:

$$0 \longrightarrow \mathrm{H}^{2r+1}_{\mathrm{dR}}((\mathcal{Y}, \mathcal{X})/\mathbb{Z}_p) \longrightarrow \mathrm{H}^{2r}_{\mathrm{dR}}(\mathcal{X}/\mathbb{Z}_p) \xrightarrow{\varsigma} \mathrm{H}^{2r+2}_{\mathrm{dR}}(\mathcal{Y}/\mathbb{Z}_p). \tag{3.2.4}$$

Corresponding to the inclusion $\iota\colon \mathcal{X} \hookrightarrow \mathcal{Y}$ we have pullback maps on cohomology

$$\iota_j^*\colon \mathrm{H}^j_{\mathrm{dR}}(\mathcal{Y}/\mathbb{Z}_p) \to \mathrm{H}^j_{\mathrm{dR}}(\mathcal{X}/\mathbb{Z}_p). \tag{3.2.5}$$

Let h be the first Chern class of the line bundle $\mathcal{O}_{\mathcal{Y}}(1)$ and h^k its k-th self product. It is easy to see that $\varsigma \circ \iota_{2r}^*(h^r) = \deg(\mathcal{X})h^{r+1}$. Since powers of h generate the cohomology of $\mathcal{Y}$, and because p does not divide the degree of $\mathcal{X}$, the composition $\varsigma \circ \iota_{2r}^*$ is an isomorphism. In particular, the last arrow in the sequence (3.2.4) is a surjection and the sequence splits.

To make the splitting equivariant with respect to the action of the Frobenius we must twist the two components. Recall that the cohomology of $\mathcal{X}$ is torsion-free. Thus each component injects into its tensor with $\mathbb{Q}_p$. Now use the discussion following Definition 2.3.3 and Proposition 2.4.1 itself in [AKR10] to conclude the proof. □

3.3 Torsion-Free Obstruction Space

Let $\mathrm{F}^\bullet \mathrm{H}^{2r}_{\mathrm{dR}}(\mathcal{X}/\mathbb{Z}_\mathfrak{p})$ denote the (decreasing) Hodge filtration on the cohomology of the smooth hypersurface $\mathcal{X} \subset \mathbb{P}^{2r+1}$.

Proposition 3.3 *For any $a, b \geq 0$ the quotient*

$$\mathrm{F}^a \mathrm{H}^{2r}_{\mathrm{dR}}(\mathcal{X}/\mathbb{Z}_\mathfrak{p})/\mathrm{F}^{a+b}\mathrm{H}^{2r}_{\mathrm{dR}}(\mathcal{X}/\mathbb{Z}_\mathfrak{p}) \tag{3.3.1}$$

is $\mathbb{Z}_\mathfrak{p}$-torsion-free.

Proof The Hodge to de Rham spectral sequence degenerates at E_1 [DI87, Ill94, Theorem 4.2.2]. Therefore, for any $s \geq 0$, the graded piece

$$\mathrm{Gr}_s := \mathrm{F}^s \mathrm{H}^{2r}_{\mathrm{dR}}(\mathcal{X}/\mathbb{Z}_\mathfrak{p})/\mathrm{F}^{s+1}\mathrm{H}^{2r}_{\mathrm{dR}}(\mathcal{X}/\mathbb{Z}_\mathfrak{p}) \tag{3.3.2}$$

is isomorphic to $\mathrm{H}^{2r-s}(\mathcal{X}, \Omega^s_{\mathcal{X}/\mathbb{Z}_\mathfrak{p}})$. Because the Hodge numbers of a smooth hypersurface depends only on degree and dimension (and not on the base field), we conclude that the graded pieces, Gr_s, are torsion-free.

The Hodge filtration on cohomology descends to a filtration on (3.3.1) whose non-zero graded pieces coincide with the graded pieces, $\mathrm{Gr}_\bullet$, of the cohomology. A filtered module with torsion-free graded pieces is torsion-free. □

Recall the set-up from Sect. 2. Let π_T be the obstruction map (1.1.3) from $T^r(X_\mathfrak{p})$ and let π_A be the obstruction map from $A^r(X_\mathfrak{p})$. Proposition 3.3 implies the following result.

Corollary 3.4 *A Tate class $v \in T^r(X_\mathfrak{p})$ is unobstructed if and only if a non-zero integral multiple of v is unobstructed. In symbols:*

$$\pi_T(v) = 0 \iff (\exists n \in \mathbb{Z}_{>0},\ \pi_T(nv) = 0)\,.$$

The equivalent formulation using π_A also holds. □

By itself, Corollary 3.4 does not imply that the cokernel of the inclusion map $A^r(\mathcal{X}) \hookrightarrow A^r(X_\mathfrak{p})$ is $\mathbb{Z}$-torsion-free. This is because the theorem of Bloch–Esnault–Kerz (Theorem 2.7) works with rational coefficients and thus implies only that a *multiple* of an obstructed class in $A^r(X_\mathfrak{p})$ can formally lift.

Nevertheless, for divisors on surfaces ($r = 1$), it is known that the cokernel of $A^1(X) \to A^1(X_\mathfrak{p})$ is torsion-free. This follows from the much stronger [Ray79, Theorem 4.1.2.1] or its generalization [EJ11b, Theorem 1.4].

3.4 Primitive Cohomology After Griffiths

In this section we will describe Griffiths' basis for the primitive cohomology $H_{\text{prim}} \subset \mathrm{H}^n_{\text{dR}}(\mathcal{X}/\mathbb{Z}_p)$ of a smooth hypersurface $\mathcal{X} \subset \mathbb{P}^{n+1}_{\mathbb{Z}_p}$ (Definition 3.9). The main result leading up to this definition is Proposition 3.8.

Let $\mathbb{P}$ denote $\mathbb{P}^{n+1}_{\mathbb{Z}_p}$ throughout this section and let d be the degree of $\mathcal{X}$. Recall that H_{prim} is naturally isomorphic to $H_{\log} := \mathrm{H}^{n+1}_{\text{dR}}((\mathbb{P}, \mathcal{X})/\mathbb{Z}_p)$. When tensored with $\mathbb{Q}_p$, this logarithmic cohomology coincides with the cohomology of the generic fiber U of the complement $\mathcal{U} := \mathbb{P} \setminus \mathcal{X}$ of our hypersurface $\mathcal{X}$. That is,

$$H_{\log} \otimes_{\mathbb{Z}_p} \mathbb{Q}_p \simeq \mathrm{H}^{n+1}_{\text{dR}}(U/\mathbb{Q}_p). \tag{3.4.1}$$

The $(n + 1)$-th de Rham cohomology of the affine variety U is readily computed. Consider the following map, induced by exterior differentiation on global sections,

$$F := \mathrm{H}^0(\mathbb{P}, \Omega^n_{\mathbb{P}/\mathbb{Z}_p}(n \cdot \mathcal{X})) \xrightarrow{\mathrm{d}} G := \mathrm{H}^0(\mathbb{P}, \Omega^{n+1}_{\mathbb{P}/\mathbb{Z}_p}((n+1) \cdot \mathcal{X})). \tag{3.4.2}$$

We will denote the quotient $G/\mathrm{d}F$ by Γ. Using characteristic 0 arguments [Voi07, §6.1], we know $\Gamma \otimes_{\mathbb{Z}_p} \mathbb{Q}_p \simeq \mathrm{H}^n_{\text{dR}}(U/\mathbb{Q}_p)$. In light of (3.4.1), we will compare $H_{\log}$ with Γ.

We have a filtration $G^\bullet$ on G induced by the pole order of forms:

$$G^j := \mathrm{im}\left(\mathrm{H}^0(\mathbb{P}, \Omega^{n+1}_{\mathbb{P}/\mathbb{Z}_p}((n+1-j) \cdot \mathcal{X})) \to \mathrm{H}^0(\mathbb{P}, \Omega^{n+1}_{\mathbb{P}/\mathbb{Z}_p}((n+1) \cdot \mathcal{X}))\right). \tag{3.4.3}$$

The induced filtration on Γ will be denoted by $\Gamma^\bullet$, so that $\Gamma^j = \mathrm{im}\left(G^j \to \Gamma\right)$.

Let $R = \mathbb{Z}_p[x_0, \dots, x_{n+1}]$ and $f \in R$ be a polynomial defining $\mathcal{X}$. Let $J = (\partial_0 f, \dots, \partial_{n+1} f)$ in R be the Jacobian ideal of f, where ∂_i denotes differentiation by x_i. We will write S for the quotient R/J. Homogeneous components of R, S, J will be denoted by subscripts. Let $N_j := (n + 1 - j)d - n - 2$ for $j = 0, \dots, n$. We have a natural isomorphism

$$R_{N_j} \xrightarrow{\sim} G^j : p \mapsto \frac{p}{f^{n+1-j}} \operatorname{vol}_{\mathbb{P}}, \tag{3.4.4}$$

here $\operatorname{vol}_{\mathbb{P}}$ is a natural generator of $\Omega^{n+1}_{\mathbb{P}/\mathbb{Z}_p}(n+2)$ given by

$$\operatorname{vol}_{\mathbb{P}} := \sum_{i=0}^{n+1} (-1)^i x_i \mathrm{d}x_0 \wedge \cdots \wedge \widehat{\mathrm{d}x_i} \wedge \cdots \wedge \mathrm{d}x_{n+1}. \tag{3.4.5}$$

Lemma 3.5 *If $p > n + 1$ and $p \nmid d$ then the map $R_{N_j} \twoheadrightarrow \Gamma^j$ induces an isomorphism*

$$(R/J)_{N_j} \xrightarrow{\sim} \Gamma^j/\Gamma^{j+1}.$$

Proof This is standard in characteristic 0 [Voi07, Gri69]. The image of $F \to G$ is generated by elements of the form

$$\left(\frac{g}{f^m} - \frac{gf}{f^{m+1}}\right) \mathrm{vol}_{\mathbb{P}} \quad \text{and} \quad \left(\frac{\partial g/\partial x_i}{f^m} - m\frac{g\partial f/\partial x_i}{f^{m+1}}\right) \mathrm{vol}_{\mathbb{P}} \tag{3.4.6}$$

each $m = 1, \dots, n+1$ and $g \in R_{N_{n+1-m}}$. Since $p \nmid d$, $f \in J$ and since $p > n+1$ we can divide by m in the relations above. It follows that, arguing as in [Gri69, §4], the pole order of an element in G can be reduced modulo $\mathrm{d}F$ if and only if it is in the image of J. □

Lemma 3.6 *If $p \nmid d$ then R/J, and in particular $(R/J)_N$ for each $N \geq 0$, is $\mathbb{Z}_{\mathrm{p}}$-torsion-free.*

Proof Let K stand for either the residue field or of the function field of $\mathbb{Z}_{\mathrm{p}}$. In either case, with $S = R/J$, $S \otimes_{\mathbb{Z}_{\mathrm{p}}} K$ is a local complete intersection ring of dimension zero. Since $J \otimes K$ is generated by $n+2$ elements forming a regular sequence each of degree $d-1$, the Hilbert series of $S \otimes K$ is independent of K. We used the smoothness of $\mathcal{X}$ here and that $f \in J$ when $p \nmid d$. Since $\dim_K S \otimes K$ is independent of K, S is torsion-free. □

Lemma 3.7 *If $p \nmid d$ and $p > n+1$ then Γ is torsion-free.*

Proof Lemmas 3.5 and 3.6 imply that the graded pieces of Γ are torsion-free. □

In light of the three lemmas above, we conclude that Γ injects into $\Gamma \otimes_{\mathbb{Z}_{\mathrm{p}}} \mathbb{Q}_{\mathrm{p}}$, which in turn is isomorphic to $H_{\log} \otimes \mathbb{Q}_{\mathrm{p}}$ and $H_{\mathrm{prim}} \otimes \mathbb{Q}_{\mathrm{p}}$. Recall that the de Rham cohomology of $\mathcal{X}$ is torsion-free and by Proposition 3.2 so is $H_{\log}$. Thus we have two injections:

$$\Gamma \hookrightarrow H_{\mathrm{prim}} \otimes \mathbb{Q}_{\mathrm{p}} \hookleftarrow H_{\log}, \quad \text{if } p \nmid d,\ p > n+1. \tag{3.4.7}$$

Proposition 3.8 *If $p \nmid d$ and $p > n+1$ then the images of Γ and $H_{\log}$ in (3.4.7) coincide. In particular, Γ is isomorphic to $H_{\log}$.*

Proof This is now a consequence of Remark 3.4.4 and Corollary 3.4.7 of [AKR10]. □

Impose the grevlex monomial ordering on R and on $R \otimes \mathsf{k} \simeq \mathbb{F}_q[x_0, \dots, x_{n+1}]$. As a result, there is a natural ordered monomial k-basis B'' for the quotient $R/J \otimes \mathsf{k}$. Let $B' \subset R$ be monomials that map to B''. By Nakayama's lemma, B' is a $\mathbb{Z}_{\mathrm{p}}$-basis for the free-module R/J. We will write $B \subset H_{\mathrm{prim}}$ for the image of $B' \cap \bigoplus_{j=0}^n (R/J)_{N_j}$ with respect to the following composition of maps:

$$\bigoplus_{j=0}^{n} (R/J)_{N_j} \twoheadrightarrow \Gamma \xrightarrow{\sim} H_{\mathrm{prim}}. \tag{3.4.8}$$

Definition 3.9 When $p \nmid d$ and $p > n + 1$ then $B \subset H_{\mathrm{prim}}$ will be called *the Griffiths basis* for the primitive cohomology of $\mathcal{X}$.

A particularly potent feature of the Griffiths basis is that it "respects the Hodge filtration" in the following sense. Write $B = (w_1, \dots, w_s)$ in increasing grevlex ordering. If $\mathrm{F}^\bullet H_{\mathrm{prim}}$ is the Hodge filtration on the primitive cohomology then

$$\forall j = 0, \dots, n, \ \exists i_j, \ \text{s.t. } \mathrm{F}^j H_{\mathrm{prim}} = \mathbb{Z}_{\mathrm{p}} \langle w_1, \dots, w_{i_j} \rangle. \tag{3.4.9}$$

This follows from the fact that pole order filtration on $\Gamma \otimes \mathbb{Q}_{\mathrm{p}}$ coincides with the Hodge filtrations on $\mathrm{H}^n_{\mathrm{dR}}(U/\mathbb{Q}_{\mathrm{p}})$ and on $H_{\mathrm{prim}} \otimes \mathbb{Q}_{\mathrm{p}}$ [Voi07].

3.4.1 Approximating the Frobenius Matrix in Terms of the Griffiths Basis

We comment briefly on how to compute an approximation of Frobenius action $\mathrm{Frob}_{\mathsf{k}}$ on the primitive cohomology H_{prim} of a hypersurface $\mathcal{X}$. We will focus on the even dimensional case, but the odd dimensional case requires minimal change. We recommend [AKR10], [Cos15] or [CHK19] for a more detailed discussion. We assume $p \nmid d$ and $p > n + 1$ throughout.

Recall $\mathcal{U} = \mathbb{P} \setminus \mathcal{X}$, $U = \mathcal{U} \otimes \mathbb{Q}_{\mathrm{p}}$ and that the de Rham cohomology $H_U := \mathrm{H}^{n+1}_{\mathrm{dR}}(U/\mathbb{Q}_{\mathrm{p}})$ of U is identified with $H_{\mathrm{prim}} \otimes \mathbb{Q}_{\mathrm{p}}$. Note also that H_{prim} is torsion-free and the natural inclusion

$$H_{\mathrm{prim}}(-1) \hookrightarrow H_{\mathrm{prim}} \otimes \mathbb{Q}_{\mathrm{p}} \xrightarrow{\sim} H_U \tag{3.4.10}$$

is equivariant with respect to the Frobenius action [AKR10, Proposition 2.4.1].

In the previous section, we represented elements of H_{prim} and of H_U using polynomials. We will work with the Griffiths basis $B \subset H_{\mathrm{prim}}$. Let $x^{\beta_1}, \dots, x^{\beta_s} \in \mathbb{Z}_{\mathrm{p}}[x_0, \dots, x_{n+1}]$ be the monomials that map to B. The image of B in H_U is given by (3.4.4). Namely, with $l_i = \big(\deg\big(x^{\beta_i}\big) + n + 2\big)/d$ the following elements give the corresponding basis in H_U:

$$\eta_i := \frac{x^{\beta_i}}{f^{\ell_i}} \mathrm{vol}_{\mathbb{P}} \bmod G, \quad i = 1, \dots, s. \tag{3.4.11}$$

Here, one switches to another cohomology theory. Since U is affine, H_U can be computed using the Monsky–Washnitzer cohomology. The advantage of Monsky–Washnitzer cohomology in this context is that it provides great flexibility in how one can choose to represent the Frobenius matrix in the chain of forms that compute the cohomology [AKR10, Definition 2.4.2].

The Frobenius action can be described on η_i and Griffiths–Dwork reduction allows one to re-cast $\mathrm{Frob}_{\mathsf{k}}(\eta_i)$ in terms of the Griffiths basis again. The catch is that $\mathrm{Frob}_{\mathsf{k}}(\eta_i)$ becomes an infinite sum, each term having coefficients with higher and higher valuation. Truncating the infinite sequence and applying the Griffiths–

Dwork reduction gives an approximation of $\mathrm{Frob}_\mathfrak{k}(\eta_i)$ in the Griffiths basis. The degree of truncation required to attain the desired precision is discussed in [AKR10, §3.4].

4 Main Algorithm

This section states and explains the steps of the main algorithm of this paper, Algorithm 4.1. The version here provides an upper bound for the geometric middle Picard number of a given smooth hypersurface. A simple variation gives bounds on the geometric Picard number of a Jacobian, see Sect. 5.1. More sophisticated variations allow for the study of non-degenerate hypersurfaces in simplicial toric varieties [CHK19].

The algorithm takes the equation $f \in K[x_0, \ldots, x_{2r+1}]$ of a smooth hypersurface $X \subset \mathbb{P}_K^{2r+1}$ as input. The output is an upper bound for the geometric middle Picard number, $\rho^r(X \times_K \overline{K})$, of X. Optionally, one may choose to specify a lower bound for the precision N used to approximate the Frobenius map. For small N, the upper bound may improve as N is increased. The bound will eventually stabilize. One may also provide a lower bound on the characteristic p of the prime $\mathfrak{p} \subset \mathcal{O}_K$ to be used for the good reduction of X.

4.1 *Clarification of the Steps in the Algorithm*

4.1.1 Pick a Good Prime

Pick the first prime number p exceeding $2r + 6$ and `char_bound`. Choose an unramified prime $\mathfrak{p}$ of K lying above p. By clearing the denominators of the polynomial f defining X, create a model $\mathcal{X}/\mathbb{Z}_\mathfrak{p}$ of X. Check if the model $\mathcal{X}/\mathbb{Z}_\mathfrak{p}$ is smooth. If not, pick another prime and repeat.

It is possible to eliminate the simple but haphazard searching method above: Compute the discriminant of X over $\mathbb{Z}$ and avoid the primes dividing this discriminant. However, discriminants tend to be huge.

4.1.2 Compute a Griffiths Basis for Primitive Cohomology

In Sect. 3.4 we described how to compute a Griffiths basis for the primitive part of $\mathrm{H}^{2r}_{\mathrm{dR}}(\mathcal{X}/\mathbb{Z}_\mathfrak{p})$. Choosing the polarization to complete the basis, we represent the arithmetic Frobenius map $\mathrm{Frob}_\mathfrak{k}$ on the cohomology $\mathrm{H}^{2r}_{\mathrm{dR}}(\mathcal{X}/\mathbb{Z}_\mathfrak{p}) \simeq \mathrm{H}^{2r}_{\mathrm{crys}}(X_\mathfrak{p}/\mathbb{Z}_\mathfrak{p})$ via the Berthelot comparison theorem, see Theorem 3.1. This square matrix with $\mathbb{Z}_\mathfrak{p}$ entries will be denoted by $[\mathrm{Frob}_\mathfrak{k}]$.

```
Input  : f ∈ O_K[x_0, x_1, ..., x_{2r+1}]; Optional: precision_bound, char_bound ∈ N
Output: An upper bound for the geometric middle Picard number of X = Z(f).
function BoundRank (f; precision_bound = 1, char_bound = 3)
    // Pick a good prime, with residue characteristic at least char_bound, Sect. 4.1.1
    p := next_good_prime (f, max (2r + 6, char_bound) )
    // Compute a Griffiths basis for primitive cohomology, Sect. 4.1.2
    B := griffiths_basis (f, p)
    // Compute a minimal working precision, Sect. 4.1.3
    q := #k, m := #B + 1
    N := max (precision_bound, log_p (2 binom(m, ⌈m/2⌉) q^{r⌈m/2⌉}))  // See Eq. (4.1.3)
    // Compute an N-digit approximation of the Frobenius matrix, Sect. 4.1.4
    [Frob_{k,N}] := frobenius (f, p, N, B)
    // Compute the characteristic polynomial of the Frobenius matrix, Sect. 4.1.5
    χ(q^{-r} Frob_k) := weil_characteristic_polynomial(Frob_{k,N}, N)
    // Represent the obstruction map, Sect. 4.1.6
    [π_X] := projection_map (B)  // π_X : H^{2r}_dR(X) → H^{2r}_dR(X)/F^r H^{2r}_dR(X)
    // Loop over the cyclotomic factors of χ(q^{-r} Frob_k), Sect. 4.1.7
    bound := 0
    for Φ_i^{γ_i} ∈ cyclotomic_factors(χ(q^{-r} Frob_k)) do
        // Approximate the space of Tate classes, Sect. 4.1.8
        B_{T_{i,N}} :=tate_basis_matrix (Φ_i, Frob_{k,N}, N)  // columns form a basis
          for T_{i,N}
        // Approximate the map π_i, Sect. 4.1.9
        [π_{i,N}] := [π_X]· column_matrix (1, [Frob_{k,N}], ··· , [Frob_{k,N}]^{deg Φ_i − 1}) ·B_{T_{i,N}}
        // Increment by the bound on the dimension of L_i := ker π_i, Sect. 4.1.10
        bound := bound + deg Φ_i^{γ_i} − bound_rank([π_{i,N}])
    return bound.
```

Algorithm 4.1: Computing an upper bound for the geometric middle Picard number

4.1.3 Compute a Minimal Working Precision

The dimension of the middle cohomology of a hypersurface of dimension $2r$ and degree d is given by the formula

$$m := \frac{(d-1)^{2r+2} + 2d - 1}{d}. \tag{4.1.1}$$

In any case, $m = \#B + 1$ where B is the Griffiths basis for the primitive cohomology.

The Weil conjectures imply that the reciprocal characteristic polynomial of $\mathrm{Frob}_{\mathrm{k}}$ on $\mathrm{H}^{2r}_{\mathrm{dR}}(\mathcal{X}/\mathbb{Z}_{\mathfrak{p}})$ has integer coefficients with constant term equal to 1, i.e.,

$$P_{2r}(X, t) := \det(1 - t\,\mathrm{Frob}_{\mathrm{k}}) \in 1 + t\mathbb{Z}[t]. \tag{4.1.2}$$

Moreover, this polynomial is completely determined by the coefficients of t^i with $i = 0, \ldots, \lceil m/2 \rceil$, and these coefficients have absolute value at most

$$\binom{m}{\lceil m/2 \rceil} q^{r\lceil m/2 \rceil}. \tag{4.1.3}$$

Thus if p^N exceeds twice this bound, then $P_{2r}(X, t)$ is determined by its reduction modulo p^N. The bound above might be significantly improved by employing Newton identities, see [Ked13, slide 8]. If the precision requested by the user is not sufficient to lift $P_{2r}(X, t)$, we increase it accordingly.

4.1.4 Compute an N-digit Approximation of the Frobenius Matrix

With N as above, we may compute a matrix $[\mathrm{Frob}_{\mathrm{k},N}]$ approximating the matrix $[\mathrm{Frob}_{\mathrm{k}}]$ to N p-adic digits as explained in [AKR10], [Cos15] or [CHK19]. We sketched the idea in Sect. 3.4.1. In practice, one may use the library `controlledreduction`.[2]

4.1.5 Compute the Characteristic Polynomial of the Frobenius Matrix

As explained above we may deduce P_{2r} from $\det(1 - t\,\mathrm{Frob}_{\mathrm{k},N})$. In practice, we compute $\chi(\mathrm{Frob}_{\mathrm{k},N})(t) \equiv \chi(\mathrm{Frob}_{\mathrm{k}})(t) \bmod p^N$ and lift each coefficient to the interval $[-p^N/2, p^N/2]$. By the discussion above, the representatives of the last coefficients of t^i for $i = \lceil m/2 \rceil, \ldots, m$ match exactly their lifts, and the remaining coefficients are deduced using the functional equation

$$\chi(\mathrm{Frob}_{\mathrm{k}})(t) = \pm \left(q^{-r} t\right)^m \chi(\mathrm{Frob}_{\mathrm{k}})\left(q^{2r}/t\right). \tag{4.1.4}$$

The sign of the functional equation is the sign of $\det(\mathrm{Frob}_k)$, which we are able to compute to one significant p-adic digit.

4.1.6 Represent the Obstruction Map

The obstruction map π_X (see Sect. 2.1.6) annihilates the polarization. Thus, it remains to describe π_X on the primitive cohomology. Since the Griffiths basis on the primitive cohomology respects the filtration (see Sect. 3.4), the map π_X on the primitive cohomology in the Griffiths basis is just the projection onto the last few coordinates. Let $[\pi_X]$ be the matrix representation of this projection.

[2] This library is made available in `SageMath` [Sag20] through the wrapper https://github.com/edgarcosta/pycontrolledreduction.

4.1.7 Extract Cyclotomic Factors from the Characteristic Polynomial

Factorize the characteristic polynomial over $\mathbb{Q}[t]$ as in (2.3.1).

4.1.8 Approximate the Space of Tate Classes

Let $\Phi_i(t)$ be a cyclotomic factor computed in the previous step, let $B_{T_{i,N}}$ be a basis for the eigenspace $T_{i,N} := \ker\left(\Phi_i\left(q^{-r}\,\mathrm{Frob}_{\mathsf{k},N}\right)\right)$. This $T_{i,N}$ approximates $T_i^r(X_{\mathfrak{p}}) := \ker\left(\Phi_i\left(q^{-r}\,\mathrm{Frob}_{\mathsf{k}}\right)\right)$ from Proposition 2.16.

The kernels are computed using standard algorithms that keep track of the p-adic precision, see [CRV18]. The computations equip $T_{i,N}$ with a basis which we record in the columns of a matrix $B_{T_{i,N}}$.

4.1.9 Approximate the Map π_i

Restricting $\pi_{\mathcal{X}}$ on to $T_{i,N}$, we obtain the approximation

$$\pi_{T_{i,N}}\colon T_{i,N} \to \mathrm{H}^{2r}_{\mathrm{dR}}(\mathcal{X})/\mathrm{F}^r\mathrm{H}^{2r}_{\mathrm{dR}}(\mathcal{X}) \tag{4.1.5}$$

of the obstruction map π_{T_i} on $T_i^r(X_{\mathfrak{p}})$. Therefore, the matrix representation $[\pi_{T_{i,N}}]$ of $\pi_{T_{i,N}}$ is then equal to $[\pi_{\mathcal{X}}] \cdot B_{T_{i,N}}$. Similarly, we may approximate the map

$$\begin{aligned}\pi_i\colon T_i^r(X_p) &\to \left(\mathrm{H}^{2r}_{\mathrm{dR}}(\mathcal{X})/\mathrm{F}^r\mathrm{H}^{2r}_{\mathrm{dR}}(\mathcal{X})\right)^{\deg\Phi_i}\\ v &\mapsto (\pi(v), \pi\circ q^{-r}\,\mathrm{Frob}_{\mathsf{k}}(v), \dots, \pi\circ (q^{-r}\,\mathrm{Frob}_{\mathsf{k}})^{\deg\Phi_i-1}(v)),\end{aligned} \tag{4.1.6}$$

from Sect. 2.4, is approximated by

$$[\pi_{i,N}] := [\pi_{\mathcal{X}}]\cdot\begin{pmatrix} 1 \\ q^{-r}[\mathrm{Frob}_{\mathsf{k},N}] \\ \vdots \\ (q^{-r}[\mathrm{Frob}_{\mathsf{k},N}])^{\deg\Phi_i-1}\end{pmatrix}\cdot B_{T_{i,N}}. \tag{4.1.7}$$

4.1.10 Bound Dimension of L_i

The approximation $\pi_{i,N}$ of π_i allows for the computation of a lower bound $b_{i,N}$ on the rank of π_i, see for example [AKR10, Algorithm 1.62]. Although we will not know when this happens, if N is large enough, then the rank $b_{i,N}$ of $\pi_{i,N}$ will match the rank of π_i. Nonetheless, as $L_i := \ker\pi_i$, we have

$$\dim L_i = \dim T_i - \mathrm{rk}\,\pi_i \le \dim T_i - b_{i,N}. \tag{4.1.8}$$

4.1.11 Return the Upper Bound on the Middle Picard Rank

From the previous argument at the end of the for loop we have $\sum_i \dim_{\mathbb{Q}_p} L_i \leq$ `bound`. Combining this with Proposition 2.22 we obtain the sought inequality:

$$\dim_{\mathbb{Q}} A^r(X_{\overline{K}}) \leq \sum_i \dim_{\mathbb{Q}_p} L_i \leq \texttt{bound}. \tag{4.1.9}$$

Remark 4.1 Let us warn once again that, even in the favorable conditions provided by K3 surfaces, the inequality (4.1.9) can be sharp. See Example 5.8 for a demonstration.

5 Examples

We now give explicit illustrations of the methods developed in this paper. We have implemented a version of Algorithm 4.1, called `crystalline_obstruction`,[3] where the prime is also given as input. We will show its basic usage below.

There are three sets of examples below. In the first set, we work with Jacobians of curves. The Hodge structure on their cohomology is inherited from the Hodge structure of the corresponding curves. The advantage is that the dimension of the cohomology is small enough that we can write the Frobenius matrices explicitly.

In the second set, we illustrate the basic usage on surfaces in projective space. We checked all of the Picard numbers in the *quartic database*[4] of [LS19]. In every case, our upper bounds agreed with the numbers listed there. We also comment on the performance gain in using the obstruction method.

In the third set, we give pathological examples. For example, the obstruction space associated to a quintic surface is four dimensional. We give an example where the image of the obstructed Tate classes span a one dimensional space although four dimensions of algebraic classes must be obstructed.

Our convention for writing a p-adic number modulo p^N is as follows: given $a \in \mathbb{Q}_p$ we write $a \equiv p^m \cdot b \bmod p^N$ where $m \in \mathbb{Z}, b \in \mathbb{Z}_{\geq 0}, p \nmid b$.

5.1 *Jacobians of Plane Curves*

The discussion on hypersurfaces above allows us to compute the Hodge decomposition on the first cohomology of a smooth curve. Since the cohomology of the

---

[3]For its implementation in `SageMath` [Sag20], see https://github.com/edgarcosta/crystalline_obstruction.

[4]One may view the database at https://pierre.lairez.fr/quarticdb/.

Jacobian is isomorphic to the natural Hodge structure on the wedge powers of the first cohomology of the curve, we can treat Jacobians explicitly. We can thus use Algorithm 4.1 with minor modifications, as illustrated below. The purpose of this change in context is to be able to display complete examples in print.

Example 5.1 We start with a genus 2 curve

$$C/\mathbb{Q}\colon y^2 - (4x^5 - 36x^4 + 56x^3 - 76x^2 + 44x - 23) = 0. \tag{5.1.1}$$

Choose the prime $\mathfrak{p} = (31)$ and write f for the equation of C. This curve has LMFDB label 1104.a.17664.1. Let J denote the Jacobian of C. We will compute the geometric Picard number of J.

When given a curve, the code `crystalline_obstruction` understands that the intention is to compute with the Jacobian of the curve. The command is simple:

```
sage: from crystalline_obstruction import crystalline
      _obstruction
sage: crystalline_obstruction(f, p=31, precision=3)
```

The first entry of the output is the upper bound (in this case sharp) on the geometric Picard number of J, and the second is a dictionary recording relevant intermediate results. The entire computation for this example takes less than a second giving the output:

```
(1, {'precision': 3, 'p': 31, 'rank T(X_Fpbar)': 2,
  'factors': [(t - 1, 2)],
      'dim Ti': [2], 'dim Li': [1]})
```

We will now walk through the intermediate steps of the computation. We first check that J has good reduction over $\mathfrak{p}$. Let $\mathcal{C}$ and $\mathcal{J}$ denote the natural models of C and J over $\mathbb{Z}_\mathfrak{p}$, respectively. As discussed in Sect. 3, we may compute an approximation of $\mathrm{Frob}_\mathfrak{k}$ on $\mathrm{H}^1_{\mathrm{dR}}(\mathcal{C}/\mathbb{Z}_\mathfrak{p}) \otimes_{\mathbb{Z}_\mathfrak{p}} \mathbb{Q}_\mathfrak{p} \simeq \mathrm{H}^1_{\mathrm{dR}}(\mathcal{J}/\mathbb{Z}_\mathfrak{p}) \otimes_{\mathbb{Z}_\mathfrak{p}} \mathbb{Q}_\mathfrak{p}$ and its characteristic polynomial:

$$\mathrm{Frob}\,|_{\mathrm{H}^1_{\mathrm{dR}}(\mathcal{C}/\mathbb{Q}_\mathfrak{p})} \equiv \begin{pmatrix} 31\cdot 482 & 31\cdot 284 & 16241 & 3075 \\ 31\cdot 386 & 31\cdot 886 & 2644 & 12126 \\ 31\cdot 284 & 31\cdot 659 & 6336 & 9750 \\ 31\cdot 194 & 31\cdot 876 & 27408 & 10841 \end{pmatrix} \pmod{31^3}, \tag{5.1.2}$$
$$\chi = 1 - 3t + 14t^2 - 93t^3 + 961t^4.$$

The natural Hodge structures on both sides of the isomorphism $\mathrm{H}^2_{\mathrm{dR}}(\mathcal{J}/\mathbb{Q}_\mathfrak{p}) \simeq \Lambda^2\mathrm{H}^1_{\mathrm{dR}}(\mathcal{J}/\mathbb{Q}_\mathfrak{p})$ agree with one another. Thus the wedge products of the Griffiths basis we used for $\mathcal{C}$ will reveal the Hodge structure on higher cohomologies of $\mathcal{J}$.

From χ and the approximate Frobenius above we can deduce

$$\det(t - \mathrm{Frob}\,|_{\mathrm{H}^2_{\mathrm{crys}}(\mathcal{J}_{\mathfrak{p}})(1)}) = (t-1)^2(31t^4 + 48t^3 + 43t^2 + 48t + 31)/31,$$

$$\mathrm{Frob}\,|_{\mathrm{H}^2_{\mathrm{dR}}(\mathcal{J}/\mathbb{Q}_{\mathfrak{p}})} \equiv \begin{pmatrix} 31^2\cdot 19 & 31\cdot 660 & 31\cdot 776 & 31\cdot 843 & 31\cdot 506 & 22499 \\ 31^2\cdot 18 & 31\cdot 250 & 31\cdot 459 & 31\cdot 270 & 31\cdot 683 & 10699 \\ 31^2\cdot 3 & 31\cdot 154 & 31\cdot 636 & 31\cdot 261 & 31^2\cdot 24 & 3010 \\ 31^2\cdot 22 & 31\cdot 557 & 31\cdot 664 & 31\cdot 392 & 31^2\cdot 23 & 10438 \\ 31^2\cdot 30 & 31\cdot 77 & 31\cdot 516 & 31^2\cdot 26 & 31\cdot 449 & 3650 \\ 31^2\cdot 7 & 31\cdot 668 & 31\cdot 509 & 31\cdot 277 & 31\cdot 513 & 17591 \end{pmatrix} \pmod{31^3}, \tag{5.1.3}$$

and $T^1(\mathcal{J}_{\mathfrak{p}}) = \mathrm{Span}(v_1, v_2)$, where, having lost a digit of precision, we have

$$\begin{aligned} v_1 &\equiv (356,\ 37,\ 831,\ 0,\ 295,\ 31) \pmod{31^2} \\ v_2 &\equiv (4,\ 957,\ 3,\ 1,\ 0,\ 0) \pmod{31^2}. \end{aligned} \tag{5.1.4}$$

The Hodge numbers for $\mathrm{H}^2_{\mathrm{dR}}(\mathcal{J}/\mathbb{Q}_{\mathfrak{p}})$ are $(1, 4, 1)$, and the projection

$$\pi_{\mathcal{J}} : \mathrm{H}^2_{\mathrm{dR}}(\mathcal{J}/\mathbb{Q}_{\mathfrak{p}}) \to \mathrm{H}^2_{\mathrm{dR}}(\mathcal{J}/\mathbb{Q}_{\mathfrak{p}})/\mathrm{F}^1\mathrm{H}^2_{\mathrm{dR}}(\mathcal{J}/\mathbb{Q}_{\mathfrak{p}}) \simeq \mathbb{Q}_{\mathfrak{p}} \tag{5.1.5}$$

corresponds to the projection onto the last coordinate in the basis chosen above. This gives

$$\pi_T \equiv \begin{pmatrix} 31 & 0 \end{pmatrix} \pmod{31^2} \tag{5.1.6}$$

in the basis v_1, v_2 for $T^1(\mathcal{J}_{\mathfrak{p}})$. Thus, the rank of the obstruction map π_T is at least 1. This is the first entry in the output of our code.

We also know that the polarization lifts and thus $\dim_{\mathbb{Q}} A^1(J_{\overline{\mathbb{Q}}})_{\mathbb{Q}} = \mathrm{rk}\,\mathrm{End}(J_{\overline{\mathbb{Q}}}) = 1$.

Example 5.2 Let us now look at an example where the rank of the Jacobian is known to be 2. In this case, an upper bound of 2 will not prove that the rank is indeed 2. However, our numerical methods has the advantage that *if* the rank was 1 then the obstruction map would be non-zero at high enough precision. Observing that the obstruction is zero to higher and higher precision would be compelling evidence that the rank is 2.

We pick the genus 2 curve $C: y^2 = x^5 - 2x^4 + 7x^3 - 5x^2 + 8x + 3$ over $\mathbb{Q}$ with LMFDB label 30976.a.495616.1. Mutatis mutandis, we may repeat Example 5.1, for $\mathfrak{p} \neq (2), (11)$. However, the Jacobian J of C will now have real multiplication by $\mathbb{Q}(\sqrt{17})$, see [CMSV19]. Therefore, $\dim_{\mathbb{Q}} A^1(J_{\overline{\mathbb{Q}}}) = 2$.

Working with finite precision, it is impossible conclude that the obstruction map is identically zero. We observe however that even over a large prime such as $p = 4999$ the obstruction map on the two-dimensional $T^1(\mathcal{J}/\mathbb{Q}_{\mathfrak{p}})$ is zero modulo p^{100}. Even with these large numbers, the computation took about 3 min.

Example 5.3 In this example, we will show how one can obstruct more than one linearly independent cycle at a single prime. Unlike [EJ11b], this method does not require an investigation of the geometry of algebraic cycles.

Consider the following genus 3 plane curve, its defining equation will be denoted by f:

$$C/\mathbb{Q}\colon xy^3 + x^3z - xy^2z + x^2z^2 + y^2z^2 - yz^3 = 0. \tag{5.1.7}$$

The following command takes less than a second to return the answer, which shows that $\dim A^1(\overline{J}) = 1$ where $J/\mathbb{Q}$ is the Jacobian of C:

```
sage: crystalline_obstruction(f, p=31, precision=3)
(1, {'precision': 3, 'p': 31, 'rank T(X_Fpbar)': 3,
  'factors': [(t - 1, 3)],
     'dim Ti': [3], 'dim Li': [1]})
```

With $\mathfrak{p} = (31)$ and $\mathcal{C}/\mathbb{Z}_\mathfrak{p}, \mathcal{J}/\mathbb{Z}_\mathfrak{p}$ as in Example 5.1, we demonstrate the intermediate steps:

$$\chi = 1 + 78t^2 + 408t^3 + 2418t^4 + 7688t^5 + 29791t^6$$

$$\mathrm{Frob}\,|_{\mathrm{H}^1_{\mathrm{dR}}(\mathcal{C}/\mathbb{Q}_\mathfrak{p})} = \begin{pmatrix} 31\cdot 104 & 31\cdot 218 & 31\cdot 7 & 27783 & 2569 & 7195 \\ 31\cdot 351 & 31\cdot 494 & 31\cdot 690 & 19524 & 8323 & 1421 \\ 31\cdot 50 & 31\cdot 237 & 31\cdot 829 & 13467 & 20050 & 19610 \\ 31\cdot 19 & 31\cdot 733 & 31\cdot 377 & 20592 & 23805 & 15085 \\ 31\cdot 482 & 31\cdot 610 & 31\cdot 793 & 12397 & 28951 & 6604 \\ 31\cdot 710 & 31\cdot 860 & 31\cdot 689 & 19294 & 18382 & 25376 \end{pmatrix} \pmod{31^3}.$$

Now the Hodge numbers for $\mathrm{H}^2_{\mathrm{dR}}(\mathcal{J}/\mathbb{Q}_\mathfrak{p})$ are $(3, 9, 3)$ and we have

$$\begin{aligned} \det(t - \mathrm{Frob}\,|_{\mathrm{H}^2_{\mathrm{crys}}(\mathcal{J}_\mathfrak{p})(1)}) &= \frac{1}{31^3}(t-1)^3(31^3 + 15\cdot 31^2 t - 642\cdot 31 t^2 \\ &\quad + 14545t^3 + 36060t^4 + 18063t^5 + 3172t^6 + 18063t^7 + \cdots). \end{aligned} \tag{5.1.8}$$

We find that the space of Tate classes, $T^1(\mathcal{J}/\mathbb{Q}_\mathfrak{p}) = \ker(\mathrm{Frob}_\mathrm{k} - 1)$, is three-dimensional. The obstruction map π_T is approximated by the following matrix with respect to a basis:

$$\pi_T \equiv \begin{pmatrix} 31\cdot 240 & 0 & 31\cdot 1 \\ 31\cdot 515 & 31\cdot 1 & 0 \\ 0 & 0 & 0 \end{pmatrix} \pmod{31^3}. \tag{5.1.9}$$

Thus, $\mathrm{rk}\,\pi_T \geq 2$ and we conclude $\dim A^1(J_{\overline{\mathbb{Q}}}) = \dim \mathrm{End}(J_{\overline{\mathbb{Q}}}) = 1$.

Example 5.4 We may also use our method to directly bound the dimension of the geometric endomorphism algebra of an abelian variety. We do this via divisorial correspondences.

If X and Y are two projective smooth varieties over a number field K, then

$$\begin{aligned}\mathrm{NS}(X \times Y) &= \mathrm{NS}(X) \oplus \mathrm{NS}(Y) \oplus \mathrm{DC}(X, Y) \\ &\subset \mathrm{H}^2_{\mathrm{dR}}(X/K) \oplus \mathrm{H}^2_{\mathrm{dR}}(Y/K) \oplus \mathrm{H}^1_{\mathrm{dR}}(X/K) \otimes \mathrm{H}^1_{\mathrm{dR}}(Y/K),\end{aligned} \tag{5.1.10}$$

where NS denotes the Néron–Severi group and DC denotes divisorial correspondences. The Néron–Severi group NS coincides with our A^1.

Furthermore, by [Lan83, VI §2 Thm 2] we have the following isomorphism

$$\mathrm{DC}(X, Y) \simeq \mathrm{Hom}(\mathrm{Alb}(X), \mathfrak{P}(Y)), \tag{5.1.11}$$

where $\mathrm{Alb}(X)$ is the Albanese variety of X and $\mathfrak{P}(Y)$ is the Picard variety of Y.

Note that the Hodge structure of the tensor $\mathrm{H}^1_{\mathrm{dR}}(\mathcal{X}/\mathbb{Q}_p) \otimes \mathrm{H}^1_{\mathrm{dR}}(\mathcal{Y}/\mathbb{Q}_p)$ is induced by the Hodge structure of its two factors. Therefore, knowing the Hodge structure on the factors allows us to apply our method to bound the dimension of $\mathrm{DC}(X_{\overline{\mathbb{Q}}}, Y_{\overline{\mathbb{Q}}})$.

For example, taking $X = Y = C$ for a curve C/K, we can bound the dimension of the endomorphism algebra $\mathrm{End}(J_{\overline{\mathbb{Q}}})$ of the Jacobian J of C. The method is analogous to the previous examples. We let $\mathfrak{J}/\mathbb{Z}_p$ denote the natural model of J and replace $\mathrm{H}^2_{\mathrm{dR}}(\mathfrak{J}/\mathbb{Q}_p)$ with $\mathrm{H}^1_{\mathrm{dR}}(\mathfrak{J}/\mathbb{Q}_p) \otimes \mathrm{H}^1_{\mathrm{dR}}(\mathfrak{J}/\mathbb{Q}_p)$ in the computations.

Our code automates this procedure with the keyword argument `tensor=True`. For example, take the smooth plane curve

$$C/\mathbb{Q}:\ -y^4 + x^3z + 2x^2z^2 - xz^3 = 0. \tag{5.1.12}$$

The following commands show that $\dim \mathrm{End}(J_{\overline{\mathbb{Q}}}) \leq 9$ and $\dim \mathrm{NS}(J_{\overline{\mathbb{Q}}}) \leq 5$, respectively.

```
sage:crystalline_obstruction(f, p=43, precision=4,
 tensor=True)
(9, {'precision': 4, 'p': 43, 'rank T(X_Fpbar)': 18,
  'factors': [(t + 1, 4), (t - 1, 6), (t^2 + 1, 4)],
  'dim Ti': [4, 6, 8], 'dim Li': [2, 3, 4]})
sage:crystalline_obstruction(f, p=43, precision=4)
(5, {'precision': 4, 'p': 43, 'rank T(X_Fpbar)': 7,
  'factors': [(t + 1, 2), (t - 1, 3), (t^2 + 1, 1)],
  'dim Ti': [2, 3, 2], 'dim Li': [1, 2, 2]})
```

5.2 *Surfaces in Projective Space*

Quartic surfaces and quintic surfaces are the prime examples in this section. We present some successful examples and some examples that demonstrate inherent weaknesses of the method.

5.2.1 K3 Surfaces

We begin with quartic smooth surfaces in $\mathbb{P}^3$. The middle cohomology of a K3 surface has dimension 22 with Hodge numbers $(1, 20, 1)$.

Example 5.5 We applied our Algorithm 4.1 to each of the 184,725 quartic surfaces in the *quartic database*[5] [LS19]. For each quartic, we found an upper bound that agreed with the Picard number listed there (and, as expected, could not find a smaller upper bound).

We will now compare the performance improvement of using the obstruction map. We distinguish between three different methods: The *Galois obstruction method* applies Algorithm 4.1 thereby using the obstruction map on each of the Galois orbits separately as in Proposition 2.22. The *vanilla obstruction method* deviates from Algorithm 4.1 only in ignoring the Galois structure on the space of Tate classes, that is, the obstruction map is considered on the entire space of Tate classes. For each surface, we start with the prime 23 and move up to the next prime if the upper bound does not match the number listed in the database or if the prime is not a good prime. We skip the primes less than 23, as the computation to deduce the Hasse–Weil zeta for these requires more precision [Cos15, Section 1.6.2], and hence more time.

In applying the *van Luijk method* [vLui07] to each surface, it would make sense to stop after *two primes* have been found where the upper bound is optimal, whereby the discriminants can be compared. This means it would require, on average, twice as many prime reductions as the vanilla obstruction method. Almost the entirety of the computation per prime is spent on computing the zeta function, a feature common to all three methods. Therefore, the van Luijk method would take roughly twice as long as the vanilla obstruction method. On Fig. 5.1, its slope would be half that of the vanilla obstruction—or about one fifth of the Galois obstruction method.

The entire computation for the Galois obstruction method took about 10 months of CPU time. The vanilla obstruction method takes about 16 months. Arguing as above, we expect the van Luijk method to take about 32 months. The Figs. 5.1 and 5.2 give more detailed information.

[5] One may explore the database at https://pierre.lairez.fr/quarticdb/.

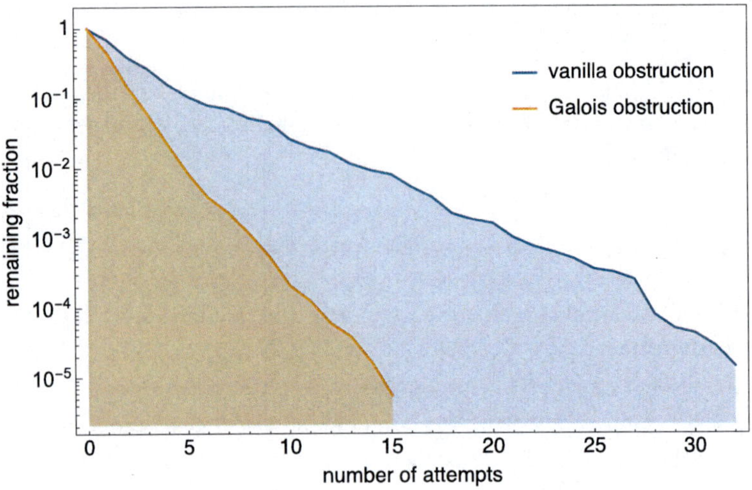

Fig. 5.1 The fraction of surfaces from the quartic database that needed a certain number of prime reductions until sharp upper bound is attained

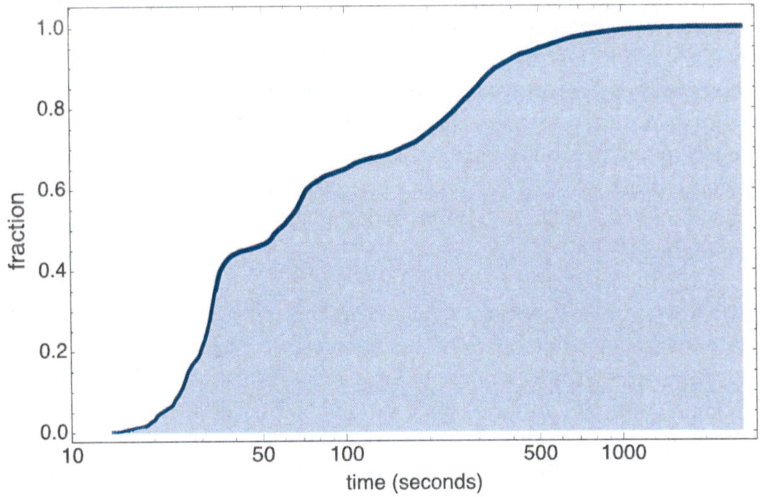

Fig. 5.2 Fraction of surfaces from the quartic database against time needed until a sharp bound is attained via Galois obstruction method

Example 5.6 In this example, we give a simple but artificial example which demonstrates that there is no *global* upper bound on how much precision must be used to reach the optimal bounds of the method.

Fix a prime p and let $\lambda = p^m$ with large $m \gg 0$. Consider the following two elements of the Dwork pencil

$$X/\mathbb{Q}\colon\; x^4 + y^4 + z^4 + w^4 + \lambda xyzw = 0, \tag{5.2.1}$$

$$Y/\mathbb{Q}\colon\; x^4 + y^4 + z^4 + w^4 = 0. \tag{5.2.2}$$

The Fermat quartic Y has Picard number 20 and it is known that Y_p has Picard number 20 when $p \equiv 1 \bmod 4$. Clearly, the two models X and Y lifting $X_p = Y_p$ are indistinguishable modulo p^m. If X has Picard number 19, then the obstruction map will be zero modulo p^m as it will not be able to make the distinction between X and Y. Although, when the p-adic precision is sufficiently large, the obstruction becomes non-zero.

Example 5.7 Take $\mathfrak{p} = (89)$ and consider the following quartic surface

$$X/\mathbb{Q}\colon y^4 - x^3 z + yz^3 + zw^3 + w^4 = 0. \tag{5.2.3}$$

The characteristic polynomial of $89^{-1}\,\mathrm{Frob}_\mathfrak{p}$ on $\mathrm{H}^2_{\mathrm{crys}}(X_\mathfrak{p}/\mathbb{Q}_\mathfrak{p})$ factors as

$$(t-1)^5 \cdot (t+1) \cdot (t^4+1) \cdot (89 + 188t^2 + 303t^4 + 316t^6 + 303t^8 + 188t^{10} + 89t^{12})/89. \tag{5.2.4}$$

Therefore, the geometric Picard number of X is at most 10. By computing an approximation of the obstruction map on the Tate classes, we would at most improve this bound by 1. Instead, we use Corollary 2.20 to drop the rank from 10 to 4.

For instance, we observe that the obstruction is non-zero on the space

$$\widetilde{T}_8^r(X_\mathfrak{p}) := \ker(\mathrm{Frob}_k^4 + p^4) \subset \mathrm{H}^2_{\mathrm{crys}}(X_\mathfrak{p}/\mathbb{Z}_\mathfrak{p}). \tag{5.2.5}$$

This already allows us to drop from 10 to 6, since $\deg(t^4+1) = 4$. Analogously, we work through the other two cyclotomic factors, observing a non-zero obstruction in each case. As these two cyclotomic polynomials are linear, we drop from 6 to 4.

All of this is automated. One may use `crystalline_obstruction` to deduce the bound 4. The following computation takes less than 4 min of CPU time.

```
sage: crystalline_obstruction(f, p=89, precision=3)
(4,
 {'precision': 3, 'p': 89, 'rank T(X_Fpbar)': 10,
  'factors': [(t - 1, 1), (t + 1, 1), (t - 1, 4),
    (t^4 + 1, 1)],
  'dim Ti': [1, 1, 4, 4], 'dim Li': [1, 0, 3, 0]})
```

Example 5.8 Consider now the K3 surface

$$X/\mathbb{Q}\colon x^4 + 2y^4 + 2yz^3 + 3z^4 - 2x^3 w - 2yw^3 = 0. \tag{5.2.6}$$

At $\mathfrak{p} = (43)$ the geometric Picard number of $X_\mathfrak{p}$ is two, and one Tate class is obstructed, thus the geometric Picard number of X is one.

Now take $\mathfrak{p} = (67)$. The characteristic polynomial of $67^{-1}\,\mathrm{Frob}_\mathfrak{p}$ factors as

$$(t-1)\cdot(t+1)^3\cdot(67-171t^1+181t^2-53t^3-71t^4+9t^5 \\ +178t^6-215t^7+54t^8+50t^9+54t^{10}\ldots)/67, \qquad (5.2.7)$$

thus the geometric Picard number of $X_{\mathfrak{p}}$ is 4.

As the polarization is unobstructed, there is only one irreducible space to consider, corresponding to $(t+1)^3$. Therefore, in contrast with the previous example, we may only obstruct the liftability of one class, as the obstruction space is one dimensional. Hence, we cannot obtain a sharp upper bound for the geometric Picard number of X by studying its specialization at $\mathfrak{p}=(67)$.

Example 5.9 Now we look at a K3 surface with real multiplication. Consider the K3 surface given by desingularizing the following double cover of the projective plane. It is given in the weighted projective plane $\mathbb{P}(1,1,3)$ and is branched over six lines.

$$X/\mathbb{Q}\colon w^2 = \left(-y^2/8+yz-z^2\right)\left(7x^2/8+5xz+7z^2\right)\left(2x^2+3xy+y^2\right). \qquad (5.2.8)$$

This is the specialization of the family $X^{(2,t)}$ in [EJ14, Theorem 6.6] with $t=1$. By *loc. cit.* it has geometric Picard number 16 and real multiplication by $\mathbb{Q}(\sqrt{2})$. In particular, every reduction of X will overestimate its geometric Picard rank by a multiple of 2 [Cha14]. In fact, the reductions must have geometric Picard rank 18 or 22 [EJ14, Corollary 4.12]. The van Luijk [vLui07] method fails in this case. We now show that the obstruction method will succeed.

Take the prime $\mathfrak{p}=(83)$. By counting points on the singular model, we deduce that the characteristic polynomial of $83^{-1}\,\mathrm{Frob}_{\mathfrak{p}}$ can be factored as $\chi_1\cdot\chi_2$ where

$$\chi_1=(t-1)^{10}(t+1)^6, \quad \chi_2=(t^2+1)(83-158t^2+83t^4)/83. \qquad (5.2.9)$$

Here, χ_1 corresponds to the action of the Frobenius on the known 16 algebraic cycles on X. The factor χ_2 corresponds to the action on their orthogonal complement (the transcendental part).

In keeping with the notation of Proposition 2.16, let $T_4^1(X_{\mathfrak{p}}) := \ker\left(\mathrm{Frob}^2+p^2\right)$. From the argument above and Proposition 2.19, we conclude that $T_4^1(X_{\mathfrak{p}})$ spans algebraic cycles in $\overline{X}_{\mathfrak{p}}$ which do not lift. In particular, by Theorem 2.7, the obstruction map must be non-zero on $T_4^1(X_{\mathfrak{p}})$.

Consequently, using sufficiently high precision, one can detect that the obstruction map π_{T_4} is non-zero. Using Corollary 2.20 one would then conclude that the geometric Picard number of X is 16. We did not implement the computation of the Frobenius on a smooth variety given as the resolution of a singular hypersurface. Therefore, we cannot comment on the precision required.

5.2.2 Quintic Surfaces

We now consider smooth quintic surfaces. For these surfaces the Hodge numbers are $(4, 45, 4)$. Now the obstruction space is four dimensional, and thus favoring richer examples.

Example 5.10 Consider the following quintic surface

$$X/\mathbb{Q}\colon\ 10x^4y - 3z^5 + 3x^4w + 3y^4w - 10xw^4 - 23x^3yz = 0. \tag{5.2.10}$$

We compute that $X_\mathfrak{p}$ has Picard number one at the prime $\mathfrak{p} = (13)$. This means the Picard number of X is one. On the other hand, the reduction $X_\mathfrak{p}$ at $\mathfrak{p} = (23)$ has 5-dimensional space of Tate classes. Fortunately, the obstruction space is 4-dimensional and we see that π_T is indeed rank 4. Thus, this method allows for the correct determination of the Picard rank of X at a prime where the jump in Picard number is four. It took about 1 day of CPU time per prime to compute the Hasse–Weil zeta function and an approximation for the Frobenius matrix.

Example 5.11 The following example suggests that we are not necessarily immune to symmetries in the transcendental lattice. Take the quintic surface

$$X/\mathbb{Q}\colon\ 3x^4z + 9xy^4 + 9y^2z^3 + z^5 + 5w^5 = 0. \tag{5.2.11}$$

The geometric Picard number of $X_\mathfrak{p}$ at $\mathfrak{p} = (31)$, and therefore of X, is one. On the other hand, the dimension of the space of eventual Tate classes $\widetilde{T}^1(X_\mathfrak{p})$ of $X_\mathfrak{p}$ at $\mathfrak{p} = (29)$ is 5, and the characteristic polynomial of the Frobenius action on $\widetilde{T}^1(X_\mathfrak{p})$ is $(t-31)^3(t+31)^2$. This time, the following result shows that there is another theoretical limit to the best possible upper bound we can compute at $\mathfrak{p} = (29)$.

We recall the decomposition $\widetilde{T}^1(X_\mathfrak{p}) = T_1^1(X_\mathfrak{p}) \bigoplus T_2^1(X_\mathfrak{p})$ as in Proposition 2.16.

Proposition 5.12 *The images of the obstruction maps π_{T_1} and π_{T_2} coincide. This image has dimension* 1*, although the codomain has dimension* 4.

Proof We use the action of the involution $\iota\colon y \mapsto -y$ on the obstruction space and on the Tate classes. By inspecting the Griffiths basis, it is immediately seen that the (-1)-eigenspace of ι^* on the obstruction space is one dimensional.

We claim that all primitive Tate classes are (-1)-eigenvectors of ι^* on cohomology. To show this, we computed an approximation of $\widetilde{T}^1(X_\mathfrak{p})$ and projected this approximation to the (-1)-eigenspace of ι^*. The minimal valuation of the 4×4 minors of a basis of the image is 0, even with 10 digits of approximation. This proves the claim.

The obstruction map is ι^* equivariant and the polarization is unobstructed. Therefore, the image of the Tate classes via $\pi_{\widetilde{T}}$ must be confined to a one dimensional space. We observe that the two ranks are at least 1 by computation. □

We note another behavior to this problem at another prime. At $\mathfrak{p} = (23)$, the characteristic polynomial of the Frobenius action on $\widetilde{T}(X_{\mathfrak{p}})$ is now $(t - 23)^2(t + 23)(t^2 + 23^2)$. Since the Galois representation $\widetilde{T}^1(X_{\mathfrak{p}})$ splits into more invariant pieces, and one piece is two dimensional, we can drop the rank to 1 using Corollary 2.20. It takes about 1 CPU day per prime to get these results.

Remark 5.13 By the computations at $\mathfrak{p} = (23), (29)$ and (31), we know that the dimension of the endomorphism algebra of the transcendental lattice is at most 4. Furthermore, we note that X has another automorphism given $w \mapsto \zeta_5 w$, thus the transcendental lattice has CM by $\mathbb{Q}(\zeta_5)$. Then, one expects the Picard number of the reductions to be congruent to 1 modulo 4 [Cha14].

Acknowledgments We thank Bjorn Poonen for his help in initiating the project and for his guidance throughout. We also benefited greatly from the constant support of Kiran Kedlaya. We thank Matthias Schütt and John Voight for their valuable comments on the first version of this text. We are grateful for the careful comments from the anonymous referees and from Avi Kulkarni to the earlier version of this article. The first author was supported by the Simons Collaboration in Arithmetic Geometry, Number Theory, and Computation via Simons Foundation grant 550033.

References

[AKR10] T. G. Abbott, K. S. Kedlaya, and D. Roe, "Bounding Picard numbers of surfaces using p-adic cohomology," in *Arithmetics, geometry, and coding theory (AGCT 2005)*, ser. Sémin. Congr. Vol. 21, Paris: Soc. Math. France, 2010, pp. 125–159.

[AMMN20] B. Antieau, A. Mathew, M. Morrow, and T. Nikolaus, "On the Beilinson fiber square," *arXiv e-prints*, Mar. 2020. arXiv: 2003.12541 [math.KT].

[BO83] P. Berthelot and A. Ogus, "F-isocrystals and de Rham cohomology. I," *Invent. Math.*, vol. 72, no. 2, pp. 159–199, 1983.

[Ber74] P. Berthelot, *Cohomologie cristalline des schemas de caractéristique p >0.*. French. Springer, Cham, 1974, vol. 407.

[Ber97] ——, "Finitude et pureté cohomologique en cohomologie rigide," *Invent. Math.*, vol. 128, no. 2, pp. 329–377, 1997, With an appendix in English by Aise Johan de Jong.

[BEK14] S. Bloch, H. Esnault, and M. Kerz, "p-adic deformation of algebraic cycle classes," *Invent. Math.*, vol. 195, no. 3, pp. 673–722, 2014.

[CRV18] X. Caruso, D. Roe, and T. Vaccon, "ZpL: A p-adic precision package," in *ISSAC'18—Proceedings of the 2018 ACM International Symposium on Symbolic and Algebraic Computation*, ACM, New York, 2018, pp. 119–126.

[Cha98] A. Chambert-Loir, "Cohomologie cristalline: Un survol." French, *Expo. Math.*, vol. 16, no. 4, pp. 333–382, 1998.

[Cha13] F. Charles, "The Tate conjecture for K3 surfaces over finite fields," *Invent. Math.*, vol. 194, no. 1, pp. 119–145, 2013.

[Cha14] ——, "On the Picard number of K3 surfaces over number fields," *Algebra Number Theory*, vol. 8, no. 1, pp. 1–17, 2014.

[Cos15] E. Costa, *Effective computations of Hasse–Weil zeta functions*. ProQuest LLC, Ann Arbor, MI, 2015, p. 78, Thesis (Ph.D.)–New York University.

[CHK19] E. Costa, D. Harvey, and K. S. Kedlaya, "Zeta functions of nondegenerate hypersurfaces in toric varieties via controlled reduction in p-adic cohomology," in *Proceedings of the Thirteenth Algorithmic Number Theory Symposium*, ser. Open Book Ser. Vol. 2, Math. Sci. Publ., Berkeley, CA, 2019, pp. 221–238.

[CMSV19] E. Costa, N. Mascot, J. Sijsling, and J. Voight, "Rigorous computation of the endomorphism ring of a Jacobian," *Math. Comp.*, vol. 88, no. 317, pp. 1303–1339, 2019.

[Del74] P. Deligne, "La conjecture de Weil. I," *Inst. Hautes Études Sci. Publ. Math.*, no. 43, pp. 273–307, 1974.

[DI87] P. Deligne and L. Illusie, "Relèvements modulo p^2 et décomposition du complexe de de Rham," *Invent. Math.*, vol. 89, no. 2, pp. 247–270, 1987.

[EJ11a] A.-S. Elsenhans and J. Jahnel, "On the computation of the Picard group for K3 surfaces," *Math. Proc. Cambridge Philos. Soc.*, vol. 151, no. 2, pp. 263–270, 2011.

[EJ11b] ——, "The Picard group of a K3 surface and its reduction modulo p," *Algebra Number Theory*, vol. 5, no. 8, pp. 1027–1040, 2011.

[EJ14] ——, "Examples of K3 surfaces with real multiplication," *LMS J. Comput. Math.*, vol. 17, no. suppl. A, pp. 14–35, 2014.

[Eme97] M. Emerton, *A p-adic variational Hodge conjecture and modular forms with complex multiplication*, http://www.math.uchicago.edu/~emerton/pdffiles/cm.pdf, [Online; accessed 8-September-2020], 1997.

[Ful98] W. Fulton, *Intersection theory*, Second, ser. Ergebnisse der Mathematik und ihrer Grenzgebiete. 3. Folge. A Series of Modern Surveys in Mathematics [Results in Mathematics and Related Areas. 3rd Series. A Series of Modern Surveys in Mathematics]. Springer-Verlag, Berlin, 1998, vol. 2, pp. xiv+470.

[GM87] H. Gillet and W. Messing, "Cycle classes and Riemann-Roch for crystalline cohomology." English, *Duke Math. J.*, vol. 55, pp. 501–538, 1987.

[Gri69] P. A. Griffiths, "On the periods of certain rational integrals. I, II," *Ann. of Math. (2) 90 (1969), 460-495; ibid. (2)*, vol. 90, pp. 496–541, 1969.

[Gro63] A. Grothendieck, "Éléments de géométrie algébrique. III. étude cohomologique des faisceaux cohérents. II," *Inst. Hautes Études Sci. Publ. Math.*, no. 17, p. 91, 1963.

[Gro66] ——, "On the De Rham cohomology of algebraic varieties." English, *Publ. Math., Inst. Hautes Étud. Sci.*, vol. 29, pp. 95–103, 1966.

[Gro68] ——, "Crystals and the de Rham cohomology of schemes," in *Dix exposés sur la cohomologie des schémas*, ser. Adv. Stud. Pure Math. Vol. 3, Notes by I. Coates and O. Jussila, North-Holland, Amsterdam, 1968, pp. 306–358.

[Har75] R. Hartshorne, "On the De Rham cohomology of algebraic varieties." English, *Publ. Math., Inst. Hautes Étud. Sci.*, vol. 45, pp. 5–99, 1975.

[Har10a] D. Harvey, *Computing zeta functions of certain varieties in larger characteristic*, http://web.maths.unsw.edu.au/~davidharvey/talks/zetasqrtp-talk.pdf, [Online; accessed 5-March-2020], 2010.

[Har10b] ——, *Computing zeta functions of projective surfaces in large characteristic*, http://web.maths.unsw.edu.au/~davidharvey/talks/zetasqrtp-talk3.pdf, [Online; accessed 5-March-2020], 2010.

[Har10c] ——, *Counting points on projective hypersurfaces*, http://web.maths.unsw.edu.au/~davidharvey/talks/zetasurface.pdf, [Online; accessed 5-March-2020], 2010.

[HKT13] B. Hassett, A. Kresch, and Y. Tschinkel, "Effective computation of Picard groups and Brauer-Manin obstructions of degree two K3 surfaces over number fields," *Rend. Circ. Mat. Palermo (2)*, vol. 62, no. 1, pp. 137–151, 2013.

[HK71] K. Hoffman and R. Kunze, *Linear algebra.*, English, Englewood Cliffs, N. J.: Prentice-Hall, Inc. VIII, 407 p. (1971). 1971.

[HM17] A. Huber and S. Müller-Stach, *Periods and Nori motives*, ser. Ergebnisse der Mathematik und ihrer Grenzgebiete. 3. Folge. A Series of Modern Surveys in Mathematics [Results in Mathematics and Related Areas. 3rd Series. A Series of Modern Surveys in Mathematics]. Springer, Cham, 2017, vol. 65, pp. xxiii+372, With contributions by Benjamin Friedrich and Jonas von Wangenheim.

[Ill94] L. Illusie, "Crystalline cohomology," in *Motives (Seattle, WA, 1991)*, ser. Proc. Sympos. Pure Math. Vol. 55, Amer. Math. Soc., Providence, RI, 1994, pp. 43–70.

[IN17] H. Inoue and K. Naito, "The shortest vector problems in p-adic lattices and simultaneous approximation problems of p-adic numbers," *Linear Nonlinear Anal.*, vol. 3, no. 2, pp. 213–224, 2017.

[Kat89] K. Kato, "Logarithmic structures of Fontaine-Illusie," in *Algebraic analysis, geometry, and number theory (Baltimore, MD, 1988)*, Johns Hopkins Univ. Press, Baltimore, MD, 1989, pp. 191–224.

[Ked08] K. S. Kedlaya, "p-adic cohomology: From theory to practice," in *p-adic geometry*, ser. Univ. Lecture Ser. Vol. 45, Amer. Math. Soc., Providence, RI, 2008, pp. 175–203.

[Ked13] ——, *Computing zeta functions of nondegenerate toric hypersurfaces via controlled reduction*, http://kskedlaya.org/slides/oxford2013.pdf, [Accessed 18-Feb-2020], 2013.

[KM16] W. Kim and K. Madapusi Pera, "2-adic integral canonical models," *Forum Math. Sigma*, vol. 4, e28, 34, 2016.

[Klo07] R. Kloosterman, "Elliptic K3 surfaces with geometric Mordell-Weil rank 15," *Canad. Math. Bull.*, vol. 50, no. 2, pp. 215–226, 2007.

[LS19] P. Lairez and E. C. Sertöz, "A numerical transcendental method in algebraic geometry: Computation of Picard groups and related invariants," *SIAM J. Appl. Algebra Geom.*, vol. 3, no. 4, pp. 559–584, 2019.

[Lan83] S. Lang, *Abelian varieties*. Springer-Verlag, New York-Berlin, 1983, pp. xii+256, Reprint of the 1959 original.

[LLL82] A. K. Lenstra, H. W. Lenstra Jr., and L. Lovász, "Factoring polynomials with rational coefficients," *Math. Ann.*, vol. 261, no. 4, pp. 515–534, 1982.

[Mad15] K. Madapusi Pera, "The Tate conjecture for K3 surfaces in odd characteristic," *Invent. Math.*, vol. 201, no. 2, pp. 625–668, 2015.

[Mil07] J. S. Milne, *The Tate conjecture over finite fields (AIM talk)*, 2007. eprint: arXiv:0709.3040.

[Mil17] ——, *Algebraic number theory (v3.07)*, www.jmilne.org/math/, 2017.

[MS] H. Movasati and E. C. Sertöz, "On reconstructing subvarieties from their periods," arXiv: 1908.03221 [math.AG].

[Pan20] X. Pan, "Automorphism and cohomology I: Fano varieties of lines and cubics," *Algebr. Geom.*, vol. 7, no. 1, pp. 1–29, 2020.

[PTvL15] B. Poonen, D. Testa, and R. van Luijk, "Computing Néron-Severi groups and cycle class groups," *Compos. Math.*, vol. 151, no. 4, pp. 713–734, 2015.

[Ray79] M. Raynaud, """p-torsion" du schéma de Picard," in *Journées de Géométrie Algébrique de Rennes (Rennes, 1978), Vol. II*, ser. Astérisque, vol. 64, Soc. Math. France, Paris, 1979, pp. 87–148.

[Sag20] T. Sage Developers, *Sagemath, the Sage Mathematics Software System (Version 9.0)*, http://www.sagemath.org, 2020.

[Sch15] M. Schütt, "Picard numbers of quintic surfaces," *Proc. Lond. Math. Soc. (3)*, vol. 110, no. 2, pp. 428–476, 2015.

[Ser19] E. C. Sertöz, "Computing periods of hypersurfaces," English, *Math. Comput.*, vol. 88, no. 320, pp. 2987–3022, 2019.

[Shi02] A. Shiho, "Crystalline fundamental groups. II. Log convergent cohomology and rigid cohomology," *J. Math. Sci. Univ. Tokyo*, vol. 9, no. 1, pp. 1–163, 2002.

[Shi72] T. Shioda, "On elliptic modular surfaces," *J. Math. Soc. Japan*, vol. 24, pp. 20–59, 1972.

[Shi86] ——, "An explicit algorithm for computing the Picard number of certain algebraic surfaces," *Amer. J. Math.*, vol. 108, no. 2, pp. 415–432, 1986.

[Shi90] ——, "On the Mordell-Weil lattices," *Comment. Math. Univ. St. Paul.*, vol. 39, no. 2, pp. 211–240, 1990.

[Tat66a] J. Tate, "Endomorphisms of abelian varieties over finite fields," *Invent. Math.*, vol. 2, pp. 134–144, 1966.

[Tat66b] ——, "On the conjectures of Birch and Swinnerton-Dyer and a geometric analog," en, in *Séminaire Bourbaki : années 1964/65 1965/66, exposés 277-312*, ser. Séminaire Bourbaki 9, talk:306, Société mathématique de France, 1966, pp. 415–440.

[Tot17] B. Totaro, "Recent progress on the Tate conjecture," *Bull. Amer. Math. Soc. (N.S.)*, vol. 54, no. 4, pp. 575–590, 2017.

[vLui07] R. van Luijk, "K3 surfaces with Picard number one and infinitely many rational points," *Algebra Number Theory*, vol. 1, no. 1, pp. 1–15, 2007.

[Voi07] C. Voisin, *Hodge theory and complex algebraic geometry. II*, English, ser. Cambridge Studies in Advanced Mathematics. Cambridge University Press, Cambridge, 2007, vol. 77, pp. x+351, Translated from the French by Leila Schneps.

Conjecture: 100% of Elliptic Surfaces Over $\mathbb{Q}$ have Rank Zero

Alex Cowan

Abstract Based on an equation for the rank of an elliptic surface over $\mathbb{Q}$ which appears in the work of Nagao, Rosen, and Silverman, we conjecture that 100% of elliptic surfaces have rank 0 when ordered by the size of the coefficients of their Weierstrass equations, and present a probabilistic heuristic to justify this conjecture. We then discuss how it would follow from either understanding of certain L-functions, or from understanding of the local behaviour of the surfaces. Finally, we make a conjecture about ranks of elliptic surfaces over finite fields, and highlight some experimental evidence supporting it.

1 Introduction

Let $\mathcal{E}$ be an elliptic surface over $\mathbb{Q}$. Fix a Weierstrass equation

$$\mathcal{E} : y^2 = x^3 + A(T)x + B(T)$$

with $A(T), B(T) \in \mathbb{Z}[T]$ and $4A(T)^3 + 27B(T)^2 \not\equiv 0$. Given an integer t such that $4A(t)^3 + 27B(t)^2 \neq 0$, let $\mathcal{E}_t$ denote the elliptic curve over $\mathbb{Q}$ with Weierstrass equation

$$\mathcal{E}_t : y^2 = x^3 + A(t)x + B(t).$$

Given an elliptic curve $E/\mathbb{Q}$ and a prime p, define $a_p(E)$ to be the pth coefficient of the L-function attached to E.

It was conjectured by Nagao [11] that

A. Cowan (✉)
Math Department, Harvard University, Cambridge, MA, USA
e-mail: cowan@math.harvard.edu

J. S. Balakrishnan et al. (eds.), *Arithmetic Geometry, Number Theory, and Computation*, Simons Symposia, https://doi.org/10.1007/978-3-030-80914-0_10

$$\operatorname{rank} \mathcal{E}(\mathbb{Q}(T)) = - \lim_{X\to\infty} \frac{1}{X} \sum_{p<X} \sum_{t=1}^{p} a_p(\mathcal{E}_t) \frac{\log p}{p}. \tag{1}$$

Rosen and Silverman [12] proved that (1) held if one assumed Tate's conjecture and that a certain L-function didn't vanish on the right edge of its critical strip. In particular, (1) holds unconditionally for rational elliptic surfaces.

In light of (1), we think it's likely that the average rank of elliptic surfaces over $\mathbb{Q}$ is 0, essentially because the average value of $a_p(\mathcal{E}_t)$ should be 0. We give a more thorough justification for this belief in Sect. 2.

There are several ways the belief that the average rank of elliptic surfaces is 0 can be formulated as a precise conjecture. For convenience, we introduce the following framework for discussing possible formulations of such a conjecture.

Let $\mathcal{S}(M)$ with $M \in \mathbb{Z}_{>0}$ be a sequence of subsets of the set of all elliptic surfaces over $\mathbb{Q}$, with the properties that $\mathcal{S}(M)$ is finite for every M, and $\mathcal{S}(M) \subseteq \mathcal{S}(M')$ whenever $M \leq M'$. We'll say that $\mathcal{S} = \cup_{M=1}^{\infty} \mathcal{S}(M)$ is a *family* of elliptic surfaces, and the filtration given by M is an *ordering* of that family. This parallels the language used when discussing statistics of ranks of elliptic curves, where, for example, people might discuss "the family of quadratic twists of a fixed elliptic curve, ordered by conductor".

Define the *average rank of the family* $\mathcal{S}$ to be the quantity

$$\lim_{M\to\infty} \frac{1}{\#\mathcal{S}(M)} \sum_{\mathcal{E}\in\mathcal{S}(M)} \operatorname{rank} \mathcal{E}(\mathbb{Q}(T)).$$

We believe that the average rank of many families $\mathcal{S}$ will be 0. In formulating Conjecture 1.1 below, we pick a specific family for the sake of concreteness, but there are many other reasonable choices.

Conversely, there are several examples in the literature of families of elliptic surfaces over $\mathbb{Q}$ which were constructed to have strictly positive rank [5, 6]. Average ranks of the elliptic curves coming from specializations of elliptic surfaces have been studied as well [1, 6, 14], and in some situations the average ranks of the specializations can exhibit rich and possibly unexpected behaviour.

To formulate Conjecture 1.1 below, we first define, for positive integers d and M, the set

$$\mathcal{P}_d(M) = \{p \in \mathbb{Z}[T] \,:\, \deg(p) = d,\ \mu(p) < M\},$$

where $\mu(p)$ is the Mahler measure of p (using the height of p would also be reasonable here). Set

$$\mathcal{S}_{m,n}(M) := \{\mathcal{E} : y^2 = x^3 + A(T)x + B(T) \,:\, A \in \mathcal{P}_m(M^2),\ B \in \mathcal{P}_n(M^3),\ 4A(T)^3 + 27B(T)^2 \not\equiv 0\}.$$

Then we believe the average rank of the family $\mathcal{S} = \mathcal{S}_{m,n}$ will be 0 whenever m and n are both positive, i.e.

Conjecture 1.1 *For any fixed positive integers m and n, we have*

$$\lim_{M\to\infty} \frac{1}{\#\mathcal{S}_{m,n}(M)} \sum_{\mathcal{E}\in\mathcal{S}_{m,n}(M)} \operatorname{rank} \mathcal{E}(\mathbb{Q}(T)) = 0.$$

In Sect. 3 we discuss the main obstacles for proving Conjecture 1.1 using the analytic framework in [12]. In Sect. 4 we outline an approach to Conjecture 1.1 based on investigating statistics of ranks of elliptic surfaces over finite fields.

2 Probabilistic Heuristics

For a fixed prime p, Birch [2] gave all moments of the distribution of $a_p(E)$ when E is chosen by selecting a (possibly singular) Weierstrass equation with coefficients in $\mathbb{F}_p$ uniformly at random. Let $(A_{p,t})$ be a sequence of independent random variables indexed by a prime p and a positive integer t, with the property that, for every t, the random variable $A_{p,t}$ has the same distribution as the values $a_p(E)$ for E a Weierstrass equation in $\mathbb{F}_p$ chosen uniformly at random, as in Birch's work. We highlight the following property of the sequence $(A_{p,t})$:

Proposition 2.1 *For any $\varepsilon > 0$, the series*

$$\frac{1}{X^{\frac{1}{2}+\varepsilon}} \sum_{p<X} \sum_{t=1}^{p} A_{p,t} \frac{\log p}{p}$$

converges to 0 *with probability* 1 *as X goes to infinity.*

Proof By construction we have

$$\operatorname{Var}\left(\frac{1}{X^{\frac{1}{2}+\frac{\varepsilon}{2}}} A_{p,t} \frac{\log p}{p}\right) \ll \frac{(\log p)^2}{X^{1+\varepsilon} p}.$$

Thus

$$\operatorname{Var}\left(\frac{1}{X^{\frac{1}{2}+\frac{\varepsilon}{2}}} \sum_{p<X} \sum_{t=1}^{p} A_{p,t} \frac{\log p}{p}\right) \ll \frac{1}{X^{1+\varepsilon}} \sum_{p<X} (\log p)^2 \ll 1.$$

By Kolmogorov's three-series theorem, as stated in [4, theorem 2.5.8] for example, it follows that the series

$$\frac{1}{X^{\frac{1}{2}+\frac{\varepsilon}{2}}} \sum_{p<X} \sum_{t=1}^{p} A_{p,t} \frac{\log p}{p}$$

converges almost surely. Hence the series

$$\frac{1}{X^{\frac{1}{2}+\varepsilon}} \sum_{p<X} \sum_{t=1}^{p} A_{p,t} \frac{\log p}{p} = \frac{1}{X^{\frac{\varepsilon}{2}}} \left(\frac{1}{X^{\frac{1}{2}+\frac{\varepsilon}{2}}} \sum_{p<X} \sum_{t=1}^{p} A_{p,t} \frac{\log p}{p} \right)$$

converges to 0 almost surely. □

Proposition 2.1 is relevant because it suggests that, if we believe that the values $a_p(\mathcal{E}_t)$ are distributed in the same way $a_p(E)$ is for elliptic curves $E/\mathbb{F}_p$ chosen uniformly at random, then we should expect that 100% of the time the quantity in (1) is 0.

There is evidence to support the belief that the values $a_p(\mathcal{E}_t)$ will be distributed in this way. If, instead of fixing p and letting t vary initially, we fix t and let p vary, then, for all $\varepsilon > 0$ the series

$$\frac{1}{X^{\varepsilon}} \sum_{p<X} A_{p,t} \frac{\log p}{p} \tag{2}$$

converges to 0 as X goes to infinity with probability 1. This suggests that, for a fixed elliptic curve $E_t/\mathbb{Q}$, we should expect that

$$\sum_{p<X} a_p(E_t) \frac{\log p}{p} \ll X^{\varepsilon} \tag{3}$$

for all $\varepsilon > 0$, because we should expect that the reductions $E_t/\mathbb{F}_p$ will be distributed uniformly at random. The bound (3) was proven by Heath-Brown [8], so the reductions $E_t/\mathbb{F}_p$ do indeed behave uniformly randomly in this case.

The popular belief that the average rank of elliptic curves over $\mathbb{Q}$ is $\frac{1}{2}$ can also support the idea that these sums of a_p values behave randomly in the way described. The Birch and Swinnerton-Dyer conjecture (BSD) connects the rank of an elliptic curve $E/\mathbb{Q}$ to the coefficients $a_p(E)$ of its L-function. The original formulation of BSD was

$$\prod_{p<X} \frac{p+1-a_p(E)}{p} \sim C_E (\log X)^{\operatorname{rank} E(\mathbb{Q})}$$

for some constant C_E which depends on E. See work of Goldfeld [7], K. Conrad [3], and Kuo-R. Murty [9] for some treatment of this specific formulation of BSD. One consequence of this form of BSD is [13]

$$\operatorname{rank} E(\mathbb{Q}) = \frac{1}{2} - \frac{1}{\log X} \int_1^X \frac{1}{x} \sum_{p<x} a_p(E) \log p \frac{dx}{x} + O\Big(\frac{1}{\log X}\Big). \tag{4}$$

It is widely believed that many families of elliptic curves over $\mathbb{Q}$ have average rank $\frac{1}{2}$. This belief suggests that the quantity

$$\frac{1}{x}\sum_{p<x} a_p(E)\log p \tag{5}$$

appearing in Eq. (4) should average to 0 in those families. However, because the error term in (4) depends on E, even knowing that the average rank of elliptic curves was $\frac{1}{2}$ wouldn't be enough to conclude that the quantity (5) averages to 0. While it would be surprising if the error terms in (4) did not average to 0 as well, controlling these error terms is difficult, and is the main obstacle in proving anything about average ranks from this perspective.

3 Obstacles for Analytic Proofs

In this section we discuss what obstacles exist which make proving Conjecture 1.1 difficult within the framework established in [12]. In proving Nagao's conjecture, Rosen and Silverman first prove an analytic version of his conjecture:

$$\operatorname*{Res}_{s=2}\sum_{p}\sum_{t=1}^{p} a_p(\mathcal{E}_t)\frac{\log p}{p^s} = -\operatorname{rank} E(\mathbb{Q}(T)). \tag{6}$$

To do this, they introduce the following notation:

- $L_2(\mathcal{E}/\mathbb{Q}, s)$ is the Hasse-Weil L-function of $\mathcal{E}/\mathbb{Q}$ attached to $H^2_{ét}(\mathcal{E}/\bar{\mathbb{Q}}, \mathbb{Q}_\ell)$.
- $\mathrm{NS}(\mathcal{E}/\bar{\mathbb{Q}})$ is the Néron-Severi group of $\mathcal{E}/\bar{\mathbb{Q}}$.
- $\mathscr{S}$ is the trivial part of $\mathrm{NS}(\mathcal{E}/\bar{\mathbb{Q}})\otimes\mathbb{Q}$, generated by the image of the zero section and by all components of all fibers.
- $\mathscr{S}_\ell(1)$ is the Tate twist $\mathscr{S}\otimes T_\ell(\bar{\mathbb{Q}}^*)$ of the $\mathrm{Gal}(\bar{\mathbb{Q}}/\mathbb{Q})$-module $\mathscr{S}$.
- $L(\mathscr{S}_\ell(1), s)$ is the Artin L-function attached to the representation $\mathscr{S}_\ell(1)$.
- $N(\mathcal{E}/\mathbb{Q}, s) = \dfrac{L_2(\mathcal{E}/\mathbb{Q}, s)}{L(\mathscr{S}_\ell(1), s)}$.

Then, assuming the Tate conjecture, Silverman and Rosen prove that both sides of (6) are equal to

$$\operatorname*{Res}_{s=2}\frac{d}{ds}\log N(\mathcal{E}/\mathbb{Q}, s).$$

The conjecture (1) then follows from standard analytic techniques applied to the series $\frac{d}{ds}\log N(\mathcal{E}/\mathbb{Q}, s)$.

We now discuss what might be involved in a proof of Conjecture 1.1 using these analytic techniques. Suppose that there exist constants $\delta_{\mathcal{E}} > 0$ and $C_{\mathcal{E}}$ such that

$$\left| \frac{1}{X} \sum_{p<X} \sum_{t=1}^{p} a_p(\mathcal{E}_t) \frac{\log p}{p} + \text{rank } \mathcal{E}(\mathbb{Q}(T)) \right| < C_{\mathcal{E}} X^{-\delta_{\mathcal{E}}},$$

as is typical in this kind of setting. To prove Conjecture 1.1, it would be sufficient to find, for every elliptic surface $\mathcal{E} \in \mathcal{S}$, a real number $X_{\mathcal{E}} > 0$ depending only on $\mathcal{E}$, such that both

$$\lim_{M\to\infty} \frac{1}{\#\mathcal{S}(M)} \sum_{\mathcal{E}\in\mathcal{S}(M)} \frac{1}{X_{\mathcal{E}}} \sum_{p<X_{\mathcal{E}}} \sum_{t=1}^{p} a_p(\mathcal{E}_t) \frac{\log p}{p} = 0 \tag{7}$$

and

$$\lim_{M\to\infty} \frac{1}{\#\mathcal{S}(M)} \sum_{\mathcal{E}\in\mathcal{S}(M)} C_{\mathcal{E}} X_{\mathcal{E}}^{-\delta_{\mathcal{E}}} = 0 \tag{8}$$

simultaneously. Condition (7) requires that M be "large" relative to the values $X_{\mathcal{E}}$ so that the averages

$$\frac{1}{\#\mathcal{S}(M)} \sum_{\mathcal{E}\in\mathcal{S}(M)} a_p(\mathcal{E}_t)$$

for fixed p and t are small. Condition (8), on the other hand, would like for M to be "small" relative to the values $X_{\mathcal{E}}$. The difficulty, then, is to find some choice of $X_{\mathcal{E}}$'s such that both of these conditions hold at once. We expect that showing that there is such a choice will be difficult. Condition (7) might require a Pólya-Vinogradov style result, where one shows that there are no "conspiracies" among the values $a_p(\mathcal{E}_t)$ that might cause them to behave differently than random variables. Condition (8) might require subconvexity bounds, and the number $\delta_{\mathcal{E}}$ will depend on the locations of the zeroes of $N(\mathcal{E}/\mathbb{Q}, s)$.

4 An Approach via Finite Fields

Let $\mathcal{S}_{\ell;m,n}$ denote the set of elliptic surfaces $\mathcal{E}/\mathbb{F}_\ell : y^2 = x^3 + A(T)x + B(T)$ with $\deg(A) = m$ and $\deg(B) = n$. This set is finite. Let $\rho_\ell(m, n)$ denote the proportion of elliptic surfaces in $\mathcal{S}_{\ell;m,n}$ which have positive rank.

Conjecture 4.1 *For every prime ℓ_0, and every pair of integers $m_0, n_0 > 0$,*

$$\lim_{\ell\to\infty} \rho_\ell(m_0, n_0) = \lim_{m\to\infty} \rho_{\ell_0}(m, n_0) = \lim_{n\to\infty} \rho_{\ell_0}(m_0, n) = \frac{1}{2}.$$

This conjecture, beyond being interesting in its own right, provides an approach for proving Conjecture 1.1. See [10] for experimental evidence towards Conjecture 4.1, where $\rho_\ell(m, n)$ is estimated computationally for $\ell = 7$, $n = 6, 12, 18, 24, 30$, and $m \leq n/2$.

Let N be a squarefree positive integer. Let $\mathcal{S}^{(N)}(M)$ denote the subset of $\mathcal{S}(M)$ for which $\gcd(N, 4A(T)^3 + 27B(T)^2) = 1$. Then, for any m, n, and M,

$$\frac{\#\{\mathcal{E} \in \mathcal{S}^{(N)}(M) \ : \ \mathcal{E}/\mathbb{F}_\ell \text{ has positive rank for all } \ell | N\}}{\#\mathcal{S}^{(N)}(M)} = \prod_{\ell | N} \rho_\ell(m, n) + \mathcal{O}(M^{-1}) \tag{9}$$

by the Chinese remainder theorem, where $\mathcal{E}/\mathbb{F}_\ell$ denotes the reduction mod ℓ of $\mathcal{E}/\mathbb{Q}$.

If $\mathcal{E}/\mathbb{Q}$ has positive rank, then either the reduction $\mathcal{E}/\mathbb{F}_\ell$ has positive rank, or the kernel of the reduction $\mathcal{E}(\mathbb{Q}(T)) \to \mathcal{E}(\mathbb{F}_\ell(T))$ is of finite index in $\mathcal{E}(\mathbb{Q}(T))$. If this kernel was never of finite index then Conjecture 1.1 would follow from Conjecture 4.1 (as well as much weaker versions of this conjecture), via observation (9). The kernel of the reduction $\mathcal{E}(\mathbb{Q}(T)) \to \mathcal{E}(\mathbb{F}_\ell(T))$ is of finite index in $\mathcal{E}(\mathbb{Q}(T))$ occasionally, but presumably not nearly enough for this approach to fail. However, proving as much seems difficult. The generators of $\mathcal{E}(\mathbb{Q}(T))$ will map to the identity of $\mathcal{E}(\mathbb{F}_\ell(T))$ if their denominators are divisible by ℓ, so one is naturally lead to investigate the dependence of the height of the generators of $\mathcal{E}(\mathbb{Q}(T))$ on the size of the coefficients of the Weierstrass model of $\mathcal{E}/\mathbb{Q}$.

Acknowledgments We thank Noam Elkies, Bjorn Poonen, and Michael Snarski for helpful discussions. This work was supported by grant 550031 from the Simons Foundation.

References

1. Sandro Bettin, Chantal David, and Christophe Delaunay. Non-isotrivial elliptic surfaces with non-zero average root number. *J. Number Theory*, 191:1–84, 2018.
2. B. J. Birch. How the number of points of an elliptic curve over a fixed prime field varies. *J. London Math. Soc.*, 43:57–60, 1968.
3. Keith Conrad. Partial Euler products on the critical line. *Canad. J. Math.*, 57(2):267–297, 2005.
4. Rick Durrett. *Probability—theory and examples*, volume 49 of *Cambridge Series in Statistical and Probabilistic Mathematics*. Cambridge University Press, Cambridge, 2019.
5. Noam D. Elkies. Three lectures on elliptic surfaces and curves of high rank. `arxiv:0709.2908`, 2007.
6. Stéfane Fermigier. Étude expérimentale du rang de familles de courbes elliptiques sur **q**. *Experiment. Math.*, 5(2):119–130, 1996.
7. Dorian Goldfeld. Sur les produits partiels eulériens attachés aux courbes elliptiques. *C. R. Acad. Sci. Paris Sér. I Math.*, 294(14):471–474, 1982.
8. D. R. Heath-Brown. The average analytic rank of elliptic curves. *Duke Math. J.*, 122(3):591–623, 2004.
9. Wentang Kuo and M. Ram Murty. On a conjecture of Birch and Swinnerton-Dyer. *Canad. J. Math.*, 57(2):328–337, 2005.

10. Alan G. B. Lauder. Ranks of elliptic curves over function fields. *LMS J. Comput. Math.*, 11:172–212, 2008.
11. Koh-ichi Nagao. $\mathbf{Q}(T)$-rank of elliptic curves and certain limit coming from the local points. *Manuscripta Math.*, 92(1):13–32, 1997. With an appendix by Nobuhiko Ishida, Tsuneo Ishikawa and the author.
12. Michael Rosen and Joseph H. Silverman. On the rank of an elliptic surface. *Invent. Math.*, 133(1):43–67, 1998.
13. Michael O. Rubinstein. Elliptic curves of high rank and the Riemann zeta function on the one line. *Exp. Math.*, 22(4):465–480, 2013.
14. Joseph H. Silverman. The average rank of an algebraic family of elliptic curves. *J. Reine Angew. Math.*, 504:227–236, 1998.

On Rational Bianchi Newforms and Abelian Surfaces with Quaternionic Multiplication

J. E. Cremona, Lassina Dembélé, Ariel Pacetti, Ciaran Schembri, and John Voight

Abstract We study the rational Bianchi newforms (weight 2, trivial character, with rational Hecke eigenvalues) in the LMFDB that are not associated to elliptic curves, but instead to abelian surfaces with quaternionic multiplication. Two of these examples exhibit a rather special kind of behaviour: we show they arise from twisted base change of a classical newform with nebentypus character of order 4 and eight inner twists.

2010 ***Mathematics Subject Classification*** Primary: 11G05, Secondary: 11G40

1 Introduction

Let f be a classical newform with weight 2 and level N. When f has rational coefficients, the Eichler–Shimura construction attaches to f an elliptic curve E_f defined over $\mathbb{Q}$, a quotient of the Jacobian of the modular curve $X_0(N)$, such that $L(f, s) = L(E_f, s)$. More generally, if f has coefficients in a number field M_f of degree d, this construction associates a simple abelian variety A_f of dimension d defined over $\mathbb{Q}$, now a quotient of the Jacobian of the modular curve $X_1(N)$,

J. E. Cremona
Warwick Mathematics Institute, Zeeman Building, University of Warwick, Coventry, UK
e-mail: j.e.cremona@warwick.ac.uk

L. Dembélé
Department of Mathematics, University of Luxembourg, Esch-sur-Alzette, Luxembourg

A. Pacetti
Center for Research and Development in Mathematics and Applications (CIDMA), Department of Mathematics, University of Aveiro, Aveiro, Portugal
e-mail: apacetti@ua.pt

C. Schembri · J. Voight (✉)
Department of Mathematics, Dartmouth College, Hanover, NH, USA
e-mail: ciaran.schembri@dartmouth.edu; jvoight@gmail.com

J. S. Balakrishnan et al. (eds.), *Arithmetic Geometry, Number Theory, and Computation*, Simons Symposia, https://doi.org/10.1007/978-3-030-80914-0_11

such that $\mathrm{End}(A_f) \otimes \mathbb{Q} \simeq M_f$ and $L(A_f, s) = \prod_\sigma L(\sigma(f), s)$, the product over all embeddings $\sigma\colon M_f \hookrightarrow \mathbb{C}$. (It may happen that A_f acquires additional endomorphisms over a number field, and in particular A_f may not be geometrically simple.)

It is natural to seek such a relationship for modular forms over other number fields. In the case of totally real number fields, in many instances (for example, over fields of odd degree) a construction analogous to the Eichler–Shimura construction is obtained by replacing the modular curve by a suitable Shimura curve. However, a general such construction over number fields is not known. So with the simplest case in mind, we are led to ask: given a Bianchi newform F of weight 2 over an imaginary quadratic field K, is there an abelian variety A_F defined over K whose L-series matches that of F (as above)?

A precise answer to this question, consistent with the predictions of the Langlands philosophy, is provided affirmatively by a conjecture of Taylor [Tay95, Conjecture 3]—but with a wrinkle. Let M_F be the number field generated by the Hecke eigenvalues of F, and let $d = [M_F : \mathbb{Q}]$. Then conjecturally there is an abelian variety A_F of dimension $2d$ defined over K with *quaternionic multiplication* (QM) by a quaternion algebra B_F over M_F (and defined over K) such that

$$L(A_F, s) = \prod_{\sigma: M_F \hookrightarrow \mathbb{C}} L(\sigma(F), s)^2. \tag{1.1}$$

In this formulation, it may happen that the quaternion algebra $B_F \simeq \mathrm{M}_2(M_F)$ is split, in which case $A_F \sim (A'_F)^2$ is isogenous over K to the square of an abelian variety A'_F of dimension d and $L(A'_F, s) = \prod_\sigma L(\sigma(F), s)$. (In general, if the base field K has a real place, then B_F is necessarily split, which explains the simplifications above.)

In particular, if F has rational coefficients (i.e., $M_F = \mathbb{Q}$), then attached to F is conjecturally a QM abelian surface that is often the square of an elliptic curve. Examples of the phenomenon where the quaternion algebra is a division algebra were exhibited by Cremona [Cre84, Cre92] in the case where F is the base-change of a classical newform with quadratic coefficients, or a quadratic twist of such a form. More recently, Schembri [Sch19] exhibited examples which are not base change. QM abelian surfaces have been called *false generalized elliptic curves* by Taylor [Tay95], *false elliptic curves* by Serre (as attributed by Deligne–Rapoport [DR73, p. DeRa-12]), and *fake elliptic curves* by others. Perhaps this terminology would be more appropriate for the non-existent geometric object whose L-function is equal to $L(F, s)$.

Given that the situation is rather unusual, in this paper we study it explicitly. We consider the Bianchi newforms F of weight 2 with rational coefficients in the *L-Functions and Modular Forms Database* (LMFDB) [LMF20]. For thousands of such forms, we can explicitly attach an elliptic curve E_F such that $L(F, s) = L(E_F, s)$, with E_F unique up to isogeny over K. For the remaining forms, we can find no such elliptic curve—in Sect. 2, we list these forms and summarize what is known about them. In all these remaining examples, we can attach a QM abelian surface A over K such that $L(A, s) = L(F, s)^2$, in agreement with the above conjecture.

Two examples in the list stand out, and because of their exotic properties they command the majority of our attention here. We consider the rational Bianchi newform F of weight 2 with LMFDB label[1] 2.0.7.1-30625.1-c over $K = \mathbb{Q}(\sqrt{-7})$, along with a certain quadratic twist G, which has label 2.0.7.1-30625.1-e. The newform F has rational coefficients and is a twist of a base-change form, but there is no *quadratic* twist of F which is base-change; in fact the simplest twist which is base-change is by a character of order 8, so has neither trivial nebentypus nor rational coefficients.

Our main result (combining Proposition 6.2, Theorem 6.9, and Proposition 7.5) is the following.

Theorem A *Let F be the rational Bianchi newform over $K := \mathbb{Q}(\sqrt{-7})$ of weight 2, level (175), and trivial character with LMFDB label 2.0.7.1-30625.1-c. Then the following statements hold.*

(a) *There exists a classical modular form f of weight 2, level 1225, character of conductor 35 and order 4, and Hecke field $\mathbb{Q}(\zeta_{24})$, such that*

$$F = f_K \otimes \psi,$$

where f_K is the base change of f to K and ψ is a character of conductor $(5\sqrt{-7})$ and order 8.

(b) *There exists an explicit genus 2 curve X_F over K whose Jacobian $J_F := \mathrm{Jac}(X_F)$ has QM by a quaternion algebra of discriminant 15 and*

$$L_{\mathfrak{p}}(X_F, T) = L_{\mathfrak{p}}(J_F, T) = L_{\mathfrak{p}}(F, T)^2$$

for all primes $\mathfrak{p} \neq (2), (5), (\sqrt{-7})$. Moreover, all endomorphisms of J_F are defined over K.

(c) *Let A_f be the abelian eightfold over $\mathbb{Q}$ attached to f. Then $\mathrm{End}(A_f) = \mathbb{Z}[\zeta_{24}]$. Let $A' := (A_f)_K^{(\psi)}$ be the base change of A_f to K twisted by ψ (relative to an automorphism of A_f of order 8). Then there is an isogeny $A' \sim J_F^2 \times J_G^2$ over K, where $G = F \otimes \psi^4$ is a quadratic twist of F (with LMFDB label 2.0.7.1-30625.1-e), and J_G the corresponding quadratic twist of J_F.*

The form is described in detail in Sect. 4, and the twists are given explicitly in Sect. 6. An equation for the curve X_F in Theorem 1(b) is given in (7.4).

[1] Here, 2.0.7.1 is the LMFDB label for the base field $K = \mathbb{Q}(\sqrt{-7})$ and 30625.1 the label for the level ideal (175), which has norm 30625. The final c is the alphabetic label for this specific newform at that level. We use either full labels such as 2.0.7.1-30625.1-c for Bianchi newforms, or the shorter version 30625.1-c which omits the field when that is clear from the context.

Contents The paper is organized as follows. In Sect. 2 we survey Bianchi newforms in the LMFDB. After quickly setting up notation for characters in Sect. 3, we describe the classical modular form f in Sect. 4; then we consider its base change and twists in Sects. 5–6. We then give a model for the curve X_F in Sect. 7 and in Sect. 8 briefly describe related Bianchi modular forms in the same level. We conclude in Sect. 9 with some connections to the Paramodular Conjecture.

2 Rational Bianchi Newforms with No Associated Elliptic Curve

As of August 2020, the LMFDB [LMF20] currently contains rational Bianchi newforms over each of the five fields $K = \mathbb{Q}(\sqrt{d})$ for $d = -3, -4, -7, -8, -11$, as computed by Cremona [Cre84], of all levels of norm up to a bound depending on the field, currently (August 2020): 150,000 for $d = -3$, 100,000 for $d = -4$, and 50,000 for $d = -7, -8, -11$ (see Table 2.1). The total number of rational newforms within these ranges is 185,581, of which 5917 arise from base change from $\mathbb{Q}$. All but 40 of the 185,581 newforms F have a candidate associated elliptic curve E defined over the field K in the sense given in the introduction, namely $L(E, s) = L(F, s)$; or equivalently that they have the same compatible system of ℓ-adic Galois representations. It is possible (in principle) to prove modularity of all these elliptic curves, for example using the Faltings–Serre–Livné method (as detailed by Dieulefait–Guerberoff–Pacetti [DGP10]). This is being carried out systematically by Mattia Sanna (Warwick) for his PhD thesis [San20], using a combination of 2-adic and 3-adic methods combined with recent modularity lifting theorems.

Of the 40 remaining rational Bianchi newforms without associated elliptic curves, six arise from base change of classical modular forms over $\mathbb{Q}$ with trivial character, real quadratic coefficients and inner twists: see [Cre92]. A further 18 are twisted base change, and for 16 of these 18, the base-change form itself and the form over $\mathbb{Q}$ have trivial character; the other two will be studied in this paper. The remaining 16 forms are associated to abelian surfaces over the base field with quaternionic multiplication (QM), and were considered by Schembri [Sch19].

There is a natural action of $\mathrm{Gal}(K \mid \mathbb{Q}) = \langle \tau \rangle$ on rational Bianchi newforms, $f \mapsto \tau(f)$, where $a_{\mathfrak{p}}(\tau(f)) = a_{\tau(\mathfrak{p})}(f)$; if $g = \tau(f)$ then we say that f, g are *Galois conjugate*. (Our forms have rational coefficients, so there should be no confusion with other Galois actions.)

Organizing according to field and level, we have the following further details about the 40 rational newforms with no associated elliptic curve. These are either base-change, or come in Galois conjugate pairs. In some cases the forms mentioned are twists of base-change, but the base-change newform itself has too large a level to be in the current database, so does not yet have a label.

Table 2.1 Summary of Bianchi modular form data in the LMFDB

Field discriminant	−4	−8	−3	−7	−11	Total
Level norm bound	100000	50000	150000	50000	50000	
Rational newform counts						
All rational newforms	40030	36843	42343	35824	30541	185581
Not twist of base change	38844	35536	40986	34830	29468	179664
Base change of rational newform	908	809	955	518	504	3694
Base change of quadratic newform	16	10	28	24	13	91
Twist of base change of rational newform	240	350	328	378	492	1788
Twist of base change of quadratic newform	22	138	46	72	64	342
Twist of base change of higher-dimensional newform	0	0	0	2	0	2
Rational newforms with no elliptic curve						
Total	4	8	21	7	0	40
Quadratic twist of base change	0	8	9	5	0	22
Higher order twist of base change	0	0	0	2	0	2
Not twist of base change	4	0	12	0	0	16

- $K = \mathbb{Q}(\sqrt{-1})$, LMFDB label 2.0.4.1: four cases, none of which are base-change (even up to twist).
  - 34225.3-a is not a base-change or twist of base-change. It is associated to a simple abelian surface with QM over K, the Jacobian of the genus 2 curve C_1 found by Schembri [Sch19, Theorem 1.1].
  - 34225.3-b is a quadratic twist of 34225.3-a.
  - 34225.7-a and 34225.7-b are Galois conjugates of the previous two.
- $K = \mathbb{Q}(\sqrt{-2})$, LMFDB label 2.0.8.1: eight cases, all base-change (up to twist).
  - 5625.1-b and its Galois conjugate 5625.3-b are both twists of base-changes of classical newforms in $S_2(28800)$.
  - 6561.5-a and its Galois conjugate 6561.5-d are both twists of a Bianchi newform at level (2592), which is a base-change from $S_2(20736)$. The classical newform has coefficients in $\mathbb{Q}(\sqrt{15})$ and attached to the Bianchi newform is an abelian surface with QM of discriminant 10, but an explicit equation is not known.
  - 21609.3-b and 21609.3-c are both twists of a Bianchi newform at level (14112), which is a base-change from $S_2(11289)$. The classical newform has coefficients in $\mathbb{Q}(\sqrt{7})$ and attached to the Bianchi newform is an abelian surface with QM of discriminant 14. Again, an explicit equation is not known.
  - 21609.1-b and 21609.1-c are Galois conjugates of the previous two.

- $K = \mathbb{Q}(\sqrt{-3})$, LMFDB label 2.0.3.1: 21 cases, of which 12 are not twists of base-change but have associated genus 2 curves with QM:
  - 61009.1-a is associated to the genus 2 curve C_2 found by Schembri [Sch19, Theorem 1.1], while 61009.1-b is a quadratic twist of this, and 61009.9-a and 61009.9-b are their Galois conjugates.
  - 67081.3-a is associated to the genus 2 curve C_3 found by Schembri [Sch19, Theorem 1.1], while 67081.3-b is a quadratic twist of this, and 67081.7-a and 67081.7-b are their Galois conjugates.
  - 123201.1-b is associated to the genus 2 curve C_4 found by Schembri [Sch19, Theorem 1.1], while 123201.1-c is a quadratic twist of this, and 123201.3-b and 123201.3-c are their Galois conjugates.

  The other nine cases are twists of base-changes of forms with trivial character, coefficients in a real quadratic field and an inner twist; in these cases, explicit equations are not known:
  - 5625.1-a is the base-change of 225.2.a.f, which has CM by -15, coefficients in $\mathbb{Q}(\sqrt{5})$, and an inner twist. This is an example of the phenomenon first described by Cremona [Cre92].
  - 6561.1-b is base-change of 243.2.a.d, with coefficients in $\mathbb{Q}(\sqrt{6})$ and inner twist, attached to an abelian surface with QM of discriminant 6.
  - 30625.1-a and its Galois conjugate 30625.3-a are twists of base-change of forms in $S_2(11025)$ with CM by -15.
  - 50625.1-c and its Galois conjugate 50625.1-d are base-changes of 675.2.a.l and 675.2.a.m respectively, with coefficients in $\mathbb{Q}(\sqrt{2})$ and inner twist, attached to abelian surfaces with QM of discriminant 6.
  - 65536.1-b and its Galois conjugate 65536.1-e are twists of a form at level (768), which is base-change of 2304.2.a.p, with CM by -6, coefficients in $\mathbb{Q}(\sqrt{2})$ and inner twist.
  - 104976.1-a is base-change of 972.2.a.e, with coefficients in $\mathbb{Q}(\sqrt{2})$ and inner twist, attached to an abelian surface with QM of discriminant 6.
- $K = \mathbb{Q}(\sqrt{-7})$, LMFDB label 2.0.7.1: seven cases. Of these, two are the forms F and G in Theorem 1 and discussed in detail in Sect. 8 below. In addition, we have
  - 30625.1-d: see Sect. 8.
  - 40000.1-b and its Galois conjugate 40000.7-b are twists of 30625.1-d.
  - 10000.1-b and its Galois conjugate 10000.5-b are twists of forms which are the base-change of forms in $S_2(19600)$ with coefficients in $\mathbb{Q}(\sqrt{5})$ and inner twist.
- $K = \mathbb{Q}(\sqrt{-11})$, LMFDB label 2.0.11.1: none.

3 Characters

We begin by setting some notation, in particular we define some characters. Let $\mathbb{Q}^{\text{al}}$ be an algebraic closure of $\mathbb{Q}$. By a *character* over $\mathbb{Q}$ we will mean either a Dirichlet character or the character of the absolute Galois group $\text{Gal}_{\mathbb{Q}} := \text{Gal}(\mathbb{Q}^{\text{al}} \mid \mathbb{Q})$ attached to it via class field theory; we define the same over a number field K allowing Hecke characters (of finite order). If χ is a character over $\mathbb{Q}$, we may talk of its *restriction* $\chi|_K$ to K thinking of it as a character of $\text{Gal}_{\mathbb{Q}}$ and restricted to $\text{Gal}_K := \text{Gal}(\mathbb{Q}^{\text{al}} \mid K)$.

Let $N := 1225 = 5^2 7^2$; we now define some Dirichlet characters of modulus N. Denote by χ_5 the mod 5 cyclotomic character of order 4 (with LMFDB label $\chi_{1225}(393, \cdot)$) and by χ_{-7} the quadratic character of conductor 7 (label $\chi_{1225}(1126, \cdot)$). As characters of $\text{Gal}_{\mathbb{Q}}$, these cut out the cyclic extensions $\mathbb{Q}(\zeta_5)$ and

$$K := \mathbb{Q}(\sqrt{-7}) \tag{3.1}$$

respectively. Define $\varepsilon := \chi_5 \chi_{-7}$ (label $\chi_{1225}(293, \cdot)$); then ε has conductor 35 and order 4. Let

$$H := \langle \chi_5 \rangle \times \langle \chi_{-7} \rangle \simeq \mathbb{Z}/4\mathbb{Z} \times \mathbb{Z}/2\mathbb{Z} \tag{3.2}$$

be the group generated by χ_{-7} and χ_5.

After restricting to K, the character $\chi_{-7}|_K$ is trivial, so $\varepsilon|_K = \chi_5|_K$. Also, while χ_5 is not a square (since $\mathbb{Q}$ has no cyclic extension of degree 8 ramified only at 5), its restriction to K is a square, since 5 is inert in K. We denote by ψ one of the two order eight characters of K such that $\psi^{-2} = \chi_5|_K = \varepsilon|_K$; the other such character is ψ^5. The conductor of ψ is $(5\sqrt{-7})$.

The reason why the character ψ must ramify at the prime $\sqrt{-7}$ can be explained as follows. The multiplicative group of the residue field of the prime ideal (5) is $\mathbb{F}_{25}^{\times}$ and so admits an order 8 character ψ_5. Such a character is odd, i.e., $\psi_5(-1) = -1$. A finite order Hecke character has infinity type (0, 0) (i.e. it is trivial at the Archimedean places), hence if it is unramified outside 5, such a character would not be trivial at -1. However, the quadratic character $\psi_{\sqrt{-7}}$ is also odd, and the character ψ corresponds to the Hecke character ramified at $\{5, \sqrt{-7}\}$ with such ramification at both primes.

4 A Classical Newform of Level 1225 and an Abelian Eightfold

In this section, we investigate a certain special classical newform; all statements below can be confirmed using MAGMA [BCP97] or PARI/GP [PAR21].

We retain notation from the previous section. Consider the newspace $S_2(N, \varepsilon)^{\text{new}}$. Its relative dimension as a vector space over $\mathbb{Q}(\varepsilon) \simeq \mathbb{Q}(i)$ is 56, and it decomposes under the action of the Hecke algebra into 7 irreducible components, of relative dimensions 2, 2, 4, 8, 12, 12, 16. Their absolute dimensions, as $\mathbb{Q}$-vector spaces, are obtained by multiplying by $[\mathbb{Q}(\varepsilon) : \mathbb{Q}] = 2$.

Let f be any one of the four Galois conjugate newforms in the unique 4-dimensional component; these four form a single orbit under $\text{Gal}(\mathbb{Q}^{\text{al}} \,|\, \mathbb{Q}(\varepsilon))$. Of the eight $\text{Gal}(\mathbb{Q}^{\text{al}} \,|\, \mathbb{Q})$-conjugates of f, four are in $S_2(N, \varepsilon)$ and the other four in $S_2(N, \overline{\varepsilon})$, where $\overline{\varepsilon} = \varepsilon^3 = \overline{\chi_5}\chi_{-7}$. The form f does not have CM and its Hecke field is $M_f := \mathbb{Q}(\zeta_{24})$.

Remark 4.1 The form f has $Nk^2 = 4900 \geq 4000$, and so is just outside the current range of forms in the LMFDB.

Proposition 4.2 *For each character $\chi \in H$, the twist $f \otimes \chi$ is a conjugate of f, i.e., there exists $\sigma \in \text{Gal}(M_f \,|\, \mathbb{Q})$ such that $f \otimes \chi = \sigma(f)$.*

Proof First observe that the level N is divisible by the square of the conductor of ε, and also that neither χ_{-7} nor χ_5 is a square. This implies that the twists have the same level as N [Shi94, Proposition 3.64].

Let $\chi \in H$. Then the nebentypus character of $f \otimes \chi$ is $\varepsilon\chi^2$, which equals either ε or $\overline{\varepsilon}$. Hence the twists are newforms in either $S_2(N, \varepsilon)^{\text{new}}$ or $S_2(N, \overline{\varepsilon})^{\text{new}}$. The $\text{Gal}_{\mathbb{Q}}$-orbit of $f \otimes \chi$ has size at most 8 (but might a priori be smaller). However, $a_2(f) = -\zeta_{24}^3$ is an eighth root of unity, while the a_2 coefficients of all the newforms of smaller dimension in $S_2(N, \varepsilon)$ (which lie in $\mathbb{Q}(\zeta_{12})$) are not roots of unity, so cannot be equal to $\chi(2)a_p(f)$. Hence $f \otimes \chi$ has the same Hecke field as f and by the uniqueness of the $\mathbb{Q}$-dimension 8 component of $S_2(N, \varepsilon)^{\text{new}}$ must be conjugates of f.

Proposition 4.2 says that the newform f has eight *inner twists* (σ, χ) (including the trivial twist, where both χ and σ are trivial). To keep track of these inner twists, we use the following notation: let $\sigma_a \in \text{Gal}(\mathbb{Q}(\zeta_{24}) \,|\, \mathbb{Q})$ be the automorphism such that $\sigma_a(\zeta_{24}) = \zeta_{24}^a$. Checking the first few Fourier coefficients $a_p(f)$ we see that the inner twists are defined by the data (σ, χ) in Table 4.1.

By the Eichler–Shimura construction, attached to (the Galois orbit of) f is a simple abelian 8-fold $A = A_f$, defined over $\mathbb{Q}$ with $\text{End}(A)_{\mathbb{Q}} := \text{End}(A) \otimes \mathbb{Q} \simeq M_f$. Write $A_K := A \times_{\mathbb{Q}} K$ for the base extension of A to $K = \mathbb{Q}(\sqrt{-7})$, and $A^{\text{al}} := A \times_{\mathbb{Q}} \mathbb{Q}^{\text{al}}$ for the base extension of A to $\mathbb{Q}^{\text{al}}$. To determine how A splits over $\mathbb{Q}^{\text{al}}$ we must determine the *geometric* endomorphism algebra $\text{End}(A_{\mathbb{Q}^{\text{al}}})_{\mathbb{Q}}$, which is generated over $\mathbb{Q}(\zeta_{24})$ by endomorphisms arising from inner twists of f,

Table 4.1 Inner twists of f

χ	1	χ_5	χ_5^2	χ_5^3	χ_{-7}	$\chi_5\chi_{-7}$	$\chi_5^2\chi_{-7}$	$\chi_5^3\chi_{-7}$
σ	σ_1	σ_{19}	σ_{13}	σ_7	σ_{17}	σ_{11}	σ_5	σ_{23}

by Ribet–Pyle [Rib80, Rib04, Pyl04] theory. We use the implementation in MAGMA [BCP97] due to Quer [Que09] and Stein.

Proposition 4.4 *The following statements hold.*

(a) *Each inner twist* (σ, χ) *of* f *gives rise to an endomorphism* $\xi_{(\sigma,\chi)}$ *of* A^{al}, *defined over the field* $\mathbb{Q}(\chi)$ *cut out by* χ *and satisfying*

$$\alpha\xi_{(\sigma,\chi)} = \xi_{(\sigma,\chi)}\sigma(\alpha) \tag{4.5}$$

for all $\alpha \in \mathbb{Q}(\zeta_{24})$. *The endomorphisms* $\xi_{(\sigma,\chi)}$ *are a* $\mathbb{Q}(\zeta_{24})$-*basis for* $\mathrm{End}(A^{\mathrm{al}})_{\mathbb{Q}}$.

(b) *We have* $\mathrm{End}(A_K)_{\mathbb{Q}} \simeq \mathrm{M}_2(\mathbb{Q}(\zeta_8))$, *with centre* $Z(\mathrm{End}(A_K)) = \mathbb{Z}[\zeta_8]$.

(c) *We have* $\mathrm{End}(A^{\mathrm{al}})_{\mathbb{Q}} \simeq \mathrm{M}_4(B)$ *where* $B \simeq \left(\dfrac{3,5}{\mathbb{Q}}\right)$ *is the division quaternion algebra of discriminant* 15 *over* $\mathbb{Q}$. *All endomorphisms of* A^{al} *are defined over* $K' := \mathbb{Q}(\zeta_5, \sqrt{-7})$.

(d) *There exists a simple abelian surface* S *defined over* K *with* $\mathrm{End}(S)_{\mathbb{Q}} \simeq B$ *and an isogeny* $A \sim S^4$ *defined over* K'.

Proof Statement (a) is an application of work of Ribet [Rib80, (5.5)] together with the computation of inner twists above (Proposition 4.2). Statement (b) follows from (a): over $K = \mathbb{Q}(\sqrt{-7})$, the abelian variety A_K obtains the endomorphism $\xi_{(\sigma_{17},\chi_{-7})}$; since the fixed field of σ_{17} is $\mathbb{Q}(\zeta_8)$ and $\mathbb{Q}(\zeta_{24}) = \mathbb{Q}(\zeta_8)(\sqrt{-3})$, we conclude that

$$\mathrm{End}(A_K)_{\mathbb{Q}} \simeq \left(\frac{-3,-7}{\mathbb{Q}(\zeta_8)}\right) \simeq \mathrm{M}_2(\mathbb{Q}(\zeta_8)). \tag{4.6}$$

We have $\zeta_8 \in \mathrm{End}(A_K)$ because already $\mathbb{Z}[\zeta_{24}] \subseteq \mathrm{End}(A)$, and $\mathbb{Q}(\zeta_8) \subseteq Z(\mathrm{End}(A_K)_{\mathbb{Q}})$.

We finish (c) and (d) following Pyle [Pyl04]. Over $\mathbb{Q}^{\mathrm{al}}$, we have $A \sim S^r$ where S (over $\mathbb{Q}^{\mathrm{al}}$) is simple, called a *building block*. We have $\dim S = 1$ or 2, and these cases are distinguished by a Brauer class; MAGMA confirms that $\dim S = 2$, $r = 4$, and therefore $\mathrm{End}(S)_{\mathbb{Q}} \simeq B$. Moreover, calculating the obstruction to descent [Pyl04, Proposition 5.2], we find that S and its endomorphism algebra may be defined, up to isogeny, over K.

The inner twist relation $f \otimes \chi = \sigma(f)$ implies, on comparing the nebentypus character on both sides, that

$$\varepsilon\chi^2 = \sigma_\chi(\varepsilon) \tag{4.7}$$

and that for all but finitely many p, we have

$$a_p(f)\chi(p) = \sigma(a_p(f)). \tag{4.8}$$

Together these imply that $a_p(f)^2/\varepsilon(p)$ is fixed by all $\sigma \in \mathrm{Gal}(\mathbb{Q}(\zeta_{24})/\mathbb{Q})$, so we conclude that $a_p(f)^2/\varepsilon(p) \in \mathbb{Q}$ for all but finitely many p (and this relation holds in generality).

The L-function of A (over $\mathbb{Q}$) is

$$L(A, s) = \prod_{\chi \in H} L(f \otimes \chi, s), \tag{4.9}$$

since the 8 Galois conjugates of f are precisely the 8 twists $f \otimes \chi$ for $\chi \in H$. After base change to $K' = \mathbb{Q}(\zeta_5, \sqrt{-7})$ to trivialize all the characters, by Proposition 4.4(d) we find that

$$L(A, K', s) = L(S, K', s)^4 = L(f_{K'}, s)^8 \tag{4.10}$$

so that

$$L(S, K', s) = L(f_{K'}, s)^2. \tag{4.11}$$

5 Base Change to $\mathbb{Q}(\sqrt{-7})$

In light of the results of the previous section, we now move to $K = \mathbb{Q}(\sqrt{-7})$. By Proposition 4.4(b), we have

$$A_K \sim C_0^2, \tag{5.1}$$

where C_0 is a simple abelian fourfold defined over K with $\mathrm{End}(C_0)_{\mathbb{Q}} \simeq \mathbb{Q}(\zeta_8)$.

We now consider the base-change f_K of the newform f to K, a Bianchi modular form of weight 2. In the classical language of automorphic forms, the base change is known as the Doi–Naganuma lift [DN70] and can be obtained using the theta function method as generalized to arbitrary weight, level, and character by Friedberg [Fri83, Theorems 3.1, 3.2]. More generally, base change can be understood in the language of automorphic representations: see Gerardin–Labesse [GL79] or Langlands [Lan80].

Although we do not use it explicitly, we recall the formula for the Hecke eigenvalues of f_K. If p splits in K and $\mathfrak{p}$ lies over p then $a_{\mathfrak{p}}(f_K) = a_p(f)$; if p is inert in K then $a_{(p)}(f_K) = a_p(f)^2 - 2\varepsilon(p)p$.

Recall the character ψ of order 8 defined in Sect. 3.

Proposition 5.2 *The base-change f_K has the following properties.*

(a) *We have $f_K \in S_2(\mathfrak{N}, \psi^{-2})$, so f_K has level $\mathfrak{N} = (175) = (5)^2(\sqrt{-7})^2$ and nebentypus character $\varepsilon|_K = \psi^{-2}$.*
(b) *The Fourier coefficients of f_K lie in $\mathbb{Q}(\zeta_8)$.*

(c) *For each inner twist* (σ, χ) *of* f, *the base change* f_K *has inner twist by* $(\sigma, \chi|_K)$; *so* f_K *has four distinct inner twists (including the trivial inner twist).*

Proof The level of f_K is determined uniquely by the properties that it is stable under Galois conjugation and of norm $N^2/\operatorname{Disc}(K)^2 = 1225^2/7^2 = 30625$ [Ţur18, Lemma 5.2], where we use the fact that the -7-twist of f has the same level by Proposition 4.2: f has inner twist by (σ_{17}, χ_{-7}). Alternatively, since $1225 = 5^2 7^2$ the level of f_K divides $(5)^2(\sqrt{-7})^2$ but this is already equal to the conductor of the character $\varepsilon|_K = \psi^{-2}$ of f_K. (This also follows from the recognition of $f_K \otimes \psi \in S_2(\Gamma_0(\mathfrak{N}))$ in Proposition 6.2 below.)

Let (σ, χ) be any of the inner twists of f, so $\sigma(f) = f \otimes \chi$. Restricting to K gives $\sigma(f_K) = f_K \otimes \chi|_K$. Taking the inner twist to be (σ_{17}, χ_{-7}) shows that $\sigma_{17}(f_K) = f_K$, so the coefficients of f_K lie in the fixed field of σ_{17}, which is $\mathbb{Q}(\zeta_8)$.

After restriction to K, the characters in H coincide in pairs to give all the even powers of ψ. Hence we have

$$L(A, K, s) = \prod_{i=1}^{4} L(f_K \otimes \psi^{2i}, s)^2, \tag{5.3}$$

and hence from (5.1) that

$$L(C_0, s) = \prod_{i=1}^{4} L(f_K \otimes \psi^{2i}, s). \tag{5.4}$$

6 Twisting

We now investigate twists of the base change form $f_K \in S_2(\mathfrak{N}, \psi^{-2})$ from the previous section. Define

$$F := f_K \otimes \psi \quad \text{and} \quad G := f_K \otimes \psi^5 = F \otimes \psi^4. \tag{6.1}$$

Proposition 6.2 *The twists* $F, G \in S_2(\mathfrak{N})$ *have rational eigenvalues; the form* F *has LMFDB label* 2.0.7.1-30625.1-c, *while* G *has label* 2.0.7.1-30625.1-e.

Proof Note that twisting leaves the level unchanged and the twisted forms have trivial nebentypus.

It is enough to show that F is fixed by all $\sigma \in \operatorname{Gal}(\mathbb{Q}(\zeta_8) \mid \mathbb{Q})$, since twisting by the quadratic character ψ^4 preserves rationality. Let (σ, χ) be an inner twist of f. Then $\sigma(f_K) = f_K \otimes \chi|_K$ and

$$\sigma(f_K \otimes \psi) = \sigma(f_K) \otimes \sigma(\psi) = f_K \otimes \chi|_K \sigma(\psi). \tag{6.3}$$

Since there is an inner twist for every $\sigma \in \mathrm{Gal}(\mathbb{Q}(\zeta_8) \mid \mathbb{Q})$, the result is equivalent to establishing the character relations

$$\psi = \chi|_K \sigma(\psi) \tag{6.4}$$

for all such χ. Inspection of Table 4.1 shows that for $a \in (\mathbb{Z}/8\mathbb{Z})^\times$,

$$(\chi_5^{(a-1)/2})|_K = (\psi^{-2})^{(a-1)/2} = \psi/\psi^a = \psi/\sigma_a(\psi) \tag{6.5}$$

verifying (6.4).

Having identified a base change, we now examine the space of Bianchi newforms of level $\mathfrak{N}$, weight 2 and trivial nebentypus character, a space with LMFDB label 2.0.7.1-30625.1. The new subspace has dimension 60 and contains 8 newforms with rational Fourier coefficients. Of these one matches F, namely 2.0.7.1-30625.1-c, while G matches 2.0.7.1-30625.1-e, and is the $\sqrt{5}$-twist of F (since $\psi^4 = \chi_5^2|_K$, the quadratic character associated to $K(\sqrt{5}) \supseteq K$).

Remark 6.6 It is straightforward to identify whether a Bianchi newform F is a twist of base change since if F and $\tau(F)$ are twist-equivalent then either F is a twist of base change or is induced from a Hecke character [LR98]. Furthermore, for two Bianchi newforms F_1 and F_2 with trivial nebentypus, if $a_\mathfrak{p}(F_1)^2 = a_\mathfrak{p}(F_2)^2$ for all but finitely many primes $\mathfrak{p}$ then F_1 and F_2 are twist-equivalent [Ram00].

Remark 6.7 In fact, all eight rational Bianchi newforms in the space $S_2(\mathfrak{N})$ (with LMFDB label 2.0.7.1-30625.1) give interesting examples: we briefly describe the other six below (Sect. 8). The LMFDB does not contain details of Bianchi newforms with nontrivial nebentypus character, such as the form f_K itself.

That the twisted forms F and G have rational eigenvalues means that one can reasonably expect the corresponding twist of the fourfold C_0 to split as a product of QM-surfaces. Recall (see after (5.1)) that $\mathrm{End}(C_0)_\mathbb{Q} \simeq \mathbb{Q}(\zeta_8)$; we choose an automorphism $\xi \in \mathrm{Aut}(C_0)$ of order 8. With this choice, we may twist C_0 via ψ (relative to ξ), to obtain a new fourfold defined over K, denoted C:

$$C := C_0^{(\psi)}. \tag{6.8}$$

Theorem 6.9 *There is an isogeny $C \sim S_F \times S_G$ defined over K, where S_F, S_G are abelian surfaces defined over K with QM by the rational quaternion algebra B of discriminant* 15, *satisfying $L(S_F, s) = L(F, s)^2$ and $L(S_G, s) = L(G, s)^2$.*

Proof Let $\ell \equiv 1 \pmod 8$ be prime and let $V_\ell C_0$ be the ℓ-adic Tate representation of C_0, a free $\mathbb{Q}_\ell[\xi]$-module of rank 2 where $\xi^4 = -1$; choosing a primitive eighth root of unity $\zeta_8 \in \mathbb{Q}_\ell$, we may write

$$V_\ell C_0 \simeq \bigoplus_{k \in (\mathbb{Z}/8\mathbb{Z})^\times} W_{\ell,k} \tag{6.10}$$

where each $W_{\ell,k}$ is a $\mathbb{Q}_\ell$-vector space of dimension 2 with ξ acting by the scalar ζ_8^k.

Now $V_\ell C_0$ is also a representation of Gal_K; comparing with the L-series decomposition (5.4), as Galois representations we have (up to the choice of ζ_8)

$$W_{\ell,k} \simeq W_{\ell,1} \otimes \psi^{k-1}. \tag{6.11}$$

On the twist, we have the Galois representation

$$V_\ell C = V_\ell C_0 \otimes_{\mathbb{Q}_\ell[\xi]} \mathbb{Q}_\ell[\xi](\psi) \tag{6.12}$$

where $\mathbb{Q}_\ell[\xi](\psi)$ is $\mathbb{Q}_\ell[\xi]$ where Gal_K acts via ψ: explicitly, on the eigenspace of $\mathbb{Q}_\ell[\xi]$ where ξ acts by ζ_8^k, now Gal_K acts via ψ^k. In terms of the splitting (6.10) above, we have

$$V_\ell C \simeq \bigoplus_{k\in(\mathbb{Z}/8\mathbb{Z})^\times} W_{\ell,k} \otimes \psi^k \simeq \bigoplus_k W_{\ell,1} \otimes \psi^{2k-1} \tag{6.13}$$

and thus

$$\begin{aligned} L(C, K, s) &= L(f_K \otimes \psi, s)L(f_K \otimes \psi^5, s)L(f_K \otimes \psi^9, s)L(f_K \otimes \psi^{13}, s) \\ &= L(F, s)^2 L(G, s)^2. \end{aligned} \tag{6.14}$$

Let $L \supseteq K' = K(\zeta_5)$ be the cyclic degree 8 extension of K cut out by ψ. By work of Kida [Kid95, p. 53] we have

$$\mathrm{Res}_{L|K} C_{0,L} \sim_K \prod_{i=0}^{7} C_0^{(\psi^i)}; \tag{6.15}$$

i.e., the restriction of scalars of the base change of A to L is isogenous over K to the product of the twists of A by powers of ψ. From the L-series decomposition (5.4) and the theorem of Faltings, we conclude that

$$\prod_{i=0}^{7} C_0^{(\psi^i)} \sim_K C_0^4 \times C^4. \tag{6.16}$$

Let S be the building block from Proposition 4.4(d); up to isogeny (over K'), we may choose S so that $\mathrm{End}(S) = \mathcal{O}$, where $\mathcal{O} \subseteq B$ is a maximal order. Then $A_L \simeq C_{0,L}^2 \simeq S_L^4$, so $C_{0,L} \sim_L S_L^2$. Since $\mathbb{Q}(\sqrt{2})$ splits B and all maximal orders in $\mathcal{O}$ are conjugate, there exists $\alpha \in \mathcal{O}$ such that $\alpha^2 = 2$; then $\begin{pmatrix} 0 & -1 \\ 1 & \alpha \end{pmatrix} \in \mathrm{M}_2(\mathcal{O})$ is an element of order 8, so we may consider twists of S^2 by powers of ψ. Applying the theorem of Kida now to S^2, with (6.15)–(6.16) we conclude there are isogenies over K

$$\prod_{i=0}^{7} (S^2)^{(\psi^i)} \sim \mathrm{Res}_{L|K} S_L^2 \sim \mathrm{Res}_{L|K} C_{0,L} \sim \prod_{i=0}^{7} C_0^{(\psi^i)} \sim C_0^4 \times C^4. \tag{6.17}$$

Since C_0 is simple, we conclude that S is an isogeny factor of C over K, and $C \sim S \times S'$ for another abelian surface S'. We have $(S^2)^{(\psi^4)} = (S^{(\psi^4)})^2$ since $S^{(\psi^4)}$ is the quadratic twist by ψ^4 (corresponding to the quadratic extension $K(\sqrt{5}) \supseteq K$), so $S^{(\psi^4)}$ is also an isogeny factor of C over K. Since S is a QM abelian surface, its L-series is a square, so by (6.14) we have $S \not\sim S'$, and so $C \sim S \times S^{(\psi^4)}$, and thus up to labelling $L(S, s) = L(F, s)^2$ and the result follows.

Remark 6.18 The argument in Theorem 6.9 can evidently be generalized to more general situations.

7 An Explicit Model for the Building Block

In light of Theorem 6.9, we now seek explicit models for the abelian surfaces S_F and S_G whose L-functions match F and G. Because G is a quadratic twist of F, it suffices to exhibit a model for S_F. We refer to Buzzard [Buz97] and Voight [Voi19, Chapter 43] as general references on abelian surfaces with quaternionic multiplication.

Let $B := \left(\dfrac{3, 5}{\mathbb{Q}}\right)$ be the quaternion algebra over $\mathbb{Q}$ of discriminant 15, and let $\mathcal{O} \subseteq B$ be a maximal order. Then there exists $\mu \in \mathcal{O}$ such that $\mu^2 + 15 = 0$, a *principal polarization* on $\mathcal{O}$. The moduli space for principally polarized abelian surfaces with QM structure by $(\mathcal{O}, \mu)$ is a Shimura curve $X_0(15, 1)$ defined over $\mathbb{Q}$ by work of Shimura [Shi67] and Deligne [Del71].

An explicit model for $X_0(15, 1)$ was computed by Jordan [Jor81, Proposition 3.2.1] (see also Jordan–Livné [JL85, Table 1]) and Elkies [Elk98, (76)]): it is described by the equation

$$X_0(15, 1)\colon (x^2 + 3)(x^2 + 243) + 3y^2 = 0. \tag{7.1}$$

In particular, $X_0(15, 1)$ is a genus 1 curve with $X_0(15, 1)(\mathbb{Q}) = X_0(15, 1)(\mathbb{R}) = \emptyset$. The Jacobian of $X_0(15, 1)$ is the elliptic curve over $\mathbb{Q}$ with LMFDB label 15.a5.

The quotient by the Atkin–Lehner involution w_{15} gives the following map.

$$\begin{aligned} X_0(15, 1) &\to X_0(15, 1)/w_{15} \simeq \mathbb{P}^1 \\ j(x, y) &= x^2/81 \end{aligned} \tag{7.2}$$

The Igusa–Clebsch invariants of weights 2, 4, 6, 10 for the 'universal' abelian surface over $X_0(15, 1)/w_{15}$ are given by $(I_2 : I_4 : I_6 : I_{10})$ where

$$\begin{aligned} I_2 &= 12(j^4 + 15j^3 + 105j^2 + 125j + 10), \\ I_4 &= 45(4j + 1)(j + 3)^2(j - 1)^4, \end{aligned} \tag{7.3}$$

$$I_6 = 9(84j^5 + 1414j^4 + 8865j^3 + 11157j^2 + 3895j + 185)(j + 3)^2(j - 1)^4,$$

$$I_{10} = 2(j + 3)^6(j - 1)^{12}.$$

The invariants (7.3) were computed by Elkies (2007) using K3 surfaces and by Guo–Yang [GY17, Table 4] using Borcherds products.

With this data in hand, we look for points on $X_0(15, 1)$ over K. The base change of its Jacobian to K has rank 1, so we will find infinitely many points. Searching in a small box, we find the points $(x, y) = (\pm 9/\sqrt{-7}, \pm 180/7)$ for which $j(x, y) = -1/7$ and normalized invariants $(648 : 23625 : 17474625 : -392000000)$. From these invariants, we construct a genus 2 curve X' over $K = \mathbb{Q}(\sqrt{-7})$ well-defined up to twist—we confirm that the Euler factors match *up to sign* in the sense that $L_{\mathfrak{p}}(X', T) = L_{\mathfrak{p}}(f, \pm T)^2$ for good primes $\mathfrak{p}$, and this gives strong evidence that we have found a match.

Next, we identify the quadratic field $K_\delta \supseteq K$ which is associated to the twist. For a good prime $\mathfrak{p}$, write $L_{\mathfrak{p}}(X', T) = (1 - b_{\mathfrak{p}}T + \mathrm{Nm}(\mathfrak{p})T^2)$ so that $b_{\mathfrak{p}} = \mu(\mathfrak{p})a_{\mathfrak{p}}(F)$ and $\mu(\mathfrak{p}) = \pm 1$. Then $\mu(\mathfrak{p}) = -1$ if and only if $\mathfrak{p}$ is inert in K_δ; searching for fields with this property supported at primes dividing the discriminant of X' gives a twist X for which the Euler factors match for all $\mathfrak{p}$ with $\mathrm{Nm}(\mathfrak{p}) \leq 200$.

In this way, we obtain a candidate model X for a curve whose Jacobian putatively corresponds to F; however, this model has *enormous* coefficients. So we seek to reduce these coefficients. We first apply the reduction algorithms of Bouyer–Streng [BS15] to reduce the discriminant. We would like to apply the Cremona–Stoll algorithm [SC03] to reduce the coefficients of X further, but the current implementation requires a real place in the base field. Nevertheless, we can apply some reductions by hand, and then iterate. In this way, we find the following model for $X = X_F \colon y^2 = f_F(x)$ where

$$
\begin{aligned}
f_F(x) := (&-1277479451106544431031637345569322244585 5\sqrt{-7} \\
&\qquad - 7695069988816582237324746836913139396789)x^6 \\
&+ (369033561110881127499984507760692059639 43\sqrt{-7} \\
&\qquad + 7476636040743270474172885336522730375581 3)x^5 \\
&+ (-321917153494381391057511777261488963972 05\sqrt{-7} \\
&\qquad - 1605941435368561319404165201637915142607 70)x^4 \\
&+ (313875730633426945520672610129421224752 0\sqrt{-7} \\
&\qquad + 1430704805457265682063815177164266973668 80)x^3 \\
&+ (872371717006802010044187864253768259583 5\sqrt{-7} \\
&\qquad - 5921319576617313055450367200522230486604 0)x^2 \\
&+ (-408065214955555266138718277062047761970 3\sqrt{-7} \\
&\qquad + 1045777357087411951262513076181384523383 3)x \\
&+ 527840106295482417952525742290110846145\sqrt{-7} \\
&\qquad - 4937131580514455349359485517406586767 43.
\end{aligned}
\tag{7.4}
$$

By construction, we know that $J_F := \mathrm{Jac}(X_F)$ has $\mathrm{End}(J_F) \supseteq \mathcal{O}$. Looking at Euler factors we see that J_F does not have CM, so $\mathrm{End}(J_F) = \mathcal{O}$, since $\mathcal{O}$ is a maximal order.

Proposition 7.5 *For all good primes* $\mathfrak{p}$ *we have* $L_\mathfrak{p}(X_F, T) = L_\mathfrak{p}(J_F, T) = L_\mathfrak{p}(F, T)^2$,

Proof The proof of this proposition is a standard application of the method of Faltings–Serre, as explained by Dieulefait–Guerberoff–Pacetti [DGP10]; we only sketch some pertinent details here. Attached to $F = f_K \otimes \psi$ is a Galois representation $\rho_{F,2}\colon \mathrm{Gal}_K \to \mathrm{GL}_2(\mathbb{Q}_2^{\mathrm{al}})$. The characteristic polynomial of Frobenius at primes above 2 and 11 are given by $1 \pm T + 2T^2$ and $1 - 4T + 11T^2$, whose roots are different and do not add zero, hence the coefficient field of $\rho_{F,2}$ may be taken as $\mathbb{Q}_2$ by Taylor [Tay94, Corollary 1]. Since J_F has QM (defined over K) by B of discriminant 15 and $2 \nmid 15$, the 2-adic Tate representation $V_2 J_F$ is free of rank 1 over $B \otimes \mathbb{Q}_2 \simeq \mathrm{M}_2(\mathbb{Q}_2)$ and so affords a Galois representation $\rho_{J_F,2}\colon \mathrm{Gal}_K \to \mathrm{GL}_2(\mathbb{Q}_2)$.

The 2-division field $K(J_F[2])$ is defined by the polynomial

$$y^{12} - 6y^{11} + 21y^{10} - 50y^9 + 69y^8 - 42y^7 - 69y^6 + 210y^5 - 190y^4 + 48y^3 + 168y^2 - 160y + 64$$

(as an absolute extension); we confirm this extension is an S_3-extension of K, so the mod 2 representation $\overline{\rho}_{J_F,2}\colon \mathrm{Gal}_K \to \mathrm{GL}_2(\mathbb{F}_2) \simeq S_3$ is surjective.

Looking at Euler factors, the mod 2 representation $\overline{\rho}_{F,2}\colon \mathrm{Gal}_K \to \mathrm{GL}_2(\mathbb{F}_2) \simeq S_3$ attached to F is also surjective. Since the conjugate of F under $\mathrm{Gal}(K \mid \mathbb{Q})$ is a twist of F, the extension $K(J_F[2])$ is Galois over $\mathbb{Q}$ with Galois group $S_3 \rtimes C_2 \simeq D_6$, and so arises as the Galois closure of a sextic extension of $\mathbb{Q}$ unramified away from 2, 5, 7. From the Jones–Roberts database [JR14], there are exactly 5 such extensions whose Galois closure contains $\mathbb{Q}(\sqrt{-7})$:

$$x^6 - 3x^5 + 2x^4 - 6x^3 + 25x^2 - 19x + 8,$$

$$x^6 - x^5 - x^4 - x^3 + 2x + 2,$$

$$x^6 - x^5 + 9x^4 - 11x^3 + 16x^2 + 14x + 4,$$

$$x^6 - 2x^5 + 11x^4 - 8x^3 + 30x^2 + 24x + 8,$$

$$x^6 - x^5 + x^4 - 29x^3 + 8x^2 - 64x + 512.$$

Looking at Euler factors, we rule out all but the first one, and then confirm that its Galois closure is isomorphic to the one for J_F; this shows that the two mod 2 representations $\overline{\rho}_{F,2}$ and $\overline{\rho}_{J_F,2}$ are equivalent.

Since both residual representations are isomorphic with absolutely irreducible image, following Dieulefait–Guerberoff–Pacetti [DGP10, Section 2.1] we then show that fully the 2-adic representations $\rho_{F,2}$ and $\rho_{J_F,2}$ are equivalent by showing there is no obstruction to lifting. Computing a set of obstructing primes (whose Frobenius classes generate the Galois group of the maximal exponent 2 extension of $K(J_F[2])$ unramified away from 2, 5, 7), we conclude it is enough to check that $\operatorname{tr} \rho_{F,2}(\mathrm{Frob}_\mathfrak{p}) = \operatorname{tr} \rho_{J_F,2}(\mathrm{Frob}_\mathfrak{p})$ for a primes $\mathfrak{p}$ of K dividing 3, 11, 13, 17, 23, 29. Having already checked this for all good primes $\mathfrak{p}$ up to norm $\mathrm{Nm}(\mathfrak{p}) \leq 200$, we conclude the proof.

8 Other Rational Bianchi Newforms at Level (175)

As well as the newforms F and G with labels 2.0.7.1-30625.1-c and 2.0.7.1-30625.1-e, there are six other rational newforms of the same level, which display a variety of phenomena. All the quadratic twists mentioned here are by $\sqrt{5}$ unless otherwise specified.

- Newform 30625.1-a is the base-change of classical modular forms in $S_2(1225)$ with LMFDB labels 1225.2.a.b and 1225.2.a.d. These are -7-twists of each other, and associated to the elliptic curves in the isogeny classes 1225.b and 1225.d. The base-change to K of the curves in both these isogeny classes all lie in the isogeny class 2.0.7.1.30625.1-a.

- Newform 30625.1-b is a base-change, but of a newform with coefficients in $\mathbb{Q}(\sqrt{2})$. The associated modular abelian variety is a surface which splits over K into the product of elliptic curves, with associated elliptic curves in the isogeny class 2.0.7.1.30625.1-b, which are $\mathbb{Q}$-curves but not base-change.
- Newform 30625.1-d is also base-change, but of a newform with coefficients in $\mathbb{Q}(\sqrt{5})$ whose associated modular abelian variety is a surface which does *not* split over K. Hence there is no elliptic curve associated to the newform, while there is an abelian surface with QM. The newform has CM by -35 (so is its own quadratic twist) and is an example of the situation described in [Cre92, p. 411].
- Newform 30625.1-f is the quadratic twist of 30625.1-a so has associated elliptic curves 2.0.7.1.30625.1-f which are base-change.
- Newform 30625.1-g is the quadratic twist of 30625.1-b so has associated elliptic curves 2.0.7.1.30625.1-g which are $\mathbb{Q}$-curves but not base-change.
- Newform 30625.1-h is the quadratic twist of 1225.1-a at level $\mathfrak{N}' = (35)$. It has associated elliptic curves in isogeny class 2.0.7.1.30625.1-h which are base-changes of those in isogeny classes 1225.h and 1225.j, and is itself the base-change of classical newforms 1225.2.a.h and 1225.2.a.j.

9 Relation with the Paramodular Conjecture

We conclude with an application to the Paramodular Conjecture. Recall the following conjecture due to Brumer–Kramer [BK14] and its corrigendum [BK19], as amended by Calegari.

Conjecture 9.1 *There is a bijection between the set of isogeny classes of QM abelian fourfolds B over $\mathbb{Q}$ of conductor N^2 and the set of cuspidal, nonlift Siegel paramodular newforms f of genus 2, weight 2, and level N with rational Hecke eigenvalues, up to nonzero scaling. Moreover, if $B \leftrightarrow f$ in this bijection, we have the equality*

$$L(B, s) = L(f, s, \mathrm{spin})^2. \tag{9.2}$$

As with Bianchi modular forms, the frequently arising case is where $\mathrm{End}(B)_{\mathbb{Q}} \simeq \mathrm{M}_2(\mathbb{Q})$, in which case $B \sim A^2$ for A an abelian surface over $\mathbb{Q}$ with $\mathrm{End}(A) = \mathbb{Z}$ with $L(A, s) = L(f, s, \mathrm{spin})$. (When $\mathrm{End}(B)_{\mathbb{Q}}$ is a division algebra, the nonexistent geometric object A' such that $L(A', s) = L(f, s, \mathrm{spin})$ could be thought of as a *fake abelian surface*.)

To our Bianchi newform F, by theta lift one can attach a Siegel paramodular form Π, with rational eigenvalues [BDPŞ15, Theorem 4.1] such that $L(F, s) = L(\Pi, s, \mathrm{spin})$. The form Π satisfies the hypothesis of the Paramodular Conjecture (Conjecture 9.1), so there should be a QM abelian fourfold attached to Π. Indeed, by Theorem 6.9 the Weil restriction of scalars $\mathrm{Res}_{K|\mathbb{Q}} S_F$ of S_F from K to $\mathbb{Q}$ gives the desired geometric object, and

$$L(\mathrm{Res}_{K|\mathbb{Q}} S_F, s) = L(S_F, K, s) = L(F, s)^2 = L(\Pi, s, \mathrm{spin})^2. \tag{9.3}$$

Remark 9.4 In work of Boxer–Calegari–Gee–Pilloni [BCGP18, Lemma 10.3.2], assuming standard conjectures a general argument is given to explain the existence of a geometric object attached to an automorphic form on GSp_4 with rational Hecke eigenvalues. In concrete examples, the veracity of the standard conjectures is very hard to check; however, when both the automorphic form and (fake) abelian surface can be explicitly given, the method of Faltings–Serre can often be applied successfully in practice [BPPTVY] to establish an equality of L-functions, such as (9.3).

Acknowledgments The authors would like to thank Frank Calegari for useful discussions and suggestions and the anonymous referees for helpful feedback. Many thanks also to David Loeffler for directing our attention to the sources in Remark 6.6. Cremona was supported by EPSRC Programme Grant EP/K034383/1 *LMF: L-Functions and Modular Forms*, and the Horizon 2020 European Research Infrastructures project *OpenDreamKit* (#676541). Pacetti was partially supported by grant PICT-2018-02073. Dembélé, Schembri, and Voight were supported by a Simons Collaboration Grant (550029; to Voight).

References

[BCGP18] George Boxer, Frank Calegari, Toby Gee, and Vincent Pilloni. Abelian surfaces over totally real fields are potentially modular, 2018. Preprint: https://arxiv.org/abs/1812.09269.

[BCP97] Wieb Bosma, John Cannon, and Catherine Playoust. The Magma algebra system. I. The user language. *J. Symbolic Comput.*, 24(3-4):235–265, 1997. Computational algebra and number theory (London, 1993).

[BDPŞ15] Tobias Berger, Lassina Dembélé, Ariel Pacetti, and Mehmet Haluk Şengün. Theta lifts of Bianchi modular forms and applications to paramodularity. *J. Lond. Math. Soc. (2)*, 92(2):353–370, 2015.

[BK14] Armand Brumer and Kenneth Kramer. Paramodular abelian varieties of odd conductor. *Trans. Amer. Math. Soc.*, 366(5):2463–2516, 2014.

[BK19] Armand Brumer and Kenneth Kramer. Corrigendum to paramodular abelian varieties of odd conductor. *Trans. Amer. Math. Soc.*, 2019. available from https://www.ams.org/journals/tran/2019-372-03/S0002-9947-2019-07792-9/home.html.

[BPPTVY] Armand Brumer, Ariel Pacetti, Cris Poor, Gonzalo Tornaría, John Voight, and David Yuen. On the paramodularity of typical abelian surfaces. Algebra & Number Theory 13(5):1145–1195, 2019.

[BS15] Florian Bouyer and Marco Streng. Examples of CM curves of genus two defined over the reflex field. *LMS J. Comput. Math.*, 18(1):507–538, 2015.

[Buz97] Kevin Buzzard. Integral models of certain Shimura curves. *Duke Math. J.*, 87(3):591–612, 1997.

[Cre84] J. E. Cremona. Hyperbolic tessellations, modular symbols, and elliptic curves over complex quadratic fields. *Compositio Mathematica*, 51:275–323, 1984.

[Cre92] J. E. Cremona. Abelian varieties with extra twist, cusp forms, and elliptic curves over imaginary quadratic fields. *J. London Math. Soc. (2)*, 45(3):404–416, 1992.

[Del71] Pierre Deligne. Travaux de Shimura. pages 123–165. Lecture Notes in Math., Vol. 244, 1971.

[DGP10] Luis Dieulefait, Lucio Guerberoff, and Ariel Pacetti. Proving modularity for a given elliptic curve over an imaginary quadratic field. *Math. Comp.*, 79(270):1145–1170, 2010.

[DN70] Koji Doi and Hidehisa Naganuma. On the functional equation of certain Dirichlet series. *Invent. Math.*, 9:1–14, 1969/1970.

[DR73] P. Deligne and M. Rapoport. Les schémas de modules de courbes elliptiques. pages 143–316. Lecture Notes in Math., Vol. 349, 1973.

[Elk98] Noam D. Elkies. Shimura curve computations. In *Algorithmic number theory (Portland, OR, 1998)*, volume 1423 of *Lecture Notes in Comput. Sci.*, pages 1–47. Springer, Berlin, 1998.

[Fri83] Solomon Friedberg. On the imaginary quadratic Doi-Naganuma lifting of modular forms of arbitrary level. *Nagoya Math. J.*, 92:1–20, 1983.

[GL79] P. Gérardin and J.-P. Labesse. The solution of a base change problem for GL(2) (following Langlands, Saito, Shintani). In *Automorphic forms, representations and L-functions (Proc. Sympos. Pure Math., Oregon State Univ., Corvallis, Ore., 1977), Part 2*, Proc. Sympos. Pure Math., XXXIII, pages 115–133. Amer. Math. Soc., Providence, R.I., 1979.

[GY17] Jia-Wei Guo and Yifan Yang. Equations of hyperelliptic Shimura curves. *Compos. Math.*, 153(1):1–40, 2017.

[JR14] John Jones and David Roberts. A database of number fields. LMS Journal of Computation and Mathematics, 17(1):595–618, 2014.

[JL85] Bruce W. Jordan and Ron A. Livné. Local Diophantine properties of Shimura curves. *Math. Ann.*, 270(2):235–248, 1985.

[Jor81] Bruce Winchester Jordan. *On The Diophantine Arithmetic of Shimura Curves*. ProQuest LLC, Ann Arbor, MI, 1981. Thesis (Ph.D.)–Harvard University.

[Kid95] Masanari Kida. Galois descent and twists of an abelian variety. *Acta Arith.*, 73(1):51–57, 1995.

[Lan80] Robert P. Langlands. *Base change for* GL(2), volume 96 of *Annals of Mathematics Studies*. Princeton University Press, Princeton, N.J.; University of Tokyo Press, Tokyo, 1980.

[LMF20] LMFDB Collaboration. The L-functions and modular forms database. http://www.lmfdb.org, 2020. [Online; accessed 3 August 2020].

[LR98] Erez Lapid and Jonathan Rogawski. On twists of cuspidal representations of GL(2). *Forum Math.*, 10(2):175–197, 1998.

[PAR21] The PARI Group, Univ. Bordeaux. *PARI/GP version* `2.13.2`, 2021. available from http://pari.math.u-bordeaux.fr/.

[Pyl04] Elisabeth E. Pyle. Abelian varieties over $\mathbb{Q}$ with large endomorphism algebras and their simple components over $\overline{\mathbb{Q}}$. In *Modular curves and abelian varieties*, volume 224 of *Progr. Math.*, pages 189–239. Birkhäuser, Basel, 2004.

[Que09] Jordi Quer. Fields of definition of building blocks. *Math. Comp.*, 78(265):537–554, 2009.

[Ram00] Dinakar Ramakrishnan. Modularity of the Rankin-Selberg L-series, and multiplicity one for SL(2). *Ann. of Math. (2)*, 152(1):45–111, 2000.

[Rib80] Kenneth A. Ribet. Twists of modular forms and endomorphisms of abelian varieties. *Math. Ann.*, 253(1):43–62, 1980.

[Rib04] Kenneth A. Ribet. Abelian varieties over **Q** and modular forms. In *Modular curves and abelian varieties*, volume 224 of *Progr. Math.*, pages 241–261. Birkhäuser, Basel, 2004.

[San20] Mattia Sanna. On the equivalence of 3-adic representations. University of Warwick PhD thesis, 2020. Available from http://wrap.warwick.ac.uk/156281/

[SC03] Michael Stoll and John E. Cremona. On the reduction theory of binary forms. *J. Reine Angew. Math.*, 565:79–99, 2003.

[Sch19] Ciaran Schembri. Examples of genuine QM abelian surfaces which are modular. *Res. Number Theory*, 5(1):Art. 11, 12, 2019.

[Shi67] Goro Shimura. Construction of class fields and zeta functions of algebraic curves. *Ann. of Math. (2)*, 85:58–159, 1967.

[Shi94] Goro Shimura. *Introduction to the arithmetic theory of automorphic functions*, volume 11 of *Publications of the Mathematical Society of Japan*. Princeton University Press, Princeton, NJ, 1994. Reprint of the 1971 original, Kanô Memorial Lectures, 1.

[Tay94] Richard Taylor. l-adic representations associated to modular forms over imaginary quadratic fields. II. *Invent. Math.*, 116(1-3):619–643, 1994.

[Tay95] Richard Taylor. Representations of Galois groups associated to modular forms. In *Proceedings of the International Congress of Mathematicians, Vol. 1, 2 (Zürich, 1994)*, pages 435–442. Birkhäuser, Basel, 1995.

[Ţur18] George C. Ţurcaş. On Fermat's equation over some quadratic imaginary number fields. *Res. Number Theory*, 4(2):Art. 24, 16, 2018.

[Voi19] John Voight. *Quaternion algebras*. May 26, 2019. available from http://quatalg.org/.

A Database of Hilbert Modular Forms

Steve Donnelly and John Voight

Abstract We describe the computation of tables of Hilbert modular forms of parallel weight 2 over totally real fields.

Tables of classical modular forms have been computed for many decades now, starting with a table of rational eigenforms (corresponding to modular elliptic curves) appearing in the proceedings of the Antwerp conference in 1972 [2]. Large tables of classical modular forms also now exist, with the most recent large-scale effort available the *L-functions and Modular Forms Database* (LMFDB) [13] described by a team of ten authors [1] along with the fascinating history.

In this article, we continue this tradition of tabulating automorphic forms on Shimura varieties by computing extensive tables of Hilbert modular cusp forms of parallel weight 2 over totally real fields. This work is in a sense complementary to work [3] on tables of (modular) elliptic curves over $\mathbb{Q}(\sqrt{5})$. Our data is available online at the LMFDB [13].

In Sect. 1, we give an overview of the basic algorithms employed. In Sect. 2, we describe in more detail the implementation in the computer algebra system Magma [4], providing some documentation and detail about computations in practice and a few auxiliary algorithms employed. In Sect. 3, we gather the data which was computed and discuss the extent of the computation. Finally, in Sect. 4 we discuss some interesting aspects of the data in detail.

S. Donnelly
School of Mathematics and Statistics F07, University of Sydney, Sydney, NSW, Australia
e-mail: donnelly@maths.usyd.edu.au

J. Voight (✉)
Department of Mathematics, Dartmouth College, Hanover, NH, USA
e-mail: jvoight@gmail.com

J. S. Balakrishnan et al. (eds.), *Arithmetic Geometry, Number Theory, and Computation*, Simons Symposia, https://doi.org/10.1007/978-3-030-80914-0_12

1 The Algorithm: An Overview

In this section, we give a basic overview of the algorithm we employ. For a motivated introduction, see the expository work by Dembélé–Voight [9]; for a more basic reference on Hilbert modular forms, see e.g. Freitag [10].

Throughout, let F be a totally real field of degree $n = [F : \mathbb{Q}]$ with ring of integers $\mathbb{Z}_F$ and let $\mathfrak{N} \subseteq \mathbb{Z}_F$ be an ideal. Let $S_k(\mathfrak{N})$ denote the space of Hilbert cusp forms of level $\mathfrak{N}$ and parallel weight $k \in \mathbb{Z}_{\geq 2}$ for F, and let $S_k^{\text{new}}(\mathfrak{N})$ be the new subspace.

The algorithm we use combines methods of Dembélé [7] and Dembélé–Donnelly [8], along with methods of Greenberg–Voight [11] and Voight [16].

Theorem 1.1 *There exists an (explicit) algorithm that, on input a totally real field F and an ideal $\mathfrak{N} \subseteq \mathbb{Z}_F$, computes as output the systems of eigenvalues for the Hecke operators $T_\mathfrak{p}$ with $\mathfrak{p} \nmid \mathfrak{N}$ and the Atkin-Lehner involutions $W_{\mathfrak{p}^e}$ with $\mathfrak{p}^e \parallel \mathfrak{N}$ that occur in the space $S_k^{\text{new}}(\mathfrak{N})$.*

In other words, there exists an explicit finite procedure which takes as input the totally real field F, an ideal $\mathfrak{N} \subseteq \mathbb{Z}_F$ (encoded in bits in the usual way [6]), and an even integer $k \geq 2$, and outputs a finite set of number fields $E_f \subseteq \overline{\mathbb{Q}}$ and sequences $(a_f(\mathfrak{p}))_\mathfrak{p}$ encoding the Hecke eigenvalues for each eigenform constituent f in $S_k(\mathfrak{N})$ and $a_f(\mathfrak{p}) \in E_f$.

In this article, we consider the first and essential case $k = 2$; we believe it would be an interesting project to extend these efforts to include tabulation of Hilbert modular forms of higher weight as well.

The general strategy of the algorithm in Theorem 1.1 is as follows. Let B be the quaternion algebra over F which is unramified at all finite places and ramified at either all or all but one real place according as $n = [F : \mathbb{Q}]$ is even or odd. Then by the Eichler–Shimizu–Jacquet–Langlands correspondence (for a precise statement, references, and discussion see Dembélé–Voight [9, Theorem 3.9]), there exists an isomorphism of Hecke modules

$$S_2^B(\mathfrak{N}) \xrightarrow{\sim} S_2(\mathfrak{N})$$

where $S_2^B(\mathfrak{N})$ denotes the space of quaternionic modular forms over B of weight k and level $\mathfrak{N}$. The description of the space $S_2^B(\mathfrak{N})$ varies accordingly as n is even or odd. Let $\mathcal{O} \subseteq B$ be a maximal order.

If n is even, we use Brandt matrices, following the method of Dembélé [7] and Dembélé–Donnelly [8]. The Hecke module $S_2^B(\mathfrak{N})$ is computed via an explicit action on maps from the set of right ideal classes in $\mathcal{O}$ to $\mathbb{P}^1(\mathbb{Z}_F/\mathfrak{N})$. We refer to this as the definite method, since the algebra B in this case is definite.

If instead n is odd, the algorithm in Theorem 1.1 is supplied by Greenberg–Voight [11] and Voight [16]. For simplicity of exposition, assume F has strict class number 1; the method works for arbitrary class number. Let $\mathcal{O}_1^\times = \{\gamma \in \mathcal{O} : \text{nrd}(\gamma) = 1\}$ denote the group of units of $\mathcal{O}$ with reduced norm 1. Suppose that the unique real split place of B corresponds to the embedding $\iota_\infty : B \hookrightarrow \mathrm{M}_2(\mathbb{R})$ and let $\Gamma(1) = \iota_\infty(\mathcal{O}_1^\times)/\{\pm 1\} \subseteq \mathrm{PSL}_2(\mathbb{R})$. Let $\mathcal{O}_0(\mathfrak{N})$ be an Eichler order of level $\mathfrak{N}$ and let $\Gamma_0(\mathfrak{N}) = \iota_\infty(\mathcal{O}_0(\mathfrak{N})_1^\times)/\{\pm 1\}$. Then by the theorem of Eichler-Shimura, and Shapiro's lemma, we have a further isomorphism of Hecke modules

$$S_2^B(\mathfrak{N}) \xrightarrow{\sim} H^1\left(\Gamma(1), \mathrm{Coind}_{\Gamma_0(\mathfrak{N})}^{\Gamma(1)} \mathbb{C}\right)^+$$

where $+$ denotes the $+1$-eigenspace for complex conjugation. The Hecke module $S_2^B(\mathfrak{N})$ is then computed via this isomorphism as group cohomology, using an explicit presentation for the group $\Gamma(1)$ based on the computation of a fundamental domain [15]. We call this the indefinite method.

Remark 1.2 More generally, one can use a quaternion algebra B over F of discriminant $\mathfrak{D}$, ramified at all or all but one real place, and $\mathfrak{M}$ an ideal coprime to $\mathfrak{D}$; then with $\mathfrak{N} = \mathfrak{D}\mathfrak{M}$, the Jacquet-Langlands correspondence yields an isomorphism

$$S_2^B(\mathfrak{M}) \xrightarrow{\sim} S_2(\mathfrak{N})^{\mathfrak{D}\text{-new}}$$

where $S_2(\mathfrak{N})^{\mathfrak{D}\text{-new}} \subseteq S_2(\mathfrak{N})$ denotes the space of Hilbert cusp forms which are new at all primes dividing $\mathfrak{D}$. Since $\mathfrak{D}$ is squarefree, these methods overlap when there is a prime $\mathfrak{p}$ which exactly divides the level $\mathfrak{N}$. We use this to verify our calculations by computing with both the definite and indefinite method when it applies; for more, see Sect. 3.

2 Computing Systems of Hecke Eigenvalues

In this section, we discuss the MAGMA [4] implementation of the above algorithms. These methods saw a first implementation over $\mathbb{Q}$ by David Kohel, and they rely heavily on code for quaternion algebras over number fields, due to Donnelly, Markus Kirschmer, Nicole Sutherland, and Voight (documented in work of Kirschmer–Voight [12]).

Our algorithm has four main steps:

1. Precomputation
2. Computing a basis
3. Hecke computation via reduction
4. Decomposition into newforms

We refer to Dembélé–Voight [9] for an explicit mathematical description; here, we provide some additional comments about the implementation.

Step 1: Precomputation The first step of the algorithm involves precomputation steps that only depend on the field F: that is to say, for all levels $\mathfrak{N}$ with $\mathfrak{p} \parallel \mathfrak{N}$, this step need only be performed once. For the definite method, the precomputation step involves determining a set of representatives for the right ideal classes in $\mathcal{O}$ up to isomorphism.

- We precompute the theta series of each left order $\mathcal{O}$ of each right ideal representative as

$$\#\{\alpha \in \mathcal{O}_L(I) : \mathrm{Tr}_{F/\mathbb{Q}} \mathrm{nrd}(\alpha) = n\}$$

for a small set of values $n \in \mathbb{Z}_{\geq 0}$. This reduces the number of isomorphism tests of ideals when computing Hecke operators below, at the marginal additional cost of requiring the computation of left orders and representation numbers.

- As computing inverses and colon ideals can also be time consuming, we compute for each representative right ideal I a small element $b \in I$ and a left ideal I' of prime power norm such that $\mathcal{O}_L(I)b = II'$: then $JI' = J(I^{-1})b$, and α generates JI' if and only if α/b generates JI^{-1}.

For the indefinite method, we compute a fundamental domain for the Shimura curve associated to the maximal order, together with a presentation for the group $\Gamma_0(1)$ [15].

Steps 2 and 3: Computing a Basis and Hecke Computation To compute a basis of forms and to compute Hecke operators, we need fast algorithms for working with $\mathbb{P}^1(\mathbb{Z}_F/\mathfrak{N})$, and our efficient implementation treats the case where $\mathfrak{N}$ is prime and composite differently: in the latter we index elements $[a : b] \in \mathbb{P}^1(\mathbb{Z}_F/\mathfrak{N})$ according to divisor $\mathfrak{a} = (a) \mid \mathfrak{N}$. With these representatives in hand, in the definite case we list orbits of $\mathbb{P}^1(\mathbb{Z}_F/\mathfrak{N})$ under the action of the stabilizer groups and in the indefinite case we record the permutation action in the coinduced module. (If we were working in higher weight, and a nontrivial representation V, we would put a copy of V on each element of $\mathbb{P}^1(\mathbb{Z}_F/\mathfrak{N})$.)

For the sake of reliability, we also verify the dimension of the space using the dimension formula (which involves class numbers and embedding numbers).

Step 4: Decomposition into Newforms With Hecke matrices in hand, we now outline the procedures used in our implementation to decompose a newspace as a direct sum of new simultaneous eigenspaces. Let $S = S_2(\mathfrak{N})$. We use the following properties.

1. The Hecke operators $T_\mathfrak{p}$ for $\mathfrak{p}$ coprime to $\mathfrak{N}$ all commute.
2. The Hecke eigenspaces in S have distinct systems of eigenvalues for the operators $T_\mathfrak{p}$ with $\mathfrak{p}$ coprime to $\mathfrak{N}$ ("multiplicity one").
3. We have $\dim S^{\mathfrak{p}-\mathrm{old}} = 2\dim S_2(\mathfrak{N}/\mathfrak{p}) - \dim S_2(\mathfrak{N}/\mathfrak{p}^2)$, with the last term omitted if $\mathfrak{p}^2$ does not divide $\mathfrak{N}$.

These properties are standard: for (2), see for instance work of Shemanske–Walling [14, Theorem 3.6], and (3) follows from standard facts about degeneracy maps: the proof is a straightforward generalization of the classical proof where $F = \mathbb{Q}$.

Using (1) and standard linear algebra techniques, we decompose the space into Hecke constituents corresponding to $\mathbb{Q}$-irreducible subspaces (a direct sum of eigenspaces). As harmless as it may seem, this step is often the bottleneck in computation.

Finally, we show how to decompose a space into old and new subspaces. For a prime $\mathfrak{p} \nmid \mathfrak{N}$, the $\mathfrak{p}$-old subspace $S^{\mathfrak{p}\text{-old}}$ of $S = S_2(\mathfrak{N})$ is the span of the images of all degeneracy maps $S_2(\mathfrak{N}/\mathfrak{p}) \to S_2(\mathfrak{N})$. The old subspace S^{old} is the sum of the spaces $S^{\mathfrak{p}\text{-old}}$ for primes $\mathfrak{p} \mid \mathfrak{N}$. The newspaces S^{new} and $S^{\mathfrak{p}\text{-new}}$ are the unique Hecke-invariant complements of S^{old} and $S^{\mathfrak{p}\text{-old}}$, respectively.

Now suppose we are given two Hecke-modules S and S_0 (e.g. $S_0 = S_2(\mathfrak{N}/\mathfrak{p})$) and we wish to compute the Hecke-submodule S^{old} of S consisting of all the irreducible Hecke-submodules of S that arise from S_0 via degeneracy maps. In addition, suppose we know a priori the dimension of S^{old}, as in (3). For a prime $\mathfrak{p} \nmid \mathfrak{N}$, the space S^{old} is contained in the kernel of the characteristic polynomial of $T_{\mathfrak{p}} \mid S_0$ evaluated at $T_{\mathfrak{p}} \mid S$. Since we know the dimension, we may compute S^{old} in this manner using finitely many primes $\mathfrak{p}$, stopping when the intersection has the right dimension. One could obtain the complement S^{new} by computing the Hecke-invariant complement, requiring extra module-splitting effort. However, for present purposes, we do not need to know S^{new} as a particular subspace of S; we merely need the Hecke action on S^{new}, which can be obtained using the quotient $S/S^{\text{old}} \cong S^{\text{new}}$.

Alternatively, we can compute the degeneracy operators themselves, and then compute the orthogonal complement.

Now suppose that S is a new subspace of dimension d. By (1), it contains d simultaneous eigenvectors having distinct systems of eigenvalues, which we wish to determine. The Hecke algebra $\mathbb{T} = \mathbb{Z}[T_{\mathfrak{n}}]_{\mathfrak{n}}$ is a commutative $\mathbb{Z}$-algebra which is generated by Hecke operators $T_{\mathfrak{p}}$ with $\mathfrak{p}$ coprime to $\mathfrak{N}$. Therefore $\mathbb{T} \otimes \mathbb{Q}$ is a direct product of number fields, an étale $\mathbb{Q}$-algebra generated by a single primitive element. For practical purposes, one may find a primitive element by trial and error (for instance taking simple combinations of the first few Hecke operators), since the set of non-primitive elements must be sparse. Then the simultaneous eigenspaces of S are simply the eigenvectors associated to the primitive element.

3 Data Computed

We compute for the totally real fields F of degree $n \leq 6$ with discriminant $\leq 2 \cdot 10^n$. Note that if $[F : \mathbb{Q}] > 6$ then $d_F > 2 \cdot 10^n$ by the Odlyzko bounds; indeed, the totally real field of degree 7 of smallest discriminant has discriminant 20134393.

In this range, the discriminant and degree of the field uniquely characterizes the field, with one exceptional pair: in degree 4, there are two fields of discriminant 16,448: one, with minimal polynomial $x^4 - 2x^3 - 6x^2 + 2$ has Galois group S_4 and the other $x^4 - 2x^3 - 7x^2 + 8x + 14$ has Galois group D_8. (The first multiple discriminant for cubic fields is 3969.)

This gives a total of 233 676 Hilbert modular forms in the database. To save space, we did not include data for conjugate levels; for consistency across the database, these were later added to the LMFDB, increasing the overall size of the data.

The computation was performed using the Vermont Advanced Computing cluster (VACC). The odd degree fields, which used the indefinite method, took approximately 4 CPU years to compute. Much of this time is spent computing Atkin-Lehner involutions and coping with higher class number issues.

The data is hosted at the LMFDB [13] and backed up in a local repository. The raw data takes 172 GB, compressed 78 GB.

Remark 3.1 As in Remark 1.2, in many situations we can employ either the definite or indefinite method. For example, if F has odd degree and $\mathfrak{p} \parallel \mathfrak{N}$, then one can apply the definite method with a quaternion algebra B ramified at all real places and the prime $\mathfrak{p}$; the Hecke module computed is then $S_2(\mathfrak{N})^{\mathfrak{p}\text{-new}}$. To compute a single space of forms, the definite method runs more quickly than the indefinite method, but for tabulation one must use the indefinite method if one wants to compute with forms of square level; it is also more efficient to use the indefinite method since for each quaternion algebra used one must repeat the precomputation step.

4 Comments on Data

Base Change Forms We detect base change forms as follows. First, a necessary condition for f to be a base change from a smaller field F_0 is that $\sigma(\mathfrak{N}) = \mathfrak{N}$ and $a_{\sigma(\mathfrak{p})} = a_{\mathfrak{p}}$ for all $\sigma \in \mathrm{Gal}(F/F_0)$. If the form fails this test (for all possible subfields F_0), it is not a base change form; if it passes this test, then is likely to be a base change. (One could detect more generally forms that are twists of base change forms.) To be certain and identify such lifts, we could look in the database for forms over the field F_0 with level supported at primes dividing the relative discriminant $\mathfrak{d}_{F|F_0}$ or $\mathfrak{N}_0 = \mathbb{Z}_{F_0} \cap \mathfrak{N}$.

CM Forms We detect CM forms as follows. Let $\mathfrak{p}$ be a prime which is trivial in the narrow class group, and let $E = \mathbb{Q}(\{a_{\mathfrak{p}}\}_{\mathfrak{p}})$ be the number field generated by the Hecke eigenvalues. Then E is totally real, and the characteristic polynomial of Frobenius $T^2 - a_{\mathfrak{p}}T + N(\mathfrak{p})$ has discriminant $d_{\mathfrak{p}} = a_{\mathfrak{p}}^2 - 4N(\mathfrak{p}) \in E$. If the form is CM and $a_{\mathfrak{p}} \notin \mathfrak{p}$, then the Frobenius belongs to the quadratic field of this discriminant. So if we find two such primes $\mathfrak{p}$ whose CM discriminants belong to different square classes in $E^{\times}/E^{\times 2}$, then the form does not have CM; if these square classes agree, then it gives strong likelihood that the form is CM with the given CM discriminant $K = E(\sqrt{d_{\mathfrak{p}}})$. One can then also verify that $a_{\mathfrak{p}} = 0$ for primes $\mathfrak{p}$ inert in K to provide further evidence. To conclude that the form really is CM, we could use the machinery for Hecke characters.

Computing the L-function Given the Hecke eigenvalues $a_{\mathfrak{p}}(f)$ of a Hilbert modular form f with trivial central character, we define its L-series via the Euler product

$$L(s, f) = \prod_{\mathfrak{p} \mid \mathfrak{N}} (1 - a_{\mathfrak{p}}(\mathrm{Nm}\ \mathfrak{p})^{-s})^{-1} \prod_{\mathfrak{p} \nmid \mathfrak{N}} (1 - a_{\mathfrak{p}}(\mathrm{Nm}\ \mathfrak{p})^{-s} + (\mathrm{Nm}\ \mathfrak{p})^{-2s})^{-1}.$$

The function $L(s, f)$ has analytic continuation to all of $\mathbb{C}$ and a functional equation

$$\Lambda(s, f) = (d_F^2 \,\mathrm{Nm}(\mathfrak{N}))^{s/2} \Gamma_{\mathbb{C}}(s + 1/2)^n L(s, f) = \Lambda(1 - s, f).$$

Factorization of New Subspaces The largest Hecke-irreducible space in our database has dimension 286, and occurs for the field $F = \mathbb{Q}(\sqrt{296})$ and level $\mathfrak{N}$ of norm 29. It would be interesting to consider the distribution of the dimensions of Hecke irreducible subspaces inside the new subspace.

Endomorphism Algebras We can also see some interesting patterns in the distribution of endomorphism algebras of modular abelian surfaces over totally real fields. First, we have the following conjecture of Coleman: see e.g. Bruin–Flynn–González–Rotger [5, Conjecture $C(e, g)$].

Conjecture 4.1 Let F be a totally real field and let $g \in \mathbb{Z}_{\geq 1}$. Then the set

$$\{E = \mathrm{End}(X)_{\mathbb{Q}} : [E : \mathbb{Q}] = g \text{ and } X \text{ an abelian variety over } F \text{ with } \dim(X) = g\}$$

is finite.

In particular, there should be only finitely many real quadratic fields E that occur as the endomorphism algebra of a GL_2-type abelian surface ($g = 2$) over a fixed totally real field F. (Our data becomes a bit too sparse to say anything for $g \geq 3$ except for a few fields.) There are a total of 37 444 Hilbert modular forms whose Hecke eigenvalue field has degree $g = 2$; corresponding to such a form f over F is an isogeny class of abelian surfaces X over F whose common endomorphism algebra $E = \mathrm{End}(X)_{\mathbb{Q}}$ is a quadratic field. We consider a refined version of Conjecture 4.1 where we consider also quadratic *orders*, and tabulate these.

Hilbert modular forms over F whose Hecke field E is an imaginary quadratic field correspond to Hecke Grossencharacters. The largest absolute discriminant of an imaginary quadratic order observed was -288, occurring for a Hilbert modular form over $F = \mathbb{Q}(\sqrt{440})$, a field with class number 2.

More interesting are the cases when the Hecke field E is a real quadratic field. The largest such discriminant of a real quadratic order observed as a Hecke algebra was 1260, occurring over the totally real quartic field F of discriminant 11025. The distribution of the quadratic Hecke fields for a fixed field F shows rapid decay by discriminant; the following histogram of the quadratic Hecke fields for $F = \mathbb{Q}(\sqrt{5})$ is typical.

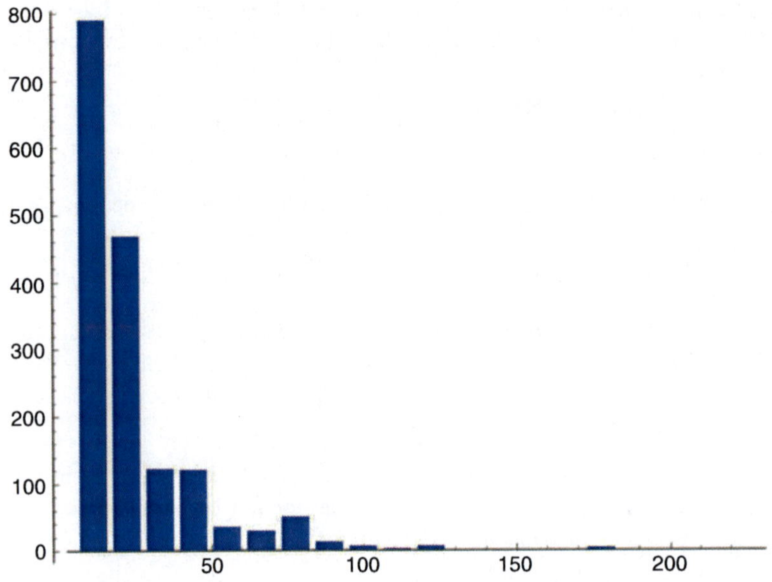

The largest discriminant for $F = \mathbb{Q}(\sqrt{5})$ is $d = 228$, and it occurs just once with level norm 3420 (up to the computed level norm 4999). We tabulate in a similar way the following maximum discriminant quadratic Hecke fields for the first few F:

d_F	$\max(d)$
5	228
8	384
12	192
13	252
17	192
21	588
24	288
28	448

Put together, this data gives some weak evidence for Conjecture 4.1—at the very least, it shows that the large discriminant fields occur with very low density.

Acknowledgments The authors would like to thank Lassina Dembélé and Matthew Greenberg for useful discussions. Particular thanks also go to the organizers (John Cremona, Nicolas Mascot, Aurel Page, and Haluk Şengün) of the LMFDB Workshop at the University of Warwick, June 12–16, 2017, and to John Cremona, Aurel Page, and Dan Yasaki and for their helpful efforts at this workshop and beyond. Voight was supported by an NSF Grant (DMS-0901971) during the time that these computations were first undertaken and by an NSF CAREER Award (DMS-1151047) while work was completed.

References

1. Alex J. Best, Jonathan Bober, Andrew R. Booker, Edgar Costa, John Cremona, Maarten Derickx, David Lowry-Duda, Min Lee, David Roe, Andrew V. Sutherland, and John Voight, *Computing classical modular forms*, Arithmetic Geometry, Number Theory, and Computation, eds. Jennifer S. Balakrishnan, Noam Elkies, Brendan Hassett, Bjorn Poonen, Andrew V. Sutherland, John Voight, Simons Symposia, Springer, Cham (2021). https://doi.org/10.1007/978-3-030-80914-0_4
2. B. J. Birch and W. Kuyk (eds.), *Modular functions of one variable IV*, Lecture Notes in Mathematics, vol. 476, Springer-Verlag, 1975.
3. Jonathan Bober, Alyson Deines, Ariah Klages-Mundt, Benjamin LeVeque, R. Andrew Ohana, Ashwath Rabindranath, Paul Sharaba, and William Stein, *A database of elliptic curves over* $\mathbb{Q}(\sqrt{5})$: *a first report*, ANTS X: Proceedings of the Tenth Algorithmic Number Theory Symposium, eds. Everett W. Howe and Kiran S. Kedlaya, Open Book Series 1, 2013, 145–166.
4. Wieb Bosma, John Cannon, and Catherine Playoust, *The Magma algebra system. I. The user language*, J. Symbolic Comput. **24** (1997), no. 3–4, 235–265.
5. Nils Bruin, E. Victor Flynn, Josep González and Victor Rotger, *On finiteness conjectures for modular quaternion algebras*, Math. Proc. Cam. Phil. Soc. **141** (2006), 383–408.
6. Henri Cohen, *A course in computational number theory*, Springer-Verlag, Berlin, 1993.
7. L. Dembélé, *Quaternionic Manin symbols, Brandt matrices and Hilbert modular forms*, Math. Comp. **76** (2007), no. 258, 1039–1057.
8. L. Dembélé and S. Donnelly, *Computing Hilbert modular forms over fields with nontrivial class group*, Algorithmic number theory (Banff, 2008), eds. Alfred van der Poorten and Andreas Stein, Lecture notes in Comput. Sci., vol. 5011, Springer, Berlin, 2008, 371–386.
9. Lassina Dembélé and John Voight, *Explicit methods for Hilbert modular forms*, Elliptic curves, Hilbert modular forms and Galois deformations, Birkhauser, Basel, 2013, 135–198.
10. E. Freitag, *Hilbert Modular Forms*, Springer-Verlag, Berlin, 1990.
11. Matthew Greenberg and John Voight, *Computing systems of Hecke eigenvalues associated to Hilbert modular forms*, Math. Comp. **80** (2011), 1071–1092.
12. Markus Kirschmer and John Voight, *Algorithmic enumeration of ideal classes for quaternion orders*, SIAM J. Comput. (SICOMP) **39** (2010), no. 5, 1714–1747.
13. The LMFDB Collaboration, *The L-functions and Modular Forms Database*, http://www.lmfdb.org, 2016.
14. T. R. Shemanske and L. Walling, *Twists of Hilbert modular forms*, Trans. Amer. Math. Soc. **338** (1993), no. 1, 375–403.
15. John Voight, *Computing fundamental domains for Fuchsian groups*, J. Théorie Nombres Bordeaux **21** (2009), no. 2, 467–489.
16. John Voight, *Computing automorphic forms on Shimura curves over fields with arbitrary class number*, Algorithmic number theory (ANTS IX, Nancy, France, 2010), eds. Guillaume Hanrot, Francois Morain, and Emmanuel Thomé, Lecture Notes in Comp. Sci., vol. 6197, Springer, Berlin, 2010, 357–371.

Isogeny Classes of Abelian Varieties over Finite Fields in the LMFDB

Taylor Dupuy, Kiran Kedlaya, David Roe, and Christelle Vincent

Abstract This document is intended to summarize the theory and methods behind `fq_isog` collection inside the `ab_var` database in the LMFDB as well as some observations gleaned from these databases. This collection consists of tables of Weil q-polynomials, which by the Honda-Tate theorem are in bijection with isogeny classes of abelian varieties over finite fields.

1 Introduction

The LMFDB (L-Functions and Modular Forms Database) includes a database of isogeny classes of abelian varieties defined over finite fields. This database can be accessed at https://www.lmfdb.org/Variety/Abelian/Fq/. The purposes of this paper are on the one hand to document the theoretical results on which these methods depend (Sect. 2) and some of the algorithms used to compile the data (Sect. 3); and on the other hand to extract some observations from the compiled data, including comparison of some statistical data with relevant heuristics (Sect. 4) and collecting some examples pertaining to theorems and conjectures in the literature (Sect. 5).

For small dimension g and prime power $q = p^r$, the database contains searchable lists of complete sets of isogeny classes of abelian varieties of dimension g over $\mathbf{F}_q$. Table 1 shows the range of g and q covered by the database as well as the total number of isogeny classes of a given dimension.

We note that by "Bound on q" we mean that the set of isogeny classes of abelian varieties over $\mathbf{F}_q$ has been computed for every prime power less than or equal to q. In dimensions $g = 1$ and 2, isogeny classes of abelian varieties defined over $\mathbf{F}_q$ for powers of 2, 3, 5, and 7 up to 1024 have also been included. The isogeny class count is the total count of isogeny classes in the database, including for q a power of 2, 3, 5, and 7 up to 1024 in dimensions 1 and 2.

T. Dupuy · K. Kedlaya · D. Roe (✉) · C. Vincent
Department of Mathematics and Statistics, University of Vermont, Burlington, VT, USA
e-mail: taylor.dupuy@uvm.edu

J. S. Balakrishnan et al. (eds.), *Arithmetic Geometry, Number Theory, and Computation*, Simons Symposia, https://doi.org/10.1007/978-3-030-80914-0_13

Table 1 The number of isogeny classes for each dimension

Dimension	Bound on q	Isogeny class count
1	499	6184
2	211	1,253,897
3	25	1,055,307
4	5	183,607
5	3	281,790
6	2	164,937

For each isogeny class, we report a variety of invariants including the following:

- L-polynomial;
- Newton polygon (and hence p-rank and ordinarity);
- endomorphism algebra;
- Frobenius angles and angle rank;
- whether the isogeny class contains a principally polarized abelian variety or even a Jacobian (when known);
- whether the isogeny class is simple;
- when it is simple, the number field defined by the L-polynomial, and the Galois group of its splitting field;
- when it is not simple, the simple isogeny factors appearing in its decomposition;
- whether the isogeny class is geometrically simple;
- the point counts over small extensions of the base field;
- if the isogeny class contains a Jacobian, the point counts for its curve over small extensions of the base field (if the isogeny class is not known to contain a Jacobian, the same computation is performed and referred to as the point counts of the "virtual curve" associated to this isogeny class);
- whether the isogeny class is a base change of an isogeny class defined over a smaller field (i.e., whether the isogeny class is primitive), and if it is not primitive, the primitive isogeny classes for which it is a base change;
- the twists of the isogeny class: the isogeny classes to which it becomes isogenous after a base change.

Some of the questions we kept in mind in our analysis of the data are:

- How does the number of objects in the database vary with g and q?
- How are these objects distributed if we sort with respect to the Galois group, Newton polygon, or angle rank?
- How is the number of points on the abelian variety distributed? Does this change if we fix other invariants?
- What are the extreme values of the number of points?

There are, of course, many more questions one can ask, but some would require further data compilation; this is especially true for questions regarding the distinction between Jacobians of curves and more general abelian varieties. We discuss possible future directions of inquiry at the very end of the paper (Sect. 6).

We conclude this introduction by directing the reader to some of the highights of the paper.

- A tabulation of Galois groups of Weil polynomials (Sect. 4.2). We attempt to explain the results using Malle's heuristics on the distribution of Galois groups of number fields, but the constants appearing in this method do not fit well with the data.
- Verification of some cases of a conjecture by Pries on the existence of abelian varieties over $\mathbf{F}_p$ with prescribed p-ranks (Problem 4.5).
- A list of possible combinations of p-rank, angle rank and Galois groups for a fixed p-rank (Sect. 4.4). This is closely related to the Tate conjecture for abelian varieties, as in the work of Lenstra and Zarhin [LZ93].
- Some analysis of the endomorphisms algebras for abelian varieties over $\mathbf{F}_q$ for small g and q, in line with a question of Oort (Sect. 4.5).
- An analysis of point counts on abelian varieties on average over isogeny classes, rather than over isomorphism classes (Sect. 4.6). The data suggests a fit not to the Sato-Tate distribution, but an alternate distribution which we have not found in the literature (called herein the isogeny Sato-Tate distribution).
- A characterization of maximal and minimal abelian varieties, including an explanation of how they are unrelated to maximality and minimality of curves, as well as some open questions regarding simple abelian varieties (Sect. 4.7).
- An example of a supersingular curve of genus 5 in characteristic 3, whose existence we were unable to infer from any general constructions (Example 5.11).
- Some counterexamples against a conjecture by Ahmadi-Sparlinski concerning the angle rank of ordinary Jacobians (Example 5.15).

2 Background

In this section, we recall various standard facts about abelian varieties and their zeta functions.

2.1 Weil Numbers, Characteristic Polynomials of Frobenius, and Zeta Functions

Here we follow [FO08, pg. 9] (up to a sign convention). Let $\mathbf{Q}^{\text{alg}}$ be an algebraic closure of $\mathbf{Q}$, $\mathbf{Z}^{\text{alg}}$ be the ring of algebraic integers in $\mathbf{Q}^{\text{alg}}$, and $w \in \mathbf{Z}$. A *Weil q-number of weight w* is an element $\alpha \in \mathbf{Q}^{\text{alg}}$ satisfying:

1. there exists $i \in \mathbf{N}$ such that $q^i\alpha \in \mathbf{Z}^{\text{alg}}$; and
2. for any embedding $\psi : \mathbf{Q}^{\text{alg}} \to \mathbf{C}$, $|\psi(\alpha)| = q^{w/2}$.

Hereafter we only consider Weil q-numbers which are *effective*, meaning that $\alpha \in \mathbf{Z}^{\text{alg}}$ (and hence $w \geq 0$).

Such numbers arise from the zeta functions of varieties over finite fields as follows: Let X be a variety over $\mathbf{F}_q$, a finite field of cardinality q and characteristic p. Then we define

$$Z(X/\mathbf{F}_q, T) := \exp\left(\sum_{n\geq 1} \#X(\mathbf{F}_{q^n})\frac{T^n}{n}\right).$$

For any Weil cohomology theory $X \mapsto H^w(X)$ (e.g., ℓ-adic étale cohomology for $\ell \neq p$, or Berthelot's p-adic rigid cohomology), we have

$$Z(X/\mathbf{F}_q, T) = \prod_{w=0}^{2\dim(X)} L_w(T)^{(-1)^{1+w}}, \qquad L_w(T) := \det(1 - FT|H^w(X)),$$

where $F\colon X \to X$ denotes the action of Frobenius.

If X is smooth and proper of dimension n, then the eigenvalues of F on $H^w(X)$ are Weil q-numbers of weight w; for étale cohomology this is Deligne's theorem [Del74], while for rigid cohomology it follows from Deligne's theorem by a result of Katz–Messing [KM74]. Also, if we define $\zeta(X, s) := Z(X/\mathbf{F}_q, q^{-s})$, then we have a functional equation

$$\zeta(X, n-s) = \pm(q^{n/2}q^{-s})^{\chi_{top}}\zeta(X, s)$$

which follows from Poincaré duality (see [FK88, §II.1] for the étale case and [Ked06a] for the p-adic case). In the displayed formula above, χ_{top} is the topological Euler characteristic; if X is the reduction of a variety over a number field then χ_{top} is the alternating sum of the Betti numbers of the cohomology of the complex manifold obtained by taking the $\mathbf{C}$-points of the variety in characteristic zero.

In the special case where $X = A$ is an abelian variety of dimension g, $\dim H^1(A) = 2g$ and $H^w(A) = \wedge^w H^1(A)$ for $w = 0, \ldots, 2g$. We summarize this second relation by saying that $L_w(T) = \wedge^w L_1(T)$. We refer to $L_1(T)$ as the *L-polynomial* of A and to its reverse $\det(T - F|H^1(X))$ as the *characteristic polynomial* of A, or the *Weil polynomial* of A. We use these two terms interchangeably throughout the text.

2.2 Weil Polynomials

Define a *Weil q-polynomial* (or Weil polynomial when q is understood) to be a monic polynomial over $\mathbf{Z}$ whose roots are all Weil q-numbers of weight 1. (Since we are exclusively interested in abelian varieties, we consider only Weil q-numbers of weight 1 in what follows.) A standard first step in classifying Weil polynomials is the following observation left as an exercise to the reader.

Proposition 2.1 *Let $f \in \mathbf{Q}[T]$ be an irreducible polynomial with all real roots. Let $\beta \in \mathbf{R}$ satisfy $f(\beta) = 0$. Choose $q = p^n$ such that for all $\psi : \mathbf{Q}(\beta) \to \mathbf{R}$ we have $\psi(\beta)^2 - 4q < 0$. If π is a zero of $T^2 - \beta T + q = 0$, then π is a Weil q-number of weight* 1.

As a converse, we have the following characterization of Weil q-numbers.

Lemma 2.2 (Weil q-number Characterization) *Let π be a Weil q-number of weight* 1.

1. *If $\mathbf{Q}(\pi)$ has a real embedding, then $\pi = \pm\sqrt{q}$.*
2. *Suppose that $\psi : \mathbf{Q}(\pi) \to \mathbf{C}$ is a non-real embedding, then*
   (a) *$\beta := \pi + q/\pi$ is totally real;*
   (b) *$\mathbf{Q}(\pi)$ is a CM-field with maximal totally real subfield $\mathbf{Q}(\beta)$, and π has no real embeddings;*
   (c) *π is a solution of $T^2 - \beta T + q$ where $\psi(\beta^2 - 4q) < 0$ for all $\psi : \mathbf{Q}(\pi) \to \mathbf{C}$.*

Proof

1. In this case, $\psi(\pi)^2 = \psi(\pi)\overline{\psi(\pi)} = q$.
2. (a) Since π has absolute value $\sqrt{q}$ in every embedding, q/π is its complex conjugate.
   (b) The fact that $\mathbf{Q}(\pi)$ is CM is [Hon68, Proposition 4], and the second part follows.
   (c) Consider the equation

$$T^2 - \beta T + q = 0.$$

Let $\gamma = \pi - q/\pi$ and note that $\beta^2 - 4q = \gamma^2$. The quadratic formula gives $\frac{1}{2}(\beta \pm \gamma)$ as the solutions, which simplify to π and q/π. The second half of the statement follows from the fact that π has no real embeddings.

□

2.3 Weil Polynomials and Isogeny Classes of Abelian Varieties

For most classes of varieties over a fixed finite field, among the possible zeta functions consistent with the Weil conjectures, it is difficult to predict in advance exactly which ones occur. However, for abelian varieties, this question is completely solved by the Honda-Tate theorem; we follow the treatment of this result given in [WM71].

Let A and B be abelian varieties over $\mathbf{F}_q$ with characteristic polynomials

$$P_A(T) := \det(T - F|H^1(A)), \qquad P_B(T) := \det(T - F|H^1(B));$$

these are Weil q-polynomials. The Honda-Tate theorem makes the following assertions:

- P_A divides P_B if and only if B is isogenous (over $\mathbf{F}_q$) to a product in which A occurs as a factor; in particular, A and B are isogenous if and only if $P_A = P_B$.
- If A is simple, then $P_A = h_A^e$ for some irreducible $h_A(T) \in \mathbf{Z}[T]$ and some positive integer $e = e_A$ which can be read off explicitly from h_A (see below).
- Every irreducible Weil q-polynomial occurs as h_A for some A (which is unique up to isogeny).

We explain the rule for computing e_A for A simple in the context of analyzing the $\mathbf{Q}$-endomorphism algebra $E := \mathrm{End}^0(A/\mathbf{F}_q)$. Let π be the class of T in $\mathbf{Q}(\pi) := \mathbf{Q}[T]/(h_A(T))$, and identify π with the action of Frobenius in E. Then E is a division algebra with center $\mathbf{Q}(\pi)$ and

$$2g = [\mathbf{Q}(\pi) : \mathbf{Q}]e_A, \qquad e_A = \sqrt{[E : \mathbf{Q}(\pi)]}.$$

The Brauer invariant $\mathrm{inv}_v(E)$ of E at a place v of $\mathbf{Q}(\pi)$ is given as follows:

- For v finite and not lying above p, $\mathrm{inv}_v(E) = 0$.
- For v finite and lying above p,

$$\mathrm{inv}_v(E) \equiv -(\log_q |\pi|_v)[K_v : \mathbf{Q}_p] \pmod{\mathbf{Z}}.$$

- For v archimedean, $\mathrm{inv}_v(E) = \frac{1}{2}$ if v is real (which by Lemma 2.2 means $\pi = \pm\sqrt{q}$) and 0 otherwise.

The exponent e_A can now be computed as

$$e_A = \mathrm{lcd}_v\{\mathrm{inv}_v(E)\},$$

where lcd denotes the least common denominator. In particular, $e_A = 1$ whenever A is ordinary or "almost ordinary" (see Sect. 2.4).

2.4 *Newton Polygons, p-rank and Ordinarity*

Let A be an abelian variety over $\mathbf{F}_q$ and define P_A as in Sect. 2.3. The *normalized Newton polygon* of P_A is the lower convex hull of the set

$$\left\{\left(i, \frac{\mathrm{ord}_p(a_i)}{\mathrm{ord}_p(q)}\right) : i = 0, \ldots, 2g\right\}$$

where $P_A(T) = T^{2g} + a_1 T^{2g-1} + \cdots + a_{2g}$. (This is invariant under base extension.) The normalized Newton polygon of P_A is the graph of a piecewise linear function on $[0, 2g]$ with changes of slope only at integer values. In particular, on each of the intervals $[0, 1], \ldots, [2g-1, 2g]$ the function has a unique slope; we call these the *slopes* of $P_A(T)$. For any place v of $K = \mathbf{Q}(\pi)$ above p, the slopes of $P_A(T)$ coincide with the ratios $v(\alpha)/v(q)$ as α varies over the roots of P_A.

The normalized Newton polygon of A satisfies the following properties:

- The left endpoint is $(0, 0)$ and the right endpoint is $(2g, g)$.
- The vertices are all lattice points with nonnegative second coordinate.
- The vertices are symmetric: (i, j) is a vertex if and only if $(2g - i, g - j)$ is a vertex. Equivalently, (i, j) lies above the polygon if and only if $(2g - i, g - j)$ does so.

We say that any convex polygon satisfying these conditions is *eligible in dimension* g. For N, N' two eligible polygons, we write $N \leq N'$ if N' lies on or above N, and $N < N'$ if $N \leq N'$ and $N \neq N'$. The eligible Newton polygons in dimension g form a partially ordered set with unique minimal and maximal elements. The minimal polygon is the one with vertices $(0, 0), (g, 0), (2g, g)$, in which the slopes are $0, 1$ (each with multiplicity g); when this occurs we say that A is *ordinary*. If we exclude the ordinary Newton polygon, then there is again a unique minimal polygon, with vertices $(0, 0), (g - 1, 0), (g + 1, 1), (2g, g)$; when this occurs we say that A is *almost ordinary*, as in [LZ93]. The maximal polygon is the one with vertices $(0, 0), (2g, g)$, in which the slopes are $\frac{1}{2}$ (with multiplicity $2g$); when this occurs we say that A is *supersingular*.

Define the *elevation* of an eligible Newton polygon N in dimension g as the number of lattice points (i, j) with $1 \leq i \leq g, j \geq 0$ which lie strictly below N. (This is not standard terminology.)

Lemma 2.3 *Let N, N' be eligible Newton polygons in dimension g with $N < N'$. Then there exists an eligible Newton polygon N'' in dimension g such that $N < N'' \leq N'$ and the elevation of N'' is one more than that of N.*

Proof Since $N < N'$, there must exist a vertex (i, j) of N lying strictly below N'; by symmetry, $(2g - i, g - j)$ also lies strictly below N'. Let N'' be the lower convex hull of the set of lattice points lying on or above N exclusive of (i, j) and $(2g - i, g - j)$; this choice has the desired form. □

Corollary 2.4 *The poset of eligible Newton polygons in dimension g is* catenary: *any two maximal chains between the same endpoint have the same length (namely the difference in elevation).*

For any positive integer d, by Katz [Kat81] the Newton polygon defines a locally closed stratification on the coarse moduli space $\mathcal{A}_{g,d}$ of g-dimensional abelian varieties equipped with a polarization of degree d^2.

Theorem 2.5 *Let N be an eligible Newton polygon of dimension g.*

1. *There is a (nonempty) stratum of $\mathcal{A}_{g,d}$ with Newton polygon N.*
2. *Each irreducible component of this stratum has codimension equal to the elevation of N.*
3. *If N is not the supersingular Newton polygon, then the corresponding stratum is geometrically irreducible.*

Proof For the supersingular Newton polygon, this is contained in [LO98, Theorem 4.9]. For general N, apply Lemma 2.3 repeatedly to construct a maximal chain $N_0 < \cdots < N_1$ containing N, where N_0 and N_1 are the ordinary and supersingular Newton polygons. By the supersingular case, the length of the chain (which is the elevation of N_1) equals the codimension of each irreducible component of the supersingular stratum.

We now appeal to the de Jong–Oort purity theorem [dJO00, Theorem 4.1], which asserts that the Newton polygon stratification on $\mathcal{A}_{g,d}$ jumps purely in codimension 1. For each Newton polygon N in the chosen chain, the union X_N of all strata corresponding to Newton polygons on or above N is a closed subscheme of $\mathcal{A}_{g,d}$; the codimension of X_N starts at 0 for $N = N_0$, increases by at most 1 at each step, and ends with the elevation of N_1. Consequently, it must increase by exactly 1 at each step; from this, the first two claims follow. The third claim is a theorem of Chai–Oort [CO11, Theorem A]. □

2.5 Galois Groups

Let π be a Weil q-number. By Lemma 2.2, either $\pi = \pm\sqrt{q}$ or $\mathbf{Q}(\pi)$ is CM; assume hereafter that the second case occurs. Write $\mathbf{Q}(\pi)^{\mathrm{gal}}$ for the Galois closure of $\mathbf{Q}(\pi)$ and set $2d = [\mathbf{Q}(\pi) : \mathbf{Q}] = \frac{2g}{e}$. Let $G = \mathrm{Gal}(\mathbf{Q}(\pi)^{\mathrm{gal}}/\mathbf{Q})$, considered as a subgroup of S_{2d} by its action on the conjugates of π.

Lemma 2.6 *Let G be the Galois group of a Weil q-number π.*

1. *G is a subgroup of $W_{2d} := C_2^d \rtimes S_d$ which acts transitively on the 2d roots, where here W_{2d} is the wreath product of C_2 by S_d.*
2. *G contains complex conjugation, which is the unique nontrivial element of the center of W_{2d}.*

Proof The fact that $G \subseteq W_{2d}$ and contains complex conjugation follows from [Dod84, Prop 1.1]. Transitivity follows from the fact that $\mathbf{Q}(\pi)$ is a field, the claim that the center of W_{2d} has order 2 follows from its presentation as a semidirect product, and that complex conjugation is central is [Hon68, Prop 1 (b)]. □

Note that we usually have $d = g$. In this case, we conjecture that Lemma 2.6 gives the only constraints on G:

Conjecture 2.7 Suppose that G is a transitive subgroup of W_{2g} containing complex conjugation. Then there is a Weil q-number π such that $\mathrm{Gal}(\mathbf{Q}(\pi)^{\mathrm{gal}}/\mathbf{Q}) \cong G$.

Note that the condition on containing complex conjugation is necessary: when $g = 4$ there is a transitive subgroup of W_8 (with transitive label `8T14` and abstractly isomorphic to S_4) that does not contain complex conjugation and thus does not arise as the Galois group of a CM-field.

2.6 Frobenius Angle Rank

For A an abelian variety of dimension g with L-polynomial $L(T) = \prod_{i=1}^{2g}(1-\alpha_i T)$, the *angle rank* of A is the quantity

$$\delta(A) = \dim_{\mathbf{Q}}(\mathrm{Span}_{\mathbf{Q}}(\{\arg(\alpha_i) : 1 \leq i \leq 2g\} \cup \{\pi\})) - 1 \in \{0, \ldots, g\}.$$

Equivalently, $\delta(A)$ equals the number of multiplicatively independent elements in the set $\{\alpha_1, \ldots, \alpha_{2g}, q^{1/2}\}$, which equals the rank of the Frobenius torus associated to A [Chi92].

The angle rank detects multiplicative relations among the roots of L; these are closely related to exceptional Hodge classes on powers of A.[1] For example, by a theorem of Zarhin [Zar94, Theorem 3.4.3], $\delta(A) = g$ if and only if there are no exceptional Hodge classes on any power of A. On the other end, $\delta(A) = 0$ if and only if A is supersingular; this follows from the fact, proven here in Example 5.1, that an abelian variety A is supersingular if and only if $\pi_A = \zeta\sqrt{q}$ for ζ a root of unity.

Another useful observation is that if A is simple and $0 < \delta(A) < g$, then the inclusion $G \subseteq W_{2d}$ from Lemma 2.6 must be strict, because the action of G preserves $\mathbf{Q}$-linear relations between π and the $\arg(\alpha_i)$. For a more refined version of this statement, and more about angle rank, see [DKZB21].

2.7 Bounds on Point Counts

Recall that if A is an abelian variety of dimension g over $\mathbf{F}_q$ with characteristic polynomial $(T - \alpha_1)\cdots(T - \alpha_{2g})$, then

$$\#A(\mathbf{F}_q) = (1 - \alpha_1)\cdots(1 - \alpha_{2g}).$$

[1] A *Hodge class* is an étale cohomology class that is invariant under the twisted action of Frobenius. A Hodge class is *exceptional* if it is not in the ring (under cup product) generated by Hodge classes of weight two. The Tate conjecture states that the space of Hodge classes is spanned by the images of algebraic cycles under the cycle class map to l-adic cohomology. This remains an open problem.

The Weil bound implies

$$(\sqrt{q}-1)^2 \leq \#A(\mathbf{F}_q)^{1/g} \leq (\sqrt{q}+1)^2;$$

this can be sharpened a bit to give

$$\lceil(\sqrt{q}-1)^2\rceil \leq \#A(\mathbf{F}_q)^{1/g} \leq \lfloor(\sqrt{q}+1)^2\rfloor$$

(see [AHL12, Théorème 1.1] or [AHL13, Corollary 2.2, Corollary 2.14]). For general A, these bounds are best possible: for $g = 1$, it follows from the Honda-Tate theorem (or an earlier theorem of Deuring) that $\#A(\mathbf{F}_q)$ can take the values $q + 1 \pm \lfloor 2\sqrt{q}\rfloor$, and the same is then true for arbitrary g by taking powers.

In light of the fact (which we just used) that

$$\#(A_1 \times_{\mathbf{F}_q} A_2)(\mathbf{F}_q) = \#A_1(\mathbf{F}_q) \times \#A_2(\mathbf{F}_q),$$

it is natural to separately consider what happens when A is required to be simple. In this case, if we exclude A of low dimension, the Weil bounds can be sharpened further.

Theorem 2.8 ([Kad19, Theorem 2.5]) *If A is simple of dimension $g > 1$, then*

$$\lfloor(\sqrt{q}-1)^2\rfloor + 1 \leq \#A(\mathbf{F}_q)^{1/g} \leq \lceil(\sqrt{q}+1)^2\rceil - 1.$$

Note that this is an improvement on the bound stated earlier when q is a square. For small q, one can make some additional improvements:

Theorem 2.9 ([Kad19, Theorem 3.2]) *For values of q listed below, one has the following lower and upper bounds on $\#A(\mathbf{F}_q)^{1/g}$ for $g \geq 4$. (More precisely, the bounds hold for all g outside of an explicit finite set of exceptional choices of A, each of which has dimension at most 3; see [Kad19, Table 2, Table 3].)*

q	2	3	4	8	9
lower bound	1	1.359	2.2750	4.635	5.47
upper bound	4.0347	5.6333	7.3818	13.05	14.303

Looking at an asymptotic version of the question, one can see that there is not much room left for potential improvement in these bounds. Building on [AHL13], Kadets has shown the following:

Theorem 2.10 ([Kad19, Proposition 1.4, Theorem 1.6]) *Define*

$$a(q) := \liminf_A \#A(\mathbf{F}_q)^{1/g}, \qquad A(q) := \limsup_A \#A(\mathbf{F}_q)^{1/g},$$

where in both cases A varies over simple abelian varieties of dimension g over $\mathbf{F}_q$. *Then*

$$\lfloor(\sqrt{q}-1)^2\rfloor+1 \leq a(q) \leq \lceil(\sqrt{q}-1)^2\rceil+2, \qquad \lfloor(\sqrt{q}+1)^2\rfloor-2-q^{-1} \leq A(q) \leq \lceil(\sqrt{q}+1)^2\rceil-1.$$

The cases where $\#A(\mathbf{F}_q) = 1$ have been classified by Madan–Pal (modulo Remark 2.12).

Theorem 2.11 ([MP77, Theorem 4]) *Suppose that A is simple of dimension g and that* $\#A(\mathbf{F}_q) = 1$.

- *We must have* $q \leq 4$. *(This is easy: if* $q \geq 5$, *then* $\sqrt{q}-1 > 1$ *and the Weil bound implies the claim.)*
- *If* $q = 3$ *or* $q = 4$, *then* $g = 1$. *(For* $q = 4$, *this is the equality case of the Weil bound. For* $q = 3$ *a more careful argument is needed.)*
- *If* $q = 2$, *then the characteristic polynomial of A belongs to an explicit (infinite) list enumerated in* op. cit.

Remark 2.12 In [MP77, Theorem 4], the case $q = 2$ of Theorem 2.11 is asserted modulo a conjecture of Robinson [Rob64]: for n a positive integer and $\zeta_n = e^{2\pi i/n}$, $\zeta_n^2 + 6\zeta_n + 1$ is not a square in $\mathbf{Q}(\zeta_n)$ unless $n = 7$ or $n = 30$. This was confirmed by Robinson [Rob77] using a method of Cassels [Cas69] and Loxton [Lox72].

Remark 2.13 A closely related question to Theorem 2.11 is to classify simple abelian varieties A of genus g over $\mathbf{F}_q$ for which $\#A(\mathbf{F}_q) = \#A(\mathbf{F}_{q^n})$ for some $n > 1$ (meaning that A acquires no new points over $\mathbf{F}_{q^n}$). In [Kad19, Corollary 3.4], it is shown (using Theorems 2.8 and 2.9) that this cannot occur for $n \geq 4$; for $n = 3$ it can only occur for $q = 2, g = 1$; and for $n = 2$ it occurs precisely when A is the quadratic isogeny twist (see Definition 3.3) of one of the cases listed in Theorem 2.11.

This in turn implies a corresponding classification for curves over $\mathbf{F}_q$ which acquire no new points over $\mathbf{F}_{q^n}$, using the classification of function fields of class number one (see Example 5.16).

3 Algorithms

We summarize some of the main algorithms used to compute data about abelian varieties in the LMFDB. The code [DKRV20] used is available at

https://github.com/LMFDB/abvar-fq.

3.1 Enumerating Weil Polynomials

We begin with the algorithm for enumerating Weil q-polynomials of fixed degree for a fixed prime power q. This algorithm originated in [AKR10] and was described in more detail in [Ked08, Section 5] and [KS16, Section 2], to which the reader is referred for more details.

Remark 3.1 The code for enumerating Weil polynomials was incorporated into `Sage-9.1.beta0` [S+20] in January 2020.

As noted earlier, identifying Weil q-polynomials of degree $2g$

$$P(T) = T^{2g} + a_1 T^{2g-1} + \cdots + a_{2g}$$

is equivalent to identifying polynomials of degree g with integer coefficients

$$Q(T) = T^g + b_1 T^{g-1} + \cdots + b_g$$

whose roots are all real and belong to the interval $[-2\sqrt{q}, 2\sqrt{q}]$, via the relation

$$P(T) = T^g Q(T + q/T).$$

In particular, for any integer $i \in \{1, \ldots, g\}$, $a_1, \ldots, a_i$ uniquely determine $b_1, \ldots, b_i$ and vice versa. The strategy is to catalog these polynomials recursively: given a putative choice of $a_1, \ldots, a_{i-1}$ (or equivalently $b_1, \ldots, b_{i-1}$), use various techniques to impose necessary conditions on a_i (or equivalently b_i), then recurse on the remaining options. Some of these necessary conditions are as follows:

- Rolle's theorem: the roots of $Q^{(g-i)}(T)$ must all be real and belong to $[-2\sqrt{q}, 2\sqrt{q}]$. This can be verified by computing a subresultant (Sturm-Habicht) sequence.
- Connectivity: the set of values of b_i satisfying the previous condition is either a closed interval or empty.
- Power sums: the i-th power sum of the roots of $P(T)$ must have absolute value at most $q^{i/2}$.
- Descartes's rule of signs: the polynomials $Q(T+2\sqrt{q})$ and $(-1)^g Q(-2\sqrt{q}-T)$ have all roots real and nonpositive, so their coefficients must be nonnegative.
- Hamburger criterion: for $s_0, s_1, \ldots$ the sequence of power sums of Q, the Hankel matrix

$$\begin{pmatrix} s_0 & s_1 & s_2 & \cdots \\ s_1 & s_2 & s_3 & \cdots \\ s_2 & s_3 & s_4 & \cdots \\ \vdots & \vdots & \vdots & \ddots \end{pmatrix}$$

is nonnegative definite (equivalently, its principal minors are nonnegative).

- Hausdorff criterion: for all i, j, the sum of $(2\sqrt{q}-x)^i(2\sqrt{q}+x)^j$ as x varies over the roots of Q is nonnegative.

A crucial but counterintuitive point is that while some subsets of these conditions together form sufficient constraints to identify Weil q-polynomials, no subset gives sufficient "on-line" criteria. That is, they cannot be applied directly to detect whether a given sequence $a_1, \ldots, a_k$ can be extended to give the coefficients of some Weil polynomial; the conditions can only be checked once a full sequence has been generated. Therefore we impose "more" conditions than are sufficient to cut down the search space as much as possible as early as possible in the computation.

One serious issue with this calculation is that, while it is trivial to check the output for false positives (e.g., Sage has a built-in function `is_weil_polynomial` to test whether a given integer polynomial is a Weil polynomial), it is rather difficult to check for false negatives other than by generating Weil polynomials using another method (e.g., by computing zeta functions of abelian varieties) and checking for their presence. One useful consistency check is to run the computation for $q = 1$, where the answer is known: by Kronecker's theorem, the irreducible Weil q-polynomials for $q = 1$ are precisely the cyclotomic polynomials. One can also attempt to use the explicit descriptions of the spaces of Weil polynomials given for $g = 3$ in [Hal10], for $g = 4$ in [HS12], and for $g = 5$ in [Soh13]. Our results mostly matched in the $g = 3$ case (we found a few extra non-simple examples), while the conditions in [HS12, Thm. 1.1] and [Soh13, Thm. 2.1] were not actually satisfied by all of the Weil polynomials that we found.

3.2 Point Counts

As noted earlier, for A an abelian variety of dimension g, we have $H^i(A) = \wedge^i H^1(A)$ for $i = 1, \ldots, 2g$, so the zeta function of A is determined completely by the L-polynomial $L(T)$. In particular, if we write $L(T) = \prod_{i=1}^{2g}(1-\alpha_i T)$, then

$$\#A(\mathbf{F}_q) = \prod_{i=1}^{2g}(1-\alpha_i) = L(1).$$

Similarly, for any positive integer r,

$$\#A(\mathbf{F}_{q^r}) = \prod_{i=1}^{2g}(1-\alpha_i^r) = \mathrm{Res}(L(T), T^r - 1).$$

In particular,

$$\#A(\mathbf{F}_{q^2}) = L(1)L(-1).$$

3.3 Curve Point Counts

If A is isogenous to the Jacobian of a curve C, then the sequence $c_n := \#C(\mathbf{F}_{q^n})$ satisfies

$$\frac{L(T)}{(1-T)(1-qT)} = \exp\left(\sum_{n=1}^{\infty} \frac{c_n}{n} T^n\right).$$

In fact, whether or not A is isogenous to a Jacobian, the sequence c_n consists of integers, which we report as the point counts of a "virtual curve" with Jacobian A. If the c_n violate any known constraints on the point counts of a curve of genus g, then A cannot be isogenous to a Jacobian; these include trivial constraints (such as $c_1 \geq 0$ and $c_{mn} \geq c_n$) and some less trivial ones (e.g., the Ihara bound).

3.4 Base Change, Primitivity and Isogeny Twists

If A is an abelian variety over $\mathbf{F}_q$, then the Weil polynomials for A and for the base change of A to $\mathbf{F}_{q^r}$ are related as follows.

Proposition 3.2 *Suppose $P(T) = \prod_{i=1}^{2g}(T - \alpha_i)$ is the Weil polynomial associated to $A/\mathbf{F}_q$ by the Honda-Tate theorem. Then the Weil polynomial associated to the base change of A to $\mathbf{F}_{q^r}$ is*

$$P_r(T) := \prod_{i=1}^{2g}(T - \alpha_i^r).$$

Proof The roots of $P(T)$ are the eigenvalues of the action of the q-Frobenius on $H^1(A)$, and the q^r-Frobenius is just the rth power of the q-Frobenius. □

Definition 3.3 We say that A is *primitive* if A is not isogenous to the base change of any abelian variety defined over a subfield of $\mathbf{F}_q$. We say that two abelian varieties A and B are *isogeny twists* if they become isogenous after some finite extension of $\mathbf{F}_q$. The *isogeny twist class* of A is the set of isogeny twists of A.

Corollary 3.4 *If A and B are simple abelian varieties of dimension g over $\mathbf{F}_q$ with associated Weil numbers α and β, then A and B are isogeny twists if and only if $\alpha = \zeta\beta$ for some root of unity ζ.*

We use various methods for computing $P_r(T)$. The simplest is to just factor $P(T)$ approximately over $\mathbf{C}$, raise each root to the rth power and then recognize the coefficients of the product as integers. In order to make this approach rigorous, one needs to use ball arithmetic in $\mathbf{C}$ and increase the precision if there are multiple integers within the error bounds for any coefficient.

The second approach is to symbolically express the coefficients of $P_r(T)$ as polynomials in the coefficients of $P(T)$ using Newton's identities to change basis between elementary symmetric polynomials and power sums. For small values of r the resulting transformation is not difficult to compute but for values of r larger than about 10 the memory footprint of the algorithm grows rapidly. We therefore only use this approach for smooth values of r, where cached transformations can be repeatedly applied to compute the overall base change.

Finally, the base change can be computed using polynomials over the cyclotomic field $\mathbf{Q}(\zeta_r)$. In particular, we can use the identity

$$P_r(T^r) = \prod_{i=0}^{r-1} P(\zeta_r^i T)$$

to determine $P_r(T)$. An equivalent approach is to compute $P_r(T)$ as the resultant of $P(U)$ and $U^r - T$.

Since we are enumerating Weil polynomials for many q, we can use these methods for computing base changes to easily determine which isogeny classes are primitive, and to find primitive models for those which are not. We simply compute all base changes from $\mathbf{F}_q$ to $\mathbf{F}_{q^r}$ when both prime powers are contained in the database.

Determining which abelian varieties are isogeny twists of each other is a little more difficult. The condition in Corollary 3.4 is difficult to determine directly from Weil polynomials. We can improve it slightly as follows.

Proposition 3.5 *Two simple abelian varieties A and B are isogenous over $\overline{\mathbf{F}}_q$ if and only if there exists a number field K containing Galois conjugates π'_A and π'_B of π_A and π_B such that $\pi'_A\mathcal{O}_K = \pi'_B\mathcal{O}_K$.*

Proof The requirement that π'_A and π'_B generate the same ideal is equivalent to $\pi'_A = u\pi'_B$ for some unit $u \in \mathcal{O}_K$. Then u has absolute value 1 at every finite place, and because $|\pi'_A|_v = |\pi'_B|_v = \sqrt{q}$ for every infinite place v, u has absolute value 1 at every infinite place as well. Therefore u is a root of unity by Kronecker and we may invoke Corollary 3.4. □

This result is not sufficient for our purposes for two reasons. First, it seems to require computations in Galois closures since we need to work with arbitrary conjugates of π'_A and π'_B. Second, as with Corollary 3.4, it only applies when both

A and B are simple, yet it is possible for a simple abelian variety to be a isogeny twist of a non-simple one.

Instead, we break up the set of isogeny classes for each g and q into clusters based on invariants, then use a pairwise test to further refine these clusters into isogeny twist classes. Isogeny twists will have the same slopes and the same geometric endomorphism algebra, whose computation we describe in the next section. These two invariants are sufficient to divide up isogeny classes into clusters whose size is already usually in the single digits, and is at most several hundred for the values of g and q that we consider. We then use the following result for each pair of isogeny classes in the cluster.

Proposition 3.6 ([CMSV19, Sec. 7.2]) *Suppose A and B are abelian varieties over $\mathbf{F}_q$ with Weil polynomials $P(T)$ and $Q(T)$. If there is an isogeny from A to B defined over $\mathbf{F}_{q^r}$ then the cyclotomic polynomial $\Phi_r(T)$ divides the resultant* $\mathrm{Res}_z(P(z), z^{2g} Q(T/z))$.

If we set m to be the least common multiple of the orders of all cyclotomic polynomials dividing this resultant, then we can determine whether A and B are isogeny twists by computing the base changes $P_m(T)$ and $Q_m(T)$ and checking for equality. Note that to find the product of all cyclotomic polynomials dividing a given polynomial, it is not necessary to factor it fully; there is a more efficient algorithm of Beukers–Smyth [BS02, §2] that finds cyclotomic factors using the Euclidean algorithm. (In Sage, a polynomial over $\mathbf{Q}$ has a method `cyclotomic_part` implementing this algorithm.)

3.5 Endomorphism Algebras

The method for determining the Brauer invariants of the endomorphism algebra $\mathrm{End}^0(A/\mathbf{F}_q)$ is described in Sect. 2.3; given an irreducible Weil polynomial h_A these invariants give the power e_A such that $h_A^{e_A}$ is the characteristic polynomial of an isogeny class of abelian varieties over a finite field. We can also use Proposition 3.6 with $B = A$ to find all possible extension degrees where the endomorphism algebra might change, and then compute the endomorphism algebra anew for each such base change. These two results allow us to, for an isogeny class defined over a finite field $\mathbf{F}_q$, give the endomorphism algebra of the base change of this isogeny class over any finite extension of $\mathbf{F}_q$.

We refer to the minimal extension over which all endomorphisms are defined as the *endomorphism field*; one can search the database by degree of the endomorphism field, which we call the *endomorphism degree*.

For the convenience of the reader, we now explain how to describe the endomorphism algebra of a simple isogeny class A over a finite field $\mathbf{F}_q$ given its Brauer invariants and its characteristic polynomial. This is enough to describe the endomorphism algebra of any isogeny class, as, if A is isogenous to an abelian variety

$$A_1^{n_1} \times A_2^{n_2} \times \cdots \times A_r^{n_r}$$

where each A_i is simple and A_i is not isogenous to A_j if $i \neq j$, then

$$\mathrm{End}^0(A/\mathbf{F}_q) \cong M_{n_1}(\mathrm{End}(A_1/\mathbf{F}_q)) \times \cdots \times M_{n_r}(\mathrm{End}(A_r/\mathbf{F}_q)).$$

Assume now that A is simple. We begin by determining the center of the endomorphism algebra of A: If the characteristic polynomial is $h_A^{e_A}$ for h_A irreducible, then the center of $\mathrm{End}^0(A/\mathbf{F}_q)$ is the field $\mathbf{Q}[x]/h_A$ generated by π_A. We then compute the degree of $\mathrm{End}^0(A/\mathbf{F}_q)$ over its center. Its square root is given by the order of the class of $\mathrm{End}^0(A/\mathbf{F}_q)$ in the Brauer group of its center, by Tate [Tat66, p. 142]. By the exact sequence

$$0 \to \mathrm{Br}(F_A) \to \bigoplus_v \mathrm{Br}(F_{A,v}) \to \mathbb{Q}/\mathbb{Z} \to 0,$$

where $F_A = \mathbf{Q}[x]/h_A$, this is the least common denominator of the Brauer invariants of $\mathrm{End}^0(A/\mathbf{F}_q)$, or more simply put, e_A.

To complete our description of $\mathrm{End}^0(A/\mathbf{F}_q)$, we note that by Tate [Tat66, Theorem 2 and its proof], if $\mathbf{Q}[x]/h_A$ has a real place, then either $\mathbf{Q}[x]/h_A = \mathbf{Q}$ and $\mathrm{End}^0(A/\mathbf{F}_q)$ is the quaternion algebra over $\mathbf{Q}$ that is ramified at p and ∞, or $\mathbf{Q}[x]/h_A = \mathbf{Q}(\sqrt{p})$ and $\mathrm{End}^0(A/\mathbf{F}_q)$ is the quaternion algebra over $\mathbf{Q}(\sqrt{p})$ ramified at both real places and nowhere else.

Otherwise, $\mathbf{Q}[x]/h_A$ is totally complex. If $e_A = 1$, then $\mathrm{End}^0(A/\mathbf{F}_q) = \mathbf{Q}[x]/h_A$. Otherwise, $\mathrm{End}^0(A/\mathbf{F}_q)$ is ramified only at places dividing p, and we identify the isomorphism class of $\mathrm{End}^0(A/\mathbf{F}_q)$ by its degree over $\mathbf{Q}[x]/h_A$ and its Brauer invariants at the places above p in $\mathbf{Q}[x]/h_A$.

3.6 Principal Polarizations

A priori, the isogeny class associated to a Weil polynomial consists of unpolarized abelian varieties. In particular, to show how polarizations may vary within an isogeny class, we remind the reader of the following standard facts:

1. For $g \geq 2$ and over an algebraically closed field, every abelian variety is isogenous to one that is principally polarizable.
2. For every field, there exist abelian varieties defined over that field which are not principally polarizable. For example, over an algebraically closed field of characteristic zero, there is a simple proof that every principally polarizable abelian variety is isogenous to an abelian variety without a principal polarization.[2] However, this last fact is not true in general, see Example 5.10.

[2]See https://mathoverflow.net/questions/16992/non-principally-polarized-complex-abelian-varieties.

3. Polarization also behaves poorly under decomposition into isogeny factors: An abelian variety can be principally polarizable without all of its isogeny factors being principally polarizable. See [How95, Example 13.8] or 3.2.ac_c_ad, which contains a Jacobian whose two-dimensional factor is not principally polarizable.

The takeaway from these facts is that the quality of admitting a principal polarization is not an isogeny invariant. Therefore in the database, when we say that an isogeny class is principally polarizable, we mean that there exists some abelian variety in the class which is principally polarizable.

Of the 2,945,722 isogeny classes in the database, there are only 3037 such that we cannot determine if the class has a principal polarization. These isogeny classes all occur in dimension greater than 3: 358 are in dimension 4, 515 are in dimension 5, and 2164 are in dimension 6.

We now discuss methods for determining when an isogeny class is principally polarizable. This code was provided by Howe; the version we use is implemented in the `has_principally_polarizable` function [DKRV20].

For simplicity, the algorithm considers only the case of simple abelian varieties. We note that since an isogeny class is principally polarizable when all of its isogeny factors are, the database does indicate that a nonsimple isogeny class is principally polarizable when all of its factors are. We remind the reader however that, as remarked above, the converse is not true, which does mean that certain non-simple cases are currently completely out of the reach of our implemented tests, and might be principally polarizable even though not all of their factors are.

Our algorithm depends on the dimension of the isogeny class: In the case $g = 1$, every isogeny class is principally polarizable. In the case $g = 2$, we apply a result of [HMNR08], which tests for principal polarizations based on a condition on the coefficients of the Weil polynomial. These conditions are that an isogeny class is *not* principally polarizable if and only if $a_1^2 - a_2 = q$, $a_2 < 0$, and every prime divisor of a_2 is 1 mod 3.

We now list the collection of tests that are implemented in the database in higher dimension. First, if g is odd and A is simple then its isogeny class is principally polarizable, which takes care of $g = 3$ and $g = 5$.

In addition, in the ordinary case, we can completely determine whether an isogeny class is principally polarizable, using Corollary 11.4 and Proposition 11.5 of [How95]. We note that the details of the proofs are given in §14 of [How93] for readers who would like to see them. The test relies on the fact that in the ordinary case, $N_{K/\mathbf{Q}}(\pi - q/\pi)$ is always a square and we can consider its positive square root N. If $q > 2$, there is a principally polarized variety in the isogeny class if and only if $N \equiv (\text{coeff of } T^g) \mod q$. If $q = 2$, we have a similar congruence condition but for a power of 2: there is a principally polarized abelian variety in the isogeny class if and only if $N \equiv (\text{coeff of } T^g) \mod 2^2$.

Remark 3.7 In [How95], Howe gives an algorithm for determining when an ordinary isogeny class (simple or not) contains a principally polarizable variety, but the non-simple case has not been implemented yet and we did not implement it for the database.

Remark 3.8 Howe has explained to us that the congruence conditions come from wanting to determine if N, the positive square root of $N_{K/\mathbf{Q}}(\pi + q/\pi)$, is the same as the square root specified by certain p-adic conditions. The result is then explained by the fact that the middle coefficient of the Weil polynomial is congruent modulo q to the square root specified by the p-adic conditions. When $q > 2$, the congruence is enough to compare the signs of the two square roots. However, when $q = 2$ we have $1 \equiv -1 \mod 2$, which explains the need to work modulo 4.

Finally, in even dimension, we verify some conditions on the field K generated by the Weil q-number π to allow us to determine if certain isogeny classes are principally polarizable (this is [How96, Theorem 1.1]). First, if K is totally real then the class is principally polarizable. (Although if $g \geq 4$ and A is simple, K is always a CM field.) Otherwise, K is a CM field. To set up the notation we will need, let $P(T)$ be the characteristic polynomial of an isogeny class of abelian varieties. Write $P(T) = h(T)^e$ for h irreducible; then h is the minimal polynomial of a Weil q-number π. Let $K = \mathbf{Q}(\pi)$ and $K_+ = \mathbf{Q}(\pi + q/\pi)$ be the CM field and its totally real subfield, respectively. Then we know that there is a principally polarizable abelian variety in the isogeny class associated to $P(T)$ if either K/K_+ is ramified at a finite prime, or there is a prime of K_+ that divides $\pi - q/\pi$ and is inert in K/K_+. We note that this second condition requires some work to test for, and we use the fact that a prime is inert in K/K_+ if and only if the prime ideal is equal to its complex conjugate to do so.

3.7 *Jacobian Testing*

Given a characteristic polynomial, to test if there exists a Jacobian with this characteristic polynomial, we apply six results from Howe and Lauter's Magma package http://ewhowe.com/Magma/IsogenyClasses.magma. (They were re-implemented for the LMFDB by Howe.) This code accompanies the paper [HL12]; see especially Sect. 6 of the article for a high-level overview of the software. In addition, the comments in the code are excellent, so rather than repeat an explanation of the tests, we refer the reader to these two excellent references. We note that currently the LMFDB does not implement positive Jacobian testing in genus 4 and higher.

3.8 *Angle Rank*

We have two algorithms to compute the angle rank of an isogeny class of abelian varieties: one that is numerical, and one that is algebraic and therefore yields a provably true answer. In the current version of the LMFDB, the angle rank $\delta(A)$ is computed numerically using lattice basis reduction.

This is done in the following way: We first approximate the roots α_i of the characteristic polynomial $P(T)$ numerically as pairs of floats (the real and imaginary parts of the number). We then pair each root α_i to its complex conjugate (which is also a root of $P(T)$) and retain only one number from each pair of complex conjugates, as we know that $\arg(\alpha_i)/2\pi$ and $\arg(\overline{\alpha}_i)/2\pi$ have a linear relation over $\mathbf{Q}$. Following this, we compute $\arg(\alpha_i)/2\pi$ numerically (using the principal branch of the logarithm) for the g remaining α_i to get values $t_1, \ldots, t_g$. Finally, to determine the dimension of the $\mathbf{Q}$-span of the appropriate set (the reader can go to Sect. 2.6 for a reminder of the definition of angle rank), we apply an LLL algorithm with a certain precision to the tuple $[t_1, \ldots, t_g, 1]$ (`PARI`'s `lindep`). Note that in these computations the branch of the logarithm chosen for the computations does not matter because we eventually take a $\mathbf{Q}$-vector space span with 1.

For the sake of completeness, we present below the algebraic algorithm yielding a provably true answer as well. Roughly speaking, the algorithm relies on expressing the roots of $P(T)$ in a common generating set for some S-units of the splitting field of $P(T)$. It would of course have been preferable to use this algorithm in the database instead, but at the time this computation was performed for the database (2015), the implementation of the software computing S-units in `Sage` was not fast enough to deploy on the full database. See Sect. 6.1 for a possible workaround.

We now present the algebraic algorithm: Let A be an abelian variety over $\mathbf{F}_q$ where $q = p^a$. Let K be the splitting field of $P(T) = P_A(T) = \prod_{i=1}^{2g}(T - \alpha_i)$. For the computation of the angle rank we consider

$$\Gamma = \langle \alpha_1, \alpha_2, \ldots, \alpha_{2g} \rangle,$$

the multiplicative subgroup of $K^\times$ generated by the roots of $P_A(T)$. In order to compute $\delta(A)$ we use the fact that

$$\mathrm{rk}(\Gamma) = \delta(A).$$

The first two authors discuss the group Γ in the sister paper [DKZB21], where some explicit relations are computed. Let $S = \{P \in \mathrm{Spec}(\mathcal{O}_K) : P|p\}$ be the collection of primes of K above p. The key observation is that Γ is a subgroup of the group of S-units of K:

$$\Gamma \leq \mathcal{O}_{K,S}^\times.$$

Using `Sage` we can then compute generators[3] for $\mathcal{O}_{K,S}^\times$:

$$\mathcal{O}_{K,S}^\times = \langle \zeta, u_1, \ldots, u_r \rangle$$

[3]The particular function we use was written by Cremona: http://doc.sagemath.org/html/en/reference/number_fields/sage/rings/number_field/unit_group.html. We also remark that there is an additional quite large bottleneck in this algorithm due to the need to compute the splitting field of the characteristic polynomial.

where ζ is a root of unity generating the torsion part of $\mathcal{O}_{K,S}^{\times}$ and $u_1, \ldots, u_r$ are of infinite order, and attempt to compute the rank of Γ in this basis, before proceeding, however, we eliminate the torsion part of Γ, if any. Let m be the cardinality of the group of roots of unity in $K^{\times}$. Then to compute the rank of Γ it suffices in fact to compute the rank of $\Gamma^m = \langle \alpha_1^m, \ldots, \alpha_{2g}^m \rangle$, since

$$\mathrm{rk}(\Gamma) = \delta(A_{\mathbf{F}_q}) = \delta(A_{\mathbf{F}_{q^m}}) = \mathrm{rk}(\Gamma^m).$$

Here, the equality of angle ranks follows from the fact that δ computes $\mathbf{Q}$-linear relations, and $\arg(\alpha)/2\pi$ and $\arg(\beta)/2\pi$ have a $\mathbf{Q}$-linear relation if and only if $\arg(\alpha^m)/2\pi$ and $\arg(\beta^m)/2\pi$ do.

We then write each α_i^m in terms of the generating elements u_j, and obtain a vector of exponents with integer coefficients. We then form a $r \times 2g$ matrix Y whose columns are these vectors, and the rank of this matrix gives us our answer:

$$\mathrm{rk}(Y) = \delta(A_{\mathbf{F}_q}) + 1.$$

We note that the explicit relations in Γ mentioned above are derived from the matrix Y; several examples of these computations can be found in [DKZB21].

4 Statistics vs. Heuristics

This database naturally invites investigation of the following motivating questions:

- How does the number of objects in the database vary with q and g?
- How about if we sort by the Galois group of $\mathbf{Q}(\pi)$, the Newton polygon, or the angle rank?
- How is the number of points on the abelian variety distributed?
- What are the extreme values of the number of points?

In this section, we gather the data available in the database and informing these questions, and compare it to the predictions given by heuristics.

4.1 The Number of Isogeny Classes

In order to set bounding boxes for our data collection, we needed to estimate

$$N(g, q) = \text{number of isogeny classes of } g\text{-dimensional abelian varieties over } \mathbf{F}_q.$$

We initially chose the limits described in Table 1 using an incorrect estimate on the growth of $N(g,q)$: Believing that $N(g,q)$ grows like $q^{g(g+1)/2}$ (rather than the correct asymptotic $q^{g(g+1)/4}$; see below), for each g we included data for q with $q^{g(g+1)/2} \leq 10^7$ to bound the number of isogeny classes per pair (g,q).[4] We later extended the bounds in order to include more fields of small characteristic in dimension up to 3, and to make the range of included q contiguous in dimension 3.[5]

We now give a more careful analysis that better models the numerical data that we have observed for $N(g,q)$. Since by Honda-Tate these isogeny classes are in bijection with characteristic polynomials, one reasonable heuristic is to count Weil q-polynomials of degree $2g$. This amounts to counting lattice points in the set of $(a_1,\ldots,a_g) \in \mathbf{R}^g$ for which the polynomial

$$T^g + a_1T^{g-1} + \cdots + a_g$$

has all roots in the interval $[-2\sqrt{q}, 2\sqrt{q}]$; it is reasonable to approximate this count by the volume of the region. This volume has been computed by DiPippo and Howe [DH98, Proposition 2.2.1]:[6] it equals

$$\left(\frac{2^g}{g!}\prod_{i=1}^{g}\left(\frac{2i}{2i-1}\right)^{g+1-i}\right) q^{g(g+1)/4}. \tag{4.1}$$

This turns out to be a good prediction in practice; see Table 2 for some examples.

For the sake of completeness, we note two facts: First, according to [DH98], for q large compared to g, the dominant contribution to the count is from ordinary abelian varieties. Secondly, to obtain the number of ordinary isogeny classes from the formula for the total number of isogeny classes, we simply multiply by a factor of $\varphi(q)/q$.

Table 2 Predicted versus actual values for the number of isogeny classes of ordinary/arbitrary abelian varieties of dimension g over $\mathbf{F}_q$

		Ordinary		Arbitrary	
g	q	Predicted	Actual	Predicted	Actual
3	25	284,444	284,740	355,556	332,166
4	5	104,025	105,600	130,032	132,839
5	3	170,796	171,180	256,194	267,465
6	2	72,362	74,122	144,724	164,937

[4]We also included (5, 3), where $q^{g(g+1)/2} \approx 1.4 \cdot 10^7$, and we only included q up to 500 for $g = 1$, rather than 10^7.

[5]Namely, we added powers of 2, 3, 5, 7 up to 1024 for $g \in \{1, 2\}$, and raised the bound for $g = 3$ from 13 to 25.

[6]We remark that the bounds used in *op. cit.* are also related to the original iterator described in [Ked08] which some reader might find enlightening.

Table 3 Predicted versus actual (least squares) values for the equation $\log N(g, q) = a\log(q) + b$

	a		b	
g	Predicted	Actual	Predicted	Actual
1	0.5	0.4971	1.3863	1.3717
2	1.5	1.4178	2.3671	2.5302
3	3	2.9135	3.1248	3.2598
4	5	4.5452	3.7283	4.2707
5	7.5	7.2188	4.2141	4.5660

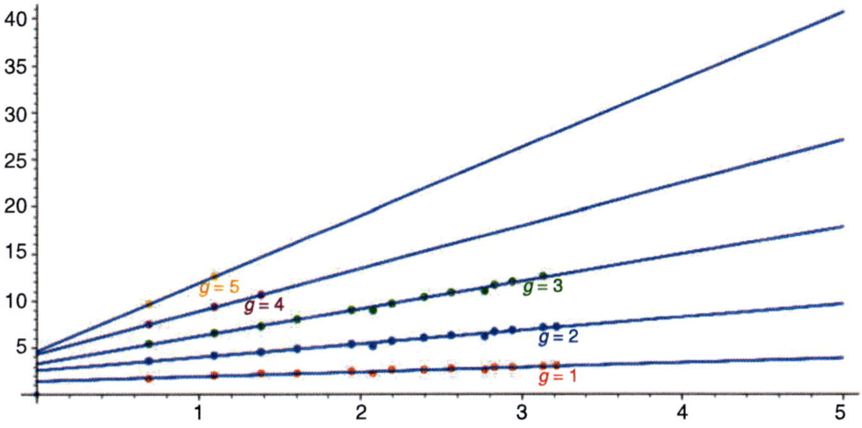

Fig. 1 Plots of $(\log q, \log N(g, q))$ for each g and $q \leq 27$ in the database. The least square values of $\log N(g, q) = a\log q + b$ are given in Table 3

Another way to evaluate our prediction for $N(g, q)$ is to make a least-squares fit for values a, b such that $\log N(g, q) \approx a\log q + b$. The results are given in Table 3. The log-log plot of q vs $N(g, q)$ is given in Fig. 1.

While we do not yet have data for isomorphism classes (see Sect. 6.3), it is worth noting that one can derive heuristics for counts of isomorphism classes within an isogeny class using Oort's theory of *isogeny leaves* in moduli spaces of abelian varieties [Oor09].

4.2 Galois Groups

Let $P(T)$ be the characteristic polynomial of an isogeny class of abelian varieties, and $\mathbf{Q}(\pi)^{\text{gal}}$ be the splitting field of P over $\mathbf{Q}$. How can one expect $\text{Gal}(\mathbf{Q}(\pi)^{\text{gal}}/\mathbf{Q})$ to be distributed as the isogeny class varies? To answer this question, we need to explain Malle's conjecture [Mal02, Mal04] and the invariant $a(G)$ for a finite group $G \subseteq S_n$. The exposition given here follows the treatment in [Mal02] (see the introduction for example). Given $g \in S_n$ we define

$$\text{index}(g) = n - \#\text{number of cycles in the cycle decomposition of } g.$$

We then define

$$a(G) = \frac{1}{\min\{\text{index}(g) : g \in G \setminus 1\}}.$$

Example 4.1

1. $\text{index}((12)(34)) = 4 - 2 = 2$
2. $a(S_4) = 1$
3. $a(D_4) = 1$
4. $a(A_4) = 1/2$
5. $a(C_4) = 1/2$

Let K be a number field, and n be an integer. We define the following counting functions for number fields L of degree n over K:

$$N_{K,n}(X) = \#\{L : [L : K] = n, |\operatorname{Disc}(L/K)| \leq X\},$$

$$N_{K,n,G}(X) = \#\{L : [L : K] = n, |\operatorname{Disc}(L/K)| \leq X, \operatorname{Gal}(L^{\text{gal}}/K) = G\}.$$

First, we have Linnik's conjecture (c.f. *op. cit.*) that there exists a constant $c_0 = c_0(K, n)$ such that

$$N_{K,n}(X) \sim c_0 X \text{ as } X \to \infty. \tag{4.2}$$

Next, the (weak) Malle conjecture states that for all $\varepsilon > 0$ there exist constants $c_1, c_{2,\varepsilon}$ such that

$$c_1 X^{a(G)} < N_{K,n,G}(X) < c_{2,\varepsilon} X^{a(G)+\varepsilon}. \tag{4.3}$$

Assuming these conjectures, given a base field K, we can describe how the proportion of extensions with Galois group G should behave on a log-log scale:

Lemma 4.2 *Assuming Eqs.* (4.2) *and* (4.3) *we have*

$$\lim_{X \to \infty} \frac{\log N_{K,G,n}(X)}{\log N_{K,n}(X)} = a(G).$$

Proof Assume the Malle and Linnik conjectures. We will prove the upper bound and omit the proof of the lower bound as it is similar and easier. Since $N_{K,n}(X) \sim c_0 X$, there exists some $\varepsilon_1(X)$, approaching zero as $X \to \infty$, such that $N_{K,n}(X) = c_0 X(1 + \varepsilon_1(X))$. Applying Malle's conjecture, we then have, for all $\varepsilon > 0$,

$$\frac{\log N_{K,n,G}(X)}{\log N_{K,n}(X)} < \frac{a(G) + \varepsilon + \frac{\log c_{2,\varepsilon}}{\log(X)}}{1 + \frac{\log c_0}{\log(X)} + \frac{\log(1+\varepsilon_1(X))}{\log(X)}} \to a(G) + \varepsilon \quad \text{as } X \to \infty.$$

This proves that for all $\varepsilon > 0$

$$\lim_{X\to\infty} \frac{\log N_{K,n,G}(X)}{\log N_{K,n}(X)} < a(G) + \varepsilon,$$

and hence

$$\lim_{X\to\infty} \frac{\log N_{K,n,G}(X)}{\log N_{K,n}(X)} \leq a(G).$$

A simpler bound using the lower end of Malle's conjecture proves

$$a(G) \leq \lim_{X\to\infty} \frac{\log N_{K,n,G}(X)}{\log N_{K,n}(X)},$$

which gives the result. □

Because of this result, we might expect that a group G should appear as $\mathrm{Gal}(\mathbf{Q}(\pi)^{\mathrm{gal}}/\mathbf{Q})$ with frequency such that the log-log ratios have a limiting value for each G as $q \to \infty$. While this does seem to be the case, the precise value of these ratios does not seem to coincide with Malle's constant $a(G)$, and seems to be more complicated. Table 4 shows them for $g = 3$. If we let $\widetilde{a}(G)$ denote these limits, to the authors it seems that

$$\widetilde{a}(\texttt{6T11}) = 1, \quad \widetilde{a}(\texttt{6T3}) = \widetilde{a}(\texttt{6T6}) \approx 3/5, \quad 1/10 \leq \widetilde{a}(\texttt{6T1}) \leq 3/10,$$

and we have no theoretical explanation for why this is the case. We do note that there seems to be some partial progress on these type of conditional Malle distributions in [BSMT17].

Remark 4.3 Alternatively, it may be the case that these distributions fit closer to a "polynomials-in-a-box" distribution. The study of the Galois group of a random polynomial goes back to van der Waerden [vdW36], and an excellent summary (and further developments) is given in [Zyw10]. One can prove (see [BSK20]) that the number of monic polynomials of degree d which are irreducible and have

Table 4 For $g = 3$ we have plotted $\log N(g, q, G)/\log N(g, q)$. Observe the phenomena for `6T1` at powers of 3. Also, note that none of these can possibly be $a(G)$ for some G, which must be of the form $1/n$ for $1 \leq n \leq 2g = 6$ (as these numbers are 1.00, 0.50, 0.33, 0.25, 0.20, 0.16)

Group\\q	2	3	4	5	7	8	9	11	13
6T1	0.31636	0.11844	0.26715	$-\infty$	0.30254	0.23744	0.14698	0.25954	0.28031
2T1	$-\infty$	$-\infty$	$-\infty$	$-\infty$	$-\infty$	0.079146	$-\infty$	$-\infty$	$-\infty$
6T3	0.60225	0.59221	0.66064	0.58783	0.60854	0.54666	0.63575	0.57555	0.60162
6T11	0.88343	0.96285	0.97157	0.98768	0.99256	0.99380	0.99424	0.99707	0.99743
6T6	0.60225	0.60257	0.58913	0.59649	0.58877	0.60836	0.58708	0.59022	0.59007

coefficients in $[-L/2, L/2]$ is asymptotic to L^d. Let's call this number $B_d(L)$. If we let $B_{d,G}(L)$ be the number of monic irreducible polynomials with coefficients in $[-L/2, L/2]$ with Galois group G, then in analogy with Malle one could naively guess that for every group G transitive on d elements, there exists some $\alpha(G) \leq 1$ such that for every $\varepsilon > 0$, there exist constants c_1 and $c_{2,\varepsilon}$ and $R > 0$ such that for every $L > R$ one has

$$c_1 L^{d\alpha(G)} < B_{d,G}(L) < c_{2,\varepsilon} L^{d(\alpha(G)+\varepsilon)}.$$

If such constants exist, is it the case that $\widetilde{a}(G) = \alpha(G)$?

4.3 Newton Polygons Data and p-rank Strata

As discussed in Sect. 2.4, for any given positive integer d, the coarse moduli space $\mathcal{A}_{g,d}$ of g-dimensional abelian varieties equipped with a polarization of degree d^2 admits a locally closed stratification by Newton polygons, in which the stratum corresponding to an individual polygon is equidimensional of codimension equal to the elevation of the polygon. A reasonable guess is that for any given eligible Newton polygon P in dimension g, the proportion of isogeny classes of abelian varieties over $\mathbf{F}_q$ with Newton polygons lying on or above P is cq^{-e} where e is the elevation of P and c is the number of geometrically irreducible components of the stratum over $\mathbf{F}_q$. By Theorem 2.5, $c = 1$ unless P is the supersingular stratum; in the supersingular case the stratum is reducible and not all irreducible components may be defined over $\mathbf{F}_q$, but it is guaranteed that $c > 0$ [Yu17]. One can even give an explicit formula for c when $q = p$ [Ibu18, Thm. 4.6].

For example, in dimension 3 the Newton polygons are linearly ordered (see Fig. A.1). In Fig. 2, we give a plot of

$$\log_q \left(\frac{N(3, q, P)}{N(3, q)} \right)$$

for each of the five possible Newton polygons, where $N(g, q, P)$ is the number of isogeny classes with Newton polygon on or above P, and $N(g, q)$ is the total number of isogeny classes. Note that the values for q prime agree quite well with the discussion above, while for non-prime q there are more isogeny classes in the smaller strata than expected. We have no explanation for this behavior. Moreover, the supersingular stratum lies consistently above -4 and $\log_q (N(3, q, P)/N(3, q))$ is increasing with q, suggesting that the number of geometrically irreducible components increases as some nonzero power of q.

Remark 4.4 In dimension 4, every Newton polygon stratum in $\mathcal{A}_{4,1}$ occurs for the Jacobian of some curve. For strata of codimension at least 2 this is implied by Shankar and Tsimerman [ST18, Proposition 2.4]; this leaves only the ordinary

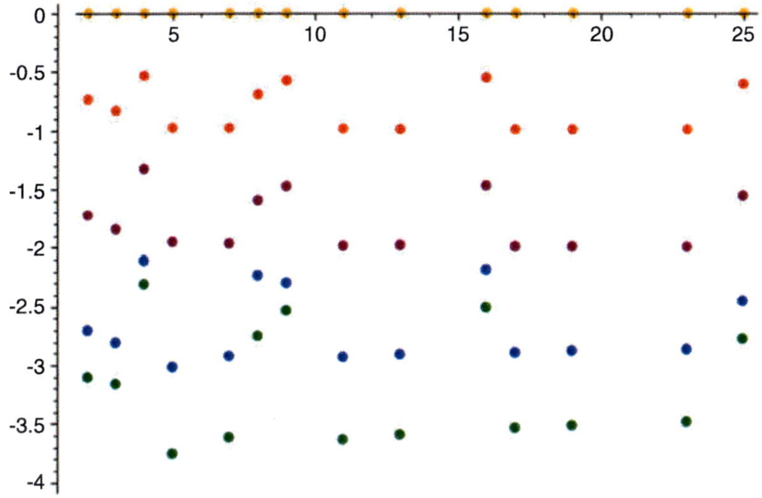

Fig. 2 $\log_q(N(3, q, P)/N(3, q))$ as a function of q, where $N(g, q, P)$ is the number of isogeny classes with Newton polygon on or above P. Colors indicate P, with orange the ordinary stratum and green the supersingular stratum

stratum which includes the generic Jacobian, and the almost ordinary stratum which includes the generic non-ordinary Jacobian.

Problem 4.5 The following question is due to Pries (see http://aimpl.org/cohomabelian/): Let p be a prime number. For a fixed dimension g and a fixed p-rank f, what is the smallest field of definition of a simple abelian variety over $\overline{\mathbf{F}}_p$ with dimension g and p-rank f?

For each pair (g, q) for which the LMFDB contains data, we have verified that each of $0, \ldots, g$ occurs as the p-rank of at least one simple abelian variety of dimension g over $\mathbf{F}_q$.

4.4 *Frobenius Angle Rank*

Table 5 shows for each g and angle rank δ with $0 \le \delta \le g$ which p-ranks do not appear in our database.

From Table 5 some simple patterns emerge.

- Angle rank 0 implies p-rank 0. This is known in general because angle rank 0 is equivalent to supersingularity.
- Up to $g = 4$, every positive angle rank can occur for ordinary abelian varieties (i.e., for p-rank g). However, please note that for $g = 5$ we only cover $q = 2, 3$, and for $g = 6$ we only cover $q = 2$.
- If $\delta = 1$ then the p-rank is 0 or g.

Table 5 Excluded p-ranks for simple abelian varieties

Dimension	Angle rank	Forbidden p-ranks
1	0	1
	1	0
2	0	1,2
	1	0,1
	2	0
3	0	1,2,3
	1	1,2
	2	0,1,3
	3	none
4	0	1,2,3,4
	1	1,2,3
	2	1,3
	3	3
	4	none
5	0	0,1,2,3,4,5
	1	1,2,3,4
	2	0,1,2,3,4,5
	3	0,1,2,3,4,5
	4	1,3,5
	5	none
6	0	1,2,3,4,5,6
	1	1,2,3,4,5
	2	1,5
	3	1,3,5
	4	0,1,3,5,6
	5	5
	6	none

- For $g \geq 3$, if $\delta = 2$ then the p-rank cannot be 1.
- The p-rank $g - 1$ only occurs when $\delta = g - 1$ or $\delta = g$. This follows from a theorem of Lenstra–Zarhin [LZ93, Theorem 5.7], who also show that the case $\delta = g - 1$ cannot occur if g is even [LZ93, Theorem 5.8].

For discussions in this direction we refer the reader to the forthcoming paper [DKZB21].

4.5 *Endomorphism Algebras*

We motivate this section with the following open problem:

Problem 4.6 ([Oor08, Open Problem 22.6]) For each $g > 0$ and p determine the possible endomorphism algebras occurring in that characteristic.

As a first step we focus on simple abelian varieties, since the general case is expressible in terms of matrix algebras over the endomorphism algebras of the simple constituents. We can then break this problem up into two parts: understanding the center of the endomorphism algebra, and understanding the endomorphism algebra as a division algebra over its center.

The center of the endomorphism algebra of an abelian variety A is just the number field $\mathbf{Q}(\pi)$ defined by $h_A(T)$, where $P_A(T) = h_A(T)^e$. In order to analyze the possible centers that can arise, we look at statistics on the discriminant Δ of $\mathbf{Q}(\pi)$. It is useful to normalize the discriminant in two ways. First, we consider the root discriminant $\mathrm{rd} = |\Delta|^{1/n}$ instead, where $n = [\mathbf{Q}(\pi) : \mathbf{Q}]$. The root discriminant is often more useful when considering number fields of different degrees. Second, we use the following result to rescale the root discriminant in a way that allows comparison across different values of q.

Theorem 4.7 *Let A be a simple abelian variety of dimension g over $\mathbf{F}_q$ with associated Weil number π. Then the root discriminant* rd *of $\mathbf{Q}(\pi)$ is bounded by*

$$\mathrm{rd} \leq 2gq^{g/2}.$$

Proof By Mordell [Mor60, Eq. 17], the maximum possible *polynomial* discriminant for a polynomial of degree $2g$ whose roots all have absolute value $\sqrt{q}$ is $(2g)^{2g}q^{g(2g-1)}$. Applied to the polynomial $h_A(T)$, this yields the bound

$$\mathrm{rd} \leq \frac{2g}{e}q^{g/e-1/2}.$$

This implies the desired bound when $e > 1$, so we may assume hereafter that $e = 1$. To improve the bound in this case, we distinguish between the discriminant Δ of $\mathbf{Q}(\pi)$ and the discriminant Δ' of $P_A(T)$; it will suffice to check that the ratio Δ'/Δ is divisible by $q^{g(g-1)}$.

Put $\beta = \pi + q/\pi$, let Δ_0 be the discriminant of $\mathbf{Q}(\beta)$, and let Δ_0' be the discriminant of the minimal polynomial of β over $\mathbf{Q}$. Let $\alpha_1, \dots, \alpha_{2g}$ be the conjugates of π in $\mathbf{Q}^{\mathrm{alg}}$, sorted so that $\alpha_i\alpha_{2g-i} = q$ for $i = 1, \dots, g$. Then on one hand,

$$\begin{aligned}
\frac{\Delta'}{(\Delta_0')^2} &= \left(\prod_{i=1}^{g}(\alpha_i - \alpha_{2g-i})^2\right)\left(\prod_{i=1}^{g}\prod_{j=i+1}^{g}\left(\frac{(\alpha_i - \alpha_j)(\alpha_i - \alpha_{2g-j})(\alpha_j - \alpha_{2g-i})(\alpha_{2g-j} - \alpha_{2g-i})}{(\alpha_i + \alpha_{2g-i} - \alpha_j - \alpha_{2g-j})^2}\right)^2\right)\\
&= \left(\prod_{i=1}^{g}(\alpha_i - q/\alpha_i)^2\right)\left(\prod_{i=1}^{g}\prod_{j=i+1}^{g}\left(\frac{(\alpha_i - \alpha_j)(\alpha_i - q/\alpha_j)(\alpha_j - q/\alpha_i)(q/\alpha_j - q/\alpha_i)}{(\alpha_i + q/\alpha_i - \alpha_j - q/\alpha_j)^2}\right)^2\right)\\
&= \left(\prod_{i=1}^{g}(\alpha_i - q/\alpha_i)^2\right)\left(\prod_{i=1}^{g}\prod_{j=i+1}^{g}\left(\frac{\alpha_i^{-2}\alpha_j^{-2}(\alpha_i - \alpha_j)(\alpha_i\alpha_j - q)(\alpha_i\alpha_j - q)(q\alpha_i - q\alpha_j)}{\alpha_i^{-2}\alpha_j^{-2}(\alpha_i - \alpha_j)^2(\alpha_i\alpha_j - q)^2}\right)^2\right)\\
&= q^{g(g-1)}\prod_{i=1}^{g}(\alpha_i - q/\alpha_i)^2.
\end{aligned}$$

On the other hand, the relative discriminant of $\mathbf{Q}(\pi)$ over $\mathbf{Q}(\beta)$ divides the polynomial discriminant of $x^2 - \beta x + q$, which is $(\pi - q/\pi)^2$. Consequently,

$$\frac{\Delta}{\Delta_0^2} \quad \text{divides} \quad \prod_{i=1}^{g} (\alpha_i - q/\alpha_i)^2.$$

By writing

$$\frac{\Delta'}{\Delta} = \frac{\Delta'}{(\Delta_0')^2} \left(\frac{\Delta_0'}{\Delta_0}\right)^2 \frac{\Delta_0^2}{\Delta}$$

and noting that Δ_0' is divisible by Δ_0, we deduce that Δ'/Δ is divisible by $q^{g(g-1)}$ as claimed. □

In Figs. A.2, A.3, A.4, A.5, A.6, and A.7 we give distributions of both the polynomial and number field root discriminants for the different values of g in the database. In the polynomial case, we divide the root discriminant by $2gq^{\frac{2g-1}{2}}$, and in the number field case we divide by $2gq^{g/2}$. The distribution in the number field case appears to be a sum of copies of the polynomial distribution after further rescaling by the appropriate roots of reciprocals of integers. This phenomenon is especially apparent in the $g = 2$ case because of the simple nature of the polynomial distribution. Large spikes can be seen at $1/2$, $1/3$, $1/4$ and $1/5$ corresponding to an extra factor of 2^4, 3^4, 4^4 or 5^4 in the index of the maximal order in the equation order of the number field; smaller spikes are visible at $1/\sqrt{2}$ and $1/\sqrt{3}$ corresponding to extra factors of 2^2 and 3^2.

Of course, while the endomorphism algebra is usually commutative, sometimes it is not. In order to give some insight into the non-commutative cases, we provide statistics on the possible Brauer invariants of the endomorphism algebra as a division algebra over its center. Table A.10 in the center summarizes the results. The length of the sequence of invariants gives the number of places above p in the center, and we have collapsed all of the commutative endomorphism algebras into a single row for each value of g.

4.6 Isogeny Sato-Tate Distribution

What is the distribution of $\#A(\mathbf{F}_q)$? What about when we restrict to ones with certain invariant types? From the Lang-Weil estimates (Sect. 2.7) we know that

$$\#A(\mathbf{F}_q) = q^g + O(q^{g-1/2}) \text{ as } q \to \infty.$$

This asymptotic suggests that the normalized error

$$E := \frac{\#A(\mathbf{F}_q) - q^g}{q^{g-1/2}}$$

will form an interesting probability distribution $P_{g,q}$ as we vary over all A's of a fixed dimension g defined over $\mathbf{F}_q$.

Let us consider what happens if we fix g and take a limit as $q \to \infty$. Writing $\alpha_1, \dots, \alpha_{2g}$ for the Frobenius eigenvalues with $\alpha_i \alpha_{g+i} = q$ for $i = 1, \dots, g$, we have

$$E = q^{-g+1/2} \left(\prod_{i=1}^{2g} (1 - \alpha_i) - \prod_{i=1}^{2g} \alpha_i \right) = \sum_{i=1}^{2g} q^{-1/2} \alpha_i + o(q^{-1/2});$$

consequently, the distribution of E will have the same limiting behavior as the distribution of the normalized Frobenius trace of A.

The philosophy of Katz–Sarnak [KS99] would predict that the distribution of the Frobenius trace should converge to the trace distribution for random matrices in the Lie group $\mathrm{USp}(2g)$. This convergence holds if we average over isomorphism classes of principally polarized abelian varieties, as this forms a geometric family with maximal monodromy [KS99, Theorem 11.0.4] to which one may apply Deligne's equidistribution theorem [KS99, Theorem 9.2.6].

However, since we do not currently have the data of how many isomorphism classes constitute a given isogeny class, we are only able to compute the average over isogeny classes. We thus predict a different distribution, given by a function whose value at a_1 is proportional to the measure of the set of $(a_2, \dots, a_g) \in \mathbf{R}^{g-1}$ for which $T^g + a_1 T^{g-1} + \cdots + a_g$ has all roots in $[-2, 2]$. We compute this distribution using the method of DiPippo–Howe (see Sect. 4.1). By computing Jacobian determinants, we see that integrating 1 over the space of coefficients $(a_1, \dots, a_g)$ is the same as integrating $1/g!$ over the space of power sums $(p_1, \dots, p_g)$, or integrating $1/g!$ times the Vandermonde determinant $V(r_1, \dots, r_g) = \prod_{1 \le i < j \le g} (r_j - r_i)$ over the space of ordered tuples of roots $(r_1, \dots, r_g)$. That is, the desired distribution is given (up to a normalizing factor) by the distribution function

$$f(x) = \int_{S \cap H_x} V(r_1, \dots, r_g) d\mu_{H_x}$$

where S denotes the simplex

$$S = \{(r_1, \dots, r_g) \in \mathbf{R}^g : -2 \le r_1 \le \cdots \le r_g \le 2\}$$

and H_x denotes the hyperplane $r_1 + \cdots + r_g = x$. Let us write this as an iterated integral over $r_1, \dots, r_{g-1}$, substituting $r_g = x - r_1 - \cdots - r_{g-1}$; the endpoints of integration of r_j are then

$$\max\{r_{j-1}, x - 2(g-j) - \sum_{k=1}^{j-1} r_k\}, \qquad \min\{2, \left(x - \sum_{k=1}^{j-1} r_k \right) / (g - j + 1)\}$$

(writing $r_0 = -2$). In particular, the distribution function is continuous, even, and piecewise polynomial: on each interval $[-2g+4(i-1), -2g+4i]$ for $i = 1, \dots, g$, it is a polynomial of degree $(g-1)(g+2)/2$ with rational coefficients. For the extreme values $i = 1$ and $i = g$, this polynomial is a scalar multiple of $(2g - |x|)^{(g-1)(g+2)/2}$.

Using Mathematica, we computed the distribution functions $f_g(x)$ for $g \leq 4$:

$$g = 1: \begin{cases} \frac{1}{4} & |x| \leq 2 \\ 0 & |x| > 2 \end{cases}$$

$$g = 2: \begin{cases} \frac{3}{2^7}(4 - |x|)^2 & |x| \leq 4 \\ 0 & |x| > 4 \end{cases}$$

$$g = 3: \begin{cases} \frac{3}{2^{13}}(15|x|^4 - 200|x|^2 + 816) & |x| \leq 2 \\ \frac{3}{2^{15}}(6 - |x|)^5 & 2 < |x| \leq 6 \\ 0 & |x| > 6 \end{cases}$$

$$g = 4: \begin{cases} \frac{5(-|x|^9 - 72|x|^8 - 2304|x|^7 + 64512|x|^6 - 516096|x|^5 + 1548288|x|^4 - 7077888|x|^2 + 24117248)}{3 \cdot 2^{27}} & |x| \leq 4 \\ \frac{5}{3 \cdot 2^{27}}(8 - |x|)^9 & 4 < |x| \leq 8 \\ 0 & |x| > 8. \end{cases}$$

See Fig. 3 for plots of the distribution for $g \leq 4$, and Figs. A.8, A.9, and A.10 for plots of this prediction against our data. A table of moments for $g = 3, 4, 5, 6$ is given in Table 6.

It is natural to ask about the limit as $g \to \infty$. While we do not have a general formula for $f_g(x)$, we can give some evidence that the limiting distribution f_∞ is a Gaussian with variance 2, which is also the limit of the Sato-Tate distributions

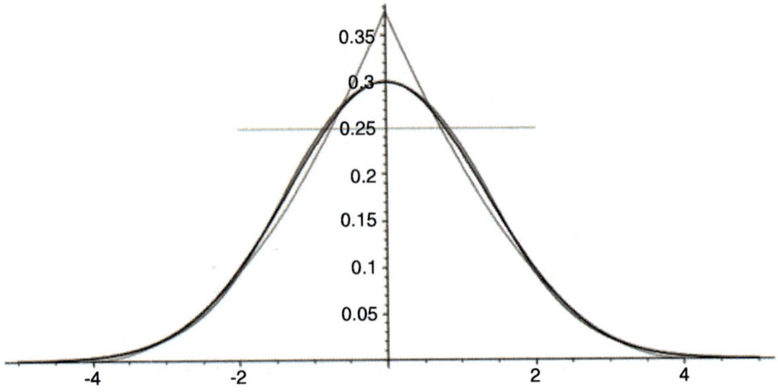

Fig. 3 Isogeny Sato-Tate distribution for $g \leq 4$

Table 6 Even moments for the isogeny Sato-Tate distributions for $g = 3, 4, 5, 6$

g	2nd	4th	6th	8th	10th	12th	14th
3	1.7142	8.6857	71.7575	796.1318	10750.4655	166954.5839	2.8786×10^6
4	1.7777	9.4199	82.5201	1001.4566	15384.2906	282674.8553	5.9748×10^6
5	1.8181	9.8834	89.1908	1121.6573	18035.9973	351973.2932	8.0435×10^6
6	1.8461	10.2041	93.7929	1203.9623	19814.7906	397315.2698	9.3803×10^6

Table 7 Predicted moments from Eq. (4.4)

r	4	6	8	10	12	14
$\widetilde{m}_6^{(r)}$	10.2041	93.7929	1203.9623	19814.7906	397315.2698	9.3803×10^6
$\left(\widetilde{m}_6^{(2)}\right)^{r/2}(r-1)!!$	10.2243	94.3750	1219.5797	20263.1948	411486.7236	9.8754×10^6

[DS94, Theorem 6].[7] Let $m_\infty^{(r)} = \int_{-\infty}^{\infty} x^r f_\infty(x)dx$ and let $\widetilde{m}_g^{(r)}$ be a numerical approximation to $\int_{-\infty}^{\infty} x^r f_g(x)dx$. If f_∞ were Gaussian, we would have

$$m_\infty^{(r)} = (m_\infty^{(2)})^{r/2}(r-1)!! \tag{4.4}$$

for even r. As Table 7 shows, this identity already holds approximately for $g = 6$.

The authors don't have enough data to determine what happens when we restrict to isogeny classes of abelian varieties with a fixed Newton polygon, endomorphism algebra, delta rank, or Galois group. We expect similar interesting distributions in these cases as $q \to \infty$ as we normalize the dimensions appropriately. We also remark that a determination of the conjectural distribution of point counts also remains open (for say, $P(T)$ of a fixed degree $2g$ and fixed Galois group G).

4.7 Maximal and Minimal Point Counts

The question of the maximum number of points on a curve of given genus over a given finite field has been studied quite extensively, due to its connection with error-correcting codes via the Goppa construction. The web site https://manypoints.org tabulates most known results on this question. In this section we investigate minimal and maximal point counts of abelian varieties.

We will say an isogeny class $[A]$ is *maximal* (resp. *minimal*) for a fixed g and q when $\#A(\mathbf{F}_q)$ is the maximum (resp. minimum) of

$$\{\#B(\mathbf{F}_q) : B \text{ an abelian variety of dim } g \text{ over } \mathbf{F}_q\}$$

[7] It may be feasible to prove that the limiting distribution is Gaussian by bounding the difference in moments directly.

Table 8 Maximal and minimal abelian varieties ($g = 3$)

	$\#A(\mathbf{F}_q)$	$P_A(T)$	Newton Polygon	Jac?
$q = 3$:	343	$(T^2 + 3T + 3)^3$	ss	No
	1	$(T^2 - 3T + 3)^3$	ss	No
$q = 4$:	729	$(T + 2)^6$	ss	No
	1	$(T - 2)^6$	ss	No
$q = 5$:	1000	$(T^2 + 4T + 5)^3$	ord	No
	8	$(T^2 - 4T + 5)^3$	ord	No
$q = 7$:	2197	$(T^2 + 5T + 7)^3$	ord	No
	27	$(T^2 - 5T + 7)^3$	ord	No
$q = 8$:	2744	$(T^2 + 5T + 8)^3$	ord	???
	64	$(T^2 - 5T + 8)^3$	ord	No
$q = 9$:	4096	$(T + 3)^6$	ss	???
	64	$(T - 3)^6$	ss	No
$q = 11$:	5832	$(T^2 + 6T + 11)^3$	ord	No
	216	$(T^2 - 6T + 11)^3$	ord	No

While studying maximal (resp. minimal) abelian varieties is formally similar to studying maximal (resp. minimal) curves, it is not directly related: an isogeny class being maximal (resp. minimal) has nothing to do with it containing the Jacobian of some maximal (resp. minimal) curve. See Example 5.8 for explicit examples.

In a similar vein one might also naively expect that, since Artin-Schreier curves are all maximal and known to be supersingular (see Example 5.5), an abelian variety A being maximal (resp. minimal) might imply that A is supersingular. Indeed, in some references (such as [KP19]) the terms *maximal* and *minimal* are used to refer exclusively to supersingular abelian varieties. However, we will see below that a maximal (resp. minimal) abelian variety in our sense need not be supersingular.

When looking at the data our first observation is that for a fixed g, q we almost always found that there were unique isogeny classes $A_{\max}$ and $A_{\min}$ with $L(A_{\max}, T) = L(A_{\min}, -T)$. This equality implies that $A_{\min}$ and $A_{\max}$ are quadratic isogeny twists.

Example 4.8 When $g = 3$ and $q = 3$ we have $P(A_{\min}, T) = T^6 - 9T^5 + 36T^4 - 81T^3, 108T^2 - 81T + 27$ which is 3.3.aj_bk_add and $P(A_{\max}, T) = T^6 + 9T^5 + 36T^4 + 81T^3 + 108T^2 + 81T + 27$, which is 3.3.j_bk_dd.

We now explain our findings. In Table 8, we report the unique minimal and maximal isogeny classes for $g = 3$ and $3 \leq q \leq 25$. These are all isogenous to cubes of elliptic curves, and the minimal and maximal examples for a given q are isogeny twists of each other; but some are ordinary and some are supersingular. We omit $q = 2$ because it is a bit anomalous: there are 7 minimal isogeny classes (and a unique maximal one). See Lemmas 4.9 and 4.10 for an explanation of these observations.

We now investigate these observations. Recall that the (sharpened) Weil bounds on an abelian variety A of dimension g over $\mathbf{F}_q$ take the form

$$\lceil(\sqrt{q}-1)^2\rceil \leq \#A(\mathbf{F}_q)^{1/g} \leq \lfloor(\sqrt{q}+1)^2\rfloor. \tag{4.5}$$

These are precisely the maximal and minimal values appearing in Table 8. We are thus led to ask whether equality in the upper (resp. lower) bounds happens only for a power of a maximal (resp. minimal) elliptic curve, or equivalently whether the inequalities become strict if we restrict to simple abelian varieties of dimension greater than 1. This is in fact claimed in both [AHL12, Théorème 1.1] and [AHL13, Corollary 2.2, Corollary 2.14], but we have already seen by an example that this is false for $q = 2$; moreover, Theorem 2.11 implies that there are infinitely many g for which there exists a simple abelian variety A of dimension g over $\mathbf{F}_2$ with $\#A(\mathbf{F}_2) = 1$.

On the other hand, by working more closely through the literature, we can recover an argument that the inequalities become strict.

Lemma 4.9 *For $q = 5, 7$, assume that $\#A(\mathbf{F}_q)^{1/g} > N_q$ for every abelian variety A of dimension g defined over $\mathbf{F}_q$, where $N_5 = 2.708$ and $N_7 = 3.970$ (reported to us by Kadets). Then for all $q > 2$, the inequalities (4.5) become strict for simple abelian varieties of dimension greater than* 1.

Proof First, by inspection of the proof of [AHL13, Proposition 2.17], one deduces that the lower bound is strict for $q \geq 8$; thus only the cases $q = 3, 4, 5, 7$ are at issue. For $q = 3$ and $q = 4$, Theorem 2.9 implies that the lower bound is strict (for $q = 4$ we may also apply Theorem 2.8). For $q = 5$ and $q = 7$, one can obtain similar lower bounds by emulating the calculation used to prove Theorem 2.9. Kadets reports that a nonrigorous version of the calculation gives the lower bounds 2.708 for $q = 5$ and 3.970 for $q = 7$, but as of this writing a rigorous calculation remains to be made. □

We raise our observations to the status of theorems below.

Lemma 4.10 *Under the hypothesis of Lemma 4.9, for every $g > 1$ and every $q > 2$ there exists unique maximal and minimal isogeny classes of dimension g. These classes are quadratic isogeny twists of each other and are a power of the unique maximal (resp. minimal) isogeny class of elliptic curves. Finally, the class is supersingular or ordinary according to whether p divides $\lfloor 2\sqrt{q}\rfloor$.*

Proof By Theorem 2.8, one always gets a maximal (resp. minimal) abelian variety of a given dimension over $\mathbf{F}_q$ by taking a power of the maximal (resp. minimal) elliptic curve over $\mathbf{F}_q$. By the discussion following Eq. (4.5), this is unique except for the minimal case over $\mathbf{F}_2$ in some genera—the isogeny class of an elliptic curve is determined by a single point count. In particular, since the minimal and maximal elliptic curve over $\mathbf{F}_q$ are isogeny twists of each other, the same is true of their powers. □

Table 9 A table of the unique simple-maximal and simple-minimal Weil polynomials for $g = 3$

	$\#A(\mathbf{F}_q)$	$P_A(T)$	NP	Jac?
$q = 5$:	631	$T^6 + 8\,T^5 + 34\,T^4 + 93\,T^3 + 170\,T^2 + 200\,T + 125$	ord	Yes
	25	$T^6 - 8\,T^5 + 32\,T^4 - 85\,T^3 + 160\,T^2 - 200\,T + 125$	ord	No
$q = 7$:	1561	$T^6 + 11\,T^5 + 59\,T^4 + 195\,T^3 + 413\,T^2 + 539\,T + 343$	ord	???
	71	$T^6 - 11\,T^5 + 59\,T^4 - 195\,T^3 + 413\,T^2 - 539\,T + 343$	ord	No
$q = 8$:	2157	$T^6 + 12\,T^5 + 69\,T^4 + 243\,T^3 + 552\,T^2 + 768\,T + 512$	ord	???
	111	$T^6 - 12\,T^5 + 69\,T^4 - 243\,T^3 + 552\,T^2 - 768\,T + 512$	ord	No
$q = 9$:	2911	$T^6 + 13\,T^5 + 81\,T^4 + 305\,T^3 + 729\,T^2 + 1053\,T + 729$	ord	???
	169	$T^6 - 13\,T^5 + 81\,T^4 - 305\,T^3 + 729\,T^2 - 1053\,T + 729$	ord	No
$q = 11$:	4861	$T^6 + 15\,T^5 + 105\,T^4 + 439\,T^3 + 1155\,T^2 + 1815\,T + 1331$	ord	???
	323	$T^6 - 15\,T^5 + 105\,T^4 - 439\,T^3 + 1155\,T^2 - 1815\,T + 1331$	ord	No
$q = 13$:	7181	$T^6 + 16\,T^5 + 122\,T^4 + 555\,T^3 + 1586\,T^2 + 2704\,T + 2197$	ord	???
	615	$T^6 - 16\,T^5 + 120\,T^4 - 543\,T^3 + 1560\,T^2 - 2704\,T + 2197$	ord	No
$q = 16$:	12649	$T^6 + 19\,T^5 + 166\,T^4 + 847\,T^3 + 2656\,T^2 + 4864\,T + 4096$	ord	???
	1189	$T^6 - 19\,T^5 + 166\,T^4 - 847\,T^3 + 2656\,T^2 - 4864\,T + 4096$	ord	No
$q = 17$:	14351	$T^6 + 19\,T^5 + 169\,T^4 + 885\,T^3 + 2873\,T^2 + 5491\,T + 4913$	ord	???
	1539	$T^6 - 19\,T^5 + 167\,T^4 - 871\,T^3 + 2839\,T^2 - 5491\,T + 4913$	ord	No
$q = 19$:	19601	$T^6 + 21\,T^5 + 201\,T^4 + 1119\,T^3 + 3819\,T^2 + 7581\,T + 6859$	ord	???
	2113	$T^6 - 21\,T^5 + 197\,T^4 - 1085\,T^3 + 3743\,T^2 - 7581\,T + 6859$	ord	No
$q = 23$:	32671	$T^6 + 24\,T^5 + 258\,T^4 + 1591\,T^3 + 5934\,T^2 + 12696\,T + 12167$	ord	???
	4049	$T^6 - 24\,T^5 + 258\,T^4 - 1591\,T^3 + 5934\,T^2 - 12696\,T + 12167$	ord	???
$q = 25$:	40391	$T^6 + 25\,T^5 + 281\,T^4 + 1809\,T^3 + 7025\,T^2 + 15625\,T + 15625$	ord	???
	5473	$T^6 - 25\,T^5 + 281\,T^4 - 1809\,T^3 + 7025\,T^2 - 15625\,T + 15625$	ord	???

Remark 4.11 As asserted in Lemma 4.10, whether or not the maximal and minimal elliptic curves over $\mathbf{F}_q$, and hence the resulting abelian varieties, are supersingular or ordinary depends on whether or not $\lfloor 2\sqrt{q} \rfloor$ is divisible by p. When $q = p$, this divisibility holds only for $p = 2, 3$ (as otherwise $0 < 2\sqrt{p} < p$). When q is a square, so $q = p^{2e}$, we have $\lfloor 2\sqrt{q} \rfloor = 2\sqrt{p^{2a}} = 2p^a$ which is obviously divisible by p. In other cases, $q = p^{2e+1}$ for some positive integer e and we are asking whether the base-p expansion of $2\sqrt{p}$ has a zero in the e-th position after the radix point; this can occur but is rather sporadic.[8]

Following the discussion in Sect. 2.7, we now restrict attention to simple abelian varieties. We say that an isogeny class $[A]$ of dimension g over $\mathbf{F}_q$ is *simple-maximal* (resp. *simple-minimal*) if $\#A(\mathbf{F}_q)$ is maximal (resp. minimal) among simple abelian varieties of dimension g over $\mathbf{F}_q$. In Table 9, we report the unique simple-maximal and simple-minimal isogeny classes for $g = 3$ and $5 \le q \le 25$. For each of $q = 2, 3, 4$, there are 2 simple-minimal isogeny classes (and a unique simple-maximal one).

[8]The *radix point* is the analogue of the decimal point for a general base expansion.

Table 10 A table of the unique simple-maximal and simple-minimal Weil polynomials for $g = 4$

	$\#A(\mathbf{F}_q)$	$P_A(T)$	NP	Jac?
$q = 3$:	979	$T^8 + 7T^7 + 27T^6 + 72T^5 + 143T^4 + 216T^3 + 243T^2 + 189T + 81$	ord	???
	5	$T^8 - 6T^7 + 13T^6 - 10T^5 + T^4 - 30T^3 + 117T^2 - 162T + 81$	ord	No
$q = 4$:	2521	$T^8 + 9T^7 + 42T^6 + 132T^5 + 305T^4 + 528T^3 + 672T^2 + 576T + 256$	ord	???
	29	$T^8 - 9T^7 + 41T^6 - 125T^5 + 285T^4 - 500T^3 + 656T^2 - 576T + 256$	ord	No
$q = 5$:	5599	$T^8 + 11T^7 + 62T^6 + 229T^5 + 601T^4 + 1145T^3 + 1550T^2 + 1375T + 625$	ord	???
	61	$T^8 - 10T^7 + 45T^6 - 130T^5 + 305T^4 - 650T^3 + 1125T^2 - 1250T + 625$	ord	No

Table 11 Bounds and extreme values for $\#A(\mathbf{F}_q)^{1/g}$ for a simple abelian variety A of dimension g over $\mathbf{F}_q$. The bounds for $q = 2, 3, 4$ are taken from [Kad19]; the bounds for $q = 5$ were computed numerically (but not rigorously) by Kadets using the same method (see Lemma 4.9)

q	g	Lower bound	Minimum	Maximum	Upper bound
2	4	1	1	3.940	4.035
2	5	1	1.149	3.717	4.035
2	6	1	1	3.697	4.035
3	4	1.359	1.495	5.594	5.634
3	5	1.359	1.670	5.423	5.634
4	4	2.275	2.321	7.086	7.382
5	4	2.708	2.795	8.615	8.938

We make a few curious observations about this data which we are unable to rigorously explain. For one, all of the simple-maximal and simple-minimal examples are ordinary. For another, the simple-maximal and simple-minimal varieties are most often isogeny twists of each other, but not always (see $q = 5, 13, 17, 19$); in any case, they have opposite sign patterns (the coefficients of $P_A(T)$ are positive for A simple-maximal and alternate in sign for A simple-minimal). Finally, note that the simple-maximal variety for $q = 5$ is a Jacobian.

In Table 10, we give similar information for $g = 4$, and in Table 11 we compare the extreme values of $\#A(\mathbf{F}_q)^{1/g}$ to the bounds given in Theorem 2.9.

5 An Isogeny Class Scavenger Hunt

In this section, we describe a number of examples related to questions or results in the literature.

Several of these examples involve Jacobians of curves, whereas the LMFDB does not currently contain complete information about Jacobians of curves of genus at least 4 (Sect. 6.2). To generate these examples, we used Sage [S+20] to exhaust over hyperelliptic curves, computing zeta functions until we found one of the desired form. (This can also be done in Magma [BCP97].)

5.1 Some Basic Examples

Example 5.1 (Supersingular Elliptic Curves) One of the first interesting examples of abelian varieties are supersingular elliptic curves defined over finite fields. If E is an elliptic curve defined over a finite field, we have

$$\zeta(E/\mathbf{F}_q, T) = \frac{1 - aT + qT^2}{(1-T)(1-qT)},$$

with $a \in \mathbf{Z}$ and $|a| < 2\sqrt{q}$ by the Hasse-Weil bound. In addition, if E is supersingular then $a \equiv 0 \mod p$, for p the characteristic of $\mathbf{F}_q$. Therefore if p is prime and $p \geq 5$, then there is a unique isogeny class of supersingular elliptic curves over $\mathbf{F}_p$, namely the one with $a = 0$. This means that

$$P_E(T) = T^2 + q = (T - i\sqrt{q})(T + i\sqrt{q});$$

a similar phenomenon holds in more generality. Indeed, when $A/\mathbf{F}_q$ is a simple abelian variety, A is supersingular if and only if $\pi_A = \zeta\sqrt{q}$ for ζ a root of unity: namely, if A is supersingular, then $\pi/\sqrt{q}$ is an algebraic number with norm 1 in every finite and infinite place, so by Kronecker–Weber it is a root of unity.

In contrast, while an elliptic curve defined over a finite field has p-rank zero if and only if it is supersingular, this does not hold in higher dimension, even for simple abelian varieties. For example, isogeny class 3.2.ac_c_ac is geometrically simple of p-rank 0, but has Newton polygon slopes $[1/3, 1/3, 1/3, 2/3, 2/3, 2/3]$ and therefore is not supersingular. (This is the smallest dimension for which p-rank 0 does not imply supersingular.) Conversely, a supersingular isogeny class always has p-rank 0.

We finally comment on the endomorphism rings of supersingular elliptic curves. While every supersingular elliptic curve has geometric endomorphism ring a maximal order in a quaternion algebra, it is not the case that every endomorphism is defined over the field of definition of the elliptic curve. For example, there are supersingular elliptic curves defined over $\mathbf{F}_p$ for each prime p (belonging to the isogeny class with $a = 0$, as above), but each of them has endomorphism ring isomorphic to an order in an imaginary quadratic field and only acquires its extra endomorphisms over an extension of $\mathbf{F}_p$. For example, isogeny class 1.3.ad has endomorphism degree 6, by which we mean that its geometric endomorphisms are defined over $\mathbf{F}_{3^6}$; isogeny class 1.2.ac has endomorphism degree 4; and isogeny class 1.2.a has endomorphism degree 2.

Example 5.2 We now give examples of particularly uncomplicated Weil polynomials for $g = 2$ and $g = 4$. These polynomials can be useful to check conjectures when one wants to produce examples quickly by hand. We first observe that by Lemma 2.2 a Weil polynomial $P_A(T) = P(T)$ of an abelian variety A of dimension g can be written as

$$P(T) = \prod_{i=1}^{g}(T^2 - \beta_i T + q) \tag{5.1}$$

where $\beta_i = \alpha_i + q/\alpha_i$ is totally real, and $\alpha_1, \ldots, \alpha_{2g}$ are the roots of the Weil polynomial, arranged so that $\alpha_i \alpha_{2g-i} = q$ for $i = 1, \ldots, g$.

When $g = 2$, we can expand this to obtain:

$$P(T) = T^4 - (\beta_1 + \beta_2)T^3 + (2q + \beta_1\beta_2)T^4 - (\beta_1 + \beta_2)qT + q^2.$$

For r a positive integer, choosing $\beta_1 = \sqrt{r}$ and $\beta_2 = -\sqrt{r}$ gives a nice family of examples. In fact, for any positive integer a not divisible by p such that $2q > |a|$, letting $r = 2q - a$ gives the characteristic polynomial

$$P(T) = T^4 + aT^2 + q^2$$

for an abelian surface. (Note that we require $p \nmid a$ to ensure that $e_A = 1$; see Sect. 2.3.) This isogeny class has Weil q-numbers

$$+\sqrt{\frac{-a + \sqrt{a^2 - 4q^2}}{2}}, +\sqrt{\frac{-a - \sqrt{a^2 - 4q^2}}{2}}, -\sqrt{\frac{-a + \sqrt{a^2 - 4q^2}}{2}}, -\sqrt{\frac{-a - \sqrt{a^2 - 4q^2}}{2}}.$$

Even more concretely, for $q = 5$ any $a \in \{-9, -8, \ldots, 8, 9\}$ is admissible. In the case $a = -1$ this is 2.5.a_ab.

One can generalize the above example by considering (5.1) where $\beta_1, \ldots, \beta_g$ are totally real algebraic integers which form a union of Galois orbits over $\mathbf{Q}$ and q is a sufficiently large prime power.

Another pleasing family of isogeny classes can be obtained by considering the case of $g = 4$ with two isogeny factors of dimension 2 each with Weil polynomial of the form $T^4 + a_i T^2 + q^2$ as above. In this case this isogeny class has Weil polynomial

$$(T^4 + a_1 T^2 + q^2)(T^4 + a_2 T^2 + q^2) = T^8 + (a_1 + a_2)T^6 + (2q^2 + a_1 a_2)T^4 + (a_1 + a_2)q^2 T + q^4,$$

which we can specialize to $a_1 = c + b\sqrt{d}$ and $a_2 = c - b\sqrt{d}$, where a_1 and a_2 are algebraic integers, and as before $2q > |a_i|$ for each i, to give

$$a_1 + a_2 = 2c, \qquad a_1 a_2 = c^2 - db^2,$$

and

$$P(T) = T^8 + 2cT^6 + (2q^2 + c^2 - db^2)T^4 + 2cq^2T + q^4.$$

In the special case $c = 1, b = 2, d = 2, q = 5$ this is 4.5.a_c_a_br.

Example 5.3 ([AP11, Example 2.1] and 4.3.a_a_a_g) Consider the hyperelliptic curve given by

$$C : y^2 = x^9 + x^7 + 2x^5 + x^4 + 2x^3 + x^2 + x.$$

The genus of this curve is 4 and

$$L(C/\mathbf{F}_3, T) = 81T^8 + 6T^4 + 1.$$

This example was computed by writing down the shape of the zeta function and equating terms in the truncated Taylor series.

Example 5.4 ([Voi05, Example 2.3] and 3.2.a_a_f) The isogeny class of the Jacobian of a curve C over $\mathbf{F}_q$ can be identified by knowing $\#C(\mathbf{F}_{q^r})$ for $1 \leq r \leq g$. In the examples where C is a projective curve defined by $x^3y + x^3z + y^3z = 0$ or $x^3y + y^3z + xz^3 = 0$ over $\mathbf{F}_2$, C has genus 3 and we can compute point counts "by hand" using a presentation of the field:

r	1	2	3	4
$\#C(\mathbf{F}_{2^r})$	3	4	24	17

This allows us to solve explicitly for the L-polynomial (see *op. cit.*)

$$L(T) = 8T^6 + 5T^3 + 1.$$

Example 5.5 Artin-Schreier curves are fan favorites. An affine Artin-Schreier curve has a model

$$U_{f,q} : y^q - y = f_d(x)$$

where $f_d(x) \in \mathbf{F}_q[x]$ is a polynomial of degree d. After a desingularization of the naive projective model one gets a proper model $X = X_{f,q}$. In the case where $q = p$, the genus of this curve is $g = (d-1)(p-1)/2$, and for $d = 5$ and $p = 3$ the curve has genus 4 so its Jacobian will be in our database. How can we find it? It turns out that the point counts of such a curve can be explicitly computed:

$$\#X_{f,q}(\mathbf{F}_{q^n}) = 1 + \left(\sum_{\psi \in \widehat{\mathbf{F}_q} \setminus \{1\}} S(\psi, n) \right) + q^n.$$

In the above expression the ψ's are additive characters and the S's are character sums given by

$$S(\psi, n) := \sum_{b \in \mathbf{F}_{q^n}} \psi(\mathrm{Tr}_{q,n}(f(b))), \qquad \mathrm{Tr}_{q,n}(a) := \sum_{i=0}^{n-1} a^{q^i}.$$

All of the additive characters of $\mathbf{F}_q$ are parametrized by $a \in \mathbf{F}_p$ and take the form $\chi_a : \mathbf{F}_q \to \mathbf{C}$ where

$$\chi_a(b) = \zeta_p^{a\mathrm{Tr}_{\mathbf{F}_q/\mathbf{F}_p}(b)}.$$

For details on this computation see [Mue13, §2.1 and §2.2].

In the special case where $f(x) = x^5$, we computed the values of these characters using Sage, and found that $\#X(\mathbf{F}_{3^n}) = 4, 10, 28, 154, 244$ for $n = 1, 2, 3, 4, 5$. Searching our database gives 4.3.a_a_a_s, as the isogeny class of the Jacobian. This isogeny class contains supersingular abelian varieties, as expected.

In the special case $f(x) = x^5 + 2x + 1$, we again compute $\#X(\mathbf{F}_{3^n}) = 1, 7, 55, 91, 271$ for $n = 1, 2, 3, 4, 5$, and find that the Jacobian belongs to 4.3.ad_d_j_abb, another non-simple supersingular isogeny class; this time the isogeny factors are not isogenous to each other.

Example 5.6 The isogeny class 2.2.a_ad provides an example of a principally polarizable isogeny class of abelian varieties of dimension 2 over $\mathbf{F}_2$ which does not contain a Jacobian (Example 5.21 is another example in higher dimension).

Also, as discussed previously, there exist principally polarizable abelian varieties which have isogeny factors whose isogeny classes are not principally polarizable. For instance, the $g = 4, q = 5$ isogeny class 4.5.ak_bp_adq_hc has a $g = 2$ isogeny factor 2.5.ac_ab which contains no principally polarizable abelian surface.

For interested readers, we point out that our methods don't allows us to determine if the particular $g = 4$ simple isogeny class 4.5.ag_o_au_bj contains a principally polarizable abelian variety. We are unable to do so whenever the isogeny class is not ordinary, and its associated CM field $K = \mathbf{Q}(\pi)$ (which in this particular case is 8.0.268960000.3) is unramified over its totally real subfield K_+ and every prime of K_+ dividing $\pi - q/\pi$ is not inert in K/K_+. In short, the condition in [How96, Theorem 1.1] summarized in Sect. 3.6 does not completely answer the question: "Does this non-ordinary isogeny class contain a principally polarizable variety?"

Remark 5.7 Howe points out to us that the 2-dimensional isogeny class 2.2.a_ad presented in the above example is one from a family that is in the Appendix to [MN02]. The theorem in *op. cit.* is that no polynomial of the form $T^4 + (1 - 2q)T^2 + q^2$ is the Weil polynomial of a Jacobian of a curve over $\mathbf{F}_q$. One can also see that such an isogeny class contains a principally polarized abelian variety. It is the restriction of scalars from $\mathbf{F}_{q^2}$ to $\mathbf{F}_q$ of an elliptic curve with Weil polynomial $T^2 + (1 - 2q)T + q^2$.

Example 5.8 In this example we show that maximality or minimality of a curve has nothing to do with the associated Jacobian being maximal or minimal as an abelian variety in the sense of Sect. 4.7. Indeed, the isogeny class 3.3.aj_bk_add is minimal for $q = 3$ and $g = 3$ with $\#A(\mathbf{F}_3) = 1$, but it does not contain a Jacobian since its virtual curve count has $\#C(\mathbf{F}_3) = -5$. Therefore, a minimal curve of genus 3 over $\mathbf{F}_3$ cannot have minimal Jacobian.

Example 5.9 When $g = 6$ and $q = 2$ there exist isogeny twists 6.2.a_ac_a_c_a_a and 6.2.a_c_a_c_a_a which have different number fields (with the same discriminant). The two number fields have a different number of places above 2 (3 in one case and 4 in the other), so the Brauer invariants of their endomorphism algebras are different: (0, 0, 0) vs (0, 0, 0, 0).

Example 5.10 (Communicated to us by Howe) Consider isogeny class 2.2.b_b, which has characteristic polynomial $P(T) = T^4 + T^3 + T^2 + 2T + 4$. The ring $\mathbf{Z}[\pi, q/\pi]$ is a maximal order in a field with class number one, and therefore this isogeny class contains a single isomorphism class of abelian varieties. Furthermore, this abelian variety must have a principal polarization (by [How95] or by the more general Theorem 1 of [How96]). This shows that while over an algebraically closed field of characteristic 0 every abelian variety is isogenous to an abelian variety that does not admit a principal polarization, this is not true in general (cf. Sect. 3.6).

5.2 *Supersingular Curves*

It is not clear whether every eligible Newton polygon can occur for curves of a given characteristic (see Remark 5.13), and the following is a related long-standing open problem.

Question 1 For every prime p and every positive integer g, does there exist a genus g curve over $\overline{\mathbf{F}}_p$ whose Jacobian is supersingular?

This is known for $g \leq 4$ [KHS19]. It is also known for all g when $p = 2$ [vdGvdV95]; however, these curves cannot always be taken to be hyperelliptic [SZ02]. Some higher-genus cases are treated in [LMPT18, LMPT19]; however, the following example is not covered by those papers (nor by [vdGvdV95], which handles some genera for $p > 2$).

Example 5.11 Let C be the hyperelliptic curve over $\mathbf{F}_3$ given by

$$y^2 = x^{11} + 2x^9 + x^5 + x^3 + x.$$

Then

$$L(C/\mathbf{F}_3, T) = 1 + 3T^2 + 81T^8 + 243T^{10},$$

so the Jacobian A of C is supersingular and belongs to isogeny class 5.3.a_d_a_a_a.

Remark 5.12 One can also consider curves whose Jacobians are *superspecial*, meaning that they are isomorphic (not just isogenous) to a product of supersingular elliptic curves. Ekedahl [Eke87] showed there do not exist superspecial genus 4 curves in characteristics 2 and 3, and asked whether conversely they exist in all characteristics at least 5; this question was answered negatively by Kudo–Harashita [KH17], who showed that none exist in characteristic 7.

Remark 5.13 The moduli space $\mathcal{M}_g$ of curves of genus g has dimension $3g - 3$, which is much smaller than the length of a maximal chain of inclusions of Newton strata in $\mathcal{A}_g$, which is asymptotic to $g^2/4$ (see Corollary 2.4). However, unlike $\mathcal{A}_g$, which has irreducible strata except for the supersingular case, Newton strata in $\mathcal{M}_g$ can be much more reducible. While it is conceivable that there are enough strata in $\mathcal{M}_g$ to account for all possible Newton polygons, it does not seem likely.

5.3 Ordinarity and Angle Ranks

The following is a conjecture of Ahmadi–Shparlinski.

Conjecture 5.14 ([AS10, §5]) Every ordinary geometrically simple Jacobian has maximal angle rank.

This is verified in the database in dimensions 2 and 3. This is a theorem in dimension 2, even without the ordinary condition: [AS10, Theorem 2]. It is also a theorem in dimension 3, but this time it requires the ordinary condition: [Zar15, Theorem 1.1].

We verified the conjecture in dimension 4 over $\mathbf{F}_2$ as follows. According to the LMFDB, there are 52 isogeny classes of ordinary, geometrically simple abelian varieties with angle rank at most 3 (in fact they are all equal to 3). Since at the time of verification the LMFDB did not yet contain full data about whether an isogeny class in dimension greater than 3 contains a Jacobian, we used the fact that every nonhyperelliptic genus 4 curve is the intersection of a quadric and a cubic in $\mathbf{P}^3$ to compute the zeta functions of all genus 4 curves over $\mathbf{F}_2$. We found 620 distinct zeta functions, none of which occur among the previous list of 52. (This result has been independently confirmed by Xarles [Xar20], and the corresponding data added to the LMFDB.)

As an aside, note that [Zar15, Theorem 1.1] implies that for each of the 52 isogeny classes in the previous paragraph, the endomorphism algebra must contain an imaginary quadratic field, which we have confirmed.

By contrast, the conjecture fails in dimension 4 over $\mathbf{F}_3$ and $\mathbf{F}_5$, as shown by the curves given in Example 5.15. We also note in passing that Conjecture 5.14 is incompatible with Conjecture 5.20 below.

Example 5.15 Let C be the hyperelliptic curve over $\mathbf{F}_3$ given by

$$y^2 = x^9 + x^8 + x^7 + 2x^5 + x.$$

Then

$$L(C/\mathbf{F}_3, T) = 1 - T + 2T^2 - 4T^3 - 2T^4 - 12T^5 + 18T^6 - 27T^7 + 81T^8,$$

so the Jacobian A of C belongs to isogeny class 4.3.ab_c_ae_ac. We see that A is ordinary, geometrically simple, and has angle rank 3. It thus constitutes a counterexample to the Ahmadi–Shparlinski conjecture (Conjecture 5.14). Consistently with Zarhin's theorem, the endomorphism algebra contains the field $\mathbf{Q}(\sqrt{-7})$.

Similarly, let C be the hyperelliptic curve over $\mathbf{F}_5$ given by

$$y^2 = x^9 + x^6 + 2x^5 + x.$$

Then

$$L(C/\mathbf{F}_5, T) = 1 - T + 2T^2 - 4T^3 + 16T^4 - 20T^5 + 50T^6 - 125T^7 + 625T^8,$$

so the Jacobian A of C belongs to isogeny class 4.5.ab_c_ae_q. Again, A is ordinary, geometrically simple, and has angle rank 3. The endomorphism algebra contains the field $\mathbf{Q}(\sqrt{-15})$.

5.4 *Function Fields of Class Number One*

Example 5.16 ([Sti14] and 4.2.ad_c_a_b) The following is the LMFDB annotation for 4.2.ad_c_a_b. In [Sti14], Stirpe exhibited an example of a genus 4 curve $C/\mathbf{F}_2$ for which

$$L(C/\mathbf{F}_2, T) = 1 - 3T + 2T^2 + T^4 + 8T^6 - 24T^7 + 16T^8.$$

The Jacobian A of this curve belongs to isogeny class 4.2.ad_c_a_b.

This example is notable because C has the largest possible genus among curves over finite fields with trivial class group, and because it refuted a published result from almost 40 years earlier. In [LMQ75], it was shown (correctly) that there are seven such curves of genus at most 3 and at most one of genus 4; it was also claimed (incorrectly) that the genus 4 case could be ruled out. Correct proofs of the classification can be found in [MS15] and [SS15].

5.5 Hypersymmetric Abelian Varieties

Let A be an abelian variety over a field $k \supset \mathbf{F}_p$. Following Chai–Oort [CO06, Definition 2.1], we say that A is *hypersymmetric* if

$$\mathrm{End}(A_{\overline{k}})_{\mathbf{Z}_p} \cong \mathrm{End}(A_{\overline{k}}[p^\infty]).$$

These are meant to provide a positive-characteristic analogue of CM points in the moduli space of abelian varieties.

According to [CO06, Theorem 3.3], for a simple abelian variety, one can read off whether it is hypersymmetric explicitly from the Frobenius polynomial.

Here is an explicit example.

Example 5.17 We exhibit a simple hypersymmetric abelian threefold over $\mathbf{F}_8$ by verifying that 3.8.ag_bk_aea satisfies the conditions of [CO06, Conclusion 3.6]. The Newton polygon has slopes $1/3$ and $2/3$ each with multiplicity 3, so it is *balanced* in the sense of [CO06, Definition 3.4]. The prime 2 splits completely in $\mathbf{Q}(\pi) = \mathbf{Q}(\sqrt{-7})$, and the Brauer invariants of the endomorphism algebra at the places above 2 are again $1/3$ and $2/3$.

The same conditions are satisfied by the quadratic isogeny twist 3.8.g_bk_ea. We have checked that other than elliptic curves (which are all hypersymmetric), these are the only examples of hypersymmetric abelian varieties currently found in the LMFDB.

5.6 Isomorphic Endomorphism Algebras and Different *p*-ranks

Example 5.18 This example is a modified version of [Gon98, Example 4.2]. (That example starts over $\mathbf{F}_3$ rather than $\mathbf{F}_2$, but we do not currently have abelian threefolds over $\mathbf{F}_{27}$ in the LMFDB.)

Let A be an abelian threefold over $\mathbf{F}_2$ in the isogeny class 3.2.ad_c_b. Then A is simple, ordinary, and of p-rank 3, and its endomorphism algebra is $\mathbf{Q}(\zeta_7)$. Although A is not geometrically simple, its base change to $\mathbf{F}_8$, which belongs to the isogeny class 3.8.ag_bd_adf, is again simple.

Let B be an abelian threefold over $\mathbf{F}_8$ in the isogeny class 3.8.ag_i_i; the Weil number for B is twice that for A. Here B is geometrically simple of p-rank 0, and its endomorphism algebra is again $\mathbf{Q}(\zeta_7)$.

Example 5.19 There are also examples where both A and B are geometrically simple and have the same endomorphism algebra but different p-ranks. For example, abelian varieties in the isogeny class 3.2.a_c_c have p-rank 0 while abelian varieties in the isogeny class 3.2.d_f_h have p-rank 3; both have endomorphism algebra isomorphic to the number field 6.0.679024.1.

5.7 Abelian Fourfolds as Jacobians

For $g \geq 4$, a generic principally polarized abelian variety is not isomorphic to a Jacobian for dimension reasons. However, when working up to isogeny, it becomes much less clear what to expect. Using ideas from the theory of unlikely intersections, Shankar–Tsimerman [ST18] have made numerous observations about this question, including the following conjecture.

Conjecture 5.20 ([ST18, Conjecture 2.5]) Every 4-dimensional abelian variety over $\overline{\mathbf{F}}_p$ is isogenous to the Jacobian of some (possibly reducible) stable curve.

Notably, however, it is not predicted that a 4-dimensional abelian variety over $\mathbf{F}_q$ is isogenous to the Jacobian of a curve over $\mathbf{F}_q$. This can fail to occur, as in the following example.

Example 5.21 Consider the isogeny class 4.2.c_c_ac_af over $\mathbf{F}_2$. This class contains a principally polarized abelian fourfold A which is not isogenous to a Jacobian: if it were, the corresponding curve would have a negative number of $\mathbf{F}_8$-points (see Sect. 3.3). However, the base change of A to $\mathbf{F}_4 = \mathbf{F}_2[\alpha]$, which belongs to the isogeny class 4.4.a_c_i_j, is isogenous to the Jacobian of the hyperelliptic curve

$$y^2 + (\alpha x^4 + x^3 + x^2 + x + \alpha)y = x^9 + x^8.$$

We can also stress-test the conjecture by considering the following example.

Example 5.22 Consider the isogeny class 4.2.a_a_b_af over $\mathbf{F}_2$. This isogeny class is geometrically simple and ordinary. It cannot contain a Jacobian over any subfield of $\mathbf{F}_{16}$: if it did, the corresponding curve would have a negative number of $\mathbf{F}_{16}$-points. In addition, Howe's criterion (see Sect. 3.6) implies that this isogeny class cannot contain a principally polarized abelian fourfold over $\mathbf{F}_{2^g}$ for any odd integer g. Consequently, if Conjecture 5.20 holds, then this isogeny class contains a Jacobian over $\mathbf{F}_{2^g}$ for some even integer $g \geq 6$.

5.8 Distinguishing Isogeny Classes by Point Counts

The Weil polynomial of an abelian variety of dimension g has g unknown coefficients, so it is expected that these can be solved for using g point counts. It turns out that we can often do better than this. This exotic phenomenon is governed by information-theoretic heuristics (see Remark 5.25).

Example 5.23 For A a 5-dimensional abelian variety over $\mathbf{F}_2$, A is determined up to isogeny by the tuple $(\#A(\mathbf{F}_{2^i}) : i = 1, \ldots, 4)$. This is best possible: for example, an abelian variety A in any of the isogeny classes

5.2.ab_b_c_d_ac, 5.2.ab_c_a_a_i, 5.2.ab_c_b_a_d

satisfies $(\#A(\mathbf{F}_2), \#A(\mathbf{F}_4), \#A(\mathbf{F}_8)) = (42, 2520, 80262)$. By contrast, over $\mathbf{F}_3$, a 5-dimensional abelian variety is determined up to isogeny by the tuple $(\#A(\mathbf{F}_{3^i}) : i = 1, \dots, 3)$.

Example 5.24 Similarly, for A a 6-dimensional abelian variety over $\mathbf{F}_2$, A is determined up to isogeny by the tuple $(\#A(\mathbf{F}_{2^i}) : i = 1, \dots, 4)$. This is best possible even if we restrict to simple abelian varieties: for example, an abelian variety A in any of the isogeny classes

6.2.ab_ab_b_g_ab_aj, 6.2.ab_b_ab_ac_f_ad, 6.2.ab_c_ab_ad_f_ap

is geometrically simple and satisfies $(\#A(\mathbf{F}_2), \#A(\mathbf{F}_4), \#A(\mathbf{F}_8))=(42, 4032, 246078)$.

Remark 5.25 It is not known exactly how many initial point counts are needed to identify a g-dimensional abelian variety over $\mathbf{F}_q$ up to isogeny (for known g and q). The fact that the complete sequence of point counts determines the Weil polynomial is already nontrivial; it follows from a theorem of Fried [Fri88] (see also [Hil05]). Using the Weil bounds, it is shown in [Ked06b] that at most $\max\{18, 2g\}$ counts suffice; however, Noam Elkies has pointed out that on information-theoretic grounds, one should expect the number of counts needed to be about $g/2$, and indeed this is consistent with these examples. Namely, we need to distinguish among $O(q^{g(g+1)/4})$ Weil polynomials (Sect. 4.1), whereas the Lang-Weil bound (Sect. 2.7) implies that the number of possible values for the tuple $(\#A(\mathbf{F}_{q^i}))_{i=1}^n$ is $O(q^{n^2/2})$.

6 Possible Generalizations and Bottlenecks

We conclude with some discussion about possible future directions for this work.

6.1 *Bottlenecks*

We first identify some bottleneck steps that limited our original work, and which we would like to overcome.

As described in Sect. 3.8, we currently present Frobenius angle ranks which were computed nonrigorously using floating-point arithmetic, because it is not feasible to run the rigorous algorithm on all cases in the database. However, since we also compute Galois groups, we can use those to certify some cases as having maximal angle rank. This covers the vast majority of cases, which might make it feasible to run the rigorous algorithm on the rest, but our present methods rely on the computation of a splitting field, which is also very costly.

As has come up on several occasions already, we do not currently implement positive Jacobian testing for abelian varieties of dimension greater than or equal to 4. See Sect. 6.2 for further discussion.

One obstruction to adding complete tables of abelian varieties for other pairs (g, q) is the overall size of the dataset. Using Eq. (4.1), including $(7, 2)$ would add about 2.2 million isogeny classes, $(6, 3)$ about 10 million, $(5, 4)$ about 2.2 million, $(4, 7)$ about 700,000, and $(3, 27)$ about 450,000. For comparison, the database currently contains about 3 million isogeny classes and takes up about 10 GB. It would certainly be feasible to extend, but a line needs to be drawn somewhere. Moreover, the computational time required per isogeny grows quickly with g, so adding data with $g \geq 5$ takes more effort than suggested by the number of classes alone.

In lieu of enlarging the tables, one could also implement the computation of the data presented in LMFDB on a case-by-case basis for individual isogeny classes. One piece of data that would be difficult to compute in this way is the isogeny twists, which we currently do by finding hash collisions across the entire table (Sect. 3.4).

It would also be useful to have a mechanism to produce random elements of the set of isogeny classes for large g and q. It should be possible to effectively simulate the uniform distribution on isogeny classes by computing analogues of the isogeny Sato-Tate distribution in which one projects onto the first k polynomial coefficients (the isogeny Sato-Tate distribution being the case $k = 1$). An alternative approach may also be to take the existing Weil polynomial iterator discussed in Remark 3.1 and replace iterations over integers in an interval with uniform random samples.

6.2 Jacobians

Currently the LMFDB does not identify any isogeny class of abelian varieties of dimension at least 4 as containing a Jacobian. For dimension 4, it may be possible to exhaust over isomorphism classes of genus 4 curves using the fact that every such curve is either hyperelliptic or the transverse intersection of a quadric and a cubic surface in $\mathbf{P}^3$; see [Sav03] for a similar calculation. For curves of genus 5, the analogous assertion is that every such curve is either hyperelliptic, trigonal, or the transverse intersection of three quadrics in $\mathbf{P}^4$. For curves of genus 6, the analogous assertion is that every such curve is either hyperelliptic, trigonal, bielliptic, isomorphic to a smooth plane quintic, or birational to a plane sextic with four double points [ACGH85, Exercises V.A]. (Beware that the previous assertions are made over an algebraically closed field.)

6.3 Isomorphism Classes

Beyond the current data in the LMFDB, it would be extremely desirable to tabulate abelian varieties up to isomorphism, not just up to isogeny. The easiest cases for this are those of ordinary abelian varieties, for which an explicit description of isomorphism classes within an isogeny class has been given by Deligne [Del69]; abelian varieties over $\mathbf{F}_p$, for which a similar description has been given by Centeleghe–Stix [CS15]; and almost ordinary abelian varieties [OS20]. The recent work of Marseglia [Mar19] has made great strides towards making these methods practical at the scale of the LMFDB, and we plan on working with him on including data on isomorphism classes into the database.

In order to handle nonordinary abelian varieties over nonprime fields, it is probably necessary to go back to the proof of Honda's theorem, by constructing CM abelian varieties over number fields and then reducing them modulo primes. Note that whereas Honda's original approach to this in [Hon68] used complex uniformization and GAGA, a more recent construction of Chai–Oort [CO15] gives a more algebraic approach that might be easier to implement as an algorithm.

6.4 K3 Surfaces and Higher Weight

It would be natural to attempt a similar compilation of other types of algebraic varieties over finite fields and their zeta functions. A strong candidate class for this is K3 surfaces, for which a weak version of the Honda-Tate theorem is known [Tae16, Ito19]. More precisely, for a given candidate Weil polynomial in this setting, one can prove there exists a corresponding K3 surface in some base change of this isogeny class. The code for tabulating Weil polynomials described in Sect. 3.1 can produce lists of possible zeta functons for K3 surfaces over $\mathbf{F}_q$ for small q (this has been tested up to $q = 5$).

This suggests the question of trying to determine exactly which zeta functions occur for K3 surfaces over a given field. An indication of the difficulties involved can be seen in [KS16], where a complete tabulation of smooth quartic surfaces in $\mathbf{P}^3$ over $\mathbf{F}_2$ and their zeta functions was made (computing the latter by enumerating points); while this search did realize every eligible zeta function that could not occur for any other type of K3 surface, not every zeta function that could have appeared did so. Furthermore, certain zeta functions can only appear for K3 surfaces of very large degree, which would be very difficult to write down explicitly (the moduli space of K3 surfaces of a given degree becomes increasingly hyperbolic as the degree increases).

A closely related case is that of cubic fourfolds. Some examples of zeta function computations to resolve specific existence questions for cubic fourfolds can be found in [AA18] and [CHK19].

In another direction, one can also identify a class of surfaces with small invariants (genus, irregularity, K^2); identify all Weil polynomials over $\mathbf{F}_q$ (for some given small values of q) which could arise from a surface with the given invariants; then exhaust over the surfaces in question to see which polynomials arise and what "isogeny classes" they fall into. (In general it is not clear what geometric conditions correspond to an equality of Weil polynomials. One nontrivial example is that surfaces which are *derived equivalent*, meaning that they have isomorphic bounded derived categories of coherent sheaves, have the same zeta function [Hon15].) In some cases, candidate polynomials can be ruled out because they would predict impossible point counts on the underlying variety; for example, this happens for K3 surfaces over $\mathbf{F}_2$ as shown in [KS16].

A Tables and Figures

Table A.1 shows how the number of distinct Galois groups and Newton polygons varies with dimension for simple isogeny classes; see Sects. 2.4, 2.5, 4.2 and 4.3 for further discussion. Table A.2 gives counts of the number of simple isogeny classes for each $q \leq 5$ with every possible Galois group; see Sect. 4.2 for the definitions of $a(G)$ and $\widetilde{a}(G)$ (maxp refers to the maximum q for which we have data). Table A.3 gives counts of the number of simple isogeny classes for each $q \leq 5$ with every possible Newton polygon; see Sect. 4.3 for a discussion of how we expect these counts to grow as a function of q and the elevation of the Newton polygon, and note that the supersingular Newton polygon does not occur among simple isogeny classes with $g = 5$ and $q \leq 3$. Tables A.4–A.9 give the counts of the number of simple isogeny classes grouped by g, q, Newton polygon, Galois group and angle rank; see Sect. 4.4 for further discussion of the relationship between angle rank and the Newton polygon. Table A.10 describes the possible Brauer invariants that can occur for endomorphism algebras of simple isogeny classes; see Sect. 4.4 and Figs. A.2–A.7 for further discussion of the centers of these algebras, which are not shown in this table.

Figure A.1 graphs the possible Newton polygons in dimension 3; see Sects. 2.4 and 4.3 for further discussion. Figures A.2–A.7 show bar charts for the distribution of normalized discriminants for the Weil polynomials of simple isogeny classes and the number fields generated by them; see Sect. 4.5 for the normalizing factor and further discussion. Figures A.8–A.10 compare the predicted isogeny Sato-Tate distribution from Sect. 4.6 with the observed point counts of $\#A(\mathbf{F}_q)$ for $g = 2, 3, 4$ and selected values of q.

Table A.1 The number of distinct Galois groups and distinct Newton polygons, by dimension

Dimension	Num of groups	Num of Newton polygons
1	2	2
2	4	3
3	5	5
4	30	8
5	9	12
6	46	20

Table A.2 Counts of isogeny classes by Galois group; $a(G)$ and $\widetilde{a}_{\mathrm{maxp}}(G)$

Dimension	G	$a(G)$	$\widetilde{a}_{\mathrm{maxp}}(G)$	$q = 2$	$q = 3$	$q = 4$	$q = 5$
1	1T1		$-\infty$			2	
1	2T1	1	1.000	5	7	7	9
2	2T1	1	0.000	1	1		1
2	4T1	1/2	0.370		2	2	2
2	4T2	1/2	0.661	11	16	14	29
2	4T3	1	0.997	8	16	30	52
3	6T1	1/3	0.199	4	2	6	
3	6T3	1/2	0.557	14	32	84	88
3	6T6	1	0.588	14	34	52	94
3	6T11	1	0.999	48	280	676	1850
4	4T2	1/2	$-\infty$			6	
4	4T3	1	$-\infty$			4	
4	8T2	1/4	0.267	13	20	14	22
4	8T3	1/4	0.259	4	13	23	20
4	8T4	1/4	0.095	1		2	3
4	8T6	1/3	0.250		8		18
4	8T9	1/2	0.519	41	112	195	405
4	8T10	1/2	0.334	12	32	34	48
4	8T11	1/2	0.155		2	2	6
4	8T12	1/4	0.267	6	12	8	22
4	8T13	1/4	0.228	2	16	8	14
4	8T17	1/2	0.199	2		8	10
4	8T18	1/2	0.581	20	116	254	832
4	8T19	1/2	$-\infty$			4	
4	8T20	1/2	0.060		2	2	2
4	8T22	1/2	0.180				8
4	8T23	1/3	0.155		8	4	6
4	8T24	1/2	0.656	92	362	630	1984
4	8T26	1/2	0.180			4	8
4	8T27	1	0.429		24	66	144
4	8T28	1/2	0.259		2	8	20
4	8T29	1/2	0.448	8	42	44	178
4	8T30	1/2	$-\infty$			2	
4	8T31	1	0.408		12	34	112
4	8T32	1/2	0.120				4
4	8T35	1	0.706	34	356	1206	3546
4	8T38	1	0.467		20	100	224
4	8T39	1/2	0.745	62	558	1214	5554
4	8T40	1/2	0.199		4	4	10
4	8T44	1	0.989	368	4986	23272	93506
5	10T1	1/5	0.114		4		
5	10T5	1/4	0.334	18	58		

(continued)

Table A.2 (continued)

Dimension	G	$a(G)$	$\widetilde{a}_{\text{maxp}}(G)$	$q = 2$	$q = 3$	$q = 4$	$q = 5$
5	10T11	1/4	0.057		2		
5	10T14	1	0.268	8	26		
5	10T22	1/2	0.562	100	932		
5	10T23	1	0.379	14	100		
5	10T29	1	0.347	4	68		
5	10T36	1	0.356	2	76		
5	10T39	1	0.999	6178	189514		
6	12T2	1/6	0.253	18			
6	12T3	1/6	0.061	2			
6	12T6	1/4	0.096	3			
6	12T10	1/4	0.326	41			
6	12T18	1/4	0.202	10			
6	12T21	1/2	0.000	1			
6	12T23	1/4	0.267	21			
6	12T24	1/4	0.141	5			
6	12T25	1/2	0.351	55			
6	12T26	1/4	0.000	1			
6	12T28	1/3	0.061	2			
6	12T37	1/4	0.157	6			
6	12T41	1/4	0.061	2			
6	12T48	1/2	0.487	259			
6	12T76	1/4	0.243	16			
6	12T77	1/2	0.382	78			
6	12T78	1/3	0.425	128			
6	12T79	1/2	0.061	2			
6	12T90	1/2	0.309	34			
6	12T101	1/2	0.400	96			
6	12T103	1/2	0.141	5			
6	12T125	1/2	0.323	40			
6	12T134	1	0.202	10			
6	12T135	1	0.061	2			
6	12T136	1/2	0.061	2			
6	12T138	1/2	0.061	2			
6	12T139	1/2	0.560	595			
6	12T148	1/2	0.061	2			
6	12T186	1/2	0.122	4			
6	12T193	1	0.243	16			
6	12T208	1	0.304	32			
6	12T219	1/2	0.656	1784			
6	12T222	1	0.415	114			
6	12T224	1	0.061	2			

(continued)

Table A.2 (continued)

Dimension	G	$a(G)$	$\tilde{a}_{\mathrm{maxp}}(G)$	$q = 2$	$q = 3$	$q = 4$	$q = 5$
6	12T226	1/2	0.122	4			
6	12T227	1	0.182	8			
6	12T236	1/2	0.441	152			
6	12T237	1/2	0.182	8			
6	12T240	1	0.061	2			
6	12T250	1	0.482	244			
6	12T255	1	0.182	8			
6	12T260	1	0.625	1240			
6	12T285	1/2	0.833	13342			
6	12T286	1	0.157	6			
6	12T287	1/2	0.182	8			
6	12T293	1	0.980	71290			

Table A.3 Counts of isogeny classes by Newton polygon

Dimension	Slopes	$q = 2$	$q = 3$	$q = 4$	$q = 5$
1	$(0, 1)$	2	4	4	8
1	$\left(\frac{1}{2}, \frac{1}{2}\right)$	3	3	5	1
2	$(0, 0, 1, 1)$	13	30	34	66
2	$\left(0, \frac{1}{2}, \frac{1}{2}, 1\right)$	2	2	8	12
2	$\left(\frac{1}{2}, \frac{1}{2}, \frac{1}{2}, \frac{1}{2}\right)$	5	3	4	6
3	$(0, 0, 0, 1, 1, 1)$	56	266	576	1696
3	$\left(0, 0, \frac{1}{2}, \frac{1}{2}, 1, 1\right)$	12	54	192	284
3	$\left(0, \frac{1}{2}, \frac{1}{2}, \frac{1}{2}, \frac{1}{2}, 1\right)$	4	16	30	36
3	$\left(\frac{1}{3}, \frac{1}{3}, \frac{1}{3}, \frac{2}{3}, \frac{2}{3}, \frac{2}{3}\right)$	8	10	18	16
3	$\left(\frac{1}{2}, \frac{1}{2}, \frac{1}{2}, \frac{1}{2}, \frac{1}{2}, \frac{1}{2}\right)$		2	2	
4	$(0, 0, 0, 0, 1, 1, 1, 1)$	453	5062	18178	87115
4	$\left(0, 0, 0, \frac{1}{2}, \frac{1}{2}, 1, 1, 1\right)$	94	1048	6966	15698
4	$\left(0, 0, \frac{1}{2}, \frac{1}{2}, \frac{1}{2}, \frac{1}{2}, 1, 1\right)$	43	381	1398	3108
4	$\left(0, \frac{1}{3}, \frac{1}{3}, \frac{1}{3}, \frac{2}{3}, \frac{2}{3}, \frac{2}{3}, 1\right)$	26	114	244	526
4	$\left(0, \frac{1}{2}, \frac{1}{2}, \frac{1}{2}, \frac{1}{2}, \frac{1}{2}, \frac{1}{2}, 1\right)$	6	12	66	70
4	$\left(\frac{1}{4}, \frac{1}{4}, \frac{1}{4}, \frac{1}{4}, \frac{3}{4}, \frac{3}{4}, \frac{3}{4}, \frac{3}{4}\right)$	36	82	275	168
4	$\left(\frac{1}{3}, \frac{1}{3}, \frac{1}{3}, \frac{1}{2}, \frac{1}{2}, \frac{2}{3}, \frac{2}{3}, \frac{2}{3}\right)$	2	2	20	16
4	$\left(\frac{1}{2}, \frac{1}{2}, \frac{1}{2}, \frac{1}{2}, \frac{1}{2}, \frac{1}{2}, \frac{1}{2}, \frac{1}{2}\right)$	5	6	5	5
5	$(0, 0, 0, 0, 0, 1, 1, 1, 1, 1)$	4034	137776		
5	$\left(0, 0, 0, 0, \frac{1}{2}, \frac{1}{2}, 1, 1, 1, 1\right)$	1248	35796		

(continued)

Table A.3 (continued)

Dimension	Slopes	$q = 2$	$q = 3$	$q = 4$	$q = 5$
5	$\left(0, 0, 0, \frac{1}{2}, \frac{1}{2}, \frac{1}{2}, \frac{1}{2}, 1, 1, 1\right)$	532	11558		
5	$\left(0, 0, \frac{1}{3}, \frac{1}{3}, \frac{1}{3}, \frac{2}{3}, \frac{2}{3}, \frac{2}{3}, 1, 1\right)$	170	3024		
5	$\left(0, 0, \frac{1}{2}, \frac{1}{2}, \frac{1}{2}, \frac{1}{2}, \frac{1}{2}, \frac{1}{2}, 1, 1\right)$	62	696		
5	$\left(0, \frac{1}{4}, \frac{1}{4}, \frac{1}{4}, \frac{1}{4}, \frac{3}{4}, \frac{3}{4}, \frac{3}{4}, \frac{3}{4}, 1\right)$	86	968		
5	$\left(0, \frac{1}{3}, \frac{1}{3}, \frac{1}{3}, \frac{1}{2}, \frac{1}{2}, \frac{2}{3}, \frac{2}{3}, \frac{2}{3}, 1\right)$	12	166		
5	$\left(0, \frac{1}{2}, \frac{1}{2}, \frac{1}{2}, \frac{1}{2}, \frac{1}{2}, \frac{1}{2}, \frac{1}{2}, \frac{1}{2}, 1\right)$	8	56		
5	$\left(\frac{1}{5}, \frac{1}{5}, \frac{1}{5}, \frac{1}{5}, \frac{1}{5}, \frac{4}{5}, \frac{4}{5}, \frac{4}{5}, \frac{4}{5}, \frac{4}{5}\right)$	140	614		
5	$\left(\frac{1}{4}, \frac{1}{4}, \frac{1}{4}, \frac{1}{4}, \frac{1}{2}, \frac{1}{2}, \frac{3}{4}, \frac{3}{4}, \frac{3}{4}, \frac{3}{4}\right)$	22	94		
5	$\left(\frac{1}{3}, \frac{1}{3}, \frac{1}{3}, \frac{1}{2}, \frac{1}{2}, \frac{1}{2}, \frac{1}{2}, \frac{2}{3}, \frac{2}{3}, \frac{2}{3}\right)$	2	20		
5	$\left(\frac{2}{5}, \frac{2}{5}, \frac{2}{5}, \frac{2}{5}, \frac{2}{5}, \frac{3}{5}, \frac{3}{5}, \frac{3}{5}, \frac{3}{5}, \frac{3}{5}\right)$	8	12		
6	$(0, 0, 0, 0, 0, 0, 1, 1, 1, 1, 1, 1)$	54730			
6	$\left(0, 0, 0, 0, 0, \frac{1}{2}, \frac{1}{2}, 1, 1, 1, 1, 1\right)$	17466			
6	$\left(0, 0, 0, 0, \frac{1}{2}, \frac{1}{2}, \frac{1}{2}, \frac{1}{2}, 1, 1, 1, 1\right)$	8570			
6	$\left(0, 0, 0, \frac{1}{3}, \frac{1}{3}, \frac{1}{3}, \frac{2}{3}, \frac{2}{3}, \frac{2}{3}, 1, 1, 1\right)$	3042			
6	$\left(0, 0, 0, \frac{1}{2}, \frac{1}{2}, \frac{1}{2}, \frac{1}{2}, \frac{1}{2}, \frac{1}{2}, 1, 1, 1\right)$	1382			
6	$\left(0, 0, \frac{1}{4}, \frac{1}{4}, \frac{1}{4}, \frac{1}{4}, \frac{3}{4}, \frac{3}{4}, \frac{3}{4}, \frac{3}{4}, 1, 1\right)$	1502			
6	$\left(0, 0, \frac{1}{3}, \frac{1}{3}, \frac{1}{3}, \frac{1}{2}, \frac{1}{2}, \frac{2}{3}, \frac{2}{3}, \frac{2}{3}, 1, 1\right)$	262			
6	$\left(0, 0, \frac{1}{2}, \frac{1}{2}, \frac{1}{2}, \frac{1}{2}, \frac{1}{2}, \frac{1}{2}, \frac{1}{2}, \frac{1}{2}, 1, 1\right)$	330			
6	$\left(0, \frac{1}{5}, \frac{1}{5}, \frac{1}{5}, \frac{1}{5}, \frac{1}{5}, \frac{4}{5}, \frac{4}{5}, \frac{4}{5}, \frac{4}{5}, \frac{4}{5}, 1\right)$	774			
6	$\left(0, \frac{1}{4}, \frac{1}{4}, \frac{1}{4}, \frac{1}{4}, \frac{1}{2}, \frac{1}{2}, \frac{3}{4}, \frac{3}{4}, \frac{3}{4}, \frac{3}{4}, 1\right)$	130			
6	$\left(0, \frac{1}{3}, \frac{1}{3}, \frac{1}{3}, \frac{1}{2}, \frac{1}{2}, \frac{1}{2}, \frac{1}{2}, \frac{2}{3}, \frac{2}{3}, \frac{2}{3}, 1\right)$	40			
6	$\left(0, \frac{2}{5}, \frac{2}{5}, \frac{2}{5}, \frac{2}{5}, \frac{2}{5}, \frac{3}{5}, \frac{3}{5}, \frac{3}{5}, \frac{3}{5}, \frac{3}{5}, 1\right)$	88			
6	$\left(0, \frac{1}{2}, \frac{1}{2}, \frac{1}{2}, \frac{1}{2}, \frac{1}{2}, \frac{1}{2}, \frac{1}{2}, \frac{1}{2}, \frac{1}{2}, \frac{1}{2}, 1\right)$	10			
6	$\left(\frac{1}{6}, \frac{1}{6}, \frac{1}{6}, \frac{1}{6}, \frac{1}{6}, \frac{1}{6}, \frac{5}{6}, \frac{5}{6}, \frac{5}{6}, \frac{5}{6}, \frac{5}{6}, \frac{5}{6}\right)$	998			
6	$\left(\frac{1}{5}, \frac{1}{5}, \frac{1}{5}, \frac{1}{5}, \frac{1}{5}, \frac{1}{2}, \frac{1}{2}, \frac{4}{5}, \frac{4}{5}, \frac{4}{5}, \frac{4}{5}, \frac{4}{5}\right)$	202			
6	$\left(\frac{1}{4}, \frac{1}{4}, \frac{1}{4}, \frac{1}{4}, \frac{1}{2}, \frac{1}{2}, \frac{1}{2}, \frac{1}{2}, \frac{3}{4}, \frac{3}{4}, \frac{3}{4}, \frac{3}{4}\right)$	66			
6	$\left(\frac{1}{3}, \frac{1}{3}, \frac{1}{3}, \frac{1}{3}, \frac{1}{3}, \frac{1}{3}, \frac{2}{3}, \frac{2}{3}, \frac{2}{3}, \frac{2}{3}, \frac{2}{3}, \frac{2}{3}\right)$	90			
6	$\left(\frac{1}{3}, \frac{1}{3}, \frac{1}{3}, \frac{1}{2}, \frac{1}{2}, \frac{1}{2}, \frac{1}{2}, \frac{1}{2}, \frac{1}{2}, \frac{2}{3}, \frac{2}{3}, \frac{2}{3}\right)$	8			
6	$\left(\frac{2}{5}, \frac{2}{5}, \frac{2}{5}, \frac{2}{5}, \frac{2}{5}, \frac{1}{2}, \frac{1}{2}, \frac{3}{5}, \frac{3}{5}, \frac{3}{5}, \frac{3}{5}, \frac{3}{5}\right)$	4			
6	$\left(\frac{1}{2}, \frac{1}{2}, \frac{1}{2}, \frac{1}{2}, \frac{1}{2}, \frac{1}{2}, \frac{1}{2}, \frac{1}{2}, \frac{1}{2}, \frac{1}{2}, \frac{1}{2}, \frac{1}{2}\right)$	8			

Table A.4 Counts of isogeny classes by Newton polygon, Galois group, and angle rank ($g = 1$)

Slopes	Group	Angle rank	$q = 2$	$q = 3$	$q = 4$	$q = 5$
(0, 1)	2T1	1	2	4	4	8
$\left(\frac{1}{2}, \frac{1}{2}\right)$	1T1	0			2	
$\left(\frac{1}{2}, \frac{1}{2}\right)$	2T1	0	3	3	3	1

Table A.5 Counts of isogeny classes by Newton polygon, Galois group, and angle rank ($g = 2$)

Slopes	Group	Angle Rank	$q = 2$	$q = 3$	$q = 4$	$q = 5$
(0, 0, 1, 1)	4T1	2		2		
(0, 0, 1, 1)	4T2	1	7	14	12	26
(0, 0, 1, 1)	4T3	2	6	14	22	40
$\left(0, \frac{1}{2}, \frac{1}{2}, 1\right)$	4T3	2	2	2	8	12
$\left(\frac{1}{2}, \frac{1}{2}, \frac{1}{2}, \frac{1}{2}\right)$	2T1	0	1	1		1
$\left(\frac{1}{2}, \frac{1}{2}, \frac{1}{2}, \frac{1}{2}\right)$	4T1	0			2	2
$\left(\frac{1}{2}, \frac{1}{2}, \frac{1}{2}, \frac{1}{2}\right)$	4T2	0	4	2	2	3

Table A.6 Counts of isogeny classes by Newton polygon, Galois group, and angle rank ($g = 3$)

Slopes	Group	Angle Rank	$q = 2$	$q = 3$	$q = 4$	$q = 5$
(0, 0, 0, 1, 1, 1)	6T1	1	4		4	
(0, 0, 0, 1, 1, 1)	6T3	1	4	10	12	28
(0, 0, 0, 1, 1, 1)	6T3	3	2	6	10	42
(0, 0, 0, 1, 1, 1)	6T6	3	14	32	48	82
(0, 0, 0, 1, 1, 1)	6T11	3	32	218	502	1544
$\left(0, 0, \frac{1}{2}, \frac{1}{2}, 1, 1\right)$	6T3	2	6	12	56	6
$\left(0, 0, \frac{1}{2}, \frac{1}{2}, 1, 1\right)$	6T6	3			2	10
$\left(0, 0, \frac{1}{2}, \frac{1}{2}, 1, 1\right)$	6T11	3	6	42	134	268
$\left(0, \frac{1}{2}, \frac{1}{2}, \frac{1}{2}, \frac{1}{2}, 1\right)$	6T3	3			2	4
$\left(0, \frac{1}{2}, \frac{1}{2}, \frac{1}{2}, \frac{1}{2}, 1\right)$	6T6	3			2	2
$\left(0, \frac{1}{2}, \frac{1}{2}, \frac{1}{2}, \frac{1}{2}, 1\right)$	6T11	3	4	16	26	30
$\left(\frac{1}{3}, \frac{1}{3}, \frac{1}{3}, \frac{2}{3}, \frac{2}{3}, \frac{2}{3}\right)$	6T3	1	2	4	4	8
$\left(\frac{1}{3}, \frac{1}{3}, \frac{1}{3}, \frac{2}{3}, \frac{2}{3}, \frac{2}{3}\right)$	6T6	3		2		
$\left(\frac{1}{3}, \frac{1}{3}, \frac{1}{3}, \frac{2}{3}, \frac{2}{3}, \frac{2}{3}\right)$	6T11	3	6	4	14	8
$\left(\frac{1}{2}, \frac{1}{2}, \frac{1}{2}, \frac{1}{2}, \frac{1}{2}, \frac{1}{2}\right)$	6T1	0		2	2	

Table A.7 Counts of isogeny classes by Newton polygon, Galois group, and angle rank ($g = 4$)

Slopes	Group	Angle Rank	$q = 2$	$q = 3$	$q = 4$	$q = 5$
(0, 0, 0, 0, 1, 1, 1, 1)	8T2	1	8	12	10	10
(0, 0, 0, 0, 1, 1, 1, 1)	8T2	2		3		
(0, 0, 0, 0, 1, 1, 1, 1)	8T3	1	4	12	9	19
(0, 0, 0, 0, 1, 1, 1, 1)	8T4	2			1	3
(0, 0, 0, 0, 1, 1, 1, 1)	8T6	4		8		8
(0, 0, 0, 0, 1, 1, 1, 1)	8T9	1	6	20	28	80
(0, 0, 0, 0, 1, 1, 1, 1)	8T9	2	19	59	100	221
(0, 0, 0, 0, 1, 1, 1, 1)	8T9	4				8
(0, 0, 0, 0, 1, 1, 1, 1)	8T10	2	10	26	28	32
(0, 0, 0, 0, 1, 1, 1, 1)	8T10	4		4	2	
(0, 0, 0, 0, 1, 1, 1, 1)	8T11	4		2	2	4
(0, 0, 0, 0, 1, 1, 1, 1)	8T12	4	6	12	8	22
(0, 0, 0, 0, 1, 1, 1, 1)	8T13	3	2	8	4	12
(0, 0, 0, 0, 1, 1, 1, 1)	8T13	4		6		2
(0, 0, 0, 0, 1, 1, 1, 1)	8T17	4	2		6	10
(0, 0, 0, 0, 1, 1, 1, 1)	8T18	2	12	84	206	688
(0, 0, 0, 0, 1, 1, 1, 1)	8T18	4	2	2	6	12
(0, 0, 0, 0, 1, 1, 1, 1)	8T19	4			4	
(0, 0, 0, 0, 1, 1, 1, 1)	8T20	4		2	2	2
(0, 0, 0, 0, 1, 1, 1, 1)	8T22	4				4
(0, 0, 0, 0, 1, 1, 1, 1)	8T23	4		6	2	4
(0, 0, 0, 0, 1, 1, 1, 1)	8T24	3	50	202	406	1292
(0, 0, 0, 0, 1, 1, 1, 1)	8T24	4	12	54	62	356
(0, 0, 0, 0, 1, 1, 1, 1)	8T26	4			4	4
(0, 0, 0, 0, 1, 1, 1, 1)	8T27	4		24	60	136
(0, 0, 0, 0, 1, 1, 1, 1)	8T28	4				4
(0, 0, 0, 0, 1, 1, 1, 1)	8T29	4	8	34	40	140
(0, 0, 0, 0, 1, 1, 1, 1)	8T30	4			2	
(0, 0, 0, 0, 1, 1, 1, 1)	8T31	4		8	28	110
(0, 0, 0, 0, 1, 1, 1, 1)	8T32	4				2
(0, 0, 0, 0, 1, 1, 1, 1)	8T35	4	26	298	926	3122
(0, 0, 0, 0, 1, 1, 1, 1)	8T38	4		10	48	172
(0, 0, 0, 0, 1, 1, 1, 1)	8T39	4	46	488	1068	4964
(0, 0, 0, 0, 1, 1, 1, 1)	8T40	4		4	4	10
(0, 0, 0, 0, 1, 1, 1, 1)	8T44	4	240	3674	15112	75662
$\left(0, 0, 0, \frac{1}{2}, \frac{1}{2}, 1, 1, 1\right)$	8T6	4				4
$\left(0, 0, 0, \frac{1}{2}, \frac{1}{2}, 1, 1, 1\right)$	8T23	4			2	2
$\left(0, 0, 0, \frac{1}{2}, \frac{1}{2}, 1, 1, 1\right)$	8T26	4				4
$\left(0, 0, 0, \frac{1}{2}, \frac{1}{2}, 1, 1, 1\right)$	8T31	4				2

(continued)

Table A.7 (continued)

Slopes	Group	Angle Rank	$q = 2$	$q = 3$	$q = 4$	$q = 5$
$\left(0, 0, 0, \frac{1}{2}, \frac{1}{2}, 1, 1, 1\right)$	8T35	4	2	32	108	260
$\left(0, 0, 0, \frac{1}{2}, \frac{1}{2}, 1, 1, 1\right)$	8T38	4		10	42	48
$\left(0, 0, 0, \frac{1}{2}, \frac{1}{2}, 1, 1, 1\right)$	8T44	4	92	1006	6814	15378
$\left(0, 0, \frac{1}{2}, \frac{1}{2}, \frac{1}{2}, \frac{1}{2}, 1, 1\right)$	8T3	2			4	
$\left(0, 0, \frac{1}{2}, \frac{1}{2}, \frac{1}{2}, \frac{1}{2}, 1, 1\right)$	8T4	2	1		1	
$\left(0, 0, \frac{1}{2}, \frac{1}{2}, \frac{1}{2}, \frac{1}{2}, 1, 1\right)$	8T6	4				4
$\left(0, 0, \frac{1}{2}, \frac{1}{2}, \frac{1}{2}, \frac{1}{2}, 1, 1\right)$	8T9	2	6	17	29	62
$\left(0, 0, \frac{1}{2}, \frac{1}{2}, \frac{1}{2}, \frac{1}{2}, 1, 1\right)$	8T9	3		4	14	22
$\left(0, 0, \frac{1}{2}, \frac{1}{2}, \frac{1}{2}, \frac{1}{2}, 1, 1\right)$	8T10	2		2	2	
$\left(0, 0, \frac{1}{2}, \frac{1}{2}, \frac{1}{2}, \frac{1}{2}, 1, 1\right)$	8T11	4				2
$\left(0, 0, \frac{1}{2}, \frac{1}{2}, \frac{1}{2}, \frac{1}{2}, 1, 1\right)$	8T13	3			2	
$\left(0, 0, \frac{1}{2}, \frac{1}{2}, \frac{1}{2}, \frac{1}{2}, 1, 1\right)$	8T17	4			2	
$\left(0, 0, \frac{1}{2}, \frac{1}{2}, \frac{1}{2}, \frac{1}{2}, 1, 1\right)$	8T18	2	4	22	30	108
$\left(0, 0, \frac{1}{2}, \frac{1}{2}, \frac{1}{2}, \frac{1}{2}, 1, 1\right)$	8T18	4				4
$\left(0, 0, \frac{1}{2}, \frac{1}{2}, \frac{1}{2}, \frac{1}{2}, 1, 1\right)$	8T22	4				4
$\left(0, 0, \frac{1}{2}, \frac{1}{2}, \frac{1}{2}, \frac{1}{2}, 1, 1\right)$	8T24	3	14	64	96	252
$\left(0, 0, \frac{1}{2}, \frac{1}{2}, \frac{1}{2}, \frac{1}{2}, 1, 1\right)$	8T24	4		2		30
$\left(0, 0, \frac{1}{2}, \frac{1}{2}, \frac{1}{2}, \frac{1}{2}, 1, 1\right)$	8T27	4			4	
$\left(0, 0, \frac{1}{2}, \frac{1}{2}, \frac{1}{2}, \frac{1}{2}, 1, 1\right)$	8T28	4			6	16
$\left(0, 0, \frac{1}{2}, \frac{1}{2}, \frac{1}{2}, \frac{1}{2}, 1, 1\right)$	8T29	4		6	4	38
$\left(0, 0, \frac{1}{2}, \frac{1}{2}, \frac{1}{2}, \frac{1}{2}, 1, 1\right)$	8T31	4		4	6	
$\left(0, 0, \frac{1}{2}, \frac{1}{2}, \frac{1}{2}, \frac{1}{2}, 1, 1\right)$	8T32	4				2
$\left(0, 0, \frac{1}{2}, \frac{1}{2}, \frac{1}{2}, \frac{1}{2}, 1, 1\right)$	8T35	4	4	22	144	150
$\left(0, 0, \frac{1}{2}, \frac{1}{2}, \frac{1}{2}, \frac{1}{2}, 1, 1\right)$	8T38	4			10	4
$\left(0, 0, \frac{1}{2}, \frac{1}{2}, \frac{1}{2}, \frac{1}{2}, 1, 1\right)$	8T39	4	2	38	68	432
$\left(0, 0, \frac{1}{2}, \frac{1}{2}, \frac{1}{2}, \frac{1}{2}, 1, 1\right)$	8T44	4	12	200	976	1978
$\left(0, \frac{1}{3}, \frac{1}{3}, \frac{1}{3}, \frac{2}{3}, \frac{2}{3}, \frac{2}{3}, 1\right)$	8T13	3		2		
$\left(0, \frac{1}{3}, \frac{1}{3}, \frac{1}{3}, \frac{2}{3}, \frac{2}{3}, \frac{2}{3}, 1\right)$	8T23	4		2		
$\left(0, \frac{1}{3}, \frac{1}{3}, \frac{1}{3}, \frac{2}{3}, \frac{2}{3}, \frac{2}{3}, 1\right)$	8T24	3	10	18	36	48
$\left(0, \frac{1}{3}, \frac{1}{3}, \frac{1}{3}, \frac{2}{3}, \frac{2}{3}, \frac{2}{3}, 1\right)$	8T24	4	2	2	6	2
$\left(0, \frac{1}{3}, \frac{1}{3}, \frac{1}{3}, \frac{2}{3}, \frac{2}{3}, \frac{2}{3}, 1\right)$	8T39	4	4	26	40	130
$\left(0, \frac{1}{3}, \frac{1}{3}, \frac{1}{3}, \frac{2}{3}, \frac{2}{3}, \frac{2}{3}, 1\right)$	8T44	4	10	64	162	346

(continued)

Table A.7 (continued)

Slopes	Group	Angle Rank	$q=2$	$q=3$	$q=4$	$q=5$
$\left(0, \frac{1}{2}, \frac{1}{2}, \frac{1}{2}, \frac{1}{2}, \frac{1}{2}, \frac{1}{2}, 1\right)$	8T6	4				2
$\left(0, \frac{1}{2}, \frac{1}{2}, \frac{1}{2}, \frac{1}{2}, \frac{1}{2}, \frac{1}{2}, 1\right)$	8T28	4		2	2	
$\left(0, \frac{1}{2}, \frac{1}{2}, \frac{1}{2}, \frac{1}{2}, \frac{1}{2}, \frac{1}{2}, 1\right)$	8T35	4				2
$\left(0, \frac{1}{2}, \frac{1}{2}, \frac{1}{2}, \frac{1}{2}, \frac{1}{2}, \frac{1}{2}, 1\right)$	8T39	4	4	2	12	12
$\left(0, \frac{1}{2}, \frac{1}{2}, \frac{1}{2}, \frac{1}{2}, \frac{1}{2}, \frac{1}{2}, 1\right)$	8T44	4	2	8	52	54
$\left(\frac{1}{4}, \frac{1}{4}, \frac{1}{4}, \frac{1}{4}, \frac{3}{4}, \frac{3}{4}, \frac{3}{4}, \frac{3}{4}\right)$	4T2	1			6	
$\left(\frac{1}{4}, \frac{1}{4}, \frac{1}{4}, \frac{1}{4}, \frac{3}{4}, \frac{3}{4}, \frac{3}{4}, \frac{3}{4}\right)$	4T3	2			4	
$\left(\frac{1}{4}, \frac{1}{4}, \frac{1}{4}, \frac{1}{4}, \frac{3}{4}, \frac{3}{4}, \frac{3}{4}, \frac{3}{4}\right)$	8T2	1				8
$\left(\frac{1}{4}, \frac{1}{4}, \frac{1}{4}, \frac{1}{4}, \frac{3}{4}, \frac{3}{4}, \frac{3}{4}, \frac{3}{4}\right)$	8T3	1			9	
$\left(\frac{1}{4}, \frac{1}{4}, \frac{1}{4}, \frac{1}{4}, \frac{3}{4}, \frac{3}{4}, \frac{3}{4}, \frac{3}{4}\right)$	8T9	1	6	12	6	12
$\left(\frac{1}{4}, \frac{1}{4}, \frac{1}{4}, \frac{1}{4}, \frac{3}{4}, \frac{3}{4}, \frac{3}{4}, \frac{3}{4}\right)$	8T9	2	4		18	
$\left(\frac{1}{4}, \frac{1}{4}, \frac{1}{4}, \frac{1}{4}, \frac{3}{4}, \frac{3}{4}, \frac{3}{4}, \frac{3}{4}\right)$	8T10	2	2		2	16
$\left(\frac{1}{4}, \frac{1}{4}, \frac{1}{4}, \frac{1}{4}, \frac{3}{4}, \frac{3}{4}, \frac{3}{4}, \frac{3}{4}\right)$	8T13	3			2	
$\left(\frac{1}{4}, \frac{1}{4}, \frac{1}{4}, \frac{1}{4}, \frac{3}{4}, \frac{3}{4}, \frac{3}{4}, \frac{3}{4}\right)$	8T18	2	2	8	12	20
$\left(\frac{1}{4}, \frac{1}{4}, \frac{1}{4}, \frac{1}{4}, \frac{3}{4}, \frac{3}{4}, \frac{3}{4}, \frac{3}{4}\right)$	8T24	3	4	20	24	4
$\left(\frac{1}{4}, \frac{1}{4}, \frac{1}{4}, \frac{1}{4}, \frac{3}{4}, \frac{3}{4}, \frac{3}{4}, \frac{3}{4}\right)$	8T27	4			2	8
$\left(\frac{1}{4}, \frac{1}{4}, \frac{1}{4}, \frac{1}{4}, \frac{3}{4}, \frac{3}{4}, \frac{3}{4}, \frac{3}{4}\right)$	8T29	4		2		
$\left(\frac{1}{4}, \frac{1}{4}, \frac{1}{4}, \frac{1}{4}, \frac{3}{4}, \frac{3}{4}, \frac{3}{4}, \frac{3}{4}\right)$	8T35	4	2	4	28	12
$\left(\frac{1}{4}, \frac{1}{4}, \frac{1}{4}, \frac{1}{4}, \frac{3}{4}, \frac{3}{4}, \frac{3}{4}, \frac{3}{4}\right)$	8T39	4	6	4	26	16
$\left(\frac{1}{4}, \frac{1}{4}, \frac{1}{4}, \frac{1}{4}, \frac{3}{4}, \frac{3}{4}, \frac{3}{4}, \frac{3}{4}\right)$	8T44	4	10	32	136	72
$\left(\frac{1}{3}, \frac{1}{3}, \frac{1}{3}, \frac{1}{2}, \frac{1}{2}, \frac{2}{3}, \frac{2}{3}, \frac{2}{3}\right)$	8T44	4	2	2	20	16
$\left(\frac{1}{2}, \frac{1}{2}, \frac{1}{2}, \frac{1}{2}, \frac{1}{2}, \frac{1}{2}, \frac{1}{2}, \frac{1}{2}\right)$	8T2	0	5	5	4	4
$\left(\frac{1}{2}, \frac{1}{2}, \frac{1}{2}, \frac{1}{2}, \frac{1}{2}, \frac{1}{2}, \frac{1}{2}, \frac{1}{2}\right)$	8T3	0		1	1	1

Table A.8 Counts of isogeny classes by Newton polygon, Galois group, and angle rank ($g = 5$)

Slopes	Group	Angle Rank	$q = 2$	$q = 3$
$(0, 0, 0, 0, 0, 1, 1, 1, 1, 1)$	`10T1`	1		4
$(0, 0, 0, 0, 0, 1, 1, 1, 1, 1)$	`10T5`	1	10	38
$(0, 0, 0, 0, 0, 1, 1, 1, 1, 1)$	`10T5`	5		2
$(0, 0, 0, 0, 0, 1, 1, 1, 1, 1)$	`10T14`	5	8	26
$(0, 0, 0, 0, 0, 1, 1, 1, 1, 1)$	`10T22`	5	34	292
$(0, 0, 0, 0, 0, 1, 1, 1, 1, 1)$	`10T23`	5	12	80
$(0, 0, 0, 0, 0, 1, 1, 1, 1, 1)$	`10T29`	5	2	50
$(0, 0, 0, 0, 0, 1, 1, 1, 1, 1)$	`10T36`	5	2	52
$(0, 0, 0, 0, 0, 1, 1, 1, 1, 1)$	`10T39`	5	3966	137232
$\left(0, 0, 0, 0, \frac{1}{2}, \frac{1}{2}, 1, 1, 1, 1\right)$	`10T22`	4	46	480
$\left(0, 0, 0, 0, \frac{1}{2}, \frac{1}{2}, 1, 1, 1, 1\right)$	`10T23`	5		8
$\left(0, 0, 0, 0, \frac{1}{2}, \frac{1}{2}, 1, 1, 1, 1\right)$	`10T29`	5	2	18
$\left(0, 0, 0, 0, \frac{1}{2}, \frac{1}{2}, 1, 1, 1, 1\right)$	`10T36`	5		12
$\left(0, 0, 0, 0, \frac{1}{2}, \frac{1}{2}, 1, 1, 1, 1\right)$	`10T39`	5	1200	35278
$\left(0, 0, 0, \frac{1}{2}, \frac{1}{2}, \frac{1}{2}, \frac{1}{2}, 1, 1, 1\right)$	`10T22`	5	4	52
$\left(0, 0, 0, \frac{1}{2}, \frac{1}{2}, \frac{1}{2}, \frac{1}{2}, 1, 1, 1\right)$	`10T23`	5	2	12
$\left(0, 0, 0, \frac{1}{2}, \frac{1}{2}, \frac{1}{2}, \frac{1}{2}, 1, 1, 1\right)$	`10T36`	5		2
$\left(0, 0, 0, \frac{1}{2}, \frac{1}{2}, \frac{1}{2}, \frac{1}{2}, 1, 1, 1\right)$	`10T39`	5	526	11492
$\left(0, 0, \frac{1}{3}, \frac{1}{3}, \frac{1}{3}, \frac{2}{3}, \frac{2}{3}, \frac{2}{3}, 1, 1\right)$	`10T11`	5		2
$\left(0, 0, \frac{1}{3}, \frac{1}{3}, \frac{1}{3}, \frac{2}{3}, \frac{2}{3}, \frac{2}{3}, 1, 1\right)$	`10T22`	5	4	32
$\left(0, 0, \frac{1}{3}, \frac{1}{3}, \frac{1}{3}, \frac{2}{3}, \frac{2}{3}, \frac{2}{3}, 1, 1\right)$	`10T36`	5		10
$\left(0, 0, \frac{1}{3}, \frac{1}{3}, \frac{1}{3}, \frac{2}{3}, \frac{2}{3}, \frac{2}{3}, 1, 1\right)$	`10T39`	5	166	2980
$\left(0, 0, \frac{1}{2}, \frac{1}{2}, \frac{1}{2}, \frac{1}{2}, \frac{1}{2}, \frac{1}{2}, 1, 1\right)$	`10T22`	4	2	48
$\left(0, 0, \frac{1}{2}, \frac{1}{2}, \frac{1}{2}, \frac{1}{2}, \frac{1}{2}, \frac{1}{2}, 1, 1\right)$	`10T39`	5	60	648
$\left(0, \frac{1}{4}, \frac{1}{4}, \frac{1}{4}, \frac{1}{4}, \frac{3}{4}, \frac{3}{4}, \frac{3}{4}, \frac{3}{4}, 1\right)$	`10T22`	5	4	4
$\left(0, \frac{1}{4}, \frac{1}{4}, \frac{1}{4}, \frac{1}{4}, \frac{3}{4}, \frac{3}{4}, \frac{3}{4}, \frac{3}{4}, 1\right)$	`10T39`	5	82	964
$\left(0, \frac{1}{3}, \frac{1}{3}, \frac{1}{3}, \frac{1}{2}, \frac{1}{2}, \frac{2}{3}, \frac{2}{3}, \frac{2}{3}, 1\right)$	`10T39`	5	12	166
$\left(0, \frac{1}{2}, \frac{1}{2}, \frac{1}{2}, \frac{1}{2}, \frac{1}{2}, \frac{1}{2}, \frac{1}{2}, \frac{1}{2}, 1\right)$	`10T39`	5	8	56
$\left(\frac{1}{5}, \frac{1}{5}, \frac{1}{5}, \frac{1}{5}, \frac{1}{5}, \frac{4}{5}, \frac{4}{5}, \frac{4}{5}, \frac{4}{5}, \frac{4}{5}\right)$	`10T5`	1	6	14
$\left(\frac{1}{5}, \frac{1}{5}, \frac{1}{5}, \frac{1}{5}, \frac{1}{5}, \frac{4}{5}, \frac{4}{5}, \frac{4}{5}, \frac{4}{5}, \frac{4}{5}\right)$	`10T39`	5	134	600
$\left(\frac{1}{4}, \frac{1}{4}, \frac{1}{4}, \frac{1}{4}, \frac{1}{2}, \frac{1}{2}, \frac{3}{4}, \frac{3}{4}, \frac{3}{4}, \frac{3}{4}\right)$	`10T22`	4	6	24
$\left(\frac{1}{4}, \frac{1}{4}, \frac{1}{4}, \frac{1}{4}, \frac{1}{2}, \frac{1}{2}, \frac{3}{4}, \frac{3}{4}, \frac{3}{4}, \frac{3}{4}\right)$	`10T39`	5	16	70
$\left(\frac{1}{3}, \frac{1}{3}, \frac{1}{3}, \frac{1}{2}, \frac{1}{2}, \frac{1}{2}, \frac{1}{2}, \frac{2}{3}, \frac{2}{3}, \frac{2}{3}\right)$	`10T39`	5	2	20
$\left(\frac{2}{5}, \frac{2}{5}, \frac{2}{5}, \frac{2}{5}, \frac{2}{5}, \frac{3}{5}, \frac{3}{5}, \frac{3}{5}, \frac{3}{5}, \frac{3}{5}\right)$	`10T5`	1	2	4
$\left(\frac{2}{5}, \frac{2}{5}, \frac{2}{5}, \frac{2}{5}, \frac{2}{5}, \frac{3}{5}, \frac{3}{5}, \frac{3}{5}, \frac{3}{5}, \frac{3}{5}\right)$	`10T39`	5	6	8

Table A.9 Counts of isogeny classes by Newton polygon, Galois group, and angle rank ($g = 6$)

Slopes	Group	Angle Rank	$q = 2$
(0, 0, 0, 0, 0, 0, 1, 1, 1, 1, 1, 1)	12T2	1	10
(0, 0, 0, 0, 0, 0, 1, 1, 1, 1, 1, 1)	12T3	1	1
(0, 0, 0, 0, 0, 0, 1, 1, 1, 1, 1, 1)	12T10	1	9
(0, 0, 0, 0, 0, 0, 1, 1, 1, 1, 1, 1)	12T10	3	3
(0, 0, 0, 0, 0, 0, 1, 1, 1, 1, 1, 1)	12T18	1	6
(0, 0, 0, 0, 0, 0, 1, 1, 1, 1, 1, 1)	12T21	3	1
(0, 0, 0, 0, 0, 0, 1, 1, 1, 1, 1, 1)	12T23	3	3
(0, 0, 0, 0, 0, 0, 1, 1, 1, 1, 1, 1)	12T23	5	6
(0, 0, 0, 0, 0, 0, 1, 1, 1, 1, 1, 1)	12T24	3	3
(0, 0, 0, 0, 0, 0, 1, 1, 1, 1, 1, 1)	12T25	3	55
(0, 0, 0, 0, 0, 0, 1, 1, 1, 1, 1, 1)	12T48	3	95
(0, 0, 0, 0, 0, 0, 1, 1, 1, 1, 1, 1)	12T48	5	12
(0, 0, 0, 0, 0, 0, 1, 1, 1, 1, 1, 1)	12T48	6	2
(0, 0, 0, 0, 0, 0, 1, 1, 1, 1, 1, 1)	12T76	6	16
(0, 0, 0, 0, 0, 0, 1, 1, 1, 1, 1, 1)	12T77	5	40
(0, 0, 0, 0, 0, 0, 1, 1, 1, 1, 1, 1)	12T78	2	74
(0, 0, 0, 0, 0, 0, 1, 1, 1, 1, 1, 1)	12T78	6	2
(0, 0, 0, 0, 0, 0, 1, 1, 1, 1, 1, 1)	12T90	3	32
(0, 0, 0, 0, 0, 0, 1, 1, 1, 1, 1, 1)	12T101	3	67
(0, 0, 0, 0, 0, 0, 1, 1, 1, 1, 1, 1)	12T103	3	4
(0, 0, 0, 0, 0, 0, 1, 1, 1, 1, 1, 1)	12T125	6	22
(0, 0, 0, 0, 0, 0, 1, 1, 1, 1, 1, 1)	12T134	6	10
(0, 0, 0, 0, 0, 0, 1, 1, 1, 1, 1, 1)	12T135	6	2
(0, 0, 0, 0, 0, 0, 1, 1, 1, 1, 1, 1)	12T136	6	2
(0, 0, 0, 0, 0, 0, 1, 1, 1, 1, 1, 1)	12T138	6	2
(0, 0, 0, 0, 0, 0, 1, 1, 1, 1, 1, 1)	12T139	3	365
(0, 0, 0, 0, 0, 0, 1, 1, 1, 1, 1, 1)	12T139	6	8
(0, 0, 0, 0, 0, 0, 1, 1, 1, 1, 1, 1)	12T148	6	2
(0, 0, 0, 0, 0, 0, 1, 1, 1, 1, 1, 1)	12T186	6	4
(0, 0, 0, 0, 0, 0, 1, 1, 1, 1, 1, 1)	12T193	6	12
(0, 0, 0, 0, 0, 0, 1, 1, 1, 1, 1, 1)	12T208	6	30
(0, 0, 0, 0, 0, 0, 1, 1, 1, 1, 1, 1)	12T219	5	1084
(0, 0, 0, 0, 0, 0, 1, 1, 1, 1, 1, 1)	12T219	6	102
(0, 0, 0, 0, 0, 0, 1, 1, 1, 1, 1, 1)	12T222	6	106
(0, 0, 0, 0, 0, 0, 1, 1, 1, 1, 1, 1)	12T224	6	2
(0, 0, 0, 0, 0, 0, 1, 1, 1, 1, 1, 1)	12T226	6	2
(0, 0, 0, 0, 0, 0, 1, 1, 1, 1, 1, 1)	12T227	6	2
(0, 0, 0, 0, 0, 0, 1, 1, 1, 1, 1, 1)	12T236	6	98
(0, 0, 0, 0, 0, 0, 1, 1, 1, 1, 1, 1)	12T237	6	4
(0, 0, 0, 0, 0, 0, 1, 1, 1, 1, 1, 1)	12T240	6	2
(0, 0, 0, 0, 0, 0, 1, 1, 1, 1, 1, 1)	12T250	6	152

(continued)

Table A.9 (continued)

Slopes	Group	Angle Rank	$q = 2$
(0, 0, 0, 0, 0, 0, 1, 1, 1, 1, 1, 1)	12T255	6	2
(0, 0, 0, 0, 0, 0, 1, 1, 1, 1, 1, 1)	12T260	6	968
(0, 0, 0, 0, 0, 0, 1, 1, 1, 1, 1, 1)	12T285	6	8810
(0, 0, 0, 0, 0, 0, 1, 1, 1, 1, 1, 1)	12T286	6	2
(0, 0, 0, 0, 0, 0, 1, 1, 1, 1, 1, 1)	12T287	6	6
(0, 0, 0, 0, 0, 0, 1, 1, 1, 1, 1, 1)	12T293	6	42488
$\left(0, 0, 0, 0, 0, \frac{1}{2}, \frac{1}{2}, 1, 1, 1, 1, 1\right)$	12T125	6	2
$\left(0, 0, 0, 0, 0, \frac{1}{2}, \frac{1}{2}, 1, 1, 1, 1, 1\right)$	12T250	6	28
$\left(0, 0, 0, 0, 0, \frac{1}{2}, \frac{1}{2}, 1, 1, 1, 1, 1\right)$	12T255	6	6
$\left(0, 0, 0, 0, 0, \frac{1}{2}, \frac{1}{2}, 1, 1, 1, 1, 1\right)$	12T260	6	34
$\left(0, 0, 0, 0, 0, \frac{1}{2}, \frac{1}{2}, 1, 1, 1, 1, 1\right)$	12T286	6	4
$\left(0, 0, 0, 0, 0, \frac{1}{2}, \frac{1}{2}, 1, 1, 1, 1, 1\right)$	12T293	6	17392
$\left(0, 0, 0, 0, \frac{1}{2}, \frac{1}{2}, \frac{1}{2}, \frac{1}{2}, 1, 1, 1, 1\right)$	12T6	3	2
$\left(0, 0, 0, 0, \frac{1}{2}, \frac{1}{2}, \frac{1}{2}, \frac{1}{2}, 1, 1, 1, 1\right)$	12T10	2	15
$\left(0, 0, 0, 0, \frac{1}{2}, \frac{1}{2}, \frac{1}{2}, \frac{1}{2}, 1, 1, 1, 1\right)$	12T23	2	6
$\left(0, 0, 0, 0, \frac{1}{2}, \frac{1}{2}, \frac{1}{2}, \frac{1}{2}, 1, 1, 1, 1\right)$	12T23	3	1
$\left(0, 0, 0, 0, \frac{1}{2}, \frac{1}{2}, \frac{1}{2}, \frac{1}{2}, 1, 1, 1, 1\right)$	12T28	4	2
$\left(0, 0, 0, 0, \frac{1}{2}, \frac{1}{2}, \frac{1}{2}, \frac{1}{2}, 1, 1, 1, 1\right)$	12T48	2	42
$\left(0, 0, 0, 0, \frac{1}{2}, \frac{1}{2}, \frac{1}{2}, \frac{1}{2}, 1, 1, 1, 1\right)$	12T48	3	36
$\left(0, 0, 0, 0, \frac{1}{2}, \frac{1}{2}, \frac{1}{2}, \frac{1}{2}, 1, 1, 1, 1\right)$	12T77	4	34
$\left(0, 0, 0, 0, \frac{1}{2}, \frac{1}{2}, \frac{1}{2}, \frac{1}{2}, 1, 1, 1, 1\right)$	12T79	4	2
$\left(0, 0, 0, 0, \frac{1}{2}, \frac{1}{2}, \frac{1}{2}, \frac{1}{2}, 1, 1, 1, 1\right)$	12T90	3	2
$\left(0, 0, 0, 0, \frac{1}{2}, \frac{1}{2}, \frac{1}{2}, \frac{1}{2}, 1, 1, 1, 1\right)$	12T101	3	19
$\left(0, 0, 0, 0, \frac{1}{2}, \frac{1}{2}, \frac{1}{2}, \frac{1}{2}, 1, 1, 1, 1\right)$	12T125	4	4
$\left(0, 0, 0, 0, \frac{1}{2}, \frac{1}{2}, \frac{1}{2}, \frac{1}{2}, 1, 1, 1, 1\right)$	12T125	6	2
$\left(0, 0, 0, 0, \frac{1}{2}, \frac{1}{2}, \frac{1}{2}, \frac{1}{2}, 1, 1, 1, 1\right)$	12T139	3	93
$\left(0, 0, 0, 0, \frac{1}{2}, \frac{1}{2}, \frac{1}{2}, \frac{1}{2}, 1, 1, 1, 1\right)$	12T193	6	2
$\left(0, 0, 0, 0, \frac{1}{2}, \frac{1}{2}, \frac{1}{2}, \frac{1}{2}, 1, 1, 1, 1\right)$	12T219	6	14
$\left(0, 0, 0, 0, \frac{1}{2}, \frac{1}{2}, \frac{1}{2}, \frac{1}{2}, 1, 1, 1, 1\right)$	12T236	6	40
$\left(0, 0, 0, 0, \frac{1}{2}, \frac{1}{2}, \frac{1}{2}, \frac{1}{2}, 1, 1, 1, 1\right)$	12T237	6	4
$\left(0, 0, 0, 0, \frac{1}{2}, \frac{1}{2}, \frac{1}{2}, \frac{1}{2}, 1, 1, 1, 1\right)$	12T250	6	34
$\left(0, 0, 0, 0, \frac{1}{2}, \frac{1}{2}, \frac{1}{2}, \frac{1}{2}, 1, 1, 1, 1\right)$	12T260	6	152
$\left(0, 0, 0, 0, \frac{1}{2}, \frac{1}{2}, \frac{1}{2}, \frac{1}{2}, 1, 1, 1, 1\right)$	12T285	6	2020
$\left(0, 0, 0, 0, \frac{1}{2}, \frac{1}{2}, \frac{1}{2}, \frac{1}{2}, 1, 1, 1, 1\right)$	12T293	6	5704

(continued)

Table A.9 (continued)

Slopes	Group	Angle Rank	$q = 2$
$\left(0, 0, 0, \frac{1}{3}, \frac{1}{3}, \frac{1}{3}, \frac{2}{3}, \frac{2}{3}, \frac{2}{3}, 1, 1, 1\right)$	12T37	2	6
$\left(0, 0, 0, \frac{1}{3}, \frac{1}{3}, \frac{1}{3}, \frac{2}{3}, \frac{2}{3}, \frac{2}{3}, 1, 1, 1\right)$	12T41	2	2
$\left(0, 0, 0, \frac{1}{3}, \frac{1}{3}, \frac{1}{3}, \frac{2}{3}, \frac{2}{3}, \frac{2}{3}, 1, 1, 1\right)$	12T78	2	12
$\left(0, 0, 0, \frac{1}{3}, \frac{1}{3}, \frac{1}{3}, \frac{2}{3}, \frac{2}{3}, \frac{2}{3}, 1, 1, 1\right)$	12T139	6	12
$\left(0, 0, 0, \frac{1}{3}, \frac{1}{3}, \frac{1}{3}, \frac{2}{3}, \frac{2}{3}, \frac{2}{3}, 1, 1, 1\right)$	12T219	5	22
$\left(0, 0, 0, \frac{1}{3}, \frac{1}{3}, \frac{1}{3}, \frac{2}{3}, \frac{2}{3}, \frac{2}{3}, 1, 1, 1\right)$	12T219	6	26
$\left(0, 0, 0, \frac{1}{3}, \frac{1}{3}, \frac{1}{3}, \frac{2}{3}, \frac{2}{3}, \frac{2}{3}, 1, 1, 1\right)$	12T227	6	2
$\left(0, 0, 0, \frac{1}{3}, \frac{1}{3}, \frac{1}{3}, \frac{2}{3}, \frac{2}{3}, \frac{2}{3}, 1, 1, 1\right)$	12T250	6	6
$\left(0, 0, 0, \frac{1}{3}, \frac{1}{3}, \frac{1}{3}, \frac{2}{3}, \frac{2}{3}, \frac{2}{3}, 1, 1, 1\right)$	12T260	6	12
$\left(0, 0, 0, \frac{1}{3}, \frac{1}{3}, \frac{1}{3}, \frac{2}{3}, \frac{2}{3}, \frac{2}{3}, 1, 1, 1\right)$	12T285	6	974
$\left(0, 0, 0, \frac{1}{3}, \frac{1}{3}, \frac{1}{3}, \frac{2}{3}, \frac{2}{3}, \frac{2}{3}, 1, 1, 1\right)$	12T293	6	1968
$\left(0, 0, 0, \frac{1}{2}, \frac{1}{2}, \frac{1}{2}, \frac{1}{2}, \frac{1}{2}, \frac{1}{2}, 1, 1, 1\right)$	12T78	2	12
$\left(0, 0, 0, \frac{1}{2}, \frac{1}{2}, \frac{1}{2}, \frac{1}{2}, \frac{1}{2}, \frac{1}{2}, 1, 1, 1\right)$	12T222	6	8
$\left(0, 0, 0, \frac{1}{2}, \frac{1}{2}, \frac{1}{2}, \frac{1}{2}, \frac{1}{2}, \frac{1}{2}, 1, 1, 1\right)$	12T226	6	2
$\left(0, 0, 0, \frac{1}{2}, \frac{1}{2}, \frac{1}{2}, \frac{1}{2}, \frac{1}{2}, \frac{1}{2}, 1, 1, 1\right)$	12T236	6	2
$\left(0, 0, 0, \frac{1}{2}, \frac{1}{2}, \frac{1}{2}, \frac{1}{2}, \frac{1}{2}, \frac{1}{2}, 1, 1, 1\right)$	12T250	6	8
$\left(0, 0, 0, \frac{1}{2}, \frac{1}{2}, \frac{1}{2}, \frac{1}{2}, \frac{1}{2}, \frac{1}{2}, 1, 1, 1\right)$	12T260	6	2
$\left(0, 0, 0, \frac{1}{2}, \frac{1}{2}, \frac{1}{2}, \frac{1}{2}, \frac{1}{2}, \frac{1}{2}, 1, 1, 1\right)$	12T285	6	504
$\left(0, 0, 0, \frac{1}{2}, \frac{1}{2}, \frac{1}{2}, \frac{1}{2}, \frac{1}{2}, \frac{1}{2}, 1, 1, 1\right)$	12T293	6	844
$\left(0, 0, \frac{1}{4}, \frac{1}{4}, \frac{1}{4}, \frac{1}{4}, \frac{3}{4}, \frac{3}{4}, \frac{3}{4}, \frac{3}{4}, 1, 1\right)$	12T24	2	2
$\left(0, 0, \frac{1}{4}, \frac{1}{4}, \frac{1}{4}, \frac{1}{4}, \frac{3}{4}, \frac{3}{4}, \frac{3}{4}, \frac{3}{4}, 1, 1\right)$	12T48	2	16
$\left(0, 0, \frac{1}{4}, \frac{1}{4}, \frac{1}{4}, \frac{1}{4}, \frac{3}{4}, \frac{3}{4}, \frac{3}{4}, \frac{3}{4}, 1, 1\right)$	12T48	5	4
$\left(0, 0, \frac{1}{4}, \frac{1}{4}, \frac{1}{4}, \frac{1}{4}, \frac{3}{4}, \frac{3}{4}, \frac{3}{4}, \frac{3}{4}, 1, 1\right)$	12T77	4	4
$\left(0, 0, \frac{1}{4}, \frac{1}{4}, \frac{1}{4}, \frac{1}{4}, \frac{3}{4}, \frac{3}{4}, \frac{3}{4}, \frac{3}{4}, 1, 1\right)$	12T78	4	2
$\left(0, 0, \frac{1}{4}, \frac{1}{4}, \frac{1}{4}, \frac{1}{4}, \frac{3}{4}, \frac{3}{4}, \frac{3}{4}, \frac{3}{4}, 1, 1\right)$	12T103	3	1
$\left(0, 0, \frac{1}{4}, \frac{1}{4}, \frac{1}{4}, \frac{1}{4}, \frac{3}{4}, \frac{3}{4}, \frac{3}{4}, \frac{3}{4}, 1, 1\right)$	12T125	4	4
$\left(0, 0, \frac{1}{4}, \frac{1}{4}, \frac{1}{4}, \frac{1}{4}, \frac{3}{4}, \frac{3}{4}, \frac{3}{4}, \frac{3}{4}, 1, 1\right)$	12T139	3	35
$\left(0, 0, \frac{1}{4}, \frac{1}{4}, \frac{1}{4}, \frac{1}{4}, \frac{3}{4}, \frac{3}{4}, \frac{3}{4}, \frac{3}{4}, 1, 1\right)$	12T219	5	58
$\left(0, 0, \frac{1}{4}, \frac{1}{4}, \frac{1}{4}, \frac{1}{4}, \frac{3}{4}, \frac{3}{4}, \frac{3}{4}, \frac{3}{4}, 1, 1\right)$	12T219	6	8
$\left(0, 0, \frac{1}{4}, \frac{1}{4}, \frac{1}{4}, \frac{1}{4}, \frac{3}{4}, \frac{3}{4}, \frac{3}{4}, \frac{3}{4}, 1, 1\right)$	12T250	6	4
$\left(0, 0, \frac{1}{4}, \frac{1}{4}, \frac{1}{4}, \frac{1}{4}, \frac{3}{4}, \frac{3}{4}, \frac{3}{4}, \frac{3}{4}, 1, 1\right)$	12T260	6	24
$\left(0, 0, \frac{1}{4}, \frac{1}{4}, \frac{1}{4}, \frac{1}{4}, \frac{3}{4}, \frac{3}{4}, \frac{3}{4}, \frac{3}{4}, 1, 1\right)$	12T285	6	372
$\left(0, 0, \frac{1}{4}, \frac{1}{4}, \frac{1}{4}, \frac{1}{4}, \frac{3}{4}, \frac{3}{4}, \frac{3}{4}, \frac{3}{4}, 1, 1\right)$	12T293	6	968

(continued)

Table A.9 (continued)

Slopes	Group	Angle Rank	$q = 2$
$\left(0, 0, \frac{1}{3}, \frac{1}{3}, \frac{1}{3}, \frac{1}{2}, \frac{1}{2}, \frac{2}{3}, \frac{2}{3}, \frac{2}{3}, 1, 1\right)$	12T125	6	2
$\left(0, 0, \frac{1}{3}, \frac{1}{3}, \frac{1}{3}, \frac{1}{2}, \frac{1}{2}, \frac{2}{3}, \frac{2}{3}, \frac{2}{3}, 1, 1\right)$	12T293	6	260
$\left(0, 0, \frac{1}{2}, \frac{1}{2}, \frac{1}{2}, \frac{1}{2}, \frac{1}{2}, \frac{1}{2}, \frac{1}{2}, \frac{1}{2}, 1, 1\right)$	12T6	3	1
$\left(0, 0, \frac{1}{2}, \frac{1}{2}, \frac{1}{2}, \frac{1}{2}, \frac{1}{2}, \frac{1}{2}, \frac{1}{2}, \frac{1}{2}, 1, 1\right)$	12T23	3	2
$\left(0, 0, \frac{1}{2}, \frac{1}{2}, \frac{1}{2}, \frac{1}{2}, \frac{1}{2}, \frac{1}{2}, \frac{1}{2}, \frac{1}{2}, 1, 1\right)$	12T26	3	1
$\left(0, 0, \frac{1}{2}, \frac{1}{2}, \frac{1}{2}, \frac{1}{2}, \frac{1}{2}, \frac{1}{2}, \frac{1}{2}, \frac{1}{2}, 1, 1\right)$	12T48	3	6
$\left(0, 0, \frac{1}{2}, \frac{1}{2}, \frac{1}{2}, \frac{1}{2}, \frac{1}{2}, \frac{1}{2}, \frac{1}{2}, \frac{1}{2}, 1, 1\right)$	12T101	3	6
$\left(0, 0, \frac{1}{2}, \frac{1}{2}, \frac{1}{2}, \frac{1}{2}, \frac{1}{2}, \frac{1}{2}, \frac{1}{2}, \frac{1}{2}, 1, 1\right)$	12T125	6	4
$\left(0, 0, \frac{1}{2}, \frac{1}{2}, \frac{1}{2}, \frac{1}{2}, \frac{1}{2}, \frac{1}{2}, \frac{1}{2}, \frac{1}{2}, 1, 1\right)$	12T139	3	16
$\left(0, 0, \frac{1}{2}, \frac{1}{2}, \frac{1}{2}, \frac{1}{2}, \frac{1}{2}, \frac{1}{2}, \frac{1}{2}, \frac{1}{2}, 1, 1\right)$	12T193	6	2
$\left(0, 0, \frac{1}{2}, \frac{1}{2}, \frac{1}{2}, \frac{1}{2}, \frac{1}{2}, \frac{1}{2}, \frac{1}{2}, \frac{1}{2}, 1, 1\right)$	12T219	5	46
$\left(0, 0, \frac{1}{2}, \frac{1}{2}, \frac{1}{2}, \frac{1}{2}, \frac{1}{2}, \frac{1}{2}, \frac{1}{2}, \frac{1}{2}, 1, 1\right)$	12T219	6	4
$\left(0, 0, \frac{1}{2}, \frac{1}{2}, \frac{1}{2}, \frac{1}{2}, \frac{1}{2}, \frac{1}{2}, \frac{1}{2}, \frac{1}{2}, 1, 1\right)$	12T236	6	10
$\left(0, 0, \frac{1}{2}, \frac{1}{2}, \frac{1}{2}, \frac{1}{2}, \frac{1}{2}, \frac{1}{2}, \frac{1}{2}, \frac{1}{2}, 1, 1\right)$	12T260	6	12
$\left(0, 0, \frac{1}{2}, \frac{1}{2}, \frac{1}{2}, \frac{1}{2}, \frac{1}{2}, \frac{1}{2}, \frac{1}{2}, \frac{1}{2}, 1, 1\right)$	12T285	6	100
$\left(0, 0, \frac{1}{2}, \frac{1}{2}, \frac{1}{2}, \frac{1}{2}, \frac{1}{2}, \frac{1}{2}, \frac{1}{2}, \frac{1}{2}, 1, 1\right)$	12T293	6	120
$\left(0, \frac{1}{5}, \frac{1}{5}, \frac{1}{5}, \frac{1}{5}, \frac{1}{5}, \frac{4}{5}, \frac{4}{5}, \frac{4}{5}, \frac{4}{5}, \frac{4}{5}, 1\right)$	12T219	6	8
$\left(0, \frac{1}{5}, \frac{1}{5}, \frac{1}{5}, \frac{1}{5}, \frac{1}{5}, \frac{4}{5}, \frac{4}{5}, \frac{4}{5}, \frac{4}{5}, \frac{4}{5}, 1\right)$	12T285	6	252
$\left(0, \frac{1}{5}, \frac{1}{5}, \frac{1}{5}, \frac{1}{5}, \frac{1}{5}, \frac{4}{5}, \frac{4}{5}, \frac{4}{5}, \frac{4}{5}, \frac{4}{5}, 1\right)$	12T293	6	514
$\left(0, \frac{1}{4}, \frac{1}{4}, \frac{1}{4}, \frac{1}{4}, \frac{1}{2}, \frac{1}{2}, \frac{3}{4}, \frac{3}{4}, \frac{3}{4}, \frac{3}{4}, 1\right)$	12T250	6	2
$\left(0, \frac{1}{4}, \frac{1}{4}, \frac{1}{4}, \frac{1}{4}, \frac{1}{2}, \frac{1}{2}, \frac{3}{4}, \frac{3}{4}, \frac{3}{4}, \frac{3}{4}, 1\right)$	12T293	6	128
$\left(0, \frac{1}{3}, \frac{1}{3}, \frac{1}{3}, \frac{1}{2}, \frac{1}{2}, \frac{1}{2}, \frac{1}{2}, \frac{2}{3}, \frac{2}{3}, \frac{2}{3}, 1\right)$	12T285	6	24
$\left(0, \frac{1}{3}, \frac{1}{3}, \frac{1}{3}, \frac{1}{2}, \frac{1}{2}, \frac{1}{2}, \frac{1}{2}, \frac{2}{3}, \frac{2}{3}, \frac{2}{3}, 1\right)$	12T293	6	16
$\left(0, \frac{2}{5}, \frac{2}{5}, \frac{2}{5}, \frac{2}{5}, \frac{2}{5}, \frac{3}{5}, \frac{3}{5}, \frac{3}{5}, \frac{3}{5}, \frac{3}{5}, 1\right)$	12T219	5	6
$\left(0, \frac{2}{5}, \frac{2}{5}, \frac{2}{5}, \frac{2}{5}, \frac{2}{5}, \frac{3}{5}, \frac{3}{5}, \frac{3}{5}, \frac{3}{5}, \frac{3}{5}, 1\right)$	12T219	6	4
$\left(0, \frac{2}{5}, \frac{2}{5}, \frac{2}{5}, \frac{2}{5}, \frac{2}{5}, \frac{3}{5}, \frac{3}{5}, \frac{3}{5}, \frac{3}{5}, \frac{3}{5}, 1\right)$	12T285	6	36
$\left(0, \frac{2}{5}, \frac{2}{5}, \frac{2}{5}, \frac{2}{5}, \frac{2}{5}, \frac{3}{5}, \frac{3}{5}, \frac{3}{5}, \frac{3}{5}, \frac{3}{5}, 1\right)$	12T293	6	42
$\left(0, \frac{1}{2}, \frac{1}{2}, \frac{1}{2}, \frac{1}{2}, \frac{1}{2}, \frac{1}{2}, \frac{1}{2}, \frac{1}{2}, \frac{1}{2}, \frac{1}{2}, 1\right)$	12T285	6	6
$\left(0, \frac{1}{2}, \frac{1}{2}, \frac{1}{2}, \frac{1}{2}, \frac{1}{2}, \frac{1}{2}, \frac{1}{2}, \frac{1}{2}, \frac{1}{2}, \frac{1}{2}, 1\right)$	12T293	6	4
$\left(\frac{1}{6}, \frac{1}{6}, \frac{1}{6}, \frac{1}{6}, \frac{1}{6}, \frac{1}{6}, \frac{5}{6}, \frac{5}{6}, \frac{5}{6}, \frac{5}{6}, \frac{5}{6}, \frac{5}{6}\right)$	12T10	1	12
$\left(\frac{1}{6}, \frac{1}{6}, \frac{1}{6}, \frac{1}{6}, \frac{1}{6}, \frac{1}{6}, \frac{5}{6}, \frac{5}{6}, \frac{5}{6}, \frac{5}{6}, \frac{5}{6}, \frac{5}{6}\right)$	12T48	3	32
$\left(\frac{1}{6}, \frac{1}{6}, \frac{1}{6}, \frac{1}{6}, \frac{1}{6}, \frac{1}{6}, \frac{5}{6}, \frac{5}{6}, \frac{5}{6}, \frac{5}{6}, \frac{5}{6}, \frac{5}{6}\right)$	12T78	2	18
$\left(\frac{1}{6}, \frac{1}{6}, \frac{1}{6}, \frac{1}{6}, \frac{1}{6}, \frac{1}{6}, \frac{5}{6}, \frac{5}{6}, \frac{5}{6}, \frac{5}{6}, \frac{5}{6}, \frac{5}{6}\right)$	12T139	3	56

(continued)

Table A.9 (continued)

Slopes	Group	Angle Rank	$q=2$
$\left(\frac{1}{6},\frac{1}{6},\frac{1}{6},\frac{1}{6},\frac{1}{6},\frac{1}{6},\frac{5}{6},\frac{5}{6},\frac{5}{6},\frac{5}{6},\frac{5}{6},\frac{5}{6}\right)$	12T219	5	48
$\left(\frac{1}{6},\frac{1}{6},\frac{1}{6},\frac{1}{6},\frac{1}{6},\frac{1}{6},\frac{5}{6},\frac{5}{6},\frac{5}{6},\frac{5}{6},\frac{5}{6},\frac{5}{6}\right)$	12T227	6	4
$\left(\frac{1}{6},\frac{1}{6},\frac{1}{6},\frac{1}{6},\frac{1}{6},\frac{1}{6},\frac{5}{6},\frac{5}{6},\frac{5}{6},\frac{5}{6},\frac{5}{6},\frac{5}{6}\right)$	12T236	6	2
$\left(\frac{1}{6},\frac{1}{6},\frac{1}{6},\frac{1}{6},\frac{1}{6},\frac{1}{6},\frac{5}{6},\frac{5}{6},\frac{5}{6},\frac{5}{6},\frac{5}{6},\frac{5}{6}\right)$	12T250	6	10
$\left(\frac{1}{6},\frac{1}{6},\frac{1}{6},\frac{1}{6},\frac{1}{6},\frac{1}{6},\frac{5}{6},\frac{5}{6},\frac{5}{6},\frac{5}{6},\frac{5}{6},\frac{5}{6}\right)$	12T260	6	32
$\left(\frac{1}{6},\frac{1}{6},\frac{1}{6},\frac{1}{6},\frac{1}{6},\frac{1}{6},\frac{5}{6},\frac{5}{6},\frac{5}{6},\frac{5}{6},\frac{5}{6},\frac{5}{6}\right)$	12T285	6	206
$\left(\frac{1}{6},\frac{1}{6},\frac{1}{6},\frac{1}{6},\frac{1}{6},\frac{1}{6},\frac{5}{6},\frac{5}{6},\frac{5}{6},\frac{5}{6},\frac{5}{6},\frac{5}{6}\right)$	12T293	6	578
$\left(\frac{1}{5},\frac{1}{5},\frac{1}{5},\frac{1}{5},\frac{1}{5},\frac{1}{2},\frac{1}{2},\frac{4}{5},\frac{4}{5},\frac{4}{5},\frac{4}{5},\frac{4}{5}\right)$	12T293	6	202
$\left(\frac{1}{4},\frac{1}{4},\frac{1}{4},\frac{1}{4},\frac{1}{2},\frac{1}{2},\frac{1}{2},\frac{1}{2},\frac{3}{4},\frac{3}{4},\frac{3}{4},\frac{3}{4}\right)$	12T23	2	3
$\left(\frac{1}{4},\frac{1}{4},\frac{1}{4},\frac{1}{4},\frac{1}{2},\frac{1}{2},\frac{1}{2},\frac{1}{2},\frac{3}{4},\frac{3}{4},\frac{3}{4},\frac{3}{4}\right)$	12T48	2	5
$\left(\frac{1}{4},\frac{1}{4},\frac{1}{4},\frac{1}{4},\frac{1}{2},\frac{1}{2},\frac{1}{2},\frac{1}{2},\frac{3}{4},\frac{3}{4},\frac{3}{4},\frac{3}{4}\right)$	12T101	3	2
$\left(\frac{1}{4},\frac{1}{4},\frac{1}{4},\frac{1}{4},\frac{1}{2},\frac{1}{2},\frac{1}{2},\frac{1}{2},\frac{3}{4},\frac{3}{4},\frac{3}{4},\frac{3}{4}\right)$	12T139	3	4
$\left(\frac{1}{4},\frac{1}{4},\frac{1}{4},\frac{1}{4},\frac{1}{2},\frac{1}{2},\frac{1}{2},\frac{1}{2},\frac{3}{4},\frac{3}{4},\frac{3}{4},\frac{3}{4}\right)$	12T260	6	2
$\left(\frac{1}{4},\frac{1}{4},\frac{1}{4},\frac{1}{4},\frac{1}{2},\frac{1}{2},\frac{1}{2},\frac{1}{2},\frac{3}{4},\frac{3}{4},\frac{3}{4},\frac{3}{4}\right)$	12T285	6	16
$\left(\frac{1}{4},\frac{1}{4},\frac{1}{4},\frac{1}{4},\frac{1}{2},\frac{1}{2},\frac{1}{2},\frac{1}{2},\frac{3}{4},\frac{3}{4},\frac{3}{4},\frac{3}{4}\right)$	12T293	6	34
$\left(\frac{1}{3},\frac{1}{3},\frac{1}{3},\frac{1}{3},\frac{1}{3},\frac{1}{3},\frac{2}{3},\frac{2}{3},\frac{2}{3},\frac{2}{3},\frac{2}{3},\frac{2}{3}\right)$	12T3	1	1
$\left(\frac{1}{3},\frac{1}{3},\frac{1}{3},\frac{1}{3},\frac{1}{3},\frac{1}{3},\frac{2}{3},\frac{2}{3},\frac{2}{3},\frac{2}{3},\frac{2}{3},\frac{2}{3}\right)$	12T10	1	2
$\left(\frac{1}{3},\frac{1}{3},\frac{1}{3},\frac{1}{3},\frac{1}{3},\frac{1}{3},\frac{2}{3},\frac{2}{3},\frac{2}{3},\frac{2}{3},\frac{2}{3},\frac{2}{3}\right)$	12T18	1	4
$\left(\frac{1}{3},\frac{1}{3},\frac{1}{3},\frac{1}{3},\frac{1}{3},\frac{1}{3},\frac{2}{3},\frac{2}{3},\frac{2}{3},\frac{2}{3},\frac{2}{3},\frac{2}{3}\right)$	12T48	3	9
$\left(\frac{1}{3},\frac{1}{3},\frac{1}{3},\frac{1}{3},\frac{1}{3},\frac{1}{3},\frac{2}{3},\frac{2}{3},\frac{2}{3},\frac{2}{3},\frac{2}{3},\frac{2}{3}\right)$	12T78	2	6
$\left(\frac{1}{3},\frac{1}{3},\frac{1}{3},\frac{1}{3},\frac{1}{3},\frac{1}{3},\frac{2}{3},\frac{2}{3},\frac{2}{3},\frac{2}{3},\frac{2}{3},\frac{2}{3}\right)$	12T101	3	2
$\left(\frac{1}{3},\frac{1}{3},\frac{1}{3},\frac{1}{3},\frac{1}{3},\frac{1}{3},\frac{2}{3},\frac{2}{3},\frac{2}{3},\frac{2}{3},\frac{2}{3},\frac{2}{3}\right)$	12T139	3	6
$\left(\frac{1}{3},\frac{1}{3},\frac{1}{3},\frac{1}{3},\frac{1}{3},\frac{1}{3},\frac{2}{3},\frac{2}{3},\frac{2}{3},\frac{2}{3},\frac{2}{3},\frac{2}{3}\right)$	12T208	6	2
$\left(\frac{1}{3},\frac{1}{3},\frac{1}{3},\frac{1}{3},\frac{1}{3},\frac{1}{3},\frac{2}{3},\frac{2}{3},\frac{2}{3},\frac{2}{3},\frac{2}{3},\frac{2}{3}\right)$	12T219	5	14
$\left(\frac{1}{3},\frac{1}{3},\frac{1}{3},\frac{1}{3},\frac{1}{3},\frac{1}{3},\frac{2}{3},\frac{2}{3},\frac{2}{3},\frac{2}{3},\frac{2}{3},\frac{2}{3}\right)$	12T260	6	2
$\left(\frac{1}{3},\frac{1}{3},\frac{1}{3},\frac{1}{3},\frac{1}{3},\frac{1}{3},\frac{2}{3},\frac{2}{3},\frac{2}{3},\frac{2}{3},\frac{2}{3},\frac{2}{3}\right)$	12T285	6	18
$\left(\frac{1}{3},\frac{1}{3},\frac{1}{3},\frac{1}{3},\frac{1}{3},\frac{1}{3},\frac{2}{3},\frac{2}{3},\frac{2}{3},\frac{2}{3},\frac{2}{3},\frac{2}{3}\right)$	12T287	6	2
$\left(\frac{1}{3},\frac{1}{3},\frac{1}{3},\frac{1}{3},\frac{1}{3},\frac{1}{3},\frac{2}{3},\frac{2}{3},\frac{2}{3},\frac{2}{3},\frac{2}{3},\frac{2}{3}\right)$	12T293	6	22
$\left(\frac{1}{3},\frac{1}{3},\frac{1}{3},\frac{1}{2},\frac{1}{2},\frac{1}{2},\frac{1}{2},\frac{1}{2},\frac{1}{2},\frac{2}{3},\frac{2}{3},\frac{2}{3}\right)$	12T78	2	2
$\left(\frac{1}{3},\frac{1}{3},\frac{1}{3},\frac{1}{2},\frac{1}{2},\frac{1}{2},\frac{1}{2},\frac{1}{2},\frac{1}{2},\frac{2}{3},\frac{2}{3},\frac{2}{3}\right)$	12T285	6	4
$\left(\frac{1}{3},\frac{1}{3},\frac{1}{3},\frac{1}{2},\frac{1}{2},\frac{1}{2},\frac{1}{2},\frac{1}{2},\frac{1}{2},\frac{2}{3},\frac{2}{3},\frac{2}{3}\right)$	12T293	6	2
$\left(\frac{2}{5},\frac{2}{5},\frac{2}{5},\frac{2}{5},\frac{2}{5},\frac{1}{2},\frac{1}{2},\frac{3}{5},\frac{3}{5},\frac{3}{5},\frac{3}{5},\frac{3}{5}\right)$	12T293	6	4
$\left(\frac{1}{2},\frac{1}{2},\frac{1}{2},\frac{1}{2},\frac{1}{2},\frac{1}{2},\frac{1}{2},\frac{1}{2},\frac{1}{2},\frac{1}{2},\frac{1}{2},\frac{1}{2}\right)$	12T2	0	8

Table A.10 Brauer invariants of endomorphism algebras. Each "zeroes" row combines all commutative endomorphism algebras for that g

g	$\mathrm{End}^0(A/\mathbf{F}_q)$ invs	$\mathrm{End}^0(A/\overline{\mathbf{F}}_q)$ invs	Count	qs	Example
1	zeroes	zeroes	6000	all	1.2.ab
	(0)	(1/2)	154	all except 169	1.2.ac
	(1/2)	(1/2)	30	squares	1.4.e
2	zeroes	zeroes	1198020	all	2.2.b_a
	(0)	(1/2)	195	all except 121	2.2.a_ae
	(0, 0)	(1/2)	80	5, 7, 9, 13, 17, 25 ...	2.5.a_a
	(1/2, 1/2)	(1/2)	7	25, 49, 169, 625	2.25.a_by
3	zeroes	zeroes	962658	all	3.2.b_b_d
	(0)	(1/2)	16	3, 4, 7, 9, 16, 25	3.3.a_a_aj
	(0, 0)	(1/3, 2/3)	140	all	3.2.a_a_ac
	(1/3, 2/3)	(1/3, 2/3)	2	8	3.8.g_bk_ea
4	zeroes	zeroes	141018	2, 3, 4, 5	4.2.ac_b_ab_d
	(0)	(1/2)	14	2, 3, 4, 5	4.2.a_c_a_e
	(0, 0)	(1/2)	7	3, 4, 5	4.3.a_a_a_aj
	(0, 0)	(1/4, 3/4)	53	2, 3, 4, 5	4.2.a_a_a_c
	(0, 0)	(1/2, 1/2)	84	2, 3, 4, 5	4.2.a_ae_a_k
	(1/2, 1/2)	(1/4, 3/4)	6	4	4.4.a_m_a_cq
	(1/2, 1/2)	(1/2, 1/2)	4	4	4.4.ae_q_abo_dw
	(0, 0, 0, 0)	(0, 0, 1/2, 1/2)	23	3, 4, 5	4.3.a_ac_a_ad
	(0, 0, 0, 0, 0, 0)	(0, 0, 1/2, 1/2)	21	3, 5	4.3.a_c_a_ad
5	zeroes	zeroes	197078	2, 3	5.2.ad_h_ao_y_abm
	(0, 0)	(1/5, 4/5)	20	2, 3	5.2.a_a_a_a_ag
	(0, 0)	(2/5, 3/5)	6	2, 3	5.2.a_a_a_a_e
6	zeroes	zeroes	89471	2	6.2.b_ab_a_ad_ad_h
	(0)	(1/2)	4	2	6.2.a_a_e_a_a_i
	(0, 0)	(1/2)	4	2	6.2.ac_c_a_ae_i_ai
	(0, 0)	(1/6, 5/6)	12	2	6.2.a_a_ac_a_a_c
	(0, 0)	(1/3, 2/3)	31	2	6.2.a_a_ac_a_a_o
	(0, 0)	(1/2, 1/2)	88	2	6.2.a_a_a_c_a_ac
	(0, 0, 0)	(0, 1/2, 1/2)	13	2	6.2.a_ac_a_g_a_ai
	(0, 0, 0, 0)	(0, 1/3, 2/3)	2	2	6.2.a_a_ac_a_a_a
	(0, 0, 0, 0)	(0, 1/2, 1/2)	1	2	6.2.a_ac_a_c_a_a
	(0, 0, 0, 0)	(0, 0, 1/2, 1/2)	42	2	6.2.a_ab_a_c_a_ag
	(0, 0, 0, 0, 0, 0)	(0, 0, 1/3, 2/3)	20	2	6.2.a_a_d_a_a_c
	(0, 0, 0, 0, 0, 0)	(0, 0, 1/2, 1/2)	14	2	6.2.a_b_a_a_a_ac

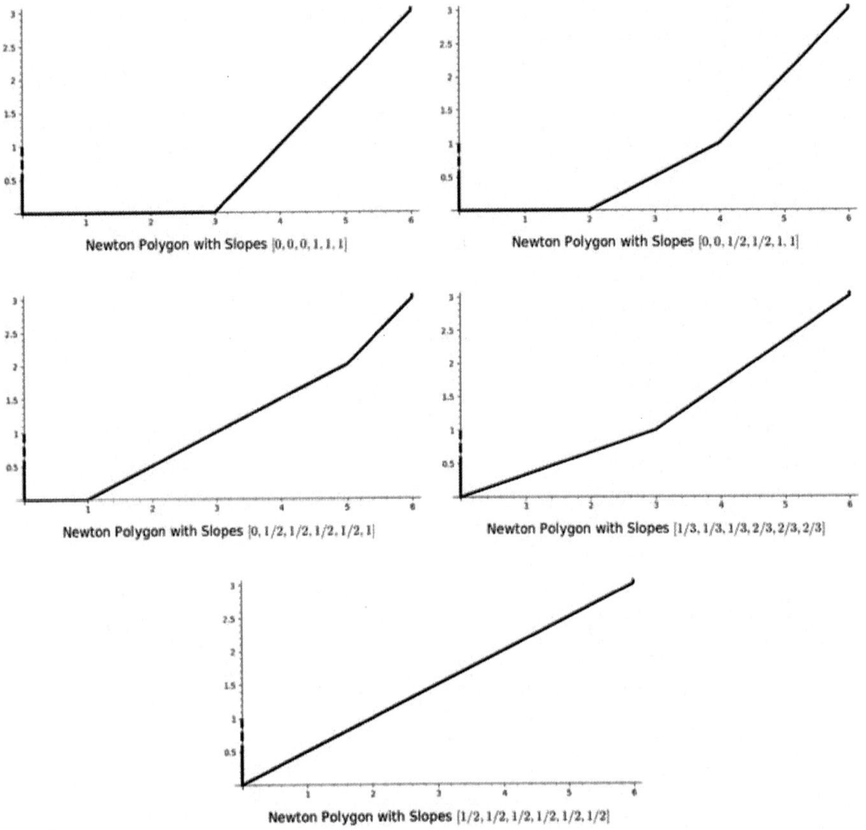

Fig. A.1 Possible Newton polygons for dimension 3 abelian varieties. The partial ordering is linear

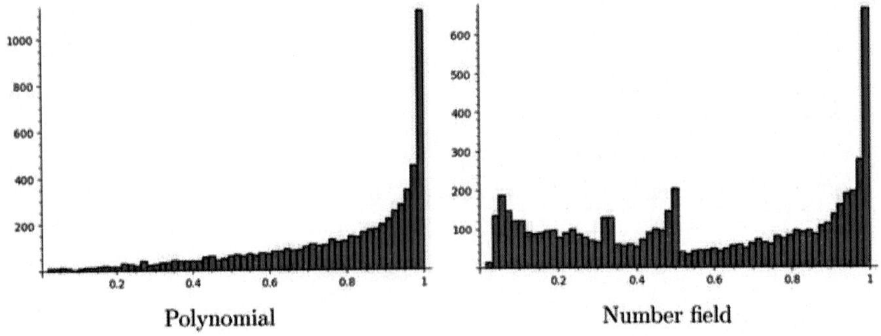

Fig. A.2 Normalized root discriminants for $g = 1$

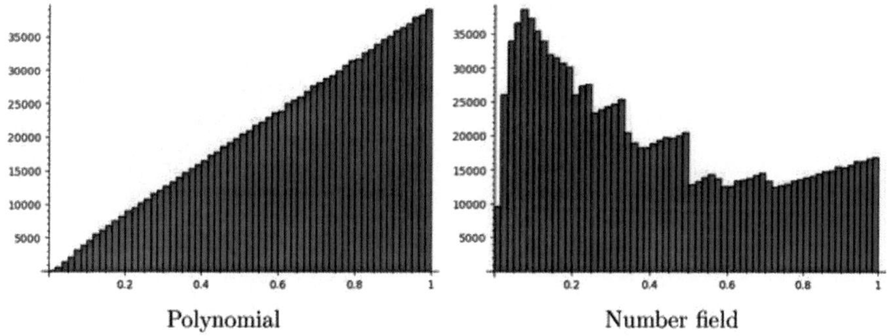

Fig. A.3 Normalized root discriminants for $g = 2$

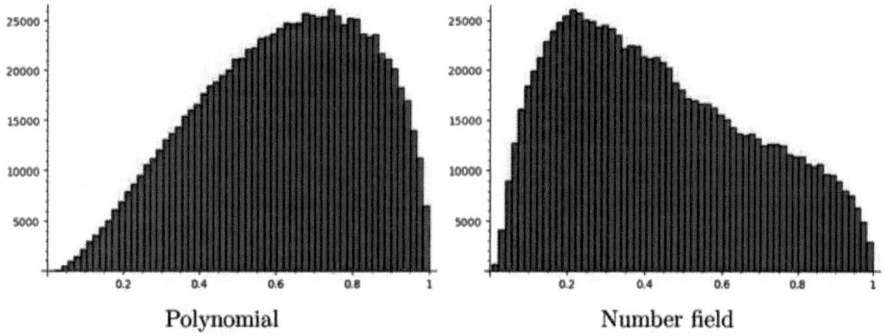

Fig. A.4 Normalized root discriminants for $g = 3$

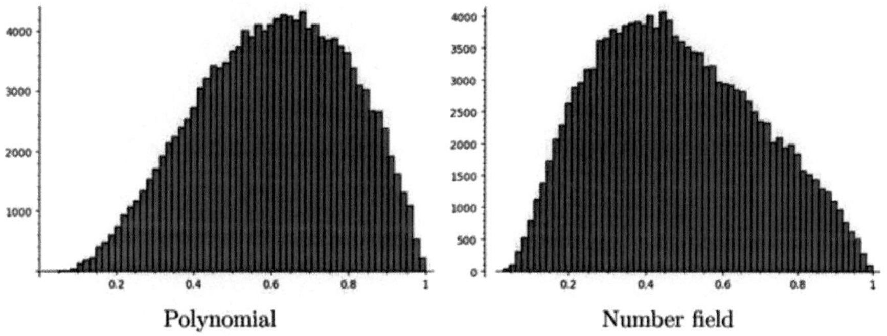

Fig. A.5 Normalized root discriminants for $g = 4$

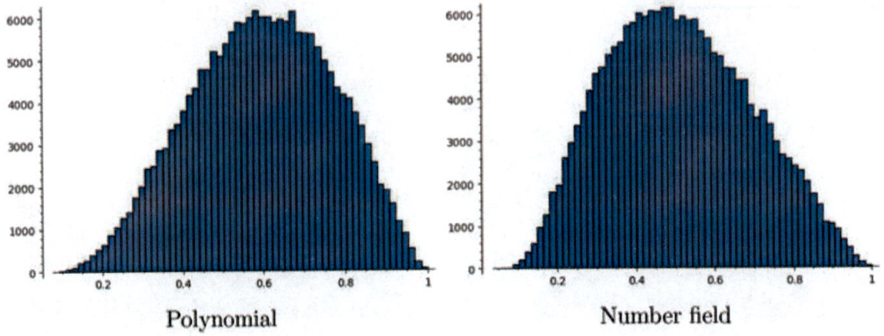

Fig. A.6 Normalized root discriminants for $g = 5$

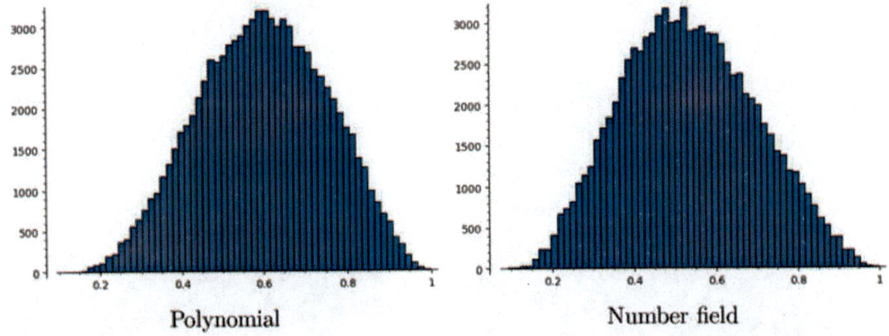

Fig. A.7 Normalized root discriminants for $g = 6$

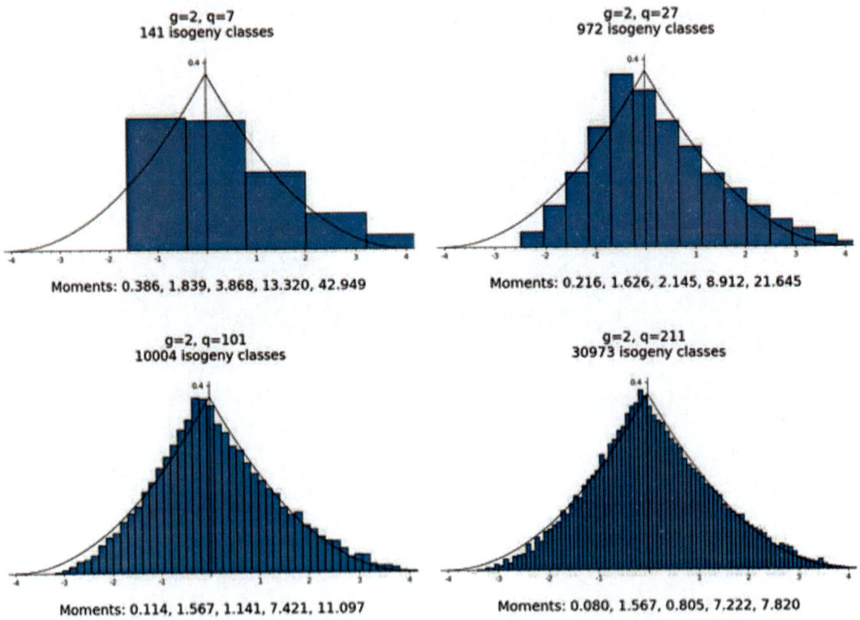

Fig. A.8 Isogeny Sato-Tate distribution for $g = 2$ with $q = 7, 27, 101, 211$

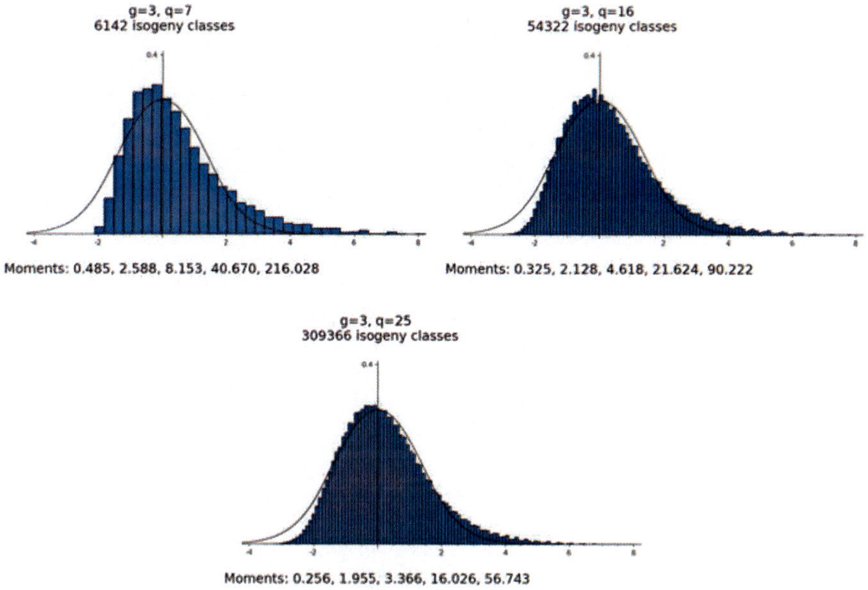

Fig. A.9 Isogeny Sato-Tate distribution for $g = 3$ with $q = 7$, 16 and 25

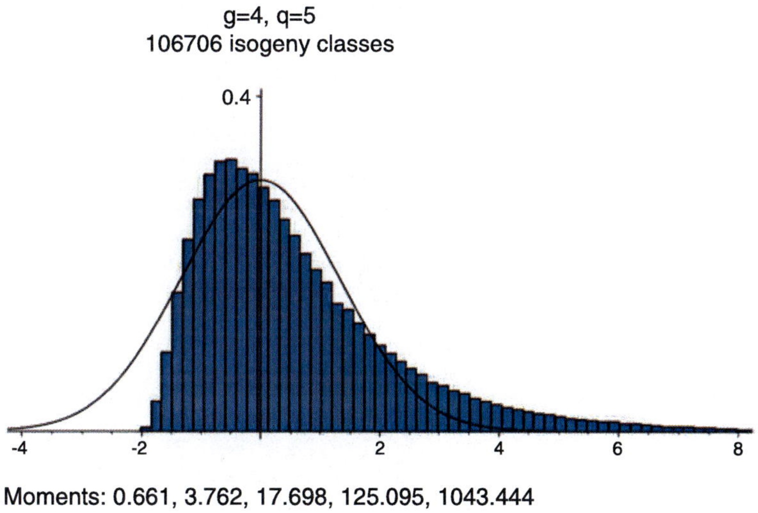

Fig. A.10 Isogeny Sato-Tate distribution for $g = 4$ and $q = 5$

Acknowledgments The authors began this project during the semester program "Computational aspects of the Langlands program" held at ICERM in fall 2015. Subsequent workshops in support of the project were sponsored by American Institute of Mathematics and the Simons Foundation. In addition Dupuy was partially supported by the European Research Council under the European Unions Seventh Framework Programme (FP7/2007–2013) / ERC Grant agreement no. 291111/ MODAG while working on this project. Kedlaya was supported by NSF (grants DMS-1501214, DMS-1802161), IAS (visiting professorship 2018–2019), UCSD (Warschawski Professorship), and a Guggenheim Fellowship (fall 2015). Roe was supported by Simons Foundation grant 550033. Vincent was partially supported by NSF (grant DMS-1802323). We would especially like to thank Everett Howe for implementing software for detecting principal polarizations in isogeny classes and for his feedback on an earlier version of this manuscript. Thanks also to Borys Kadets for computing bounds on point counts of abelian varieties over $\mathbf{F}_5$ and $\mathbf{F}_7$.

References

[AA18] Nicolas Addington and Asher Auel, *Some non-special cubic fourfolds* Doc. Math. **23** (2018), 637–651. MR3846048

[ACGH85] Enrico Arbarello, Maurizio Cornalba, Phillip A. Griffiths, and Joseph Harris, *Geometry of algebraic curves. Vol. I*, Grundlehren der Mathematischen Wissenschaften [Fundamental Principles of Mathematical Sciences], vol. 267, Springer-Verlag, New York, 1985. MR770932

[AHL12] Yves Aubry, Safia Haloui, and Gilles Lachaud, *Sur le nombre de points rationnels des variétés abéliennes et des Jacobiennes sur les corps finis*, C. R. Math. Acad. Sci. Paris **350** (2012), no. 19-20, 907–910. MR2990901

[AHL13] ——, *On the number of points on abelian and Jacobian varieties over finite fields*, Acta Arith. **160** (2013), no. 3, 201–241. MR3106095

[AKR10] Timothy G. Abbott, Kiran S. Kedlaya, and David Roe, *Bounding Picard numbers of surfaces using p-adic cohomology*, Arithmetic, Geometry, and Coding Theory (AGCT 2005), 2010, pp. 125–159. MR2856564

[AP11] Jeffrey D. Achter and Rachel Pries, *The p-rank strata of the moduli space of hyperelliptic curves*, Adv. Math. **227** (2011), no. 5, 1846–1872. MR2803789

[AS10] Omran Ahmadi and Igor E. Shparlinski, *On the distribution of the number of points on algebraic curves in extensions of finite fields*, Math. Res. Lett. **17** (2010), no. 4, 689–699. MR2661173

[BCP97] Wieb Bosma, John Cannon, and Catherine Playoust, *The Magma algebra system. I. The user language*, J. Symbolic Comput. **24** (1997), no. 3-4, 235–265. Computational algebra and number theory (London, 1993). MR1484478

[BS02] Frits Beukers and Christopher J. Smyth, *Cyclotomic points on curves*, Number Theory for the Millennium, I (Urbana, IL, 2000), 2002, pp. 67–85. MR1956219

[BSK20] Lior Bary-Soroker and Gady Kozma, *Irreducible polynomials of bounded height*, Duke Math. J. **169** (2020), no. 4, 579–598. MR4072635

[BSMT17] Adrian Barquero-Sanchez, Riad Masri, and Frank Thorne, *The distribution of G-Weyl CM fields and the Colmez conjecture*, arXiv preprint arXiv:1708.00044 (2017).

[Cas69] J. W. S. Cassels, *On a conjecture of R. M. Robinson about sums of roots of unity*, J. Reine Angew. Math. **238** (1969), 112–131. MR246852

[Chi92] Wenchen Chi, *l-adic and λ-adic representations associated to abelian varieties defined over number fields*, Amer. J. Math. **114** (1992), no. 2, 315–353. MR1156568

[CHK19] Edgar Costa, David Harvey, and Kiran S. Kedlaya, *Zeta functions of nondegenerate hypersurfaces in toric varieties via controlled reduction in p-adic cohomology*, Proceedings of the Thirteenth Algorithmic Number Theory Symposium, 2019, pp. 221–238. MR3952014

[CMSV19] Edgar Costa, Nicolas Mascot, Jeroen Sijsling, and John Voight, *Rigorous computation of the endomorphism ring of a Jacobian*, Math. Comp. **88** (2019), no. 317, 1303–1339. MR3904148

[CO06] Ching-Li Chai and Frans Oort, *Hypersymmetric abelian varieties*, Pure Appl. Math. Q. **2** (2006), no. 1, Special Issue: In honor of John H. Coates. Part 1, 1–27. MR2217565

[CO11] ——, *Monodromy and irreducibility of leaves*, Ann. of Math. (2) **173** (2011), no. 3, 1359–1396. MR2800716

[CO15] ——, *An algebraic construction of an abelian variety with a given Weil number*, Algebr. Geom. **2** (2015), no. 5, 654–663. MR3421786

[CS15] Tommaso Giorgio Centeleghe and Jakob Stix, *Categories of abelian varieties over finite fields, I: Abelian varieties over* $\mathbb{F}_p$, Algebra Number Theory **9** (2015), no. 1, 225–265. MR3317765

[Del69] Pierre Deligne, *Variétés abéliennes ordinaires sur un corps fini*, Invent. Math. **8** (1969), 238–243. MR254059

[Del74] ——, *La conjecture de Weil. I*, Inst. Hautes Études Sci. Publ. Math. **43** (1974), 273–307. MR340258

[DH98] Stephen A. DiPippo and Everett W. Howe, *Real polynomials with all roots on the unit circle and abelian varieties over finite fields*, J. Number Theory **73** (1998), no. 2, 426–450. MR1657992

[dJO00] Aise Johan de Jong and Frans Oort, *Purity of the stratification by Newton polygons*, J. Amer. Math. Soc. **13** (2000), no. 1, 209–241. MR1703336

[DKRV20] Taylor Dupuy, Kiran S. Kedlaya, David Roe, and Christelle Vincent, *Abvar-Fq*, 2020. https://github.com/LMFDB/abvar-fq.

[DKZB21] Taylor Dupuy, Kiran S. Kedlaya, and David Zureick-Brown, *Angle ranks of abelian varieties*, arXiv:2112.02455v1 (2021).

[Dod84] Bruce Dodson, *The structure of Galois groups of CM-fields*, Trans. Amer. Math. Soc. **283** (1984), no. 1, 1–32. MR735406

[DS94] Persi Diaconis and Mehrdad Shahshahani, *On the eigenvalues of random matrices*, 1994, pp. 49–62. Studies in applied probability. MR1274717

[Eke87] Torsten Ekedahl, *On supersingular curves and abelian varieties*, Math. Scand. **60** (1987), no. 2, 151–178. MR914332

[FK88] Eberhard Freitag and Reinhardt Kiehl, *Étale cohomology and the Weil conjecture*, Ergebnisse der Mathematik und ihrer Grenzgebiete (3) [Results in Mathematics and Related Areas (3)], vol. 13, Springer-Verlag, Berlin, 1988. Translated from the German by Betty S. Waterhouse and William C. Waterhouse, with an historical introduction by J. A. Dieudonné. MR926276

[FO08] Jean-Marc Fontaine and Yi Ouyang, *Theory of p-adic Galois representations*, preprint (2008).

[Fri88] David Fried, *Cyclic resultants of reciprocal polynomials*, Holomorphic dynamics (Mexico, 1986), 1988, pp. 124–128. MR980956

[Gon98] Josep González, *On the p-rank of an abelian variety and its endomorphism algebra*, Publ. Mat. **42** (1998), no. 1, 119–130. MR1628150

[Hal10] Safia Haloui, *The characteristic polynomials of abelian varieties of dimension 3 over finite fields*, J. Number Theory **130** (2010), no. 12, 2745–2752. MR2684495

[Hil05] Christopher J. Hillar, *Cyclic resultants*, J. Symbolic Comput. **39** (2005), no. 6, 653–669. MR2167674

[HL12] Everett W. Howe and Kristin E. Lauter, *New methods for bounding the number of points on curves over finite fields*, Geometry and arithmetic, 2012, pp. 173–212. MR2987661

[HMNR08] Everett W. Howe, Daniel Maisner, Enric Nart, and Christophe Ritzenthaler, *Principally polarizable isogeny classes of abelian surfaces over finite fields*, Math. Res. Lett. **15** (2008), no. 1, 121–127. MR2367179

[Hon15] Katrina Honigs, *Derived equivalent surfaces and abelian varieties, and their zeta functions*, Proc. Amer. Math. Soc. **143** (2015), no. 10, 4161–4166. MR3373916

[Hon68] Taira Honda, *Isogeny classes of abelian varieties over finite fields*, J. Math. Soc. Japan **20** (1968), 83–95. MR229642

[How93] Everett W. Howe, *Elliptic curves and ordinary abelian varieties over finite fields*, Ph.D. Thesis, University of California Berkeley, 1993. MR2690210

[How95] ——, *Principally polarized ordinary abelian varieties over finite fields*, Trans. Amer. Math. Soc. **347** (1995), no. 7, 2361–2401. MR1297531

[How96] ——, *Kernels of polarizations of abelian varieties over finite fields*, J. Algebraic Geom. **5** (1996), no. 3, 583–608. MR1382738

[HS12] Safia Haloui and Vijaykumar Singh, *The characteristic polynomials of abelian varieties of dimension 4 over finite fields*, Arithmetic, Geometry, Cryptography and Coding Theory, 2012, pp. 59–68. MR2961400

[Ibu18] Tomoyoshi Ibukiyama, *Type numbers of quaternion hermitian forms and supersingular abelian varieties*, Osaka J. Math. **55** (2018), no. 2, 369–384. MR3787750

[Iha81] Yasutaka Ihara, *Some remarks on the number of rational points of algebraic curves over finite fields*, J. Fac. Sci. Univ. Tokyo Sect. IA Math. **28** (1981), no. 3, 721–724 (1982). MR656048

[Ito19] Kazuhiro Ito, *Unconditional construction of $K3$ surfaces over finite fields with given L-function in large characteristic*, Manuscripta Math. **159** (2019), no. 3-4, 281–300. MR3959263

[Kad19] Borys Kadets, *Estimates for the number of rational points on simple abelian varieties over finite fields*, Mathematische Zeitschrift **297** (2021), 465–473, https://doi.org/10.1007/s00209-020-02520-w

[Kat81] Nicholas M. Katz, *Crystalline cohomology, Dieudonné modules, and Jacobi sums*, Automorphic Forms, Representation Theory and Arithmetic (Bombay, 1979), 1981, pp. 165–246. MR633662

[Ked06a] Kiran S. Kedlaya, *Finiteness of rigid cohomology with coefficients*, Duke Math. J. **134** (2006), no. 1, 15–97. MR2239343

[Ked06b] ——, *Quantum computation of zeta functions of curves*, Comput. Complexity **15** (2006), no. 1, 1–19. MR2226067

[Ked08] ——, *Search techniques for root-unitary polynomials*, Computational Arithmetic Geometry, 2008, pp. 71–81. MR2459990

[KH17] Momonari Kudo and Shushi Harashita, *Superspecial curves of genus 4 in small characteristic*, Finite Fields Appl. **45** (2017), 131–169. MR3631358

[KHS19] Momonari Kudo, Shushi Harashita, and Hayato Senda, *The existence of supersingular curves of genus 4 in arbitrary characteristic*, arXiv preprint arXiv:1903.08095 (2019).

[KM74] Nicholas M. Katz and William Messing, *Some consequences of the Riemann hypothesis for varieties over finite fields*, Invent. Math. **23** (1974), 73–77. MR332791

[KP19] Valentijn Karemaker and Rachel Pries, *Fully maximal and fully minimal abelian varieties*, J. Pure Appl. Algebra **223** (2019), no. 7, 3031–3056. MR3912957

[KS16] Kiran S. Kedlaya and Andrew V. Sutherland, *A census of zeta functions of quartic K3 surfaces over $\mathbb{F}_2$*, LMS J. Comput. Math. **19** (2016), no. suppl. A, 1–11. MR3540942

[KS99] Nicholas M. Katz and Peter Sarnak, *Random matrices, Frobenius eigenvalues, and monodromy*, American Mathematical Society Colloquium Publications, vol. 45, American Mathematical Society, Providence, RI, 1999. MR1659828

[LMPT18] Wanlin Li, Elena Mantovan, Rachel Pries, and Yunqing Tang, *Newton polygons of cyclic covers of the projective line branched at three points*, arXiv preprint arXiv:1805.04598 (2018).

[LMPT19] ——, *Newton polygons arising from special families of cyclic covers of the projective line*, Res. Number Theory **5** (2019), no. 1, Art. 12, 31. MR3897613

[LMQ75] James R. C. Leitzel, Manohar L. Madan, and Clifford S. Queen, *Algebraic function fields with small class number*, J. Number Theory **7** (1975), 11–27. MR369326

[LO98] Ke-Zheng Li and Frans Oort, *Moduli of supersingular abelian varieties*, Lecture Notes in Mathematics, vol. 1680, Springer-Verlag, Berlin, 1998. MR1611305

[Lox72] J. H. Loxton, *On the maximum modulus of cyclotomic integers*, Acta Arith. **22** (1972), 69–85. MR309896

[LZ93] Hendrik W. Lenstra Jr. and Yuri G. Zarhin, *The Tate conjecture for almost ordinary abelian varieties over finite fields*, Advances in number theory (Kingston, ON, 1991), 1993, pp. 179–194. MR1368419

[Mal02] Gunter Malle, *On the distribution of Galois groups*, J. Number Theory **92** (2002), no. 2, 315–329. MR1884706

[Mal04] ——, *On the distribution of Galois groups. II*, Experiment. Math. **13** (2004), no. 2, 129–135. MR2068887

[Mar19] Stefano Marseglia, *Computing abelian varieties over finite fields isogenous to a power*, Res. Number Theory **5** (2019), no. 4, Paper No. 35. MR4030241

[MN02] Daniel Maisner and Enric Nart, *Abelian surfaces over finite fields as Jacobians*, Experiment. Math. **11** (2002), no. 3, 321–337. With an appendix by Everett W. Howe. MR1959745

[Mor60] Louis Joel Mordell, *On a discriminant inequality*, Canadian J. Math. **12** (1960), 699–704. MR125493

[MP77] Manohar L. Madan and Sat Pal, *Abelian varieties and a conjecture of R. M. Robinson*, J. Reine Angew. Math. **291** (1977), 78–91. MR439848

[MS15] Pietro Mercuri and Claudio Stirpe, *Classification of algebraic function fields with class number one*, J. Number Theory **154** (2015), 365–374. MR3339577

[Mue13] Alexander Mueller, *Applications of generalized Fermat varieties to zeta functions of Artin-Schreier curves*, Ph.D. Thesis, University of Michigan, 2013.

[Oor08] Frans Oort, *Abelian varieties over finite fields*, Higher-dimensional geometry over finite fields, 2008, pp. 123–188. MR2484079

[Oor09] ——, *Foliations in moduli spaces of abelian varieties and dimension of leaves*, Algebra, arithmetic, and geometry: in honor of Yu. I. Manin. Vol. II, 2009, pp. 465–501. MR2641199

[OS20] Abhishek Oswal and Ananth N. Shankar, *Almost ordinary abelian varieties over finite fields*, J. Lond. Math. Soc. (2) **101** (2020), no. 3, 923–937. MR4111929

[Rob64] Raphael M. Robinson, *Intervals containing infinitely many sets of conjugate algebraic units*, Ann. of Math. (2) **80** (1964), 411–428. MR175881

[Rob77] ——, *Conjugate algebraic units in a special interval*, Math. Z. **154** (1977), no. 1, 31–40. MR447175

[Sav03] David Savitt, *The maximum number of points on a curve of genus 4 over* $\mathbb{F}_8$ *is 25*, Canad. J. Math. **55** (2003), no. 2, 331–352. With an appendix by Kristin Lauter. MR1969795

[Soh13] Gyoyong Sohn, *The bounds of the coefficients of the characteristic polynomials for abelian varieties of dimension 5 over finite fields*, Adv. Stud. Contemp. Math. (Kyungshang) **23** (2013), no. 3, 415–421. MR3100101

[SS15] Qibin Shen and Shuhui Shi, *Function fields of class number one*, J. Number Theory **154** (2015), 375–379. MR3339578

[ST18] Ananth N. Shankar and Jacob Tsimerman, *Unlikely intersections in finite characteristic*, Forum Math. Sigma **6** (2018), e13, 17. MR3841493

[S$^+$20] W. A. Stein et al., *Sage Mathematics Software (Version 9.1.beta0)*, The Sage Development Team, 2020. http://www.sagemath.org.

[Sti14] Claudio Stirpe, *A counterexample to 'Algebraic function fields with small class number'*, J. Number Theory **143** (2014), 402–404. MR3227356

[SZ02] Jasper Scholten and Hui June Zhu, *Hyperelliptic curves in characteristic 2*, Int. Math. Res. Not. **17** (2002), 905–917. MR1899907

[Tae16] Lenny Taelman, *K3 surfaces over finite fields with given L-function*, Algebra Number Theory **10** (2016), no. 5, 1133–1146. MR3531364

[Tat66] John Tate, *Endomorphisms of abelian varieties over finite fields*, Invent. Math. **2** (1966), 134–144. MR206004 [vdGvdV95] Gerard van der Geer and Marcel van der Vlugt, *On the existence of supersingular curves of given genus*, J. Reine Angew. Math. **458** (1995), 53–61. MR1310953

[vdW36] B. L. van der Waerden, *Die Seltenheit der reduziblen Gleichungen und der Gleichungen mit Affekt*, Monatsh. Math. Phys. **43** (1936), no. 1, 133–147. MR1550517

[Voi05] John Voight, *Curves over finite fields with many points: an introduction*, Computational aspects of algebraic curves, 2005, pp. 124–144. MR2182038

[WM71] William C. Waterhouse and James S. Milne, *Abelian varieties over finite fields*, 1969 Number Theory Institute (Proc. Sympos. Pure Math., Vol. XX, State Univ. New York, Stony Brook, N.Y., 1969), 1971, pp. 53–64. MR0314847

[Xar20] Xavier Xarles, *A census of all genus 4 curves over the field with 2 elements*, arXiv preprint arXiv:2007.07822 (2020).

[Yu17] Chia-Fu Yu, *On fields of definition of components of the Siegel supersingular locus*, Proc. Amer. Math. Soc. **145** (2017), no. 12, 5053–5058. MR3717936

[Zar15] Yuri G. Zarhin, *Eigenvalues of Frobenius endomorphisms of abelian varieties of low dimension*, J. Pure Appl. Algebra **219** (2015), no. 6, 2076–2098. MR3299720

[Zar94] Yu. G. Zarhin, *The Tate conjecture for nonsimple abelian varieties over finite fields*, Algebra and Number Theory (Essen, 1992), 1994, pp. 267–296. MR1285371

[Zyw10] David Zywina, *Hilbert's irreducibility theorem and the larger sieve*, arXiv preprint arXiv:1011.6465 (2010).

Computing Rational Points on Rank 0 Genus 3 Hyperelliptic Curves

María Inés de Frutos-Fernández and Sachi Hashimoto

Abstract We compute rational points on genus 3 odd degree hyperelliptic curves C over $\mathbb{Q}$ that have Jacobians of Mordell–Weil rank 0. The computation applies the Chabauty–Coleman method to find the zero set of a certain system of p-adic integrals, which is known to be finite and include the set of rational points $C(\mathbb{Q})$. We implemented an algorithm in Sage to carry out the Chabauty–Coleman method on a database of 5870 curves.

1 Introduction

Given a curve C of genus $g \geq 2$ defined over $\mathbb{Q}$, Mordell's conjecture, proved by Faltings [10] implies that the set of rational points $C(\mathbb{Q})$ is finite. Our goal is to explicitly compute this finite set of points in the case of a genus 3 hyperelliptic curve with Jacobian having Mordell–Weil rank 0. For a general curve, computing the set of its rational points is an unsolved problem. However, there are techniques available when the Mordell–Weil rank r of the Jacobian is smaller than the genus. In this case it is often possible, for a specific curve, to provably find its rational points.

The rational point-finding techniques applied here for curves of small Mordell–Weil rank date back to an idea of Chabauty [6], who proved in 1941 that when the rank r is less than the genus g, fixing a prime of good reduction p, the set $C(\mathbb{Q}_p) \cap \overline{J(\mathbb{Q})}$ is finite, and hence $C(\mathbb{Q})$ is, where $\overline{J(\mathbb{Q})}$ denotes the closure in the p-adic topology of $J(\mathbb{Q})$. In the 1980s, Coleman made Chabauty's idea effective: he gave an upper bound on the number of rational points [7], now known as the Chabauty–Coleman bound. Coleman's upper bound comes from counting the zeros of p-adic (Coleman) integrals on $J(\mathbb{Q}_p)$ that vanish on $\overline{J(\mathbb{Q})}$. This bound was refined by Stoll [13]. We exhibit some curves that reach the Stoll bound. In 2010, Balakrishnan, Bradshaw, and Kedlaya [3] gave a practical algorithm to compute

M. I. de Frutos-Fernández · S. Hashimoto (✉)
Department of Mathematics and Statistics, Boston University, Boston, MA, USA
e-mail: svh@bu.edu

J. S. Balakrishnan et al. (eds.), *Arithmetic Geometry, Number Theory, and Computation*, Simons Symposia, https://doi.org/10.1007/978-3-030-80914-0_14

these p-adic Coleman integrals on odd degree hyperelliptic curves and implemented the algorithm in Sage [16].

Recently, Sutherland [15] constructed a database of genus 3 hyperelliptic curves of discriminant up to 10^7 using the approach of [4]. Armed with a decent upper bound on the number of rational points on a curve, a practical algorithm for computing these Coleman integrals that vanish on the rational points, and a large number of curves to experiment on, one might ask: how effective is Coleman's method for determining rational points?

For example, Balakrishnan, Bianchi, Cantoral-Farfán, Çiperiani, and Etropolski [2] carried out computations to find rational points on 16,977 genus 3 odd degree hyperelliptic curves with rank 1 Jacobian. They found that in most cases, the Chabauty–Coleman method picks up exactly the rational points, or the rational points and the Weierstrass points that are defined over $\mathbb{Q}_p$. In the remaining cases, they were able to explain the existence of the extra points either by torsion, linearity of the integral, or by extra endomorphisms of the Jacobian.

In order to satisfy the Chabauty–Coleman hypothesis that $r < g$ we need $r = 0, 1$, or 2, but in the case of $r = g - 1$, one will likely find many extra p-adic but non-rational points when applying the Chabauty–Coleman method, and in general the Chabauty–Coleman set will be larger than the set of rational points. More work would be needed to determine the set of rational points; for example, one could implement a Mordell–Weil sieve. For these reasons, we focused on the case of $g = 3$ and $r = 0$. In this case, we computed the rational points in Magma [5], up to height 10^5 and verified these points using Chabauty–Coleman calculations. Most of the time, in 3070 curves, the only extra points in the Chabauty–Coleman set were Weierstrass points that were defined over $\mathbb{Q}_p \setminus \mathbb{Q}$. In 17 cases, we picked up points on the curve giving rise to torsion points of higher order on the Jacobian. These points are defined over quadratic fields where the prime p splits.

In Sect. 2, we present background on Coleman integration: we define the Coleman integral and list its properties, and we also discuss explicit Coleman integration. In Sect. 3, we discuss our implementation of the Chabauty–Coleman method for computing rational points on hyperelliptic curves of genus 3 with rank 0 Jacobian. Finally, in Sect. 4, we present our results and give an overview of the rational points and extra $\mathbb{Q}_p$-points found on the database of curves. We discuss several interesting examples in detail, including a curve with an 18-torsion point on its Jacobian and a curve that is sharp for Stoll's bound.

2 Background on Coleman Integration

In this section we will define the basics of Coleman integration and necessary background for the rest of the paper. For a more thorough introduction to the rigid analytic geometry and theory behind Coleman integrals, we refer the reader to [8], whose exposition we have loosely followed herein. Other good background on the Chabauty–Coleman method can be found in [11] and [17].

Fix K a field that is complete with respect to a non-archimedean absolute value and fix some embedding of K into $\mathbb{C}_p$. Let R be the ring of integers of K.

Let C/R be a smooth connected affinoid curve over K and set $C_K = C \times_R K$. Let F be the residue field of K, and $\widetilde{C}_K$ be $C \times_R F$. There is a reduction map red : $C_K \to \widetilde{C}_K$. We say that C *has good reduction over* K if $\widetilde{C}_K$ is smooth. We call the preimage of a point under the reduction map $\mathrm{red}^{-1}(x)$ for $x \in \widetilde{C}_K$ the *residue disc of* x. These residue discs partition $C_K(K)$.

Let C have good reduction, ω be a holomorphic one-form on C, and $P, Q \in C(K)$ be K-rational points. Coleman defined an integral $\int_P^Q \omega$ in the following way. Fix a Frobenius lift $\phi : C_K \to C_K$, that is, a morphism of rigid analytic varieties which reduces to (relative) Frobenius on $\widetilde{C}_K$.

Theorem 1 ([8, Theorem 2.1]) *Suppose there is a polynomial $\mathcal{P}(T) \in \mathbb{C}_p[T]$ which does not vanish on any root of unity, and that*

$$\mathcal{P}(\phi^*)\omega \text{ is exact.}$$

Then there is a function f_ω on $C(\mathbb{C}_p)$ which is analytic on each residue disc such that $df_\omega = \omega$ and $\mathcal{P}(\phi^*)(f_\omega)$ is analytic. The function f_ω is unique up to a constant and is independent of ϕ and $\mathcal{P}$.

In this way, Coleman provides a way to compute a locally analytic antiderivative of ω. Coleman proves, in several corollaries, properties of the antiderivative, which we collect here in one theorem:

Theorem 2 ([8, Theorem 2.3, Proposition 2.4]) *The integral satisfies the following properties:*

1. *Linearity:* $\int_P^Q (\alpha\omega + \beta\omega') = \alpha \int_P^Q \omega + \beta \int_P^Q \omega'$.
2. *Additivity:* $\int_P^R \omega = \int_P^Q \omega + \int_Q^R \omega$.
3. *Change of variables: if C' is a curve and $\phi : C \to C'$ a rigid analytic map between wide opens then* $\int_P^Q \phi^*\omega = \int_{\phi(P)}^{\phi(Q)} \omega$. *For example, ϕ can be taken to be a lift of the pth power Frobenius.*
4. *Fundamental theorem of calculus:* $\int_P^Q df = f(Q) - f(P)$.
5. *For any divisor $D := \sum_i Q_i - P_i$ of degree zero on C, the integral $\int_D \omega = \sum_i \int_{P_i}^{Q_i} \omega$ is well-defined, and $\int_D \omega = 0$ when D is principal.*
6. *Galois equivariance: if σ is an automorphism of $\mathbb{C}_p$, then* $\left(\int_P^Q \omega\right)^\sigma = \int_{\sigma(P)}^{\sigma(Q)} \omega^\sigma$.

Then Coleman shows that we can use this integration theory to obtain arithmetic information about points.

Suppose now that C_K is the base change of a curve C defined over $\mathbb{Q}$ and that the Jacobian J of $C/\mathbb{Q}$ has Mordell–Weil rank $r < g$. Fix $\iota : C \to J$ to be an embedding given by choosing a $\mathbb{Q}$-rational basepoint $b \in C(\mathbb{Q})$ and mapping P to $[P - b]$. Denote by J_K the base change of J to K. The embedding ι induces an isomorphism of vector spaces $\iota^* : H^0(J_K, \Omega^1) \to H^0(C_K, \Omega^1)$ which is independent of the chosen basepoint b (Proposition 2.2 of [12]).

We have functionals

$$\lambda_\omega(D) = \int_D \omega, \text{ for } \omega \in H^0(J_K, \Omega^1).$$

If $D = \sum_i [Q_i - P_i]$, then λ_ω can also be thought of as functionals on C_K by $\lambda_\omega(D) = \sum_i \int_{P_i}^{Q_i} \iota^*\omega$.

Note that $H^0(J_K, \Omega^1)$ is g-dimensional but the Jacobian $J(\mathbb{Q})$ has rank $r < g$. Fix a basis $\omega_0, \ldots, \omega_{g-1}$ for the space of holomorphic differentials on J_K. Define functionals

$$\lambda_i(D) = \int_D \omega_i : J(\mathbb{Q}) \to K.$$

Consider all linear relations between the restrictions of the λ_i to $J(\mathbb{Q})$: there are at least $g - r$ independent linear relations and they are of the form

$$\sum_{i=0}^{g-1} \alpha_i \int_D \omega_i = 0, \text{ with } \alpha_i \in K.$$

Let S be the set of functionals $\mu(D) := \sum_{i=0}^{g-1} \alpha_i \int_D \omega_i$ from J_K to K such that $\mu(D) = 0$ for all $D \in J(\mathbb{Q})$. Via the map ι, we can define $\mu(z) := \sum_{i=0}^{g-1} \alpha_i \int_{\iota(z)} \iota^*\omega_i$ for any $z \in C(K)$. Note that in particular $\mu(z) = 0$ for every z in $C(\mathbb{Q})$.

By Theorem 1, the functionals μ in S are locally analytic, and we can use bounds on the numbers of zeros of p-adic power series to get control over the number of zeros for μ on each residue disc. In particular, there are at most finitely many common zeros of any non-zero linear combination of functionals. Hence the set of rational points $C(\mathbb{Q})$ is contained in the finite set of points satisfying $\mu(z) = 0$ for every μ in S.

Coleman's proof suggests an effective method for computing $C(\mathbb{Q})$: first, compute the linear relations that the functionals satisfy. Then, compute the finite zero set of the functionals on each residue disc. This second step requires an algorithm for computing the Coleman integral to some p-adic precision.

Balakrishnan, Bradshaw, and Kedlaya give a practical algorithm for computing $\int_P^Q \omega$, on an odd-degree hyperelliptic curve C, where P and Q are in $C(\mathbb{Q}_p)$ and ω is a one-form of the second kind. We give a brief outline here, and for more details refer the reader to [3].

To evaluate Coleman integrals, we work locally in each residue disc, where the Coleman integral can be expressed as a convergent power series in a parameter, called a *local coordinate*. If $K(C)$ is the function field of C, then our local coordinate is simply a uniformizing parameter t for $K(C)_P$, that is, an element $t \in K(C)_P$ such that the valuation of t is one.

Our residue discs come in two flavors: those that are the residue disc of a Weierstrass point, the *Weierstrass residue discs*, and those that are not, the *non-Weierstrass* residue discs. The *infinity disc* which contains the point at infinity is a Weierstrass residue disc in the odd degree hyperelliptic case.

For a hyperelliptic curve C/K we have differentials $\omega_i = (x^i/2y)\,\mathrm{d}x$, for $i = 0, \ldots, 2g-1$ such that their classes $[\omega_i]$ generate $H^1_{dR}(C)$. For any one-form ω, we can write $\omega = \sum_{i=0}^{2g-1} a_i\omega_i + df$ for some $a_i \in K$ and exact differential *df*, and then use linearity to break up an integral.

Assume that P and Q are in the same residue disc, and that this disc does not contain a pole of ω. To compute a Coleman integral between P and Q, called a *tiny integral*, we simply write P, Q, and ω in a local coordinate t, $P(t)$, $Q(t)$, $\omega(t)$, and integrate formally:

$$\int_P^Q \omega = \sum_{i=0}^{2g-1} a_i \int_P^Q \omega_i + \int_P^Q df = \sum_{i=0}^{2g-1} a_i \int_{P(t)}^{Q(t)} \frac{x(t)^i}{2y(t)} \frac{dx(t)}{dt} dt + f(Q) - f(P).$$

To compute a Coleman integral between P and Q in different non-Weierstrass residue discs we fix Teichmüller points P' in the residue disc of P and Q' in the residue disc of Q such that $\phi(P') = P'$ and $\phi(Q') = Q'$. By the previous algorithm we can compute tiny integrals $\int_P^{P'} \omega$ and $\int_Q^{Q'} \omega$ between our desired points and the Teichmüller points. Applying the change of variables property of the integral, $\int_{P'}^{Q'} \phi^*\omega = \int_{\phi(P')}^{\phi(Q')} \omega$. Thus we compute $\phi^*\omega_i$ for each basis vector ω_i. Then ϕ^* acts linearly on cohomology, so we can represent ϕ^* as a matrix M. However, $\phi^*\omega_i$ is only cohomologous to a linear combination of ω_i, so to represent the action of ϕ^* as a matrix, we write

$$\phi^*\omega_i = \sum_{j=0}^{2g-1} M_{ij}\omega_j + df_i.$$

This gives us a linear system, and we can solve

$$(M - I)\begin{bmatrix} \vdots \\ \int_{P'}^{Q'} \omega_i \\ \vdots \end{bmatrix} = \begin{bmatrix} \vdots \\ \sum_{i=0}^{2g-1} f_i(P') - f_i(Q') \\ \vdots \end{bmatrix}$$

for the value of $\int_{P'}^{Q'} \omega_i$. By linearity, we can compute $\int_{P'}^{Q'} \omega$. The value of the original integral is simply the sum $\int_P^{P'} \omega + \int_{P'}^{Q'} \omega + \int_{Q'}^{Q} \omega$. Finally, for the case of P or Q in a Weierstrass disc, we use the fact that if P is a Weierstrass point $\int_P^Q \omega = 1/2 \int_{\iota(Q)}^{Q} \omega$ where ι is the hyperelliptic involution (Lemma 16 of [3]).

3 Algorithm

In this section we describe our implementation of the Chabauty–Coleman method for our case of interest. For a more general description of this method, see for instance [11]. Our code is available at [9].

We require C to be a hyperelliptic curve of genus 3 given by an odd degree model

$$C : y^2 = F(x),$$

where $F(x) \in \mathbb{Q}[x]$ is a monic polynomial of degree 7. Every such model has a unique rational point at infinity, that we denote ∞ and take as our basepoint for the Abel–Jacobi embedding $\iota : C \hookrightarrow J$. Moreover, we assume that the Jacobian J of C has Mordell–Weil rank 0 over $\mathbb{Q}$. Let $p \geq 7$ be the smallest prime such that C has good reduction mod p (note that p does not divide the leading coefficient of F since we assume that F is monic).

Our goal is to provably compute the set of rational points of C. We first use Magma to compute $C(\mathbb{Q})_{\text{known}}$ to be the set of all rational points in C of naive height bounded by 10^5. Having this set reduces the number of Coleman integrations required by the algorithm.

The input of our algorithm consists of the hyperelliptic curve C, the prime p and the set $C(\mathbb{Q})_{\text{known}}$. The output is a finite subset of $C(\mathbb{Q}_p)$ containing $C(\mathbb{Q})$, which is returned as three separate subsets:

- The set $C(\mathbb{Q})$ of rational points of C.
- The set of points Q in $C(\mathbb{Q}_p) \setminus C(\mathbb{Q})$ such that $[Q - \infty] \in J(\mathbb{Q}_p)$ is 2-torsion.
- The set of points Q in $C(\mathbb{Q}_p) \setminus C(\mathbb{Q})$ such that $[Q - \infty] \in J(\mathbb{Q}_p)$ is an n-torsion point for some $n > 2$.

Note that since we are assuming that J has rank 0, no point in C can give rise to a point of infinite order in J. Moreover, we remark that under this rank hypothesis we automatically know that the annihilator of $J(\mathbb{Q})$ under the integration pairing is spanned by $\omega_i = x^i/2y\, dx$ for $i = 0, 1, 2$, which simplifies Step 2 of our algorithm.

We use essentially the same precision analysis as in [2]. Proposition 3.11 in [2] can be easily adapted to show that it is enough to set $N = 2p + 4$ to be the p-adic precision and $M = 2p + 1$ to be the t-adic precision. We give a basic outline of the adaptations needed here (the reader should replace the statements in their paper about the annihilating differentials α, β with the annihilating differentials $\omega_0, \omega_1, \omega_2$).

First note that assuming [2, Proposition 3.11], $N = 2p + 4$ and $M = 2p + 1$ is sufficient for p prime $p > 2g = 6$ (this follows from Riemann–Roch, see [2, Remark 3.12]). The proof of this proposition depends on two earlier statements: [2, Lemmas 3.10 and 3.9], which do not depend on the rank of the curve or the exact set of annihilating differentials.

While the lemmas and propositions in [2, Section 3.1] that bound the number of zeros in each residue disc of the curve remain true with minimal modification to the proofs, we did not use these bounds to rule out residue discs in our implementation.

An outline of our algorithm is given in Algorithm 3. Below we explain its main steps.

---

function CHABAUTY–COLEMAN(C, p, $C(\mathbb{Q})_{\text{known}}$)
    Set the p-adic precision N to $2p+4$.
    Set the t-adic precision M to $2p+1$.
    Initialize found-points := empty list.
    for each $\overline{P} \in \overline{C}(\mathbb{F}_p)$ up to the hyperelliptic involution ι **do**
        Compute f_0, f_1 and f_2 in local coordinates.
        for each point $Q \in C(\mathbb{Q}_p)$ corresponding to a common zero of the f_i **do**
            Add Q and $\iota(Q)$ to found-points.
        end for
    end for
    Classify found-points into three lists: $\mathbb{Q}$-points, non-rational 2-torsion-points, and non-rational higher-torsion-points.
    return $\mathbb{Q}$-points, non-rational 2-torsion-points, and non-rational higher-torsion-points.
end function

Algorithm 3 Chabauty–Coleman method for a genus 3 hyperelliptic curve with rank 0 Jacobian

Step 1 (Required Precision) We need to choose the p-adic precision N and the t-adic precision M to guarantee that, in Step 3, we will obtain all the roots of $f_i(pt)$ in $\mathbb{Z}_p$. Set $N = 2p+4$ and $M = 2p+1$.

Step 2 (Annihilator) A basis of the space of differentials $H^0(C_{\mathbb{Q}_p}, \Omega^1)$ is given by $\{\omega_0, \omega_1, \omega_2\}$, where $\omega_i = (x^i/2y)\,dx$. For each $i = 0, 1, 2$, define

$$f_i(z) = \int_{\infty}^{z} \omega_i,$$

where ∞ denotes the point at infinity. The functions $f_i(z)$ are zero on all rational points of C, but not identically zero.

Step 3 (Searching in Residue Discs) For each point $\overline{P}$ in $\overline{C}(\mathbb{F}_p)$, we compute the set of $\mathbb{Q}_p$-rational points P reducing to $\overline{P}$ such that $f_i(P) = 0$ for $i = 0, 1, 2$. To perform this computation, we consider two different cases:

(i) If there is a point $P \in C(\mathbb{Q})_{\text{known}}$ reducing to $\overline{P}$, then we expand each ω_i in terms of a uniformizer t at P and we formally integrate to obtain three power series $f_i(t)$, that parametrize the integrals of the ω_i between P and any other point in the residue disc.

(ii) Otherwise, we start by finding a $\mathbb{Q}_p$-point P reducing to $\overline{P} = (\overline{x_0}, \overline{y_0})$ (note that $\overline{P}$ cannot be $\overline{\infty}$ in this case). If $\overline{y_0} = 0$ we can take $P = (x_0, 0)$ where x_0 is the Hensel lift of $\overline{x_0}$ to a root of $F(x)$. Otherwise, we can take $P = (x_0, y_0)$ where x_0 is any lift of $\overline{x_0}$ to $\mathbb{Z}_p$ and y_0 is obtained from $\overline{y_0}$ by applying Hensel's Lemma to $y^2 = F(x_0)$. Then we set $f_i(t) = \tilde{f}_i(t) + \int_\infty^P \omega_i$, where each $\tilde{f}_i(t)$ parametrizes the integral of ω_i between P and any other point in the residue disc.

To provably compute the set of common zeros of f_0, f_1 and f_2 to a desired precision, we need to impose the condition that at least one of the f_i has only simple roots. We check this requirement by computing the discriminant of the truncations of the power series to t-adic precision M. If all the f_i have multiple roots, we run the algorithm using a bigger prime p for which they do not.

Say f_{i_0} has no double roots. Then we use the PARI/GP function polrootspadic to compute its roots, truncating it first to t-adic precision $M = 2p + 1$ and computing the coefficients up to p-adic precision $N = 2p + 4$. For each root that lies in $p\mathbb{Z}_p$, we check whether it is also a root of the other two f_i; if so, it corresponds to a point $P \in C(\mathbb{Q}_p)$ lying over $\overline{P}$.

Step 4 (Identifying the Rational Points) Now, for each of the points Q found in Step 3, we attempt to reconstruct Q as a $\mathbb{Q}$-rational point, using Sage. If this is not possible, then $[Q-\infty]$ must be a torsion point in $J(\mathbb{Q}_p)$, because J has rank 0. If Q is a Weierstrass point, it will give rise to a 2-torsion point in the Jacobian; otherwise we classify it as a higher order torsion point.

Step 5 (Identifying Higher Order Torsion Points) For each of the points Q that is not identified as a rational or Weierstrass point in Step 4, we use Sage's algebraic_dependency function to attempt to find the minimal polynomial of the x-coordinate and reconstruct the point as a point with coordinates defined over an extension of $\mathbb{Q}$. We then use Magma to compute the torsion order of $[Q - \infty]$.

4 Analysis and Examples

We ran the Sage implementation of our algorithm on a database of 5870 hyperelliptic curves of genus 3 with Jacobian of rank 0 taken from the list [1]. This list [1] is a subset of Sutherland's database of genus 3 hyperelliptic curves of discriminant bounded by 10^7 [15]. It was created by first filtering for curves with odd-degree models and then filtering for curves that are provably rank 0. The rank computation was done with the Magma function `RankBounds`, which returns an upper and lower bound for the rank of the curve through descent computations. For curves in [1], the upper bound and the lower bound returned by `RankBounds` agreed and equaled 0. We computed a set of monic models for each curve in [1] for the purpose of our computations.

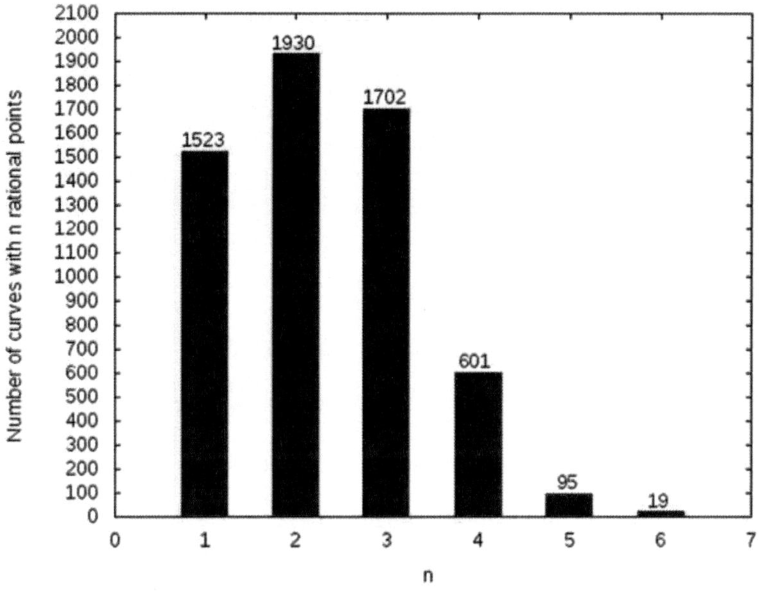

Fig. 1 Number of curves in the database with n rational points

In 23 of these curves, there was at least one $\mathbb{Q}_p$-point for which the three power series $f_0(z)$, $f_1(z)$ and $f_2(z)$ had double roots at the first prime $p \geq 7$ of good reduction. To analyze these curves, we replaced p by a bigger prime for which at least one of the power series had only simple roots.

Our implementation proves that, for each of the studied curves, the entire set of rational points equals the set of rational points of naive height less than 10^5. In fact, we proved that, with respect to the odd-degree monic models used in the computation, the maximum among the global heights of rational points in those models is 30.7611440827071, reached only by points at three curves. Or, if we translate this to the odd degree models in [1], those three curves have points of global height 4.39444915467244, while the heights of the rational points in the rest of the curves is bounded above by 3.[1]

Next, we show in Fig. 1 how many of the curves in our database have a certain number of rational points. We observe that the maximum number of points is six, and that a vast majority of the curves have three or fewer rational points.

As mentioned on the introduction, the Chabauty–Coleman sets of these curves agreed with the set of rational points on 2783 cases, contained only rational points and Weierstrass points on 3070 cases, and contained points on the curve giving

[1]The global height is the absolute logarithmic height of the point, which is the maximum of the absolute logarithmic heights of its coordinates. For a rational point $\frac{n}{d}$, this height is computed as $\max(\log(|n|), \log(|d|))$.

rise to higher order torsion points on the Jacobian on 17 cases, 13 of which also presented non-rational Weierstrass points. In each of these 17 curves, there was exactly one pair of points corresponding to higher order torsion points, of order 8 in ten cases, order 12 in six cases, and order 18 in one case, which we present in Example 2. We made these 17 higher torsion examples available on [9].

We conclude this section by showing how the algorithm works on a particular curve, giving an example where the algorithm detects a torsion point defined over a number field, and exhibiting an example curve where the Stoll bound is sharp.

Example 1 Consider the hyperelliptic curve

$$C : y^2 = x^7 - 37024x^6 + 3134464x^5 - 101220352x^4 + 1613758464x^3 - \\ - 13656653824x^2 + 59055800320x - 103079215104,$$

with minimal discriminant 2168084 and conductor 1084042.

Using Magma, we find that the set of rational points of C with height bounded by 10^5 is

$$C(\mathbb{Q})_{\text{known}} = \{\infty, (32, 0)\}.$$

Since C has good reduction modulo 7, we run our code using this prime. The points of $\overline{C}(\mathbb{F}_7)$ are as follows:

$$\{\overline{\infty}, \overline{(0, \pm 4)}, \overline{(1, \pm 5)}, \overline{(2, \pm 6)}, \overline{(4, 0)}, \overline{(6, \pm 2)}\}.$$

After Hensel lifting each of these points to a point of $C(\mathbb{Q}_7)$, we write $f_0(z)$, $f_1(z)$, and $f_2(z)$ in local coordinates and find their common zeros. This yields the following results:

Disc	Common roots of $f_0(z)$, $f_1(z)$ and $f_2(z)$
$\overline{\infty}$	∞
$\overline{(0, \pm 4)}$	No common roots
$\overline{(1, \pm 5)}$	No common roots
$\overline{(2, \pm 6)}$	No common roots
$\overline{(4, 0)}$	(32, 0)
$\overline{(6, \pm 2)}$	No common roots

Therefore, we have shown that $C(\mathbb{Q}) = \{\infty, (32, 0)\}$.

Example 2 Let

$$C : y^2 = x^7 + \frac{5}{64}x^6 - \frac{51}{256}x^5 - \frac{243}{4096}x^4 - \frac{53}{8192}x^3 - \frac{27}{65536}x^2 - \frac{1}{65536}x - \frac{1}{4194304}$$

be the hyperelliptic curve with minimal discriminant 6856704 and conductor 6856704. Using the algorithm described in the paper, we can determine that this curve has only one rational point, the point at infinity. However, we also detect a pair of 7-adic points with x-coordinate

$$6 + 6 \cdot 7^2 + 6 \cdot 7^4 + 6 \cdot 7^6 + 6 \cdot 7^8 + 6 \cdot 7^{10} + 6 \cdot 7^{12} + 6 \cdot 7^{14} + 6 \cdot 7^{16} + O(7^{18}).$$

Using sage's algebraic_dependency function, we identify this pair of points as $(-1/8, \pm\sqrt{-3}/2^{11})$. The points $[(-1/8, \pm\sqrt{-3}/2^{11}) - \infty]$ have order 18 on the Jacobian over $\mathbb{Q}(\sqrt{-3})$. The Chabauty–Coleman algorithm detects these torsion points at the prime $p = 7$ because 7 splits in this field.

Example 3 When $p > 2g$ is a prime of good reduction and $r < g - 1$, then Stoll [14, Corollary 6.7] has improved the bound given by Coleman to give $\#X(\mathbb{Q}) \leq \#X(\mathbb{F}_p) + 2r$. In our case, this shows that $\#X(\mathbb{Q}) \leq \#X(\mathbb{F}_p)$. In our database of 5870 curves, 94 achieve equality for Stoll's bound.

For example, the curve

$$y^2 = x^7 - \frac{1}{2^5}x^5 - \frac{7}{2^{12}}x^4 + \frac{1}{2^{13}}x^3 + \frac{3}{2^{17}}x^2 + \frac{1}{2^{19}}x + \frac{1}{2^{24}},$$

which has minimal discriminant 4089600 and conductor 170400, has 6 rational points of height less than 10^5:

$$\left\{\infty, \left(0, \frac{1}{4096}\right), \left(0, -\frac{1}{4096}\right), \left(\frac{1}{8}, 0\right), \left(-\frac{1}{8}, 0\right), \left(-\frac{1}{16}, 0\right)\right\}$$

as well as 6 points over $\mathbb{F}_7$:

$$\{\infty, (0, 1), (0, 6), (1, 0), (6, 0), (3, 0)\}.$$

By Stoll's theorem, we know that there cannot be any more rational points.

Acknowledgments We would like to thank Jennifer Balakrishnan for suggesting these computations, and for many helpful discussions throughout. We would also like to thank Alex Best for several productive conversations about Coleman integration. The second author was supported by a Clare Boothe Luce grant while working on this project.

References

1. J. S. Balakrishnan, *Genus 3 hyperelliptic curves with Mordell–Weil rank 0*, 2018, http://math.bu.edu/people/jbala/GWAG/g3oddr0.txt.
2. J. S. Balakrishnan, F. Bianchi, V. Cantoral-Farfán, M. Çiperiani, and A. Etropolski, "Chabauty-Coleman experiments for genus 3 hyperelliptic curves," in *Research Directions in Number Theory*, ser. Association for Women in Mathematics Series. Springer, 2019, vol. 19, pp. 67–90.

3. J. S. Balakrishnan, R. W. Bradshaw, and K. S. Kedlaya, "Explicit Coleman integration for hyperelliptic curves," in *Algorithmic Number Theory. ANTS 2010*, Lecture Notes in Computer Science. Springer, Berlin, 2010, vol. 6197, pp. 16–31.
4. A. R. Booker, J. Sijsling, A. V. Sutherland, J. Voight, and D. Yasaki, "A database of genus-2 curves over the rational numbers," *LMS J. Comput. Math.*, vol. 19, no. suppl. A, pp. 235–254, 2016.
5. W. Bosma, J. Cannon, and C. Playoust, "The Magma Algebra System. I. The User Language," *J. Symbolic Comput.*, vol. 24, no. 3-4, pp. 235–265, 1997.
6. C. Chabauty, "Sur les points rationnels des courbes algébriques de genre supérieur à l'unité," *C. R. Acad. Sci. Paris*, vol. 212, pp. 882–885, 1941.
7. R. F. Coleman, "Effective Chabauty," *Duke Math. J.*, vol. 52, no. 3, pp. 765–770, 1985.
8. ——, "Torsion points on curves and p-adic abelian integrals," *Ann. of Math. (2)*, vol. 121, no. 1, pp. 111–168, 1985.
9. María Inés de Frutos-Fernández and S. Hashimoto, *Sage code*, https://github.com/sachihashimoto/rational-points-hyperelliptic.
10. G. Faltings, "Endlichkeitssätze für abelsche Varietäten über Zahlkörpern," *Invent. Math.*, vol. 73, no. 3, pp. 349–366, 1983.
11. W. McCallum and B. Poonen, "The method of Chabauty and Coleman," *Explicit Methods in Number Theory, Rational Points and Diophantine Equations, Panor. Synthèses*, vol. 36, pp. 99–117, 2012.
12. J. S. Milne, "Jacobian varieties," in *Arithmetic geometry (Storrs, Conn., 1984)*. Springer, New York, 1986, pp. 167–212.
13. M. Stoll, "On the height constant for curves of genus two. II," *Acta Arith.*, vol. 104, no. 2, pp. 165–182, 2002.
14. ——, "Independence of rational points on twists of a given curve," *Compos. Math.*, vol. 142, no. 5, pp. 1201–1214, 2006.
15. A. V. Sutherland, *Genus 3 hyperelliptic curves*, https://math.mit.edu/~drew/genus3curves.html.
16. The Sage Developers, *Sagemath, the Sage Mathematics Software System (Version 8.0)*, 2017, https://www.sagemath.org.
17. J. L. Wetherell, *Bounding the Number of Rational Points on Certain Curves of High Rank*, 1997, Ph.D. thesis, University of California at Berkeley.

Curves with Sharp Chabauty-Coleman Bound

Stevan Gajović

Abstract We construct curves of each genus $g \geq 2$ for which Coleman's effective Chabauty bound is sharp and Coleman's theorem can be applied to determine rational points if the rank condition is satisfied. We give numerous examples of genus two and rank one curves for which Coleman's bound is sharp. Based on one of those curves, we construct an example of a curve of genus five whose rational points are determined using the descent method together with Coleman's theorem.

1 Introduction

One of the fundamental problems in the arithmetic of curves and abelian varieties is to find ways to use p-adic information on curves to determine rational points. It is known that p-adic points have a richer structure than rational points, and, in principle, are easier to determine. Thus, it is natural to try to use the knowledge of p-adic points in studying rational points. There are several such approaches, e.g., the Hasse principle and its generalizations, or bounding ranks of elliptic curves by Selmer groups.

The method of Chabauty and Coleman, nowadays also called "abelian Chabauty," is a powerful p-adic method, and it is often very successful in determining rational points on curves whose Jaconians satisfy a certain rank condition. Coleman's "effective Chabauty" theorem (Theorem 2) gives an upper bound on the number of rational points on these curves. However, it is rare that the direct application of Coleman's theorem succeeds in determining all rational points because the bound from Coleman's theorem is not sharp in most cases.

In fact, there are only a few curves known for which Coleman's bound is sharp. Hence it is of interest to study the question highlighted by Coleman's result: Regardless of the rank of its Jacobian, when does the number of rational points on C meet or exceed Coleman's bound at p? We will call curves for which the

S. Gajović (✉)
University of Groningen, Bernoulli Institute, Department of Algebra, Groningen, Netherlands

J. S. Balakrishnan et al. (eds.), *Arithmetic Geometry, Number Theory, and Computation*, Simons Symposia, https://doi.org/10.1007/978-3-030-80914-0_15

rank condition is satisfied, and for which Coleman's bound is sharp for some prime number p of good reduction, *sharp curves at* p. The curves such that the number of their known rational points meets Coleman's bound regardless of the rank condition will be called *potentially sharp curves at* p. Note that we cannot use the method of Chabauty and Coleman to determine the set $C(\mathbb{Q})$ for potentially sharp curves C because we did not check the rank condition. However, if the rank condition holds for a potentially sharp curve, then the curve is sharp; hence, we determined its rational points. Curves that exceed the bound will be called *excessive at* p. Curves for which there is a prime p such that the curve is sharp at p, potentially sharp at p, or excessive at p are called sharp, potentially sharp, and excessive, respectively. We note that the rank of the Jacobians of excessive curves is at least g.

We concisely introduce the method in Sect. 1.1, and the computations of the rank of Jacobians of hyperelliptic curves in Sect. 1.2. In Sect. 1.3, we briefly present the descent method for determining rational points on curves. The descent method is usually used to descend to curves of genus zero and one. Thus, in Sect. 2.3, we construct a curve whose rational points are determined by using descent to curves of genus two, where one of them is sharp.

After mentioning two known sharp curves in Sect. 2.1, we give various examples of sharp curves of genus two in Sect. 2.2, and one with the smallest number of rational points in Sect. 2.4. In Sect. 2.5 we list two curves that violate Coleman's bound, and we use this information to determine their ranks.

In the third section, we start by presenting sharp curves of genus three, four, and five in Sect. 3.1. Using results on the existence of primes in short intervals, we construct infinitely many potentially sharp curves of each genus $g \geq 2$ in Sect. 3.3. We give examples of sharp curves of genus $g = 4$ and $g = 5$ based on this construction.

All computations were done in Magma [BCP97].

1.1 The Method of Chabauty and Coleman and Its History

In 1922, in [Mor22], Louis J. Mordell proved that the group of rational points of an elliptic curve $E/\mathbb{Q}$ is finitely generated. At the end of the article, Mordell observes that curves of genus $g \geq 2$ have only finitely many rational points. The statement that for a nice (smooth, projective, and geometrically irreducible) curve C of genus $g \geq 2$ and a number field K it holds that $\#C(K)$ is finite was named the "Mordell conjecture". After Mordell's theorem it was a natural question whether the group $A(K)$ is finitely generated for all abelian varieties A/K, where K is any number field. In 1928, André Weil proved in [Wei29] that for all abelian varieties A over the number field K we have

$$A(K) \cong \mathbb{Z}^r \oplus A(K)_{\text{tors}},$$

where $A(K)_{\text{tors}}$ is the torsion subgroup of $A(K)$ and $r \in \mathbb{N}_0$ is called the rank of $A(K)$.

In the 1940s, Claude Chabauty proved the Mordell conjecture for curves that satisfy a certain condition.

Theorem 1 (Chabauty 1941, [Cha41]) *Let C be a nice curve of genus $g \geq 2$. Let $J(C)$ be its Jacobian, and let r be the rank of $J(C)$ over $\mathbb{Q}$. If $r < g$, then $C(\mathbb{Q})$ is finite.*

We briefly discuss the idea behind the proof. We may assume that $C(\mathbb{Q}) \neq \emptyset$ and embed $C(\mathbb{Q})$ via an Abel-Jacobi map into $J(C)(\mathbb{Q})$, and do the same over the field $\mathbb{Q}_p$, where p is a prime of good reduction for C. That means that $\overline{C}$, the curve that is obtained by reduction of C modulo p, remains a smooth curve over $\mathbb{F}_p$. We have the following commutative diagram.

$$\begin{array}{ccc} C(\mathbb{Q}) & \hookrightarrow & J(C)(\mathbb{Q}) \\ \downarrow & & \downarrow \\ C(\mathbb{Q}_p) & \hookrightarrow & J(C)(\mathbb{Q}_p) \end{array}$$

The set $J(C)(\mathbb{Q}_p)$ has the structure of a g-dimensional p-adic manifold, containing $C(\mathbb{Q}_p)$ as a 1-dimensional submanifold. The discrete group $J(C)(\mathbb{Q})$ also sits inside $J(C)(\mathbb{Q}_p)$, and we take its closure in the p-adic topology of $J(C)(\mathbb{Q}_p)$, denoted by $\overline{J(C)(\mathbb{Q})}$, to make it a submanifold of $J(C)(\mathbb{Q}_p)$. Then the dimension of $\overline{J(C)(\mathbb{Q})}$ is not greater than r, so heuristically for dimensional reasons, it should have finite intersection with $C(\mathbb{Q}_p)$, which was what Chabauty proved. This intersection contains the set $C(\mathbb{Q})$ (more precisely, $C(\mathbb{Q})$ injects into it), implying the finiteness of $C(\mathbb{Q})$.

This theorem remained the most significant result towards the Mordell conjecture until Gerd Faltings proved the conjecture unconditionally, i.e. for all nice curves of genus $g \geq 2$, in [Fal83]. It seemed that Chabauty's theorem lost its value. But, that is not true because Robert Coleman in 1985 found a way to use Chabauty's approach to state and prove an effective version of the theorem.

Theorem 2 *(Coleman 1985, [Col85a]) Let C be a nice curve of genus $g \geq 2$. Let $J(C)$ be its Jacobian, and let r be the rank of $J(C)(\mathbb{Q})$. Let p be a prime of good reduction for C. If $r < g$, and $p > 2g$, then*

$$\#C(\mathbb{Q}) \leq \#\overline{C}(\mathbb{F}_p) + 2g - 2.$$

Coleman invented a theory of p-adic integration on curves in [Col85b], which has properties that we would expect from integration, such as linearity in integrands and additivity in endpoints. When two endpoints are in the same residue disc, i.e., when both have the same reduction modulo p, we can compute the integral in the expected way by expressing the integrand as a power series in a local parameter and then integrating term by term. There are more important properties, such as Frobenius equivariance, and unexpectedly, path-independence, unlike the real and complex case. The last two properties turn out to be very convenient, and all properties together make Coleman integrals practically computable, which makes Chabauty's method effective. Using Coleman integration, we can construct a locally analytic function $\rho : C(\mathbb{Q}_p) \longrightarrow \mathbb{Q}_p$, which vanishes on $C(\mathbb{Q})$. We can estimate the number of zeros in each residue disc, and summing all estimates gives us the bound in Theorem 2.

We can furthermore investigate the function ρ in each residue disc (and sometimes combine results with some other methods) to provably determine $C(\mathbb{Q})$. This locally analytic function ρ is the main ingredient in determining rational points on curves, and it is a main object of research nowadays to extend the method to the complementary case $r \geq g$. This is an explicit version of a non-abelian extension of this method, a program initiated by Kim in [Kim05] and [Kim09] when $r \geq g$. Quadratic Chabauty is possible when $r = g$, because then, instead of linear functionals (integrals), we use quadratic functions (p-adic heights) to determine rational points. The first explicit examples can be found in [BBM16].

By a careful investigation inside each residue disc, Michael Stoll in [Sto06] proved a theorem that improves Coleman's bound.

Theorem 3 (Stoll 2006, [Sto06]) *Let C be a nice curve of genus $g \geq 2$. Let $J(C)$ be its Jacobian, and let r be the rank of $J(C)$ over $\mathbb{Q}$. Let p be a prime of good reduction for C. If $r < g - 1$, and $p > 2r + 2$, then*

$$\#C(\mathbb{Q}) \leq \#\overline{C}(\mathbb{F}_p) + 2r.$$

As a corollary, we have the following rank statement for sharp curves. We know that $r < g$. If $r < g - 1$, then Coleman's bound cannot be sharp.

Corollary 4 *Let C be a sharp curve of genus $g \geq 2$. Then the rank of the Jacobian of C over $\mathbb{Q}$ is $r = g - 1$.*

If a curve satisfying the rank condition is not sharp, this means that looking at the local information at p is not sufficient to determine its rational points. Nevertheless, it is often possible to compute the rational points by combining the Chabauty information at p with v-adic information for other primes v, for instance, obtained through the Mordell-Weil sieve. The Mordell-Weil sieve was introduced in [Sch99], and well explored in the paper [BS10]. We also note that the Mordell-Weil sieve can be successfully combined with the Quadratic Chabauty method, see, e.g., [BBM17].

One reference for more details about the method of Chabauty and Coleman is [MP12]. The notes [Sik15] contain some very explicit examples, including the combination of the method with the Mordell-Weil sieve.

1.2 On the Computation of the Rank of the Jacobian of Hyperelliptic Curves

The problem of computing the rank of abelian varieties is extremely difficult in general and still open. It is related to one of the most important conjectures in arithmetic geometry. This is the Birch and Swinnerton-Dyer conjecture, which states that the rank of an abelian variety can be computed analytically, namely, as the order of vanishing of its L-function at $s = 1$ (for more details see, e.g., [Tat66]). The latter quantity is called the analytic rank. Thus, if one is willing to assume the Birch and Swinnerton-Dyer conjecture, then the rank of the Jacobian of a hyperelliptic curve can be computed as an analytic rank. We can compute the L-function of hyperelliptic curves in Magma using the algorithm by Dokchitser [Dok04], and we can evaluate it at $s = 1$ as well as its derivatives. We need to be careful that we can compute these values only up to some precision. In general, we cannot prove that some value is zero, although we can verify that it is very close to zero.

Algebraically, there has been a lot of progress since Weil's theorem, and there are methods to compute the rank that work in many cases, but none of them is guaranteed to work, even for the simplest case of elliptic curves. One of the first explicit computations for dimension greater than one appeared in the paper [GG93], followed by [Sch95]. It was later generalized and implemented for hyperelliptic curves in Magma by Stoll, [Sto01].

Let J be the Jacobian of a hyperelliptic curve $C : y^2 = f(x)$. From the Kummer exact sequence $0 \longrightarrow J[2] \longrightarrow J \xrightarrow{[2]} J \longrightarrow 0$, and the long exact sequence of Galois cohomology, we get the following commutative diagram ($M_{\mathbb{Q}}$ is the set of places of $\mathbb{Q}$):

$$
\begin{array}{ccccccccc}
0 & \longrightarrow & J(\mathbb{Q})/2J(\mathbb{Q}) & \longrightarrow & H^1(\mathbb{Q}, J[2]) & \longrightarrow & H^1(\mathbb{Q}, J)[2] & \longrightarrow & 0 \\
 & & \downarrow & & \downarrow & \searrow^{\alpha} & \downarrow & & \\
0 & \longrightarrow & \prod_{v \in M_{\mathbb{Q}}} J(\mathbb{Q}_v)/2J(\mathbb{Q}_v) & \longrightarrow & \prod_{v \in M_{\mathbb{Q}}} H^1(\mathbb{Q}_v, J[2]) & \longrightarrow & \prod_{v \in M_{\mathbb{Q}}} H^1(\mathbb{Q}_v, J)[2] & \longrightarrow & 0
\end{array}
$$
% the diagonal arrow α goes from H^1(Q, J[2]) to ∏ H^1(Q_v, J)[2]

The kernel of the map α is *the Selmer group* $\mathrm{Sel}^{(2)}(J/\mathbb{Q})$, and the kernel

$$Ш(\mathbb{Q}, J) := \ker\left(H^1(\mathbb{Q}, J) \longrightarrow \prod_{v \in M_{\mathbb{Q}}} H^1(\mathbb{Q}_v, J)\right)$$

is called *the Shafarevich-Tate group*.

We have the following exact sequence from the diagram above

$$0 \longrightarrow J(\mathbb{Q})/2J(\mathbb{Q}) \longrightarrow \mathrm{Sel}^{(2)}(J/\mathbb{Q}) \longrightarrow Ш(\mathbb{Q}, J)[2] \longrightarrow 0.$$

It is a well-known fact that $\mathrm{Sel}^{(2)}(J/\mathbb{Q})$ is a finite group. The idea of the proof, which works for more general Selmer groups and can be found in, e.g., [HS00, C.4], is to construct an injection into a finite commutative group indexed by bad and infinite primes. In [HS00, C.4], we also see that the Selmer group is, in principle, effectively computable. We have

$$\dim_{\mathbb{F}_2} J(\mathbb{Q})/2J(\mathbb{Q}) = r + \dim_{\mathbb{F}_2} J(\mathbb{Q})[2] \leq \dim_{\mathbb{F}_2} \mathrm{Sel}^{(2)}(J/\mathbb{Q}).$$

The difference between the two sides is exactly $\dim_{\mathbb{F}_2} \text{Ш}(\mathbb{Q}, J)[2]$.

We can easily determine $J(\mathbb{Q})[2]$; it depends on the factorization of f over $\mathbb{Q}$. Thus, we can obtain an upper bound on the rank of $J(\mathbb{Q})$.

For the lower bound, we search for rational points on J, up to some bounded height, and hope that eventually, we will find the same number of independent points as the upper bound for the rank. If the curve has many rational points, then we might be able to find a lower bound on the rank, see the remark in Sect. 2.5. The idea to use Coleman's bound to obtain the lower bound for the rank with the infinite families of curves as the examples and the improvements on the lower bound can be found in [JT99].

From the discussion above, we easily see why the algorithm is not guaranteed to work, but it still works very well in practice. We also see that the (partial) knowledge of $\text{Ш}(\mathbb{Q}, J)$ can improve upper bounds, e.g., as in [Sto01, Chapter 8].

1.3 Introduction to the Descent Method

The descent method is one of the first methods ever used for determining rational points on curves. It is still very applicable, especially for hyperelliptic curves. The basics of this method are explained in more detail in [Sto15]. For explicit descent on hyperelliptic curves, see [BS09]. The descent method is based on the property of unique factorization in $\mathbb{Z}$. In a sense, it is a generalization of the fact that if a product of two numbers is a square, then these two numbers have the same squarefree part. We can bound the possible greatest common divisor of these numbers, which gives only finitely many possibilities for their squarefree parts. In principle, we have a strategy to reduce all possible factorizations to a finite number.

Let us make it more precise. Suppose that $C/\mathbb{Q}$ is a hyperelliptic curve given by the equation

$$C : y^2 = f_1(x) f_2(x),$$

where f_1 and f_2 are non-constant coprime integral polynomials, with at least one of $\deg(f_1)$ and $\deg(f_2)$ even. Then, for every point $(x_0, y_0) \in C(\mathbb{Q})$, we have $f_1(x_0) \neq 0$ or $f_2(x_0) \neq 0$, and there is a unique squarefree $d \in \mathbb{Z}$, and $z_0, t_0 \in \mathbb{Q}$, such that

$$f_1(x_0) = dz_0^2, \quad f_2(x_0) = dt_0^2.$$

Denote by C_d the curve given by equations

$$C_d : f_1(x) = dz^2, \quad f_2(x) = dt^2;$$

then there is a covering $\pi_d : C_d \longrightarrow C$ given by

$$(x, z, t) \mapsto (x, dzt).$$

It follows that

$$C(\mathbb{Q}) = \bigcup_{d \in S} \pi_d(C_d),$$

where $S \subset \mathbb{Z}$ denotes the set of all squarefree numbers. This set is infinite, but we can reduce it to a finite set. Let us assume that f_1 and f_2 are monic polynomials (otherwise, we have to include primes that divide their leading coefficients). As these polynomials are coprime, their resultant is a non-zero integer R. The resultant gives information on possible common roots. So, for all primes p such that $p \nmid R$, the resultant of f_1 and f_2 over $\mathbb{F}_p$ is not zero, implying that f_1 and f_2 do not have a common root in $\mathbb{F}_p$. It means that f_1 and f_2 are "coprime at p", i.e., that it is not possible that p divides the squarefree part d. So, the number of possible primes that can divide the squarefree part d is finite, and we need to investigate a finite number of curves C_d (which, in principle, should be easier than doing so for C) to find all rational points on C.

2 Examples of Sharp Curves of Genus Two

2.1 Two Known Sharp Curves in Genus Two

We first mention two known examples.

Example 1 ([Gra94]) This is the first known example of a sharp curve, so we see that it took almost 10 years after Coleman's paper [Col85a] until the first example appeared. Let C be the curve given by the equation

$$C : y^2 = x(x-1)(x-2)(x-5)(x-6).$$

This curve has good reduction at $p = 7$, and it has eight $\mathbb{F}_7$-rational points

$$\overline{C}(\mathbb{F}_7) = \{(0,0), (1,0), (2,0), (3,\pm 1), (5,0), (6,0), \infty\}.$$

In [GG93], it is computed that $J(C)(\mathbb{Q}) \cong \mathbb{Z} \times (\mathbb{Z}/2\mathbb{Z})^4$. We can apply Theorem 2 to obtain that $C(\mathbb{Q}) \leq 8 + 2 = 10$. We can find ten rational points, so

$$C(\mathbb{Q}) = \{(0,0), (1,0), (2,0), (3,\pm 6), (5,0), (6,0), (10,\pm 120), \infty\}.$$

Example 2 ([HM19]) This example is miraculous because it arises from a geometric problem. Namely, this curve occurs in analysing the problem whether there exists a pair of one right triangle and one isosceles triangle, both with rational sides, having the same area and perimeter. More details about the problem can be found in [HM19]. The curve C is given by the equation

$$C : y^2 = (x^3 - x + 6)^2 - 32.$$

We compute that the rank of its Jacobian over $\mathbb{Q}$ is equal to $r = 1$, that C has good reduction at $p = 5$, and that its reduction modulo 5 has exactly eight points. All conditions for Theorem 2 are satisfied; thus we conclude

$$C(\mathbb{Q}) = \left\{(0,\pm 2), (1,\pm 2), (-1,\pm 2), \left(\frac{5}{6}, \pm\frac{217}{216}\right), \infty_{\pm}\right\}.$$

2.2 A Finite Family of Sharp Curves of Genus Two

We construct new examples of sharp curves of genus two. Let C be a curve. We focus on genus two curves because the computation of the rank becomes more difficult for larger genus. Thus, the bound is given by

$$\#C(\mathbb{Q}) \leq \#\overline{C}(\mathbb{F}_p) + 2.$$

Since we want this bound to be sharp, we would like that two residue discs have extra points (this condition is natural due to the hyperelliptic involution), and other ones have exactly one rational point. The easiest way to achieve this is to require a small number of residue discs. We will construct a suitable monic polynomial $f(x)$ of degree 5. One natural choice for a prime p to consider is $p = 11$, because $x^5 \equiv 0, \pm 1 \pmod{11}$. If we want to include other monomials in x, we want them to have a coefficient divisible by 11, to control the polynomial $f(x)$ modulo 11 easily. It remains to pick the constant term. We want some constant for which we can find rational points, but not too many modulo 11. Quadratic residues modulo 11 are in the set $\{0, 1, 3, 4, 5, 9\}$, so if we pick the constant term to be 9 modulo 11, we have that

$$f(x) \equiv x^5 + 9 \equiv \{8, 9, 10\} \pmod{11}.$$

This implies that

$$\overline{C}(\mathbb{F}_{11}) = \{(0, \pm 3), \infty\}.$$

When the rank of the Jacobian of C over $\mathbb{Q}$ is less than two, by Coleman's bound, C can have at most 5 rational points.

Note, as in [Gra94], that we should check if a curve C has absolutely irreducible Jacobian. If this were not the case, then $J(C)$ would be isomorphic to the product of two elliptic curves over some number field K, and one of them would have rank zero, providing an easier way to determine rational points on C. In [Gra94], David Grant together with Jaap Top, proved absolute simplicity of the Jacobian of the curve from Example 1 by proving that its L-function cannot be a product of two L-functions of elliptic curves. In the meantime, in [HZ02], Everett W. Howe and Hui J. Zhu found a criterion when an abelian variety over a finite field is absolutely simple.

Lemma 5 ([HZ02, Lemma 8]) *Let q be a prime power and $n > 2$ an integer. Suppose that π is an ordinary Weil q-number (the minimal polynomial for π is a characteristic polynomial of Frobenius of an ordinary abelian variety). Let $K = \mathbb{Q}(\pi)$, K^+ its maximal real subfield, and $n = [K^+ : \mathbb{Q}]$. Suppose that*

(1) the minimal polynomial of π is not of the form $x^{2n} + ax^n + q^n$,
(2) the field K^+ has no proper subfields other than $\mathbb{Q}$,
(3) the field K^+ is not the maximal real subfield of a cyclotomic field.

Then the isogeny class corresponding to π consists of absolutely simple abelian varieties.

We finally give a number of examples.

Proposition 6 *Let $C_k : y^2 = f(x)$ be a hyperelliptic curve over $\mathbb{Q}$, where*

$$f(x) = x^5 + 11x^4 + (11k + 3)^2, \quad k \in \{0, 1, 2, 3, 7, 10, 11, 12, 15, 21, 22, 31, 40, 42, 44, 47, 50\}$$

or

$$f(x) = x^5 + 11x^4 + (11k - 3)^2, \quad k \in \{1, 4, 9, 15, 16, 17, 19, 27, 28, 31, 40, 41, 42, 43\}.$$

(1) Then $\#C_k(\mathbb{Q}) = 5$, and in the first case

$$C_k(\mathbb{Q}) = \{(0, \pm(11k + 3)), (-11, \pm(11k + 3)), \infty\},$$

whereas in the second case

$$C_k(\mathbb{Q}) = \{(0, \pm(11k - 3)), (-11, \pm(11k - 3)), \infty\}.$$

(2) All curves C_k from (1) have absolutely simple Jacobian.

Proof

(1) In both cases, we run the Magma command `RankBound` to compute an upper bound on the rank of the Jacobian of curves C_k for $0 \leq k \leq 50$. We listed above all k's in both cases for which the upper bound is less than two. As we saw, $p = 11$ is a prime of good reduction of all curves C_k and $\#\overline{C_k}(\mathbb{F}_{11}) = 3$, so we conclude by Theorem 2 that $\#C_k(\mathbb{Q}) \leq 5$. The fact that we have five obvious rational points on each C_k finishes the proof. All listed curves C_k are sharp and have rank $r = 1$.

(2) Using Magma and Lemma 5 we check that all $J(C_k)$ are absolutely simple by finding a prime p such that the reduction of $J(C_k)$ modulo p is absolutely simple.

□

2.3 An Example with Descent and a Sharp Curve

There are many examples of applying descent from curves of higher genus to curves of genus zero or one in the literature. Lack of examples of curves whose rational points can be determined by applying descent to curves of genus at least two motivated us to construct one such curve.

Proposition 7 *Let $C/\mathbb{Q}$ be the hyperelliptic curve given by the equation*

$$C : y^2 = f(x) := (x^6 + 11x^5 + 64x + 729)(x^5 + 11x^4 + 64).$$

The set of rational points on C is

$$C(\mathbb{Q}) = \{(0, \pm 216), (-11, \pm 40), \infty\}.$$

Proof As we can see, the polynomial f is monic, so when applying the descent method we need to compute the resultant of two polynomials, whose product is f, which is

$$R := \mathrm{Res}(x^6 + 11x^5 + 64x + 729, x^5 + 11x^4 + 64)) = 3^{30}.$$

We are only interested in the radical of the resultant, since it is enough to consider squarefree numbers, and $\mathrm{rad}(R) = 3$. Thus, after applying the descent, we know that x corresponds to an x-coordinate of a rational point on one of curves

$$x^5 + 11x^4 + 64 = dz^2, \quad x^6 + 11x^5 + 64x + 729 = dt^2$$

for some $d \in \{-3, -1, 1, 3\}$.

We first prove that $d < 0$ is impossible. If $x^5 + 11x^4 + 64 < 0$, then $x < 0$ and so

$$x(x^5 + 11x^4 + 64) + 729 > 0,$$

giving a contradiction.

If $d = 3$, then we can use the linear change of coordinates $X = 3x$, $Y = 3^3 z$, and then the equation of the curve

$$C_3 : x^5 + 11x^4 + 64 = 3z^2$$

transforms to

$$C_3' : X^5 + 33X^4 + 64 \cdot 3^5 = Y^2.$$

For the curve C_3', we compute in Magma that $J(C_3')(\mathbb{Q}) = \langle 0 \rangle$. We know that $C_3'(\mathbb{Q})$ embeds into $J(C_3')(\mathbb{Q})$, so C_3' has at most one rational point: this is the one at infinity.

If $d = 1$, then we consider the hyperelliptic curve

$$C_1 : x^5 + 11x^4 + 64 = z^2,$$

which is one of the curves in Proposition 6, and we already know

$$C_1(\mathbb{Q}) = \{(0, \pm 8), (-11, \pm 8), \infty\}.$$

We see that all of these points give rise to the rational points of C, and that

$$C(\mathbb{Q}) = \{(0, \pm 216), (-11, \pm 40), \infty\}.$$

□

Comment Let r be the rank of the Jacobian of C over $\mathbb{Q}$. Assuming the Generalized Riemann Hypothesis, Magma gives that $2 \leq r \leq 4$. So, in principle, we would be able to apply the abelian Chabauty method to C. However, this seems to be more difficult than the approach in the proof, which is furthermore unconditional. As we can check in Magma, the Jacobian of C is absolutely simple.

2.4 A Sharp Curve with the Smallest Possible Number of Points

We give an example of a sharp curve that satisfies the property that it has the smallest number of rational points amongst all sharp curves of genus $g \geq 2$. For a sharp curve C of genus $g \geq 2$ it holds $\#C(\mathbb{Q}) \geq 3$. Indeed, we know that in that case

$$\#C(\mathbb{Q}) = \#\overline{C}(\mathbb{F}_p) + 2g - 2 \geq \#\overline{C}(\mathbb{F}_p) + 2 \geq 2.$$

Therefore, the set $C(\mathbb{Q})$ is non-empty, implying that also $\overline{C}(\mathbb{F}_p)$ is non-empty, giving the desired conclusion. If we want to construct such a curve, we want a curve of genus two with only one residue disc. If we consider a hyperelliptic curve $C : y^2 = f(x)$, where f is a monic polynomial of degree 5, we already know that we have one residue disc at infinity. So, we want to construct a curve having no other residue discs, meaning that the reduction over $\mathbb{F}_p$ can have only the point at infinity. We will use a similar strategy as before, so that

$$\overline{C} : y^2 = x^5 + c/\mathbb{F}_{11}.$$

If $c \equiv 7 \pmod{11}$, then $\overline{C}$ has only one rational point, $\overline{C}(\mathbb{F}_{11}) = \{\infty\}$ because 6, 7, and 8 are not quadratic residues modulo 11, and C has good reduction at $p = 11$. There should be two more rational points on C. Let one of them be $P = (a, b)$. Then P maps to infinity modulo 11. Furthermore, we know that the denominator of a is divisible by 11^2 since otherwise $f(a)$ is not a square of a rational number. Let us try to find an example such that a is already a square of a rational number and that a is a zero of a polynomial $f(x) - x^5$, with f satisfying the conditions above.

Let C be the hyperelliptic curve over $\mathbb{Q}$ defined by

$$C : y^2 = x^5 + 121x - 4.$$

Magma gives us that r, the rank of the Jacobian of C over $\mathbb{Q}$, is zero or one. We can apply Theorem 2, and by the previous discussion,

$$C(\mathbb{Q}) = \left\{ \left(\frac{4}{121}, \pm\frac{2^5}{11^5} \right), \infty \right\}.$$

We also know that $r = 1$ in this case by Corollary 4. We could also conclude this in another way: Using Magma we see that the torsion subgroup is trivial, which is impossible when $r = 0$ and $\#C(\mathbb{Q}) > 1$.

2.5 Examples on Improving Lower Rank Bounds in Magma

We now present two examples of excessive curves, for which we can improve Magma's lower bounds for the rank. In both examples, we can determine the rank of the curve, although Magma gives inconclusive information.

Again, it is desirable to have a small number of residue discs, and in particular, we would like only one residue disc. This time, we will consider $p = 5$ and the curve

$$C : y^2 = 8x^6 - 314x^5 + 3250x^4 - 10000x^3 + 64x = x(2(x-25)(4x-25)(x^3-8x^2)+64).$$

The reduction of C modulo 5 is

$$\overline{C} : y^2 = x(3x^5 + x^4 - 1)/\mathbb{F}_5$$

We easily see that C has good reduction at $p = 5$ and that $\overline{C}(\mathbb{F}_5) = \{(0, 0)\}$. We can find at least five points on this curve:

$$\left\{(0,0), \left(\frac{25}{4}, \pm 20\right), (25, \pm 40)\right\} \subset C(\mathbb{Q}),$$

so C is excessive. Magma gives us that the rank r of the Jacobian of C satisfies $0 \leq r \leq 2$. Hence, we conclude that $r = 2$.

Note that we do not find the set $C(\mathbb{Q})$ here because Chabauty's condition $r < g$ is violated. It might be possible to compute $C(\mathbb{Q})$ using Quadratic Chabauty.

The following example is conditional, i.e., for the rank computations in Magma we need to assume the Generalized Riemann Hypothesis. We can do the same for the following curve

$$C : y^2 = x^5 - 12(121x - 1)(121x - 4)$$

looking at the good prime $p = 11$. This curve has at least five rational points, so it is excessive. Magma gives us that the rank of its Jacobian over $\mathbb{Q}$ is between zero and two, thus, it is two.

Remark We see that a curve $C/\mathbb{Q}$ with many rational points (at least $2g - 2$) might be potentially sharp or excessive. There is an algorithm to check that. We need to determine all prime numbers p for which C can be potentially sharp or excessive at p. We use the Hasse-Weil bounds and the number of known rational points to give an upper bound for primes p, so we need to check finitely many of them. Therefore, if we find that the curve is potentially sharp or excessive, we obtain a lower bound for the rank, $r \geq g - 1$, or $r \geq g$, respectively.

Note that here we only considered primes of good reduction. There are results similar to Coleman's bound when we consider primes of bad reduction, and these can be found in [LT02] and improved, with a bound in Stoll's style, in [KZB13]. It would be interesting to investigate the case of bad reduction in the future. In this case, we might be able to construct infinite families of examples with sharp bounds. We could then allow multiple Weierstrass points in the same residue discs, which makes computations of ranks of curves easier and hope that we can determine the ranks of Jacobians for an infinite number of curves.

3 Examples of Sharp Curves of Higher Genus

We present a few sharp curves of genus $g > 2$. There is a way to construct potentially sharp curves, which works for any genus g and the construction is contained in Sect. 3.3, including examples of sharp curves of genus $g = 4$ and $g = 5$, one of each. In some cases, there is an easier way to find examples of sharp curves, and it is presented in the following subsection.

3.1 *Concrete Examples of Sharp Curves of Genus Three, Four and Five*

Let g be a positive integer. In the case where $2g + 1$ or $2g + 3$ is a prime number, we can construct examples of potentially sharp curves in the following ways.

If $2g+1 = p$, where p is some prime number, then $x^{2g+1} - x \equiv 0 \pmod{p}$ for all $x \in \mathbb{F}_p$. If c is any quadratic nonresidue modulo p, then for the curve

$$\overline{C} : y^2 = x^{2g+1} - x + c/\mathbb{F}_p$$

we have $\overline{C}(\mathbb{F}_p) = \{\infty\}$. So, all possible rational points on the curve C reduce modulo p to ∞, thus are not integral and have denominators divisible by p. Let $a_1, \dots, a_{g-1}$ be integers not divisible by p, with distinct absolute values. Let $b \in \mathbb{Z}$ be any inverse of $(a_1 \dots a_{g-1})^2$ modulo p. The following curve is potentially sharp:

$$C : y^2 = x^{2g+1} + b(a_1^2 - p^2x) \cdots (a_{g-1}^2 - p^2x)(c - x).$$

If $2g + 3 = p > 3$ is prime, then we can use the property that there are two consecutive quadratic nonresidues modulo p. If $p \equiv 1 \pmod 4$, then between 2 and $p - 2$ there are $\frac{p-1}{2}$ quadratic nonresidues implying that two of them must be consecutive. If $p \equiv 3 \pmod 4$, -1 and -4 are quadratic nonresidues, so if -2 or -3 is a quadratic nonresidue, we have the conclusion. If not, then 2 and 3 are consecutive quadratic nonresidues. Denote by c and $c+1$ two consecutive quadratic nonresidues modulo p. Then the curve

$$\overline{C} : y^2 = x^{2g+2} + c/\mathbb{F}_p$$

has two $\mathbb{F}_p$-points, both at infinity because, for $x \in \mathbb{F}_p$, we have that

$$x^{2g+2} + c \equiv \{c, c+1\} \pmod{p}.$$

We construct examples from this curve. Let $a_1, \dots, a_{g-1}$ be distinct integers not divisible by p. Let $b \in \mathbb{Z}$ be any inverse of $a_1 \cdots a_{g-1}$ modulo p. The following curve is potentially sharp:

$$C' : y^2 = x^{2g+2} + b(a_1 - px) \cdots (a_{g-1} - px)c.$$

We summarize the conclusions above into the theorem.

Theorem 8 *Let $g \geq 2$ be a positive integer such that $2g+1$ or $2g+3$ is a prime number. Define C and C' as above in each case. If r, the rank of the Jacobian of C over $\mathbb{Q}$, satisfies $r < g$, then $r = g-1$, and*

$$C(\mathbb{Q}) = \left\{ \left(\frac{a_1^2}{p^2}, \pm \frac{a_1^{2g+1}}{p^{2g+1}} \right), \ldots, \left(\frac{a_{g-1}^2}{p^2}, \pm \frac{a_{g-1}^{2g+1}}{p^{2g+1}} \right), \infty \right\}.$$

If r', the rank of the Jacobian of C' over $\mathbb{Q}$, satisfies $r' < g$, then $r' = g-1$, and

$$C'(\mathbb{Q}) = \left\{ \left(\frac{a_1}{p}, \pm \frac{a_1^{g+1}}{p^{g+1}} \right), \ldots, \left(\frac{a_{g-1}}{p}, \pm \frac{a_{g-1}^{g+1}}{p^{g+1}} \right), \infty_\pm \right\}.$$

Proof Follows from the above and Corollary 4. □

We give one example of a sharp curve in each genus $g \in \{3, 4, 5\}$. For these curves we are able to check the rank condition $r < g$. However, we need to assume the Generalized Riemann Hypothesis for the rank computations in Magma. The curves are as follows:

$$C_3 : y^2 = x^7 - (49x-1)(49x-36)(x+1), \quad r(J(C_3)(\mathbb{Q})) = 2,$$

$$C_3(\mathbb{Q}) = \left\{ \left(\frac{1}{49}, \pm \frac{1}{7^7} \right), \left(\frac{36}{49}, \pm \frac{6^7}{7^7} \right), \infty \right\};$$

$$C_4 : y^2 = x^{10} - (11x-3)(11x-4)(11x-6), \quad r(J(C_4)(\mathbb{Q})) = 3,$$

$$C_4(\mathbb{Q}) = \left\{ \left(\frac{3}{11}, \pm \frac{3^5}{11^5} \right), \left(\frac{4}{11}, \pm \frac{4^5}{11^5} \right), \left(\frac{6}{11}, \pm \frac{6^5}{11^5} \right), \infty_\pm \right\};$$

$$C_5 : y^2 = x^{12} - (13x-1)(13x-2)(13x-3)(13x-12), \quad r(J(C_5)(\mathbb{Q})) = 4,$$

$$C_5(\mathbb{Q}) = \left\{ \left(\frac{1}{13}, \pm \frac{1}{13^6} \right), \left(\frac{2}{13}, \pm \frac{2^6}{13^6} \right), \left(\frac{3}{13}, \pm \frac{3^6}{13^6} \right), \left(\frac{12}{13}, \pm \frac{12^6}{13^6} \right), \infty_\pm \right\}.$$

3.2 Bertrand's Postulate for Primes Modulo 8

For our construction, we will need a stronger version of Bertrand's Postulate. We formulate one version first, and then we cover smaller cases.

Proposition 9 *Let $n \geq 15$ be a positive integer. In the interval $[n, 2n)$ there is a prime number p such that $p \equiv 5 \pmod 8$.*

Proof Let $\mathbb{P}$ denote the set of prime numbers in $\mathbb{Z}$. For any coprime integers k and l, we define the function

$$\theta(x; k, l) = \sum_{p \in \mathbb{P},\ p \leq x,\ p \equiv l \pmod k} \log(p).$$

Theorem 1, for $k \leq 13$, from [RR96], states that

$$\max_{1 \leq y \leq x} \left| \theta(y; k, l) - \frac{y}{\varphi(k)} \right| \leq \varepsilon \frac{x}{\varphi(k)},$$

where one can take $\varepsilon = 0.00456$ for $x \geq 10^{10}$. It implies that when $y = x$

$$\left| \theta(x; 8, 5) - \frac{x}{4} \right| \leq 0.00456\frac{x}{4} \leq 0.005\frac{x}{4} \implies 0.995\frac{x}{4} \leq \theta(x; 8, 5) \leq 1.005\frac{x}{4}$$

for $x \geq 10^{10}$. If $x \geq 10^{10}$, then also

$$\theta(x; 8, 5) \leq 1.005\frac{x}{4} < 0.995\frac{2x}{4} \leq \theta(2x; 8, 5)$$

giving that $\theta(2x; 8, 5) > \theta(x; 8, 5)$ for $x \geq 10^{10}$. It follows that there is a prime congruent to 5 modulo 8 between x and $2x$. We need to prove the statement for the remaining interval $[15, 10^{10}]$. We can easily find a list of primes congruent to 5 modulo 8 such that the next one is larger than half of the previous one (e.g., we can use Magma for it), which completes the proof. For instance, we can take the list to be

$$10000000061, 5000000141, 2500000117, 1250000077, 625000069, 312500077, 156250093,$$

$$78125141, 39062581, 19531381, 9765757, 4882957, 2441573, 1220797, 610429, 305237,$$

$$152629, 76333, 38189, 19141, 9613, 4813, 2437, 1229, 653, 349, 181, 101, 53, 29.$$

□

Proposition 10 *Let $n \geq 2$ be a positive integer. Then in the interval $[n, 2n)$ there is a prime number p such that $p \equiv 3 \pmod 8$ or $p \equiv 5 \pmod 8$.*

Proof By the previous proposition, we only need to check small cases, but the statement is trivially true since 3, 5, 11, 19 are primes that fill in the missing gaps. □

As a corollary, in each interval $[n, 2n)$, where $n > 1$ is a positive integer, there is a prime p such that 2 is not a quadratic residue modulo p, which is the property we will use in our construction of curves in the following subsection.

3.3 Potentially Sharp Curves of Genus $g \geq 2$

Let $g \geq 2$ be a positive integer. We will construct potentially sharp hyperelliptic curves $C : y^2 = f(x)$ of genus g, where f is a monic polynomial. For simplicity, we present the construction and ideas in a simpler way. We add a comment on how we can slightly generalize it using the same ideas after Theorem 12 and use these generalized ideas to construct concrete examples.

By our version of Bertrand's postulate, Proposition 10, there is a prime number p, congruent to 3 or 5 modulo 8, that is contained in the interval $(2g + 2, 4g + 4)$. Define the polynomial

$$Q(x) = x^{2g+2} - x^{\frac{p-1}{2}} + x^{2g+2-\frac{p-1}{2}} + 1.$$

Then we have the following congruence modulo p:

$$Q(x) \equiv \begin{cases} 2x^{2g+2} \pmod{p}, \text{ if } \left(\dfrac{x}{p}\right) = 1, \\ 2 \pmod{p}, \text{ if } \left(\dfrac{x}{p}\right) = -1, \\ 1 \pmod{p}, \text{ if } p \mid x. \end{cases}$$

Since p is chosen so that 2 is quadratic nonresidue modulo p, it follows that

$$\overline{C}(\mathbb{F}_p) = \{(0, \pm 1), \infty_{\pm}\},$$

if $\overline{C}$ is the hyperelliptic curve over $\mathbb{F}_p$, defined by $y^2 = Q(x)$.

For each $s \leq g$, $s \in \mathbb{N}$, we will construct curves $C_s : y^2 = H_s$, whose reduction modulo p is $\overline{C}$ and which will have at least $4 + 2s$ rational points. Four rational points on C_s will be from the set $\{(0, \pm 1), \infty_{\pm}\}$, and we need to add $2s$ points, i.e. s values of x for which $H_s(x)$ is a square of a rational number. We will construct H_s such that new rational points on C_s are integral, so that all x-coordinates will be integers divisible by p. We note that then C_{g-1} is a potentially sharp curve and C_g excessive.

Let $a_1, \ldots, a_s \in \mathbb{Z}$ be distinct and non-zero. We will construct the polynomial H_s such that $H_s(pa_1)$, $H_s(pa_2)$, …, $H_s(pa_s)$ are squares of integers. The strategy is as follows. All monomials x^k with $k > s$ will be replaced by

$$t_s(x^k) := x^{k-s}(x - pa_1)(x - pa_2)\cdots(x - pa_s).$$

We extend this transformation linearly so that we can compute $t_s(P) \in \mathbb{Z}[x]$ for all polynomials $P \in \mathbb{Z}[x]$ that have only monomials in x of degree greater than s. This transformation has two important properties, namely for all such $P \in \mathbb{Z}[x]$, we have

$$t_s(P) \equiv P \pmod{p},$$

$$t_s(P)(0) = t_s(P)(pa_1) = \cdots = t_s(P)(pa_s) = 0.$$

We can apply this transformation to the higher degree part of $Q(x)$, e.g. to the polynomial $x^{2g+2} - x^{\frac{p-1}{2}}$ (note that $\frac{p-1}{2} > g \geq s$). We need to deal with the remaining part $1 + x^l$, where $l := 2g + 2 - \frac{p-1}{2}$. If $l > s$, then we can apply t_s to x^l and define a polynomial

$$Z_s(x) := t_s(x^{2g+2} - x^{\frac{p-1}{2}} + x^l) + 1.$$

If $s \geq l$, the polynomial $1 + x^l$ will be transformed into a part of a square of some polynomial. In this way, we can assure that $H_s(pa_i)$ is a square for $i = 1, \ldots, s$. Denote $m \in \mathbb{N}$ the number for which $ml \leq s < (m+1)l$. We use the following lemma.

Lemma 11 *There exist integers $c_1, c_2, \ldots, c_m$ such that*

$$(1 + c_1x^l + c_2x^{2l} + \cdots + c_mx^{ml})^2 \equiv 1 + x^l \pmod{p, x^s},$$

meaning that these two polynomials agree modulo p up to degree s.

Proof We prove this lemma constructively. We can start by taking $c_1 = \frac{p+1}{2}$ (or any other number congruent modulo p to it). Then, looking at coefficients with x^{jl}, where $j \in \{2, \ldots, m\}$ we have

$$2c_j + S_j \equiv 0 \pmod{p},$$

where S_j is some polynomial in the previous coefficients $c_1, \ldots, c_{j-1}$. Since p is odd, 2 is invertible and we can determine c_j modulo p. □

If c_j are the coefficients from the previous lemma, then

$$(1 + c_1x^l + c_2x^{2l} + \cdots + c_mx^{ml})^2 = G_s(x) + L_s(x),$$

where $\deg(L_s) \leq s$ and $L_s(x) \equiv 1+x^l \pmod{p}$, and all monomials in x in G_s have degree at least $s + 1$. Note $\deg(G_s) = 2ml \leq 2s < 2g + 2$. Define the polynomial Z_s as

$$Z_s(x) = t_s(x^{2g+2} - x^{\frac{p-1}{2}} - G_s(x)) + G_s(x) + L_s(x).$$

In both cases ($l > s$ and $s \geq l$) for all $i = 1, \ldots, s$, we have

$$Z_s(pa_i) = G_s(pa_i)+L_s(pa_i) = (1+c_1p^la_i^l+c_2p^{2l}a_i^{2l}+\ldots+c_mp^{ml}a_i^{ml})^2 =: b_i^2,$$

where, if $l > s$, we have $c_1 = \cdots = c_m = 0$, i.e. $b_1 = \cdots = b_s = 1$, $Z_s(0) = 1$, and

$$Z_s(x) \equiv x^{2g+2} - x^{\frac{p-1}{2}} + x^l + 1 = Q(x) \pmod{p}.$$

We can change Z_s by any polynomial $px(x - pa_1)\cdots(x - pa_s)R(x)$, where $R \in \mathbb{Z}[x]$ has degree at most $2g - s$, and the new curve will still have all known rational points and the same reduction modulo p. So, for any choice as above, define

$$C_s : y^2 = H_s(x), \quad H_s(x) := Z_s(x) + px(x - pa_1)\cdots(x - pa_s)R(x).$$

Note that we can take $b_i \equiv 1 \pmod{p}$, so $b_i \neq 0$, and we indeed have at least $4 + 2s$ rational points on C_s:

$$\{\infty_\pm, (0, \pm 1), (pa_1, \pm b_1), \ldots, (pa_s, \pm b_s)\} \subset C_s(\mathbb{Q}).$$

Therefore, we constructed many examples of potentially sharp curves C_{g-1} and curves C_g of large rank. Let r_s be the rank of the Mordell-Weil group of the Jacobian of C_s over $\mathbb{Q}$. We summarize this in the following theorem.

Theorem 12

(1) Let $s = g - 1$, i.e., let C_{g-1} be a curve defined as above. If $r_{g-1} < g$, then $r_{g-1} = g - 1$, and

$$C_{g-1}(\mathbb{Q}) = \{\infty_\pm, (0, \pm 1), (pa_1, \pm b_1), \ldots, (pa_{g-1}, \pm b_{g-1})\}.$$

(2) Let $s = g$, i.e., let C_g be a curve defined as above. Then $r_g \geq g$.

Proof

(1) By construction, C_{g-1} is potentially sharp. If the rank condition is satisfied, then C_{g-1} is sharp, hence has exactly $2g + 2$ points and rank $r_{g-1} = g - 1$.
(2) The curve C_g is constructed to be excessive, so $r_g \geq g$.

□

Comment In the spirit of the ideas in the construction, we can change the curves in the following ways.

(1) The transformation t_s can be slightly changed to ($k > e_1 + \cdots + e_s$)

$$t_{e_1,\dots,e_s}(x^k) = x^{k-e_1-\dots-e_s}(x-pa_1)^{e_1}(x-pa_2)^{e_2}\cdots(x-pa_s)^{e_s}.$$

(2) The numbers $a_1, \dots, a_s$ do not have to be integers; we can allow them to be rational numbers such that $a_i = \frac{b_i}{c_i}$, $\mathrm{GCD}(b_i, c_i) = 1$, $p \mid b_i$ for all $i = 1, \dots, s$. But we need to ensure that the curve still has two points at infinity. Note that the resulting polynomial, after a linear change of variables, might not be monic anymore.

(3) We can change anything modulo p, so that the reduction of the curve stays the same, as long as we do not change the number of known rational points. For example, the constant term of the polynomial can be any square of an integer congruent to 1 modulo p, or we can change the polynomials constructed in Lemma 11, or increase their degree until it is not greater than g, and similar constructions.

We now want to construct some concrete examples of sharp curves from the previous family. To this end we need to check is whether the rank assumption is satisfied. In a sense, we want to find curves with many rational points and small rank of its Jacobian.

Remark There is a program started by Noam Elkies, and continued by Michael Stoll and Andreas Kühn dealing with a very similar question: Can we construct curves of genus 2 with many rational points whose differences are contained in the same arithmetic progression inside the Jacobian? Such curves have small rank despite having many rational points. In particular, we can apply the method of Chabauty and Coleman for these curves, at least with respect to the subgroup of the Jacobian, which contains the arithmetic progression. In many cases, this subgroup is a finite index subgroup of the Jacobian. Some examples can be found in [Elk] and [Sto].

We now look at some of their examples and we determine whether they are sharp. In [Elk], we can, e.g., find the family

$$C(b) : y^2 = (x^3 - x^2 + bx + (b+1))^2 - 4b(b+1)x.$$

The curve $C(b)$ has rational points $\infty_\pm$, $(0, \pm(b+1))$, $(1, \pm 1)$, and $(-1, \pm(2b+1))$. We denote $q := [\infty_+ - \infty_-]$. Then $[(0, -(b+1)) - (0, b+1)] = 5q$, $[(1,1) - (1,-1)] = 13q$, $[(-1, 2b+1) - (-1, -(2b+1)] = 29q$. One special curve, for $b = -\frac{5}{2}$, has six pairs of points in the arithmetic progression. The curve $C(-\frac{5}{2})$ has to more pairs of rational points, $(3, \pm 6)$, and $(\frac{1}{2}, \pm\frac{7}{8})$, and $[(3,6) - (3,-6)] = 61q$, $[(\frac{1}{2}, \frac{7}{8}) - (\frac{1}{2}, -\frac{7}{8})] = 83q$.

The rank of $J(C(-\frac{5}{2}))(\mathbb{Q})$ is equal to 1, and it is almost sharp, since it can be proven using the method of Chabauty and Coleman combined with the Mordell-Weil sieve that $\#C(\mathbb{Q}) = 12$, whereas $\#\overline{C}(\mathbb{F}_7) = 11$.

There is another curve, found by Stoll, $C : y^2 = (x+2)(x^2 - 3x + 6)(4x^3 + 8x^2 - 3x + 3)$, that is sharp and is interesting because it has 13 rational points and still rank 1. In this case, the differences of non-Weierstrass points and their images under the hyperelliptic involution do not form an arithmetic progression, but an arithmetic progression modulo the torsion subgroup of the Jacobian.

In the end, we mention the curve from [Sto], $C : y^2 = 25x^6 + 20x^5 - 76x^4 - 134x^3 + 124x^2 + 96x + 9$. This curve is interesting because its Jacobian contains a rational point of canonical height ≈ 0.000298, which is the smallest known positive canonical height on a Jacobian surface over $\mathbb{Q}$. It turns out that this curve is sharp.

Based on the construction of curves C_{g-1} from Theorem 12, we searched for examples of sharp curves using Magma. We were unable to find sharp examples of genus two or three. For these genera, the generic rank of the subgroup generated by differences of the rational points on the curve appears to be at least g. We give one example of a sharp curve of genus four and one example of a sharp curve of genus five. For both examples, we assume the Generalized Riemann Hypothesis to compute the rank. We did not check the rank conditions of curves of genus at least six, since this is an extremely computationally difficult task.

We start with a curve of genus four, and we choose the prime $p = 11$. Then the reduction modulo 11 is the curve $\overline{C} : y^2 = x^{10} + 1$. Since there are no small powers of x, we only need to add more rational points using some convenient t transformation. We consider the curve

$$C : y^2 = x^4(x - 11)^2(x - 22)^2(x - 33)^2 + 1.$$

This curve has the property that the element $[\infty_- - \infty_+]$ of the Jacobian has finite order 5, increasing the chances to obtain smaller rank. Indeed, it is verified in Magma that this curve has rank 3, so it is a sharp curve of genus four, and

$$C(\mathbb{Q}) = \{\infty_\pm, (0, \pm 1), (11, \pm 1), (22, \pm 1), (33, \pm 1)\}.$$

We use a similar strategy to construct a sharp curve of genus five. We choose $p = 13$ and start with the reduction curve $\overline{C} : y^2 = x^{12} + 1$. Then we consider the curve

$$C' : y^2 = x^4\left(x^2 - \frac{13^2}{3^2}\right)^2\left(x^2 - \frac{13^2}{4^2}\right)^2 + 1,$$

or, after the linear change of variables

$$C : y^2 = x^4(9x^2 - 169)^2(16x^2 - 169)^2 + 144^2.$$

This curve is potentially sharp, as $\#\overline{C}(\mathbb{F}_{13}) = 4$, and $\#C(\mathbb{Q}) \geq 12$. Hence, the rank of its Jacobian satisfies $r \geq 4$.

The group of automorphisms of C contains the group $(\mathbb{Z}/2\mathbb{Z})^2$, so by [Pau08, 3.1.1., Theorem 5], C has two quotients of degree two, C_1, and C_2, such that $J(C)$ is isogenous to the product $J(C_1) \times J(C_2)$. We conclude that the rank of $J(C)$ is equal to the sum of ranks of $J(C_1)$ and $J(C_2)$. Using Magma, we compute that C_1 and C_2 are given by the equations

C_1:$y^2+(x^2+x)y=5184x^6+37944x^5-802992x^4-3580910x^3+35285157x^2+32655315x+7268209$,
C_2:$y^2=-20736x^7-55584x^6+4461023x^5+5168390x^4-294280801x^3-78257128x^2+4786378304x+8098228736$.

Furthermore, for both curves C_1 and C_2, Magma gives the lower rank bound 0, and the upper bound 2. We conclude that both $J(C_1)$ and $J(C_2)$ have rank 2, because the sum of their ranks is at least 4. So the Jacobian of C has rank 4. Hence the curve is sharp and we have

$$C(\mathbb{Q}) = \left\{\infty_\pm, (0, \pm 144), \left(\pm\frac{13}{3}, \pm 12\right), \left(\pm\frac{13}{4}, \pm 12\right)\right\}.$$

We note that it might be possible to use the method of Chabuaty and Coleman to find the rational points on C_2, and to use that information to find $C(\mathbb{Q})$. However, the curve C_2 is not sharp, so it would be more involved to compute the set $C_2(\mathbb{Q})$.

Acknowledgments The author is funded by the DFG-Grant MU 4110/1-1. Part of this research was done during the "Summer school in computational number theory" in Bristol and during a visit to Boston University where the author was partially supported by the Diamant PhD Travel Grant. The author thanks Steffen Müller for his support, checking examples and numerous comments and corrections on the text, Jennifer Balakrishnan, Francesca Bianchi, Céline Maistret, Michael Stoll, and Jaap Top for useful discussions, Pieter Moree for significant help on obtaining the special case of Bertrand's postulate used here, Lazar Radičević for many suggestions and checking examples, Alex Best for helpful suggestions, Oana Adascalitei for help and support when creating examples, and Boston University for their hospitality and providing Magma on their computer. The author thanks Noam Elkies and Kirti Joshi for bringing attention to examples related to the work and ideas mentioned in this paper. The author thanks the anonymous referees for carefully reading the article and many useful comments.

References

[BBM16] Jennifer S. Balakrishnan, Amnon Besser, and J. Steffen Müller. Quadratic Chabauty: p-adic heights and integral points on hyperelliptic curves. *Journal für die reine und angewandte Mathematik*, 2016(720):51–79, 01 Nov. 2016.

[BBM17] Jennifer Balakrishnan, Amnon Besser, and J. Steffen Müller. Computing integral points on hyperelliptic curves using quadratic Chabauty. *Mathematics of Computation*, 86:1403–1434, 2017.

[BCP97] Wieb Bosma, John Cannon, and Catherine Playoust. The Magma algebra system. I. The user language. *J. Symbolic Comput.*, 24(3–4):235–265, 1997.

[BS09] Nils Bruin and Michael Stoll. Two-cover descent on hyperelliptic curves. *Mathematics of Computation*, 78(268):2347–2370, 2009.

[BS10] Nils Bruin and Michael Stoll. The Mordell–Weil sieve: proving non-existence of rational points on curves. *LMS Journal of Computation and Mathematics*, 13:272–306, 2010.

[Cha41] Claude Chabauty. Sur les points rationnels des courbes algébriques de genre supérieur à l'unité. *C. R. Acad. Sci. Paris*, 212:882–885, 1941.

[Col85a] Robert F. Coleman. Effective Chabauty. *Duke Math. J.*, 52(3):765–770, 09 1985.

[Col85b] Robert F. Coleman. Torsion points on curves and p-adic abelian integrals. *Annals of Mathematics*, 121(1):111–168, 1985.

[Dok04] Tim Dokchitser. Computing special values of motivic L-functions. *Experimental Mathematics*, 13(2):137–149, 2004.

[Elk] Noam Elkies. Curves of genus 2 over $\mathbb{Q}$ whose Jacobians are absolutely simple abelian surfaces with torsion points of high order. http://people.math.harvard.edu/~elkies/g2_tors.html#more.

[Fal83] Gerd Faltings. Endlichkeitssätze für abelsche Varietäten über Zahlkörpern. *Inventiones mathematicae*, 73(3):349–366, 1983.

[GG93] Daniel M. Gordon and David J. W. Grant. Computing the Mordell-Weil rank of Jacobians of curves of genus two. *Transactions of the American Mathematical Society*, 337(2):807–824, 1993.

[Gra94] David Grant. A curve for which Coleman's effective Chabauty bound is sharp. *Proceedings of the American Mathematical Society*, 122(1):317–319, 1994.

[HM19] Yoshinosuke Hirakawa and Hideki Matsumura. A unique pair of triangles. *Journal of Number Theory*, 194:297–302, 2019.

[HS00] Marc Hindry and Joseph H. Silverman. *Diophantine geometry. An Introduction*, volume 201 of *Graduate Texts in Mathematics*. Springer-Verlag, New York, 2000.

[HZ02] Everett W. Howe and Hui June Zhu. On the existence of absolutely simple abelian varieties of a given dimension over an arbitrary field. *Journal of Number Theory*, 92(1):139–163, 2002.

[JT99] Kirti Joshi and Pavlos Tzermias. On the Coleman-Chabauty bound. *C. R. Acad. Sci. Paris Sér. I Math.*, 329(6):459–463, 1999.

[Kim05] M. Kim. The motivic fundamental group of $\mathbb{P}^1\backslash\{0, 1, \infty\}$ and the theorem of Siegel. *Inventiones mathematicae*, 161(3):629–656, 2005.

[Kim09] Minhyong Kim. The unipotent Albanese map and Selmer varieties for curves. *Publ. Res. Inst. Math. Sci.*, 45(1):89–133, 2009.

[KZB13] Eric Katz and David Zureick-Brown. The Chabauty–Coleman bound at a prime of bad reduction and Clifford bounds for geometric rank functions. *Compositio Mathematica*, 149(11):1818–1838, 2013.

[LT02] Dino Lorenzini and Thomas J. Tucker. Thue equations and the method of Chabauty-Coleman. *Inventiones mathematicae*, 148(1):47–77, 2002.

[Mor22] Louis J. Mordell. On the rational solutions of the indeterminate equations of the third and fourth degrees. *Proc. Camb. Phil. Soc.*, 21:179–192, 1922.

[MP12] William McCallum and Bjorn Poonen. The method of Chabauty and Coleman. *Explicit Methods in Number Theory, Panor. Synthèses*, 36:99–117, 2012.

[Pau08] Jennifer Paulhus. Decomposing Jacobians of curves with extra automorphisms. *Acta Arith.*, 132(3):231–244, 2008.

[RR96] Olivier Ramaré and Robert Rumely. Primes in arithmetic progressions. *Mathematics of Computation*, 65(213):397–425, 1996.

[Sch95] Edward F. Schaefer. 2-descent on the Jacobians of hyperelliptic curves. *Journal of Number Theory*, 51(2):219–232, 1995.

[Sch99] Victor Scharaschkin. Local-global problems and the Brauer-Manin obstruction. *ProQuest LLC, Ann Arbor, MI, Thesis (Ph.D.)–University of Michigan*, 1999.

[Sik15] Samir Siksek. Chabauty and the Mordell-Weil sieve. In *Advances on superelliptic curves and their applications*, volume 41 of *NATO Sci. Peace Secur. Ser. D Inf. Commun. Secur.*, pages 194–224. IOS, Amsterdam, 2015.

[Sto] Michael Stoll. Genus 2 curves with several points contained in an arithmetic progression. http://www.mathe2.uni-bayreuth.de/stoll/talks/MIT2018-print.pdf.

[Sto01] Michael Stoll. Implementing 2-descent for Jacobians of hyperelliptic curves. *Acta Arith*, 98(3):245–277, 2001.

[Sto06] Michael Stoll. Independence of rational points on twists of a given curve. *Compositio Mathematica*, 142(5), 2006.

[Sto15] Michael Stoll. Descent and covering collections. In *Advances on superelliptic curves and their applications*, volume 41 of *NATO Sci. Peace Secur. Ser. D Inf. Commun. Secur.*, pages 176–193. IOS, Amsterdam, 2015.

[Tat66] John Tate. On the conjectures of Birch and Swinnerton-Dyer and a geometric analog. Séminaire N. Bourbaki Exp. No. 306, 415–440, 1966.

[Wei29] André Weil. L'arithmétique sur les courbes algébriques. *Acta Math.*, 52(1):281–315, 1929.

Chabauty–Coleman Computations on Rank 1 Picard Curves

Sachi Hashimoto and Travis Morrison

Abstract We provably compute the full set of rational points on 1403 Picard curves defined over $\mathbb{Q}$ with Jacobians of Mordell–Weil rank 1 using the Chabauty–Coleman method. To carry out this computation, we extend Magma code of Balakrishnan and Tuitman for Coleman integration. The new code computes p-adic (Coleman) integrals on curves to points defined over number fields where the prime p splits completely and implements effective Chabauty for curves whose Jacobians have infinite order points that are not the image of a rational point under the Abel–Jacobi map. We discuss several interesting examples of curves where the Chabauty–Coleman set contains points defined over number fields.

1 Introduction

Let $X/\mathbb{Q}$ be a smooth projective curve of genus at least 2. Faltings's theorem [Fal83], proved in 1983, establishes that the set of rational points $X(\mathbb{Q})$ of X is finite, but falls short of giving an explicit method for computing this set. Our goal in this paper is to compute this set explicitly for certain curves of genus 3.

Let p be a prime of good reduction for X. We focus on curves amenable to Chabauty–Coleman calculations, which use p-adic (Coleman) integrals to carve out

SH was supported by a Clare Boothe Luce Fellowship (Henry Luce Foundation) and NSF Graduate Research Fellowship grant number DGE-1840990. TM was supported by funding from the Natural Sciences and Engineering Research Council of Canada, the Canada First Research Excellence Fund, CryptoWorks21, Public Works and Government Services Canada, and the Royal Bank of Canada.

S. Hashimoto (✉)
Department of Mathematics and Statistics, Boston University, Boston, MA, USA
e-mail: svh@bu.edu

T. Morrison
Mathematics Department, Virginia Tech, Blacksburg, VA, USA
e-mail: tmo@vt.edu

J. S. Balakrishnan et al. (eds.), *Arithmetic Geometry, Number Theory, and Computation*, Simons Symposia, https://doi.org/10.1007/978-3-030-80914-0_16

a finite set of p-adic points containing the rational points. The Chabauty–Coleman method is one attempt at an effective Faltings's theorem. The method dates back to work of Chabauty in 1941 [Cha41], who showed finiteness of $X(\mathbb{Q})$ for certain curves X using the embedding of $X(\mathbb{Q})$ into the p-adic closure of $J(\mathbb{Q})$, where J is the Jacobian variety of X. In the 1980s, Coleman [Col85a, Col85b] made Chabauty's approach effective with the development of the Coleman integral. With this computational tool, he was able to bound the number of rational points on some curves.

For a given curve X, the Chabauty–Coleman method requires that the Mordell–Weil rank of its Jacobian be computable and smaller than the genus of X. In this paper we apply the Chabauty–Coleman method to genus 3 non-hyperelliptic curves. While there exist techniques to compute the rank of Jacobians of general smooth plane quartics [BPS16], practical implementations of rank-bounding techniques only exist for superelliptic curves, based on work of Poonen and Schaefer [Poo97]. A Picard curve is a smooth plane quartic $X/\mathbb{Q}$ that admits a Galois cover of degree 3 of $\mathbb{P}^1$. Every Picard curve has an affine equation of the form $y^3 = f(x)$ where $f(x) \in \mathbb{Q}[x]$ is degree 4 and without multiple roots in $\overline{\mathbb{Q}}$. Therefore we focus on Picard curves. While for some particular genus 3 curves X of Mordell–Weil rank 2 it is possible to determine the rational points using the Chabauty–Coleman method, and one can always extend the method using the Mordell–Weil sieve [Sik15, BS10], which combines p-adic information for multiple primes p, when the rank is one less than the genus we expect the Chabauty–Coleman method alone to fail to determine $X(\mathbb{Q})$ for most X. So, in particular, we focus on the rank 1 case, where we can work with at least two linearly independent functionals on $J(\mathbb{Q}_p)$.

Recently, Balakrishnan and Tuitman [BT20] have made Coleman integration explicit and practical, opening up the ability to compute rational points on entire databases of curves satisfying Chabauty's rank hypothesis.

Sutherland has computed a database of genus 3 curves of bounded discriminant [Sut19, Sut], providing the necessary data for large-scale computational experiments in genus 3. Balakrishnan–Bianchi–Cantoral-Farfán–Çiperiani–Etropolski [BBCF$^+$] computed rational points on a database of genus 3 hyperelliptic curves with Mordell–Weil rank 1 using Chabauty–Coleman methods and de Frutos-Fernández–Hashimoto [dFFH19] computed rational points on genus 3 hyperelliptic curves with Mordell–Weil rank 0. This work is the first instance of such computations being carried out on non-hyperelliptic curves.

The Chabauty–Coleman method cuts out a finite set of p-adic points $X(\mathbb{Q}_p)_1$ on a curve X that contains the rational points $X(\mathbb{Q})$; sometimes this set is strictly larger than the set of rational points.

For every Picard curve X we consider, we classify the set of extra points $X(\mathbb{Q}_p)_1 \setminus X(\mathbb{Q})$ in the Chabauty–Coleman set. We show that each extra point P is defined over a number field, that is, we can find a number field K where p is divisible by a prime $\mathfrak{p}$ of degree one dividing p, and $P \in X(K) \cap X(K_{\mathfrak{p}})$. Let $\infty \in X(\mathbb{Q})$ be the unique rational point at infinity. These extra points fall into one of three categories:

- $P \in X(K)$ such that $[P - \infty]$ is in $J(K)_{\text{tors}}$;
- $P \in X(K)$ such that there exists $n \in \mathbb{Z}$ so that $n[P - \infty]$ is infinite order in $J(\mathbb{Q})$;
- $P \in X(K)$ that appear due to the existence of extra automorphisms of $X_{\overline{\mathbb{Q}}}$.

We exhibit examples of all three possibilities in Sect. 4.1.

Our Magma code extending the functionality of Balakrishnan and Tuitman's Coleman integration code as well as our code for computing $X(\mathbb{Q}_p)_1$ on the database of Picard curves is available at [HMb, HMa].

2 The Chabauty–Coleman Method

2.1 *Chabauty–Coleman*

In this section we introduce the Coleman integral as a p-adic tool for computing the rational points on a curve. We follow the exposition of the construction of the Coleman integral in [Wet] and [MP12]. Let X be a smooth projective curve of genus $g > 1$ over $\mathbb{Q}$ with Jacobian J. Let B be a nonzero rational divisor on X inducing an Abel–Jacobi map

$$\begin{aligned} \iota\colon X &\to J \\ Q &\mapsto [\deg(B)Q - B] \end{aligned}$$

(for example we can always take a canonical divisor or we can take a rational point if one is known). Let p be a prime of good reduction for X. We introduce two ways to construct the p-adic Coleman integral of a regular one-form on $X_{\mathbb{Q}_p}$: one by pulling back an integral on $J_{\mathbb{Q}_p}$ thought of as a p-adic Lie group and the other directly on $X_{\mathbb{Q}_p}$ as a rigid analytic space. We first describe the p-adic Lie group approach and do not address the equivalence here; Coleman discusses the details in [Col85b].

Let $\omega \in H^0(J_{\mathbb{Q}_p}, \Omega^1)$. Since $J(\mathbb{Q}_p)$ is a p-adic Lie group, the differential ω is translation-invariant, and we can define an antiderivative homomorphism $\int_0 : J(\mathbb{Q}_p) \to \mathbb{Q}_p$. This homomorphism sends a point $Q \in J(\mathbb{Q}_p)$ to $\int_0^Q \omega$. For any open subgroup $U \subset J(\mathbb{Q}_p)$, the antiderivative of ω can be computed by expanding ω in local coordinates at $Q \in U$ and taking formal antiderivative in the power series ring; this yields a convergent p-adic power series on small enough open subgroups.

The pullback map

$$\iota^*\colon H^0(J_{\mathbb{Q}_p}, \Omega^1) \to H^0(X_{\mathbb{Q}_p}, \Omega^1)$$

induced by ι is an isomorphism, independent of the choice of B [Mil86, Proposition 2.2]. This gives us a way to think of the integral on the curve $X_{\mathbb{Q}_p}$: for any $\omega \in H^0(X_{\mathbb{Q}_p}, \Omega^1)$ and any $P, Q \in X(\mathbb{Q}_p)$, we can define $\int_P^Q \omega$ by $\int_0^{[Q-P]} (\iota^*)^{-1}\omega$.

Conversely, we can write an integral on $J_{\mathbb{Q}_p}$ in terms of integrals on $X_{\mathbb{Q}_p}$: for any $D \in J(\mathbb{Q}_p)$ and $\omega \in H^0(J_{\mathbb{Q}_p}, \Omega^1)$, we can write $D = \sum_i [Q_i - P_i]$ where $P_i \in X(L)$ for $L/\mathbb{Q}_p$ a finite extension. Then

$$\int_0^D \omega = \sum_i \int_{P_i}^{Q_i} \iota^* \omega.$$

Note $\int_0^D \omega = 0$ when $D \in J(\mathbb{Q}_p)_{\mathrm{tors}}$ because the antiderivative is a homomorphism.

Coleman gives a definition of the integral $\int_P^Q \eta$ on the rigid analytic space $X_{\mathbb{Q}_p}$; to do so, we work over a wide open space $V \subset X_{\mathbb{Q}_p}$ constructed by deleting a finite number of closed disks from $X_{\mathbb{Q}_p}$ of radius less than 1. He defines an integral for $P, Q \in V$ and η any differential ω of the second kind on V [Col85b] (a differential ω is of the second kind if it is everywhere meromorphic and the residue at every pole is zero). Coleman shows his definition agrees with the one induced by the p-adic Lie group structure of $J_{\mathbb{Q}_p}$ when $\eta \in H^0(X_{\mathbb{Q}_p}, \Omega^1)$. Fix an algebraic closure $\overline{\mathbb{Q}}_p$ of $\mathbb{Q}_p$. For all points $P, Q, R \in V(\overline{\mathbb{Q}}_p)$ and one-forms ω, η of the second kind on $X_{\mathbb{Q}_p}$, the Coleman integral enjoys the following properties:

Theorem 2.1 (Coleman)

(1) Linearity: $\int_P^Q (\alpha\omega + \beta\eta) = \alpha \int_P^Q \omega + \beta \int_P^Q \eta$, *for all* $\alpha, \beta \in \mathbb{Q}_p$.
(2) Additivity: $\int_P^Q \omega = \int_P^R \omega + \int_R^Q \omega$.
(3) Fundamental theorem of calculus: $\int_P^Q df = f(Q) - f(P)$ *for* df *a rigid analytic function on* V.
(4) Change of variables: if $V' \subset X'$ *is another wide open subspace of a rigid analytic space* X', *for any rigid analytic map* $\phi\colon V \to V'$, *we have* $\int_{\phi(P)}^{\phi(Q)} \omega = \int_P^Q \phi^*(\omega)$.
(5) For any divisor $D := \sum_i Q_i - P_i$ *of degree zero on* X *defined over* $\overline{\mathbb{Q}}_p$, *and for* ω *a regular one-form, then* $\int_D \omega := \sum_i \int_{P_i}^{Q_i} \omega$ *is well-defined, and* $\int_D \omega = 0$ *when* D *is principal.*
(6) Galois equivariance: if $\sigma \in \mathrm{Gal}(\overline{\mathbb{Q}}_p/\mathbb{Q}_p)$, *then* $\left(\int_P^Q \omega\right)^\sigma = \int_{\sigma(P)}^{\sigma(Q)} \omega^\sigma$.

Remark 2.2 Let $Q_i, P_i \in \overline{\mathbb{Q}}_p$ such that $D := \sum_i Q_i - P_i$ is a divisor of degree zero on X defined over $\mathbb{Q}_p$. Let $\omega \in H^0(X_{\mathbb{Q}_p}, \Omega^1)$. By (5) and (6) of Theorem 2.1, the integral

$$\int_D \omega = \sum_i \int_{P_i}^{Q_i} \omega$$

is well-defined, only depends on $[D]$, and belongs to $\mathbb{Q}_p$, even though the individual integrals in the sum $\int_{P_i}^{Q_i} \omega$ may not belong to $\mathbb{Q}_p$.

When the rank r of $J(\mathbb{Q})$ is less than the genus g of X, the space of vanishing differentials

$$\mathrm{Van}(X(\mathbb{Q})) := \left\{\omega \in H^0(X_{\mathbb{Q}_p}, \Omega^1) \colon \int_D \omega = 0 \text{ for all } D \in J(\mathbb{Q})\right\}$$

is at least $(g - r)$-dimensional, and thus nonzero. In particular, when $r = 1$, the space of vanishing differentials is exactly $(g - 1)$-dimensional: for $D \in J(\mathbb{Q})$ the integrals $\int_0^D \omega_i$ on a basis of regular differentials $i = 1, \ldots, g$ are zero if and only if D is torsion [Bou89, III, §7.6, Proposition 12]. Since we are guaranteed some $D \in J(\mathbb{Q})$ which is non-torsion, $\mathrm{Van}(X(\mathbb{Q}))$ cannot be all of $H^0(X_{\mathbb{Q}_p}, \Omega^1)$.

Remark 2.3 We will think of the p-adic integral of ω in $\mathrm{Van}(X(\mathbb{Q}))$ as the Coleman integral $\int_D \omega := \sum_i \int_{P_i}^{Q_i} \omega$ taking place directly on the curve $X_{\mathbb{Q}_p}$ as a rigid analytic space (as in Theorem 2.1 part 6). While we could consider both the p-adic Lie group integral $\int_0^D (\iota^*)^{-1}\omega$ on the Jacobian and the Coleman integral when $\omega \in H^0(X_{\mathbb{Q}_p}, \Omega^1)$, in the process of explicitly computing the Coleman integral between two points in different residue disks, we compute integrals of differentials of the second kind that are not everywhere regular (see Sect. 2.2). Therefore throughout the paper we consider all p-adic integrals to take place on the curve as a rigid analytic space.

Coleman [Col85a] proves a bound on the number of rational points $\#X(\mathbb{Q}) \leq \#X(\mathbb{F}_p) + 2g - 2$ by bounding the number of zeros of a nonzero differential in $\mathrm{Van}(X(\mathbb{Q}))$, assuming $p > 2g$.

Coleman's proof suggests an algorithm for computing $X(\mathbb{Q})$ by computing the set

$$X(\mathbb{Q}_p)_1 := \left\{Q \in X(\mathbb{Q}_p) : \int_B^{\deg(D)Q} \omega = 0 \text{ for all } \omega \in \mathrm{Van}(X(\mathbb{Q}))\right\}.$$

Since the Coleman integral is locally analytic, it can be expressed as the integral of a convergent p-adic power series on each residue disk. Computing $X(\mathbb{Q}_p)_1$ yields finitely many solutions in each disk, and so finitely many in total. Therefore $X(\mathbb{Q}) \subseteq X(\mathbb{Q}_p)_1$ is finite.

2.2 *Explicit Coleman Integration After Balakrishnan and Tuitman*

To carry out the Chabauty–Coleman method in practice, we need an explicit method for computing the set $X(\mathbb{Q}_p)_1$, which is defined as the vanishing locus of some p-adic integrals. In this section, we give an overview of a practical algorithm to compute p-adic integrals on curves, following the method of Balakrishnan and

Tuitman [BT20]. For a more detailed description of the algorithm, which applies in more generality than described here, we refer the reader to their paper, as well as the zeta function algorithm of Tuitman described in [Tui16, Tui17] which develops the reduction in cohomology, extending an earlier algorithm for hyperelliptic curves due to Kedlaya [Ked01, Ked03].

Let $X/\mathbb{Q}$ be a smooth projective geometrically integral curve with (possibly singular) affine plane model given by the equation $Q(x, y) = 0$ with $Q(x, y) \in \mathbb{Z}[x, y]$ and let p be a prime of good reduction. Tuitman's algorithm exploits the existence of a low-degree map $x : X \to \mathbb{P}^1$ of degree d_x. After removing the ramification locus of this map, the algorithm chooses a lift of Frobenius which sends $x \mapsto x^p$ and Hensel lifts y. Then, we compute the action of Frobenius on a certain basis of differentials whose classes generate $H^1_{rig}(X_{\mathbb{Q}_p})$, and reduce using Lauder's fibration method which is given by explicit and fast linear algebra.

To specify a point P on X, Balakrishnan and Tuitman's algorithm uses the following data:

- `P'x` the x-coordinate of P (or 0 if x is infinite).
- `P'b` the values of $b^0 = [b^0_0, \ldots, b^0_{d_x-1}]$ that form an integral basis for the function field $\mathbb{Q}(X)/\mathbb{Q}[x]$ at x, ($b^\infty = [b^\infty_0, \ldots, b^\infty_{d_x-1}]$ an integral basis for $\mathbb{Q}(X)/\mathbb{Q}[1/x]$ if P is infinite).
- `P'inf` a boolean value specifying whether P is infinite.

We must impose some additional assumptions on $Q(x, y)$, p, b^0, and b^∞, which are listed in [BT20, Assumption 2.6], in order to compute p-adic Coleman integrals on $Q(x, y) = 0$ using Balakrishnan and Tuitman's algorithm. These hypotheses can be verified directly for our case of Picard curves from the computation of b^0 and b^∞ provided here.

In the case where X is a monic Picard curve $y^3 = f(x)$, the map $x : X \to \mathbb{P}^1$ is projection to the x-coordinate, $b^0 = [1, y, y^2]$, $b^\infty = [y, y/x^2, y/x^3]$. The regular differentials $\omega_1 = ydx/f, \omega_2 = xydx/f$ and $\omega_3 = y^2dx/f$ form a basis of $H^0(X_{\mathbb{Q}_p}, \Omega^1)$. We extend this basis by $\omega_4 = x^3ydx/f, \omega_5 = xy^2dx/f, \omega_6 = x^2y^2dx/f$ to obtain classes $[\omega_1], \ldots, [\omega_6]$ that generate $H^1_{rig}(X_{\mathbb{Q}_p})$.

The p-adic points $X(\mathbb{Q}_p)$ are a disjoint union of residue disks, which are the preimages under reduction modulo p of points in $X(\mathbb{F}_p)$. After Balakrishnan and Tuitman, we classify the residue disks based on their ramification with respect to the map $x : X \to \mathbb{P}^1$. Let $\Delta_y(x)$ be the discriminant of Q thought of as a polynomial in y, and $r(x)$ be the squarefree polynomial with the same zeros as $\Delta_y(x)$ given by $r(x) = \Delta_y(x)/\gcd(\Delta_y(x), \frac{d\Delta_y(x)}{dx})$. Then our residue disks partition as follows:

- Bad residue disks are residue disks that contain a point whose x-coordinate is a zero of $r(x)$ or is infinite. Such points are called very bad points. If the very bad point of a disk has infinite x-coordinate, the disk is also an infinite residue disk, and the point is also a very infinite point. Otherwise the disk is a finite residue disk.
- Good residue disks comprise the remaining residue disks. All good residue disks are finite.

Note that each residue disk contains at most one very bad point $P = (x, y)$ that will be defined over an unramified extension of $\mathbb{Q}_p$ [BT20, Remark 2.9]. We denote by e_P the ramification index of the map by x at a very bad point P. If $Q(x, y) = y^3 - f(x)$ is a Picard curve, then $r(x) = -27f(x)$, the very bad points are the ramification points (where $y = 0$), the point at infinity ∞ is $[0 : 1 : 0]$, the residue disks containing ramification points are the finite bad disks, and all very bad points have ramification index 3.

There are two kinds of Coleman integrals on X: integrals between two points P and Q in $X(\mathbb{Q}_p)$ in the same residue disk, called tiny Coleman integrals and integrals between two points that are not in the same residue disk. Tiny integrals are computed using uniformizing parameters. A uniformizing parameter gives an isomorphism from a residue disk to $p\mathbb{Z}_p$ and transforms a Coleman integral into a power series integral.

Algorithm 2.4 *Compute a tiny integral $\int_P^Q \omega$ and uniformizing parameter t*

Input: P and $Q \in X(\mathbb{Q}_p)$ in the same residue disk, and ω a differential of the second kind on X such that ω does not have a pole in the residue disk of P and Q.

Output: $\int_P^Q \omega$ and t a uniformizing parameter in the residue disk of P and Q

(1) If P, Q are in a bad disk, find P', the very bad point in the disk, and break up the integral as $\int_P^{P'} \omega + \int_{P'}^Q \omega$. Otherwise proceed with $P = P'$.
(2) Compute a uniformizing parameter t at P': in a good disk, t is given by $x - x(P')$; at a very bad point $t = b_i^0 - b_i^0(P')$ for some i (replace with b_i^∞ for very infinite P').
(3) Compute x and the b^0 vector (or b^∞) as a function of t to desired precision, and using these expansions compute ω as a power series in t.
(4) Integrate

$$\int_P^Q \omega = \int_{t(P)}^{t(Q)} \omega(t)$$

as a power series.

For more details on the computations of the tiny integral and the uniformizing parameters see [BT20, Algorithm 3.7, Proposition 3.2].

For the second kind of Coleman integral, between points in different residue disks, we turn to the full machinery of Tuitman's zeta function algorithm to compute the action of Frobenius on cohomology. Let P and Q in $X(\mathbb{Q}_p)$ be points in different good residue disks. We use Tuitman's lift ϕ of the Frobenius morphism on $X_{\mathbb{F}_p}$. Let ω_i be Tuitman's basis differentials such that $[\omega_i]$ generate $H^1_{rig}(X_{\mathbb{Q}_p})$ for $i = 1, \ldots, 2g$. By applying Frobenius to P and Q, we obtain

$$\int_P^Q \omega_i = \int_P^{\phi(P)} \omega_i + \int_{\phi(P)}^{\phi(Q)} \omega_i + \int_{\phi(Q)}^Q \omega_i = \int_P^{\phi(P)} \omega_i + \int_P^Q \phi^*\omega_i + \int_{\phi(Q)}^Q \omega_i$$

where P and $\phi(P)$ are in the same residue disk, so define a tiny integral, and similarly for Q and $\phi(Q)$.

So, we expressed the unknown integral as a linear combination of two computable quantities and the integral of $\phi^*\omega_i$. This motivates computing $\phi^*\omega_i$; Balakrishnan and Tuitman compute the reduction in cohomology of $\phi^*\omega_i$ as a sum of basis elements $\sum_j M_{ij}\omega_j$ plus an exact differential df_i. We compute the reductions of all $\phi^*\omega_i$ and consider the linear system of equations to compute the Coleman integral. For more details see [BT20, Algorithm 3.7, Algorithm 3.12] as well as [Tui17, Section 4.1] which discusses how to construct the $2g$ basis differentials whose classes generate $H^1_{rig}(X_{\mathbb{Q}_p})$. If P lies in a bad disk, some f_i might not converge at P, but near the boundary of the disk of P they do; we compute the integral $\int_P^Q \omega$ by taking $S \in X(\mathbb{Q}_p(p^{1/e}))$ on the boundary for e large enough and breaking up the integral as $\int_P^S \omega + \int_S^Q \omega$.

In practice, to compute $X(\mathbb{Q}_p)_1$, we integrate to arbitrary endpoints $\int_b^z \omega$; to do this we iterate over residue disks, breaking up the integral by fixing P in each residue disk and integrating $\int_b^P \omega + \int_P^z \omega$ where $\int_b^P \omega$ sets the constant of integration and $\int_P^z \omega$ is a tiny integral which can be expanded as a p-adic power series to arbitrary precision and whose zeros can be found using analytic techniques. Balakrishnan and Tuitman implement the Coleman integral as well as the computation of the Chabauty–Coleman set $X(\mathbb{Q}_p)_1$ in Magma [BT] under the assumption that we can find rational points $P_1, P_2, \ldots, P_r \in X(\mathbb{Q})$ such that the classes $[P_i - b]$ are infinite order and linearly independent in $J(\mathbb{Q})$, where r is the rank of $J(\mathbb{Q})$. However, their implementation does not check the precision (for example, when solving for roots of p-adic power series the code does not check whether all of the roots obtained are single roots) and needs a few modifications to produce provably correct results.

3 The Algorithm

In this section, we describe the algorithm we implemented to compute $X(\mathbb{Q})_1$ for each X a rank 1 Picard curve in our database. These curves come from Sutherland's database of genus 3 curves of small discriminant [Sut19, Sut] and have discriminant Δ bounded by 10^{12}. See [Sut19] for a definition of the discriminant of a smooth plane curve; in particular, $p|\Delta$ if and only if p is a prime of bad reduction. Each curve in the database is given by the data of an affine model $y^3 = f(x)$ with $f(x)$ degree four, along with its discriminant Δ.

We began by filtering the database for those curves whose Jacobians have Mordell–Weil rank 1. We ran the Magma function `RankBounds(f,3:Return Generators:=true)` which computes the rank based on work of Poonen and Schaefer [Poo97] and Creutz [Cre13] and is implemented in Magma by Creutz. It gives a lower and upper bound for the rank of the Jacobian of the cyclic cover of $\mathbb{P}^1$

defined by $y^3 = f(x)$ along with a list of polynomials $g(T)$ which define divisors $\mathrm{div}(g)$ on $\mathbb{P}^1$ that lift to rational divisors on X. The image of such a divisor in the Jacobian is either a 3-torsion point or a point of infinite order.

If `RankBounds` returns upper and lower bounds that prove J is rank 1, we proceed. Otherwise, we discard the curve and move on to the next one.[1] For J rank 1, we consider the non-torsion generators $g(T)$; if there is some non-torsion $g(T)$ of degree one, then we have a rational point $P \in X(\mathbb{Q})$ such that $[P - \infty]$ is infinite order in $J(\mathbb{Q})$. Otherwise, we are left with $g(T)$ of higher degree, which correspond to rational divisors that are supported on Galois conjugate points defined over a number field K. To handle this second case, which comprises 544 of the 1403 Picard curves in our rank 1 database, we extended functionality in Balakrishnan and Tuitman's code to handle rational divisors on curves when running effective Chabauty computations. This is done by extending the Coleman integral functionality to compute integrals to points on curves defined over number fields when p is totally split in the field.

3.1 With a Rational Point Whose Class in the Jacobian Is of Infinite Order

Algorithm 3.1 *Computing $X(\mathbb{Q}_p)_1$ with $P \in X(\mathbb{Q})$ such that $[P - \infty]$ is infinite order*
Input: a Picard curve X with affine equation $y^3 = f(x)$ such that df is monic, with discriminant Δ, and Jacobian J such that $J(\mathbb{Q})$ is rank 1, along with a point $P \in X(\mathbb{Q})$ such that $[P - \infty] \in J(\mathbb{Q})$ has infinite order, and parameters (N, e), where p^N is the p-adic precision and $\mathbb{Q}_p(p^{1/e})$ is the totally ramified extension over which we work in the bad disks (see Sect. 2.2).[2]
Output: $S \subseteq X(\mathbb{Q})$ and $T \subseteq X(\mathbb{Q}_p)_1 \setminus S$ or "Failure". If the algorithm does not return "Failure", and $T = \emptyset$, then $S = X(\mathbb{Q})$.

(1) *Find the first*[3] *prime $p > 3$ such that $p \nmid \Delta$. (We discard $p = 3$ because 3 always divides the discriminant of a Picard curve* [BBW17, Theorem 1.1]. *We avoid $p = 2$ for simplicity when Hensel lifting.)*
(2) *Compute a list of points $S \subseteq X(\mathbb{Q})$ by searching in a box up to naive height 1000.*

[1]For practical purposes, we terminated RankBounds if it ran for 120 s without returning an answer. We discarded 520 curves from the total 5335 curves in the database provided by Sutherland either from terminating RankBounds or due to a "Runtime error".

[2]For some heuristics for choosing e given a fixed N see Appendix 4.1.

[3]Due to a minor bug in the code of Balakrishnan and Tuitman for computing local coordinates at very infinite points, which in certain cases yields a Runtime Error instead of computing the coordinates, we sometimes choose the next largest possible prime.

(3) *Compute a basis* v_1, v_2 *of vanishing differentials* $Van(X(\mathbb{Q}))$ *by computing a basis for the kernel of the matrix of integrals* $[\int_{\infty}^{P} \omega_i]$ *for* $i = 1, 2, 3$ *regular basis differentials. (We compute* v_1, v_2 *to finite precision: let* M *be the matrix of Frobenius and* I *the identity matrix. Then* v_1, v_2 *are computed to precision* $N - \mathrm{ord}_p(\det(M - I)) - \delta$ *as described in* [BT20, Section 4] *where* δ *is a precision loss defined in* [Tui17, Definition 4.4]. *If* $N - \mathrm{ord}_p(\det(M - I)) - \delta < 0$, *return "Failure".)*
(4) *Solve for* $X(\mathbb{Q}_p)_1 = \{Q \in X(\mathbb{Q}_p) | \int_{\infty}^{Q} v_i = 0, i = 1, 2\}$ *using Algorithm 3.7. If Algorithm 3.7 returns "Failure" (due to a bad choice of parameters* N *and* e*), return "Failure".*
(5) *Let* $T := X(\mathbb{Q}_p)_1 \setminus S$.
(6) *Run the Magma function* `PowerRelation` *to see if we can recognize the points of* T *as defined over a number field. If any point* Q *of* T *is defined over* $\mathbb{Q}$, *append it to* S *and remove it from* T.

   i. *First check if* $f(x(Q)) = 0$; *if so,* Q *is a ramification point, and therefore* $[Q - \infty]$ *is 3-torsion. Remove* Q *from* T.
   ii. *Compute* $\int_{\infty}^{Q} \omega_i, i = 1, 2, 3$ *on regular basis differentials. If these are* 0 *to our* p*-adic precision, then* $[Q - \infty]$ *is potentially torsion. We verify it is torsion by finding* $n \in \mathbb{Z}$ *such that* $n[Q - \infty] = 0 \in J(\mathbb{Q})$ *using Jacobian arithmetic implemented in Magma. If we verify it is torsion, remove* Q *from* T.
   iii. *Search in a box for relations with coefficients of small height between* $[Q - \infty]$ *and known rational divisors on* $J(\mathbb{Q})$. *If we succeed, remove* Q *from* T. *We have explained* $Q \in X(\mathbb{Q}_p)_1$ *by linearity.*

(7) *Return* S *and* T. *If* $T = \emptyset$, *then* $S = X(\mathbb{Q})$. *If* $T \neq \emptyset$, *then more work must be done to explain the presence of the remaining points in* $T \subseteq X(\mathbb{Q}_p)_1 \setminus S$ *and provably determine* $X(\mathbb{Q})$.

3.2 Precision Bounds

In Algorithm 3.1 Step 4 we solve for the zeros of the integral of a regular differential v on X. To do this, we solve for the zeros in each residue disk of $\overline{Q} \in X(\mathbb{F}_p)$, expanding the integral $\int_b^z v = \int_b^Q v + \int_{t(Q)}^{z(t)} v(t)dt$ for some $Q \in X(\mathbb{Q}_p)$ reducing to $\overline{Q}$, and where t is a uniformizing parameter at Q computed using Algorithm 2.4. In practice, we must truncate the p-adic power series to be a polynomial with finite p-adic precision. Therefore we need to determine how much p-adic and t-adic precision we need to determine all of the zeros of $\int_b^Q v + \int_{t(Q)}^{z(t)} v(t)dt$ on the residue disk of $\overline{Q}$.

First we define a *normalizing* process for p-adic power series.

Algorithm 3.2 *Normalize a* p*-adic power series* f
Input: $f(t) \in \mathbb{Q}_p[[t]]$, *a* p*-adic power series such that* $f'(t) = \sum_{i \geq 0} a_i t^i \in \mathbb{Z}_p[[t]]$, *and* df *has constant term in* $c \in \mathbb{Z}_p$.

Output: $F(x) \in \mathbb{Z}_p[[x]] \setminus p\mathbb{Z}_p[[x]]$ *such that the roots of* $f(t)$ *are in bijection with those of* F *by the map* $r \mapsto pr$.

(1) *Set* $F_0(x) := f(px)$. *Then* $F_0(x) = c + \sum_{i \geq 1} \frac{a_{i-1} p^i x^i}{i}$.
(2) *Set* $\lambda := \min\left(\left\{v_p\left(\frac{a_{i-1}p^i}{i}\right)\right\}_{i \geq 0} \cup \{v_p(c)\}\right)$.
(3) *Return* $F(x) := F_0(x)/p^\lambda$.

First, it is easy to see $F(x)$ is in $\mathbb{Z}_p[[x]]$. For $i > 0$, the coefficient of x^i is $\frac{a_{i-1}p^{i-\lambda}}{i}$, which has positive valuation for all $i > 0$ (by assumption the constant term has nonnegative valuation). In Step 2, we ensure that $F(x) \in \mathbb{Z}_p[[x]] \setminus p\mathbb{Z}_p[[x]]$.

Lemma 3.3 *Let* $f(t) \in \mathbb{Q}_p[[t]]$ *be a* p*-adic power series such that* $f'(t) \in \mathbb{Z}_p[[t]]$. *Write* $f'(t) = a_0 + a_1 t + a_2 t^2 + \ldots$ *and* $f(t) = c + a_0 t + a_1 t^2/2 + a_2 t^3/3 + \cdots$. *Let* $F(x)$ *be the output of Algorithm 3.2 on input* $f(t)$, *so that*

$$F(x) = \sum_{i=1}^{\infty} \frac{a_{i-1} x^i p^{i-\lambda}}{i} + c$$

where we assume $c \in \mathbb{Z}_p$.

Fix a positive integer $N \in \mathbb{N}$. *Define*

$$M(N, \lambda) := \min_m \{m - \lambda - \log_p(m) > N\}$$

and let $M = M(N, \lambda)$. *Write* $F(x) = F_M(x) + F_\infty(x)$ *where* $F_M(x)$ *contains the terms* F *of degree at most* M, *and* $F_\infty(x)$ *contains higher order terms.*

Then $F_M(x) \equiv F(x) \mod p^N$.

Proof Let $b_i := \frac{a_{i-1}p^{i-\lambda}}{i}$ be the coefficient of x^i.

To show $F_M(x) \equiv F(x) \mod p^N$, we show that for $i \geq M$, $v_p(b_i) > N$. Then $v_p(b_i) \geq v_p(p^{i-\lambda}/i) \geq i - \lambda - \log_p(i)$, and for $i \geq M$, this is greater than N. □

Note that if $c \in \mathbb{Q}_p$ and $v_p(c) < 0$, then $F(x)$ and therefore $f(t)$ has no zeros, so we will not need to consider this case.

Remark 3.4 In order to work with the prime $p = 5 < 2g$, we provide a more detailed analysis of the precision here than in [BBCF$^+$, Section 3.3]. Their algorithm assumes $p > 2g$ in order to obtain explicit formulas for the p-adic and t-adic precision. Let $b \in X(\mathbb{Q})$.

Definition 3.5 Let N be a positive integer $G \in \mathbb{Z}_p[[x]]$. Let $\overline{G}$ denote the reduction of G to $\mathbb{Z}/p^N\mathbb{Z}$. We say G satisfies Hensel's lemma for $r \in \mathbb{Z}/p^N\mathbb{Z}$ if

- $\overline{G}(r) = 0$ modulo p^N and
- $\overline{G}'(r)^2 \neq 0$ modulo p^N.

If G satisfies Hensel's lemma for r then by Hensel's Lemma [Con, Theorem 4.1] there exists a unique $\tilde{r} \in \mathbb{Z}_p$ such that $G(\tilde{r}) = 0$ and $|r - \tilde{r}|_p < |\overline{G}'(r)|_p$. Furthermore,

1. $|r - \tilde{r}|_p = |\overline{G}(r)|_p / |\overline{G}'(r)|_p$ and
2. $|\overline{G}'(\tilde{r})|_p = |\overline{G}'(r)|_p$.

Lemma 3.6 *Let $F(x)$ be a normalized power series, obtained as output from Algorithm 3.2, which converges on $\mathbb{Z}_p$. Fix N a positive integer, $M = M(N, \lambda)$, and write $F(x) = F_M(x) + F_\infty(x)$ as in Lemma 3.3.*

Let $r \in \mathbb{Z}/p^N\mathbb{Z}$ such that $F(x)$ satisfies Hensel's lemma for r. Let $\tilde{r} \in \mathbb{Z}_p$ be the unique lift of r satisfying $|r - \tilde{r}|_p < |F'_M(r)|_p$.

Then $\tilde{r} \equiv r \mod p^{N - v_p(F'_M(r))}$.

Proof Since F satisfies Hensel's lemma for $r \in \mathbb{Z}/p^N\mathbb{Z}$, we have $F_M(r) = 0$ modulo p^N. Furthermore, by (1), letting $\tilde{r}$ be the unique lift of r satisfying $|F'_M(r)|_p > |r - \tilde{r}|_p$, we have

$$p^{-N + v_p(F'(r))} \geq |F_M(r)|_p / |F'_M(r)|_p = |r - \tilde{r}|_p.$$

So $\tilde{r} = r$ modulo $p^{N - v_p(F'_M(r))}$. □

Every root of $F(x)$ reduces to a root of $F_M(x)$. Conversely, suppose we start with a set of roots R in $\mathbb{Z}/p^N\mathbb{Z}$ satisfying Hensel's Lemma for F_M, so that for any $r, r' \in R$, we have

$$r \not\equiv r' \mod p^{N - v_p(F'_M(r))}. \qquad (\dagger)$$

Then Lemma 3.6 implies that the roots $r \in R$ lift uniquely to simple roots $\tilde{r}$ of F in $\mathbb{Z}_p$ if F_M satisfies Hensel's lemma for all $r \in R$.

We call a set R with property (†) a system of roots for F_M modulo p^N. We can compute a system of roots iteratively, lifting roots of F_M modulo p to higher precision, as described in Algorithm B.0.1.

Algorithm 3.7 *Solving for the zeros of a tiny integral in a residue disk*

Input: $b \in X(\mathbb{Q})$, $\omega \in Van(X(\mathbb{Q}))$, $Q \in X(\mathbb{Q}_p)$ with uniformizer t computed as in Algorithm 2.4, and a precision N.
Output: The zeros in the residue disk of Q of the antiderivative of ω with constant term $\int_b^Q \omega$, or "Failure".

(1) Compute $c = \int_b^Q \omega$ the constant term (if Q is a bad point, $c = \int_b^S \omega + \int_S^Q \omega$ where $S \in X(\mathbb{Q}_p(p^{1/e}))$ is a point near the boundary of the residue disk of Q with e large enough so that the integral from b to S converges).
(2) Compute $f'(t) := \omega(t)$ to precision N using Algorithm 2.4.

(3) Compute $f(t)$ the antiderivative of $f'(t)$ with constant term c, accurate to precision $N' = N - \delta$, where δ is the precision loss when integrating discussed in [BT20, Section 4]. If $N' \leq 0$, return "Failure".
(4) Compute $M(N', \lambda)$ and let $F_M(x)$ be as in Lemma 3.3.
(5) Compute a system of roots R of $F_M(x)$ modulo $p^{N'}$, as described in Algorithm B.0.1.
(6) For each $r \in R$, check if $F'_M(r)^2 \not\equiv 0$ modulo $p^{N'}$. If this is true for every r, these correspond uniquely to the zeros of $f(t)$, as per Lemmas 3.3 and 3.6. If not, return "Failure".

Remark 3.8 If $f(t)$ has double roots, then $F_M(x)$ always has double roots. In Algorithm 3.1, we solve for common roots of two power series; we check the roots are single roots of at least one of the truncations. In the 1403 Picard curves in our database, at $N = 15$ we found common double roots for a single curve. By using a larger prime, we were able to provably compute that curve's rational points.

Remark 3.9 According to [BT20, Remark 4.2] the precision M in Balakrishnan and Tuitman's implementation used to truncate $F(x)$ t-adically is chosen so that for all $i \geq M$, $v_p(b_i) \geq N$ for b_i the coefficient of x^i. The chosen $M \geq M(\lambda, N')$ is sufficient.

3.3 Beyond Rational Points

If we cannot find $P \in X(\mathbb{Q})$ such that $[P - \infty]$ is infinite order in $J(\mathbb{Q})$, then we are must instead use a rational divisor on X whose image is infinite order in $J(\mathbb{Q})$. Because X is a genus 3 curve with a rational basepoint $\infty \in X(\mathbb{Q})$, we have a surjection $\mathrm{Sym}^3(X) \twoheadrightarrow J$ [Mil86]. Because we assume the rank of $J(\mathbb{Q})$ is 1, such a divisor exists, and the points in the support of the divisor are defined over an extension of degree at most 3.

Using our extension of Balakrishnan and Tuitman's code, it is possible to integrate over any rational divisor of the form $D - \deg(D)b \in X(\mathbb{Q})$ for rational primes p split completely in L, the smallest number field over which the points in the support of D are defined, and also to points defined over a number field K on a curve $X/\mathbb{Q}$ for p completely split in K. The curve $X/\mathbb{Q}$ should be smooth projective and geometrically integral with (not necessarily smooth) affine model $Q(x, y)$ satisfying the same assumptions as in [BT20, Assumption 2.6]. However, we focus our discussion on the situation for superelliptic curves where it is practical to compute rank bounds. Our code is tailored to this case. For non-hyperelliptic superelliptic curves, when computing infinite order points in the Mordell–Weil group $J(\mathbb{Q})$, `RankBounds` outputs polynomials[4] $g(T)$ such that the divisor $\mathrm{div}(g)$

[4]For a given curve there are often multiple choices of g, and we pick the one with the smallest first split prime.

on $\mathbb{P}^1$ lifts to a degree $\deg(g(T))$ rational divisor on X under the map $X \to \mathbb{P}^1$. Then $\deg(g(T))$ is at most the genus of X, which is 3 in our case of Picard curves. (For odd degree hyperelliptic curves `RankBounds` returns the Mumford representation of the divisor, and one can recover g by taking the polynomial for the x-coordinate. Our code does not handle the even degree hyperelliptic case.)

Let D be the degree $d = \deg(g(T))$ rational divisor obtained by lifting $\operatorname{div}(g)$ to X. Then $D = P_1 + \cdots + P_d - d\infty$. We use L to denote the splitting field of g, which is the smallest number field over which the points P_j are defined. To compute the Chabauty–Coleman set for X, we need to compute the set of vanishing differentials, which is done by taking a basis for the kernel of the 1×3 matrix of integrals of the regular differentials over D. The integral $\int_D \omega_i$ is equal to $\sum_{j=1}^{d} \int_{\infty}^{P_j} \omega_i$. By choosing a prime p that splits completely in L, for any $\mathfrak{p}|p$ in L we have $L_{\mathfrak{p}} \cong \mathbb{Q}_p$. Therefore, we can proceed as before and compute these integrals in $L_{\mathfrak{p}}$.

There are no theoretical innovations needed extend the code to use inert primes p in L, but it would require extensive changes to the base code. To integrate using inert primes, one would need to implement the q-power Frobenius for residue fields $\mathbb{F}_q$ where q is a power of p. The code directly extending Balakrishnan and Tuitman's code for Coleman integrals and the code for batch computations of rational points and testing the algebricity of points and Jacobian relations are available on GitHub [HMb, HMa].

4 The Data

In this section we discuss the overall data and give in-depth explanations on examples of interest. For the rest of the section, we fix $\infty \in X(\mathbb{Q})$ as our basepoint for the Abel–Jacobi embedding.

We computed the Chabauty–Coleman sets $X(\mathbb{Q}_p)_1$ of 1403 rank 1 Picard curves of discriminant bounded by 10^{12} and provably determined the set of rational points on these curves. In 859 cases, we found a rational point $P \in X(\mathbb{Q})$ whose class $[P - \infty]$ under the Abel–Jacobi map is infinite order. For the other 544 curves, we used infinite order divisor classes not given by the image of a rational point under the Abel–Jacobi map.[5]

On all curves, the naive search for points of height at most 1000 yielded all of the rational points. In Fig. 1 we display the number of Picard curves with n rational points. Every Picard curve has exactly one point at infinity and it is rational; the curves in our database had at most six rational points. In Fig. 2 we show the primes

[5]For the remaining 544 curves, it can be quite computationally expensive to compute $X(\mathbb{Q})$ depending on the parameters N, e, and p. Individual curves can require up to several hours. This computation was run on a single core of a 28-core 2.2 GHz Intel 2 Xeon Gold server with 256 GB RAM.

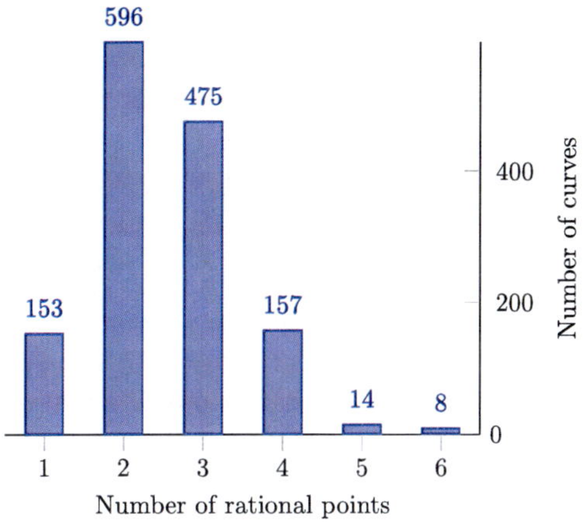

Fig. 1 Graph of number of rational points

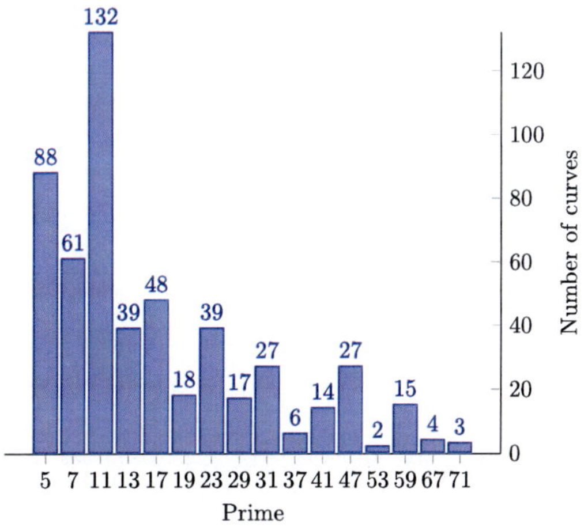

Fig. 2 Primes used to determine $X(\mathbb{Q}_p)_1$ for $\deg(g) > 1$

used for the curves with divisors not given by the image of a rational point under the Abel–Jacobi map. These are the first split primes in the splitting field of g, where the divisor is the rational lift of $\mathrm{div}(g)$, and g is chosen from a list of possible g returned by `RankBounds`. For ease of reading, we leave out the primes used once: those primes are 43, 61, 83, and 107.

For all but 14 curves, the only extra points in the Chabauty–Coleman set were the ramification points whose image in J is 3-torsion. For six curves, we found points $P_{\mathrm{tor}} \in X(K)$ such that $[P_{\mathrm{tor}} - \infty] \in J(K)$ is a torsion point of higher order defined over a number field K. These torsion points were 9-torsion and 4-torsion; the corresponding points P_{tor} are all of the form $(a, b^{1/3})$ with $a, b \in \mathbb{Z}$. For five curves, we found extra points $P_{\text{lin}} \in X(K)$ defined over a number field K such that there exists $n \in \mathbb{Z}$ so that $n[P_{\text{lin}} - \infty]$ is infinite order in $J(\mathbb{Q})$. For three curves, we found an extra point in the Chabauty–Coleman set whose presence is explained by an automorphism of the curve.

4.1 Examples

For some Picard curves X and choices of prime p, we found interesting global points in $X(\mathbb{Q}_p)_1$ that were not ramification points or rational points. We discuss four notable examples here.

Example 4.3 Consider the curve $X : y^3 = x^4 + 6x^3 - 48x - 64$ with discriminant 1289945088. Then J has the infinite order point $[(-3, -1) - \infty]$ so we compute the Chabauty–Coleman set at $p = 5$ and find an extra point $T = (t, 1/2t^2 - 4)$ defined over $\mathbb{Q}[t]/(t^3 - 24t - 48)$ (LMFDB 3.1.108.1). However, if we instead compute the Chabauty–Coleman set at the prime $p = 17$ we find $X(\mathbb{Q}_{17})_1 = X(\mathbb{Q}) \cup W \cup \{T, S\}$ where W is the pair of 17-adic ramification points. The point S is a different point $S = (s, s^2 + 6s + 8)$ defined over a different number field $\mathbb{Q}[s]/(s^3 + 9s^2 + 24s + 24)$ (LMFDB 3.1.216.1).

In the Jacobian of X, we have that $18[T - \infty] = 9[S - \infty]$. Furthermore, relating this back to the known point of infinite order in $J(\mathbb{Q})$, $[(-3, -1) - \infty]$, we can show that $18[T - \infty] = 3[(-3, -1) - \infty]$. So S and T appear in the Chabauty–Coleman set by linearity of the Coleman integral.

Example 4.4 The curve $X : y^3 = x^4 + 25x^3 - 78x^2 + 76x - 24$ with discriminant 411210307584 has Mordell–Weil rank 1 and possesses the infinite order point $D \in J(\mathbb{Q})$ that is a lift of $\text{div}(x^2 - 6x + 4)$.

The first prime that splits completely in $\mathbb{Q}[x]/(x^2 - 6x + 4)$ is 11, so we compute $X(\mathbb{Q}_{11})_1$. We obtain the rational points, one ramification point, as well as the point $(2, 32^{1/3})$, whose image in $J(\mathbb{Q})$ under the Abel–Jacobi map is 9-torsion.

Example 4.5 In their Chabauty–Coleman experiments on genus 3 hyperelliptic curves with Mordell–Weil rank 1, Balakrishnan–Bianchi–Cantoral-Farfán–Çiperiani–Etropolski [BBCF+] find a novel reason for extra points in the Chabauty–Coleman set. They exhibit a curve with two extra points coming from the presence of extra endomorphisms of the Jacobian, which are not explained by linearity or torsion. Motivated by their example, we searched Sutherland's Picard curve database for similar examples and found several. This first example of a curve with an extra point in the Chabauty–Coleman set explained by the automorphism group of the curve comes from searching outside of our rank 1 database.

Let $X : y^3 = x^4 - 2$ be the Picard curve with discriminant 10319560704000000. The Jacobian of this curve has rank 2 and has linearly independent non-torsion classes $[(2, 3) - \infty]$ and $[(-2, 3) - \infty]$ in $J(\mathbb{Q})$.

Table 1 Points in $X(\mathbb{Q}_{13})_1$

Set	Points
$X(\mathbb{Q})$	$(2, 3), (-2, 3), \infty$
W	Four Galois conjugates of $((5/2)^{1/4}, 0)$
T	Twelve Galois conjugates of $((45/2)^{1/4}, 40^{1/3})$, Three Galois conjugates of $(0, (-5)^{1/3})$
A	Two Galois conjugates of $((-4)^{1/2}, 3)$

The $\overline{\mathbb{Q}}$-automorphism group of X is a group of order 48 defined over the degree 12 number field $L := \mathbb{Q}[x]/(x^{12} - 6x^{11} + 21x^{10} - 50x^9 + 90x^8 - 126x^7 + 191x^6 - 276x^5 - 285x^4 + 950x^3 - 354x^2 - 156x + 676)$ (LMFDB 12.0.7652750400000000.2). The group is isomorphic to $\mathrm{SL}_2(\mathbb{F}_3) \rtimes_\rho C_2$ where the image of $\rho : C_2 \to \mathrm{Aut}(\mathrm{SL}_2(\mathbb{F}_3))$ is given by conjugation by an element of order 2. Let $p = 13$ be the first prime that splits completely in L. Computing the Chabauty–Coleman set at p, we find $X(\mathbb{Q}_{13})_1 = X(\mathbb{Q}) \cup W \cup T \cup A$ where W is the set of ramification points, T is the set of points whose image in $J(\mathbb{Q})$ is torsion, and A is a set of points that will be explained only by automorphisms of the curve, as shown in Table 1.

To verify that the points in T give rise to torsion classes in the Jacobian, while those in A do not, we compute the Coleman integrals $\int_\infty^P \omega_i$, $i = 1, 2, 3$ on regular differentials for each P in T and A using the new functionality for integrating to points defined over number fields. These integrals are 0 on points in T to our p-adic precision, but not zero on points in A. We then algebraically check that for $P \in T$, the class $n[P - \infty] = 0 \in J(\mathbb{Q})$ for some $n \in \mathbb{Z}$ using Jacobian arithmetic in Magma. The image of each point in T of the form $((45/2)^{1/4}, 40^{1/3})$ under the Abel–Jacobi map is a 12-torsion point on the Jacobian, while those of the form $(0, (-5)^{1/3})$ are 4-torsion points in J.

On the other hand, we show that the classes $A_1 := [((-4)^{1/2}, 3) - \infty]$ and $A_2 := [(-(-4)^{1/2}, 3) - \infty]$ do not have $\mathbb{Z}$-linear relations with $P_1 := [(2, 3) - \infty]$ and $P_2 := [(-2, 3) - \infty]$ in $J(\mathbb{Q})$. Suppose $nA_1 = aP_1 + bP_2$ for some $n, a, b \in \mathbb{Z}$ with $n \neq 0$. Then $nA_1^c = aP_1 + bP_2$ where c is complex conjugation. But $A_1^c = A_2$, and subtracting we would have $n(A_1 - A_2) = 0$ in $J(\mathbb{Q})$. However, computing the Coleman integrals to A_1 minus the Coleman integrals to A_2 we get a nonzero value, so $A_1 - A_2$ cannot be torsion.

Therefore we need a new reason to explain the presence of the points in A in the Chabauty–Coleman set. This reason will come from examining the automorphisms of X. When the vanishing differential is an eigenvector for a nontrivial automorphism of the curve, there is potential for extra points to appear in the Chabauty–Coleman set. Recall that $\omega_1 = ydx/f, \omega_2 = xydx/f$, and $\omega_3 = y^2dx/f$. Note X has an order 4 automorphism given by $\varphi : x \mapsto ix$ that acts by pullback on the differentials: $\varphi^*\omega_1 = i\omega_1, \varphi^*\omega_2 = -\omega_2, \varphi^*\omega_3 = i\omega_3$. The set of vanishing differentials $\mathrm{Van}(X(\mathbb{Q}))$ has dimension 1; let $v \in \mathrm{Van}(X(\mathbb{Q}))$ be a nonzero element. We will show v is an eigenvector for φ^* with eigenvalue i.

First we will explain this phenomenon by showing why the points of A are in the Chabauty–Coleman set, assuming v is an eigenvector for φ^* with eigenvalue i. It is enough to show that integrating A_j against the vanishing differential v is zero. Note that choosing square roots, $\varphi((2,3)) = ((-4)^{1/2}, 3)$ and $\varphi((-2,3)) = (-(-4)^{1/2}, 3)$. Using change of variables and applying the fact that our vanishing differential is an eigenvector, we have

$$\int_{A_1} v = \int_{\varphi(\infty)}^{\varphi((2,3))} v = \int_{P_1} \varphi^* v = i \int_{P_1} v = 0$$

and similarly for A_2. Thus the points of A are in the Chabauty–Coleman set because of φ.

Now we prove that v is an eigenvector for φ^*. Write $v = A\omega_1 + B\omega_2 + C\omega_3$, with $A, B, C \in \mathbb{Q}_p$. The vanishing differential is defined (up to a scalar multiple) by the equation $\int_{P_j} v = 0$ for $j = 1, 2$. Acting by φ^2 we have

$$0 = \int_{P_2} v = \int_{\varphi^2(\infty)}^{\varphi^2((2,3))} v = \int_{P_1} (\varphi^*)^2 v = \int_{P_1} -A\omega_1 + B\omega_2 - C\omega_3.$$

Adding this to the fact that $\int_{P_1} v = \int_{P_1} A\omega_1 + B\omega_2 + C\omega_3 = 0$ we see that

$$2 \int_{P_1} B\omega_2 = 0.$$

We compute that the integral over P_1 of ω_2 is not zero, so B must be zero, and $v = A\omega_1 + C\omega_3$ is indeed an eigenvector for φ^* with eigenvalue i.

Example 4.6 In addition to the previous example, we found three examples of extra points coming from automorphisms of the curve that arose during the computation of the Chabauty–Coleman sets in our rank 1 database. We describe one here. Let $X : y^3 = x^4 + 2x^3 + 6x^2 + 5x + 2$ be the Picard curve with discriminant 277826509467 and Mordell–Weil rank 1. There are no small affine rational points on this curve, so we use an infinite order point $D \in J(\mathbb{Q})$ that is a lift of $\mathrm{div}(x^2 + x - 1)$, and we compute the Chabauty–Coleman set for X at $p = 11$, the first split prime in $\mathbb{Q}[x]/(x^2 + x - 1)$. We obtain:

$$X(\mathbb{Q}_{11})_1 = \{\infty, (-1/2, \sqrt[3]{13/16})\}.$$

Let $R := (-1/2, \sqrt[3]{13/16})$. To check if R is torsion, we compute the Coleman integrals on the regular basis differentials $\int_\infty^R \omega_i$, $i = 1, 2, 3$, which are not all zero. The presence of R in $X(\mathbb{Q}_{11})_1$ also cannot be explained by relations in $J(\mathbb{Q})$: suppose $n[R - \infty] = mD$ for some $n, m \in \mathbb{Z} \setminus \{0\}$. Let ζ_3 be a primitive third root of unity. Acting pointwise by $\sigma : (x, y) \mapsto (x, \zeta_3 y)$ on each divisor, we have $nR^\sigma - n\infty^\sigma = mD^\sigma$. Subtracting from the original relation, we would have $n(R -$

$R^\sigma) = 0$ in $J(\mathbb{Q})$, but a computation with Coleman integrals confirms that $R - R^\sigma$ is not torsion.

By computing the automorphism group of X over $\mathbb{Q}[x]/(16x^3 - 13)$, we find X has an order 2 automorphism φ sending $x \mapsto -x - 1$ and fixing y. The pullback φ^* acts on Tuitman's basis of regular differentials by $\varphi^*\omega_1 = -\omega_1$, $\varphi^*\omega_2 = \omega_2$, and $\varphi^*\omega_3 = -\omega_3$. By a similar method used in Example 4.5, comparing the Coleman integral over D against v, and the Coleman integral over D against φ^*v, we can show that the vanishing differentials are of the form $v = A\omega_1 + C\omega_3$, and are eigenvectors for φ^* with eigenvalue -1. Furthermore, $\varphi(R) = R$ and $\varphi(\infty) = \infty$ and therefore

$$\int_\infty^R v = \int_{\varphi(\infty)}^{\varphi(R)} v = \int_\infty^R \varphi^* v = -\int_\infty^R v.$$

So $2\int_\infty^R v = 0$, showing $R \in X(\mathbb{Q}_{11})_1$.

Acknowledgments We are very grateful to Jennifer Balakrishnan for suggesting this project and for numerous helpful discussions and suggestions. We are also grateful to Drew Sutherland for providing us with data, and to Alex Best and Borys Kadets for their generous help and advice. We are also very thankful to the anonymous referees for their helpful suggestions for improving the paper.

Appendix 1: Precision Heuristics

Recall that, in the process of computing (non-tiny) Coleman integrals $\int_P^Q \omega$ where one endpoint Q is in a bad disk, we must break the integral up into $\int_P^S \omega + \int_S^Q \omega$ such that S is near the boundary of the disk. In other words, we find a point S defined over a totally ramified extension $\mathbb{Q}_p(p^{1/e})$. Balakrishnan and Tuitman's code takes as an input this e, but it can be a balancing act to decide what value to enter: too small and the integral will not converge, but the larger e is, the longer the runtime (Tables B.1 and B.2).

This section gives tables of the smallest e needed for `effective_chabauty` rounded up to the nearest multiple of five, fixing p and N, for 10 arbitrarily chosen curves of rank 1 with a rational point whose class in the Jacobian is of infinite order taken from Sutherland's database. Entries "bad" denote that a curve has bad reduction at that prime. For the tables, after small primes $p \leq 13$ we chose a sample of primes less than 100 spaced roughly by 10.

For every curve X in the full database of 1403 Picard curves, we also recorded the e value used to compute $X(\mathbb{Q}_p)_1$ and to select a non-torsion divisor when $N = 15$. We initialized e at 40 and then incremented by 20 until the integrals to compute $X(\mathbb{Q}_p)_1$ converged. These values can be found in the master list of Chabauty data on the 1403 Picard curves `masteralldata.txt` available in the repository [HMa].

Table B.1 Table for $p = 5, 7, 11, 13$, $N = 10$

Discriminant : Curve	5	7	11	13
$31492800 : y^3 = x^4 + 3x^3 - 3x + 1$	bad	50	85	100
$47258883 : y^3 = x^4 + 2x^3 - x^2 - x$	40	bad	85	100
$70858800 : y^3 = x^4 + 3x^3 - x^2 - 4x + 2$	bad	50	85	100
$106288200 : y^3 = x^4 + x^3 - 66x^2 - 324x - 432$	bad	30	50	55
$151322904 : y^3 = x^4 + 11x^3 - 32x^2 + 28x - 8$	30	35	50	100
$212891328 : y^3 = x^4 + 5x^3 + 4x^2 - 5x + 1$	10	50	85	bad
$105303439827 : y^3 = x^4 + 4x^3 + x^2 - 3x - 1$	30	50	85	95
$108095630841 : y^3 = x^4 + 4x^3 + 6x^2 - 9x$	35	40	bad	55
$113232992256 : y^3 = x^4 + x^3 - 4x^2 - 8x$	30	50	80	100
$988868881152 : y^3 = x^4 + x^2 - 2x + 3$	10	55	85	100

Table B.2 Table for select $p \geq 23$, $N = 10$

Curve	23	31	41	53	61	71	83	97
$y^3 = x^4 + 3x^3 - 3x + 1$	180	245	325	420	485	565	660	770
$y^3 = x^4 + 2x^3 - x^2 - x$	180	245	325	420	485	565	660	770
$y^3 = x^4 + 3x^3 - x^2 - 4x + 2$	180	245	325	420	485	565	660	770
$y^3 = x^4 + x^3 - 66x^2 - 324x - 432$	95	105	155	200	205	240	280	325
$y^3 = x^4 + 11x^3 - 32x^2 + 28x - 8$	180	bad	325	420	485	565	660	770
$y^3 = x^4 + 5x^3 + 4x^2 - 5x + 1$	180	240	325	420	485	565	660	770
$y^3 = x^4 + 4x^3 + x^2 - 3x - 1$	180	245	325	420	485	565	660	770
$y^3 = x^4 + 4x^3 + 6x^2 - 9x$	95	120	bad	200	230	270	315	365
$y^3 = x^4 + x^3 - 4x^2 - 8x$	180	245	320	bad	485	565	660	770
$y^3 = x^4 + x^2 - 2x + 3$	180	245	325	420	485	565	660	770

Appendix 2: Iterative Hensel Lifting

In this section we give an algorithm for computing approximate roots in $\mathbb{Z}_p$ of a polynomial with coefficients in $\mathbb{Z}_p$. Concretely, the elements of $\mathbb{Z}_p$ are represented by integers in $[0, p^N - 1]$ for some positive integer N, the coefficient precision (here, $[a, b]$ is the set of all integers k such that $a \leq k \leq b$). Let $f \in \mathbb{Z}[x]$ be a polynomial whose coefficients are in $[0, p^N - 1]$. We compute elements $r \in [0, p^N - 1]$ such that

(1) $f(r) = 0 \pmod{p^N}$;
(2) there exists an integer $0 \leq k_r \leq N$ such that $r \in [0, p^{k_r} - 1]$ and $f(r + p^{k_r} s)$ is identically zero in $\mathbb{Z}/p^N[s]$;
(3) if $r' \in [0, p^{k_r - 1} - 1]$ satisfies $r' = r \pmod{p^{k_r - 1}}$ then $f(r' + p^{k_r - 1} s)$ is not identically zero in $\mathbb{Z}/p^N[s]$; and
(4) if $2v_p(f'(r)) < N$ then there exists $\tilde{r} \in \mathbb{Z}_p$ such that $f(\tilde{r}) = 0$ and $|\tilde{r} - r| \leq p^{-k_r}$ (in other words, the approximation of $\tilde{r}$ by r is correct to precision k_r).

These properties follow in directly from Algorithm B.0.1, except (4) which is an application of Hensel's lemma (see Definition 3.5).

From these properties, we see the algorithm computes a system of roots as described in Sect. 3.2, with the additional property (2) which keeps track of the precision of the roots. To compute a system of roots, we inductively lift roots of df mod p^i to roots mod p^{i+1} in what is essentially a depth-first-search through the elements of $\mathbb{Z}/p^N$.

Algorithm B.0.1 *Solve for a system of roots R of $f \in \mathbb{Z}/p^N\mathbb{Z}[x]$.*
Input: Prime p, precision N, and $f \in \mathbb{Z}/p^N\mathbb{Z}[x]$ such that $f \mod p$ is not identically zero.
Output: A set R of roots in $\mathbb{Z}/p^N\mathbb{Z}$ such that if $r, r' \in R$, then $r \neq r' \mod p^{N-v_p(f'(r))}$.

(1) Let $A = []$.
(2) Let $B = \{a \in \mathbb{F}_p : f(a) = 0 \mod p\}$ be the $\mathbb{F}_p$-roots of f.
(3) Update $A = [(b, 1) : b \in B]$.
(4) Let $Z = []$.
(5) While A is nonempty:

(a) Let (a, i) be the first element of A. Remove (a, i) from A.
(b) Let $g_0(s) := f(a + p^i s)$.
(c) If $g_0(s) \neq 0 \pmod{p^N}$:

(i) Let v be the minimum valuation of the coefficients of g_0 and set $g := g_0/p^v$.
(ii) Let B be the roots of g in $\mathbb{F}_p$ and prepend $[(b, i + 1) : b \in B]$ to A.

(d) Else:

(i) Append (a, i) to Z.

(6) Return Z.

References

[BBCF+] J. S. Balakrishnan, F. Bianchi, V. Cantoral-Farfán, M. Çiperiani, and A. Etropolski. Chabauty-Coleman experiments for genus 3 hyperelliptic curves. *Research Directions in Number Theory, Association for Women in Mathematics Series*, 19:67–90.

[BT20] J. S. Balakrishnan and J. Tuitman. Explicit Coleman integration for curves. *Math. Comp.*, 89(326):2965–2984, 2020.

[BT] J. S. Balakrishnan and J. Tuitman. *Magma code.* https://github.com/jtuitman/Coleman.

[BBW17] M. Börner, I. Bouw, and S. Wewers. Picard curves with small conductor. In *Algorithmic and experimental methods in algebra, geometry, and number theory*, pages 97–122. Springer, Cham, 2017.

[Bou89] N. Bourbaki. *Lie groups and Lie algebras. Chapters 1–3.* Elements of Mathematics (Berlin). Springer-Verlag, Berlin, 1989. Translated from the French, Reprint of the 1975 edition.

[BPS16] N. Bruin, B. Poonen, and M. Stoll. Generalized explicit descent and its application to curves of genus 3. *Forum Math. Sigma*, 4:e6, 80, 2016.

[BS10] N. Bruin and M. Stoll. The Mordell-Weil sieve: proving non-existence of rational points on curves. *LMS J. Comput. Math.*, 13:272–306, 2010.

[Cha41] C. Chabauty. Sur les points rationnels des courbes algébriques de genre supérieure à l'unité. *C. R. Acad. Sci. Paris*, 212:882–885, 1941.

[Col85a] R. F. Coleman. Effective Chabauty. *Duke Math. J.*, 52(3):765–770, 1985.

[Col85b] R. F. Coleman. Torsion points on curves and p-adic abelian integrals. *Ann. of Math. (2)*, 121(1):111–168, 1985.

[Con] K. Conrad. *Hensel's Lemma*. https://kconrad.math.uconn.edu/blurbs/gradnumthy/hensel.pdf.

[Cre13] B. Creutz. Explicit descent in the Picard group of a cyclic cover of the projective line. In *ANTS X—Proceedings of the Tenth Algorithmic Number Theory Symposium*, volume 1 of *Open Book Ser.*, pages 295–315. Math. Sci. Publ., Berkeley, CA, 2013.

[Fal83] G. Faltings. Endlichkeitssätze für abelsche Varietäten über Zahlkörpern. *Invent. Math.*, 73(3):349–366, 1983.

[dFFH19] M. I. de Frutos-Fernández and S. Hashimoto. Computing rational points on rank 0 genus 3 hyperelliptic curves, 2021. arXiv:1909.04808.

[HMa] S. Hashimoto and T. Morrison. *Coleman experiments, Magma code*. https://github.com/travismo/picard.

[HMb] S. Hashimoto and T. Morrison. *Coleman extensions, Magma code*. https://github.com/travismo/Coleman.

[Ked01] K. S. Kedlaya. Counting points on hyperelliptic curves using Monsky-Washnitzer cohomology. *J. Ramanujan Math. Soc.*, 16(4):323–338, 2001.

[Ked03] K. S. Kedlaya. Errata for: "Counting points on hyperelliptic curves using Monsky-Washnitzer cohomology" [J. Ramanujan Math. Soc. **16** (2001), no. 4, 323–338; MR1877805]. volume 18, pages 417–418. 2003. Dedicated to Professor K. S. Padmanabhan.

[MP12] W. McCallum and B. Poonen. The method of Chabauty and Coleman. *Explicit Methods in Number Theory, Panor. Synthèses*, 36:99–117, 01 2012.

[Mil86] J. S. Milne. Jacobian varieties. In *Arithmetic Geometry (Storrs, Conn., 1984)*, pages 167–212. Springer, New York, 1986.

[Poo97] B. Poonen and E. Schaefer. Explicit descent for Jacobians of cyclic covers of the projective line. *J. Reine Angew. Math.*, 488:141–188, 1997.

[Sik15] S. Siksek. Chabauty and the Mordell-Weil sieve. In *Advances on superelliptic curves and their applications*, volume 41 of *NATO Sci. Peace Secur. Ser. D Inf. Commun. Secur.*, pages 194–224. IOS, Amsterdam, 2015.

[Sut] A. V. Sutherland. *Genus 3 non-hyperelliptic curves*. https://math.mit.edu/~drew/genus3curves.html.

[Sut19] A. V. Sutherland. A database of nonhyperelliptic genus-3 curves over $\mathbf{Q}$. In *Proceedings of the Thirteenth Algorithmic Number Theory Symposium*, volume 2 of *Open Book Ser.*, pages 443–459. Math. Sci. Publ., Berkeley, CA, 2019.

[Tui16] J. Tuitman. Counting points on curves using a map to $\mathbf{P}^1$. *Math. Comp.*, 85(298):961–981, 2016.

[Tui17] J. Tuitman. Counting points on curves using a map to $\mathbf{P}^1$, II. *Finite Fields Appl.*, 45:301–322, 2017.

[Wet] J. L. Wetherell. Bounding the number of rational points on certain curves of high rank. PhD Thesis, 1997. UC Berkeley.

Linear Dependence Among Hecke Eigenvalues

Dohyeong Kim

Abstract We prove an explicit upper bound on the absolute value of the coefficients of a non-trivial integral linear relation among Hecke eigenvalues of a given cuspidal eigenform. Our motivation lies in its algorithmic application. For any fixed positive integer n, the bound established here yields an algorithm that computes cuspidal Hecke eigenforms with a given weight k whose Hecke eigenvalues generate a number field of degree n. The resulting algorithm reduces to Cremona's when $n = 1$ and $k = 2$.

1 Introduction

Hecke showed that, if

$$f(q) = \sum_{m=1}^{\infty} a_m q^m, \ a_m \in \mathbb{C}$$

is the Fourier expansion of a primitive normalized cuspidal modular form of weight $k \geq 2$, then its coefficients satisfy

$$a_m = O(m^{\frac{k}{2}})$$

where the implied constant only depends on the form. Deligne, using his deep work confirming the Weil conjecture, lowered the exponent by one half. Furthermore, the implied constant was made absolute. More precisely, he showed that the inequality

$$|a_m| \leq d(m) m^{\frac{k-1}{2}}$$

D. Kim (✉)
Department of Mathematical Sciences and RIM, Seoul National University, Seoul, Republic of Korea
e-mail: dohyeongkim@snu.ac.kr

J. S. Balakrishnan et al. (eds.), *Arithmetic Geometry, Number Theory, and Computation*, Simons Symposia, https://doi.org/10.1007/978-3-030-80914-0_17

holds true for any $m \geq 1$, where $d(m)$ denotes the number of positive divisors of m. See Ch. 9 of [2] for a treatment of the topic.

On the other hand, the Hecke field of $f(q)$, the subfield of $\mathbb{C}$ generated by all a_m's is of finite degree, say n, over $\mathbb{Q}$. Then, Deligne's bound allows one to bound the coefficients of the minimal polynomial of a_m in terms of k, m, and n. This cuts out a finite collection of polynomials that can be potentially satisfied by the m-th eigenvalue a_m of some $f(q)$.

Our aim in this article is to construct a similar collection of linear relations among multiple a_m's. Concretely, for a given choice $(m_1, m_2, \ldots, m_{n+1})$ of $n+1$ distinct indices, we want to produce positive integers $b_1, b_2, \ldots, b_{n+1}$ which shall guarantee a non-trivial relation

$$\sum_{i=1}^{n+1} \lambda_i a_{m_i} = 0, \quad \lambda_i \in \mathbb{Z} \tag{1.1}$$

such that $|\lambda_i| \leq b_i$ for all i. More generally, we will consider the possibility of a subcollection $a_{m_1}, \ldots, a_{m_{r+1}}$ for some $r \leq n$ admitting a non-trivial relation among themselves. This motivates us to consider the relations of the form

$$\sum_{i=1}^{r+1} \lambda_i a_{m_i} = 0, \quad \lambda_i \in \mathbb{Z} \tag{1.2}$$

where $a_{m_1}, \ldots, a_{m_{r+1}}$ generate a $\mathbb{Z}$-module of rank exactly r. Of course, if there is a non-trivial relation of the form (1.1), then there exists some $r \leq n$ for which (1.2) holds true after reordering indices. Therefore, it suffices to consider relations of the form (1.2) for each $r = 0, 1, 2, \ldots, n$ instead of (1.1), maintaining throughout the assumption that $a_{m_1}, \ldots, a_{m_{r+1}}$ generate a $\mathbb{Z}$-module of rank r. This assumption will be exploited repeatedly in subsequent arguments.

We first introduce a result which applies to all r. For integers $k \geq 1$ and $m \geq 1$, let

$$\rho_{m,k} = d(m)m^{\frac{k-1}{2}} \tag{1.3}$$

be the Deligne bound. The inequality (1.5) below is what we call the trivial bound.

Theorem 1.4 *Let $f(q) = \sum_{m \geq 1} a_m q^m$ be a normalized cuspidal Hecke eigenform of weight $k \geq 1$. If $m_1, \ldots, m_{r+1}$ are chosen so that $a_{m_1}, \ldots, a_{m_{r+1}}$ span a $\mathbb{Z}$-module of rank r, then there is a non-trivial relation $\sum_{i=1}^{r+1} \lambda_i a_{m_i} = 0$ with $\lambda_i \in \mathbb{Z}$ such that*

$$|\lambda_i| \leq r^r \left(\prod_{j_1 \neq i} \rho_{m_{j_1},k} \right) \left(\sum_{j_2 \neq i} \rho^2_{m_{j_2},k} \right)^{\frac{1}{2}} \tag{1.5}$$

where the product and the sum are taken over $1 \le j_1, j_2 \le r+1$ *and* $j_1, j_2 \ne i$.

Note that the trivial bound does not involve the degree of the Hecke field.

We move on to our main result whose aim is to refine the trivial bound. Note that if $\chi(m) = 1$, where χ is the character of $f(q)$, then the corresponding eigenvalue a_m is totally real. Therefore, such eigenvalues will belong to the maximal totally real subfield F of the Hecke field. From now on, we will denote the degree of F by n, and the main result applies to the case $r = n$. To formulate it, we will need two positive quantities $\eta(n)$ and $\theta(n)$. We require that $\eta(n)$ is a lower bound for the root discriminant of a totally real number field of degree n. Explicit values of $\eta(n)$ have been obtained by Odlyzko [9]. For $\theta(n)$, we require that any lattice of dimension n inside $\mathbb{R}^n$ admits a basis whose orthogonality defect is at most $\theta(n)$. Various forms of $\theta(n)$ are available [7, 10]. Asymptotically, $\eta(n) = M - o(n^{-2/3})$ with $M = 60.8395\ldots$ is due to [9], and $\log(\theta(n)) = O(n \log n)$ due to [7]. In any case, we define

$$\rho(n) = \frac{\theta(n)}{\eta(n)^{n/2}}$$

to be the ratio. Note that $\rho(n)$ is independent of k. We list some of its values in Table 1.

Here is our main result.

Theorem 1.6 *Let* $f(q) = \sum_{m \ge 1} a_m q^m$ *be a normalized cuspidal Hecke eigenform with character* χ *and weight* $k \ge 1$. *Let* n *be the degree of the maximal totally real subfield of the Hecke field of* $f(q)$. *If* $m_1, \ldots, m_{n+1}$ *are chosen so that* $\chi(m_i) = 1$ *for all* i *and that* $a_{m_1}, \ldots, a_{m_{n+1}}$ *generate a* $\mathbb{Z}$*-module of rank* n, *then there is a non-trivial relation* $\sum_{i=1}^{n+1} \lambda_i a_{m_i} = 0$ *with* $\lambda_i \in \mathbb{Z}$ *such that*

$$|\lambda_i| \le n^{n/2} \rho(n) \left(\sum_{j \ne i} \rho_{m_j,k}^2 \right)^{\frac{1}{2}} \tag{1.7}$$

where the sum is taken over $1 \le j \le n+1$ *and* $j \ne i$.

Since the right hand side of (1.7) depends on the choice of $m_1, m_2, \ldots, m_{n+1}$, one is naturally interested in making a good choice. We largely ignore the problem to make the best possible choice, and contend ourselves with the fact that some choice can be made without recourse to the form $f(q)$. For example, given k and χ, one may choose first n values of m_i's with $\chi(m_i) = 1$.

We compare the trivial bound (1.5) and the main theorem (1.7). Since the last factor is common to both, we may compare $r^r \prod_{j_1} \rho_{m_{j_1},k}$ and $n^{n/2}\rho(n)$ when $r = n$. The latter is, first of all, independent of the weight k of the form in consideration. Therefore, (1.7) is an improvement of (1.5) in the weight aspect. On the other hand, we observe that $\rho(n)$ decreases at least up to $n = 8$. We give some approximate values of $\rho(n)$ in Table 1. Asymptotically, as we noted earlier, we know from [7]

Table 1 Some approximate values of $\rho(n)$

n	2	3	4	5	6	7	8
$\rho(n)$	0.52	0.21	0.078	0.030	0.013	0.0064	0.0039

that $\rho(n)$, in logarithm, belongs to $O(n \log n)$. Therefore, (1.7) is not worse than (1.5) as n grows to infinity.

Refinements of the trivial bound are relevant in the following computational problem. Let n and k be fixed positive integers. If V is a vector space of modular forms of weight k on which a set of Hecke operators act, then the problem asks one to determine whether V has an irreducible subquotient of dimension n. It was studied by Cremona [5, 6] in the case $n = 1$ and $k = 2$, giving rise to an influential database of elliptic curves. We will explain in Sect. 7 how our work generalizes a part of it.

The organization of the paper is as follows. In Sect. 2, we will prove Lemma 2.2, a key technical tool, about linear relation among algebraic integers. Then, we will continue to review bounds for discriminant in Sect. 3 and orthogonality defects in Sect. 4, both of which we employ to bound $\rho(n)$. In Sect. 5, we apply Deligne's bound, and note a connection to Kronecker's theorem. After reviewing these results, the trivial bound and the main theorem will follow by straightforward bootstrapping in Sect. 6. In Sect. 7, we conclude with a description of our algorithm and a comparison to Cremona's work.

Acknowledgments The author is grateful for helpful comments by J. Cremona, A. Sutherland, and J. Voight. This work is partially supported by (1) Simons Foundation grant 550033, (2) Research Resettlement Fund for the new faculty of Seoul National University, and (3) the National Research Foundation of Korea (NRF) grants funded by the Korea government (MSIP) (No. 2020R1C1C1A01006819 and No. 2019R1A6A1A10073437).

2 Linear Relations Among Algebraic Integers in a Given Number Field

Suppose that F is a number field of degree n with its ring of integers O_F. Consider $n + 1$ elements $\alpha_1, \ldots, \alpha_{n+1} \in O_F$ which generate a $\mathbb{Z}$-submodule of rank n.

Using α_i's, form an additive map

$$A\colon \mathbb{Z}^{n+1} \to O_F$$

which sends the i-th standard basis to α_i, for $i = 1, 2, \ldots, n + 1$.

Choose an identification $O_F \cong \mathbb{Z}^n$ using a $\mathbb{Z}$-basis of O_F. Then, A is represented by an $n \times (n+1)$ matrix $\tilde{A}$. Denote by $\tilde{A}_i$ the submatrix of $\tilde{A}$ obtained by removing the i-th column for $i = 1, 2, \ldots, n+1$. It is a square matrix whose determinant is independent of the basis we have chosen for O_F. Put $\lambda_i = (-1)^i \det(\tilde{A}_i)$ and $\underline{\lambda} = (\lambda_1, \ldots, \lambda_{n+1})$.

Lemma 2.1 *We have* $\underline{\lambda} \neq 0$ *and* $A(\underline{\lambda}) = 0$.

Proof The first assertion follows from the assumption that $\alpha_1, \ldots, \alpha_{n+1}$ span a $\mathbb{Z}$-module of rank n. The second assertion follows from Cramer's rule. □

Our aim in this section is to bound $|\lambda_i|$'s. For each $i = 1, \ldots, n+1$, let L_i be the submodule of O_F generated by the columns of $\tilde{A}_i$. A basic property of determinants tells us that $|\lambda_i| = [O_F : L_i]$, where we use the convention that $[O_F : L_i] = 0$ when L_i is of infinite index in O_F. We conclude that the problem to bound λ_i's reduce to bounding $[O_F : L_i]$. We describe below an elementary method for it.

Without loss of generality, we may assume $i = n+1$ and L_{n+1} has rank n. Put $L = L_{n+1}$ for simplicity of notation. We would like to bound $[O_F : L]$ in terms of $\sigma(\alpha_j)$ for $\sigma \in I_F$, the set of archimedean places of F. Let

$$\mathcal{M} \colon F \to \mathbb{R}$$

be the function which assigns to each $\alpha \in F$ the average of $|\sigma(\alpha)|^2$ taken over $\sigma \in I_F$. In fact, $\mathcal{M}(\alpha)$ is absolute in that it does not depend on the choice of a number field that contains α. For example, $\mathcal{M}(1+\sqrt{2}) = \frac{1}{2}(3+2\sqrt{2}) + \frac{1}{2}(3-2\sqrt{2}) = 3$ and $\mathcal{M}(1+\sqrt{-3}) = 4$. We also define

$$\mu(\alpha) = (\mathcal{M}(\alpha))^{1/2}.$$

Lemma 2.2 *We have*

$$[O_F : L] \leq \frac{n^n}{|D_F|} \left(\prod_{i=1}^{n} \mu(\beta_i) \right) \cdot \left(\sum_{j=1}^{n} \mu(\alpha_j)^2 \right)^{\frac{1}{2}}$$

for any $\mathbb{Z}$*-basis* $\beta_1, \ldots, \beta_n$ *of* O_F.

Proof Let $\tau \colon F \to \mathbb{Q}$ be the trace map. We rewrite

$$[O_F : L] = \frac{|\det(\tau(\beta_i \alpha_j))|}{|\det(\tau(\beta_i \beta_j))|}$$

whose denominator is $|D_F|$ by definition. Thus, it suffices to prove the inequality

$$|\det(\tau(\beta_i \alpha_j))| \leq n^n \left(\prod_{i=1}^{n} \mu(\beta_i) \right) \cdot \left(\sum_{j=1}^{n} \mu(\alpha_j)^2 \right)^{\frac{1}{2}}. \tag{2.3}$$

By the Hadamard inequality, the magnitude of the determinant of a matrix is at most the product of the Euclidean lengths of its columns. On the other hand, $|\tau(\alpha\beta)| \leq n\mu(\alpha)\mu(\beta)$ for any $\alpha, \beta \in F$ by the Cauchy-Schwartz inequality. Applying these two, one obtains

$$
\begin{aligned}
|\det\left(\tau(\beta_i\alpha_j)\right)| &\leq \prod_{i=1}^{n}\left(\sum_{j=1}^{n}\tau(\beta_i\alpha_j)^2\right)^{1/2} \\
&\leq \prod_{i=1}^{n}\left(\sum_{j=1}^{n}\left(n\mu(\beta_i)\mu(\alpha_j)\right)^2\right)^{\frac{1}{2}} \\
&\leq n^n\left(\prod_{i=1}^{n}\mu(\beta_i)\right)\cdot\left(\sum_{j=1}^{n}\mu(\alpha_j)^2\right)^{\frac{1}{2}}
\end{aligned}
$$

as desired. □

We rewrite the inequality in Lemma 2.2 in the form

$$
[O_F : L] \leq \frac{n^{n/2}}{|D_F|^{1/2}} \cdot \frac{n^{n/2}}{|D_F|^{1/2}} \prod_{i=1}^{n}\mu(\beta_i) \cdot \left(\sum_{j=1}^{n}\mu(\alpha_j)^2\right)^{\frac{1}{2}}. \tag{2.4}
$$

Since this holds for any choice of $(\beta_1, \ldots, \beta_n)$, it becomes

$$
[O_F : L] \leq \frac{n^{n/2}}{|D_F|^{1/2}} \cdot \left(\inf_{\beta_1,\ldots,\beta_n} \frac{n^{n/2}}{|D_F|^{1/2}} \prod_{i=1}^{n}\mu(\beta_i)\right) \cdot \left(\sum_{j=1}^{n}\mu(\alpha_j)^2\right)^{\frac{1}{2}}, \tag{2.5}
$$

where the infimum is taken over the set of $\mathbb{Z}$-bases of O_F. The problem to bound $[O_F : L]$ is reduced to three subproblems corresponding to each factor. First, we need a lower bound for $|D_F|$. Second, we need to find an upper bound for the factor in the middle. Lastly, we need an upper bound for each $\mu(\alpha_j)$. In subsequent sections, we will discuss bounds for these factors. Note that only the last part depends on the lattice L, while the first two are determined by O_F. After that, in Sect. 6, we summarize the results and prove Theorems 1.4 and 1.6.

3 Discriminants and Odlyzko Bound

We review the work of Odlyzko on lower bounds for discriminants of number fields. Let F be a number field of degree n. Let D_F be the discriminant of F, and $\delta_F :=$

Table 2 Root discriminant bounds, totally real fields

n	2	3	4	5	6	7	8	9
$\eta(n)$	2.223	3.610	5.067	6.523	7.941	9.301	10.596	11.823
$\eta^*(n)$	2.225	3.630	5.124	6.640	8.143	9.611	11.036	12.410

$|D_F|^{1/n}$ be its root discriminant. Our main concern is to bound δ_F when F is totally real.

By a root discriminant bound, we shall mean an inequality of the form

$$\delta_F \geq \eta(n)$$

where $\eta(n)$ is a function of n. Table 2 contains some of the root discriminant bounds, extracted from Table 4 of [9], where $\eta(n)$ denotes an unconditional bound while $\eta^*(n)$ is conditional on a generalized Riemann hypothesis.

This allows us to bound the first factor in (2.5).

4 Lattice Reduction and Orthogonality Defect

Let $\Lambda \subset \mathbb{R}^n$ be a lattice of rank n. Let $v(\Lambda)$ be the covolume of Λ. For a basis $\mathbf{x} = (x_1, \ldots, x_n)$ of Λ, define the orthogonality defect of it to be

$$od(\mathbf{x}) = \frac{\prod_{i=1}^{n} ||x_i||}{v(\Lambda)},$$

where $||\cdot||$ denotes the Euclidean norm. The Hadamard inequality tells us that the orthogonality defect is at least one, and is equal to one if and only if the basis is orthogonal. Thus, it provides a way to measure the extent to which a basis fails to be orthogonal. It is natural to consider

$$od(\Lambda) := \inf_{\mathbf{x}} od(\mathbf{x}),$$

where the infimum is taken over all bases of Λ. We will call it the orthogonality defect of the lattice Λ.

We explain the relevance of the orthogonality defect in (2.5). We view O_F as a lattice in $F \otimes_{\mathbb{Q}} \mathbb{R}$, which we identify with the Euclidean space $\mathbb{R}^n$ via field embeddings $F \to \mathbb{R}$.

Proposition 4.1 *The infimum*

$$\inf_{\beta_1,\ldots,\beta_n} \frac{n^{n/2}}{|D_F|^{1/2}} \prod_{i=1}^{n} \mu(\beta_i)$$

taken over all $\mathbb{Z}$-bases $(\beta_1, \ldots, \beta_n)$ is equal to $od(O_F)$.

Proof It suffices to note that $v(O_F)^2 = |D_F|$ and that $\sqrt{n} \cdot \mu(\beta)$ is the Euclidean length of β for $\beta \in O_F$. □

Now we explain how to bound orthogonality defect of a lattice $\Lambda \subset \mathbb{R}^n$. An important quantity associated to Λ is the length of shortest non-zero vector in it, often denoted by $\lambda_1(\Lambda)$. The Hermite constant γ_n is defined to be

$$\gamma_n = \sup_{\Lambda} \lambda_1(\Lambda),$$

where the supremum is taken over the set of all lattices with unit covolume. Clearly, $\gamma_1 = 1$, and other known values of γ_n range over $2 \leq n \leq 8$ and $n = 24$. On the other hand, we have an upper bound

$$\gamma_n \leq \frac{2}{\pi} \Gamma\left(2 + \frac{n}{2}\right)^{2/n}$$

due to Blichfeldt [1]. Better upper bounds for γ_n were established in [3], and computed for $n \leq 36$ using the relation between lattice packing density and Hermite constants [4, p. 20] (Table 3).

Various reduction algorithms provide upper bounds for $od(\Lambda)$ in the form

$$od(\Lambda) \leq \theta(n)$$

where $\theta(n)$ only depends on the lattice dimension n. Whenever such $\theta(n)$ is established, we can apply it to the particular lattice $\Lambda = O_F$ associated to the number field and deduce

$$\inf_{\beta_1,\ldots,\beta_n} \frac{n^{n/2}}{|D_F|^{1/2}} \prod_{i=1}^{n} \mu(\beta_i) = od(O_F) \leq \theta(n) \tag{4.2}$$

We will recall those arising from the Minkowski reduction and the Korkin-Zolotarev reduction. The Minkowski reduction implies [10]

$$od(\Lambda) \leq \begin{cases} \gamma_n^{n/2} & \text{if } n \leq 4, \\ \gamma_n^{n/2} \left(\frac{5}{4}\right)^{(n-3)(n-4)/4} & \text{if } n > 4. \end{cases} \tag{4.3}$$

The Lagarias-Lenstra-Schnorr bound established in [7], which is based on the Korkin-Zolotarev reduction, is

Table 3 Some known values of powers of Hermite constants

n	2	3	4	5	6	7	8	24
γ_n^n	$\frac{4}{3}$	2	4	8	$\frac{64}{3}$	64	2^8	4^{24}

Table 4 Orthogonality defect bounds

n	2	3	4	5	6	7	8	12	16	24
$\theta(n)$	1.155	1.415	2	3.163	6.455	15.63	48.83	1.4×10^4	1.6×10^7	4.4×10^{13}

$$od(\Lambda) \le \gamma_n^{n/2} \prod_{i=1}^{n} \left(\frac{i+3}{4} \right)^{1/2}. \tag{4.4}$$

Lastly, Nguyen and Vallée [8, p. 47] obtained

$$od(\Lambda) \le \left(\frac{4}{3} \right)^{n(n-1)/4} \tag{4.5}$$

based on the Hermite reduction.

For $n = 1, 2$, (4.3) and (4.5) agree and they are better than (4.4). For $3 \le n \le 10$, (4.3) is the best. For $11 \le n \le 14$, (4.5) is the best. For $n \ge 15$, (4.4) is the best. Table 4 summarizes in numbers the best among these three methods for $2 \le n \le 8$, and $n = 12, 16, 24$.

5 Kronecker's Theorem

We come to the last factor in (2.5). Let α be an algebraic integer in a number field F, and M a positive real number such that $|\sigma(\alpha)| \le M$ for any embedding $\sigma : F \to \mathbb{C}$. For example, when $\alpha = a_m$ for some cuspidal eigenform $f(q) = \sum_{m \ge 1} a_m q^m$, M may be taken to be the corresponding Deligne bound. The trivial bound for $\mu(\alpha)$ is

$$\mu(\alpha) \le M. \tag{5.1}$$

For the proof of Theorem 1.6, all we will need is (5.1), but we would like to note in this section that we are often able to refine (5.1).

Let α be an irrational totally real algebraic integer. Kronecker proved that the conjugates of α are contained in the interval $[-2, 2]$ if and only if α can be written as $\alpha = \zeta + \zeta^{-1}$ where ζ is a root of unity.

Proposition 5.2 *Let α be a totally real algebraic integer of degree n such that α^2 is irrational. Choose a positive real number M with $|\sigma(\alpha)| \le M$ for any embedding of F into $\mathbb{R}$. Also assume that $M^2 - \frac{4}{n} > 4$. Then*

$$\mu(\alpha) < \left(M^2 - \frac{4}{n} \right)^{1/2}. \tag{5.3}$$

Proof If α is of the form $\alpha = \zeta + \zeta^{-1}$ for a root of unity ζ, then $\mu(\alpha) \leq 2$ holds trivially and the claimed inequality follows from the assumption $M^2 - \frac{4}{n} > 4$. Otherwise, Kronecker's theorem forces a conjugate of α^2 to be strictly smaller than $M^2 - 4$. This yields the claimed inequality. □

Returning to (2.5), let us temporarily assume that the field F has an odd degree. Since F cannot have any quadratic subfield, it follows that an irrational element $\alpha \in F$ never satisfies $\alpha^2 \in \mathbb{Q}$. This allows us to use (5.3) instead of (5.1).

6 The Proof of the Main Theorem and the Trivial Bound

The proof of the main theorem requires little more than adding up results from previous sections. Let $a_{m_1}, \ldots, a_{m_{n+1}}$ be Hecke eigenvalues of $f(q)$ where $m_1, \ldots, m_{n+1}$ are appropriately chosen so that each a_{m_i} is totally real. Apply Lemma 2.1 to $\alpha_i = a_{m_i}$, $i = 1, 2, \ldots, n+1$, and obtain a non-trivial relation $\sum_i \lambda_i \alpha_i = 0$. Each $|\lambda_i|$ is the index $[O_F : L_i]$, where F is the totally real subfield of the Hecke field of $f(q)$ and L_i is the lattice generated by $\alpha_1, \ldots, \alpha_{i-1}, \alpha_{i+1}, \ldots, \alpha_{n+1}$. As explained below Lemma 2.1, we have the equality $|\lambda_i| = [O_F : L_i]$. Note that when L_i has rank less than that of O_F both sides of the equality are regarded as zero. Then, we apply the inequality (2.5) and apply the bounds from Sects. 3 and 4 in order. For the sum $\sum_{j \neq i} \mu(\alpha_i)$, we apply trivial bound (5.1) together with the Deligne bound. This completes the proof of Theorem 1.6.

We continue to prove Theorem 1.4, the trivial bound. Suppose that the Hecke eigenvalues $a_{m_1}, \ldots, a_{m_{r+1}}$ generate a lattice L of rank $r > 0$. For simplicity of notation, put $\alpha_i = a_{m_i}$. We will apply a method similar to what we used in Sect. 2. The argument for Lemma 2.1 and the following explanation implies that there is a non-trivial relation $\sum_{i=1}^{r+1} \lambda_i \alpha_i = 0$ where λ_i is, up to sign, the index in L of the $\mathbb{Z}$-submodule generated by $\alpha_1, \ldots, \alpha_{i-1}, \alpha_{i+1}, \ldots, \alpha_{r+1}$. Without loss of generality, we assume $i = r+1$. For a given basis $\beta_1, \ldots, \beta_r$ for L, we have

$$|\lambda_{r+1}| = \frac{\det(\tau(\alpha_i \beta_j))}{\det(\tau(\beta_i \beta_j))}, \qquad i, j = 1, 2, \ldots, r.$$

Because both $\tau(\alpha_i \alpha_j)$ and $\tau(\alpha_i \beta_j)$ are integral and $|\det(\tau(\alpha_i \beta_j))| \leq |\det(\tau(\alpha_i \alpha_j))|$, the above has a rather trivial upper bound $|\det(\tau(\alpha_i \alpha_j)|$. Then, the Hadamard inequality together with Deligne's bound implies

$$\begin{aligned} |\det(\tau(\alpha_i \alpha_j))| &\leq |\det(r \rho_{m_i,k} \rho_{m_j,k})| \\ &\leq r^r \left(\prod_{i=1}^{r} \rho_{m_i,k} \right) \left(\sum_{j=1}^{r} \rho_{m_j,k}^2 \right)^{\frac{1}{2}}. \end{aligned}$$

This completes the proof of the trivial bound.

7 Application: An Algorithm for Modular Forms

A typical computational problem in the theory of modular forms is to decompose a module V over a Hecke algebra into irreducible subquotients. For example, V can be a space of modular forms stable under a set of Hecke operators. We will focus on the case when V is cuspidal, defined over $\mathbb{Q}$, and the Hecke algebra is generated over $\mathbb{Q}$ by a suitable set of commuting Hecke operators.

We describe a general and elementary method for attacking the problem. If V and some Hecke operators on it are given, one may start by choosing a Hecke operator T and computing its characteristic polynomial, say $R_T(X)$. If $R_T(X)$ is irreducible, then V must be irreducible and the decomposition is complete. If not, there must be a non-trivial factor of $R_T(X)$, say $\tilde{R}_T(X)$, and the kernel and the image of $\tilde{R}_T(T)$ are non-trivial subquotients of V, stable under other Hecke operators as well. One repeats this until all Hecke modules are irreducible.

Often, all one wants to know about V is whether it has a subquotient of particular type. For example, one might wonder if V has an eigenform with trivial character, weight two, and rational Hecke eigenvalues. In this case, it suffices to compute the kernels of $T_2 - \lambda$ with $\lambda = -2, -1, 0, 1, 2$. An attentive reader may wonder how this is possible, given the fact that there do exist eigenforms with irrational Hecke fields whose T_2-eigenvalues happen to be rational. Let us temporarily call them fake rational eigenforms. With the presence of fake rational eigenforms, one would have a non-zero kernel of $T_2 - \lambda$, say V_λ, which does not contain any rational eigenform. However, V_λ is almost always much smaller in dimension than V, and it is much less demanding to handle V_λ. This is what Cremona employed in order to generate his database [5, 6] without fully decomposing relevant spaces of modular forms.

Suppose that one is instead interested in the type of irreducible subquotients of V. If W is an irreducible subquotient, then the Hecke algebra $H(W)$ acting on W itself is a number field[1] and its degree is equal to the dimension of W. Let $n(W)$ be the degree of the maximal totally real subfield in $H(W)$, and we shall call $n(W)$ the essential degree of W. We consider the problem to find all irreducible subquotients of V with essential degree n. Since $n(W)$ is at most the dimension of $H(W)$, it subsumes the problem to find all irreducible subquotients of dimension n. A natural question is, especially when V has such a large dimension that the characteristic polynomials of Hecke operators are difficult to compute and factor, whether one can obtain all the desired subquotients without fully decomposing V. It is indeed possible, as we explained in the introduction, that one can use Deligne's bound to produce a finite family of polynomials with integer coefficients in a Hecke operator, which plays the role of $\{T_2 - \lambda \colon |\lambda| \leq 2\}$ in Cremona's strategy. Our

[1]Indeed, it is isomorphic to the Hecke field of any eigenform in $W \otimes_{\mathbb{Q}} \mathbb{C}$.

results provide an alternative answer to this question. If W is such a subquotient, then there must be some non-trivial linear relation among the restrictions to W of $T_{m_1}, \ldots, T_{m_{n+1}}$, where $m_1, \ldots, m_{n+1}$ are chosen so that each T_{m_i} has totally real eigenvalues. Reordering indices, for some $0 \le r \le n$, $\{T_{m_1}, \ldots, T_{m_{r+1}}\}$ is a minimal set admitting a non-trivial linear relation. Then, $T_\lambda := \sum_{i=1}^{r+1} \lambda_i T_{m_i}$ would annihilate[2] W for some vector $\lambda = (\lambda_1, \ldots, \lambda_{r+1}) \neq 0$ with integer entries. If $r = 0$, then some T_{m_i} annihilates W. If $0 < r < n$, then the trivial bound (1.5) applies. If $r = n$, then (1.7) applies. We note that if $n = 1$ and the space V consists of forms with trivial character, then our method reduces to Cremona's.

We illustrate the algorithm by reporting two numerical examples which we ran in SageMath. Using Brandt matrices, we computed the Hecke operators T_2 and T_3 acting on the Brandt module V of level $N = 13{,}499$. The module V is defined as a vector space over $\mathbb{Q}$ and its dimension is 1126. Elements in V correspond to modular forms of weight two. The problem is to determine whether V has any two-dimensional irreducible subquotients. The finite set of linear combinations of 1, T_2, and T_3, as constrained by Theorems 1.4 and 1.6, turn out to have trivial kernels. This shows that V has no irreducible subquotients of dimension two. For the second example, we take $N = 12{,}553$, whose associated Brandt module has dimension 1046. Then, $1 + T_2 - T_3$ turns out to be the unique linear combination with a non-trivial kernel. Its dimension is two, on which the action of T_3 is irreducible. We conclude that the Brandt module of level 12553 has a unique irreducible subquotient of dimension two.

References

[1] H. F. Blichfeldt. The minimum value of quadratic forms, and the closest packing of spheres. *Math. Ann.*, 101(1):605–608, 1929.

[2] Henri Cohen and Fredrik Strömberg. *Modular forms*, volume 179 of *Graduate Studies in Mathematics*. American Mathematical Society, Providence, RI, 2017. A classical approach.

[3] Henry Cohn and Noam Elkies. New upper bounds on sphere packings. I. *Ann. of Math. (2)*, 157(2):689–714, 2003.

[4] J. H. Conway and N. J. A. Sloane. *Sphere packings, lattices and groups*, volume 290 of *Grundlehren der Mathematischen Wissenschaften [Fundamental Principles of Mathematical Sciences]*. Springer-Verlag, New York, third edition, 1999. With additional contributions by E. Bannai, R. E. Borcherds, J. Leech, S. P. Norton, A. M. Odlyzko, R. A. Parker, L. Queen and B. B. Venkov.

[5] J. E. Cremona. *Algorithms for modular elliptic curves*. Cambridge University Press, Cambridge, second edition, 1997.

[6] John Cremona. The elliptic curve database for conductors to 130000. In *Algorithmic number theory*, volume 4076 of *Lecture Notes in Comput. Sci.*, pages 11–29. Springer, Berlin, 2006.

[7] J. C. Lagarias, H. W. Lenstra, Jr., and C.-P. Schnorr. Korkin-Zolotarev bases and successive minima of a lattice and its reciprocal lattice. *Combinatorica*, 10(4):333–348, 1990.

[2]As explained in the preceding paragraph, the kernel of T_λ may be strictly bigger than W, while being harmless to the purpose.

[8] Phong Q. Nguyen and Brigitte Valle. *The LLL Algorithm: Survey and Applications*. Springer Publishing Company, Incorporated, 1st edition, 2009.
[9] A. M. Odlyzko. Bounds for discriminants and related estimates for class numbers, regulators and zeros of zeta functions: a survey of recent results. *Sém. Théor. Nombres Bordeaux (2)*, 2(1):119–141, 1990.
[10] B. L. van der Waerden. Die Reduktionstheorie der positiven quadratischen Formen. *Acta Math.*, 96:265–309, 1956.

Congruent Number Triangles with the Same Hypotenuse

David Lowry-Duda, with an appendix by Brendan Hassett

Abstract In this article, we discuss whether a single congruent number t can have two (or more) distinct corresponding triangles with the same hypotenuse. We describe and carry out computational experimentation providing evidence that this does not occur.

1 Introduction and Motivation

A *congruent* number is a rational number that appears as the area of a rational right triangle. The *congruent number classification problem* is the problem of determining whether a given number t is congruent. By scaling the triangle, one may consider only the classification of squarefree integers t; we do this throughout the article. The study of congruent numbers is ancient and has a long history; see the survey article of Conrad [Con08] for more.

We consider whether a single congruent number t can have two distinct rational right triangles with the same hypotenuse. This question was first motivated by an observation in [HKLDW18].

In that paper, Hulse, Kuan, Walker, and the author consider a convolution sum whose main term exists only when t is congruent. Let $\tau(n)$ denote the square

The author was supported by the Simons Collaboration in Arithmetic Geometry, Number Theory, and Computation via the Simons Foundation grant 546235.

The author wants to thank John Cremona, Brendan Hassett, Joe Silverman, and Damiano Testa for the encouraging remarks and discussion. The author also wants to thank an anonymous Referee for several comments, and for directing the author to the paper [WDE+14].

BH was supported by National Science Foundation grant 1701659 and Simons Foundation award 546235.

D. Lowry-Duda (✉)
Institute of Computational and Experimental Research in Mathematics, Providence, RI, USA
e-mail: david@lowryduda.com

J. S. Balakrishnan et al. (eds.), *Arithmetic Geometry, Number Theory, and Computation*, Simons Symposia, https://doi.org/10.1007/978-3-030-80914-0_18

indicator function, taking 1 if n is a square and 0 otherwise. The primary theorem of [HKLDW18] is that

$$\sum_{n \leq X} \sum_{m \leq X} \tau(m+n)\tau(m)\tau(m-n)\tau(nt) = C_t\sqrt{X} + O_t((\log X)^{r/2}), \qquad (1.1)$$

where r is the rank of the elliptic curve $E_t : Y^2 = X^3 - t^2X$ and

$$C_t = \sum_{h \in \mathcal{H}(t)} \frac{1}{h} \qquad (1.2)$$

is the (convergent) sum over the set of hypotenuses $\mathcal{H}(t)$ of dissimilar primitive right integer triangles with area equal to tu^2 (i.e. the squarefree part of the area is t).

Although there is implicit dependence on t in the error term and poorly understood dependence on t in the main term, heuristically one should expect that for large X, the main term "quickly" dominates.

In [HKLDW18], it is also shown that the set of hypotenuses $\mathcal{H}(t)$ is only logarithmically dense in $\mathbb{Z}_{\geq 1}$. Thus heuristically the coefficient C_t of the main term is well-estimated by its first term m/h, where h is the smallest hypotenuse in $\mathcal{H}(t)$ and m is the multiplicity of this hypotenuse. Further, the main term should heuristically grow larger than the error term quickly once $X > h/m$.

The question of the size of the smallest hypotenuse in $\mathcal{H}(t)$ is closely related to the least height of a non-torsion point on E_t (we make this more precise in Sect. 3), but the multiplicity m is more mysterious.

In this article, we describe numerical experimentation suggesting that the multiplicity is always exactly 1. We also propose that this is always the case as a conjecture.

Conjecture *There do not exist two dissimilar primitive right triangles with the same hypotenuse and whose squarefree parts of the areas are equal to each other.*

In Sect. 2, we describe why this problem might be hard to fully resolve. Hassett describes this more completely in the Appendix.

We then describe and carry out numerical experimentation using the free and open source math software SageMath [Sag20]. Our experimentation comes in two forms. Firstly, we do a deep investigation for those congruent numbers below 1000. For each congruent number, we generate many different triangles by generating rational points $E_t(\mathbb{Q})$ and the corresponding right triangles.

Secondly, we do a broad investigation for several congruent numbers corresponding to curves E_t of high rank. These curves will have more rational points up to a bounding height, heuristically leading to more triangles with hypotenuses up to some bound. In this investigation, we make heavy use of [WDE$^+$14], which gives several congruent number curves E_t of rank 6 and 7. In total, we investigate 1513 curves of ranks 6 and 7.

We describe these approaches and the results in Sects. 3 and 4, respectively.

2 Algebraic Formulation

One can parametrize right triangles with two variables (s, t) such that the sides are given by $(s^2 - t^2, 2st, s^2 + t^2)$. We follow the convention that the hypotenuse is written last in any triple (a, b, c) giving a right triangle, and that $s > t$. In this correspondence, primitive integer right triangles correspond to relatively prime integers s and t.

Thus finding two right triangles with the same hypotenuse and whose areas are the same up to multiplication by squares can be reformulated as finding integers (s, t, S, T) such that

$$s^2 + t^2 = S^2 + T^2$$

and where

$$\frac{st(s^2 - t^2)}{ST(S^2 - T^2)} \text{ is a square.}$$

This last equation can be rewritten as

$$u^2 st(s^2 - t^2) = v^2 ST(S^2 - T^2)$$

for some positive integers u, v.

These two equations define a surface

$$X \subset \mathbb{P}^3_{[s,t,S,T]} \times \mathbb{P}^1_{[u,v]}.$$

If we could understand all points on this surface, we could almost certainly resolve the conjecture. Unfortunately, this understanding appears to be beyond current algebraic techniques. This is considered by Hassett in the Appendix, where it is shown that the surface X admits a resolution of singularities that is of general type and simply connected (see Proposition A).

Thus we expect that a complete resolution of the conjecture is not within reach.

3 Description of Methodology

With a purely theoretical resolution likely out of reach, we turn now to computational experimentation. There is a well-known correspondence between right triangles (a, b, c) with $ab/2 = t$ (which we may assume without loss of generality is a squarefree integer) and $\mathbb{Q}$-rational points on the elliptic curve $E_t : Y^2 = X^3 - t^2 X$ where $Y \neq 0$. The inverse maps of this correspondence are given by

$$(a, b, c) \mapsto \left(\frac{tb}{c-a}, \frac{2t^2}{c-a}\right) = (X, Y),$$

$$(X, Y) \mapsto \left(\frac{X^2 - t^2}{Y}, \frac{2tX}{Y}, \frac{X^2 + t^2}{Y}\right) = (a, b, c).$$

(See [Kob93] for a historical overview and description of the relationship between congruent numbers and elliptic curves.)

3.1 Enumeration Through Elliptic Curves

This provides an explicit and computable way of enumerating all rational right triangles with area t: find generators for $E_t(\mathbb{Q})$ and enumerate their linear combinations. This method of enumeration is of course vastly superior to naive enumeration and is at the core of the organization of the numerical experimentation.

In principle, it is necessary to also include the torsion subgroup of $E_t(\mathbb{Q})$. However for these curves, the torsion subgroup is completely understood and can be ignored. In particular, the torsion subgroup is isomorphic to $\mathbb{Z}/2\mathbb{Z} \times \mathbb{Z}/2\mathbb{Z}$, generated by two points T_1 and T_2. If the triangle (a, b, c) corresponds to the point $P \in E_t(\mathbb{Q})$, the eight equivalent triangles $(\pm a, \pm b, \pm c)$ correspond to the points $\pm P + \epsilon_1 T_1 + \epsilon_2 T_2$, where $\epsilon_i \in \{0, 1\}$. Thus we omit no triangles (up to similarity) by omitting consideration of torsion points.

3.2 Hypotenuses, and Heights, Grow Exponentially

Given an elliptic curve with known rank and generators, the primary obstruction to collecting numerical evidence is the size of the triangles and hypotenuses. There is a general theory that the heights of points on elliptic curves grow very rapidly under repeated addition. In Corollary 1 of [HKLDW18], it is shown that

$$|\{h \in \mathcal{H}(t) : h \leq X/2t\}| = O_t((\log X)^{r/2}),$$

where r is the rank of the corresponding elliptic curve E_t. In particular, letting $H(P)$ denote the naive height of the X-coordinate of P, it is shown that the hypotenuse h corresponding to a point P on E_t satisfies $h \geq H(P)/2t$, and thus hypotenuses grow at least as quickly as the heights.

Example 1 The elliptic curve $E_6 : Y^2 = X^3 - 36X$ has rank 1 and the free part of $E_6(\mathbb{Q})$ is generated by $g = (-3, 9)$. This point corresponds to the primitive right triangle $(3, 4, 5)$. The hypotenuse is 3 bits long. The point $2 * g = (25/4, -35/8)$ corresponds to the primitive right triangle $(49, 1200, 1201)$, whose hypotenuse is 11 bits long. The point $4 * g$ has hypotenuse 2094350404801, which is 29 bits long.

The point $400 * g$ (the largest multiple we considered on this curve) corresponds to a right triangle whose hypotenuse requires 410426 bits to store.

As heights grow exponentially, we see that hypotenuses grow exponentially. If we assume that the hypotenuses behave like exponentially growing random variables, then heuristically we would expect that it is very unlikely that two points of large height correspond to the same hypotenuse. Informally, we should expect that if a pair of points yields two triangles with hypotenuse that are close in size, then both points are probably of small height.

For curves of rank 1, it is impractical to compute more than a few hundred triangles—they simply grow too quickly. For curves of higher rank, however, it is possible to compute millions of triangles before the individual sides of the triangles become prohibitively large.

3.3 *Two Strategies of Experimentation*

We investigate two potential sources for a counterexample to the conjecture: when all numbers are small and coincidental collisions are most likely, and when there are many triangles up to a certain size due to high rank. Thus we organize our experimentation into two strategies: one for elliptic curves E_t with small t, and one for high rank elliptic curves.

First, we study the curves E_t for all squarefree $t \leq 1000$. Each of these curves has rank 0, 1, or 2. In total, this includes 361 curves corresponding to congruent numbers. These are the curves with the smallest t and with the simplest triangles.

As these curves have low rank, we can carry out "deep" investigation on each curve. More specifically, if $\{g_i\}$ denote generators for the free part of $E_t(\mathbb{Q})$, then we can study triangles coming from $\sum a_i g_i$ for all coefficients $\{a_i\}$ in a large box $|a_i| \leq K$ for a constant $K = K(t)$. For rank 1 curves, we choose $K(t) \geq 300$, and for rank 2 curves we choose $K(t) \geq 75$. As the points P and $-P$ generate the same triangle, this translates to considering at least 300 triangles from each rank 1 curve and at least $(151^2 - 1)/2 = 11{,}400$ triangles from each rank 2 curve.

Second, we study curves E_t of known high rank. To get these curves, we use the large-scale project detailed in [WDE$^+$14], which builds off of earlier work of Rogers [Rog00, Rog04]. In this project, Watkins et al. investigated ranks of congruent number curves. They identify 1486 curves E_t of rank at least 6 and 27 curves of rank 7, giving 1513 curves in total of high rank. We examine rational points on each of these 1513 curves.

We note that no curves of higher rank are known, and it is conjectured that ranks in the family of congruent number curves are bounded [Hon61]. In §11 of [WDE$^+$14], it is further demonstrated that a heuristic of Granville's suggests that the maximum rank of a congruent number elliptic curve is 7.

As noted in Sect. 3.2, we should heuristically expect that it is more likely for points of small height to give two triangles with the same hypotenuse. For this reason, we carry out a "wide but shallow" investigation of these curves. Again letting $\{g_i\}$ denote a set of generators for the free part of $E_t(\mathbb{Q})$, we study triangles coming from points $\sum a_i g_i$, where all coefficients $\{a_i\}$ are in the small box $|a_i| \leq 4$. This gives 267520 triangles from each rank 6 curve and 2391484 triangles from each rank 7 curve.

Remark 2 It is likely that using the box $|a_i| \leq 2$ would be sufficient. In Sect. 4, we observe that $|a_i| \leq 2$ finds all closest pairs of hypotenuses on the curves investigated. We use the larger box to conduct a more robust investigation. Computing and manipulating these points is by far the most time intensive portion of this experiment. With the larger coefficient boxes $|a_i| \leq 4$, we spend approximately 32 min on average for each rank 6 curve, and approximately 5 h on average for each rank 7 curve. In total, we used approximately 940 CPU hours on this part of the experimentation.

3.4 Computing Rank and Generators

In order to generate triangles for a congruent number t, we require that we can compute generators for (the free part of) $E_t(\mathbb{Q})$. We perform computations for two different sets of curves: those curves E_t with $t < 1000$ and for several chosen curves of higher rank.

For the 1513 curves of higher rank, we use ranks and generators available as an electronic supplement to [WDE+14].[1] We note that not all of these curves have explicitly known rank: all have rank at least 6, and 27 have rank 7, but Watkins et al. couldn't verify the ranks of each curve. We do not try to complete this verification here, and instead use the 6 known generators for each curve of presumed rank 6. Curves corresponding to $t < 1000$ can be handled within SageMath [Sag20]. The methods *EllipticCurve.rank* and *EllipticCurve.gens* from SageMath works for all but 84 cases. The smallest problematic example is 113. We note that the curve E_{113} is contained within the L-function and Modular Form Database (LMFDB) [LMF19]. However, it is also possible to directly use Tunnell's criterion [Tun83] to quickly confirm that 113 is not congruent. Applying Tunnell's criterion on the remaining 83 cases to remove non-congruent numbers reduces this to 54 cases. The smallest remaining example is 157.

The number 157 is congruent and is somewhat famous for its late classification by Zagier [Zag89]; the simplest rational triangle with area 157 is extremely complicated.

The techniques SageMath uses to compute the rank and generators for elliptic curves are based on John Cremona's MWRANK [CMP+19]. SageMath includes

[1]The supplement to their paper made this data available in Magma code. We have converted it to a SageMath-friendly format and made it available at https://github.com/davidlowryduda/notebooks/blob/master/Papers/largetdata.sage.

a copy of MWRANK and a partial interface to MWRANK's functionality. Calling MWRANK directly and allowing up to 2000 seconds of computation time per curve provides generators and ranks for an additional 20 cases (including 157). This leaves 34 congruent numbers up to 1000, the smallest of which is 277.

To understand the remaining 34 curves, we use functionality from PARI-GP [BBB+00], which is also packaged with SageMath. Although MWRANK struggles to exactly compute the rank and generators, it can compute that the upper bound on the rank is 1 for each of these curves. For the curve E_{277} for example, calling MWRANK through Sagemath with the command *EllipticCurve([-277*277,0]).rank_bound()* shows an upper bound of 1 for the rank.

Thus one can use the Heegner point method [Elk94] to find a generator for the rational points on these curves. The existence of such a generator confirms that the rank is exactly 1. This is implemented remarkably well in PARI. Calling *pari(EllipticCurve([-277*277,0])).ellheegner()* from SageMath quickly returns a generator for $E_{277}(\mathbb{Q})$.

Remark 3 The CAS Magma [BCP97] is also capable of generating ranks and generators for these elliptic curves. An anonymous Referee noted that *Generators(EllipticCurve([-277*277,0]))* returns a set of generators almost immediately, and functions almost as quickly for all congruent numbers up to 1000.

4 Experimental Results

We performed the strategies outlined in Sect. 3 for each squarefree congruent number t with $t \leq 1000$ and for the 1513 curves of higher rank from [WDE+14]. (See the author's demonstration github [LD] for a reference implementation to obtain and manipulate generators when $t \leq 1000$; for higher t, we do similar manipulations with the generators found by Watkins et al.)

4.1 Results for $t \leq 1000$

Out of the 608 squarefree numbers up to 1000, we verified that 327 of them are congruent numbers and produced several triangles for each one. Of these, 274 corresponding to curves with rank 1 and 53 to curves with rank 2. We note that this is consistent with the celebrated conjecture of Goldfeld [Gol79], which implies that although there may be infinitely many such curves of rank at least 2, these should be sparse and correspond to 0% of all congruent numbers in natural density.

Among all these computed triangles, we found no counterexample to the conjecture. There were no two dissimilar right triangles with the same hypotenuse (after scaling to a primitive right triangle) corresponding to the same congruent number.

Recall that we computed at least 300 triangles from each curve of rank 1 and at least 11,400 from each curve of rank 2. Per the heuristic in Sect. 3.2, we would expect that pairs of points corresponding to triangles with the close hypotenuses should come from points of small height.

This is true for all the curves tested for $t \leq 1000$. Further, the nearest pairs of hypotenuses among triangles associated to the same curve were always among the smallest hypotenuses. In terms of our generators $\{g_i\}$, the smallest pair of hypotenuses always came from triangles corresponding to the points $\sum a_i g_i$ with $|a_i| \leq 2$. (Although there are not canonical choices of generators, we note that MWRANK returns generators with small naive height [Cre97].)

Example 4 On $E_6(\mathbb{Q})$, whose free part is generated by $(-3, 9)$, the two nearest hypotenuses come from the triangles $(3, 4, 5)$ and $(49, 1200, 1201)$, which come from g and $2g$ respectively.

For $t = 34$, we have that $\text{rank}\big(E_{34}(\mathbb{Q})\big) = 2$. The free part of $E_{34}(\mathbb{Q})$ is generated by the points $g_1 = (-16, 120)$ and $g_2 = (-2, 48)$. The two closest hypotenuses come from the triangles $(225, 272, 353)$ and $(17, 144, 145)$, corresponding to g_1 and g_2 respectively.

This adds support for our choice of box $|a_i| \leq 4$ for points on higher rank elliptic curves.

Although no pair of hypotenuses exactly matched, we can ask about how close two hypotenuses could be. Of those congruent numbers $t \leq 1000$, we determined that 5 have pairs of dissimilar triangles with hypotenuses that differ by less than t: 210, 330, 546, 609, 915. All five of these come from rank 2 curves.

The curve E_{210} has a particularly close pair of triangles whose hypotenuses differ by 8. This is the pair of triangles $(12, 35, 37)$ and $(20, 21, 29)$, which correspond to the two generators for the free part of $E_{210}(\mathbb{Q})$.

Smallest Hypotenuses Further, we took this opportunity to collect data on the size of the smallest hypotenuse for each congruent number ≤ 1000. In Fig. 1, we plot squarefree congruent numbers t against the log of the smallest hypotenuse of a primitive right triangle with squarefree part of the area equal to t.

Within the plot, the first congruent number t with smallest hypotenuse greater than e^{100} is Zagier's number, 157. The *largest* smallest hypotenuse we computed corresponds to the congruent number $t = 997$, and is approximately e^{449}.

Many smallest-hypotenuses appear to be moderate in size, but no consistent pattern emerges.

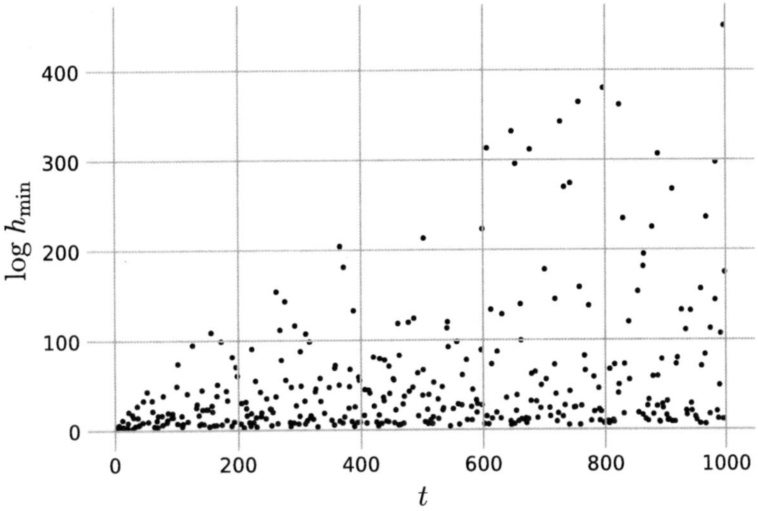

Fig. 1 Smallest hypotenuses for congruent numbers $t \leq 1000$

4.2 Results for Curves of Rank 6 and 7

We produced hundreds of thousands of triangles for each of the 1513 curves of high rank. We found no counterexample to the conjecture.

As above, we tracked closest pairs of hypotenuses for each curve. Even though we computed all rational points $\sum a_i g_i$ in the coefficient box where each coefficient satisfies the bound $|a_i| \leq 4$, we again observed that for each t, the closest pair of hypotenuses came from points in the smaller box $|a_i| \leq 2$. The heuristic from Sect. 3.2 appears to continue to hold, indicating that nearest pairs of hypotenuses occur among the points of the least height on the curve.

We consider how close hypotenuses from two different triangles could be. Of all the 1513 higher rank curves investigated, 375 has two hypotenuses that differed by less than the corresponding congruent number t.

In absolute terms, the closest pair came from $t = 6611719866 \approx 6.611 \times 10^9$, which has triangles with hypotenuses 30544225 and 67119265, differing by about 3.657×10^7. This is very large. We also noted a pair for the congruent number $t = 1902736244939034 \approx 1.9027 \times 10^{15}$, which has triangles with hypotenuses given approximately by 1.061×10^{10} and 1.109×10^{10}, differing by about 4.769×10^8. These are also the two smallest hypotenuses of triangles for t. Although this is also very large, this is significantly smaller than both t and the smallest hypotenuse.

Both of these "near-miss" examples came from rank 6 elliptic curves.

Smallest Hypotenuses We again took this opportunity to collect data on the size of the smallest hypotenuse coming from each of the 1513 curves of higher rank. As the numbers are encompass a vastly larger domain, we present this in two log-log plots.

In Fig. 2, we show smallest hypotenuses for each t from curves of (presumed) rank 6. We first explain the lack of points in the upper-right portion of the plot. This is caused by the methodology used to generate the 1513 curves in [WDE$^+$14]—they performed an almost-exhaustive search for high rank elliptic curves for all $t \leq 2^{50}$, while for higher t they only looked for curves with points of relatively low height. As $2^{50} \approx e^{34.5}$ and as the hypotenuse h corresponding to a point P satisfies the bound $h > H(p)/2t$, this explains the missing region of data. (Although we found it surprising that this plot reveals the choices of bounds used to find the elliptic curves so clearly.)

Comparing Fig. 2 with Fig. 1, we note the disparity in size. Every hypotenuse coming from a higher rank elliptic curve was smaller than the smallest hypotenuse on E_{157}. On the other hand, each smallest hypotenuse was larger than e^{15}.

In Fig. 3, we show smallest hypotenuses coming from curves of rank 7. As with the previous plot, there are patterns among the curves that result from the way in which the curves were initially found. In [WDE$^+$14], they performed an extensive search for curves of rank 7 for $t \leq 2^{60} \approx e^{41.6}$, and for higher t they only looked for curves with rational points of low height. Although it is less obvious in this plot, we note that we should expect that there is a lot of missing data in the upper right portion of this plot.

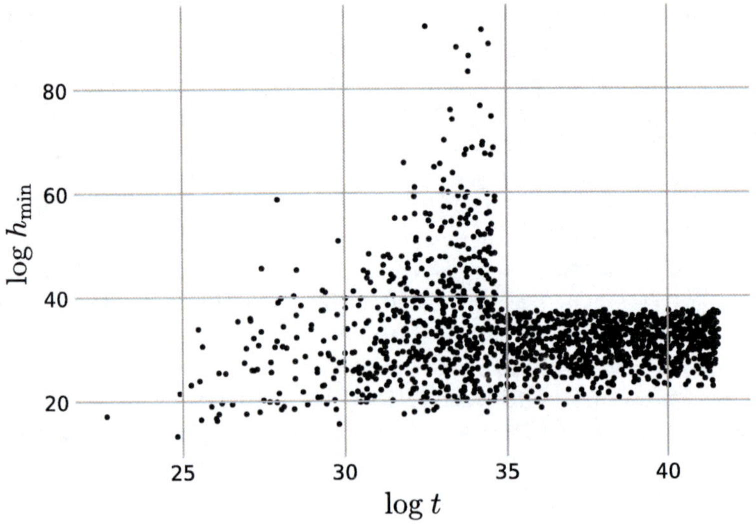

Fig. 2 Smallest hypotenuses from (presumed) rank 6 curves

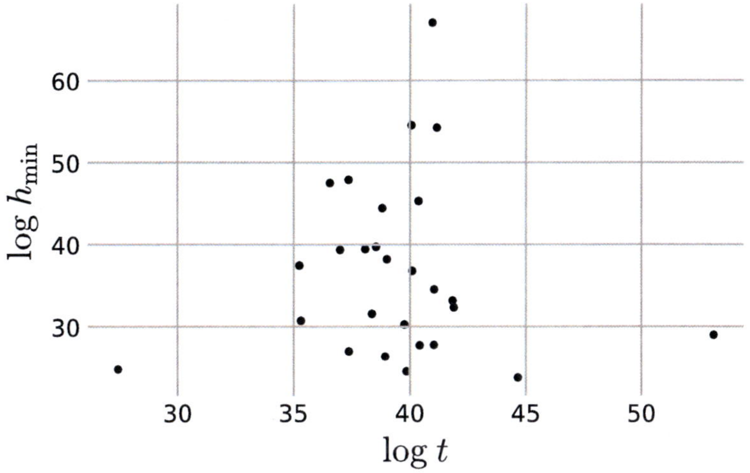

Fig. 3 Smallest hypotenuses from rank 7 curves

Appendix: Invariants of the Lowry-Duda Moduli Space

by Brendan Hassett

We verify that the moduli space X introduced by Lowry-Duda admits a resolution of singularities $\widetilde{X} \to X$ that is of general type and simply connected. Since $\widetilde{X}(\mathbb{Q}) \neq \emptyset$, existing Diophantine techniques (like Faltings' Theorem [Fal94]) shed little light on the structure of the rational points on X.

Double Cover Realization

Let

$$X \subset \mathbb{P}^3_{[s,t,S,T]} \times \mathbb{P}^1_{[u,v]}$$

be the surface over $\mathbb{Q}$ defined by the equations

$$s^2 + t^2 = S^2 + T^2$$
$$u^2 st(s^2 - t^2) = v^2 ST(S^2 - T^2).$$

The first equation defines a surface $Y \subset \mathbb{P}^3$ isomorphic to $\mathbb{P}^1 \times \mathbb{P}^1$. Projection onto the first factor induces a morphism $\varpi : X \to Y$ that is generically finite of degree two. The branch curve

$$B := \{st(s^2 - t^2)ST(S^2 - T^2) = 0\} \subset Y$$

is a union of eight planar sections of Y, each a smooth conic curve defined over $\mathbb{Q}$.

The morphism ϖ fails to be flat over the locus

$$Z := \{st(s^2 - t^2) = ST(S^2 - T^2) = 0\} \subset B \subset Y$$

consisting of the following 32 points:

$$\begin{aligned}
[0, 1, 0, \pm 1] &= \{s = 0\} \cap \{S = 0\} \\
[0, 1, \pm 1, 0] &= \{s = 0\} \cap \{T = 0\} \\
[0, \pm\sqrt{2}, 1, 1] &= \{s = 0\} \cap \{S - T = 0\} \\
[0, \pm\sqrt{2}, 1, -1] &= \{s = 0\} \cap \{S + T = 0\} \\
[1, 0, 0, \pm 1] &= \{t = 0\} \cap \{S = 0\} \\
[1, 0, \pm 1, 0] &= \{t = 0\} \cap \{T = 0\} \\
[\pm\sqrt{2}, 0, 1, 1] &= \{t = 0\} \cap \{S - T = 0\} \\
[\pm\sqrt{2}, 0, 1, -1] &= \{t = 0\} \cap \{S + T = 0\} \\
[1, 1, 0, \pm\sqrt{2}] &= \{s - t = 0\} \cap \{S = 0\} \\
[1, 1, \pm\sqrt{2}, 0] &= \{s - t = 0\} \cap \{T = 0\} \\
[1, 1, \pm 1, \pm 1] &= \{s - t = 0\} \cap \{S - T = 0\} \\
[1, 1, \pm 1, \mp 1] &= \{s - t = 0\} \cap \{S + T = 0\} \\
[1, -1, 0, \pm\sqrt{2}] &= \{s + t = 0\} \cap \{S = 0\} \\
[1, -1, \pm\sqrt{2}, 0] &= \{s + t = 0\} \cap \{T = 0\} \\
[1, -1, \pm 1, \pm 1] &= \{s + t = 0\} \cap \{S - T = 0\} \\
[1, -1, \pm 1, \mp 1] &= \{s + t = 0\} \cap \{S + T = 0\}
\end{aligned}$$

Points of Z are nodes of B. The double cover of Y branched over B has A_1 singularities over these 32 nodes which are automatically resolved in X.

The remaining points of intersection among components of B are:

$$\begin{aligned}
[0, 0, 1, \pm i] &= \{s = 0\} \cap \{t = 0\} \cap \{s - t = 0\} \cap \{s + t = 0\} \\
[1, \pm i, 0, 0] &= \{S = 0\} \cap \{T = 0\} \cap \{S - T = 0\} \cap \{S + T = 0\}
\end{aligned}$$

Write

$$W := \{[0, 0, 1, \pm i], [1, \pm i, 0, 0]\},$$

where B has multiplicity-four singularities. There are $\binom{8}{2} = 28$ pairs of conics altogether so all intersections are accounted for.

Resolving the Bad Singularities

We briefly review background on surface singularities: A double cover of a smooth surface branched along a curve has ADE singularities if and only if the curve has simple (ADE) singularities. An isolated hypersurface singularity $s \in S$ is *Du Val* if there exists a resolution $\beta : \widetilde{S} \to S$ such that $K_{\widetilde{S}} - \beta^* K_S$ is effective. Complex Du Val singularities coincide with the ADE surface singularities [Rei97, §4.20]. Given a projective surface S with Du Val singularities, pluricanonical differentials on S are the same as pluricanonical differentials on a resolution $\widetilde{S}$. In particular, if K_S is ample then $\widetilde{S}$ is of general type.

Returning to X, we seek to resolve the singularities that are not Du Val, which are associated with the multiplicity four singularities of B along W. We observe that

$$W \subset \{s^2 + t^2 = S^2 + T^2 = 0\},$$

i.e., the vertices of a cycle of four rational curves contained in Y, whose complement is a torus. The blow-up

$$\widetilde{Y} := \mathrm{Bl}_W(Y) \to Y$$

is therefore toric with boundary consisting of an octagon of $\mathbb{P}^1$'s with self-intersections alternating between -1 and -2. The induced blow-up

$$\beta : \widetilde{X} \to X$$

resolves the singularities of X.

Canonical Class of X

First, we choose a basis for the Picard group of $\widetilde{Y}$: Let f_1 and f_2 be the rulings of $Y \simeq \mathbb{P}^1 \times \mathbb{P}^1$ and $E_1, \ldots, E_4$ the exceptional divisors over W so

$$\mathrm{Pic}(\widetilde{Y}) = \langle f_1, f_2, E_1, E_2, E_3, E_4 \rangle .$$

We order so that the boundary octagon has sides with classes

$$E_1, f_1 - E_1 - E_2, E_2, f_2 - E_2 - E_3, E_3, f_1 - E_3 - E_4, E_4, f_2 - E_4 - E_1.$$

The canonical class

$$K_{\widetilde{Y}} \equiv -2f_1 - 2f_2 + E_1 + E_2 + E_3 + E_4.$$

The proper transform $\widetilde{B}$ of B—noting the multiplicity four singularities along W—has class

$$8(f_1 + f_2) - 4(E_1 + E_2 + E_3 + E_4);$$

$\widetilde{B}$ is disjoint from the four (-2) curves in the octagon. Blowing down these four curves induces a birational morphism

$$\widetilde{Y} \to \Sigma$$

to a quartic del Pezzo surface with four A_1 singularities. This morphism is induced by the linear series $|2f_1 + 2f_2 - E_1 - E_2 - E_3 - E_4|$ corresponding to the quadratic equations for W modulo the defining equation for Y.

We refer the reader to [BHPVdV04, IV.22] for a discussion of the invariants of double covers of surfaces branched along curves with simple singularities.

The canonical class of $\widetilde{X}$ is obtained by pulling back

$$K_{\widetilde{Y}} + \frac{1}{2}\widetilde{B} \equiv 2(f_1 + f_2) - (E_1 + \ldots + E_4)$$

through the induced morphism $\widetilde{X} \to \widetilde{Y}$. The image of $\widetilde{Y}$ and $\widetilde{X}$ under this linear series equals Σ. We summarize this in the following proposition.

Proposition A *The surface $\widetilde{X}$ is of general type with $K^2_{\widetilde{X}} = 8$, and the linear series $|K_{\widetilde{X}}|$ induces a morphism*

$$\widetilde{X} \to \Sigma$$

that is generically finite of degree two.

In particular, we find

$$\dim \Gamma(K_{\widetilde{X}}) = \dim \Gamma(\widetilde{Y}, K_{\widetilde{Y}} + \frac{1}{2}\widetilde{B}) = 5.$$

$\widetilde{X}$ *Is Simply Connected*

To compute the other invariants of $\widetilde{X}$, we use observations following immediately from simultaneous resolution of ADE singularities [Bri68]: $\widetilde{X}$ is deformation equivalent to a double cover of $\widetilde{Y}$ branched along a *generic* member of

$$|\widetilde{B}| = |8(f_1 + f_2) - 4(E_1 + E_2 + E_3 + E_4)|,$$

disjoint from the (-2)-curves. Smoothing the A_1 singularities of Σ and noting that the branch divisor lies in $|-4K_\Sigma|$, we find that $\widetilde{X}$ is also deformation equivalent to a double cover of a quartic del Pezzo surface Σ' branched over a divisor $A \in |-4K_{\Sigma'}|$. Such surfaces are simply connected by the Lefschetz hyperplane theorem, hence

$$\pi_1(\widetilde{X}) = \{1\}.$$

The Remaining Invariants

The branch curve A satisfies

$$\deg K_A = 48$$

by adjunction on Σ'. Thus we conclude

$$\chi(\widetilde{X}) = 2(\chi(\Sigma') - \chi(A)) + \chi(A) = 2(8 + 48) - 48 = 64,$$

$b_2(\widetilde{X}) = 62$, and $h^1(\Omega^1_{\widetilde{X}}) = 52$.

References

[BBB+00] Christian Batut, Karim Belabas, Dominique Bernardi, Henri Cohen, and Michel Olivier. *User's Guide to PARI-GP*. Université de Bordeaux I, 2000.

[BCP97] Wieb Bosma, John Cannon, and Catherine Playoust. The Magma algebra system. I. The user language. *J. Symbolic Comput.*, 24(3–4):235–265, 1997. Computational algebra and number theory (London, 1993).

[BHPVdV04] Wolf P. Barth, Klaus Hulek, Chris A. M. Peters, and Antonius Van de Ven. *Compact complex surfaces*, volume 4 of *Ergebnisse der Mathematik und ihrer Grenzgebiete. 3. Folge. A Series of Modern Surveys in Mathematics [Results in Mathematics and Related Areas. 3rd Series. A Series of Modern Surveys in Mathematics]*. Springer-Verlag, Berlin, second edition, 2004.

[Bri68] Egbert Brieskorn. Die Auflösung der rationalen Singularitäten holomorpher Abbildungen. *Math. Ann.*, 178:255–270, 1968.

[CMP+19] John Cremona, Marcus Mo, Julien Puydt, Fernando Perez, François Bissey, Isuru Fernando, William Stein, qed777, Giovanni Mascellani, Keno Fischer, Jerry James, Dima Pasechnik, abergeron, and Antonio Rojas. *JohnCremona/eclib v20190226*, February 2019. Also at http://homepages.warwick.ac.uk/~masgaj/mwrank/.

[Con08] Keith Conrad. The congruent number problem. *Harvard College Mathematical Review*, 2:58–74, 2008. http://www.math.harvard.edu/hcmr/issues/2a.pdf.

[Cre97] John E Cremona. *Algorithms for Modular Elliptic Curves*. Cambridge University Press, 1997.

[Elk94] Noam D Elkies. Heegner point computations. In *International Algorithmic Number Theory Symposium*, pages 122–133. Springer, 1994.

[Fal94] Gerd Faltings. The general case of S. Lang's conjecture. In *Barsotti Symposium in Algebraic Geometry (Abano Terme, 1991)*, volume 15 of *Perspect. Math.*, pages 175–182. Academic Press, San Diego, CA, 1994.

[Gol79] Dorian Goldfeld. Conjectures on elliptic curves over quadratic fields. In *Number Theory Carbondale 1979*, pages 108–118. Springer, 1979.

[HKLDW18] Thomas A. Hulse, Chan Ieong Kuan, David Lowry-Duda, and Alexander Walker. A shifted sum for the congruent number problem, 2018.

[Hon61] Taira Honda. Isogenies, rational points and section points of group varieties. In *Japanese journal of mathematics: transactions and abstracts*, volume 30, pages 84–101. The Mathematical Society of Japan, 1961.

[Kob93] Neal Koblitz. *Introduction to elliptic curves and modular forms*, volume 97 of *Graduate Texts in Mathematics*. Springer-Verlag, New York, second edition, 1993. https://doi.org/10.1007/978-1-4612-0909-6.

[LD] David Lowry-Duda. Sage implementation for congruent number triangles with the same hypotenuse. https://github.com/davidlowryduda/notebooks/blob/master/Papers/CongruentNumberTrianglesSameHypotenuse.sage. Accessed: 2020-01-31.

[LMF19] The LMFDB Collaboration. The L-functions and modular forms database. http://www.lmfdb.org, 2019. [Online; accessed 30 October 2019].

[Rei97] Miles Reid. Chapters on algebraic surfaces. In *Complex algebraic geometry (Park City, UT, 1993)*, volume 3 of *IAS/Park City Math. Ser.*, pages 3–159. Amer. Math. Soc., Providence, RI, 1997.

[Rog00] Nicholas F Rogers. Rank computations for the congruent number elliptic curves. *Experimental Mathematics*, 9(4):591–594, 2000.

[Rog04] Nicholas F. Rogers. *Elliptic curves x(3) + y(3) = k with high rank*. PhD thesis, 2004. Accessed through ProQuest Dissertations and Theses; Last updated 2020-05-06.

[Sag20] Sage Developers. *SageMath, the Sage Mathematics Software System (Version 8.8)*, 2020. https://www.sagemath.org.

[Tun83] Jerrold B. Tunnell. A classical Diophantine problem and modular forms of weight 3/2. *Invent. Math.*, 72(2):323–334, 1983. https://dx.doi.org//10.1007/BF01389327.

[WDE+14] Mark Watkins, Stephen Donnelly, Noam D Elkies, Tom Fisher, Andrew Granville, and Nicholas F Rogers. Ranks of quadratic twists of elliptic curves. *Publications Mathématiques de Besançon*, (2):63–98, 2014.

[Zag89] Don Zagier. *Elliptische Kurven: Fortschritte und Anwendungen*. Max-Planck-Institut für Mathematik, 1989.

Visualizing Modular Forms

David Lowry-Duda

Abstract We examine several currently used techniques for visualizing complex-valued functions applied to modular forms. We plot several examples and study the benefits and limitations of each technique. We then introduce a method of visualization that can take advantage of colormaps in Python's `matplotlib` library, describe an implementation, and give more examples. Much of this discussion applies to general visualizations of complex-valued functions in the plane.

1 Introduction

1.1 Motivation

Graphs of real-valued functions are ubiquitous and commonly serve as a source of mathematical insight. But graphs of complex functions are simultaneously less common and more challenging to make. The reason is that the graph of a complex function is naturally a surface in four dimensions, and there are not many intuitive embeddings available to capture this surface within a 2d plot.

In this article, we examine different methods for visualizing plots of modular forms on congruent subgroups of $\mathrm{SL}(2, \mathbb{Z})$. These forms are highly symmetric functions and we should expect their plots to capture many distinctive, highly symmetric features.

This work was supported by the Simons Collaboration in Arithmetic Geometry, Number Theory, and Computation via the Simons Foundation grant 546235.

The author benefitted from many conversations at the fall 2019 ICERM program on Illustrating Mathematics (where this project was born) and at the special number theory day organized at Bowdoin College by Naomi Tanabe.

D. Lowry-Duda (✉)
Institute of Computational and Experimental Research in Mathematics, Providence, RI, USA
e-mail: david@lowryduda.com

J. S. Balakrishnan et al. (eds.), *Arithmetic Geometry, Number Theory, and Computation*, Simons Symposia, https://doi.org/10.1007/978-3-030-80914-0_19

In addition, we wish to take advantage of the broader capabilities that exist in the Python/SageMath data visualization ecosystem. There are a vast number of color choices and colormaps implemented in terms of Python's `matplotlib` library [Hun07].

In `matplotlib`, a colormap is a Python data type that associates a color to an input real number. While many colormaps are purely aesthetic, others are chosen to allow the most accurate perception by as many viewers as possible. By using established `matplotlib` colormaps, it is possible to choose colors to support different types of color perception, including for people who see color differently. We describe this further in Sect. 3.

Although our emphasis is on plots of modular forms, many of the techniques described apply more generally to graphing complex-valued functions. We hope this paper will be of help to any reader wanting to visualize complex functions.

1.2 Broad Overview of Complex Function Plotting

Over the last 20 years, different approaches towards representing graphs of complex functions have emerged. The most commonly used approach is one first introduced by Frank Farris in a review [Far98] of Needham's *Visual Complex Analysis*.

The idea is to represent the output of a complex function though color. Frequently, one uses hue to represent argument and lightness to represent magnitude. Or more generally, one can associate a color to each point in the complex plane and then color the domain of a complex function f by the color of $f(z)$. This technique is now called *domain coloring*. Using hue and lightness to represent argument and magnitude is the default complex plotting method in SageMath [Sag20], as well as other common complex plotting libraries.

In principle, such a graph gives perfect information about the function. But in practice, it is often difficult to distinguish between hue and lightness — and in particular it is challenging to determine if two points with different hues have the same lightness. Thus some variations omit a dimension from the graph and plot only the magnitude or phase. Even though the resulting graphs do not perfectly capture f, the omission of a dimension can be beneficial. Other variations specially color the complex plane in other ways.

Wegert [WS10] describes an approach using domain coloring, but emphasizing phase plots as a visual tool. He also produces phase plots with small brightness adjustments near certain magnitude thresholds. The effect is similar to looking at a contour map of a landscape, except that each pixel's color carries meaning (the phase of f evaluated at that point).

1.3 Paper Overview

In Sect. 2, we give a short survey of common existing complex plotting techniques applied to modular forms. We include images coming from the ideas of Farris and Wegert, built using SageMath. We also reproduce the plots associated to modular forms on the L-Function and Modular Form Database (LMFDB [LMF19]).

In Sect. 3, we describe how one can incorporate pre-established colormaps into modular form visualizations. And in Sect. 3.1, we discuss one such implementation in terms of `matplotlib`. This is a new implementation.

Throughout, we revisit the same basic figures with each visualization technique. We plot four different modular forms on three different domains in each technique. Each plot reveals different behaviors of the underlying forms.

2 Survey of Visualizations

2.1 Halfplane and Disk Models

Plots of modular forms typically represent the form either on the upper half-plane or on the Poincaré disk. We first describe the relationship between these two types of plots, and then survey different visualizations of forms in each plot type.

Let $\mathcal{H} = \{x + iy : y > 0\}$ denote the complex upper halfplane. A modular form of weight k on a congruence subgroup $\Gamma \subseteq \mathrm{SL}(2, \mathbb{Z})$ is a holomorphic function $f : \mathcal{H} \longrightarrow \mathbb{C}$ satisfying

$$f(\gamma z) = (cz + d)^k f(z) \qquad \forall \gamma = \left(\begin{smallmatrix} a & b \\ c & d \end{smallmatrix}\right) \in \Gamma, \tag{1}$$

as well as certain growth conditions at the cusps of $\Gamma \backslash \mathcal{H}$. See [DS05] for complete definitions.

To the congruence subgroup Γ is associated a positive integer N called the level. The automorphy condition (1) implies strong symmetry conditions on f in terms of N. In particular, $\left(\begin{smallmatrix} 1 & N \\ 0 & 1 \end{smallmatrix}\right) \in \Gamma$ implies that f will be periodic with period N. Thus one can constrain a plot to a vertical strip of width N without omitting any information.

In practice, one would cut off the vertical strip at some height H. If f is a cuspform, so that $\lim_{y \to \infty} f(x + iy) = 0$, then heuristically most interesting behavior of the form will lie within a well-chosen box $[0, N] \times [0, H]$. In practice, N is typically much larger than H, and it's necessary to investigate some smaller box. We refer to this type of plot as a plot of f in the upper halfplane.

Alternately, the upper halfplane is conformally equivalent to the Poincaré disk $\mathbb{D} = \{x + iy : x^2 + y^2 < 1\}$. There are infinitely many such maps taking $\mathbb{D}$ to $\mathcal{H}$, but we choose the möbius transform

$$\phi(z) = \frac{1 - iz}{z - i}. \tag{2}$$

Under ϕ, the points $-i, 0, i$ in $\mathbb{D}$ are mapped to $0, i, \infty$, respectively, in $\mathcal{H}$. Thus the apparent vertical orientation remains fixed in both models.

Plotting a modular form f on $\mathbb{D}$ is a complete picture; values of f at every point of $\mathcal{H}$ will be represented through such a plot. We refer this type of plot as a plot of f in the disk.

2.2 *Visualizations*

Let $\Delta(z)$ denote the Ramanujan Delta function, the unique holomorphic cuspform of weight 12 on $\mathrm{SL}(2, \mathbb{Z})$. And let $g(z)$ denote the unique holomorphic cuspform of weight 4 on $\Gamma(5)$. We think of g as a "simple" modular form, and Δ is perhaps the most widely-recognized modular form. Each of these forms is catalogued in the LMFDB [LMF19], referenced by a label. The Δ function has label 1.12.a.a.1.1 and g has label 5.4.a.a.1.1.

We also consider two more complicated modular forms: f_{105}, a weight 2 modular form on $\Gamma(105)$, identified by label 105.2.a.a.1.1 and $f_{10}(z)$, a weight 20 modular form on $\Gamma(10)$, identified by label 10.20.a.a.1.1.

We should expect forms of similar weight to share characteristics, and we should expect forms whose levels have similar numbers of cusps to share some characteristics. Here, we have chosen levels that correlate strongly with the number of cusps. Thus the two forms of smaller level have fewer distinct cusps than the two forms of larger level.

Note that $g(z)$ has small level and small weight, $\Delta(z)$ has small level and moderately large weight, f_{105} has very large level and very small weight, and f_{10} has moderately large level and high weight,

In the plots that follow, we will show these four forms in the order $(g, \Delta, f_{105}, f_{10})$ from top to bottom. To produce the plots, we approximate each modular form by its first 400 Fourier coefficients and evaluate the function on a 600×600 grid, and store it as a PNG. We plot each form on the Poincaré disk, on a natural box in $\mathcal{H}$, and then on a "zoomed-in" portion of $\mathcal{H}$. Specifically, all "zoomed-in" plots on $\mathcal{H}$ are of the region $[0.1, 0.4] \times [0, 0.25]$. For the other plots on $\mathcal{H}$, we plot Δ on $[-1, 1] \times [0, 2]$, g on $[-2.5, 2.5] \times [0, 2]$, f_{105} on $[-1, 1] \times [0, 1]$, and f_{10} on $[-1, 1] \times [0, 2]$.

The region choices for Δ and g are natural: these are each wide enough to include at least one period, and tall enough to capture most interesting properties. (This will be apparent in the graphs.) For f_{105} and f_{10}, it is not practical to plot a box sufficiently wide to capture a period. Instead we plot a smaller region.

We plot each of these four forms on each of the three regions with a variety of different visualizations and discuss each in turn.

Remark 1 We do not intend to emphasize one type of visualization as being universally better than all the others. As we will see, each type has flaws and only presents a single view. Nonetheless, at the end of paper we give our recommendations for producing visualizations.

2.2.1 Standard Domain Coloring

In Fig. 1, we present default plots (as created by `complex_plot` in SageMath). In these plots, the $\arg z$ is represented by hue and $|z|$ is represented by brightness, where 0 is black and ∞ is white.

The fact that each form plotted is a cuspform is immediately clear as each plot on the disk is dominated by the color black (corresponding to the overall size being small). Further, a closer look will reveal that each cusp appears to be visibly dark. This is most obvious in the two lower level forms, g and Δ, where each cusp has a relatively large dark region around it. As the number of cusps increases, the behavior around cusps becomes more varied. Very near the boundary, our approximation for the form is less accurate. This is one factor contributing to why the cusps of f_{10} are far less pronounced.

The weight of each form manifests primarily in how quickly the form can go from very bright to very dark. Here, the two forms of higher weight, Δ and f_{10}, have regions of near-white brightness near the boundary that very rapidly darken towards the top of the disk. The two forms of lower weight are much more even in brightness. The high weight of f_{10} is another factor contributing to why its cusps are less pronounced: there are smaller regions of darkness around each cusp.

It is possible to adjust the mapping from magnitude to brightness, but in practice these plots look roughly the same for a wide choice of maps and it takes quite a bit of tuning to choose an appropriate map.[1]

[1]Adjusting the map from magnitude to brightness for SageMath's `complex_plot` is possible, but nontrivial; the map is defined internally and the current plotting interface doesn't give any option to choose or alter this map. In order to adjust this map, it is necessary to modify the source for SageMath's plotting routines directly.

Fig. 1 Default plots using `complex_plot` in SageMath. From top to bottom, the forms are g, Δ, f_{105}, and f_{10}

2.2.2 Magnitude Plot Without Color

We now produce plots after ignoring the phase of $f(z)$ and instead only using the magnitude $|f(z)|$. There remains a choice of how to map magnitude to brightness;

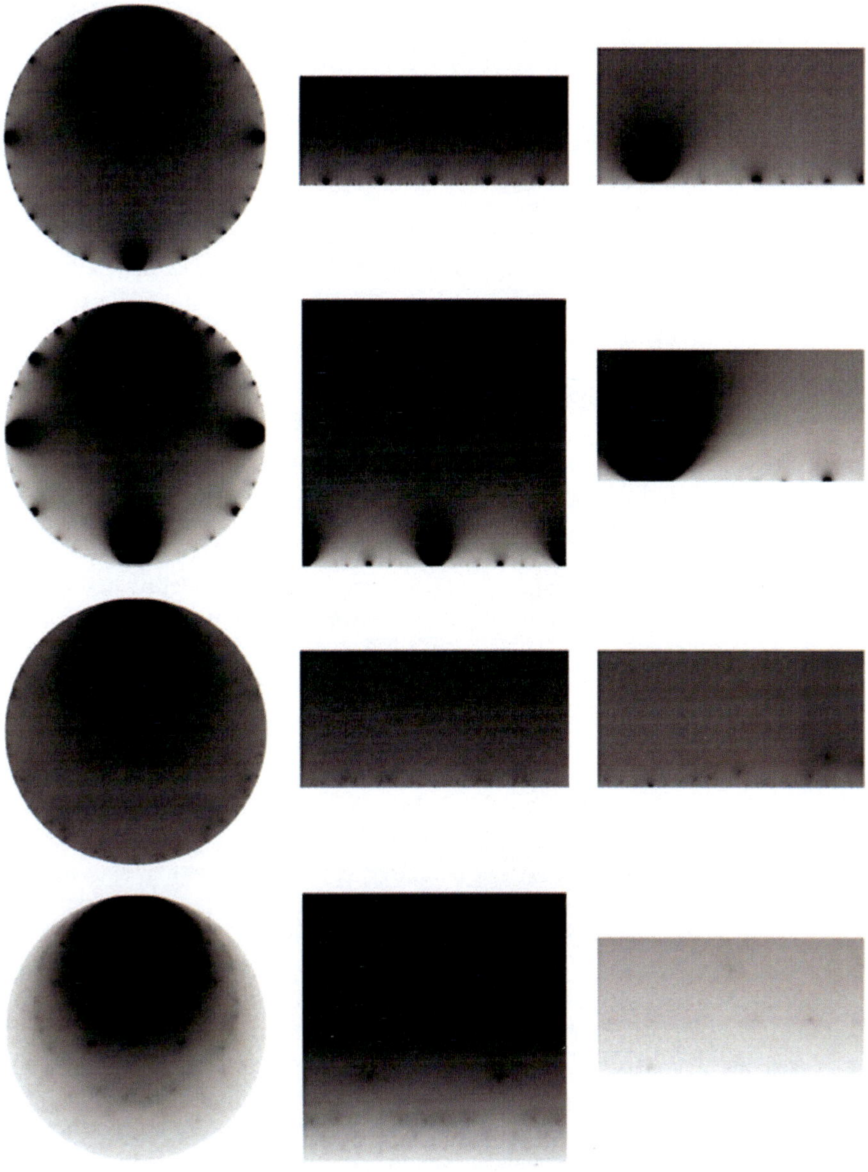

Fig. 2 Magnitude plots where the magnitude corresponds to darkness. Black corresponds to zero magnitude. From top to bottom, the forms are g, Δ, f_{105}, and f_{10}

it is common to use maps of the form $\arctan(\log(|f(z)|^{\alpha}+1))$ for some $0<\alpha<1$. Taking $\alpha=0.25$ yields the plots in Fig. 2. We note that adjusting the power α is the most natural and clearest way to adjust the overall darkness/lightness scale of this style of plot: larger alpha causes points near zeros to be much darker and points near poles to be much lighter; smaller alpha decreases this effect.

Qualitatively, each of these pictures offers very similar information as those in Fig. 1. The lack of phase information makes it slightly more challenging to distinguish between the images of Δ and g in the disk. But note that it is very easy to determine forms of low weight and high weight, as the forms with low weight have much slower changes in brightness.

The complicated and varied behavior of f_{10} near the boundary is more evident in this plot. The presence of many more points of small size is made far more obvious. It is just barely possible to observe the region of decreasing size around the 0 cusp.

2.2.3 Magnitude Plot with Periodic Linear Color

In the LMFDB, magnitude is mapped periodically to color.[2] That is, hue represents the magnitude modulo 1, where blue is zero, and increases through purple, red, orange, yellow, and so on. In Fig. 3, we mimic the plots in the LMFDB attained by ignoring the phase of mapping the magnitude mod 1 to hue.

Recall from Sect. 2.2 that each of the four forms we are considering has a label on the LMFDB, and we gave links to their homepages. Each page has a medallion with a plot similar to the plots on the Poincaré disk given here.

The large bluish blob corresponds to a region where the magnitude of the form is never more than 1. The larger weights of $\Delta(z)$ and f_{10} correspond to more rapid growth away from cusps, leading both to smaller bluish regions and more tightly packed lines; f_{10} has larger weight, and both of these traits are even more prominent in its graphs.

For the lower weight forms g and f_{105}, this is a very accurate plot. But $\Delta(z)$ and f_{10} change too rapidly near the boundary, causing a static-like appearance. Increasing resolution helps, but in practice it is not feasible to produce sufficiently high resolution images to completely make sense of the boundary with this type of plot.

This sort of plot is very well-suited for low weight forms of any level. The small weight means that even at moderate resolution, the image will appear well-defined even near the boundary. Then the cusps associated to the level will be visually apparent by the varying types of behaviors near the boundary. This is most obvious in the detail images for g and f_{105}, where we can clearly see the greater variety in behavior near the cusps at the boundary in plots of f_{105}.

[2]In the period between writing and publishing this paper, the LMFDB changed its method for plotting modular forms to incorporate methods described in this paper. Thus in the rest of this paper, LMFDB-style plots refer to the style of plots in the LMFDB on October 1, 2020.

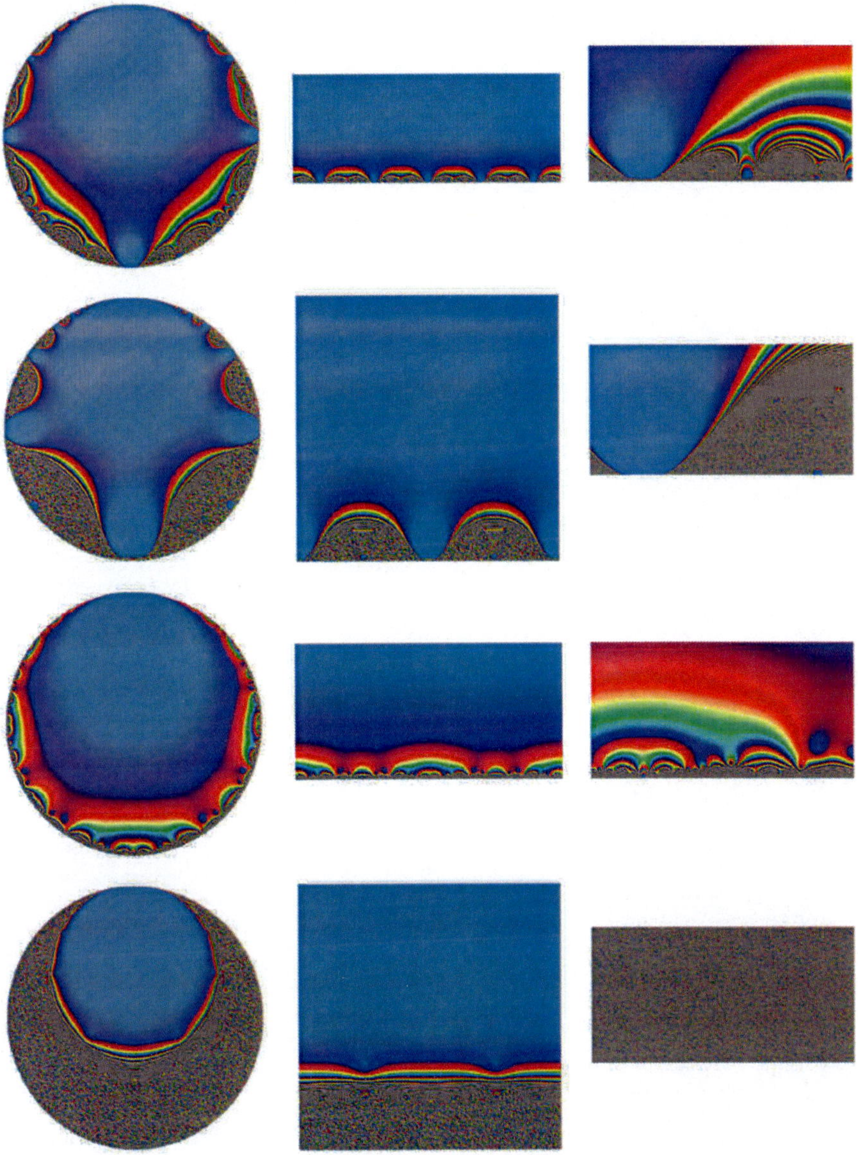

Fig. 3 Magnitude plots mimicking the style of the LMFDB. The phase is ignored; the hue represents the magnitude modulo 1. From top to bottom, the forms are g, Δ, f_{105}, and f_{10}

2.2.4 Magnitude Plot with Periodic Logarithmic Spacing

In Fig. 4, we change yet again the map from magnitude to color. We now allow the hue to represent $\log_\alpha(|f(z)|) \bmod 1$ for some base α. Thus two consecutive bands

Fig. 4 Magnitude plots. The phase is ignored; hue represents the log of the magnitude, modulo 1. From top to bottom, the forms are g, Δ, f_{105}, and f_{10}

of red will correspond to the points in one band being α times the size of the points in the other.

We note that adjusting the factor between consecutive bands (i.e. choosing the base of the logarithm in the map from magnitude to color) controls the number of spacing of bands, and thus it is possible to make the plots visually simpler or more complex by experimenting with this parameter. The plots in Fig. 4 were produced with $\alpha = 7$.

The effect in these plots is similar to giving a higher resolution of Fig. 3. A large amount of detail and data is visible within these plots; however this is double-edged, and the plots are now busier. The high growth rate in the forms of higher weight is indicated by many more tightly packed bands of color near the boundary.

Note that there is a region where the resolution appears insufficient in the detail plot for $\Delta(z)$. This is actually caused by small errors in the approximation for $\Delta(z)$ interfering with the extremely tiny values of $\Delta(z)$ there. As can be seen from the nearly black portions of Figs. 1 and 2, $\Delta(z)$ is extremely small in this region, and thus even small errors in the approximation produce noticeable artifacts in the graph.

These plots are particularly useful for plots on $\mathcal{H}$, where even for high weight the plots appear mostly well-formed. In the disk, the rapid decay at the top of the disk (at the ∞ cusp) causes tight concentric circles of color and attracts a lot of visual attention, even though the actual behavior is mostly regular and relatively uninteresting there.

2.2.5 Pure Phase Plots

Following the ideas of Wegert [WS10], we now consider phase plots. In Fig. 5, we ignore magnitude completely and only consider the phase. There is a natural mapping from phase to hue: we allow hue to represent $\arg(z)/2\pi$, and the natural periodicity of phase works well with the natural periodicity of the color wheel.

Qualitatively, less behavior of each modular form is evident in its phase plot in comparison to its magnitude plot. This is partly due to the lack of poles and infrequent zeros away from the boundary. Poles and zeros are singularities in phase plots, and the behavior of the colors around these points describes the order and nature of these points.

In our opinion, pure phase plots should not be used on their own to study or describe modular forms. However, pairing each image in Fig. 5 with its corresponding image in Fig. 3 or Fig. 4 completely represents the modular form on the indicated domain.

2.2.6 Phase Plots with Magnitude Contours

Again following Wegert [WS10], we create phase plots with contours representing magnitude in Fig. 6. We indicate the contours by adjusting the lightness: below each contour we adjust the lightness to be darker and above each contour we adjust the lightness to be lighter. The sudden change from darker to lighter indicates the contour itself.

The phase naturally maps to hue as in the pure phase plots in Fig. 5. Contours could be chosen with spacing defined linearly as in Sect. 2.2.3 or logarithmically as in Sect. 2.2.4. In these plots, we have chosen logarithmic spacing with respect to

Fig. 5 Phase plots. Magnitude is ignored, and hue represents the phase. From top to bottom, the forms are g, Δ, f_{105}, and f_{10}

$\log_2$. That is, two consecutive contour lines indicate that the points along one have twice the magnitude of the points along the other.

Remark 2 We emphasize that it is not necessary to compute the contours themselves. Instead, one defines a coloring of the complex plane that includes brightness

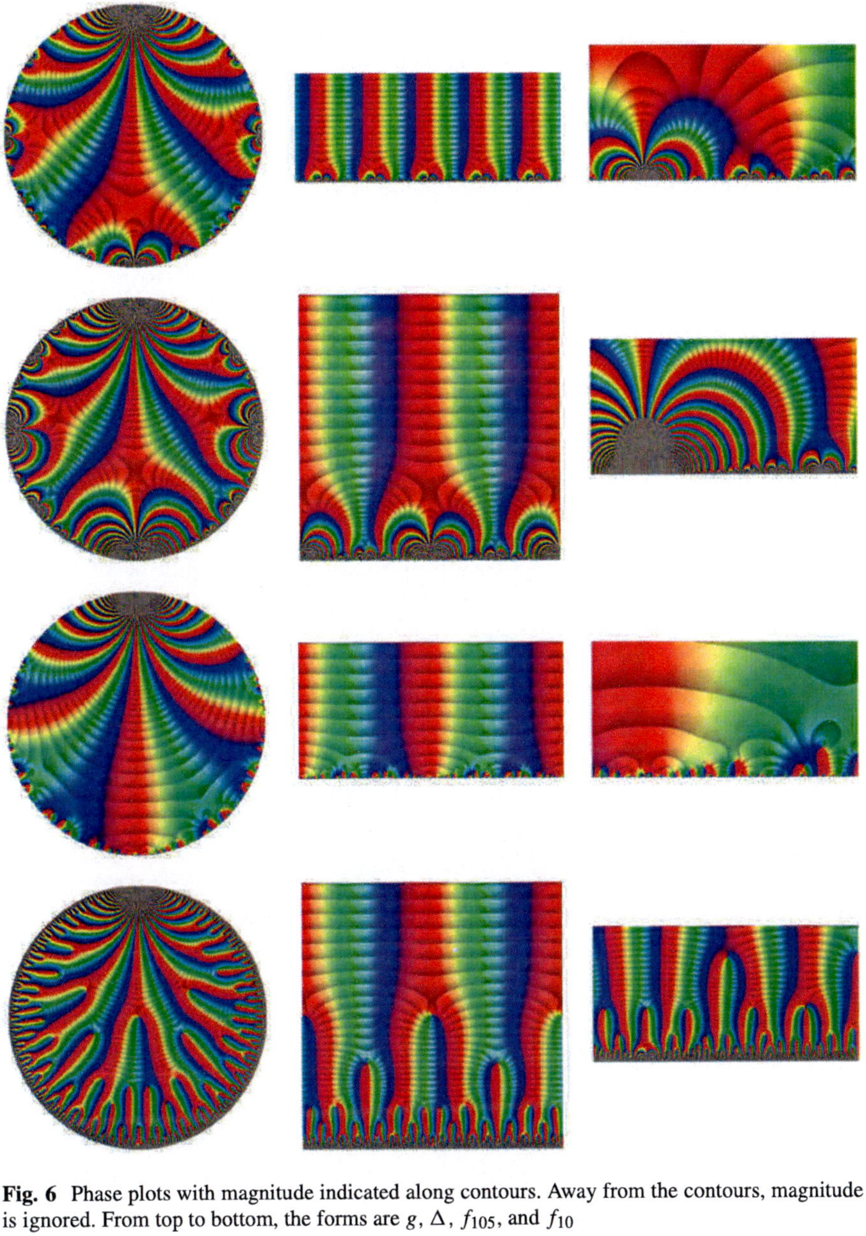

Fig. 6 Phase plots with magnitude indicated along contours. Away from the contours, magnitude is ignored. From top to bottom, the forms are g, Δ, f_{105}, and f_{10}

adjustments along the desired contours. In this case, we defined brightness adjustments around points of magnitude 2^n for integers n. Then simply plotting the function with respect to this domain coloring causes the contours to appear. This is a key idea why approaches based on domain coloring are simple to compute.

The plotting program used to produce these plots is available at [LD20].

These plots carry all the information visible in the pure phase plots of Sect. 2.2.5 and some of the information contained within the logarithmically spaced magnitude plots of Sect. 2.2.4, but these two sets of information are not displayed equally. The colors grab attention far more than the brightness adjustments around contours—analogous to the lightness difficulties for the standard domain coloring plots of Sect. 2.2.1. These plots emphasize argument more than magnitude.

3 Choosing Color

We surveyed many different visualizations in Sect. 2, but one thing remained constant: the choices of color. In most complex function visualization packages, there is only one color choice available. All visualizations in Sect. 2 were created using the only color scheme for `complex_plot` available in SageMath.[3]

Unfortunately, this color scheme has flaws. For one example, it has uneven perceptual contrast: it has points of locally high color contrast (giving the impression of additional local variation) and points of locally low color contrast (hiding local variation). Thus in Fig. 5, for instance, green is particularly visible and the yellow-green border is particularly striking, even though there is no underlying feature that this identifies. Further, in the plots in the upper halfplane, the vertical bands of color seem to favor green-yellow at the expense of purple-blue.

In his excellent article [Kov15], Kovesi describes this problem and details how to create colormaps that are perceptually uniform — that don't have uneven perceptual contrast. His work led to the creation of new default colormap, called *viridis*, for Python's `matplotlib` plotting library.

More recently, Nuñez, Anderton, and Renslow wrote an article [NAR18] describing their Python module `cmaputil`, which helps make perceptually uniform colormaps that additionally support perception by those with various forms of color vision deficiency. They created the Python colormap *cividis* (obtained by modifying *viridis*, hence the name), which they show is a good colormap for those with reduced sensitivity to red or green light. It is possible to make other colormaps suitable for other types of perception using this tool.

We have incorporated `matplotlib` colormaps into our plotting routines. In particular, we implemented phase plots and phase plots with contours using these colormaps. In this section, we describe this implementation and give examples of the resulting plots.

Remark 3 SageMath and `matplotlib` share many plotting routines, and newer versions of SageMath will default to these new perceptually uniform colormaps

[3]The behavior in Matlab and Maple appears similar, though it is possible to define a color scheme. On the other hand Mathematica has an extensive library of complex plotting color schemes.

for real-valued plotting. But SageMath's `complex_plot` does not directly use `matplotlib`. We are working towards making these new plotting routines easy to use and publicly available, and possibly incorporating them into SageMath.

3.1 Implementation Overview

Some of the details of this implementation are specific to Python and SageMath, but the overall idea readily generalizes.

(1) Begin with a desired complex function f, a region $\Omega \subset \mathbb{C}$, and a desired colormap cm.
(2) Evaluate $\arg(f(z))$ and $|f(z)|$ for each point in an $M \times N$ grid containing Ω. (Optionally mask points (m, n) in this grid outside of Ω to avoid unnecessary computation. For example, to produce the plots of Δ and g in the disk, we mask points z with $|z| \geq 1$.) This gives an $M \times N \times 2$ array.
(3) Map each phase to $[0, 1]$. This is most often done through the natural map $\theta \mapsto (\theta/2\pi) \bmod 1$, but note that it is possible to add an offset Θ through $\theta \mapsto (\theta/2\pi + \Theta) \bmod 1$. This might be done to choose which phase corresponds to the endpoints of the colormap. (See plots in Fig. 9 for examples).
(4) Each `matplotlib` colormap is implemented as a function $\text{cm} : [0, 1] \to \text{RGBA}$, where $\text{RGBA} \cong [0, 1]^4$ is red-green-blue-alpha colorspace. We don't currently use the alpha channel. Thus apply cm to each mapped phase and discard the alpha channel to get RGB.

If the intention is to produce a pure phase plot, then (after discarding the magnitude) one now has an $M \times N \times \text{RGB}$ array, and one now produces a plot using this data in this grid, either directly or using a plotting library. For example, `matplotlib`'s method `matplotlib.pyplot.imshow` will directly plot this array.

However, if the intention is to produce a phase plot with (logarithmically spaced) magnitude contours, then additional work is necessary. We would like to adjust the brightness as done in Sect. 2.2.6, but the brightness of an RGB pixel is nontrivial to work with directly.

Instead we convert each pixel to HSL colorspace (hue-saturation-lightness) and adjust lightness. As most plotting libraries use RGB (as does the underlying hardware), we convert back to RGB prior to plotting. Thus to produce a phase plot with magnitude contours we perform these additional steps.

(5) Map each RGB element to HSL colorspace, where HSL refers to "hue-saturation-lightness". In HSL, it is possible to directly manipulate the lightness. This gives an $M \times N \times 3$ array.
(6) Map each magnitude to a lightness adjustment in the range $[-1, 1]$. For example, to have logarithmic contours, one could use a map of the form $(\log_2(|f(z)|) \bmod 1)/2$ with a fixed lightness adjustment in a neighborhood

of 0. This has the effect of creating logarithmically spaced contours. This gives an $M \times N \times 1$ array.

We note that we refer to $x \bmod 1$ to mean the least nonnegative number of the form $x + \ell$ for $\ell \in \mathbb{Z}$, i.e. the "computer science" modulo operation.

(7) For each (m, n), apply the corresponding lightness adjustment to each HSL value. (It may be necessary to renormalize lightness at this step.) This gives a single $M \times N \times 3$ array.

(8) Map each HSL back to RGB to get the final $M \times N \times 3$ array.

And now one can directly plot this array or use a library like `matplotlib` to plot this array, as noted above.

Remark 4 Each step can be done very quickly if one uses vectorized array operations as in NumPy [Oli06]. We also note that Python's `colorsys` library implements NumPy-vectorized conversions between RGB and HSL. Then the most time-consuming step is then the evaluation of f on the $M \times N$ grid, as should be expected.

3.2 *Example Visualizations with Colormaps*

We now give example visualizations with `matplotlib` colormaps. The three mostly commonly used `matplotlib` colormaps are currently *twilight*, *viridis*, and *cividis*. Each of these maps is perceptually uniform and designed for clean presentation of data.

The *twilight* colormap is also cyclic. This means that the two ends of the colormap line up perfectly, and thus it is a good colormap to represent phase in phase plots. The *viridis* and *cividis* colormaps are not cyclic. There is a sharp difference in colors at the two ends of the colormap. Plotting a modular form using either *viridis* or *cividis* thus highlights the points where the phase crosses over the end of the colormap.

In some cases, this may be undesirable. For exploratory visualization, using a cyclic perceptually uniform colormap often seems superior. But more generally it is sometimes desirable, or even advantageous, to emphasize certain aspects of a complex plot, and using a non-cyclic colormap to highlight phase-changes is just one possible method.

In Fig. 7, we plot the modular form g in four different color schemes: the default in SageMath, and the three `matplotlib` colormaps *twilight*, *viridis*, and *cividis*. In contrast to the default color scheme, these three colormaps produce a plot that is more balanced in appearance. In particular, the perceptually uniform plots don't overemphasize regions like the bright green and yellow portions in the default color scheme.

Of course, perceptually uniform doesn't mean perfect. In the *twilight* colormap, there is a sensation of peaks and troughs resulting from the light and dark areas of the plots; and as both *viridis* and *cividis* aren't cyclic, the sharp color change at the

Fig. 7 Phase plots of g with four color schemes. The top left is the default color scheme of SageMath. The top right is *twilight*, bottom left is *viridis*, and the bottom right is *cividis*

edges of the colormap gives the impression of contours. These are features of the colormap, not of the underlying function.

In Fig. 8, we give phase plots in the *cividis* colormap. For half of the plots, we include logarithmically spaced contours, implemented as described above. These plots should compared with the phase plots in Fig. 5 and the contoured phase plots in Fig. 6.

In these images, we have mapped the arguments to the colormap in such a way that the sharp change in color from light yellow to dark blue occurs when the phase is π, or equivalently when the value of the modular form is a negative real number. On the upper halfplane, a modular form with real coefficients will take real values on vertical lines $z = \frac{n}{2} + it$ for $n \in \mathbb{Z}, t > 0$. This explains the regularly spaced lines in the plots.

On the one hand, this gives a strong and simple visual indicator of phase change. The regularity of this phase change is more obvious in these plots than the previous phase plots in the default color scheme. But on the other hand, this strongly emphasizes a small subset of phases.

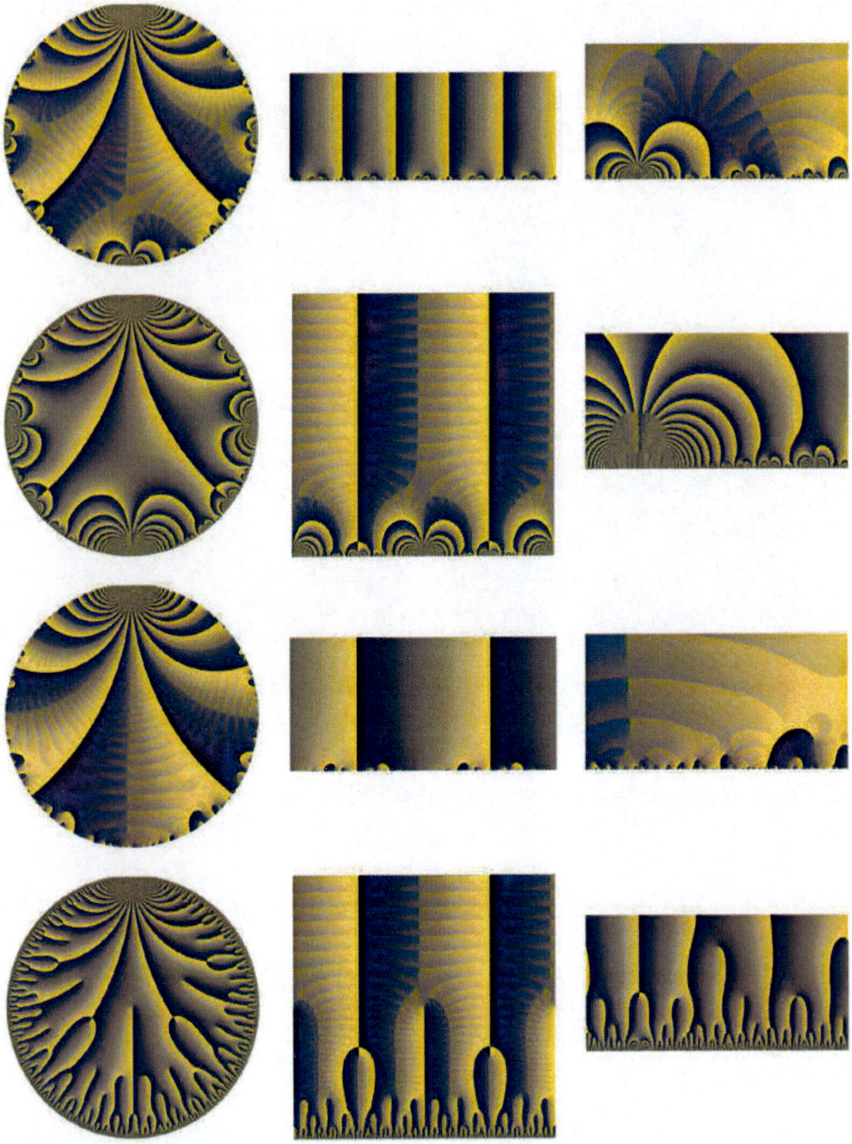

Fig. 8 Phase plots with the `matplotlib` perceptually uniform colormap *cividis* applied. In half of the plots, logarithmically spaced contours are included. From top to bottom, the forms are g, Δ, f_{105}, and f_{10}

We note also that the colormap *cividis* is both visually appealing and widely accessible.

In Fig. 9 we have included plots in a variety of other `matplotlib` colormaps. For the top six plots, we used the three colormaps *viridis*, *twilight*, and *coolwarm*.

Fig. 9 Phase plots with a variety of colormaps from `matplotlib` applied. From the top left and going row by row, the (function, colormap) are (Δ, *viridis*), (Δ, *twilight*), (Δ, an offset *coolwarm*); (f_{10}, *viridis*), (f_{10}, *plasma*), (g, *twilight*); (f_{105}, *coolwarm*), (f_{105}, *inferno*), (g, *Paired*)

Fig. 10 Two visualizations of Δ

Note that in the middle plot in the second row, we have normalized the non-cyclic colormap *coolwarm* so that the sharp distinction between red and blue occurs when the argument of Δ is $3\pi/2$ instead of π. In the bottom row, we applied the colormaps *inferno*, *plasma*, and *Paired*. Both *inferno* and *plasma* are also perceptually uniform, but are very orange. The colormap *Paired* is a discrete colormap, which has the effect of coloring different regions of the complex plane based on discrete argument sectors.

While each plot in Fig. 9 may not be appropriate for initial exploratory visualization, they demonstrate the ability to choose colormap to highlight certain behaviors or areas of the function. And most importantly, they demonstrate that implementing `matplotlib` colormaps gives access to an existing, very large library of color schemes.

Remark 5 Choosing color schemes to highlight certain behaviors is not a new concept. For example, Frank Farris [Far15] has produced several pieces of art based on domain colorings.

Finally, we note that it is possible to apply these colormaps to the previous styles of visualization. For example, we can produce a colormapped version of the default SageMath plot (as in Fig. 1) by changing steps (5)–(8) in the implementation described in Sect. 3.1; map the phase to RGB as described there, but map magnitude to brightness as in Sect. 2.2.1 or 2.2.2.

A Strategy for Making Visualizations Having given examples of several different types of complex plot, we feel we should describe our current preferences for making plots. If we were to choose a single plotting method to present a visualization, we would use contoured phase plots as in Fig. 8.

We have found that for initial exploration, it is very effective to pair a contoured phase plot as in Fig. 8 with a plot showing the absolute magnitude, such as a colormapped version of the default Sagemath plot.

In Fig. 10, we do this for Δ on the disk in the *cividis* colormap. The strengths and weaknesses of these two plots complement each other. Specifically, the relative magnitudes are very clearly given in the (colormapped) standard domain coloring plot on the right and the phase information is more apparent in the contoured phase plot on the left. The presence of phase information in the standard domain coloring helps to identify the same location in the two plots.

After this initial investigation, it is possible to choose regions and colormaps to highlight different behaviors of the function.

But overall, the idea of choosing colormaps for exploratory complex function visual analysis is young, and there remains much to be discovered.

References

[DS05] Fred Diamond and Jerry Shurman. *A first course in modular forms*, volume 228. Springer Verlag, 2005.

[Far98] Frank A Farris. Visual complex analysis by Tristan Needham. *The American Mathematical Monthly*, 105(6):570–576, 1998.

[Far15] Frank A Farris. *Creating Symmetry: The artful mathematics of wallpaper patterns*. Princeton University Press, 2015.

[Hun07] J. D. Hunter. Matplotlib: A 2d graphics environment. *Computing in Science & Engineering*, 9(3):90–95, 2007.

[Kov15] Peter Kovesi. Good colour maps: How to design them. *arXiv preprint arXiv:1509.03700*, 2015.

[LD20] David Lowry-Duda. phase_mag_plot. https://github.com/davidlowryduda/phase_mag_plot/, September 2020. [Online; Reference version at https://doi.org/10.5281/zenodo.4035117].

[LMF19] The LMFDB Collaboration. The L-functions and modular forms database. http://www.lmfdb.org, 2019. [Online; accessed 30 October 2019].

[NAR18] Jamie R Nuñez, Christopher R Anderton, and Ryan S Renslow. Optimizing colormaps with consideration for color vision deficiency to enable accurate interpretation of scientific data. *PloS one*, 13(7):e0199239, 2018.

[Oli06] Travis E Oliphant. *A guide to NumPy*, volume 1. Trelgol Publishing USA, 2006.

[Sag20] Sage Developers. *SageMath, the Sage Mathematics Software System (Version 8.8)*, 2020. https://www.sagemath.org.

[WS10] Elias Wegert and Gunter Semmler. Phase plots of complex functions: a journey in illustration. *Notices AMS*, 58:768–780, 2010.

A Prym Variety with Everywhere Good Reduction over $\mathbb{Q}(\sqrt{61})$

Nicolas Mascot, Jeroen Sijsling, and John Voight

Abstract We compute an equation for a modular abelian surface A that has everywhere good reduction over the quadratic field $K = \mathbb{Q}(\sqrt{61})$ and that does not admit a principal polarization over K.

1 Introduction

A foundational theorem of Abrashkin [1] and Fontaine [17] asserts that there is no abelian variety over $\mathbb{Q}$ with everywhere good reduction; this theorem is an analogue in arithmetic geometry of Minkowski's theorem that every number field has a ramified prime. More generally, the finiteness of the set of isomorphism classes of abelian varieties over a given number field of given dimension and with good reduction outside a given finite set is a key ingredient in the proof of Faltings [15] that a curve of genus at least two defined over a number field has finitely many rational points. For this reason and many others, it remains an important task to explicitly study abelian varieties over number fields with controlled ramification [3, 34].

Moving beyond elliptic curves, Dembélé–Kumar [12] have given examples of abelian surfaces with everywhere good reduction over real quadratic fields. More recently, Dembélé [11] gave a classification of abelian varieties with everywhere

N. Mascot
School of Mathematics, Trinity College, Dublin, Ireland
e-mail: mascotn@tcd.ie

J. Sijsling
Universität Ulm, Institut für Algebra und Zahlentheorie, Ulm, Germany
e-mail: jeroen.sijsling@uni-ulm.de

J. Voight (✉)
Department of Mathematics, Dartmouth College, Hanover, NH, USA
e-mail: jvoight@gmail.com

J. S. Balakrishnan et al. (eds.), *Arithmetic Geometry, Number Theory, and Computation*, Simons Symposia, https://doi.org/10.1007/978-3-030-80914-0_20

good reduction over $\mathbb{Q}(\sqrt{d})$ with $d = 53, 61, 73$: assuming the generalized Riemann hypothesis (GRH), he proves that every such abelian variety is isogenous to the power of a unique abelian surface A'_d. Moreover, for $d = 53, 73$ he was able to find an explicit equation for a genus 2 curve whose Jacobian is isogenous to A'_d. In this article, we finish off the case $d = 61$ as follows.

Theorem 1.1 *Let $K := \mathbb{Q}(\sqrt{61})$, let $\nu := (1+\sqrt{61})/2$, and let X be the projective plane curve over K defined by the equation $F(x, y, z) = 0$ where*

$$\begin{aligned} F(x,y,z) := {} & (5\nu+17)x^4 - (14\nu+48)x^2y^2 + (8\nu+28)x^2yz \\ & - (4\nu+14)x^2z^2 + (10\nu+33)y^4 - (12\nu+44)y^3z \\ & + (2\nu+26)y^2z^2 + (4\nu-16)yz^3 + (\nu-6)z^4 \end{aligned} \tag{1.2}$$

Then the following statements hold.

(a) *The curve X is smooth of genus 3 and has bad reduction at the prime (2) only.*
(b) *The curve X admits an involution $\iota\colon (x : y : z) \mapsto (-x : y : z)$ over K, and the quotient of X by ι defines a map from X onto an elliptic curve E of conductor (64). An equation for E is given by*

$$-(10\nu+34)y^2 = x^3 - \nu x^2 + (-\nu+1)x - 1.$$

(c) *Up to isogeny over K, we have* $\mathrm{Jac}(X) \sim A \times E$, *where A is the Prym variety of the natural map* $\mathrm{Jac}(X) \to \mathrm{Jac}(E)$, *an abelian surface over K with everywhere good reduction.*
(d) *All geometric endomorphisms of A are defined over K, and* $\mathrm{End}(A) \simeq \mathbb{Z}[\sqrt{3}]$ *so A is of* GL_2*-type over K. Moreover, A is isogenous to A'_{61} over K.*
(e) *The abelian surface A has a polarization over K of type* $(1, 2)$, *but no abelian surface isogenous to A over K admits a principal polarization over K.*

Although the full classification results of Dembélé are conditional on the GRH, the realization in Theorem 1.1 of the particular abelian surface A uses a direct computation, making it unconditional.

The abelian surface A'_{61} arises naturally as an isogeny factor over $\mathbb{Q}(\sqrt{61})$ of a modular abelian fourfold A_{61} associated via the Eichler–Shimura construction to the Hecke irreducible space $S_2(\Gamma_0(61), \chi)$ of classical modular forms of level 61 with quadratic character χ, a space having nontrivial inner twist. Shimura [35, p. 210] conjectured and Casselman [9, Theorem 1.1] proved that A'_{61} has everywhere good reduction (more generally, for good reduction properties of the Jacobians of modular curves, see Katz–Mazur [23, Theorem 14.7.2]). Using the Atkin–Lehner involution and explicit period computations, Dembélé–Kumar [12, Remark 3] observed that A'_{61} is naturally (1, 2)-polarized but not principally polarized (using a general criterion of González–Guàrdia–Rotger [18]); consequently, this surface was an outlier for their method (the “sole exception” in the range of their computations).

We construct X as follows. First, we compute the period lattice of the modular abelian fourfold A_{61} (attached to $S_2(\Gamma_0(61), \chi)$) to large precision. Adapting the methods of Costa–Mascot–Sijsling–Voight [10], we then numerically split A_{61} as the square of a simple abelian surface with a $(1, 2)$-polarization—a *building block* in the language of Ribet and Pyle [32, 33]—providing a numerical candidate for A. We then glue with the elliptic curve E (in Theorem 1.1(b)) via analytic methods to find a curve of genus three over K, but with a terrible equation. Finally, we simplify the cover to obtain $X \to E$ as in Theorem 1.1 and confirm that the Prym of this cover has everywhere good reduction as claimed. Code that was used to obtain the results in this article can be found at [31].

Contents The paper is organized as follows. In Sect. 2, we explain explicit methods to compute the period matrix of a modular abelian variety to large precision. In Sect. 3, we illustrate our adapted methods to numerically decompose a complex abelian variety with a focus on our example. In Sect. 4, we realize the resulting building block as a Prym explicitly; then in Sect. 5 we simplify the equations for the curve X and verify the properties announced in Theorem 1.1.

Acknowledgments The authors would like to thank Lassina Dembélé and the anonymous referees for comments. Sijsling was supported by a Juniorprofessurprogramm "Endomorphismen algebraischer Kurven" of the Science Ministry of Baden-Württemberg. Voight was supported by a Simons Collaboration grant (550029). This work was sparked during the conference *Arithmetic of Low-Dimensional Abelian Varieties* at the Institute for Computational and Experimental Research in Mathematics (ICERM) in June 2019; the authors would like to thank ICERM for their hospitality and support.

2 Computing the Period Matrix

In this section, we show how to compute the period matrix of our modular abelian variety to large precision. We begin by recalling how this variety shows up as a factor in the decomposition up to isogeny of the Jacobian of the modular curve $X_1(61)$. After this, we explain how to determine the periods of a newform of weight 2 in terms of its q-expansion. Finally, we show how to compute many coefficients of this q-expansion in reasonable time, so as to be able to compute the period matrix with very high accuracy. (The current implementation in MAGMA [6] allows computations to low precision; this is not enough for our purposes.)

Conventions Before we begin, we set up a few conventions we use throughout. Let K be a field with algebraic closure K^{al}. For an abelian variety A over K, we denote by $\mathrm{End}(A)$ the endomorphisms of A defined over K; if we wish to consider endomorphisms over a field extension $L \supseteq K$, we write A_L for the base change of A to L and write $\mathrm{End}(A_L)$ for endomorphisms over L.

Setup Let $k \in \mathbb{Z}_{\geq 0}$ and $n, N \in \mathbb{Z}_{>0}$ be integers. We denote by $\sigma_k(n) := \sum_{d|n} d^k$ the sum of the k-th powers of the positive divisors of n. In particular, $\sigma_0(n)$ is the number of divisors of n.

Given a subgroup $H \leq (\mathbb{Z}/N\mathbb{Z})^\times$, let

$$\Gamma_H(N) := \left\{ \begin{pmatrix} a & b \\ c & d \end{pmatrix} \in \mathrm{SL}_2(\mathbb{Z}) \ : \ c \equiv 0 \pmod{N} \ \text{ and } \ a, d \bmod N \in H \right\}.$$

By a newform of level $\Gamma_H(N)$, we mean a newform of level $\Gamma_1(N)$ whose nebentypus character $\varepsilon\colon (\mathbb{Z}/N\mathbb{Z})^\times \to \mathbb{C}^\times$ satisfies $H \leq \ker \varepsilon$. Let $\mathcal{N}_k(\Gamma_H(N))$ denote the finite set of newforms of weight k and level $\Gamma_H(N)$. This set is acted on (the left) by $\mathrm{Gal}_\mathbb{Q}$, the absolute Galois group of $\mathbb{Q}$, via the coefficients of q-expansions at the cusp ∞: if $f(q) = \sum_{n=1}^\infty a_n(f) q^n$ lies in $\mathcal{N}_k(\Gamma_H(N))$, so that in particular $a_1 = 1$, then the coefficients $\{a_n(f)\}_n$ are algebraic integers, and for $\tau \in \mathrm{Gal}_\mathbb{Q}$ we have $\tau f \in \mathcal{N}_k(\Gamma_H(N))$ with q-expansion $(\tau f)(q) = \sum_n \tau(a_n(f)) q^n$. Let $\mathrm{Gal}_\mathbb{Q} \backslash \mathcal{N}_k(\Gamma_H(N))$ be the set of Galois orbits of $\mathcal{N}_k(\Gamma_H(N))$. For each $f \in \mathcal{N}_k(\Gamma_H(N))$, we denote by $K_f := \mathbb{Q}(\{a_n(f)\}_n)$ the number field generated by the coefficients of f. The field K_f contains the values of the nebentypus character ε_f of f; furthermore the nebentypus of $\tau(f)$ is $\varepsilon_{\tau(f)} = \tau \circ \varepsilon_f$ for all $\tau \in G_\mathbb{Q}$.

Let $\mathcal{H}^* := \mathcal{H} \cup \mathbb{P}^1(\mathbb{Q})$ denote the completed upper half-plane. Recall that for $N \in \mathbb{Z}_{>0}$ and $H \leq (\mathbb{Z}/N\mathbb{Z})^\times$, the modular curve $X_H(N) := \Gamma_H(N) \backslash \mathcal{H}^*$ attached to the congruence subgroup $\Gamma_H(N)$ is defined over $\mathbb{Q}$, and its Jacobian $J_H(N)$ decomposes up to isogeny over $\mathbb{Q}$ as

$$J_H(N) \sim \prod_{M \mid N} \prod_{f \in \mathcal{N}_2(\mathrm{Gal}_\mathbb{Q} \backslash \Gamma_{H_M}(M))} A_f^{\sigma_0(N/M)}, \tag{2.1}$$

where $H_M \leq (\mathbb{Z}/M\mathbb{Z})^\times$ denotes the image of H in $(\mathbb{Z}/M\mathbb{Z})^\times$. Moreover, for each Galois orbit $f \in \mathrm{Gal}_\mathbb{Q} \backslash \mathcal{N}_2(\Gamma_H(N))$, the abelian variety A_f is simple over $\mathbb{Q}$ with $\dim A_f = [K_f : \mathbb{Q}]$, and A_f has endomorphism algebra $\mathrm{End}(A_f) \otimes \mathbb{Q} \simeq K_f$.

Let $\mathbb{T}$ be the Hecke algebra of weight 2 and level $\Gamma_H(N)$. Roughly speaking, A_f can be thought of as the piece of $J_H(N)$ on which $\mathbb{T}$ acts with the same eigenvalue system as f. More precisely, A_f is defined by

$$A_f := J_1(N)/I_f J_1(N), \tag{2.2}$$

where $I_f := \{T \in \mathbb{T} : Tf = 0\}$ is the annihilator of f under $\mathbb{T}$, so that for instance the image of $T_n|_{A_f} \in \mathrm{End}(A_f)$ under the isomorphism $\mathrm{End}(A_f) \otimes \mathbb{Q} \simeq K_f$ is the coefficient $a_n(f) \in K_f$. In particular, up to isogeny, the matrix

$$\begin{pmatrix} & \vdots & \\ \cdots & \displaystyle\int_{\gamma_j} \tau(f)(z)\,\mathrm{d}z & \cdots \\ & \vdots & \end{pmatrix}$$

whose rows are indexed by the newforms $\tau(f)$ in the Galois orbit of f and whose columns are indexed by a $\mathbb{Z}$-basis $(\gamma_j)_j$ of

$$H_1(X_H(N), \mathbb{Z})[I_f] = \{\gamma \in H_1(X_H(N), \mathbb{Z}) : T\gamma = 0 \text{ for all } T \in I_f\}$$

is a period matrix of A_f.

It is therefore possible to extract a period matrix of A_f out of a period matrix of the modular curve $X_H(N)$. While it may seem excessive to work with the periods of the whole Jacobian $J_H(N)$ instead of A_f, this approach presents the advantage of making it easy to compute explicitly in the $\mathbb{T}$-module $H_1(X_H(N), \mathbb{Z})$ using modular symbols.

Our Modular Jacobian Factor In what follows, we use LMFDB [26] labels for our classical modular forms. In our case, we observe that the Galois orbit of newforms attached to A_{61} is

$$f := \text{61.2.b.a} = q + \alpha q^2 + (\alpha^2 + 3)q^3 + O(q^4)$$

where $\alpha^4 + 8\alpha^2 + 13 = 0$; the four newforms in this orbit correspond to the four embeddings of $\mathbb{Q}(\alpha) \hookrightarrow \mathbb{C}$, and their nebentypus is the quadratic character χ_{61}. Thus $N = 61$ and $H = \ker \chi_{61}$, the subgroup of squares of $(\mathbb{Z}/61\mathbb{Z})^\times$.

This Galois orbit 61.2.b.a happens to be the only one with quadratic nebentypus, whereas there are two Galois orbits with trivial nebentypus, namely

$$\text{61.2.a.a} = q - q^2 - 2q^3 + O(q^4)$$

whose coefficient field is $\mathbb{Q}$, and

$$\text{61.2.a.b} = q + \beta q^2 + (-\beta^2 + 3)q^3 + O(q^4)$$

whose coefficient field is $\mathbb{Q}(\beta)$ where $\beta^3 - \beta^2 - 3\beta + 1 = 0$. Since 61 is prime and since there are no cuspforms of weight 2 and level 1, the decomposition (2.1) informs us that $J_H(61)$ is isogenous to the product of an elliptic curve corresponding to 61.2.a.a, an abelian threefold corresponding to 61.2.a.b, and our fourfold A_{61} corresponding to 61.2.b.a. It also follows that the modular curve $X_H(61)$ has genus 8, which is reasonable for practical computations.

In order to compute the period lattice of A_{61}, we must first isolate the rank-8 sublattice $H_1(X_H(61), \mathbb{Z})[I]$ of the rank-16 lattice $H_1(X_H(61), \mathbb{Z})$, where $I \leq \mathbb{T}$ is the annihilator of 61.2.b.a. We can obtain this sublattice as

$$H_1(X_H(61), \mathbb{Z})[I] = \bigcap_{T \in I} \ker(T|_{H_1(X_H(61), \mathbb{Z})}).$$

We observe that the Hecke operator $T_2^4 + 8T_2^2 + 13$ lies in I, since $T_2 f = a_2(f) f = \alpha f$ and $\alpha^4 + 8\alpha^2 + 13 = 0$; furthermore, the kernel of this operator acting on $H_1(X_H(61), \mathbb{Z})$ happens to have rank 8, so that it agrees with the sublattice $H_1(X_H(61), \mathbb{Z})[I]$.

Remark 2.3 The fact that the hunt for this sublattice was so short-lived is not a coincidence. Indeed, since 61 is prime and since there is no cuspform of weight 2 and level 1, all the eigenforms of level 61 are new, so that $\mathbb{T}\otimes\mathbb{Q}$ is a semi-simple $\mathbb{Q}$-algebra by the multiplicity-one theorem. It is therefore étale, so that most of its elements generate it as a $\mathbb{Q}$-algebra; in fact, an element of $\mathbb{T}\otimes\mathbb{Q}$ is a generator if and only if its eigenvalues on each of the Galois conjugates of 61.2.a.a, 61.2.a.b, and 61.2.b.a are all distinct. In particular, it is not surprising to find that $\mathbb{T}\otimes\mathbb{Q}=\mathbb{Q}[T_2]$, so that $I\otimes\mathbb{Q}$ is generated by $T_2^4+8T_2^2+13$, which explains why taking the kernel of this operator led us directly to the correct sublattice.

Twisted Winding Elements We now recall a formula that we will use to compute the periods $\int_\gamma f(z)\,\mathrm{d}z$ of a newform $f\in\mathcal{N}_2(\Gamma_H(N))$ for γ ranging over a $\mathbb{Z}$-basis of $H_1(X_H(N),\mathbb{Z})$.

Since f is an eigenform, it is actually enough to compute these integrals for γ ranging over a generating set of $H_1(X_H(N),\mathbb{Z})$ as a $\mathbb{T}$-module, since

$$\int_{T\gamma} f(z)\,\mathrm{d}z=\int_\gamma (Tf)(z)\,\mathrm{d}z=\lambda_T(f)\int_\gamma f(z)\,\mathrm{d}z$$

for all $T\in\mathbb{T}$, where $\lambda_T(f)$ is such that $Tf=\lambda_T(f)f$.

Our strategy consists in integrating term-by-term the q-expansion of f along modular symbols chosen so that the resulting series has good convergence properties. We choose to use the so-called *twisted winding elements*, that is to say modular symbols of the form

$$s_\chi=\sum_{a \bmod m}\overline{\chi}(-a)\{\infty,a/m\}$$

where χ is a primitive Dirichlet character whose modulus m is coprime to N.

Indeed, let ε_f be the nebentypus of f, and $\lambda_f\in\mathbb{C}^\times$ be the Fricke pseudo-eigenvalue of f, i.e.

$$f\big|\left(\begin{smallmatrix}0&-1\\N&0\end{smallmatrix}\right)=\lambda_f\sum_{n=1}^{\infty}\overline{a_n(f)}q^n; \tag{2.4}$$

if N is squarefree, the exact value of λ_f can be read off the coefficients $a_n(f)$ of f [30, Theorem 2.2.1]; in any case, a numerical approximation of λ_f can be obtained by evaluating (2.4) at a particular point of the upper half plane. A computation [30, Proposition 3.2.3] based on the fact that the twist $f\otimes\chi:=\sum_{n=1}^\infty a_n(f)\chi(n)q^n$ of f by χ is modular of level m^2N and on the identity

$$\int_\infty^0=\int_\infty^{\frac{i}{m\sqrt{N}}}+\int_{\frac{i}{m\sqrt{N}}}^0$$

yields the formula

$$\int_{s_\chi} f(z)\,\mathrm{d}z = \frac{m}{2\pi i}\sum_{n=1}^{\infty}\frac{1}{n}\left(\frac{\chi(n)a_n(f)}{\mathfrak{g}(\chi)} - \chi(-N)\varepsilon_f(m)\lambda_f\frac{\overline{\chi(n)a_n(f)}}{\mathfrak{g}(\overline{\chi})}\right)R^n \tag{2.5}$$

where $R := e^{-2\pi/m\sqrt{N}}$ and $\mathfrak{g}(\chi)$ denotes the Gauss sum

$$\mathfrak{g}(\chi) := \sum_{a\in\mathbb{Z}/m\mathbb{Z}} \chi(a)e^{2a\pi i/m}$$

of χ (and the same with $\overline{\chi}$).

In view of the Deligne bound $a_n(f) \ll n^{\frac{1}{2}+\varepsilon}$, it is necessary to sum numerically the series (2.5) up to $n \approx \frac{m\sqrt{N}\log 10}{2\pi}D$ in order to compute this integral with D correct decimal digits. For instance, the homology of the curve $X_H(61)$ introduced above happens to be generated as a $\mathbb{T}$-module by the s_χ for χ ranging over the set of primitive characters of modulus $m \leq 7$; therefore, in order to compute the period lattice of A_{61} with 1000 digits of accuracy, we need to compute the coefficients $a_n(f)$ for n up to approximately 20000. We show how this can be done in reasonable time below.

Remark 2.6 For safety, we actually computed the coefficients $a_n(f)$ and summed the series up to $n \leq 24000$. This 20% increase is quite arbitrary, but at this point we do not need to have a very tight control over the accuracy of our computations. Note that since we sum the series (2.5) using floating-point arithmetic and since its terms are roughly geometrically decreasing in magnitude, summing them in reverse (from $n = 24000$ to $n = 1$) increases numerical stability.

High-Accuracy q-Expansions Classical algorithms, such as that based on modular symbols or that based on trace formulas for Hecke operators, compute the coefficients $a_n(f)$ of a newform for $n \leq B$ in complexity which is at least quadratic in B. Since we are aiming for $B = 24000$, these algorithms are not suitable for our purpose. We therefore turn to the method presented by Mascot [29, 3.1], which can compute the first B coefficients $a_n(f)$ in complexity which is quasi-linear in B.

This method is based on the observation that any two rational functions u and v on an algebraic curve X are necessarily algebraically dependent, since the transcendence degree of the function field of X is only 1. There must therefore exist a polynomial $\Phi(x, y)$ satisfying $\Phi(u, v) = 0$. Besides, this polynomial can be taken to have degree at most $\deg v$ in x and $\deg u$ in y; indeed, a given value of u generically corresponds to $\deg u$ points of X, which in turn yield at most $\deg u$ values of v.

Let $f \in \mathcal{N}_k(\Gamma_H(N))$, let K_f be its coefficient field, and let $\mathcal{O}_f$ be the ring of integers of K_f. Our method thus consists in using such a relation between the inverse of the j-invariant

$$u := 1/j = \frac{E_4^3 - E_6^2}{(12E_4)^3} \in q\mathbb{Z}[[q]] \tag{2.7}$$

where

$$E_4 = 1 + 240 \sum_{n=1}^{\infty} \sigma_3(n) q^n, \quad E_6 = 1 - 504 \sum_{n=1}^{\infty} \sigma_5(n) q^n,$$

and the rational function $v := f/u' \in \mathcal{O}_f[[q]]$ where $u' := q\,du/dq$.

The first B coefficients of u can be computed in quasi-linear complexity in B using the sieve of Eratosthenes and fast series arithmetic, and we can then determine the first B coefficients of v and hence of $f = vu'$ by Newton iteration using $\Phi(u, v) = 0$.

In order to increase the efficiency of this computation, we run it modulo a large enough prime number p which splits completely in the coefficient field K_f of f. More precisely, let $(\omega_j)_{j \le [K_f:\mathbb{Q}]}$ be a $\mathbb{Z}$-basis of $\mathcal{O}_f$, so that each element of $\mathcal{O}_f$ may be uniquely written as $\sum_j \lambda_j \omega_j$ with $\lambda_j \in \mathbb{Z}$ for all j. The matrix

$$\begin{pmatrix} & \vdots & \\ \cdots & \tau(\omega_j) & \cdots \\ & \vdots & \end{pmatrix}$$

whose rows (resp. columns) are indexed by the complex embeddings $\tau : K_f \hookrightarrow \mathbb{C}$ (resp. the integers $1 \le j \le [K_f : \mathbb{Q}]$) is square, and invertible since the square of its determinant is the discriminant of K_f. Let $(c_{j,\tau})_{j,\tau}$ be the coefficients of its inverse, and define

$$C := \max_j \sum_\tau |c_{j,\tau}|.$$

Then for all $r > 0$ and for any element $\theta = \sum_j \lambda_j \omega_j \in \mathcal{O}_f$, we have

$$|\tau(\theta)| \le r \text{ for all } \tau : K_f \hookrightarrow \mathbb{C} \implies |\lambda_j| \le Cr \text{ for all } j. \tag{2.8}$$

Since for all $n \in \mathbb{Z}_{\ge 1}$, the coefficient $a_n(f) \in \mathcal{O}_f$ has absolute value at most $\sigma_0(n)\sqrt{n}$ under every embedding $\tau : K_f \hookrightarrow \mathbb{C}$, if we require that

$$p \ge 1 + 2C \max_{n \le B} \sigma_0(n)\sqrt{n}, \tag{2.9}$$

then for all $n \le B$ we can recover the exact value of $a_n(f) = \sum_j \lambda_{j,n} \omega_j$ from the $\lambda_{j,n}$ mod p, since we will always have $p \ge 2|\lambda_{j,n}| + 1$.

Our method is summarized in Algorithm 2.10; for more detail, see Mascot [29, 3.1], [30, 3.3].

Remark 2.11 Although the complexity of this method is quasilinear in B, the initial computation of the $a_n(f)$ for $n \le b$ and the linear algebra required to find the coefficients $a_{m,n}$ of $\Phi_{\mathfrak{p}}(x, y)$ represent a significant overhead if the degrees of u and v are large. In our case, it is more efficient to proceed as follows:

(i) Compute the first B coefficients of the newform $f_0 = $ 61.2.a.a $= q - q^2 - 2q^3 + O(q^4)$ introduced above using Algorithm 2.10. This is faster than for f, since f_0 has trivial nebentypus, so that instead of $X_H(61)$, we can work on the curve $X_0(61)$, where the degree of $u = 1/j$ is smaller.
(ii) Compute the first B coefficients of f using a variant of Algorithm 2.10 based on a relation between the functions $u = 1/j$ and $v = (f/f_0)^2$. Since the nebentypus of f has order 2, this can again be done on the curve $X_0(61)$.

Algorithm 2.10 High-accuracy q-expansion in quasi-linear time

---

Input: A newform f and $B \in \mathbb{Z}_{\geq 1}$.
Output: The coefficients $a_n(f)$ for $n \leq B$.
Algorithm:

1. Evaluate the degrees of the rational functions $u = 1/j$ and $v = f/u'$ (where $u' = q\,dq/du$) on the modular curve $X_H(N)$ in terms of the index $[\mathrm{SL}_2(\mathbb{Z}) : \Gamma_H(N)]$ and of the ramification of the j-invariant map.
2. Use Riemann-Roch to find a bound $b \in \mathbb{Z}_{\geq 1}$ such that any linear combination of the $u^m v^n$ with $m \leq \deg v$ and $n \leq \deg u$ whose first b q-expansion coefficients vanish is necessarily the 0 function.
3. Compute the coefficients $a_n(f)$ for $n \leq b$ using one of the quadratic algorithms mentioned above.
4. Pick a prime p satisfying (2.9) and which splits completely in K_f, and compute the first B coefficients of u mod p using fast series arithmetic in $\mathbb{F}_p[[q]]$.
5. For each prime $\mathfrak{p} \mid p$ of K_f, in parallel:

    a. By linear algebra, find coefficients $a_{m,n} \in \mathbb{F}_{\mathfrak{p}}$ such that the first b coefficients of

    $$\sum_{m \leq \deg v} \sum_{n \leq \deg u} a_{m,n} u^m v^n$$

    are all 0 mod $\mathfrak{p}$, and define

    $$\Phi_{\mathfrak{p}}(x, y) = \sum_{m \leq \deg v} \sum_{n \leq \deg u} a_{m,n} x^m y^n \in \mathbb{F}_p[x, y].$$

    b. Use Newton iteration in $\mathbb{F}_p[[q]]$ over the second variable of the equation $\Phi_p(u, v) = 0$ mod $\mathfrak{p}$ to find the first B coefficients of v (mod $\mathfrak{p}$).

6. Use CRT over the primes $\mathfrak{p}$ to determine the first B coefficients of v (mod $p\mathcal{O}_f$).
7. Express the first B coefficients of $f = vu'$ (mod $p\mathcal{O}_f$) on the $\mathbb{Z}/p\mathbb{Z}$-basis $(\omega_j)_{j \leq [K_f:\mathbb{Q}]}$ of $\mathcal{O}_f/p\mathcal{O}_f$, and deduce their exact value thanks to Deligne's bounds.

In general, if we wanted the coefficients of more modular forms (e.g., the other newform 61.2.a.b of trivial nebentypus), it is better to proceed inductively by replacing the j-invariant with forms that have already been computed (such as f_0 and f), since this leads to polynomials $\Phi_{\mathfrak{p}}(x, y)$ of much lower degree [30, 3.3].

Using this method, we can obtain the coefficients $a_n(f)$ for $n \leq 24{,}000$ as elements of $K_f = \mathbb{Q}(\alpha)$, where $\alpha^4 + 8\alpha^2 + 13 = 0$. We then deduce the complex coefficients of each of the newforms $\tau(f)$ in the Galois orbit of f by applying each of the complex embeddings of K_f to these coefficients. It then remains to evaluate the integrals $\int_{s_\chi} \tau(f)(z)\,dz$ using (2.5), and then to extract the periods of our abelian fourfold A_{61}, as explained at the beginning of this section.

3 Isolating the Building Block

The result of the algorithms in the previous sections is a 4×8 period matrix $\Pi \in \mathrm{M}_{4,8}(\mathbb{C})$ with respect to some basis of the homology of the corresponding abelian fourfold A_{61}. In this section, we find the period sublattice corresponding to an abelian surface as isogeny factor.

Finding a Polarization The homology basis used for the period computations in the previous section need not, in general, be symplectic. Indeed, the computation of the intersection pairing on the homology of a modular curve is currently implemented only for the case of $X_0(N)$. Therefore, we start this section by briefly describing how to numerically recover a polarization for Π.

Remark 3.1 Dembélé–Kumar [12, §4.1.1, Method 1] showed that when the character is quadratic (as in our case), the Atkin–Lehner involution splits the modular abelian variety, giving a polarization whose degree is a power of 2. In cases where the resulting polarization is not principal, they search (in the Hecke algebra) for a suitable principal polarization, so the method we give can be seen as a more general version of this approach.

More generally, let $\Pi \in \mathrm{M}_{g,2g}(\mathbb{C})$ be a period matrix, with corresponding abelian variety $A = V/\Lambda$, where $V = \mathbb{C}^g$ and $\Lambda = \Pi\mathbb{Z}^{2g}$. Having Π at our disposal yields an identification of the homology group $H_1(A, \mathbb{Z})$ with $\mathbb{Z}^{2g}$. Multiplication by i on A yields an endomorphism J of $H_1(A, \mathbb{R})$, which we similarly identify with the unique matrix $J \in \mathrm{M}_{2g}(\mathbb{R})$ obtained from the equality [2, §6.1]

$$i\Pi = \Pi J. \tag{3.2}$$

Via the above identification and the Riemann relations [5, 4.2.1], finding a polarization for A comes down to finding an alternating matrix $E \in \mathrm{M}_{2g}(\mathbb{Z})$ satisfying $J^t E J = E$ (where ^{t} denotes transpose) and the positive definiteness condition $i\Pi E^{-1}\Pi^* > 0$ (where * denotes the conjugate transpose).

We first focus on the linear equations

$$\begin{aligned} E^t &= -E \\ J^t E J &= E. \end{aligned} \tag{3.3}$$

Let $m = g(2g-1)$. We choose a $\mathbb{Z}$-basis of size m of alternating matrices $\mathrm{M}_{2g}(\mathbb{Z})_{\mathrm{alt}}$ (e.g., indexed by entries above the diagonal). Numerically solving the equations (3.3) then reduces to finding common approximate zeros of a number of linear forms $\ell_1, \ldots, \ell_m : \mathbb{Z}^m \to \mathbb{R}$. We determine a $\mathbb{Z}$-basis of vectors v for which all $\ell_i(v)$ are numerically negligible by LLL-reducing the lattice spanned by the rows of the matrix

$$\begin{pmatrix} & | & & | \\ I_m & C\ell_1 & \cdots & C\ell_m \\ & | & & | \end{pmatrix}. \tag{3.4}$$

Here I_m is the m-by-m identity matrix, and the other columns correspond to the linear forms ℓ_i multiplied by a sufficiently large scalar $C \in \mathbb{R}_{\gg 0}$. The first elements of the resulting LLL-reduced basis will give approximate solutions of (3.3), and we preserve those that do so to large enough precision to get a $\mathbb{Z}$-basis of numerical solutions for (3.3).

Having found a $\mathbb{Z}$-basis of solutions of (3.3), we take small linear combinations until we find an element E of determinant 1 such that the Hermitian matrix $i\Pi E^{-1}\Pi^*$ is positive definite. We summarize our steps in Algorithm 3.5.

Algorithm 3.5 Numerically recovering principal polarizations

Input: A period matrix $\Pi \in \mathrm{M}_{g,2g}(\mathbb{C})$, not necessarily with respect to a symplectic basis.

Output: An alternating matrix $E \in \mathrm{M}_{2g}(\mathbb{Z})$ defining a principal polarization for the abelian variety corresponding to Π.

Algorithm:

1. Use LLL to find a $\mathbb{Z}$-basis $\{E_1, \ldots, E_d\}$ of solutions of (3.3) in $\mathrm{M}_{2g}(\mathbb{Z})_{\mathrm{alt}}$.
2. Take a (random) $\mathbb{Z}$-linear combination $E = \sum_{i=1}^d c_i E_i$, with $c_i \in \mathbb{Z}$.
3. If $\det(E) = 1$ and $i\Pi E^{-1}\Pi^*$ is positive definite, return E; else, return to Step 2.

In this way, we quickly recover a principal polarization E on Π.

Remark 3.6 More generally, this method can be used to search for polarizations of any desired type.

For our modular abelian fourfold $A = A_{61}$, whose period matrix $\Pi \in \mathrm{M}_{4,8}(\mathbb{C})$ was computed in the previous section, we carry out Algorithm 3.5 and recover (numerically) a principal polarization on A.

Finding Small Idempotents With a polarization in hand, we numerically decompose the abelian variety $A = A_{61}$, following Costa–Mascot–Sijsling–Voight [10]: we compute numerically that $\mathrm{End}(A_{\mathbb{C}})$ is an order of index 13 in $\mathrm{M}_2(\mathbb{Z}[\sqrt{3}])$. Using methods similar to those above, we find a $\mathbb{Z}$-basis $\{(T_1, R_1), \ldots, (T_8, R_8)\}$ for the pairs (T, R) with $T \in \mathrm{M}_4(\mathbb{C})$ and $R \in \mathrm{M}_8(\mathbb{Z})$ such that numerically

$$T\Pi = \Pi R. \tag{3.7}$$

Specifically, $R_1 = 1$, and the remaining matrices R_i are

$$\begin{pmatrix} -1 & 0 & 0 & 1 & 1 & 0 & 0 & 0 \\ -1 & -1 & -1 & 0 & 1 & -1 & 2 & 0 \\ 1 & 0 & 0 & 0 & -1 & 1 & 0 & 2 \\ -1 & 0 & -1 & 0 & -1 & -1 & -1 & 0 \\ 1 & 0 & 1 & 0 & 1 & 1 & 1 & 0 \\ -1 & 0 & -1 & -1 & -1 & -2 & -1 & -2 \\ -1 & 0 & -1 & 0 & -1 & -1 & -1 & 0 \\ 1 & 0 & 1 & 0 & 1 & 1 & 1 & 0 \end{pmatrix}, \begin{pmatrix} -2 & 0 & -1 & 0 & 0 & -2 & -1 & -2 \\ 1 & 0 & 0 & 1 & 1 & 0 & 1 & 2 \\ -1 & -1 & -1 & -2 & -1 & 0 & 0 & 0 \\ 0 & -1 & -2 & 0 & 0 & 0 & 0 & 0 \\ 1 & 1 & 1 & 1 & -1 & 2 & 1 & 2 \\ -1 & -1 & -1 & 0 & -1 & 0 & -1 & 0 \\ -1 & -1 & 0 & 1 & 1 & -2 & 0 & 0 \\ 1 & 2 & 2 & 0 & 1 & 1 & 1 & 0 \end{pmatrix}, \begin{pmatrix} 0 & 0 & 0 & 1 & 1 & 0 & 0 & 0 \\ 0 & 0 & 0 & 0 & 2 & 0 & 2 & 0 \\ 0 & 0 & 0 & 1 & 0 & 0 & 1 & 2 \\ 0 & 0 & 0 & 1 & 0 & 0 & 1 & 2 \\ 0 & 0 & 0 & -2 & -1 & 0 & -1 & -2 \\ -1 & 0 & -1 & 0 & -1 & -1 & 0 & 0 \\ 0 & 0 & 0 & 2 & 0 & 0 & 0 & 2 \\ 0 & 0 & 0 & -2 & 0 & 0 & -1 & -3 \end{pmatrix},$$

$$\begin{pmatrix} -1 & 0 & 0 & -1 & 0 & -2 & -1 & -2 \\ 0 & 0 & -1 & -2 & -1 & 1 & -2 & -1 \\ 0 & -1 & -1 & 1 & 0 & 0 & 1 & 1 \\ -1 & -1 & 0 & 0 & 0 & 0 & 1 & 1 \\ 1 & 1 & 0 & -1 & -1 & 0 & -1 & -1 \\ -1 & 0 & 0 & 1 & 0 & 1 & 1 & 2 \\ -1 & -1 & 0 & 1 & 0 & 0 & 0 & 1 \\ 1 & 1 & 0 & -1 & 0 & 0 & -1 & -2 \end{pmatrix}, \begin{pmatrix} -1 & 0 & 0 & 0 & 0 & 0 & 0 & 0 \\ -1 & -1 & -1 & -1 & -2 & -1 & -1 & -1 \\ 0 & 0 & -1 & 1 & 0 & 0 & 0 & 1 \\ 0 & 0 & 0 & -1 & -1 & 0 & 0 & 1 \\ 2 & 0 & 2 & 1 & 2 & 2 & 1 & 1 \\ 0 & 0 & 0 & 0 & 1 & -1 & 1 & 0 \\ -2 & 0 & -2 & -2 & -1 & -2 & -1 & -3 \\ 1 & 0 & 1 & 1 & 0 & 1 & -1 & 0 \end{pmatrix},$$

$$\begin{pmatrix} 1 & 0 & 1 & -1 & 0 & 0 & 0 & 0 \\ 2 & 1 & 1 & -2 & 1 & 1 & -2 & -3 \\ -3 & -1 & -2 & -1 & -1 & -2 & -2 & -3 \\ 2 & -1 & 0 & -1 & -1 & 2 & 0 & 1 \\ -1 & 1 & -1 & 1 & -1 & 0 & 1 & 1 \\ -1 & -1 & -1 & 0 & -2 & 1 & 0 & 2 \\ 1 & -1 & 2 & 1 & 0 & 0 & -1 & 1 \\ -1 & 2 & 0 & 0 & 2 & -1 & 1 & -2 \end{pmatrix}, \begin{pmatrix} 1 & 0 & 0 & 0 & 0 & 0 & 0 & 0 \\ 1 & 1 & 0 & -2 & 0 & -2 & -2 & -3 \\ 1 & 1 & 1 & 1 & 1 & 1 & 1 & 1 \\ 0 & 1 & 1 & 0 & 1 & -1 & 0 & -1 \\ -3 & -1 & -1 & 1 & -1 & 3 & 1 & 3 \\ 0 & 0 & 1 & 2 & 1 & 1 & 2 & 2 \\ 2 & -1 & 1 & 1 & 1 & -3 & 1 & -1 \\ -1 & 0 & -2 & -2 & -2 & 2 & -2 & 0 \end{pmatrix}.$$

The matrices R_i are a $\mathbb{Z}$-basis for the numerical endomorphism ring $\mathrm{End}(A_{\mathbb{C}}) \subset \mathrm{M}_8(\mathbb{Z})$; we look for small idempotents therein as follows. Inspired by algorithms for computing an isomorphism with the matrix ring [16, 22], we equip $\mathrm{M}_8(\mathbb{Z})$ with the Frobenius norm (sum of squares of entries) $N\colon \mathrm{M}_8(\mathbb{Z}) \to \mathbb{Z}$, a positive definite quadratic form; then idempotents (and, more generally, zerodivisors) are likely to be short vectors with respect to this norm. This very quickly identifies scores of idempotents

$$R_2 + 1,\ R_4 + 1,\ R_2 - R_3 + 1,\ R_8, \ldots \tag{3.8}$$

and we select one that gives us the simplest looking decomposition.

Inducing the Polarization Let (T, R) be a small idempotent in $\mathrm{End}(A_{\mathbb{C}}) \otimes \mathbb{Q}$ as constructed above. We can then find an isogeny factor of A in two ways:

(i) Taking the kernel of T yields a linear subspace of W of $\mathbb{C}^g$ in which $W \cap \Pi\mathbb{Z}^{2g}$ is a lattice. Choosing bases, we get a new period matrix Π' as well as a pair abusively still denoted (T, R) such that $T\Pi' = \Pi R$. The matrix T represents the *inclusion* of $W \cong \mathbb{C}^{\dim(W)}$ into $V \cong \mathbb{C}^g$.
(ii) Alternatively, taking the image of T yields a linear subspace of W of $\mathbb{C}^g$ in which $T\Pi\mathbb{Z}^{2g}$ is a lattice. Choosing bases, we get a new period matrix Π' as well as a pair abusively still denoted (T, R) such that $T\Pi = \Pi' R$. The matrix T represents the *projection* from $V \cong \mathbb{C}^g$ to $W \cong \mathbb{C}^{\dim(W)}$.

Note that by the dimensions of the abelian varieties W in approaches (i) and (ii) above are complementary in the sense that they sum to g.

Proposition 3.9 *In approach* (i), *the restriction of E to W is represented by the matrix $E' := R^t E R$, and E' defines a polarization for Π'.*

Proof It is clear that E' is alternating. The complex structure J restricts to give the complex structure on W, in the sense that because $T\Pi' = \Pi R$ we have

$$\Pi R J' = T\Pi' J' = iT\Pi' = i\Pi R = \Pi J R, \tag{3.10}$$

so that $RJ' = JR$. Therefore E', as the restriction of E under R, is again compatible with J', or more precisely

$$J'^t E' J' = J'^t R^t E R J' = R^t J^t E J R = R^t E R = E'. \tag{3.11}$$

Similarly, since E' is the restriction of E under R, the corresponding Hermitian form on W remains positive definite, which is equivalent to $i\Pi' E'^{-1} \Pi'^* > 0$ (as in Birkenhake–Lange [5, Proof of Lemma 4.2.3]). □

Proposition 3.12 *In approach* (ii), *an integer multiple of $E' := (RE^{-1}R^t)^{-1}$ induces a polarization on Π'.*

Proof This follows from the previous proposition by double duality. The projection (T, R) induces a dual map $(T^\vee, R^\vee) : (A')^\vee \to A^\vee$ on the corresponding dual abelian varieties. Recall that we can identify $A^\vee = V^\vee/\Lambda^\vee$, where $V^\vee := \mathrm{Hom}_{\overline{\mathbb{C}}}(V, \mathbb{C})$ is the set of antilinear maps from V to $\mathbb{C}$ and where $\Lambda^\vee$ is the dual of Λ under the canonical pairing between V and $V^\vee$. The original pairing E gives rise to a map $V \to V^\vee$ sending the lattice Λ into its antilinear dual Λ^*. In other words, we have

$$E(\Lambda \otimes \mathbb{Q}) = \Lambda^\vee \otimes \mathbb{Q}. \tag{3.13}$$

Therefore a suitable multiple of the inverse map $E^{-1} : V^\vee \to V$ maps $\Lambda^\vee$ into its dual Λ. This defines a polarization on $A^\vee$; it is compatible with the complex structure because E is, and it is positive definite for the same reason. We can transfer this polarization to $A'^\vee$ by the previous proposition, using the inclusion $(T^\vee, R^\vee)$ dual to the projection (T, R). Note that $R^\vee = R^t$ by duality. Dualizing a second time, we get our induced polarization on A' after again taking a suitable integral multiple if needed. □

Remark 3.14 If the morphism (T, R) in the previous proposition corresponds to a map of curves of degree d under the Torelli map, then E' is in fact d times the principal polarization on Π' by Birkenhake–Lange [5, Lemma 12.3.1]. However, the above notion of an induced polarization applies whether (T, R) comes from a map of curves or not.

We apply approach (ii) to the chosen polarization E and idempotent (T, R) for our period matrix $\Pi \in M_{4,8}(\mathbb{C})$ and obtain a matrix $\Pi_2 \in M_{2,4}(\mathbb{C})$ with a polarization E_2 of type $(1, 2)$, which defines an abelian surface A_2. Note that the above approaches are essentially equivalent by duality; we slightly preferred the second as it comes with a projection map $A \to A_2$.

4 Computing an Equation for the Cover

The $(1, 2)$-polarized abelian surface A_2 defined by Π_2 and E_2 in the previous section was obtained from a numerical idempotent of Π defined over the base field $K = \mathbb{Q}(\sqrt{61})$. Therefore the abelian surface A_2 is also defined over K. However, the polarization E_2 is not principal, so we do not obtain in this way a Jacobian of a genus 2 curve over K. In this section, we show how to realize A_2 as the Prym variety associated to a map $X \to A_1$ from a genus 3 curve to a genus 1 curve.

Along the way, we will continue our numerical computations and retain our sensitivity to fields of definition; at the end, we will rigorously certify our construction.

To obtain a principal polarization, one can first try to take a maximal isotropic subgroup of the kernel of the isogeny $A \to A^\vee$ defined by E_2. This kernel is isomorphic to $\mathbb{Z}/2\mathbb{Z} \times \mathbb{Z}/2\mathbb{Z}$ and carries a non-trivial Weil pairing; there are therefore 3 such maximal isotropic groups. Calculating the corresponding principally polarized quotients of A_2, one obtains that all of them give rise to indecomposable polarizations and therefore correspond to algebraic curves under Torelli. However, calculating the corresponding Igusa invariants, or alternatively reconstructing after Guàrdia [19] (implemented at [36]) shows that all of them define the same cubic Galois extension L of K, namely that defined by the polynomial $x^3 - \sqrt{61}x^2 + 4x + 2$.

This cubic polynomial also shows the way out. We will glue A_2 with an elliptic curve A_1 over K to realize A_2 as the Prym of genus-3 double cover of A_1 over K. This is a slight generalization of the same gluing operation when both A_2 and A_1 have a principal polarization, which will be considered in detail in the upcoming work by Hanselman [20] (for code, see Hanselman–Sijsling [21]).

Via the aforementioned cubic polynomial, we consider the elliptic curve

$$A_1 \colon y^2 = x^3 - \sqrt{61}x^2 + 4x + 2. \qquad (4.1)$$

Let Π_2 be a period matrix for the 2-dimensional abelian variety A_2 with respect to which the polarization E_2 is represented by the alternating form

$$E_2 = \begin{pmatrix} 0 & -1 & 0 & 0 \\ 1 & 0 & 0 & 0 \\ 0 & 0 & 0 & -2 \\ 0 & 0 & 2 & 0 \end{pmatrix}. \tag{4.2}$$

The elliptic curve A_1 will have a principal polarization represented by the standard symplectic matrix $E_1 = \left(\begin{smallmatrix} 0 & -1 \\ 1 & 0 \end{smallmatrix}\right)$, and with corresponding period matrix Π_1 say.

Consider the period matrix

$$\Pi_3' = \begin{pmatrix} \Pi_2 & 0 \\ 0 & \Pi_1 \end{pmatrix}, \tag{4.3}$$

with corresponding abelian variety A_3'; it admits a polarization represented by

$$E_3' = \begin{pmatrix} E_2 & 0 \\ 0 & 2E_1 \end{pmatrix} = \begin{pmatrix} 0 & -1 & 0 & 0 & 0 & 0 \\ 1 & 0 & 0 & 0 & 0 & 0 \\ 0 & 0 & 0 & -2 & 0 & 0 \\ 0 & 0 & 2 & 0 & 0 & 0 \\ 0 & 0 & 0 & 0 & 0 & -2 \\ 0 & 0 & 0 & 0 & 2 & 0 \end{pmatrix}. \tag{4.4}$$

Let $G := \ker(E_3') < \frac{1}{2}\Lambda'/\Lambda'$, where $\Lambda' = H_1(A_3', \mathbb{Z}) \cong \Pi_3'\mathbb{Z}^6$. Then G is equipped with the basis corresponding to the final four columns of E_3', and under the corresponding identification

$$G \simeq (\mathbb{Z}/2\mathbb{Z})^2 \times (\mathbb{Z}/2\mathbb{Z})^2 \tag{4.5}$$

the Weil pairing on G is the product pairing. By Birkenhake–Lange [5, 2.4.4], the quotients of A_3' on which the polarization E_3' induces a principal polarization are in bijection with the maximal isotropic subgroups $H < G$—there are 15 such subgroups.

Proposition 4.6 *Suppose that A_3' and G are defined over K. Then there exists a maximal isotropic subgroup $H < G$ that is defined over K. In particular, the quotient A_3'/H is defined over K.*

Proof The first factor in (4.5) corresponds to the kernel of the polarization E_2 on A_2. By the discussion at the beginning of this section, we know the structure of this group as a Galois module: the elements of order 2 are cyclically permuted by a generator of the Galois extension L of K defined by $x^3 - \sqrt{61}x^2 + 4x + 2$. This is the very same Galois module structure as that on the second factor in (4.5), which corresponds to $A_1[2]$.

We can therefore obtain a Galois-stable isotropic subgroup $H < G$ by starting with a pair (x_1, x_2) with $x_1 \neq 0$ and $x_2 \neq 0$ and taking the image under Galois, after which we adjoin 0:

$$H := \{(0,0), (x_1, x_2), (\sigma(x_1), \sigma(x_2)), \\ (\sigma^2(x_1), \sigma^2(x_2)) = (x_1 + \sigma(x_1), x_2 + \sigma(x_2))\}. \quad (4.7)$$

Since $\sigma_i(x_1)$ and $\sigma_j(x_1)$ pair to 1 for $i \neq j$ and similarly for x_2, this subgroup is indeed isotropic, because the product pairing on G assumes the values $0 + 0 = 0$ and $1 + 1 \equiv 0 \pmod 2$. By construction, H is defined over K.

The second part of the proposition follows immediately. □

Given a maximal isotropic subgroup H of G, say with generators h_1 and h_2, we take inverse images $\widetilde{h}_1$ and $\widetilde{h}_2$ in $\frac{1}{2}\Lambda'$. This yields a lattice

$$\Lambda = \Lambda' + \mathbb{Z}\widetilde{h}_1 + \mathbb{Z}\widetilde{h}_2. \quad (4.8)$$

that contains Λ'. Choosing a basis of Λ, and keeping the basis for Λ' used heretofore, we find a matrix $R_3 \in M_6(\mathbb{Z})$ such that $\Lambda R_3 = \Lambda'$. By construction, E_3' induces a principal polarization on the abelian variety corresponding to the period matrix

$$\Pi_3 = \Pi_3' R_3^{-1}. \quad (4.9)$$

Note that we have a projection map corresponding to the pair (I_3, R_3) in the equality $I_3 \Pi_3' = \Pi_3 R_3$. Under this map, the induced principal polarization on Π_3 corresponds to the polarization class from Proposition 3.12.

While we can in principle identify the subgroup H in terms of the columns of E_3' by using the Abel–Jacobi map from [27], we just tried all 15 subgroups, as in Algorithm 4.10.

Invoking this calculation using 1000 decimal digits, only one isotropic subgroup numerically gave rise to invariants over K; we obtain in this way a period matrix Π_3 whose abelian variety $A_3 := \mathbb{C}^3/\Pi_3\mathbb{Z}^6$ is equipped with the principal polarization E_3.

Finally, using the main result in Lercier–Ritzenthaler–Sijsling [25] (recognizing invariants), we constructed a plane quartic equation defining a genus 3 curve whose Jacobian (numerically) matches the abelian surface A_3.

5 Simplifying, Twisting, and Verifying the Cover

The methods from the previous section furnish us with an equation $F(x, y, z) = 0$ defining a smooth plane quartic curve $X \subset \mathbb{P}^2$ over K whose Jacobian putatively

Algorithm 4.10 Finding a principally polarized quotient over K that is a Jacobian

Input: The period matrices Π_2 and Π_1 for the factors A_2 and A_1 over K, with polarizations E_2 and E_1 of type $(1, 2)$ and (1) respectively, along with the associated matrices Π_3' and E_3' as in (4.3) and (4.4).

Output: A matrix $\Pi_3 = \Pi_3' R_3^{-1}$ on which E_3' induces a principal polarization E_3 such that Π_3 numerically corresponds to a genus-3 curve X over K, along with E_3 and the invariants of X.

Algorithm:

1. Determine the 15 isotropic subgroups H of $G = \ker(E_3') \cong (\mathbb{Z}/2\mathbb{Z})^2 \times (\mathbb{Z}/2\mathbb{Z})^2$ with the product Weil pairing.
2. For a given H as in Step 1, lift generators h_1 and h_2 to elements $\widetilde{h}_1$ and $\widetilde{h}_1$ of $\frac{1}{2}\Lambda'/\Lambda'$.
3. Construct $\Lambda = \Lambda' + \mathbb{Z}\widetilde{h}_1 + \mathbb{Z}\widetilde{h}_2$, choose a basis of Λ, and find $R_3 \in M_6(\mathbb{Z})$ such that $\Lambda R_3 = \Lambda'$.
4. Let $\Pi_3 = \Pi_3' R_3^{-1}$, and let $E_3 = R_3^{-t} E_3' R_3^{-1}$.
5. Using the methods of [24, §2] and LLL, check if the principally polarized abelian variety defined by Π_3 and E_3 corresponds to an algebraic curve X, and if so, whether the invariants of X are numerically in K. If so, return Π_3, E_3, and the corresponding invariants. Otherwise proceed to the next H in Step 1.

contains the abelian surface $A = A_2$ and the elliptic curve $E = A_1$ in Theorem 1.1 as isogeny factors over $\mathbb{Q}^{\text{al}}$. Since this reconstruction from invariants is only up to $\mathbb{Q}^{\text{al}}$-isomorphism, this Jacobian may not yet contain the given factors over K. Moreover, the implementation of these methods does not yet optimize the equations involved when working over proper extensions of $\mathbb{Q}$; indeed, the first homogeneous defining polynomial F is of a rather terrible form, whose coefficients are thousands upon thousands of decimal digits long. In this section, we simplify, twist, and then verify that the cover is as claimed.

Simplification Since the ring of integers of K has class number 1, a first step is to remove superfluous primes from the discriminant by generalizing methods of Elsenhans [14]. This approach makes the equation somewhat smaller, but its size is still thousands of digits.

Although we constructed X from a period matrix Π, the reconstruction methods involved only uses invariants and does not furnish the period matrix of $\text{Jac}(X)$ in terms of Π. Fortunately, using integration algorithms due to Neurohr [28], we can simply recompute the period matrix of $\text{Jac}(X)$ from scratch. Then, with the techniques of Bruin–Sijsling–Zotine [7] one can find the numerical geometric automorphism group $\text{Aut}(X_{\mathbb{C}})$ of the curve X along with its tangent representation on the ambient $\mathbb{P}^2$. It turns out that $\text{Aut}(X_{\mathbb{C}}) \simeq \mathbb{Z}/2\mathbb{Z}$, and in particular the nontrivial involution ι generating $\text{Aut}(X_{\mathbb{C}})$ is putatively defined over K itself.

Let $T \in \text{GL}_3(K)$ be a matrix representing ι, i.e., $F \cdot T = \lambda F$ under the substitution acting on the right. Since X is canonically embedded, we can read off the matrix T from its tangent representation. Moreover, we can write $T = UDU^{-1}$, where D is a diagonal matrix with diagonal entries $(-\mu, \mu, \mu)$ say. Then for $H = F \cdot U$ we have $H \cdot D = F \cdot UD = \lambda F \cdot U = \lambda H$, so that $H(-x, y, z) = \lambda H(x, y, z)$. In this way, we recognize an exact automorphism and reduce to the case where

$$F(x, y, z) = G(x^2, y, z), \tag{5.1}$$

with G homogeneous in x, y, z of weights 2, 1, 1. Now the equation $G(x, y, z) = 0$ defines a genus-1 curve E, which is the quotient of X by its involution ι and which is still defined over K. The map $X \to E$ is ramified over the divisor of zeros D of x on E.

We then calculate the j-invariant of the binary quartic $q(y, z)$ corresponding to the genus-1 curve E. An elliptic curve E_0 with the same j-invariant over K is defined by

$$E_0 \colon u^2 = p_0(t) \tag{5.2}$$

over K, where

$$p_0(t) := (10\nu + 34)(t^3 + \nu t^2 + (1 - \nu)t + 1) \tag{5.3}$$

where $\nu := (1+\sqrt{61})/2 \in K$ satisfies $\nu^2 = \nu+15$. The conductor of E_0 is (64). As luck would have it, the binary quartic q is equivalent over K to the quartic $z^4 p_0(y/z)$ corresponding to the polynomial p_0, yielding an isomorphism $E \xrightarrow{\sim} E_0$. Pushing the divisor D along this isomorphism, one obtains a divisor D_0 on E_0. Again luck is on our side and the divisor $D_0 - 4\infty$ on E_0 is principal; the corresponding degree-2 cover X_0 of E_0 is defined by the affine equations

$$X_0 \colon \begin{cases} (10\nu + 34)u^2 = t^3 + \nu t^2 + (1 - \nu)t + 1 \\ (-2\nu + 9)w^2 = 2u + (-7\nu - 24)t^2 + (4\nu + 14)t - 2\nu - 7 \end{cases} \tag{5.4}$$

Eliminating u using the second equation and substituting in the first, replacing $w \leftarrow x$ and $t \leftarrow y$, and homogenizing yields the defining homogeneous quartic

$$\begin{aligned} F_0(x, y, z) = {} & (-195\nu + 859)x^4 + (-64\nu + 282)x^2y^2 + (-5\nu + 24)y^4 \\ & + (148\nu - 652)x^2yz + (-16\nu + 68)y^3z + (-74\nu + 326)x^2z^2 \\ & + (96\nu - 422)y^2z^2 + (-148\nu + 652)yz^3 + (-47\nu + 207)z^4. \end{aligned} \tag{5.5}$$

We are blessed for one final time because F_0 defines a non-singular curve $X_0 \subset \mathbb{P}^2$ with the same Dixmier–Ohno invariants as the original terrible curve X. (This is not automatic, as a degree 2 cover of a genus-1 curve is not determined by its ramification locus only, in contrast to the case of genus-0 curves.) In any event, we have finally obtained a simplified equation X_0.

Twisting It remains to find the correct twist of X_0 (with respect to the involution on X_0) that matches A_2 as an isogeny factor *over* K (instead of merely over $\mathbb{Q}^{\mathrm{al}}$). Again, this step is needed because the algorithms in Lercier–Ritzenthaler–Sijsling [25] only yield the correct geometric isomorphism class. We proceed as follows.

First, we considering the scheme over $\mathbb{Z}$ cut out by (5.5) and the equation for ν. By a Gröbner basis computation, we find that X_0 has bad reduction only at the prime (2); so our desired twist will be by a 2-unit (in $K^\times$). We begin by computing generators for the group $U := \mathbb{Z}_K[1/2]^\times/\mathbb{Z}_K[1/2]^{\times 2}$ of 2-units of K modulo squares, a group with 2-rank 3.

To find the right twist, we loop over primes $p \geq 5$ that split in K and by counting points compare the L-polynomial $L_p(X, T) \in 1 + T\mathbb{Z}[T]$ of X and the absolute L-polynomial of f

$$L_p(f, T) := (1 - a_p T + \chi(p) p T^2)(1 - \tau(a_p) T + \chi(p) T^2) \in 1 + T\mathbb{Z}[T],$$

where $\langle \tau \rangle = \mathrm{Gal}(K \mid \mathbb{Q})$. We find always that $L_p(f, \pm T) \mid L_p(X, T)$ for a unique choice of sign ε_p.

We then look for an element $\delta \in U$ such that $\left(\dfrac{\delta}{\mathfrak{p}}\right) = \varepsilon_p$ for each prime $\mathfrak{p}$ above p: we find a unique match, namely $\delta = -5\nu + 22$, an element of norm -1. Twisting by this element (replacing $x^2 \leftarrow \delta^{-1} x^2$) gives an equation all of whose good Euler factors match:

$$\begin{aligned} F(x, y, z) := {} & (5\nu + 17)x^4 - (14\nu + 48)x^2y^2 + (8\nu + 28)x^2yz \\ & - (4\nu + 14)x^2z^2 + (10\nu + 33)y^4 - (12\nu + 44)y^3z \\ & + (2\nu + 26)y^2z^2 + (4\nu - 16)yz^3 + (\nu - 6)z^4. \end{aligned} \tag{5.6}$$

Verification We now conclude with the proof of our main result (Theorem 1.1); for convenience, we reproduce the statement.

Theorem 5.7 *Let X be the projective plane curve over K defined by the equation $F(x, y, z) = 0$ as in (5.6). Then the following statements hold.*

(a) *The curve X is smooth of genus 3 and has bad reduction at the prime (2) only.*
(b) *The curve X admits an involution $\iota\colon (x : y : z) \mapsto (-x : y : z)$ over K, and the quotient of X by ι defines a map from X onto an elliptic curve E of conductor (64). An equation for E is given by*

$$-(10\nu + 34)y^2 = x^3 - \nu x^2 + (-\nu + 1)x - 1.$$

(c) *Up to isogeny over K, we have $\mathrm{Jac}(X) \sim A \times E$, where A is the Prym variety of the natural map $\mathrm{Jac}(X) \to \mathrm{Jac}(E)$, an abelian surface over K with everywhere good reduction.*

(d) *All geometric endomorphisms of A are defined over K, and* $\mathrm{End}(A) \simeq \mathbb{Z}[\sqrt{3}]$ *so A is of* GL_2*-type over K. Moreover, A is isogenous to* A'_{61} *over K.*
(e) *The abelian surface A has a polarization over K of type* $(1, 2)$, *but no abelian surface isogenous to A over K admits a principal polarization over K.*

Proof Part (a) was verified directly via a Gröbner basis computation as above: more precisely, we compute that the ideal N generated by F and its derivatives in $\mathbb{Z}[\nu][x, y, z]$, saturated at the irrelevant ideal (x, y, z), has $N \cap \mathbb{Z}[\nu] = (2)^{14}$.

Part (b) is because $F(x, y, z) = G(x^2, y, z)$, and $G(x, y, z)$ defines an elliptic curve, equipped with the point at infinity $(\nu - 3 : 1 : 0)$; we then compute the given equation (arising up to quadratic twist by construction). For the first part of (c), the splitting over K comes from the map in (b).

Next, we prove part (d) using the methods of Costa–Mascot–Sijsling–Voight [10], which rigorously certifies that $\mathrm{End}(\mathrm{Jac}(X)) = \mathrm{End}(\mathrm{Jac}(X^{\mathrm{al}})) \simeq \mathbb{Z} \times \mathbb{Z}[\sqrt{3}]$, where X^{al} denotes the base change to an algebraic closure K^{al} of K. More precisely, with respect to a K-basis of differentials, we certify that there is an endomorphism with tangent representation

$$\begin{pmatrix} 1 & 0 & 0 \\ 0 & 3/2 & (\nu - 4)/2 \\ 0 & (\nu + 3)/2 & -3/2 \end{pmatrix} \tag{5.8}$$

representing multiplication by $\sqrt{3}$.

To finish the proof of (c), frustratingly it is not clear how to algorithmically verify that the Prym has everywhere good reduction directly. (We could not find an implementation for the computation of regular models over number fields.) Instead, we look at Galois representations and apply the now standard method of Faltings–Serre, the relevant details of which we now sketch: see Dieulefait–Guerberoff–Pacetti [13, §4] and more generally Brumer–Pacetti–Poor–Tornaría–Voight–Yuen [4, §2].

The 2-adic Tate module of the abelian surface factor yields a Galois representation

$$\rho_{A,2}\colon \mathrm{Gal}_K \to \mathrm{GL}_2(\mathbb{Z}_2[\sqrt{3}]), \tag{5.9}$$

where $\mathrm{Gal}_K := \mathrm{Gal}(K^{\mathrm{al}} \mid K)$ is the absolute Galois group of K; since $(2) = \mathfrak{l}^2$ ramifies in $\mathbb{Z}[\sqrt{3}]$, reducing modulo $\mathfrak{l} = (1 + \sqrt{3})$ gives the mod $\mathfrak{l}$ representation

$$\overline{\rho}_{A,\mathfrak{l}}\colon \mathrm{Gal}_K \to \mathrm{GL}_2(\mathbb{F}_2).$$

Counting points, we find that the conjugacy classes of Frobenius elements for either prime above 5 has order 3, so the image of $\overline{\rho}_{A,\mathfrak{l}}$ has order 3 or 6, and correspondingly the fixed field of $\ker \overline{\rho}_2$ is an extension of K of degree 3 or 6 ramified only at (2). By computational class field theory, we verify there is in fact a *unique* such extension:

it is the normal closure of the extension of K defined by the polynomial $p_0(t)$ in (5.3), whose absolute field is defined by

$$x^6 + x^5 - 7x^4 + 14x^2 - 16x + 4; \tag{5.10}$$

the full normal extension L of $\mathbb{Q}$ is defined by

$$\begin{aligned} &x^{12} - 4x^{11} + 10x^{10} - 44x^9 + 174x^8 - 422x^7 + 618x^6 - 512x^5 \\ &\quad +170x^4 + 56x^3 - 58x^2 + 6x + 9. \end{aligned} \tag{5.11}$$

We repeat the same calculation for the newform $f \in S_2(\Gamma_0(61), \chi)$: its base change f_K to K has level (1) and so gives a Galois representation $\rho_{f_K,\mathfrak{l}}\colon \mathrm{Gal}_K \to \mathrm{GL}_2(\mathbb{Z}_2^{\mathrm{al}})$ for which the characteristic polynomial of Frobenius lies in $\mathbb{Z}_2[\sqrt{3}]$. Reducing modulo $\mathfrak{l}$ and semisimplifying gives $\overline{\rho}^{\mathrm{ss}}_{f_K,\mathfrak{l}}\colon \mathrm{Gal}_K \to \mathrm{GL}_2(\mathbb{F}_2^{\mathrm{al}})$; since the traces belong to $\mathbb{F}_2$, up to equivalence the representation takes values in $\mathbb{F}_2$ [37, Corollary 1], and as in the previous paragraph we confirm that it is surjective so absolutely irreducible so equal to its semisimplification: indeed, we must have $\overline{\rho}_{A,\mathfrak{l}} \simeq \overline{\rho}_{f_K,\mathfrak{l}}$ by uniqueness. By a result of Carayol [8, Théorème 2], up to equivalence the representation takes values in $\mathbb{Z}_2[\sqrt{3}]$, i.e., we have $\rho_{f_K,\mathfrak{l}}\colon \mathrm{Gal}_K \to \mathrm{GL}_2(\mathbb{Z}_2[\sqrt{3}])$.

We conclude by showing that $\rho_{A,\mathfrak{l}} \simeq \rho_{f_K,\mathfrak{l}}$, from which the remainder of part (c) follows as f_K has level (1). We compute $\mathbb{Z}_L[1/2]^\times/\mathbb{Z}_L[1/2]^{\times 2} \simeq (\mathbb{Z}/2\mathbb{Z})^9$ and that the primes of L above $3, 5, 61, 97$ generate the corresponding elementary 2-abelian extension of L. The method of Faltings–Serre then provides that $\rho_{A,\mathfrak{l}} \simeq \rho_{f_K,\mathfrak{l}}$ if and only if we have the equality of traces $\operatorname{tr} \rho_{A,\mathfrak{l}}(\mathrm{Frob}_\mathfrak{p}) = \operatorname{tr} \rho_{f_K,\mathfrak{l}}(\mathrm{Frob}_\mathfrak{p})$ for the primes $\mathfrak{p}$ of K above $3, 5, 61, 97$, where $\mathrm{Frob}_\mathfrak{p} \in \mathrm{Gal}_K$ denotes (a representative of) the conjugacy class of the $\mathfrak{p}$-Frobenius automorphism. Having checked equality of L-polynomials for these primes (indeed, we checked all primes of norm up to 200), the result follows.

Finally, having proven in the previous paragraph that A is modular, part (e) was proven for a particular abelian surface in the isogeny class of A by Dembélé–Kumar [12, Remark 3] (using a criterion of González–Guàrdia–Rotger [18]). We now show it extends to all abelian surfaces in the isogeny class. We have exhibited a (1, 2)-polarization $\lambda\colon A \to A^\vee$, where $A^\vee$ denotes the dual abelian variety. Assume for purposes of contradiction that there exists an isogeny $\varphi\colon A' \to A$ and a *principal* polarization λ' on A', both over K. Then

$$\mu' := \varphi^\vee \circ \lambda \circ \varphi\colon A' \to (A')^\vee$$

is another polarization, with $\deg(\mu') = 2\deg(\varphi)^2$. From work of González–Guàrdia–Rotger [18, Theorem 2.3, Theorem 2.10(ii)], we conclude that μ' corresponds to a *totally positive* element of $\mathrm{End}(A')$ of norm $\deg(\mu')$. But $\mathrm{End}(A')$ is an order in $\mathrm{End}(A') \otimes \mathbb{Q} = \mathbb{Q}(\sqrt{3})$, and 2 is not a norm from this field—equivalently, the conic $x^2 - 3y^2 = 2z^2$ has no rational points over $\mathbb{Q}$, or also equivalently, the corresponding quaternion algebra $(2, 3 \mid \mathbb{Q}) \not\simeq M_2(\mathbb{Q})$ is not split, having discriminant 6—a contradiction. (The element $1 + \sqrt{3} \in \mathbb{Z}[\sqrt{3}]$ has norm -2, and (2) represents the nontrivial element of the narrow class group of $\mathbb{Z}[\sqrt{3}]$.) □

References

1. V. A. Abrashkin, *Galois modules of group schemes of period p over the ring of Witt vectors*, translated from Izv. Akad. Nauk SSSR Ser. Mat. **51** (1987), no. 4, 691–736, 910, *Math. USSR-Izv.* **31** (1988), no. 1, 1–46.
2. Andrew R. Booker, Jeroen Sijsling, Andrew V. Sutherland, John Voight, and Dan Yasaki, *A database of genus-2 curves over the rational numbers,* LMS J. Comput. Math. **19** (Special Issue A) (2016), 235–254.
3. Armand Brumer and Kenneth Kramer, *Certain abelian varieties bad at only one prime*, Algebra Number Theory **12** (2018), no. 5, 1027–1071.
4. Armand Brumer, Ariel Pacetti, Cris Poor, Gonzalo Tornaría, John Voight, and David S. Yuen, *On the paramodularity of typical abelian surfaces*, Algebra & Number Theory **13** (2019), no. 5, 1145–1195.
5. Christina Birkenhake and Herbert Lange, *Complex abelian varieties*, volume 302 of *Grundlehren der Mathematischen Wissenschaften [Fundamental Principles of Mathematical Sciences]*, Springer-Verlag, Berlin, second edition, 2004.
6. Wieb Bosma, John Cannon, and Catherine Playoust, *The Magma algebra system. I. The user language*, J. Symbolic Comput. **24** (1997), vol. 3–4, 235–265.
7. Nils Bruin, Jeroen Sijsling, and Alexandre Zotine, *Numerical computation of endomorphism rings of Jacobians*, Proceedings of the Thirteenth Algorithmic Number Theory Symposium, Open Book Series 2 (2019), 155–171.
8. Henri Carayol, *Formes modulaires et représentations galoisiennes à valeurs dans un anneau local complet*, *p*-adic monodromy and the Birch and Swinnerton-Dyer conjecture (Boston, MA, 1991), Contemp. Math., vol. 165, Amer. Math. Soc., Providence, RI, 1994, 213–237.
9. W. Casselman, *On abelian varieties with many endomorphisms and a conjecture of Shimura's*, Invent. Math. **12** (1971), 225–236.
10. Edgar Costa, Nicolas Mascot, Jeroen Sijsling, and John Voight, *Rigorous computation of the endomorphism ring of a Jacobian*, Math. Comp. **88** (2019), 1303–1339.
11. Lassina Dembélé, *Abelian varieties with everywhere good reduction over certain real quadratic fields with small discriminant*, 2019, preprint.
12. Lassina Dembélé and Abhinav Kumar, *Examples of abelian surfaces with everywhere good reduction*, Math. Ann. **364** (2016), no. 3–4, 1365–1392.
13. Luis Dieulefait, Lucio Guerberoff, and Ariel Pacetti, *Proving modularity for a given elliptic curve over an imaginary quadratic field*, Math. Comp. **79** (2010), no. 270, 1145–1170.
14. Andreas-Stephan Elsenhans, *Good models for cubic surfaces*, preprint at https://math.uni-paderborn.de/fileadmin/mathematik/AG-Computeralgebra/Preprints-elsenhans/red_5.pdf.
15. Gerd Faltings, *Finiteness theorems for abelian varieties over number fields*, translated from Invent. Math. **73** (1983), no. 3, 349–366; ibid. **75** (1984), no. 2, 381, Arithmetic geometry (Storrs, Conn., 1984), Springer, New York, 1986, 9–27.
16. Tom Fisher, *How to trivialise an algebra*, 2007, http://www.faculty.jacobs-university.de/mstoll/workshop2007/fisher2.pdf.
17. Jean-Marc Fontaine, *Il n'y a pas de variété abélienne sur* $\mathbb{Z}$, Invent. Math. **81** (1985), no. 3, 515–538.
18. Josep González, Jordi Guàrdia, and Victor Rotger, *Abelian surfaces of* GL_2*-type as Jacobians of curves*, Acta Arith. **116** (2005), no. 3, 263–287.

19. Jordi Guàrdia, *Jacobian nullwerte and algebraic equations*, J. Algebra **253** (2002), no. 1, 112–132.
20. Jeroen Hanselman, *Gluing curves along their torsion*, Ph.D. thesis, Universität Ulm, in progress, (2021) available at https://oparu.uni-ulm.de/xmlui/bitstream/handle/123456789/33412/Hanselman_Thesis.pdf
21. Jeroen Hanselman and Jeroen Sijsling, `gluing`: *Gluing curves of low genus along their torsion*, https://github.com/JRSijsling/gluing.
22. Ivanyos, Gábor, Lajos Rónyai, and Josef Schicho, *Splitting full matrix algebras over algebraic number fields*, J. Algebra **354** (2012), 211–223.
23. Nicholas M. Katz and Barry Mazur, *Arithmetic moduli of elliptic curves*, Annals Math. Studies, vol. 108, Princeton University Press, 1985.
24. Pınar Kılıçer, Hugo Labrande, Reynald Lercier, Christophe Ritzenthaler, Jeroen Sijsling, and Marco Streng, *Plane quartics over* $\mathbb{Q}$ *with complex multiplication*, Acta Arith. **185** (2018), no. 2, 127–156.
25. Reynald Lercier, Christophe Ritzenthaler, and Jeroen Sijsling, *Reconstructing plane quartics from their invariants*, Discrete Comput. Geom. (2018), https://doi.org/10.1007/s00454-018-0047-4.
26. The LMFDB Collaboration, *The L-functions and Modular Forms Database*, http://www.lmfdb.org, 2019, [Online; accessed 30 July 2019].
27. Pascal Molin and Christian Neurohr, *Computing period matrices and the Abel-Jacobi map of superelliptic curves*, Math. Comp. **88** (2019), no. 316, 847–888.
28. Christian Neurohr, *Efficient integration on Riemann surfaces & applications*, Ph.D. thesis, Carl-von-Ossietzky-Universität Oldenburg, 2018.
29. Nicolas Mascot, *Computing modular Galois representations*, Rendiconti Circ. Mat. Palermo, vol. 62 no 3 (2013), 451–476.
30. Nicolas Mascot, *Companion forms and explicit computation of* PGL_2*-number fields with very little ramification*, Journal of Algebra, vol. 509 (2018) 476–506.
31. Nicolas Mascot, Jeroen Sijsling, and John Voight, *Prym calculation code*, https://github.com/JRSijsling/prym.
32. Elisabeth E. Pyle, *Abelian varieties over* $\mathbb{Q}$ *with large endomorphism algebras and their simple components over* $\overline{\mathbb{Q}}$, Modular curves and abelian varieties, Progr. Math., vol. 224, Birkhäuser, Basel, 2004, 189–239.
33. Kenneth A. Ribet, *Twists of modular forms and endomorphisms of abelian varieties*, Math. Ann. **253** (1980), no. 1, 143–162.
34. René Schoof, *Abelian varieties over cyclotomic fields with good reduction everywhere*, Math. Ann. **325** (2003), no. 3, 413–448.
35. Goro Shimura, *Introduction to the arithmetic theory of automorphic functions*, Kanô Memorial Lectures, no. 1, Princeton University Press, Princeton, 1971.
36. Jeroen Sijsling, `curve_reconstruction`: *Geometric and arithmetic reconstruction of curves from their period matrices*, https://github.com/JRSijsling/curve_reconstruction.
37. Richard Taylor, *Representations of Galois groups associated to modular forms*, Proc. ICM (Zürich, 1994), vol. 2, Birkhäuser, Basel, 1995, 435–442.

The S-Integral Points on the Projective Line Minus Three Points via Finite Covers and Skolem's Method

Bjorn Poonen

Abstract We describe a p-adic proof of the finiteness of $(\mathbb{P}^1 - \{0, 1, \infty\})(\mathbb{Z}[S^{-1}])$ using only Skolem's method applied to finite covers.

2020 Mathematics Subject Classification Primary 11G30

1 Introduction

Let S be a finite set of primes of $\mathbb{Q}$. Let $R = \mathbb{Z}[S^{-1}]$. Let $X = \mathbb{P}^1_R - \{0, 1, \infty\} = \operatorname{Spec} R[x, x^{-1}, (1-x)^{-1}]$. The set $X(R)$ is in bijection with $\{(x, y) \in R^\times \times R^\times : x + y = 1\}$.

Siegel [Sie26] proved that $X(R)$ is finite. Kim [Kim05] gave a new, p-adic proof of this fact, as an application of his nonabelian analogue of the Skolem–Chabauty method. Inspired by Kim's proof, we give a different p-adic proof, using only Skolem's method applied to finite covers. (In fact, the proof we present dates from an April 23, 2005 email to Kim following a talk he gave on his method, but we have not published our proof before now.)

2 Review of the Skolem–Chabauty Method

Let k be a finite extension of $\mathbb{Q}$. Let S be a finite set of places of k containing all the archimedean places. The *ring of S-integers in k* is

This research was supported in part by National Science Foundation grant DMS-1601946 and Simons Foundation grants #402472 (to Bjorn Poonen) and #550033.

B. Poonen (✉)
Department of Mathematics, Massachusetts Institute of Technology, Cambridge, MA, USA
e-mail: poonen@math.mit.edu

J. S. Balakrishnan et al. (eds.), *Arithmetic Geometry, Number Theory, and Computation*, Simons Symposia, https://doi.org/10.1007/978-3-030-80914-0_21

$$R := \{x \in k : v(x) \geq 0 \text{ for all } v \notin S\}.$$

Skolem devised a method that, in modern terms, for some subvarieties X in an algebraic torus over R, could prove finiteness of $X(R)$ or even determine it explicitly [Sko34]. His method was generalized by Chabauty [Cha38], who also adapted Skolem's method to study rational points on a curve X in an abelian variety [Cha41]; see [MP12] for an introduction to the latter. Although we need only the torus case, it is not much extra work to describe the method in a more general setting, so we will do so.

A semiabelian variety J is a commutative group variety fitting in an exact sequence $1 \to T \to J \to A \to 1$ with T a torus and A an abelian variety.

Proposition 2.1 *Let R be a ring of S-integers in a number field k. Let J be a finite-type group scheme over R whose generic fiber J_k is a semiabelian variety. Then the abelian group $J(R)$ is finitely generated.*

Proof By replacing k by a finite extension and enlarging S, we may assume that J fits in an exact sequence of R-group schemes

$$1 \longrightarrow \mathbb{G}_m^n \longrightarrow J \longrightarrow A \longrightarrow 1$$

for some $n \geq 0$ and abelian scheme A over R. Taking R-points yields an exact sequence

$$1 \longrightarrow (R^\times)^n \longrightarrow J(R) \longrightarrow A(R). \tag{1}$$

The group $R^\times$ is finitely generated by the Dirichlet S-unit theorem. By the valuative criterion for properness, $A(R) = A(K)$, which is finitely generated by the Mordell–Weil theorem. Now (1) shows that $J(R)$ is finitely generated. □

For a group variety J over a field, say that a subvariety $X \subset J$ generates J if, for some n, the addition morphism $X^n \to J$ is surjective (which amounts to requiring that it gives a surjective map on points over an algebraically closed field).

Proposition 2.2 *Let J be a semiabelian variety over $\mathbb{Q}_p$. Equip $J(\mathbb{Q}_p)$ with the p-adic topology. Let Γ be a finitely generated subgroup of a compact subgroup $G \leq J(\mathbb{Q}_p)$. Let X be an irreducible curve over $\mathbb{Q}_p$. Let $\iota\colon X \to J$ be a morphism whose image generates J. If* $\operatorname{rank}\Gamma < \dim J$, *then $\{x \in X(\mathbb{Q}_p) : \iota(x) \in \Gamma\}$ is finite.*

Proof (The details are analogous to those in [MP12, §4].) We may assume that ι is proper. Since $J(\mathbb{Q}_p)$ has a basis of neighborhoods of 1 consisting of compact open subgroups K, we may replace G by $G + K$ for any such K to assume that G is open in $J(\mathbb{Q}_p)$. By [Bou98, III.§7.6], there is a canonical homomorphism

$$\log\colon G \to \operatorname{Lie} G = \operatorname{Lie} J.$$

The group $\log \Gamma$ is generated by at most $\operatorname{rk} \Gamma$ elements, and $\operatorname{rk} \Gamma < \dim J = \dim(\operatorname{Lie} J)$, so there is a linear functional $\lambda \colon \operatorname{Lie} J \to \mathbb{Q}_p$ that vanishes on $\log \Gamma$ and even on its closure $\log \overline{\Gamma}$. Pulling λ back to the compact subset $\{x \in X(\mathbb{Q}_p) : \iota(x) \in G\}$ yields an analytic function η that is locally the integral of a nonzero 1-form on X, so the zero locus of η is discrete, and hence finite. Finally, $\{x \in X(\mathbb{Q}_p) : \iota(x) \in \Gamma\}$ is contained in the zero locus of η, by definition of λ. □

Theorem 2.3 (The Skolem–Chabauty Method) *Let R be a ring of S-integers in a number field k. Let X be a finite-type separated R-scheme such that X_k is an irreducible curve. Let J be a finite-type separated R-group scheme such that J_k is a semiabelian variety. Let $\iota \colon X_k \to J_k$ be a morphism whose image generates J_k. If* $\operatorname{rk} J(R) < \dim J_k$*, then $X(R)$ is finite.*

Proof Let $R_v \subset k_v$ denote the completions of $R \subset k$ at v. Let $\widehat{R} = \prod_{\ell \notin S} R_v$. Let $\mathbf{A}$ be the restricted product $\prod'_{v \notin S}(k_v, R_v)$, so the R-algebra $\widehat{R}$ is open in the k-algebra $\mathbf{A}$, and $\widehat{R} \cap k = R$. Since $X(\widehat{R})$ is compact and $J(\widehat{R})$ is open in $J(\mathbf{A})$, the map $\iota \colon X(\mathbf{A}) \to J(\mathbf{A})$ maps $X(\widehat{R})$ into a finite union of cosets of $J(\widehat{R})$. Intersecting with $J(k)$ shows that ι maps $X(R)$ into a finite union of cosets of $J(R)$ in $J(k)$.

Choose $\mathfrak{p} \notin S$ that is unramified of degree 1 over a prime p of $\mathbb{Q}$. Then $R_{\mathfrak{p}} \simeq \mathbb{Z}_p$ and $k_{\mathfrak{p}} \simeq \mathbb{Q}_p$. Proposition 2.2 applied to $\iota_{\mathbb{Q}_p}$ with $\Gamma = J(R)$ (finitely generated by Proposition 2.1) and $G = J(\mathbb{Z}_p)$ shows that $\{x \in X(R) : \iota(x) \in J(R)\}$ is finite. The same argument with ι composed with a translation shows that $\{x \in X(R) : \iota(x) \in j + J(R)\}$ is finite for each $j \in J(\mathbb{Q})$. By the first paragraph, $X(R)$ is contained in a finite union of these. □

3 Proof of Siegel's Theorem

Let $R = \mathbb{Z}[S^{-1}]$ and $X = \mathbb{P}^1_R - \{0, 1, \infty\}$. Let ℓ be a prime. Let A be a (finite) set of representatives for $R^\times / R^{\times \ell}$. For each $a \in A$, let $\pi_a \colon Y_a \to X$ be the finite cover obtained as the inverse image of X under the morphism

$$\begin{aligned} \mathbb{P}^1_R &\longrightarrow \mathbb{P}^1_R \\ y &\longmapsto ay^\ell. \end{aligned}$$

Any element of $X(R) \subset \mathbb{G}_m(R) = R^\times$ is ay^ℓ for some $a \in A$ and $y \in R^\times \subset \mathbb{P}^1(R)$, and then $y \in Y_a(R)$ by definition of Y_a. Thus $X(R) = \bigcup_{a \in A} \pi_a(Y_a(R))$. It remains to prove that each set $Y_a(R)$ is finite.

Let $\mathcal{O} = R[t]/(at^\ell - 1)$ and $K = \mathbb{Q}[t]/(at^\ell - 1)$. If a represents the trivial class in $R^\times / R^{\times \ell}$, then $K \simeq \mathbb{Q} \times K_0$ for a number field K_0; otherwise define $K_0 := K$, which is already a number field. Let $\mathcal{O}_0$ be the integral closure of R in K_0.

We have $Y_a = \mathbb{P}^1_R - \{0, \infty, \text{zeros of } ay^\ell - 1\}$. Over $\overline{\mathbb{Q}}$, the generalized Jacobian of Y_a (i.e., of $\mathbb{P}^1$ with modulus consisting of the $\ell + 2$ removed points) is a torus of dimension $\ell + 1$, and it has a natural model over R, namely $J := \mathbb{G}_{m,R} \times$

$\mathrm{Res}_{\mathcal{O}/R}\,\mathbb{G}_{m,\mathcal{O}}$, where Res denotes restriction of scalars. The usual morphism from $(Y_a)_{\mathbb{Q}}$ to its generalized Jacobian $J_{\mathbb{Q}}$, up to translation in $J_{\mathbb{Q}}$, is $y \mapsto (y, y-t)$, and its image generates $J_{\mathbb{Q}}$. Skolem's method (Theorem 2.3) applies if we can prove that $\operatorname{rk} J(R) < \ell + 1$.

We will show that $\operatorname{rk} J(R) < \ell + 1$ holds when ℓ is large. Below, $O(1)$ denotes a quantity whose size depends only on X and S, not on ℓ or a. We have

$$\operatorname{rk} J(R) = \operatorname{rk} R^\times + \operatorname{rk} \mathcal{O}^\times \le 2 \operatorname{rk} R^\times + \operatorname{rk} \mathcal{O}_0^\times = O(1) + \operatorname{rk} \mathcal{O}_0^\times. \tag{2}$$

Since K_0 is of degree $\ell + O(1)$ and has at most one real place, the Dirichlet S-unit theorem implies that

$$\operatorname{rk} \mathcal{O}_0^\times = \ell/2 + \#\{\text{primes in } K_0 \text{ above } S\} + O(1). \tag{3}$$

Let $s = \#S$. For fixed $p \in S$, the number of primes of K_0 above p having degree $< 3s$ is at most p^{3s}, because these primes correspond to distinct irreducible factors of $ax^\ell - 1 \bmod p$ of degree $< 3s$; on the other hand, there are at most $\ell/(3s)$ primes of K_0 above p having degree $\ge 3s$, because their degrees sum to at most $[K_0 : \mathbb{Q}] \le \ell$. Thus

$$\#\{\text{primes in } K_0 \text{ above } S\} = \sum_{p \in S} \left(p^{3s} + \ell/(3s)\right) = \ell/3 + O(1). \tag{4}$$

Substituting (4) into (3) and then (3) into (2) yields

$$\operatorname{rk} J(R) \le \ell/2 + \ell/3 + O(1) < \ell + 1$$

if ℓ is sufficiently large in terms of S.

Remark 3.1 The proof does not seem to generalize readily to prove the analogue for rings of S-integers in number fields other than $\mathbb{Q}$. The limitations of this approach are investigated thoroughly in [Tri19].

Acknowledgments I thank the referees for encouraging me to include an exposition of the Skolem–Chabauty method.

References

[Bou98] Nicolas Bourbaki, Lie groups and Lie algebras. Chapters 1–3, Elements of Mathematics (Berlin), Springer-Verlag, Berlin, 1998. Translated from the French; Reprint of the 1989 English translation. MR1728312 (2001g:17006)

[Cha38] Claude Chabauty, *Sur les équations diophantiennes liées aux unités d'un corps de nombres algébriques fini*, Ann. Mat. Pura Appl. **17** (1938), no. 1, 127–168, https://doi.org/10.1007/BF02410698 (French). MR1553308

[Cha41] Claude Chabauty, *Sur les points rationnels des courbes algébriques de genre supérieur à l'unité*, C. R. Acad. Sci. Paris **212** (1941), 882–885 (French). MR0004484 (3,14d)
[Kim05] Minhyong Kim, *The motivic fundamental group of* $\mathbf{P}^1\backslash\{0, 1, \infty\}$ *and the theorem of Siegel*, Invent. Math. **161** (2005), no. 3, 629–656. MR2181717 (2006k:11119)
[MP12] William McCallum and Bjorn Poonen, *The method of Chabauty and Coleman*, Explicit Methods in Number Theory: Rational Points and Diophantine Equations, Panoramas et Synthèses, vol. 36, Société Mathématique de France, Paris, 2012, pp. 99–117.
[Sie26] Carl Ludwig Siegel, *The integer solutions of the equation* $y^2 = ax^n + bx^{n-1} + \ldots + k$, J. London Math. Soc. **1** (1926), 66–68.
[Sko34] Th. Skolem, *Ein Verfahren zur Behandlung gewisser exponentialer Gleichungen und diophantischer Gleichungen*, 8. Skand. Mat.-Kongr., Stockholm, 1934, pp. 163–188 (German).
[Tri19] Nicholas George Triantafillou, *Restriction of scalars, the Chabauty-Coleman method, and* $\mathbb{P}^1\backslash\{0, 1, \infty\}$, June 2019. Ph.D. thesis, Massachusetts Institute of Technology.